Graphs of Common Functions

The function $f(x) = x$

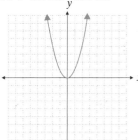
The function $f(x) = x^2$

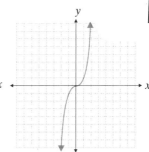
The function $f(x) = x^3$

The function $f(x) = \dfrac{1}{x}$

The functions e^x and $\ln(x)$

The function $f(x) = \sqrt{x}$ or $x^{\frac{1}{2}}$

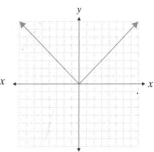
The function $f(x) = |x|$

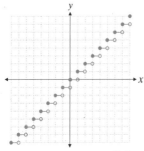
The function $f(x) = [\![x]\!]$

Conic Sections

ellipse

parabola

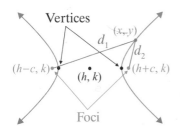
hyperbola

$a, b > 0$ with $a > b$. The standard form of the equation for the ellipse centered at (h, k) with major axis of length $2a$ and minor axis of length $2b$ is:

1. $\dfrac{(x-h)^2}{a^2} + \dfrac{(y-k)^2}{b^2} = 1$
(major axis is horizontal)

2. $\dfrac{(x-h)^2}{b^2} + \dfrac{(y-k)^2}{a^2} = 1$
(major axis is vertical)

The foci are located c units away from the center of the ellipse where $c^2 = a^2 - b^2$.
The eccentricity is defined as:
$$e = \dfrac{c}{a} = \dfrac{\sqrt{a^2 - b^2}}{a}$$

Note: an ellipse with $e = 1$ is a circle (see inside back cover).

p is a nonzero real number. The standard form of the equation for the parabola with vertex (h, k) is:

1. $(x-h)^2 = 4p(y-k)$
(vertically-oriented)
The focus is at $(h, k+p)$.
The equation for the directrix is:
$y = k - p$

2. $(y-k)^2 = 4p(x-h)$
(horizontally-oriented)
The focus is at $(h+p, k)$.
The equation for the directrix is:
$x = h - p$

The standard form of the equation for the hyperbola with center at (h, k) is:

1. $\dfrac{(x-h)^2}{a^2} - \dfrac{(y-k)^2}{b^2} = 1$
(foci are aligned horizontally)
Asymptotes: $y - k = \dfrac{b}{a}(x-h)$
and $y - k = -\dfrac{b}{a}(x-h)$

2. $\dfrac{(y-k)^2}{a^2} - \dfrac{(x-h)^2}{b^2} = 1$
(foci are aligned vertically)
Asymptotes: $y - k = \dfrac{a}{b}(x-h)$
and $y - k = -\dfrac{a}{b}(x-h)$

The foci are located c units away from the center where $c^2 = a^2 + b^2$.
The vertices are a units away from the center.

Index of Symbols

Symbol	Meaning
$\mathbb{N}$	set of all natural numbers $\{1, 2, 3, 4, 5, ...\}$
$\mathbb{Z}$	set of all integers $\{..., -4, -3, -2, -1, 0, 1, 2, 3, 4, ...\}$
$\mathbb{Q}$	set of all rational numbers, that is the set of all numbers that can be represented as a ratio
$\mathbb{R}$	set of all real numbers
π	the ratio of the circumference to the diameter of a circle
E	calculator notation for scientific notation (ex. '$a\text{E}b$' means '$a \times 10^b$')
∞	infinity
i	imaginary unit defined as $\sqrt{-1}$
$\Leftrightarrow$	"is equivalent to"
$\Rightarrow$	"implies"
$\cup$	the union of two sets (i.e. the set of all elements found in either set)
$\cap$	the intersection of two sets (i.e. the set of all elements found in both sets)
$\in$	"is an element of"
$\emptyset$	empty set or the set containing no elements
$\rightarrow$	"approaches"
$\approx$	"approximately equal to"
Δ	"change in"
$\mathbb{R}^2$	the plane of all real x-values by all real y-values (the Cartesian plane)
$\sum_{i=a}^{n}$	summation notation (or sigma notation) used to express the sum of a sequence from a to n
e	base of the natural logarithm
$n!$	"n factorial" stands for the product of all the integers from 1 to n ($0! = 1$ and $1! = 1$)
$\binom{n}{k}$	"n choose k"; combination of n objects taken k at a time

SECOND EDITION

COLLEGE ALGEBRA

HAWKES
LEARNING
SYSTEMS

PAUL SISSON

Editor: Bethany Loftis
Development Director: Marcel Prevuznak
Production Editors: Harding Brumby, Phillip Bushkar, Cynthia Ellison, Mandy Glover, Nina Miller, Kim Scott
Editorial Assistants: K.V.S. Anil, Kelly Epperson, William Epperson, D. Kanthi, Amy Riesel, Brian Smith, B. Syamprasad, Tristan Vogler, Chris Yount
Layout: QSI (Pvt.) Ltd.: E. Jeevan Kumar, U. Nagesh
Art: Ayvin Samonte

Copyright © 2008 by Hawkes Learning Systems/Quant Systems, Inc. All rights reserved.
1023 Wappoo Rd. Suite A-6/ Charleston, SC 29407
www.hawkeslearning.com

No part of this publication may be reproduced, stored in a retrieval system, or transmitted in any form or by any means, electronic, mechanical, photocopying, recording, or otherwise, without the prior written consent of the publisher.

Library of Congress Control Number: 2006933712

Printed in the United States of America

ISBN (Student): 978-1-932628-27-2
(Student Bundle): 978-1-932628-29-6

contents

preface vii

chapter one 1 Number Systems and Fundamental Concepts of Algebra

1.1	The Real Number System	3
1.2	The Arithmetic of Algebraic Expressions	16
1.3	Properties of Exponents	28
1.4	Properties of Radicals	45
1.5	Polynomials and Factoring	61
1.6	The Complex Number System	77
	Chapter Project	86
	Chapter Summary	87
	Chapter Review	91
	Chapter Test	93

chapter two 97 Equations and Inequalities of One Variable

2.1	Linear Equations in One Variable	99
2.2	Linear Inequalities in One Variable	112
2.3	Quadratic Equations in One Variable	122
2.4	Higher Degree Polynomial Equations	137
2.5	Rational Expressions and Equations	143
2.6	Radical Equations	158
	Chapter Project	164
	Chapter Summary	165
	Chapter Review	169
	Chapter Test	171

chapter three 175 Linear Equations and Inequalities of Two Variables

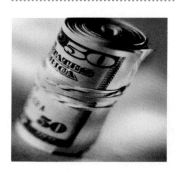

- 3.1 The Cartesian Coordinate System 177
- 3.2 Linear Equations in Two Variables 189
- 3.3 Forms of Linear Equations 197
- 3.4 Parallel and Perpendicular Lines 215
- 3.5 Linear Inequalities in Two Variables 223
- 3.6 Introduction to Circles 235
- Chapter Project 242
- Chapter Summary 243
- Chapter Review 246
- Chapter Test 248

chapter four 253 Relations, Functions, and Their Graphs

- 4.1 Relations and Functions 255
- 4.2 Linear and Quadratic Functions 271
- 4.3 Other Common Functions 287
- 4.4 Transformations of Functions 304
- 4.5 Combining Functions 322
- 4.6 Inverses of Functions 336
- Chapter Project 349
- Chapter Summary 350
- Chapter Review 353
- Chapter Test 356

chapter five 361 Polynomial Functions

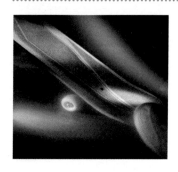

- 5.1 Introduction to Polynomial Equations and Graphs 363
- 5.2 Polynomial Division and the Division Algorithm 377
- 5.3 Locating Real Zeros of Polynomials 390
- 5.4 The Fundamental Theorem of Algebra 405
- Chapter Project 419
- Chapter Summary 420
- Chapter Review 423
- Chapter Test 426

chapter six 429   Rational Functions and Conic Sections

- 6.1   Rational Functions and Rational Inequalities   431
- 6.2   The Ellipse   449
- 6.3   The Parabola   467
- 6.4   The Hyperbola   478
- Chapter Project   493
- Chapter Summary   495
- Chapter Review   498
- Chapter Test   500

chapter seven 505   Exponential and Logarithmic Functions

- 7.1   Exponential Functions and Their Graphs   507
- 7.2   Applications of Exponential Functions   517
- 7.3   Logarithmic Functions and Their Graphs   533
- 7.4   Properties and Applications of Logarithms   545
- 7.5   Exponential and Logarithmic Equations   561
- Chapter Project   577
- Chapter Summary   578
- Chapter Review   581
- Chapter Test   583

chapter eight 587   Systems of Equations

- 8.1   Solving Systems by Substitution and Elimination   589
- 8.2   Matrix Notation and Gaussian Elimination   605
- 8.3   Determinants and Cramer's Rule   619
- 8.4   The Algebra of Matrices   635
- 8.5   Inverses of Matrices   650
- 8.6   Linear Programming   664
- 8.7   Nonlinear Systems of Equations   675
- Chapter Project   687
- Chapter Summary   688
- Chapter Review   691
- Chapter Test   695

chapter nine   701   **An Introduction to Sequences, Series, Combinatorics, and Probability**

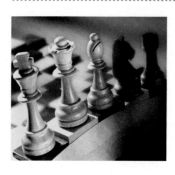

9.1	Sequences and Series	703
9.2	Arithmetic Sequences and Series	720
9.3	Geometric Sequences and Series	732
9.4	Mathematical Induction	750
9.5	An Introduction to Combinatorics	760
9.6	An Introduction to Probability	783

Chapter Project   798
Chapter Summary   799
Chapter Review   802
Chapter Test   805

appendix a   **Fundamentals of *Mathematica*®**   809

answers   **Odd Numbered Exercises**   815

index of applications   865

technology index   869

index   870

preface

Introduction

As the title indicates, this book is written for college students studying algebra. This audience is a diverse one, and includes students for whom algebra is a necessary precursor to calculus and other topics, as well as students for whom algebra is (they think) the last math class they will ever take.

Both of the preceding sentences warrant elaboration. First, it is important to note what this book is not. It is not written for high school students, nor is it a mere repackaging of a high school text in a college cover. Given the range in quality of secondary education in this country, students and teachers will find varying degrees of overlap between the material here and that in the high school classes of their experience. That being said, it is assumed that students using this book have already taken an elementary or intermediate algebra course (either in high school or college) and are proficient with fundamental algebraic skills and notions. As a rough guide, the material in the first two chapters of this book should be largely review. It is included here for reference purposes and to fill any gaps that might exist in a student's prior education.

Second, this book will certainly help to lay a solid algebraic foundation for those students who already know they'll be using the tools learned here in later courses. But I have also tried to make this book of lasting value for those on the opposite end of the spectrum. One reason for this is that college is a formative experience, and teachers and especially students must always keep in mind that plans change. Some students will find their educational paths unexpectedly veering in a "mathward" direction at some point, and a college algebra text must be robust enough to prepare them for such a shift. Finally, even for those who remain determined to resist the charms of math, I have tried to include numerous examples of the applications of algebra that affect all of us in everyday life.

What has changed in the second edition? I have tried to keep the aspects of this book that are unique, and indeed hope to have enhanced them. Rigor and a comprehensive coverage of the material are important to me, and I have tried to achieve both while maintaining a less stuffy delivery than is commonly found. More particular changes include the addition of many new exercises and application problems throughout the text, an updated and expanded collection of exercises that make use of graphing utilities, and the inclusion of such topics as Linear Regression, Introduction to Circles, Linear Programming, and Mathematical Induction. Many of these additions came at the suggestion of users of the first edition, and such suggestions are always welcome.

preface

Features

Topics:

Each section begins with a list of topics. These concise objectives are a helpful guide for both reference and class preparation.

section 1.1   The Real Number System

1.1 The Real Number System

TOPICS

1. Common subsets of real numbers
2. The real number line
3. Order on the real number line
4. Set-builder notation and interval notation
5. Absolute value and distance on the real number line
6. Interlude: working with repeating digits

Topic 1:   **Common Subsets of Real Numbers**

Certain types of numbers occur so frequently in mathematics that they have been given special names and symbols. These names will be used throughout this book and in later classes when referring to members of the following sets:

mbers

atural (or Counting) Numbers: This is the set of numbers $\mathbb{N} = \{1, 2, 3, 4, 5, \ldots\}$. et is infinite, so in list form we can write only the first few numbers.

hole Numbers: This is the set of natural numbers and 0: $\{0, 1, 2, 3, 4, 5, \ldots\}$. , we can list only the first few members of this set. No special symbol will be ed to this set in this text.

ntegers: This is the set of natural numbers, their negatives, and 0. As a list, the set $\mathbb{Z} = \{\ldots, -4, -3, -2, -1, 0, 1, 2, 3, 4, \ldots\}$. Note that the list continues nitely in both directions.

Rational Numbers: This is the set, with symbol $\mathbb{Q}$ for quotient, of *ratios* of rs (hence the name). That is, any rational number can be written in the $\frac{p}{q}$, where p and q are both integers and $q \neq 0$. When written in decimal form, al numbers either terminate or repeat a pattern of digits past some point.

continued on next page...

3

chapter

OVERVIEW

1.1   The Real Number System

1.2   The Arithmetic of Algebraic Expressions

1.3   Properties of Exponents

1.4   Properties of Radicals

1.5   Polynomials and Factoring

1.6   The Complex Number System

ONE

NUMBER SYSTEMS AND FUNDAMENTAL CONCEPTS OF ALGEBRA

By the end of this chapter you should be able to work with and manipulate expressions involving exponents. One common task involving exponents is building geometric formulas. On page 37 you will find a problem involving a soup can. You will develop a formula for its surface area by mentally taking it apart and combining simpler formulas for its top, bottom, and sides. You will master this type of problem, making use of tools such as the table of *Properties of Exponents* on page 32.

Chapter Openers:

Each chapter begins with a list of sections and an engaging preview of an application appearing later in the chapter.

1

viii

preface

Definitions and Theorems:

All of the primary definitions and theorems are clearly identified and set off in yellow boxes that are highly visible and easily found again. All important terms appear in bold print when first formally defined, and other useful terms appear in italic font when informally defined.

Cautions:

Common pitfalls are pointed out, and examples of common errors and their corrections abound.

---

chapter one   Number Systems and Fundamental Concepts of Algebra

We could describe such sets with set-builder notation, but intervals occur frequently enough that special notation has been devised to define them succinctly.

Interval Notation

Interval Notation	Set-Builder Notation	Meaning
(a,b)	$\{x \mid a < x < b\}$	all real numbers strictly between a and b
$[a,b]$	$\{x \mid a \leq x \leq b\}$	all real numbers between a and b, including both a and b
$(a,b]$	$\{x \mid a < x \leq b\}$	all real numbers between a and b, including b but not a
$(-\infty, b)$	$\{x \mid x < b\}$	all real numbers less than b
$[a, \infty)$	$\{x \mid x \geq a\}$	all real numbers greater than or equal to a

In interval notation, the left endpoint is always written first. That is, as we read the notation from left to right, the left endpoint of the interval appears before the comma. Intervals of the form (a,b) are called **open** intervals, while those of the form $[a,b]$ are **closed** intervals. The interval $(a,b]$ is **half-open** (or **half-closed**). Of course, a half-open interval may be open at either endpoint, as long as it is closed at the other. The symbols $-\infty$ and ∞ indicate that the interval extends indefinitely in the left and right directions, respectively. Note that $(-\infty, b)$ excludes the endpoint b, while $[a, \infty)$ includes the endpoint a.

caution

The symbols $-\infty$ and ∞ are just that: symbols! They are not real numbers, and so they cannot, for instance, be solutions to a given equation. The fact that they are symbols, and not numbers, also means that as endpoints of intervals they can never be included. For this reason, a parenthesis always appears next to either $-\infty$ or ∞; a bracket should never appear next to either infinity symbol.

example 5

a. The notation $[-3, 10)$ stands for the set $\{x \mid -3 \leq x < 10\}$. This interval is closed at the left endpoint, -3, and open at the right endpoint, 10.
b. The interval $(-\infty, \infty)$ is just another way of describing the entire set of real numbers.

Numerous Examples:

Each section contains many examples that illustrate the concepts presented and the skills to be mastered. The examples are clearly set off from the accompanying text, and the exercises in each section refer the student to the relevant examples to study.

preface

Historical Contexts:

Each chapter begins with a brief introduction setting the historical context of the math that follows. Mathematics is a human endeavor, and knowledge of how and why a particular idea developed is of great help in understanding it. Too often, math is presented in cold abstract chunks completely divorced from the rest of reality. While a (very) few students may be able to master material this way, most benefit from an explanation of how math ties into the rest of what people were doing at the time it was created.

Technology Notes and Exercises:

Where appropriate, the use of technology (in the form of graphing calculators or computer software) is illustrated. Such notes are then followed by specially identified problems in the Exercises for which technology is helpful. *See the Technology Index on page 869 for a complete listing of these notes and exercises.*

preface

chapter one Number Systems and Fundamental Concepts of Algebra

chapter one project

Polynomials

A chemistry professor calculates final grades for her class using the polynomial

$$A = 0.3f + 0.15h + 0.4t + 0.15p,$$

where A is the final grade, f is the final exam, h is the homework average, t is the chapter test average, and p is the semester project.

The following is a table containing the grades for various students in the class:

Name	Final Exam	Homework Avg.	Test Avg.	Project
Alex	77	95	79	85
Ashley	91	95	88	90
Barron	82	85	81	75
Elizabeth	75	100	84	80
Gabe	94	90	90	85
Lynn	88	85	80	75

1. Find the course average for each student, rounded to the nearest tenth.
2. Who has the highest total score?
3. Why is the total grade raised more with a grade of 100 on the final exam than with a grade of 100 on the semester project?
4. Assume you are a student in this class. With 1 week until the final exam, you have a homework average of 85, a test average of 85, and a 95 on the semester project. What score must you make on the final exam to achieve at least a 90.0 overall? (Round to the nearest tenth.)

86

Chapter Projects:

Each project describes a plausible scenario and related constraints, and is suitable for individual or group assignments.

---

section 1.3 Properties of Exponents

Topic 4: Scientific Notation

Scientific notation is an important application of exponents. Very large and very small numbers arise naturally in a variety of situations, and working with them without scientific notation is an unwieldy and error-prone process. Scientific notation uses the properties of exponents to rewrite very large and very small numbers in a less clumsy form. It applies to those numbers which contain a long string of 0's before the decimal point (very large numbers) or to those which contain a long string of 0's between the decimal point and the non-zero digits (very small numbers).

Scientific Notation

A number is in scientific notation when it is written in the form

$$a \times 10^n$$

where $1 \leq |a| < 10$ and n is an integer. If n is a positive integer, the number is large in magnitude, and if n is a negative integer, the number is small in magnitude (close to 0). The number a itself can be either positive or negative, and the sign of a determines the sign of the number as a whole.

caution

The sign of the exponent n in scientific notation does *not* determine the sign of the number as a whole. The sign of n only determines if the number is large (positive n) or small (negative n) in magnitude.

example 4

a. The distance from Earth to the sun is approximately 93,000,000 miles. Scientific notation takes advantage of the observation that multiplication of a number by 10 moves the decimal point one place to the right, and we can repeat this process as many times as necessary. Thus 9.3×10^7 is equal to 93,000,000, as 93,000,000 is obtained from 9.3 by moving the decimal point 7 places to the right:

$$9.3 \times 10^7 = 9\underbrace{3000000}_{7 \text{ places}}$$

Note: In a calculator, scientific notation is often expressed with an E. To represent this on the TI-84 PLUS, press 2ND and , .

93,000,000 miles

continued on next page...

35

Graphing Calculators:

Throughout the text, graphing calculators (with an emphasis on the TI-84 Plus) are integrated into the text to demonstrate usage, with helpful operating hints where appropriate.

Applications:

Many exercises and examples illustrate practical applications, keeping students engaged. *See the Index of Applications on page 865 for a complete list of application problems and examples.*

xi

preface

Mathematical Interludes:

Many sections end with an Interlude which ties together some of the ideas just presented, extends the math slightly, or showcases some concept in an important application. Depending on the class in which this book is used, Interludes may be integral to the course or may serve as nothing more than extracurricular reading for interested students.

Exercises:

Each section concludes with a selection of exercises designed to allow the student to practice skills and master concepts. References to appropriate chapter examples are clearly labeled for those who desire assistance. Many levels of difficulty exist within each exercise set, allowing teachers to adapt the exercises as necessary and allowing students to practice elementary skills or stretch themselves, as appropriate.

---

section 1.3 Properties of Exponents

Topic 5: Interlude: Working with Geometric Formulas

Exponents occur in a very natural way when geometric formulas are considered. Some of the problems that you encounter in this book (and elsewhere) will require nothing more than using one of the basic geometric formulas found on the inside front cover of the text, but others will require a bit more work. Often, the exact geometric formula that you need to solve a given problem can be derived from simpler formulas.

We will look at several examples of how a new geometric formula is built up from known formulas. The general rule of thumb in each case is to break down the task at hand into smaller pieces that can be easily handled.

example 6

Find formulas for each of the following:
 a. the surface area of a box
 b. the surface area of a soup can
 c. the volume of a birdbath in the shape of half of a sphere
 d. the volume of a gold ingot whose shape is a right trapezoidal cylinder

Solutions:

a. A box whose six faces are all rectangular is characterized by its length l, its width w, and its height h. The area of a rectangle is one of the basic formulas that you should be familiar with, and can be found on the inside front cover of the text. The formula for the surface area of a box, then, is just the sum of the areas of the six sides. If we let S stand for the total surface area, we obtain the formula $S = lw + lw + lh + lh + hw + hw$ or $S = 2lw + 2lh + 2hw$.

b. A soup can, an example of a right circular cylinder, is characterized by its height h and the radius r of the circle that makes up the base (or the top).

---

section 1.3 Properties of Exponents

exercises

Simplify each of the following expressions, writing your answer with only positive exponents. See Example 1.

1. $(-2)^4$ 2. -2^4 3. -3^2 4. $(-3)^2$ 5. $3^2 \cdot 3^2$
6. $2^3 \cdot 3^2$ 7. $4 \cdot 4^2$ 8. $(-3)^3$ 9. $\dfrac{8^2}{4^3}$ 10. $2^2 \cdot 2^3$
11. $\dfrac{7^4}{7^5}$ 12. $n^2 \cdot n^5$ 13. $\dfrac{x^5}{x^2}$ 14. $\dfrac{y^3 \cdot y^8}{y^2}$ 15. $\dfrac{3^7}{3^4 \cdot s^{-10}}$

Use the properties of exponents to simplify each of the following expressions, writing your answer with only positive exponents. See Examples 1, 2, and 3.

16. $\dfrac{3t^{-2}}{t^3}$ 17. $-2y^0$ 18. $\dfrac{1}{7x^{-5}}$ 19. $9^0 x^3 y^0$ 20. $\dfrac{2n^3}{n^{-5}}$
21. $\dfrac{11^{21}}{11^{19} x^{-7}}$ 22. $\dfrac{x^7 y^{-3} z^{12}}{x^{-1} z^9}$ 23. $\dfrac{x^4(-x^{-3})}{-y^0}$ 24. $\dfrac{s^3}{s^{-2}}$ 25. $\dfrac{x^{-1}}{x}$
26. $x^{(y^0)} \cdot x^9$ 27. $\dfrac{x^2 y^{-2}}{x^{-1} y^{-5}}$ 28. $\dfrac{s^5 y^{-5} z^{-11}}{s^8 y^{-7}}$ 29. $\dfrac{2^7 s^{-3}}{2^3}$ 30. $\dfrac{3^{-5}}{(3^{-4} x^5 y^4)^2}$
31. $\dfrac{-9^0 (x^2 y^{-2})^{-3}}{3x^{-4} y}$ 32. $\left[(2x^{-1} z^3)^{-2}\right]^{-1}$ 33. $\dfrac{(3yz^{-2})^0}{3y^2 z}$
34. $(12a^2 - 3b^4)^0$ 35. $\dfrac{3^{-1}}{(3^2 xy^2)^{-2}}$ 36. $\left[9m^2 - (2n^2)^3\right]^{-1}$
37. $\left[(12x^{-6} y^4 z^3)^5\right]^0$ 38. $\dfrac{x(x^{-2} y^3)^3}{(2x^4)^{-2} y}$ 39. $\dfrac{(-3a)^{-2}(bc^{-2})^{-3}}{a^5 c^4}$
40. $\left[(5m^4 n^{-2})^{-1}\right]^{-2}$ 41. $(9x^{-1} z)^2 (2xy^{-3})^{-1}$ 42. $(4^{-2} x^5 y^{-3} z^4)^{-2}$
43. $\left[(4a^2 b^{-5})^{-1}\right]^3$ 44. $\left[(2^{-3} m^{-6} n^3)^{-1}\right]^{-1}$ 45. $\left[(3^{-1} x^{-1} y)(x^2 y)^{-1}\right]^{-3}$
46. $\left[\dfrac{100^0 (x^{-1} y^3)^{-1}}{x^2 y}\right]^{-3}$ 47. $(5z^6 - (3x^3)^4)^{-1}$ 48. $\left[\dfrac{y^6 (xy^2)^{-3}}{3x^{-3} z}\right]^{-2}$

preface

Chapter Summary:

Each chapter ends with a concise summary of the concepts learned and the skills acquired, arranged by section and topic.

---

chapter one summary

CHAPTER SUMMARY

A summary of concepts and skills follows each chapter. Refer to these summaries to make sure you feel comfortable with the material in the chapter. The concepts and skills are organized according to the section title and topic title in which the material is first discussed.

1.1 — The Real Number System

topics	pages
Common Subsets of Real Numbers	p. 3 – 4
• The sets $\mathbb{N}, \mathbb{Z}, \mathbb{Q}$, and $\mathbb{R}$, as well as *whole* numbers and *irrational* numbers	
• Identifying numbers as elements of one or more of the common sets	
The Real Number Line	p. 5 – 6
• Plotting numbers on the real number line	
• The *origin* of the real line, and its relation to nega numbers	
Order on the Real Number Line	
• The four inequality symbols $<, \leq, >,$ and $\geq$	
• Distinguishing between strict and non-strict inequ	
Set-Builder Notation and Interval Notation	
• Using set-builder notation to define sets	
• The empty set and its two common symbols	
• Interval notation and its relation to inequality sta	

Chapter Review & Test:

Immediately following each Chapter Summary is a Chapter Review and a Chapter Test. Each presents several more problems pertaining to the major ideas of the chapter.

---

chapter one review

chapter review

Section 1.1

Write the following rational numbers as ratios of integers.

1. $7.\overline{6}$ 2. $-2.0\overline{42}$

te the absolute value expressions.

$-|11-2|$ 4. $-|-4-3|$ 5. $\left|\sqrt{5}-\sqrt{11}\right|$

ction 1.2

te each expression for the given values of the variables.

$x^2z^3 + 5\sqrt{3x-2y}$ for $x=2, y=1,$ and $z=-1$

$7y^2 - \frac{1}{3}\pi xy + 8x^3$ for $x=-2$ and $y=2$

y the property that justifies each of the following statements. If one of the cancellation ies is being used to transform an equation, identify the quantity that is being added sides or the quantity that is multiplied by both sides.

$2a^2 = 8b \Leftrightarrow 3a^2 = 2b$ 9. $(x-3)(z-2)=0 \Rightarrow x-3=0$ or $z-2=0$

ction 1.3

properties of exponents to simplify each of the following expressions, writing your with only positive exponents.

$\dfrac{-4t^0\left(s^2t^{-2}\right)^{-3}}{2^3st^{-3}}$ 11. $\left[\left(3y^{-2}z\right)^{-1}\right]^{-3}$

---

chapter one test

chapter test

1. Determine if the following statement is true or false and explain your answer: If $|x|=-x$, then $x-4<0$ for all x.

2. If $|a|<6$, what is the smallest possible integer value of a?

Evaluate each of the following expressions.

3. $\left(-3^2-2^3\right)\left(1-2^4\right)+\left[-\left(-4^3\right)\cdot\left(-2\right)^2\right]$

4. $\left[2^3-3^2\cdot(-3)^3-\left(2-5^0\right)\right]\left[-4^2+(-2)^3\right]$

5. $\dfrac{2-2^{-1}}{2+2^{-1}} \div \dfrac{1+2^{-1}}{1-2^{-1}}$ 6. $3+4^2\left(4^{-2}\right)^{-1}-259$

7. $(-1)^{-10}\cdot(-1)^5-(-2)^2\cdot\left(-2^2\right)(-2)^3$ 8. $\dfrac{3^{-3}\cdot(-3)^{-2}\left(-3^2\right)}{\left(1-2^2\right)(-3)\left(2^2-1\right)}$

Simplify each of the following expressions, writing your answer with only positive exponents.

9. $-2^3\left(-2^2\right)\cdot\left(-2^{-2}\right)$ 10. $\dfrac{2^{-3}\left(b^2\right)^{-2}}{\left(2^{-2}\right)^{-3}\left(b^{-1}\right)^5}$ 11. $\left(2x^3y^{-1}\right)^2\cdot\left(-4xy^3\right)^2$

Evaluate each expression using the properties of exponents and express your answer using scientific notation.

12. $\left(3.46\times10^8\right)\left(1.2\times10^4\right)$ 13. $\left(-3\times10^{-7}\right)\left(1.2\times10^4\right)$

14. $\dfrac{2.4\times10^{-12}}{(1.2)\times10^{-4}}$ 15. $\dfrac{1.6\times10^{-4}}{8.0\times10^8}$

16. $\dfrac{2{,}000{,}000}{5{,}000{,}000{,}000}$

17. Arrange the following numbers from largest to smallest: $\sqrt{3}, \sqrt[3]{4}, \sqrt[5]{15}$

Evaluate each of the following radical expressions.

18. $-\sqrt{121}$ 19. $\sqrt[3]{-125}$ 20. $\sqrt{3^2+4^2}$

21. $\sqrt{-81}$ 22. $\sqrt{8\sqrt{4}}$ 23. $\sqrt{\sqrt{81}}$

xiii

preface

A Note on Technology

The purpose of this book is to help the reader learn the mathematics contained within. To the extent that technology (in whatever form) can aid in pursuit of that goal, it should be enthusiastically embraced by both student and teacher. Aside from masochism, there is no excuse for not using a calculator or computer program when faced with tedious, pedagogically unenlightening tasks. But technology can be terribly abused as well, and the risks of relying too heavily on technological aids when attempting to understand a concept can't be stressed too much. There is great temptation, on the part of students and teachers alike, to let calculators and computers turn every problem into a mindless "plug and chug" operation.

Experienced educators have long realized the dangers of teaching students rote algorithmic procedures for solving problems; chief among them is the fact that if a problem is not presented in "just the right way" a student is unable to solve it. A textbook that takes a technology-based cookbook approach to mathematics has exactly the same problem, and far too many students suffer later in life because they were taught how to use a calculator, but not mathematics.

The above should by no means be taken as a dismissal of technology, however. Once the mathematics has been learned, calculators and (especially) computers are essential everyday tools for many, many people. The important thing is to make the distinction between learning mathematics and using mathematics. In general, technology is much more useful in using mathematics than in learning it.

I have tried to take great care in this book to note the exceptions. For example, the software package *Mathematica*® is featured in many of the Technology Notes and Exercises, and an Appendix on its elementary use is included. For many of the Technology Exercises a graphing calculator can also be substituted. A further very important example of how technology can be used as an aid to learning mathematics is seen in the multimedia courseware that has been designed as a companion to this text. See the section entitled *Hawkes Learning Systems: College Algebra* for more details on installing and using this tutorial and testing package.

Supplements

The **Instructor's Annotated Edition** of this text contains, in the margins, solutions to all of the exercises.

Hawkes Learning Systems: College Algebra is a comprehensive tutoring and testing software package that has been specifically designed to complement this book. While the book stands alone and can be used independently of the software, the computer component provides an effective second approach to learning and to developing problem-solving skills.

preface

To the Student: Learning Mathematics

There is a saying among math teachers that you might have heard: "Math is not a spectator sport." While trite, it is undeniably true (and while trite, at least it's concise). Mathematics is not something you can learn by watching someone else do it. You have probably already had the experience of watching a teacher solve a problem and marveling at how easy it seems, only to discover that a nearly identical problem is much harder to solve at home or (far worse) on a test.

The lesson to be learned is that you have to practice mathematics in order to learn mathematics. Make sure you do the homework problems your teacher assigns you, not only because that's how mathematics is learned but also because it gives you insight into the sort of problems your teacher thinks is important.

Beyond that, there are some other key ideas to keep in mind. One is that very, very few people will fully grasp a mathematical concept or master a skill the first time they are presented with it. If you find yourself lost after the first reading of a section of this book (or any math book), don't despair. Just remember what is puzzling you, take a break, and try it again when you're fresh. Most math is learned in a cyclic process of plowing ahead until lost, backing up and re-reading, and then plowing ahead a bit further. A math book is not like a novel: you shouldn't expect to be able to read it quickly in a purely linear fashion.

A third point related to the first two is that math must be read with a pencil (okay, the writing instrument of your choice). Mathematical reading is slow, and must be broken up with your own writing as you make sure you understand the worked examples and maybe do a few homework problems. Take your time as you read, and work carefully and deliberately.

Finally, make sure you take advantage of your teacher and peers. Go to class and pay attention to what your teacher emphasizes and learn from his or her unique insight into the material. Work with friends when possible, and ask others for help if they understand something you haven't gotten yet. And when you have the opportunity to explain some math to someone else, take advantage of that too. Teaching mathematics to others is an amazingly effective way to improve your own understanding.

preface

Acknowledgements

I am very grateful to all the people at Hawkes Learning Systems for their support, their dedication, and their indefatigable cheerfulness. In particular, many thanks to Marcel Prevuznak (development director), Bethany Loftis (editor), Emily Cook (marketing), and James Hawkes. Thanks also to Will Pridgen, who did a superb job of reviewing the text and concocting additional exercises for the first edition.

I am also very grateful to the following for their many insightful comments and reviews:

Brenda Cates, *Mount Olive College*
Robert Crise, *Crafton Hills College*
Michelle DeDeo, *University of North Florida*
Pamela Duck, *NorthWest Arkansas Community College*
Rebecca Heiskell, *Mountain View College*
Theresa Hert, *Mount San Jacinto College–San Jacinto Campus*
Christopher Imm, *Johnson County Community College*
Abdusamad Kabir, *Bowie State University*
Jeff Lewis, *Johnson County Community College*
Russell May, *Morehead State University*
Ron Palcic, *Johnson County Community College*
Chris Schroeder, *Morehead State University*
Kevin Sukanek, *University of Mississippi*
Tracy Vaughan, *NorthWest Arkansas Community College*
Raymond Williams, *Mississippi Valley State University*

Thanks also to Drs. Richard Mabry, Carlos Spaht, and Al Vekovius of LSU-Shreveport, and to their college algebra students, for giving the original text a trial run. My thanks also to Drs. Larry Anderson and Mary Ellen Foley of LSUS for their reading of Chapter 1.

I would also like to thank John Scofield, Charlie Hunter, Miles Davis, Herbie Hancock, the Dave Matthews Band, Groove Collective, Carlos Santana, Medeski, Martin & Wood, Lee Ritenour, and Miriam Maria among others, for their unwitting but much-appreciated assistance.

Finally, my continued profound thanks to my wife Cindy for putting up with many more early-morning hours of textbook writing.

preface

Hawkes Learning Systems: College Algebra

Overview:

This multimedia courseware allows students to become better problem-solvers by creating a mastery level of learning in the classroom. The software includes an "Instruct," "Practice," "Tutor," and "Certify" mode in each lesson, allowing students to learn through step-by-step interactions with the software. This automated homework system's tutorial and assessment modes extend instructional influence beyond the classroom. Intelligence is what makes the tutorials so unique. By offering intelligent tutoring and mastery level testing to measure what has been learned, the software extends the instructor's ability to influence students to solve problems. This courseware can be ordered either separately or bundled together with this text.

Minimum Requirements:

In order to run **HLS: College Algebra**, you will need:
Intel® Pentium® 1 GHz or faster processor
Windows® 2000 or later
256 MB RAM
200 MB hard drive space
256 color display (1024x768, 16-bit color recommended)
Internet Explorer 6.0 or later
CD-ROM drive

Getting Started:

Before you can run **HLS: College Algebra**, you will need an access code. This 30 character code is your personal access code. To obtain an access code, go to http://www.hawkeslearning.com and follow the links to the access code request page (unless directed otherwise by your instructor.)

Installation:

1. Place CD #1 in the CD-ROM drive. (CD#1 is the only CD needed for the installation.)
2. Double-click on the **My Computer** icon. (If the installation automatically begins, skip to step 5.)
3. Double-click on the CD-ROM drive (has a picture of a CD-ROM disk).
4. Double-click on Setup.exe.
5. Follow the on-screen instructions.
6. You will be prompted for a Course ID.

If you have internet access, select "Yes, the Course ID is:" and enter the Hawkes Course ID provided by your instructor in the box provided. If you do not have internet access, select the option that says "No, I will not be accessing an online progress report from this computer."

Starting the Courseware:

After you have installed **HLS: College Algebra** on your computer, to run the courseware select Start/Programs/Hawkes Learning Systems/College Algebra.

You will be prompted to enter your access code with a message box similar to the following:

Type your entire access code in the box. No spaces or dashes are necessary; they will be supplied automatically. When you are finished, press OK.

If you typed in your access code correctly, you will be prompted to save the code to a disk. If you choose to save your code to disk, typing in the authorization code each time you run **HLS: College Algebra** will not be necessary. Instead, select the [Load from File] button when prompted to enter your access code and choose the path to your saved access code.

Now that you have entered your authorization code and saved it to disk, you are ready to run a lesson. From the table of contents screen, choose the appropriate chapter and then choose the lesson you wish to run.

preface

Features:

Each lesson in ***HLS: College Algebra*** has four modes: Instruct, Practice, Tutor, and Certify.

Instruct: Instruct provides an exposition on the material covered in the lesson in a multimedia environment. This same instruct mode can be accessed via the tutor mode.

Practice: Practice allows you to hone your problem-solving skills. It provides an unlimited number of randomly generated problems. Practice also provides access to the Tutor mode by selecting the Tutor button located by the Submit button.

Tutor: Tutor mode is broken up into several parts: Instruct, Explain Error, Step by Step, and Solution.

1. **Instruct**, which can also be selected directly from Practice mode, contains a multimedia lecture of the material covered in a lesson.

2. **Explain Error** is active whenever a problem is incorrectly answered. It will attempt to explain the error that caused you to incorrectly answer the problem.

3. **Step by Step** is an interactive "step through" of the problem. It breaks each problem into several steps, explains to you each step in solving the problem, and asks you a question about the step. After you answer the last step correctly, you have solved the problem.

4. **Solution** will provide you with a detailed "worked-out" solution to the problem.

Throughout the Tutor, you will see words or phrases colored green with a dashed underline. These are called Hot Words. Clicking on a Hot Word will provide you with more information on these word(s) or phrases.

Certify: Certify is the testing mode. You are given a finite number of problems and a certain number of strikes (problems you can get wrong). If you correctly answer the required number of questions, you will receive a certification code and a certificate. Write down your certification code and/or print out your certificate. The certification code will be used by your instructor to update your records. The Tutor is not available in Certify.

Integration of Courseware and Textbook:

HLS: College Algebra courseware has been carefully designed to complement this textbook. Each section of this text has a corresponding lesson in the courseware which provides additional instruction, examples and problems.

Support:

If you have questions about ***HLS: College Algebra*** or are having technical difficulties, we can be contacted as follows:

Phone: (843) 571-2825
Email: support@hawkeslearning.com
Web: www.hawkeslearning.com

Support hours are 8:30 a.m. to 5:30 p.m., Eastern Time, Monday through Friday.

chapter ONE

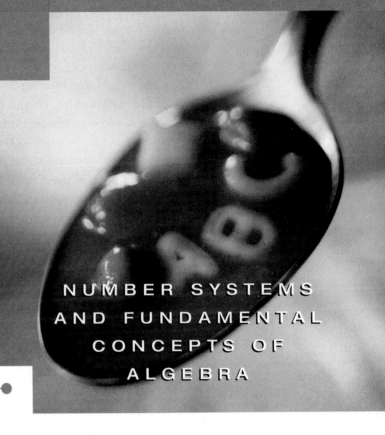

NUMBER SYSTEMS AND FUNDAMENTAL CONCEPTS OF ALGEBRA

OVERVIEW

1.1　The Real Number System

1.2　The Arithmetic of Algebraic Expressions

1.3　Properties of Exponents

1.4　Properties of Radicals

1.5　Polynomials and Factoring

1.6　The Complex Number System

By the end of this chapter you should be able to work with and manipulate expressions involving exponents. One common task involving exponents is building geometric formulas. On page 37 you will find a problem involving a soup can. You will develop a formula for its surface area by mentally taking it apart and combining simpler formulas for its top, bottom, and sides. You will master this type of problem, making use of tools such as the table of *Properties of Exponents* on page 32.

chapter one — Number Systems and Fundamental Concepts of Algebra

Introduction

In this chapter, we review the terminology, the notation and properties of the real number system frequently encountered in algebra, and the extension of the real number system to the larger set of complex numbers.

We begin with a discussion of common subsets of the set of real numbers. Certain types of numbers are important from both a historical and a mathematical perspective. There is archaeological evidence that people used the simplest sort of numbers, the *counting* or *natural numbers*, as far back as 50,000 years ago. Over time, many cultures discovered needs for various refinements to the number system, resulting in the development of such classes of numbers as the *integers*, the *rational numbers*, the *irrational numbers*, and ultimately, the *complex numbers*, a number system which contains the real numbers.

Many of the ideas in this chapter have a history dating as far back as Egyptian and Babylonian civilizations of around 3000 BC, with later developments and additions due to Greek, Hindu, and Arabic mathematicians. Much of the material was also developed independently by Chinese mathematicians. It is a tribute to the necessity, utility, and objectivity of mathematics that so many civilizations adopted so much mathematics from abroad, and that different cultures operating independently developed identical mathematical concepts so frequently.

Pythagoras

As an example of the historical development of just one concept, consider the notion of an irrational number. The very idea that a real number could be irrational (which simply means not rational, or not the ratio of two integers) is fairly sophisticated, and it took some time for mathematicians to come to this realization. The Pythagoreans, members of a school founded by the Greek philosopher Pythagoras in southern Italy around 540 BC, discovered that the square root of 2 was such a number, and there is evidence that for a long time $\sqrt{2}$ was the only known irrational number. A member of the Pythagorean school, Theodorus of Cyrene, later showed (c. 425 BC) that $\sqrt{3}, \sqrt{5}, \sqrt{6}, \sqrt{7}, \sqrt{8}, \sqrt{10}, \sqrt{11}, \sqrt{12}, \sqrt{13}, \sqrt{14}, \sqrt{15}$, and $\sqrt{17}$ also are irrational. It wasn't until 1767 that a European mathematician, Johann Lambert, showed that the famous number π is irrational, and the modern rigorous mathematical description of irrational numbers is due to work by Richard Dedekind in 1872.

As you review the concepts in Chapter 1, keep the larger picture firmly in mind. All of the material presented in this chapter was developed over long periods of time by many different cultures with the aim of solving problems important to them. We will encounter some examples of these problems in Chapter 2.

1.1 The Real Number System

TOPICS

1. Common subsets of real numbers
2. The real number line
3. Order on the real number line
4. Set-builder notation and interval notation
5. Absolute value and distance on the real number line
6. Interlude: working with repeating digits

Topic 1: Common Subsets of Real Numbers

Certain types of numbers occur so frequently in mathematics that they have been given special names and symbols. These names will be used throughout this book and in later math classes when referring to members of the following sets:

Types of Real Numbers

Examples of rational numbers:

The Natural (or Counting) Numbers: This is the set of numbers $\mathbb{N} = \{1, 2, 3, 4, 5, ...\}$. The set is infinite, so in list form we can write only the first few numbers.

The Whole Numbers: This is the set of natural numbers and 0: $\{0, 1, 2, 3, 4, 5, ...\}$. Again, we can list only the first few members of this set. No special symbol will be assigned to this set in this text.

The Integers: This is the set of natural numbers, their negatives, and 0. As a list, this is the set $\mathbb{Z} = \{..., -4, -3, -2, -1, 0, 1, 2, 3, 4, ...\}$. Note that the list continues indefinitely in both directions.

The Rational Numbers: This is the set, with symbol $\mathbb{Q}$ for quotient, of *ratios* of integers (hence the name). That is, any rational number can be written in the form $\frac{p}{q}$, where p and q are both integers and $q \neq 0$. When written in decimal form, rational numbers either terminate or repeat a pattern of digits past some point.

continued on next page...

Types of Real Numbers (continued)

Examples of irrational numbers.

The Irrational Numbers: Every real number that is not rational is, by definition, irrational. In decimal form, irrational numbers are non-terminating and non-repeating. No special symbol will be assigned to this set in this text.

The Real Numbers: Every set above is a subset of the set of real numbers, which is denoted $\mathbb{R}$. Every real number is either rational or irrational, and no real number is both.

example 1

Consider the set $S = \left\{-15, -7.5, -\dfrac{7}{3}, 0, \sqrt{2}, 1.87\overline{35}, \sqrt{9}, \pi, 10^{17}\right\}$.

a. The natural numbers in S are $\sqrt{9}$ and 10^{17}. $\sqrt{9}$ is a natural number since $\sqrt{9} = 3$.
b. The whole numbers in S are 0, $\sqrt{9}$, and 10^{17}.
c. The integers in S are $-15, 0, \sqrt{9}$, and 10^{17}.
d. The rational numbers in S are $-15, -7.5, -\dfrac{7}{3}, 0, 1.87\overline{35}, \sqrt{9}$, and 10^{17}. The numbers -7.5 and $1.87\overline{35}$ qualify as rational numbers since $-7.5 = \dfrac{-15}{2}$ and $1.87\overline{35} = \dfrac{4637}{2475}$ (the bar over the last two digits indicates that these digits constitute a pattern that is repeated indefinitely). Any integer p automatically qualifies as a rational number since it can be written as $\dfrac{p}{1}$.
e. The only irrational numbers in S are $\sqrt{2}$ and π. Although well known now, the irrationality of $\sqrt{2}$ came as a bit of a surprise to the early Greek mathematicians who discovered this fact. The irrationality of π was not proven until 1767 (see p. 2).

The following figure shows the relationships among the subsets of $\mathbb{R}$ defined above and on the previous page. This figure indicates, for example, that every natural number is automatically a whole number, and also an integer, and also a rational number.

section 1.1   The Real Number System

Real Numbers ($\mathbb{R}$)

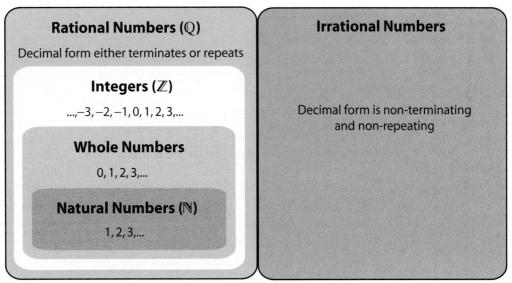

Figure 1: The Real Numbers

Topic 2:   The Real Number Line

It is often convenient and instructive to depict the set of real numbers as a horizontal line, with each point on the line representing a unique real number and each real number associated with a unique point on the line. The real number corresponding to a given point is called the **coordinate** of that point. Thus one (and only one) point on the line represents the number 0, and this point is called the **origin** of the real number line. Points to the right of the origin represent positive real numbers, while points to the left of the origin represent negative real numbers. Figure 2 is an illustration of the real number line with several points plotted. Note that two irrational numbers are plotted, though their locations on the line are necessarily approximations.

Figure 2: The Real Number Line

example 2

We choose which portion of the real number line to depict and the physical length that represents one unit based on the numbers that we wish to plot.

a. If we want to plot the numbers 101, 106, and 107, we might construct the diagram below.

continued on next page...

chapter one Number Systems and Fundamental Concepts of Algebra

b. If we want to plot the numbers $-\frac{3}{4}$, $-\frac{1}{2}$, and $\frac{1}{4}$, we might make the unit interval longer.

Topic 3: Order on the Real Number Line

The representation of the real numbers as a line leads naturally to the idea of *ordering* the real numbers. We say that the real number a is **less than** the real number b (in symbols, $a < b$) if a lies to the left of b on the real number line. This is equivalent to saying that b is **greater than** a (in symbols, $b > a$). The following definition gives the meaning of these and two other symbols indicating order.

Inequality Symbols (Order)

Symbol	Meaning
$a < b$ (read "a is less than b")	a lies to the left of b on the number line.
$a \leq b$ (read "a is less than or equal to b")	a lies to the left of b or is equal to b.
$b > a$ (read "b is greater than a")	b lies to the right of a on the number line.
$b \geq a$ (read "b is greater than or equal to a")	b lies to the right of a or is equal to a.

The two symbols $<$ and $>$ are called *strict* inequality signs, while the symbols $\leq$ and $\geq$ are *non-strict* inequality signs.

caution

Remember that order is defined by placement of real numbers on the number line, *not* by magnitude (its distance from zero). For instance, $-36 < 5$ because -36 lies to the left of 5 on the number line.

example 3

a. $5 \leq 9$, since 5 lies to the left of 9.
b. $5 \leq 5$, since 5 is equal to 5. Note that for every real number a, we have $a \leq a$ and $a \geq a$.
c. $-7 > -163$, since -7 lies to the right of -163.
d. The statement "5 is greater than -2" can be written $5 > -2$.
e. The statement "a is less than or equal to $b + c$" can be written $a \leq b + c$.
f. The statement "x is strictly less than y" can be written $x < y$.
g. The negation of the statement $a \leq b$ is the statement $a > b$.
h. If $a \leq b$ and $a \geq b$, then it must be the case that $a = b$.

Topic 4: Set-Builder Notation and Interval Notation

In describing the solutions to equations and inequalities, we will often need a precise way of defining sets of real numbers. **Set-builder notation** is a general method of describing the elements that belong to a given set. **Interval notation** is a way of describing certain subsets of the real line.

Set-Builder Notation

The notation $\{x \mid x \text{ has property } P\}$ is used to describe a set of real numbers, all of which have the property P. This can be read "the set of all real numbers x having property P."

example 4

a. $\{x \mid x \text{ is an even integer}\}$ is a more precise way of describing the set $\{\ldots, -4, -2, 0, 2, 4, \ldots\}$. We could also describe this set as $\{2n \mid n \text{ is an integer}\}$, since every even integer is a multiple of 2.
b. $\{x \mid x \text{ is an integer such that } -3 \leq x < 2\}$ describes the set $\{-3, -2, -1, 0, 1\}$.
c. $\{y \mid y > 1 \text{ and } y \leq -4\}$ describes the **empty set**, which is denoted $\varnothing$ or $\{\ \}$, as no real number satisfies the stated property.

Sets that consist of all real numbers bounded by two endpoints, possibly including those endpoints, are called **intervals**. Intervals can also consist of a portion of the real line extending indefinitely in either direction from just one endpoint.

Chapter one — Number Systems and Fundamental Concepts of Algebra

We could describe such sets with set-builder notation, but intervals occur frequently enough that special notation has been devised to define them succinctly.

Interval Notation

Interval Notation	Set-Builder Notation	Meaning
(a,b)	$\{x \mid a < x < b\}$	all real numbers strictly between a and b
$[a,b]$	$\{x \mid a \leq x \leq b\}$	all real numbers between a and b, including both a and b
$(a,b]$	$\{x \mid a < x \leq b\}$	all real numbers between a and b, including b but not a
$(-\infty, b)$	$\{x \mid x < b\}$	all real numbers less than b
$[a, \infty)$	$\{x \mid x \geq a\}$	all real numbers greater than or equal to a

In interval notation, the left endpoint is always written first. That is, as we read the notation from left to right, the left endpoint of the interval appears before the comma. Intervals of the form (a,b) are called **open** intervals, while those of the form $[a,b]$ are **closed** intervals. The interval $(a,b]$ is **half-open** (or **half-closed**). Of course, a half-open interval may be open at either endpoint, as long as it is closed at the other. The symbols $-\infty$ and ∞ indicate that the interval extends indefinitely in the left and the right directions, respectively. Note that $(-\infty, b)$ excludes the endpoint b, while $[a, \infty)$ includes the endpoint a.

caution

The symbols $-\infty$ and ∞ are just that: symbols! They are not real numbers, and so they cannot, for instance, be solutions to a given equation. The fact that they are symbols, and not numbers, also means that as endpoints of intervals they can never be included. For this reason, a parenthesis always appears next to either $-\infty$ or ∞; a bracket should never appear next to either infinity symbol.

example 5

a. The notation $[-3, 10)$ stands for the set $\{x \mid -3 \leq x < 10\}$. This interval is closed at the left endpoint, -3, and open at the right endpoint, 10.

b. The interval $(-\infty, \infty)$ is just another way of describing the entire set of real numbers.

Topic 5: Absolute Value and Distance on the Real Number Line

In addition to order, the depiction of the set of real numbers as a line leads to the notion of *distance*. Physically, distance is a well-understood concept: the distance between two objects is a non-negative number, dependent on a choice of measuring system, indicating how close the objects are to one another. The mathematical idea of **absolute value** gives us a means of defining distance in a mathematical setting.

Absolute Value

The **absolute value** of a real number a, denoted as $|a|$, is defined by:

$$|a| = \begin{cases} a & \text{if } a \geq 0 \\ -a & \text{if } a < 0 \end{cases}$$

The absolute value of a number is also referred to as its **magnitude**; it is the non-negative number corresponding to its distance from the origin.

Note that this definition implicitly gives us a system of measurement; 1 and −1 are the two real numbers which have a magnitude of 1, and so the distance between 0 and 1 (or between −1 and 0) is one unit. Note also that 0 is the only real number whose absolute value is 0.

Distance on the Real Number Line

Given two real numbers a and b, the distance between them is defined to be $|a-b|$. In particular, the distance between a and 0 is $|a-0|$ or just $|a|$.

Of course, distance should be symmetric. That is, the distance from a to b should be the same as the distance from b to a. Also, no mention was made in the above definition of which of the two numbers a and b is smaller, and our intuition suggests that this is immaterial as far as distance is concerned. These two concerns are really the same, and happily (not by chance) our mathematical definition of distance coincides with our intuition and is indeed symmetric. Given two distinct real numbers a and b, exactly one of the two differences $a-b$ and $b-a$ will be negative, and since $b-a = -(a-b)$ the definition of absolute value makes it clear that these two differences have the same magnitude. That is, $|a-b| = |b-a|$.

example 6

A few examples should make the previous points clear.

a. $|17-3|=|3-17|=14$. 17 and 3 are 14 units apart.

b. $|-\pi|=|\pi|=\pi$. Both $-\pi$ and π are π units from 0.

c. $\dfrac{|7|}{7}=\dfrac{7}{7}=1$.

d. $\dfrac{|-7|}{-7}=\dfrac{7}{-7}=-1$.

e. $-|-5|=-5$. Note that the negative sign outside the absolute value symbol is not affected by the absolute value. Compare this with the fact that $-(-5)=5$.

f. $|\sqrt{7}-2|=\sqrt{7}-2$. Even without a calculator, we know $\sqrt{7}$ is larger than 2 (since $2=\sqrt{4}$), so $\sqrt{7}-2$ is positive and hence $|\sqrt{7}-2|=\sqrt{7}-2$.

g. $|\sqrt{7}-19|=19-\sqrt{7}$. In contrast to the last example, we know $\sqrt{7}-19$ is negative, so its absolute value is $-(\sqrt{7}-19)=19-\sqrt{7}$.

The examples above illustrate some of the basic properties of absolute value. The list of properties below can all be derived from the definition of absolute value.

Properties of Absolute Value

For all real numbers a and b:

1. $|a| \geq 0$
2. $|-a|=|a|$
3. $a \leq |a|$
4. $|ab|=|a||b|$
5. $\left|\dfrac{a}{b}\right|=\dfrac{|a|}{|b|}$, $b \neq 0$
6. $|a+b| \leq |a|+|b|$ (This is called the **triangle inequality**, as it is a reflection of the fact that one side of a triangle is never longer than the sum of the other two sides.)

section 1.1 The Real Number System

example 7

a. $|(-3)(5)| = |-15| = 15 = |-3||5|$

b. $1 = |-3+4| \le |-3| + |4| = 7$

c. $7 = |-3-4| \le |-3| + |-4| = 7$

d. $\left|\dfrac{-3}{7}\right| = \dfrac{|-3|}{|7|} = \dfrac{3}{7}$

Topic 6: Interlude: Working with Repeating Digits

A rational number that appears as a decimal with a repeating pattern of digits can be written as a ratio of integers by following the procedure outlined below.

example 8

As a sum, the decimal number $1.8\overline{735}$ stands for $1 + \dfrac{87}{100} + 0.00\overline{35}$. The first two terms pose no problem, but we will have to work to write the third term as a ratio of integers.

Let $x = 0.00\overline{35}$. Then $100x = 0.\overline{35}$, so $100x = 0.35 + 0.00\overline{35}$ or $100x = 0.35 + x$. Thus $99x = 0.35$ or $99x = \dfrac{35}{100}$. Solving for x, we have $x = \dfrac{35}{9900}$. (We have just solved a linear equation; we will study these in detail in Section 2.1.) Altogether, we have $1.8\overline{735} = 1 + \dfrac{87}{100} + \dfrac{35}{9900}$ or, after adding the three terms, $x = \dfrac{18548}{9900}$ which is equal to $\dfrac{4637}{2475}$ when reduced.

exercises

1-57 odd, 60

Which elements of the following sets are (**a**) natural numbers, (**b**) whole numbers, (**c**) integers, (**d**) rational numbers, (**e**) irrational numbers, (**f**) real numbers, (**g**) undefined? See Example 1.

1. $\left\{19, -4.3, -\sqrt{3}, \dfrac{15}{0}, \dfrac{0}{15}, 2^5, -33\right\}$ **2.** $\left\{5\sqrt{7}, 4\pi, \sqrt{16}, 3.\overline{3}, -1, \dfrac{22}{7}, |-8|\right\}$

3. $\left\{5.41, |-16|, \dfrac{12}{3}, 0, \sqrt{4}, 2.\overline{145}, \dfrac{1}{4}\right\}$ **4.** $\left\{2\sqrt{25}, -4, 0.125, |32|, 2.1563, 6, \sqrt[3]{8}\right\}$

chapter one — Number Systems and Fundamental Concepts of Algebra

Plot the real numbers in the following sets on a number line. Choose the unit length appropriately for each set. See Example 2.

5. $\{-4.5, -1, 2.5\}$
6. $\{5.1, 5.2, 5.8\}$
7. $\{-24, 2, 15\}$
8. $\left\{0, \dfrac{1}{2}, \dfrac{5}{6}\right\}$

Select all of the symbols from the set $\{<, \leq, >, \geq\}$ that can be placed in the blank to make each statement true. See Example 3.

9. -3.4 ___ -3.5
10. -102 ___ 9
11. 3 ___ 3
12. -50 ___ -45
13. $\dfrac{-1}{4}$ ___ $\dfrac{-1}{3}$
14. 0.0087 ___ -42.9
15. $\dfrac{2}{16}$ ___ 0.125
16. -7 ___ -9
17. -8 ___ 2

Write each statement as an inequality, using the appropriate inequality symbol. See Example 3.

18. "$2a + b$ is strictly greater than c"
19. "2 is less than or equal to x"
20. "9 is greater than or equal to 7"
21. "7 is less than or equal to 9"
22. "$x + 5$ is strictly less than 3"
23. "$2c$ is no more than $3d$"
24. "9 is no less than 8"
25. "$6 + x$ is greater than or equal to $4x$"

Describe each of the following sets using set-builder notation. There may be more than one correct way to do this. See Example 4.

26. $\{-6, -3, 0, 3, 6, 9\}$
27. $\{5, 6, 7, \ldots, 105\}$
28. $\{2, 3, 5, 7, 11, 13, 17, \ldots\}$
29. $\{1, 2, 4, 8, 16, 32, \ldots\}$
30. $\left\{\ldots, \dfrac{1}{3}, \dfrac{1}{5}, \dfrac{1}{7}, \dfrac{1}{9}, \ldots\right\}$
31. $\{0, 1, 2, 3, 4, 5, \ldots\}$

Write each set as an interval using interval notation. See Example 5.

32. $x < 15$
33. $-9 \leq x \leq 6$
34. $2.5 < x \leq 3.7$
35. $\{x \mid -3 \leq x < 19\}$
36. $\{x \mid x < 4\}$
37. The positive real numbers
38. $\left\{x \mid -\dfrac{1}{2} < x < \dfrac{2}{5}\right\}$
39. $\{x \mid 1 \leq x \leq 2\}$
40. The non-negative real numbers

section 1.1 The Real Number System

Evaluate the absolute value expressions. See Examples 6 and 7.

41. $-|-11|$ **42.** $|3-7|$ **43.** $-|4-9|$

44. $|\sqrt{3}-\sqrt{5}|$ **45.** $\sqrt{|-4|}$ **46.** $-|-4-|-11||$

47. $|-\sqrt{2}|$ **48.** $\dfrac{|-x|}{|x|}\ (x \neq 0)$ **49.** $|(-7)(-5)|$

50. $-|\sqrt{16}-5|$ **51.** $|2-\sqrt{7}|$ **52.** $-|-\sqrt{|-9|}-|-9||$

Find the distance on the real number line between each pair of numbers given. See Example 6.

53. $a = 8, b = 3$ **54.** $a = 6, b = 14$ **55.** $a = 5, b = 5$

56. $a = 4, b = -2$ **57.** $a = -7, b = 7$ **58.** $a = -12, b = -1$

Write the following rational numbers as ratios of integers. See Example 8.

59. $2.\overline{3}$ **60.** $-5.0\overline{82}$ **61.** $0.\overline{41836}$

62. $0.\overline{9}$ **63.** $-1.\overline{01}$ **64.** $7.\overline{421}$

Write the following rational numbers in decimal form.

65. $\dfrac{-11}{3}$ **66.** $\dfrac{13}{12}$ **67.** $\dfrac{6}{7}$

Use your knowledge of the real numbers to answer the following questions.

68. Jess, Stan, Nina, and Michele are in a marathon. Twenty-five minutes after beginning, Jess has run 3.4 miles, Stan has run 4 miles, Nina has run 2.25 miles, and Michele has walked 1.6 miles. Using 0 as the beginning point, plot each competitor's location on a real number line using an appropriate interval.

69. Freddie, Sarah, Elizabeth, JR, and Aubrey are trying to line up by height for a photo shoot. JR is the tallest and Elizabeth is the shortest. Freddie is taller than Sarah, and Sarah is taller than Aubrey. Express their line-up using appropriate inequality symbols.

70. Sue boards an east-bound train in Center Station at the same time Joy boards a west-bound train in Center Station. After riding the Straight Line for 20 minutes, Sue's train has traveled 13 miles east, while Joy's train (also on the Straight Line) has traveled 7 miles west. Find the distance between the two trains at this time. (Assume the Straight Line is true to its name and that the tracks lie literally along a straight line.)

71. The admission prices at the local zoo are as follows:

Express the age range for each of these prices in set-builder notation and interval notation.

72. A particular fudge recipe calls for at least 3 but no more than 4 cups of sugar and at least $\frac{1}{2}$ but no more than $\frac{2}{3}$ of a cup of walnuts. Express the amount of sugar and nuts needed in both set-builder and interval notation.

73. Can a natural number be irrational? Explain.

74. Are all whole numbers also integers? Are all integers also whole numbers? Explain your answers.

75. In your own words, define absolute value.

76. Write a short paragraph explaining the similarities and differences between $>$ and $\geq$.

section 1.1 The Real Number System

technology exercises

A computer algebra system such as *Mathematica* can be used to check many of your answers. As an example, *Mathematica* can be used to verify the truth or falsehood of inequality statements. In the image below, the lines appearing in bold font have been typed and executed within *Mathematica*; the response (either True or False) appears in normal font. *Mathematica* has changed the appearance below; the relation ≤ is entered as <=. (Similarly, ≥ is entered as >=.)

If available, check your answers to the following problems with a computer algebra system. (See Appendix A for guidance on using computer algebra systems like Mathematica.)

77. −2.9 ___ −3.1   **78.** 2.1 ___ −5.5   **79.** 100 ___ −4

80. .001 ___ −99.8   **81.** $\dfrac{1}{3}$ ___ $\dfrac{1}{4}$ **82.** $\dfrac{-1}{5}$ ___ $\dfrac{-3}{4}$

chapter one Number Systems and Fundamental Concepts of Algebra

1.2 The Arithmetic of Algebraic Expressions

TOPICS

1. Components and terminology of algebraic expressions
2. The field properties and their use in algebra
3. The order of mathematical operations
4. Basic set operations and Venn diagrams

Topic 1: Components and Terminology of Algebraic Expressions

Algebraic expressions are made up of constants and variables, combined by the operations of addition, subtraction, multiplication, division, exponentiation and the taking of roots. The basic operations of addition, subtraction, multiplication and division are familiar to you, and exponents and roots will be discussed in the following two sections. **Constants** are fixed numbers, while **variables** are usually letters that represent unspecified numbers. To *evaluate* a given expression means to replace the variables (if there are any) with specific numbers and then to perform the indicated mathematical operations and simplify the result.

The **terms** of an algebraic expression are those parts joined by addition, while the **factors** of a term are the individual parts of the term that are joined by multiplication. In this context, as is commonly the case, addition also covers subtraction (as $a - b$ can be thought of as $a + (-b)$) and multiplication covers division (as $\frac{a}{b}$ can be thought of as $a \cdot \frac{1}{b}$). The **coefficient** of a term is the constant factor of the term, while the remaining part of the term constitutes the **variable factor**.

section 1.2   The Arithmetic of Algebraic Expressions

example 1

a. The algebraic expression $-17x^3(x^2+4y)+5\sqrt{x}-13(x-y)$ consists of the three terms $-17x^3(x^2+4y)$, $5\sqrt{x}$, and $-13(x-y)$. The factors of the term $-17x^3(x^2+4y)$ are -17, x^3, and (x^2+4y). The coefficient of $-17x^3(x^2+4y)$ is -17, and the variable part is $x^3(x^2+4y)$. Note that the factor (x^2+4y) itself consists of the two terms x^2 and $4y$, but these two terms are *not* terms of the original algebraic expression $-17x^3(x^2+4y)+5\sqrt{x}-13(x-y)$.

b. If we were asked to evaluate the expression $-3x^2-2(x+y)$ for $x=-2$ and $y=3$ we would replace x with -2 and y with 3 and then simplify $-3(-2)^2-2(-2+3)$ to obtain $(-3)(4)-2(1)=-12-2=-14$.

Topic 2: The Field Properties and Their Use in Algebra

The following properties of addition and multiplication on the set of real numbers are probably very familiar. You have likely used many of them in the past, though you may not have known the technical names of the properties. These properties and the few that follow them form the basis of the logical steps we use to solve equations and inequalities in algebra.

The set of real numbers forms what is known mathematically as a *field*, and consequently, the properties below are called *field properties*. These properties also apply to the set of complex numbers, which is a larger field containing the real numbers. Complex numbers will be discussed in Section 6 of this chapter.

Field Properties

In this table, a, b, and c represent arbitrary real numbers. The first five properties apply to addition and multiplication, while the last combines the two.

Name of Property	Additive Version	Multiplicative Version
Closure	$a+b$ is a real number	ab is a real number
Commutative	$a+b=b+a$	$ab=ba$
Associative	$a+(b+c)=(a+b)+c$	$a(bc)=(ab)c$
Identity	$a+0=0+a=a$	$a\cdot 1=1\cdot a=a$
Inverse	$a+(-a)=0$	$a\cdot\dfrac{1}{a}=1$ (for $a\neq 0$)
Distributive		$a(b+c)=ab+ac$

chapter one   Number Systems and Fundamental Concepts of Algebra

It is important to remember that variables represent numbers, so the previous properties apply to algebraic expressions as well as single variables.

example 2

a. $3(2+(-8)) = 3 \cdot (-6) = -18$ and $3 \cdot 2 + 3 \cdot (-8) = 6 - 24 = -18$. This demonstrates the distributive property: $3(2+(-8)) = 3 \cdot 2 + 3 \cdot (-8)$.

b. $3 - 4 = 3 + (-4) = -4 + 3 = -1$. While subtraction is not commutative, we can rewrite any difference as a sum (with the sign changed on the second term) and then apply the commutative property.

c. $(-x)\left(\dfrac{1}{y}\right) = \dfrac{-x}{y} = \left(\dfrac{1}{y}\right)(-x)$. Division can be restated as multiplication by the reciprocal of the denominator, and multiplication is commutative.

d. $(x^2+y)\left(\dfrac{1}{x^2+y}\right) = 1$, provided that $x^2+y \neq 0$. Any non-zero expression, when multiplied by its reciprocal, yields the multiplicative identity 1. The expression $\dfrac{1}{x^2+y}$ is the multiplicative inverse of x^2+y.

While the field properties are of fundamental importance to algebra, they imply further properties that are often of more immediate use.

Cancellation Properties

Throughout this table, *A*, *B*, and *C* represent algebraic expressions. The symbol ⇔ can be read as "if and only if" or "is equivalent to."

Property	Description
$A = B \Leftrightarrow A + C = B + C$	**Additive Cancellation** Adding the same quantity to both sides of an equation results in an equivalent equation.
For $C \neq 0$, $A = B \Leftrightarrow A \cdot C = B \cdot C$	**Multiplicative Cancellation** Multiplying both sides of an equation by the same non-zero quantity results in an equivalent equation.

section 1.2   **The Arithmetic of Algebraic Expressions**

Zero-Factor Property

Let A and B represent algebraic expressions. If the product of A and B is 0, then at least one of A and B is itself 0. Using the symbol $\Rightarrow$ for "implies," we write

$$AB = 0 \Rightarrow A = 0 \text{ or } B = 0.$$

example 3

a. The equation $x^5 + y = 3xy^3 + y$ is equivalent to the equation $x^5 = 3xy^3$. Using additive cancellation, we add the quantity $-y$ to both sides of the first equation which leads to the second equation. Note that this process can also be thought of as subtracting y from both sides.

b. The equation $-6\left(x - \frac{y}{2}\right)^2 = 18$ is equivalent to the equation $\left(x - \frac{y}{2}\right)^2 = -3$. Using multiplicative cancellation, we have multiplied both sides of the original equation by $-\frac{1}{6}$ to obtain the second. Alternatively, dividing both sides of the first equation by -6 leads to the second equation.

c. Multiplying both sides of the equation $x^2 - x = 2$ by 0 leads to the equation $0 = 0$ which, while trivial, is undeniably true. The problem is that these two equations are *not* equivalent. Convince yourself that replacement of x in the first equation by -1 or 2 leads to a true statement, while any other value for x leads to a false statement. But in the equation $0 = 0$ *any* value for x leads to a true statement, as there is no x in the equation to replace with a number. The equation $0 = 0$ is an example of an *identity*, a statement that is always true, while the equation $x^2 - x = 2$ is only true if x is replaced by -1 or 2.

d. The equation $(x - y)(x + y) = 0$ means that either $x - y = 0$ or $x + y = 0$. Remember that the only way for a product of two (or more) factors to be 0 is for at least one of the factors to be 0 itself. It is also important to remember that in mathematics, the word *or* is not exclusive. For instance, in this example it might be the case that both $x - y = 0$ and $x + y = 0$. If $x = 0$ and $y = 0$, this is indeed the case.

chapter one — Number Systems and Fundamental Concepts of Algebra

Topic 3: The Order of Mathematical Operations

Consider the two arithmetic expressions $4 - \dfrac{6}{2}$ and -3^2. Both expressions contain two operations: subtraction and division in the first one and multiplication and exponentiation in the second. Two reasonable people, lacking any indication of which operation is to be performed first, might very well proceed to simplify these expressions in two different ways and consequently arrive at different answers. It is important, therefore, that we decide beforehand the order in which the various mathematical operations are to be performed. The following list is the order which has evolved over time and which is assumed to be understood by both writers and readers of mathematics.

Order of Operations

1. If the expression is a fraction, simplify the numerator and denominator individually, according to the guidelines in the following steps.
2. Parentheses, braces and brackets are all used as grouping symbols. Simplify expressions within each set of grouping symbols, if any are present, working from the innermost outward.
3. Simplify all powers (exponents) and roots.
4. Perform all multiplications and divisions in the expression in the order they occur, working from left to right.
5. Perform all additions and subtractions in the expression in the order they occur, working from left to right.

example 4

a. Following this order of operations, the meaning of $4 - \dfrac{6}{2}$ is unambiguous. The division is to be performed first, followed by the subtraction, and the correct result is 1. Note that if 6 were subtracted from 4 first, and that result divided by 2, the (incorrect) answer would be -1.

b. The expression -3^2 when simplified, is equal to -9. Exponentiation will be discussed in the next section, but for the moment all we need to know is that $3^2 = 9$ and that this power is evaluated before the result is multiplied by -1. If we had incorrectly squared -3, we would have gotten the wrong answer of 9.

section 1.2   The Arithmetic of Algebraic Expressions

Make sure to enter parentheses around the numerator and denominator.

c. The expression $\dfrac{3-\left(3\sqrt{4}-2^2\right)\left(\dfrac{6}{-2}\right)}{3+2(-2)}$ is correctly simplified by following the steps below:

$$\dfrac{3-\left(3\sqrt{4}-2^2\right)\left(\dfrac{6}{-2}\right)}{3+2(-2)} = \dfrac{3-(3\cdot 2-4)(-3)}{3-4}$$

$$= \dfrac{3-(2)(-3)}{-1}$$

$$= \dfrac{3+6}{-1}$$

$$= -9$$

First, we simplify the numerator and denominator separately. We then simplify the expressions inside the two sets of parentheses in the numerator.

Remember to multiply 2 by –3 and subtract the result from 3.

Topic 4:   Basic Set Operations and Venn Diagrams

The sets that arise most frequently in algebra are sets of real numbers, and these sets are often the solutions of equations or inequalities. We will, on occasion, find it necessary to combine two or more such sets through the set operations of **union** and **intersection**. These operations are defined on sets in general, not just sets of real numbers, and are perhaps best illustrated by means of Venn diagrams.

A **Venn diagram** is a pictorial representation of a set or sets, and its aim is to indicate, through shading, the outcome of set operations such as union and intersection. In the definition below, these two operations are first defined with set-builder notation and then demonstrated with a Venn diagram. The symbol $\in$ is read "is an element of".

Union

In this definition, A and B denote two sets, and are represented in the Venn diagrams by circles. The operation of union is demonstrated in the diagram by means of shading.

The **union** of A and B, denoted $A \cup B$, is the set $\{x \mid x \in A \text{ or } x \in B\}$. That is, an element x is in $A \cup B$ if it is in the set A, the set B, or both. Note that the union of A and B contains both individual sets.

Intersection

In this definition, *A* and *B* denote two sets, and are represented in the Venn diagrams by circles. The operation of intersection is demonstrated in the diagram by means of shading.

The **intersection** of A and B, denoted $A \cap B$, is the set $\{x \mid x \in A \text{ and } x \in B\}$. That is, an element x is in $A \cap B$ if it is in both A and B. Note that the intersection of A and B is contained in each individual set.

In Chapter 2, we will apply these two set operations to intervals. Recall that intervals are sets of real numbers and that they are described with the interval notation defined in Section 1 of this chapter. The following example illustrates the sort of operations we will perform frequently in Chapter 2.

example 5

Simplify each of the following set expressions, if possible.

a. $(-2,4] \cup [0,9]$ **b.** $(-2,4] \cap [0,9]$ **c.** $[3,4) \cap (4,9)$ **d.** $(-\infty,4] \cup (-1,\infty)$

Solutions:

a. $(-2,4] \cup [0,9] = (-2,9]$ Since these two intervals overlap, their union is best described with a single interval.

b. $(-2,4] \cap [0,9] = [0,4]$ This intersection of two intervals can also be described with a single interval.

c. $[3,4) \cap (4,9) = \emptyset$ These two intervals have no elements in common, and their intersection is the empty set.

d. $(-\infty,4] \cup (-1,\infty) = (-\infty,\infty)$ The union of these two intervals constitutes the entire set of real numbers.

section 1.2 The Arithmetic of Algebraic Expressions

exercises

Identify the components of the algebraic expressions, as indicated. See Example 1.

1. Identify the terms in the expression $3x^2y^3 - 2\sqrt{x+y} + 7z$.
2. Identify the coefficients in the expression $3x^2y^3 - 2\sqrt{x+y} + 7z$.
3. Identify the factors in the term $-2\sqrt{x+y}$.
4. Identify the terms in the expression $x^2 + 8.5x - 14y^3$.
5. Identify the coefficients in the expression $x^2 + 8.5x - 14y^3$.
6. Identify the factors in the term $8.5x$.
7. Identify the terms in the expression $\dfrac{-5x}{2yz} - 8x^5y^3 + 6.9z$.
8. Identify the coefficients in the expression $\dfrac{-5x}{2yz} - 8x^5y^3 + 6.9z$.
9. Identify the factors in the term $\dfrac{-5x}{2yz}$.

Evaluate each expression for the given values of the variables. See Example 1.

10. $3x^2y^3 - 2\sqrt{x+y} + 7z$ for $x = -1, y = 2,$ and $z = -2$.
11. $-3\pi y + 8x + y^3$ for $x = 2$ and $y = -2$.
12. $\dfrac{|x|\sqrt{2}}{x^3 y^2} - \dfrac{3y}{x}$ for $x = -3$ and $y = 2$.
13. $y\sqrt{x^3 - 2} + \sqrt{x - 2y} - 3y$ for $x = 3$ and $y = -\dfrac{1}{2}$.
14. $\left| -x^2 + 2xy - y^2 \right|$ for $x = -3$ and $y = -5$.
15. $\dfrac{1}{32}x^2y^3 + y\sqrt{x} - 7y$ for $x = 4$ and $y = 2$.
16. $6x^2 + 3\pi y + y^2$ for $x = 3$ and $y = 2$.
17. $|x - 9y| - (8z - 8)$ for $x = -3, y = 1,$ and $z = 5$.
18. $\dfrac{x^2 y^3}{8z} - \dfrac{|2xy|}{8z}$ for $x = 2, y = -1,$ and $z = 3$.
19. $5\sqrt{x+6} - 8y^2$ for $x = 10$ and $y = -2$.

23

chapter one Number Systems and Fundamental Concepts of Algebra

Identify the property that justifies each of the following statements. See Example 2.

20. $(x-y)(z^2) = (z^2)(x-y)$
21. $3 - 7 = -7 + 3$
22. $(3x+2) + z = 3x + (2+z)$
23. $4(y-3) = 4y - 12$
24. $-3(4x^6 z) = (-3)(4)(x^6 z) = -12x^6 z$
25. $4 + (-3+x) = (4-3) + x = 1 + x$
26. $-2(4-x) = -8 + 2x$
27. $(x+y)\left(\dfrac{1}{x+y}\right) = 1$
28. $(-5+1)(7^7) = (7^7)(-5+1)$
29. $-5(-7x^8 y^4 z) = [(-5)(-7)](x^8 y^4 z)$

Identify the property that justifies each of the following statements. If one of the cancellation properties is being used to transform an equation, identify the quantity that is being added to both sides or the quantity by which both sides are being multiplied. See Example 3.

30. $25x^3 = 10y \Leftrightarrow 5x^3 = 2y$
31. $-14y = 7 \Leftrightarrow y = -\dfrac{1}{2}$
32. $14 - x = 2x \Leftrightarrow 14 = 3x$
33. $5 + 3x - y = 2x - y \Leftrightarrow 5 + x = 0$
34. $x^2 z = 0 \Rightarrow x^2 = 0 \text{ or } z = 0$
35. $(a+b)(x) = 0 \Rightarrow a+b = 0 \text{ or } x = 0$
36. $\dfrac{x}{6} + \dfrac{y}{3} - 2 = 0 \Leftrightarrow x + 2y - 12 = 0$
37. $(x-3)(x+2) = 0 \Rightarrow x - 3 = 0 \text{ or } x + 2 = 0$
38. $21x^4 = 15y^4 z \Leftrightarrow 7x^4 = 5y^4 z$
39. $6x + \dfrac{25}{4} y^9 - z = \dfrac{1}{4} y^9 - z \Leftrightarrow 6x + 6y^9 = 0$

Evaluate each of the following expressions. Be sure to use the correct order of operations. See Example 4.

40. $2 + 3 - 4 \div 8 + (-1)^2$
41. $\dfrac{-2(13 - \sqrt{9} + 2)}{14 - 4 \div 2}$
42. $-3^2 - 2 \div 2$
43. $(-3^2 - 2) \div 2$
44. $\dfrac{\sqrt{\sqrt{81} + 4^2}}{10(4 - 7 \div 2)}$
45. $4\pi + 6^{\sqrt{5 - \frac{2}{2}}} - 3\pi[8 - 15 \div (2+3)]$
46. $4 - 10 \cdot (-1) \div 5 + (-8)^2$
47. $-3^2 + 2 \cdot \sqrt{2 + 1 \cdot 2} - 7\pi$
48. $1 \div 6 + 3^{\sqrt{2^2}} - (-4 \cdot 2)$
49. $\dfrac{8 - 9 \cdot 5 - 7}{-4(-9 - 5 \div (2+4))}$
50. $-3 + 6 \cdot 1 \div 5 + (-3)^3$
51. $-5^2 + 4 \cdot \sqrt{2 + 7 \cdot 2} - 2\pi$
52. $9 \div 2 + 2^{\sqrt{2^4}} - (1 \cdot 2)$
53. $\dfrac{4 + 3 \cdot 8 - 6}{-5(3 - 8 \div (2+5))}$

section 1.2   The Arithmetic of Algebraic Expressions

Use a calculator to evaluate each of the following expressions. Be sure to use the correct order of operations. Round your answers to the nearest hundredth. See Example 4.

54. $(-3.28)^2 + 4 \cdot \sqrt{2 + 7 \cdot 3} - 2\pi$

55. $2.66 - 7 \cdot 4 \div 5 + (2 \div 3)^2$

56. $\dfrac{7.6 - 5.2 \cdot 9.8 - 8.1}{-3.22(11 - 6 \div (-1.45 + 6.32))}$

57. $7 \div 4.6 + 2.4^{\sqrt{3}} - (1.23 \cdot 2)^4$

Translate the following directions into an arithmetic expression.

58. Begin with 3. Add 7, and multiply the result by 3. Subtract 5. Take the square root, raise the result to the 3rd power, and then multiply by $-\dfrac{1}{5}$.

59. Begin with –6. Add 4, raise the result to the 3rd power, multiply by –2, and take the fourth root of the result.

60. Begin with x. Subtract 4, and take the third root of the result. Divide by 2, and square the result.

Simplify the following unions and intersections of intervals. See Example 5.

61. $(-5, 2] \cup (2, 4]$
62. $(-5, 2] \cap (2, 4]$
63. $[3, 5] \cap [2, 4]$
64. $(-\infty, 4] \cup (0, \infty)$
65. $\mathbb{Q} \cap \mathbb{Z}$
66. $\mathbb{N} \cup \mathbb{R}$
67. $(-\infty, \infty) \cap [-\pi, 21)$
68. $[2, \infty) \cap (-4, 7) \cap (-3, 2]$
69. $(3, 5] \cup [5, 9]$
70. $[-\pi, 2\pi) \cap [0, 4\pi]$
71. $\mathbb{N} \cup \mathbb{Z} \cap \mathbb{Q}$
72. $(-4.8, -3.5) \cap \mathbb{Z}$

Use your knowledge of the arithmetic of algebraic expressions to answer the following questions.

73. At the beginning of the month, your checking account contains $128. For your birthday, your mother deposits $50 and your grandmother deposits $25. After you write three checks for $17, $23, and $62, you make a deposit of $41. At the end of the month, your bank removes half of the balance to put in your savings account and then charges you a $5 fee for doing so. How much do you have remaining in your checking account?

74. A particular liquid boils at 268° F. Given the formula $C = \dfrac{5}{9}(F - 32)$ for converting temperatures from Celsius (C) to Fahrenheit (F), find the boiling point of this liquid in the Celsius scale. Round your answer to the nearest hundredth.

75. Stephen received $75 as a gift from his aunt. With this money, he decided to start saving to buy the newest gaming console, which costs $398 after tax. After working two weeks at his part-time job, he got one check for $123 and a second check for $98. How much more does Stephen need to save to buy his gaming console?

76. Body mass index, abbreviated BMI, is one way doctors determine an adult's weight status. A BMI below 18.5 is considered underweight, the range 18.5 – 24.9 is normal, the range 25.0 – 29.9 is overweight, and a BMI above 30.0 indicates obesity. The formula used to determine BMI is $BMI = 703 \left(\dfrac{weight\ in\ pounds}{(height\ in\ inches)^2} \right)$. Derek weighs 180 lbs and is 73 inches tall. Use this formula to determine Derek's BMI and weight status. Round your answer to the nearest tenth.

77. The Du Bois Method provides a formula used to estimate your body's surface area in meters squared: $BSA = 0.007184 (height)^{0.725} (weight)^{0.425}$, where *height* is in centimeters and *weight* is in kilograms. Assume Juan is 193 cm tall and weighs 88 kg. Use the Du Bois Method to estimate his body's surface area in square meters. Round your answer to the nearest hundredth.

78. Samantha drops a tennis ball from the top of the mathematics building. If it takes the ball 3.42 seconds to hit the ground, use the formula $distance = \dfrac{1}{2}(acceleration)(time)^2$ to find the height of the building, which is equivalent to the distance the ball falls. Use the value of 32 ft/sec² for the acceleration of a falling object. Round your answer to the nearest foot.

79. Choose a number. Multiply it by 3 and then add 4. Now multiply by 2 and subtract 8. Finally divide by 6. What do you notice about your final answer? Explain why you got this as a result.

80. Use your knowledge of the order of operations to check the following problem for accuracy. Explain any errors you find.

$$-8 \div 4 + 2^3 - (3 \cdot 2) = -8 \div 4 + 2^3 - (6)$$
$$= -8 \div 4 + 8 - 6$$
$$= -8 \div 4 + 2$$
$$= -8 \div 6$$
$$= \dfrac{-4}{3}$$

section 1.2 Arithmetic of Algebraic Expressions

81. A mnemonic is a device used to recall particular information. For example, "*My Very Educated Mother Just Served Us Nachos*" is often used to recall the order of the planets in our solar system: *My* = Mercury, *Very* = Venus, *Educated* = Earth, and so on. Come up with your own mnemonic for remembering the order of operations.

82. After taking a poll in her town, Sally began grouping the citizens into various sets. One set contained all the citizens with brown hair and another set contained all the citizens with blue eyes. What do you know about the citizens who would be listed in the union of these two sets? What do you know about the citizens who would be listed in the intersection of these two sets?

83. In your own words, describe the difference between a union and an intersection of two sets.

technology exercises

In *Mathematica*, for example, the expression $\sqrt{x^3 y^2} + 5z - 14$ can be evaluated for the values $x = 2$, $y = -1$, and $z = -5$ by typing and executing the lines shown in bold below; the output appears in normal font.

Use a calculator or computer software to assign the indicated values to the variables and evaluate the corresponding expressions. (See Appendix A for guidance on using computer algebra systems like Mathematica.)

```
Section 1-2 Technology Exercise.nb
In[1]:= x = 2;
       y = -1;
       z = -5;
       Sqrt[x^3 * y^2] + 5 z - 14
Out[4]= -39 + 2 √2
```

84. $\sqrt{x^4 y - z} + \dfrac{x - y^3}{z^2}$ for $x = -3$, $y = 2$, and $z = -2$.

85. $\dfrac{(x - pq^2)^3}{2q^3}$ for $x = -5$, $p = 2$, and $q = -3$.

86. $\dfrac{|x^2 - y^3| - 4x}{3y^5}$ for $x = 2$ and $y = 3$.

87. $\sqrt{p^3 q - q^3} - |p + q^2|$ for $p = -5$ and $q = 2$.

chapter one — Number Systems and Fundamental Concepts of Algebra

1.3 Properties of Exponents

TOPICS

1. Natural number exponents
2. Integer exponents
3. Properties of exponents and their use
4. Scientific notation
5. Interlude: working with geometric formulas

Topic 1: Natural Number Exponents

As we progress in Chapter 1, we will encounter a variety of algebraic expressions. As discussed in Section 1.2, algebraic expressions consist of constants and variables combined by the basic operations of addition, subtraction, multiplication, and division, along with exponentiation and the taking of roots. In this section, we will explore the meaning of exponentiation and the properties of exponents.

We will begin with the most basic type of exponent: an exponent consisting of a natural number.

Natural Number Exponents

If a is any real number and if n is any natural number, then $a^n = \underbrace{a \cdot a \cdot \ldots \cdot a}_{n \text{ factors}}$. That is, a^n is merely a shorter, and more precise, way of denoting the product of n factors of a. In the expression a^n, a is called the **base**, and n is the **exponent**. The process of multiplying n factors of a is called "raising a to the nth power," and the expression a^n may be referred to as "the nth power of a" or "a to the nth power." Note that $a^1 = a$.

section 1.3   Properties of Exponents

Other phrases are also commonly used to denote the raising of something to a power, especially when the power is 2 or 3. For instance, a^2 is often referred to as "a squared" and a^3 is often referred to as "a cubed." These phrases have their basis in geometry, as the area of a square region with side length a is a^2 and the volume of a three-dimensional cube, again with side length a, is a^3.

If the base and exponent are known constants, the expression a^n may be evaluated and written as a simple number. Even if an expression with one or more natural number exponents contains a variable, some simplification may be possible based only on the above definition.

example 1

a. $4^3 = 4 \cdot 4 \cdot 4 = 64$. Thus "four cubed is sixty-four."

b. $(-3)^2 = (-3)(-3) = 9$. Thus "negative three, squared, is nine."

c. $-3^4 = -(3 \cdot 3 \cdot 3 \cdot 3) = -81$. Note that, by our order of operations from the previous section, the exponent of 4 applies only to the number 3. After raising 3 to the 4th power, the result is multiplied by -1.

d. $-(-2)^3 \cdot 5^2 = -(-2 \cdot -2 \cdot -2)(5 \cdot 5) = -(-8)(25) = 200$.

e. $x^3 \cdot x^4 = (x \cdot x \cdot x)(x \cdot x \cdot x \cdot x) = x^7$. Even though x is a variable, preventing us from writing the expression as simply a number, we can use the definition of natural number exponents to write the product in a simpler way.

f. $\dfrac{7^6}{7^4} = \dfrac{7 \cdot 7 \cdot \cancel{7} \cdot \cancel{7} \cdot \cancel{7} \cdot \cancel{7}}{\cancel{7} \cdot \cancel{7} \cdot \cancel{7} \cdot \cancel{7}} = 7 \cdot 7 = 49$. We can cancel four factors of 7 from the numerator and denominator of the original fraction, leaving us with 7^2 or 49.

Example 1c illustrates an important point that we discussed in Section 1.2: if we have not decided beforehand on the order of carrying out such operations as exponentiation and multiplication, it is possible for two people to interpret the same expression in two different ways. However, if we use the rules for orders of operation, there is no ambiguity.

Examples 1e and *1f* are illustrations of two basic properties of exponents that we will shortly state more generally. For the moment, however, we will use similar examples to motivate our extension of the definition of exponents to include integer exponents.

chapter one Number Systems and Fundamental Concepts of Algebra

Topic 2: Integer Exponents

The ultimate goal is to give meaning to the expression a^n for any real number n and to do so in such a way that certain properties of exponents hold consistently. For example, analysis of *Example 1e* on the previous page leads to the observation that if n and m are natural numbers, then

$$a^n \cdot a^m = \underbrace{a \cdot a \cdot \ldots \cdot a}_{n \text{ factors}} \underbrace{a \cdot a \cdot \ldots \cdot a}_{m \text{ factors}} = \underbrace{a \cdot a \cdot \ldots \cdot a}_{n+m \text{ factors}} = a^{n+m}.$$

To extend the meaning of a^n to the case where $n = 0$, we might start by noting that we would like the following to hold:

$$a^0 \cdot a^m = a^{0+m} = a^m$$

In order to make this specific property hold in the case when one exponent is 0, we want $a^0 \cdot a^m$ to be equal to a^m. This suggests the following:

0 as an Exponent

> For any real number $a \neq 0$, we define $a^0 = 1$. 0^0 is undefined, just as division by 0 is undefined.

Note that this definition means $a^n \cdot a^m = a^{n+m}$ whenever n and m are whole numbers. With this small extension of the meaning of exponents as inspiration, let us continue. Consider the following table:

The exponent is decreased by one at each step.

$3^3 = 27$
$3^2 = 9$
$3^1 = 3$
$3^0 = 1$
$3^{-1} = ?$
$3^{-2} = ?$
$3^{-3} = ?$

The result is $\frac{1}{3}$ of the result from the previous line.

section 1.3   Properties of Exponents

In order to maintain the pattern that has begun to emerge, we are led to complete the table with $3^{-1} = \frac{1}{3}$, $3^{-2} = \frac{1}{9}$, and $3^{-3} = \frac{1}{27}$. In general, we define negative integer exponents as follows:

Negative Integer Exponents

> For any real number $a \neq 0$ and for any natural number n, $a^{-n} = \frac{1}{a^n}$. (We don't allow a to be 0 simply to avoid the possibility of division by 0.) Since any negative integer is the negative of a natural number, this defines exponentiation by negative integers.

We have now given meaning to a^n in the case where n is any integer, and we have done so by noting certain properties or patterns that we have hoped to maintain. Of course, there are many other properties and patterns that could have been used as inspiration equally well, as will be seen in some of the following examples. What we have not done yet is define a^n in the case where n is not an integer. This idea will be revisited in Section 1.4.

example 2

a. $\dfrac{y^2}{y^7} = \dfrac{\cancel{y} \cdot \cancel{y}}{y \cdot y \cdot y \cdot y \cdot y \cdot \cancel{y} \cdot \cancel{y}} = \dfrac{1}{y \cdot y \cdot y \cdot y \cdot y} = \dfrac{1}{y^5} = y^{-5}$.

b. $\dfrac{6x^2}{-3x^2} = \dfrac{6}{-3} = -2$. Note that the variable x cancels out entirely. Recall, as in *Example 1c*, $-3x^2$ means -3 times x^2, not the quantity $(-3x)$ squared.

c. $5^0 5^{-3} = 5^{0-3} = 5^{-3} = \dfrac{1}{125}$. Note that $5^0 = 1$, as does a^0 for any $a \neq 0$.

d. $\dfrac{1}{t^{-3}} = \dfrac{1}{\frac{1}{t^3}} = 1 \cdot \dfrac{t^3}{1} = t^3$.

e. $(x^2 y)^3 = (x^2 y)(x^2 y)(x^2 y) = x^2 \cdot x^2 \cdot x^2 \cdot y \cdot y \cdot y = x^6 y^3$.

chapter one   Number Systems and Fundamental Concepts of Algebra

Topic 3:   Properties of Exponents and Their Use

The table below lists the properties of exponents that are used frequently in algebra. Most of these properties have been illustrated already in *Examples 1* and *2*. All of them can be readily demonstrated by applying the definition of integer exponents.

Properties of Exponents

Throughout this table, *a* and *b* may be taken to represent constants, variables, or more complicated algebraic expressions. The letters *n* and *m* represent integers.

	Property	Example
1.	$a^n \cdot a^m = a^{n+m}$	$3^3 \cdot 3^{-1} = 3^{3+(-1)} = 3^2 = 9$
2.	$\dfrac{a^n}{a^m} = a^{n-m}$	$\dfrac{7^9}{7^{10}} = 7^{9-10} = 7^{-1}$
3.	$a^{-n} = \dfrac{1}{a^n}$	$5^{-2} = \dfrac{1}{5^2} = \dfrac{1}{25}$ and $x^3 = \dfrac{1}{x^{-3}}$
4.	$\left(a^n\right)^m = a^{nm}$	$\left(2^3\right)^2 = 2^{3 \cdot 2} = 2^6 = 64$
5.	$(ab)^n = a^n b^n$	$(7x)^3 = 7^3 x^3 = 343 x^3$ and $\left(-2x^5\right)^2 = (-2)^2 \left(x^5\right)^2 = 4x^{10}$
6.	$\left(\dfrac{a}{b}\right)^n = \dfrac{a^n}{b^n}$	$\left(\dfrac{3}{x}\right)^2 = \dfrac{3^2}{x^2} = \dfrac{9}{x^2}$ and $\left(\dfrac{1}{3z}\right)^2 = \dfrac{1^2}{(3z)^2} = \dfrac{1}{9z^2}$

Here it is assumed that every expression is defined. That is, if an exponent is 0, then the base is non-zero, and if an expression appears in the denominator of a fraction, then that expression is non-zero. Remember that $a^0 = 1$ for every $a \neq 0$.

section 1.3   Properties of Exponents

example 3

Simplify the following expressions by using the properties of exponents. Write the final answers with only positive exponents. (As in the table on the previous page, it is assumed that every expression is defined.)

a. $(17x^4 - 5x^2 + 2)^0$ **b.** $\dfrac{(x^2y^3)^{-1} z^{-2}}{x^3 z^{-3}}$ **c.** $\dfrac{(-2x^3 y^{-1})^{-3}}{(18x^{-3})^0 (xy)^{-2}}$

d. $(7xz^{-2})^2 (5x^2y)^{-1}$

Solutions:

a. $(17x^4 - 5x^2 + 2)^0 = 1$

Any non-zero expression with an exponent of 0 is 1.

b. $\dfrac{(x^2y^3)^{-1} z^{-2}}{x^3 z^{-3}} = \dfrac{x^{-2} y^{-3} z^{-2}}{x^3 z^{-3}}$

$= \dfrac{z^3}{x^3 x^2 y^3 z^2}$

$= \dfrac{z}{x^5 y^3}$

Note that we have used several properties in this example. We could have used the applicable properties in many different orders to achieve the same result. Also note that the final answer contains only positive exponents. If we had not been told to write the answer in this way, we could have written the result as $zx^{-5}y^{-3}$.

c. $\dfrac{(-2x^3 y^{-1})^{-3}}{(18x^{-3})^0 (xy)^{-2}} = \dfrac{(-2)^{-3} x^{-9} y^3}{x^{-2} y^{-2}}$

$= (-2)^{-3} x^{-9-(-2)} y^{3-(-2)}$

$= (-2)^{-3} x^{-7} y^5$

$= \dfrac{y^5}{-8x^7}$

$= -\dfrac{y^5}{8x^7}$

We have chosen to apply the appropriate properties in a slightly different order than in the previous example, just to illustrate an alternative way to go about the task of simplifying such an expression.

d. $(7xz^{-2})^2 (5x^2y)^{-1} = \dfrac{49x^2 z^{-4}}{5x^2 y}$

$= \dfrac{49}{5yz^4}$

Note that the variable x disappeared entirely from the expression. If we had simplified the expression in a slightly different order, we would have obtained a factor of x^{2-2} which is 1.

chapter one   Number Systems and Fundamental Concepts of Algebra

technology note

Computer algebra systems, as well as some calculators, are programmed to apply the properties of exponents and can demonstrate the result of simplifying expressions. In *Mathematica*, the command to do this is `Simplify`, illustrated with the first expression below. In many cases, *Mathematica* will simplify a given expression automatically upon evaluation, making the `Simplify` command unnecessary. This is illustrated with the second and third expressions below. *(See Appendix A for guidance on using computer algebra systems like Mathematica.)*

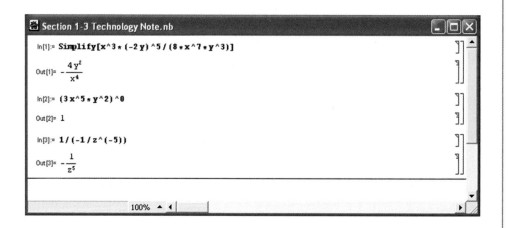

caution

Many errors are commonly made in applying the properties of exponents and are a result of forgetting the exact form of the properties. The first column in the table below contains examples of some common errors. The second column contains the corrected statements.

Incorrect Statements	Corrected Statements
$x^2 x^5 = x^{10}$	$x^2 x^5 = x^{2+5} = x^7$
$2^4 2^3 = 4^7$	$2^4 2^3 = 2^{4+3} = 2^7$
$(3+4)^2 = 3^2 + 4^2$	$(3+4)^2 = 7^2$
$\left(x^2 + 3y\right)^{-1} = \dfrac{1}{x^2} + \dfrac{1}{3y}$	$\left(x^2 + 3y\right)^{-1} = \dfrac{1}{x^2 + 3y}$
$(3x)^2 = 3x^2$	$(3x)^2 = 3^2 x^2 = 9x^2$
$\dfrac{x^5}{x^{-2}} = x^3$	$\dfrac{x^5}{x^{-2}} = x^{5-(-2)} = x^7$

section 1.3   Properties of Exponents

Topic 4:   Scientific Notation

Scientific notation is an important application of exponents. Very large and very small numbers arise naturally in a variety of situations, and working with them without scientific notation is an unwieldy and error-prone process. Scientific notation uses the properties of exponents to rewrite very large and very small numbers in a less clumsy form. It applies to those numbers which contain a long string of 0's before the decimal point (very large numbers) or to those which contain a long string of 0's between the decimal point and the non-zero digits (very small numbers).

Scientific Notation

A number is in scientific notation when it is written in the form
$$a \times 10^n$$
where $1 \le |a| < 10$ and n is an integer. If n is a positive integer, the number is large in magnitude, and if n is a negative integer, the number is small in magnitude (close to 0). The number a itself can be either positive or negative, and the sign of a determines the sign of the number as a whole.

caution

The sign of the exponent n in scientific notation does *not* determine the sign of the number as a whole. The sign of n only determines if the number is large (positive n) or small (negative n) in magnitude.

example 4

a. The distance from Earth to the sun is approximately 93,000,000 miles. Scientific notation takes advantage of the observation that multiplication of a number by 10 moves the decimal point one place to the right, and we can repeat this process as many times as necessary. Thus 9.3×10^7 is equal to 93,000,000, as 93,000,000 is obtained from 9.3 by moving the decimal point 7 places to the right:

$$9.3 \times 10^7 = 9\underbrace{3000000}_{\text{7 places}}$$

Note: In a calculator, scientific notation is often expressed with an E. To represent this on the TI-84 PLUS, press **2ND** and **,**.

93,000,000 miles

continued on next page...

chapter one   Number Systems and Fundamental Concepts of Algebra

b. The mass of an electron, in kilograms, is approximately
$$0.00000000000000000000000000000911,$$
clearly not a convenient number to work with. Scientific notation takes advantage of the observation that multiplication of a number by 10^{-1} (which is equivalent to division by 10) moves the decimal point one place to the left. Again, we can repeat this process as many times as necessary. Thus in scientific notation,
$$0.00000000000000000000000000000911 = 9.11 \times 10^{-31}.$$

electron

c. The speed of light in a vacuum is approximately 3×10^8 meters/second. In standard (non-scientific) notation, this number is written as 300,000,000.

We can also use the properties of exponents to simplify computations involving two or more numbers that are large or small in magnitude, as illustrated by the next set of examples.

example 5

Simplify the following expressions, writing your answer either in scientific or standard notation, as appropriate.

a. $\dfrac{(3.6 \times 10^{-12})(-6 \times 10^4)}{1.8 \times 10^{-6}}$

b. $\dfrac{(7 \times 10^{34})(3 \times 10^{-12})}{6 \times 10^{-7}}$

Solutions:

a. $\dfrac{(3.6 \times 10^{-12})(-6 \times 10^4)}{1.8 \times 10^{-6}} = \dfrac{(3.6)(-6)}{1.8} \times 10^{-12+4-(-6)}$
$= -12 \times 10^{-2}$
$= -0.12$

We have written the final answer in standard notation (as it is not inconvenient to do so). Note that in scientific notation, it would be written as -1.2×10^{-1}.

b. $\dfrac{(7 \times 10^{34})(3 \times 10^{-12})}{6 \times 10^{-7}} = \dfrac{(7)(3)}{6} \times 10^{34+(-12)-(-7)}$
$= 3.5 \times 10^{29}$

This answer is best written in scientific notation.

section 1.3   **Properties of Exponents**

Topic 5:   Interlude: Working with Geometric Formulas

Exponents occur in a very natural way when geometric formulas are considered. Some of the problems that you encounter in this book (and elsewhere) will require nothing more than using one of the basic geometric formulas found on the inside front cover of the text, but others will require a bit more work. Often, the exact geometric formula that you need to solve a given problem can be derived from simpler formulas.

We will look at several examples of how a new geometric formula is built up from known formulas. The general rule of thumb in each case is to break down the task at hand into smaller pieces that can be easily handled.

example 6

Find formulas for each of the following:
  a. the surface area of a box
  b. the surface area of a soup can
  c. the volume of a birdbath in the shape of half of a sphere
  d. the volume of a gold ingot whose shape is a right trapezoidal cylinder

Solutions:

a. A box whose six faces are all rectangular is characterized by its length l, its width w, and its height h. The area of a rectangle is one of the basic formulas that you should be familiar with, and can be found on the inside front cover of this text. The formula for the surface area of a box, then, is just the sum of the areas of the six sides. If we let S stand for the total surface area, we obtain the formula $S = lw + lw + lh + lh + hw + hw$ or $S = 2lw + 2lh + 2hw$.

b. A soup can, an example of a right circular cylinder, is characterized by its height h and the radius r of the circle that makes up the base (or the top).

continued on next page...

To determine the surface area of such a shape, imagine removing the top and bottom surfaces, cutting the soup can along the dotted line as shown, and flattening out the curved piece of metal making up the side. The flattened piece of metal would be a rectangle with height h and width $2\pi r$. Do you see why? The width of the rectangle would be the same as the circumference of the circular top and base, and the circumference of a circle is $2\pi r$. So the surface area of the curved side is $2\pi rh$. We also know that the area of a circle is πr^2, so if we let S stand for the surface area of the entire can, we have $S = \pi r^2 + \pi r^2 + 2\pi rh$, or $S = 2\pi r^2 + 2\pi rh$.

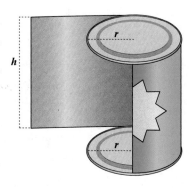

c. The volume of a sphere of radius r is $\frac{4}{3}\pi r^3$, and the birdbath of which we are to find the volume has the shape of half a sphere. So if we let V stand for the birdbath's volume,

$$V = \left(\frac{1}{2}\right)\left(\frac{4}{3}\pi r^3\right), \text{ or } V = \frac{2}{3}\pi r^3.$$

d. A *right cylinder* is the three-dimensional object generated by extending a plane region along an axis perpendicular to itself for a certain distance. (Such objects are often called prisms when the plane region is a polygon.) As indicated on the inside front cover, the volume of any right cylinder is the product of the area of the plane region and the distance that region has been extended perpendicular to itself. The gold ingot under consideration in this example is a right cylinder based on a trapezoid, as shown. The area of the trapezoid is $\frac{1}{2}(B+b)h$ and the ingot has length l, so its volume is $V = \frac{1}{2}(B+b)hl$. This could also be written as $V = \frac{(B+b)hl}{2}$.

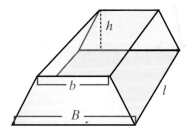

section 1.3   Properties of Exponents

exercises

Simplify each of the following expressions, writing your answer with only positive exponents. See Example 1.

1. $(-2)^4$
2. -2^4
3. -3^2
4. $(-3)^2$
5. $3^2 \cdot 3^2$
6. $2^3 \cdot 3^2$
7. $4 \cdot 4^2$
8. $(-3)^3$
9. $\dfrac{8^2}{4^3}$
10. $2^2 \cdot 2^3$
11. $\dfrac{7^4}{7^5}$
12. $n^2 \cdot n^5$
13. $\dfrac{x^5}{x^2}$
14. $\dfrac{y^3 \cdot y^8}{y^2}$
15. $\dfrac{3^7}{3^4 s^{-10}}$

Use the properties of exponents to simplify each of the following expressions, writing your answer with only positive exponents. See Examples 1, 2, and 3.

16. $\dfrac{3t^{-2}}{t^3}$
17. $-2y^0$
18. $\dfrac{1}{7x^{-5}}$
19. $9^0 x^3 y^0$
20. $\dfrac{2n^3}{n^{-5}}$
21. $\dfrac{11^{21}}{11^{19} x^{-7}}$
22. $\dfrac{x^7 y^{-3} z^{12}}{x^{-1} z^9}$
23. $\dfrac{x^4(-x^{-3})}{-y^0}$
24. $\dfrac{s^3}{s^{-2}}$
25. $\dfrac{x^{-1}}{x}$
26. $x^{(y^0)} \cdot x^9$
27. $\dfrac{x^2 y^{-2}}{x^{-1} y^{-5}}$
28. $\dfrac{s^5 y^{-5} z^{-11}}{s^8 y^{-7}}$
29. $\dfrac{2^7 s^{-3}}{2^3}$
30. $\dfrac{3^{-5}}{(3^{-4} x^5 y^4)^2}$
31. $\dfrac{-9^0 (x^2 y^{-2})^{-3}}{3x^{-4} y}$
32. $\left[(2x^{-1} z^3)^{-2}\right]^{-1}$
33. $\dfrac{(3yz^{-2})^0}{3y^2 z}$
34. $(12a^2 - 3b^4)^0$
35. $\dfrac{3^{-1}}{(3^2 xy^2)^{-2}}$
36. $\left[9m^2 - (2n^2)^3\right]^{-1}$
37. $\left[(12x^{-6} y^4 z^3)^5\right]^0$
38. $\dfrac{x(x^{-2} y^3)^3}{(2x^4)^{-2} y}$
39. $\dfrac{(-3a)^{-2} (bc^{-2})^{-3}}{a^5 c^4}$
40. $\left[(5m^4 n^{-2})^{-1}\right]^{-2}$
41. $(9x^{-1} z)^2 (2xy^{-3})^{-1}$
42. $(4^{-2} x^5 y^{-3} z^4)^{-2}$
43. $\left[(4a^2 b^{-5})^{-1}\right]^{-3}$
44. $\left[(2^{-3} m^{-6} n^3)^3\right]^{-1}$
45. $\left[(3^{-1} x^{-1} y)(x^2 y)^{-1}\right]^{-3}$
46. $\left[\dfrac{100^0 (x^{-1} y^3)^{-1}}{x^2 y}\right]^{-3}$
47. $(5z^6 - (3x^3)^4)^{-1}$
48. $\left[\dfrac{y^6 (xy^2)^{-3}}{3x^{-3} z}\right]^{-2}$

Convert each number from scientific notation to standard notation, or vice versa, as indicated. See Example 4.

49. -1.76×10^{-5}; convert to standard

50. $-912{,}000{,}000$; convert to scientific

51. 0.00000021; convert to scientific

52. 3.2×10^{7}; convert to standard

53. 5100; convert to scientific

54. -0.000187; convert to scientific

55. 3.1212×10^{2}; convert to standard

56. 1.934×10^{-4}; convert to standard

57. 0.00000002587; convert to scientific

58. -8.039×10^{6}; convert to standard

59. There are approximately 31,536,000 seconds in a calendar year. Express the number of seconds in scientific notation.

60. Together, the 46 human chromosomes are estimated to contain some 3.0×10^{9} base pairs of DNA. Express the number of pairs of DNA in standard notation.

61. A particular Italian sports car can be bought new for $675,000. Express this price in scientific notation.

62. A white blood cell is approximately 3.937×10^{-4} inches in diameter. Express this diameter in standard notation.

63. The probability of winning the lottery with one dollar is approximately 0.0000002605. Express this probability in scientific notation.

Evaluate each expression, using the properties of exponents. Use a calculator only to check your final answer. See Example 5.

64. $(2.3 \times 10^{13})(2 \times 10^{12})$

65. $\dfrac{(8 \times 10^{-3})(3 \times 10^{-2})}{2 \times 10^{5}}$

66. $(2 \times 10^{-13})(5.5 \times 10^{10})(-1 \times 10^{3})$

67. $\dfrac{(4 \times 10^{34})(3 \times 10^{-32})}{24}$

68. $(6 \times 10^{21})(5 \times 10^{-19})(5 \times 10^{4})$

69. $(3.2 \times 10^{7})(5 \times 10^{-4})(2 \times 10^{-10})$

70. $\dfrac{4 \times 10^{-6}}{(5 \times 10^{4})(8 \times 10^{-3})}$

71. $\dfrac{(4.6 \times 10^{12})(9 \times 10^{3})}{(1.5 \times 10^{8})(2.3 \times 10^{-5})}$

Apply the definition of integer exponents to demonstrate the following properties.

72. $a^{n} \cdot a^{m} = a^{n+m}$

73. $(a^{n})^{m} = a^{nm}$

74. $(ab)^{n} = a^{n}b^{n}$

section 1.3   Properties of Exponents

75. A farmer fences in three square garden plots that are situated along a road, as shown. Each square plot has a side-length of s, and he doesn't put fence along the road-side. Find an expression, in the variable s, for the amount of fencing used.

76. The prism shown below is a right triangular cylinder, where the triangular base is a right triangle. Find the volume of the prism in terms of b, h, and l.

77. Determine the volume of the right circular cylinder shown, in terms of r and h.

78. Matt wants to let people in the future know what life is like today, so he goes shopping for a time capsule. Capacity, along with price and quality, is an important consideration for him. One time capsule he looks at is a right circular cylinder with a hemisphere on each end. Find the volume of the time capsule, given that the height h is 16 inches and the radius r is 3 inches.

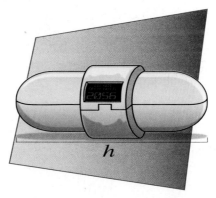

79. Bill and Dee are buying a new house. The house is a right cylinder based on a trapezoid atop a rectangular prism. The bases of the trapezoid are $B = 10$ m and $b = 8$ m, and the length of the house is $l = 15$ m. The height of the house up to the bottom of the roof is $H = 3$ m, and the height of the roof is $h = 1$ m. Find the volume of the house.

80. Construct the expression for the volume of water contained in a rectangular swimming pool of length l feet and width w feet, assuming the water has a uniform depth of 6 feet.

81. Construct the expression for the volume of water contained in an above-ground circular swimming pool that has a diameter of 18 feet, assuming the water has a uniform depth of d feet.

82. The floor of a rectangular bedroom measures N feet wide and M feet long. The height of the walls is 7 feet. Construct an expression for the number of square feet of wallpaper needed to cover all the walls. (Ignore the presence of doors and windows.)

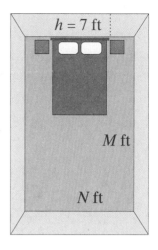

83. The interior surface of the birdbath in *Example 6c* needs to be painted with a waterproof (and nontoxic) coating. Construct the expression for the interior surface area.

84. Give a few examples of instances in which it would be more useful to use scientific notation rather than standard.

85. In February of 2006, the US national debt was approximately 8.2 trillion dollars. How is saying 8.2 trillion similar to scientific notation? How is it different?

86. In your own words, explain why $a^0 = 1$.

chapter one   Number Systems and Fundamental Concepts of Algebra

technology exercises

Simplify each of the following expressions, writing your answer with only positive exponents. Then use a calculator (if possible) or a computer algebra system to check your work. See the Technology Note after Example 3 for guidance.

87. $\left[\left(5xy^0z^{-2}\right)^{-1}\left(25x^4z\right)^2\right]^{-2}$

88. $\dfrac{\left(3pq^{-1}r^{-2}\right)^{-3}}{(pqr)^3}$

89. $\left(\left(2x^{-2}yz^3\right)^{-1}\right)^2$

90. $\left(\dfrac{3x^{-2}}{(2y)^{-3}}\right)^{-1}$

91. $\dfrac{3^{-5}}{\left(3^4z^{-3}x^{-4}\right)^{-2}}$

92. $\left(\left(3y^{-3}x^2\right)^5\right)^{-1}$

93. $\left(6y^3 - \left(3x^3\right)^5\right)^{-1}$

94. $\dfrac{y^{-16}\left(y^4x\right)^4}{4zx^4}$

1.4 Properties of Radicals

TOPICS

1. Roots and radical notation
2. Simplifying radical expressions
3. Combining radical expressions
4. Rational number exponents

Topic 1: Roots and Radical Notation

The nth root of an expression, where n is a natural number, is related to exponentiation by n in the sense of being the opposite operation. For example, the statement $4^3 = 64$ (which can be read as "four cubed is sixty-four") implies that the *cube root* of 64 is 4 (written $\sqrt[3]{64} = 4$). By analogy, the statement $2^4 = 16$ would lead us to write $\sqrt[4]{16} = 2$ (read "the *fourth root* of sixteen is two").

At first glance, then, the formal definition of nth roots would appear to be a simple matter, but a few subtle and important difficulties intervene. For instance, the statement $(-2)^4 = 16$ is also true, which might lead us to write $\sqrt[4]{16} = -2$. Also, since any real number a can be raised to any natural number power, we might be led to think that for any natural number n and any real number a, the nth root of a should be defined. But then how should we interpret $\sqrt[4]{-16}$? In order to evaluate $\sqrt[4]{-16}$ we seek a number whose fourth power is -16. Is there such a number?

The complete resolution of such issues will have to wait until Section 1.6, but we can begin now by defining $\sqrt[n]{a}$ in the cases where $\sqrt[n]{a}$ is a real number.

chapter one   Number Systems and Fundamental Concepts of Algebra

n^{th} Roots and Radical Notation

Case 1: n is an even natural number. If a is a non-negative real number and n is an even natural number, $\sqrt[n]{a}$ is the non-negative real number b with the property that $b^n = a$. That is, $\sqrt[n]{a} = b \Leftrightarrow a = b^n$. In this case, note that $\left(\sqrt[n]{a}\right)^n = a$ and $\sqrt[n]{a^n} = a$.

Case 2: n is an odd natural number. If a is any real number and n is an odd natural number, $\sqrt[n]{a}$ is the real number b (whose sign will be the same as the sign of a) with the property that $b^n = a$. Again, $\sqrt[n]{a} = b \Leftrightarrow a = b^n$, $\left(\sqrt[n]{a}\right)^n = a$, and $\sqrt[n]{a^n} = a$.

The expression $\sqrt[n]{a}$ expresses the nth root of a in **radical notation**. The natural number n is called the **index**, a is the **radicand**, and $\sqrt{}$ is called a **radical sign**. By convention, $\sqrt[2]{}$ is usually simply written as $\sqrt{}$.

The important distinction between the two cases is that when n is even, $\sqrt[n]{a}$ is defined only when a is non-negative, whereas if n is odd, $\sqrt[n]{a}$ is defined for all real numbers a. This asymmetrical situation will be remedied in Section 1.6, but to do so requires the introduction of *complex numbers*.

Another point to note is that we prevent any ambiguity in the meaning of $\sqrt[n]{a}$ when n is even and a is non-negative by *defining* $\sqrt[n]{a}$ to be the **non-negative** number whose n^{th} power is a. For example, $\sqrt{4}$ is equal to 2, *not* -2.

Finally, a word on nomenclature: just as in the case of powers, alternative phrases are commonly used when the index is 2 or 3. We usually read $\sqrt{a}$ as the "square root of a" and $\sqrt[3]{a}$ as the "cube root of a."

example 1

a. $\sqrt[5]{-32} = -2$ because $(-2)^5 = -32$.

b. $\sqrt[4]{-16}$ is not a real number, as no real number raised to the fourth power is -16.

c. $-\sqrt[4]{16} = -2$. Note that the fourth root of 16 is a real number, which in this expression is then multiplied by -1.

d. $\sqrt{0} = 0$. In fact, $\sqrt[n]{0} = 0$ for any natural number n.

e. $\sqrt[n]{1} = 1$ for any natural number n. $\sqrt[n]{-1} = -1$ for any odd natural number n.

f. $\sqrt[3]{-\dfrac{27}{64}} = -\dfrac{3}{4}$ because $\left(-\dfrac{3}{4}\right)^3 = -\dfrac{27}{64}$.

g. $\sqrt[5]{-\pi^5} = -\pi$ because $(-\pi)^5 = (-1)^5 \pi^5 = -\pi^5$.

h. $\sqrt[4]{(-3)^4} = \sqrt[4]{81} = 3$. In general, if n is an even natural number, $\sqrt[n]{a^n} = |a|$ for any real number a. Remember, though, that $\sqrt[n]{a^n} = a$ if n is an odd natural number.

Topic 2: Simplifying Radical Expressions

When solving equations that contain radical expressions, it is often helpful to simplify the expressions first. The word *simplify* is one which arises frequently in mathematics, and its meaning depends on the context.

Simplified Radical Form

A radical expression is in **simplified form** when:

1. The radicand contains no factor with an exponent greater than or equal to the index of the radical.
2. The radicand contains no fractions.
3. The denominator, if there is one, contains no radical.
4. The greatest common factor of the index and any exponent occurring in the radicand is 1. That is, the index and any exponent in the radicand have no common factor other than 1.

These four conditions deserve some explanation. The first condition is a reflection of the fact that roots and powers undo one another. Specifically,

$$\sqrt[n]{a^n} = \begin{cases} |a|, & \text{if } n \text{ is even} \\ a, & \text{if } n \text{ is odd} \end{cases}.$$

So if the radicand contains an exponent greater than or equal to the index, a factor can be brought out from under the radical. The second and third conditions both aim to remove radicals from the denominators of fractions.

chapter one   Number Systems and Fundamental Concepts of Algebra

The process of removing radicals from the denominator is called *rationalizing the denominator*, and is useful when solving equations that contain radicals. The fourth condition is again a reflection of the fact that roots and powers undo one another, and is perhaps best motivated by an example:

$$\sqrt[3]{a^6} = \sqrt[3]{a^2 a^2 a^2} = \sqrt[3]{\left(a^2\right)^3} = a^2$$

This simplification was possible because the exponent in the radicand and the index had a common factor of 3. This sort of simplification must be approached with caution, however, as will be seen in the following examples.

Before illustrating the process of simplifying radicals, a few useful properties of radicals will be listed. These properties can all be proved using nothing more than the definition of roots, but their validity will be more clear once we have discussed rational exponents later in this section.

Properties of Radicals

Throughout the following table, a and b may be taken to represent constants, variables, or more complicated algebraic expressions. The letters n and m represent natural numbers. It is assumed that all expressions are defined and are real numbers.

Property	Example
1. $\sqrt[n]{ab} = \sqrt[n]{a} \cdot \sqrt[n]{b}$	$\sqrt[3]{3x^6 y^2} = \sqrt[3]{3} \cdot \sqrt[3]{x^3} \cdot \sqrt[3]{x^3} \cdot \sqrt[3]{y^2} = \sqrt[3]{3} \cdot x \cdot x \cdot \sqrt[3]{y^2} = x^2 \sqrt[3]{3y^2}$
2. $\sqrt[n]{\dfrac{a}{b}} = \dfrac{\sqrt[n]{a}}{\sqrt[n]{b}}$	$\sqrt[4]{\dfrac{x^4}{16}} = \dfrac{\sqrt[4]{x^4}}{\sqrt[4]{16}} = \dfrac{|x|}{2}$
3. $\sqrt[m]{\sqrt[n]{a}} = \sqrt[mn]{a}$	$\sqrt[3]{\sqrt{64}} = \sqrt[3]{\sqrt[2]{64}} = \sqrt[6]{64} = 2$

section 1.4   Properties of Radicals

example 2

Simplify the following radical expressions:

a. $\sqrt[3]{-16x^8y^4}$ **b.** $\sqrt{8z^6}$ **c.** $\sqrt[3]{\dfrac{72x^2}{y^3}}$

Solutions:

a. $\sqrt[3]{-16x^8y^4} = \sqrt[3]{(-2)^3 \cdot 2 \cdot x^3 \cdot x^3 \cdot x^2 \cdot y^3 \cdot y}$

$\qquad\qquad\quad = -2x^2y\sqrt[3]{2x^2y}$

Note that since the index is 3, we look for all of the *perfect cubes* in the radicand.

b. $\sqrt{8z^6} = \sqrt{2^2 \cdot 2 \cdot (z^3)^2}$

$\qquad\quad = |2z^3|\sqrt{2}$

$\qquad\quad = 2|z^3|\sqrt{2}$

Remember that $\sqrt[n]{a^n} = |a|$ if n is even, so absolute value signs are necessary around the factor of z^3.

c. $\sqrt[3]{\dfrac{72x^2}{y^3}} = \dfrac{\sqrt[3]{8 \cdot 9 \cdot x^2}}{\sqrt[3]{y^3}}$

$\qquad\qquad = \dfrac{2\sqrt[3]{9x^2}}{y}$

All perfect cubes have been brought out from under the radical, and the denominator has been rationalized.

caution

Notice that the expressions are not equal.

As with the properties of exponents, many mistakes arise from forgetting the properties of radicals. One common error, for instance, is to rewrite $\sqrt{a+b}$ as $\sqrt{a} + \sqrt{b}$. These two expressions are not equal! To convince yourself of this, evaluate the two expressions with actual constants in place of a and b. For example, note that $5 = \sqrt{9+16} \neq \sqrt{9} + \sqrt{16} = 7$.

Rationalizing denominators sometimes requires more effort than in *Example 2c*, and in fact is sometimes impossible. The following methods will, however, take care of two common cases.

chapter one   Number Systems and Fundamental Concepts of Algebra

Case 1: Denominator is a single term containing a root.

If the denominator is a single term containing a factor of $\sqrt[n]{a^m}$, we will take advantage of the fact that $\sqrt[n]{a^m} \cdot \sqrt[n]{a^{n-m}} = \sqrt[n]{a^m \cdot a^{n-m}} = \sqrt[n]{a^n}$ and this last expression is either a or $|a|$, depending on whether n is odd or even. Of course, we can't multiply the denominator by a factor of $\sqrt[n]{a^{n-m}}$ without multiplying the numerator by the same factor, as we otherwise change the expression. The method in this case is thus to multiply the fraction by $\dfrac{\sqrt[n]{a^{n-m}}}{\sqrt[n]{a^{n-m}}}$.

Case 2: Denominator consists of two terms, one or both of which are square roots.

To discuss the method used in this case, let $A+B$ represent the denominator of the fraction under consideration, where at least one of A and B stands for a square root term. We will take advantage of the fact that $(A+B)(A-B) = A(A-B) + B(A-B) = A^2 - AB + BA - B^2 = A^2 - B^2$ and the exponents of 2 in the end result negate the square root (or roots) initially in the denominator. Just as in case 1, though, we can't multiply the denominator by $A-B$ unless we multiply the numerator by this same factor. The method is thus to multiply the fraction by $\dfrac{A-B}{A-B}$. The factor $A-B$ is called the **conjugate radical expression** of $A+B$.

example 3

Simplify the following radical expressions:

a. $\sqrt[5]{\dfrac{-4x^6}{8y^2}}$

b. $\dfrac{4}{\sqrt{7}+\sqrt{3}}$

c. $\dfrac{-\sqrt{5x}}{5-\sqrt{x}}$

Solutions:

a. $\sqrt[5]{\dfrac{-4x^6}{8y^2}} = \dfrac{\sqrt[5]{-4x \cdot x^5}}{\sqrt[5]{8y^2}}$

$= \dfrac{-x\sqrt[5]{4x}}{\sqrt[5]{2^3 y^2}} \cdot \dfrac{\sqrt[5]{2^2 y^3}}{\sqrt[5]{2^2 y^3}}$

$= \dfrac{-x\sqrt[5]{16xy^3}}{\sqrt[5]{2^5 y^5}}$

$= \dfrac{-x\sqrt[5]{16xy^3}}{2y}$

The first step is to simplify the numerator and denominator and determine what the denominator must be multiplied by in order to eliminate the radical. Remember to multiply the numerator by the same factor.

b. $\dfrac{4}{\sqrt{7}+\sqrt{3}} = \left(\dfrac{4}{\sqrt{7}+\sqrt{3}}\right)\left(\dfrac{\sqrt{7}-\sqrt{3}}{\sqrt{7}-\sqrt{3}}\right)$

$= \dfrac{4(\sqrt{7}-\sqrt{3})}{7-3}$

$= \dfrac{4(\sqrt{7}-\sqrt{3})}{4}$

$= \sqrt{7}-\sqrt{3}$

Again, we have multiplied the fraction by 1, but this time we have had to multiply the numerator and denominator by the *conjugate radical* of the original denominator.

c. $\dfrac{-\sqrt{5x}}{5-\sqrt{x}} = \left(\dfrac{-\sqrt{5x}}{5-\sqrt{x}}\right)\left(\dfrac{5+\sqrt{x}}{5+\sqrt{x}}\right)$

$= \dfrac{-5\sqrt{5x}-\sqrt{5x^2}}{25-x}$

$= \dfrac{-5\sqrt{5x}-x\sqrt{5}}{25-x}$

In this example, the original denominator is the sum of 5 and $-\sqrt{x}$, so the conjugate radical of the denominator is $5+\sqrt{x}$. Note that in simplifying the term $-\sqrt{5x^2}$, we can simply write $-x\sqrt{5}$ instead of $-|x|\sqrt{5}$, since the original expression is not real if x is negative.

As an aside, there are occasions when rationalizing the numerator is desirable. For instance, some problems in Calculus (which the author encourages all college students to take as a consciousness-raising experience!) are much easier to solve after rationalizing the numerator of a given fraction. This is accomplished by the same method, as seen in the following example.

example 4

Rationalize the numerator of the fraction $\dfrac{\sqrt{4x}-\sqrt{6y}}{2x-3y}$.

Solution:

$\dfrac{\sqrt{4x}-\sqrt{6y}}{2x-3y} = \left(\dfrac{\sqrt{4x}-\sqrt{6y}}{2x-3y}\right)\left(\dfrac{\sqrt{4x}+\sqrt{6y}}{\sqrt{4x}+\sqrt{6y}}\right)$

$= \dfrac{4x-6y}{(2x-3y)(\sqrt{4x}+\sqrt{6y})}$

$= \dfrac{2(2x-3y)}{(2x-3y)(2\sqrt{x}+\sqrt{6y})}$

$= \dfrac{2}{2\sqrt{x}+\sqrt{6y}}$

We begin by multiplying the numerator and denominator by the conjugate radical of the *numerator*. Note that we could have simplified the term $\sqrt{4x}$ first, but this would not have altered the work in any substantial way, and the final answer would be the same.

chapter one   Number Systems and Fundamental Concepts of Algebra

Topic 3:   **Combining Radical Expressions**

Often, a sum of two or more radical expressions can be combined into one. This can be done if the radical expressions are **like radicals**, meaning that they have the same index and the same radicand. It is frequently necessary to simplify the radical expressions before it can be determined if they are like or not.

example 5

Combine the radical expressions, if possible.

a. $-3\sqrt{8x^5} + \sqrt{18x}$

b. $\sqrt[3]{54x^3} + \sqrt{50x^2}$

c. $\sqrt{\dfrac{1}{12}} - \sqrt{\dfrac{25}{48}}$

Solutions:

a. $-3\sqrt{8x^5} + \sqrt{18x} = -3\sqrt{2^2 \cdot 2 \cdot x^4 \cdot x} + \sqrt{2 \cdot 3^2 \cdot x}$
$= -6|x^2|\sqrt{2x} + 3\sqrt{2x}$
$= -6x^2\sqrt{2x} + 3\sqrt{2x}$
$= (-6x^2 + 3)\sqrt{2x}$

We begin by simplifying the radicals. Note that upon simplification, the two radicals have the same index and the same radicand. Also note that the absolute value bars around the factor of x^2 are unnecessary, since x^2 is always non-negative.

b. $\sqrt[3]{54x^3} + \sqrt{50x^2} = \sqrt[3]{2 \cdot 3^3 \cdot x^3} + \sqrt{2 \cdot 5^2 \cdot x^2}$
$= 3x\sqrt[3]{2} + 5|x|\sqrt{2}$

Upon simplification, the radicands are the same, but the indices are not. We have written the radicals in simplest form, but they cannot be combined.

c. $\sqrt{\dfrac{1}{12}} - \sqrt{\dfrac{25}{48}} = \dfrac{1}{\sqrt{2^2 \cdot 3}} - \dfrac{\sqrt{5^2}}{\sqrt{4^2 \cdot 3}}$
$= \dfrac{1}{2\sqrt{3}} \cdot \dfrac{\sqrt{3}}{\sqrt{3}} - \dfrac{5}{4\sqrt{3}} \cdot \dfrac{\sqrt{3}}{\sqrt{3}}$
$= \dfrac{2\sqrt{3}}{4 \cdot 3} - \dfrac{5 \cdot \sqrt{3}}{4 \cdot 3}$
$= -\dfrac{3\sqrt{3}}{12} = -\dfrac{\sqrt{3}}{4}$

The first step, again, is to simplify the radicals. Note that both denominators are being rationalized in the second step. The first fraction must be multiplied by $\dfrac{2}{2}$ in order to get a common denominator. After subtracting, the fraction can be reduced.

section 1.4 Properties of Radicals

Topic 4: Rational Number Exponents

We can now return to the task of defining exponentiation, and give meaning to a^r when r is a rational number.

Rational Number Exponents

Meaning of $a^{\frac{1}{n}}$: If n is a natural number and if $\sqrt[n]{a}$ is a real number, then $a^{\frac{1}{n}} = \sqrt[n]{a}$.

Meaning of $a^{\frac{m}{n}}$: If m and n are natural numbers with $n \neq 0$, if m and n have no common factors greater than 1, and if $\sqrt[n]{a}$ is a real number, then $a^{\frac{m}{n}} = \sqrt[n]{a^m} = \left(\sqrt[n]{a}\right)^m$. Either $\sqrt[n]{a^m}$ or $\left(\sqrt[n]{a}\right)^m$ can be used to evaluate $a^{\frac{m}{n}}$, as they are equal. $a^{\frac{-m}{n}}$ is defined to be $\frac{1}{a^{\frac{m}{n}}}$.

In addition to giving meaning to rational exponentiation, the above definition provides us with the means of converting between *radical notation* and *exponential notation*. In some cases, one notation or the other is more convenient, and the choice of notation should be based on which makes the task at hand easier.

Although originally stated only for integer exponents, the properties of exponents listed in Section 1.3 also hold for rational exponents (and for real exponents as well, though we don't require that fact at the moment). And now that rational exponentiation has been defined, we are better able to demonstrate the properties of radicals mentioned earlier. For instance,

$$\sqrt[m]{\sqrt[n]{a}} = \left(a^{\frac{1}{n}}\right)^{\frac{1}{m}} = a^{\frac{1}{n} \cdot \frac{1}{m}} = a^{\frac{1}{mn}} = \sqrt[mn]{a}.$$

In fact, the desire to extend the properties of exponents from integers to rationals forces us to define rational exponents as we did above. For instance, since we want

$$\left(a^{\frac{1}{2}}\right)^2 = a^{\frac{1}{2}} a^{\frac{1}{2}} = a^1 = a,$$

it must be the case that $a^{\frac{1}{2}} = \sqrt{a}$.

We close this section with a variety of examples illustrating radical notation, exponential notation, and the properties of both.

chapter one   Number Systems and Fundamental Concepts of Algebra

example 6

Simplify each of the following expressions, writing your answer using the same notation as the original expression.

a. $27^{\frac{-2}{3}}$ **b.** $\sqrt[9]{-8x^6}$ **c.** $(5x^2+3)^{\frac{8}{3}}(5x^2+3)^{\frac{-2}{3}}$

d. $\sqrt[5]{\sqrt[3]{x^2}}$ **e.** $\dfrac{5x-y}{(5x-y)^{\frac{-1}{3}}}$

Solutions:

a. $27^{\frac{-2}{3}} = (27^{\frac{1}{3}})^{-2}$
$= 3^{-2}$
$= \dfrac{1}{3^2}$
$= \dfrac{1}{9}$

The only task here is to evaluate the expression. We could also have begun by noting that $27^{\frac{-2}{3}} = (27^{-2})^{\frac{1}{3}}$, but this would have made the calculations much more tedious.

b. $\sqrt[9]{-8x^6} = -\sqrt[9]{2^3 x^6}$
$= -2^{\frac{3}{9}} x^{\frac{6}{9}}$
$= -2^{\frac{1}{3}} x^{\frac{2}{3}}$
$= -\sqrt[3]{2x^2}$

To begin, the only factor that can be brought out from under the radical is –1, but we can still simplify the expression by reducing the exponents and index. Note that we switched to exponential form temporarily in doing this.

c. $(5x^2+3)^{\frac{8}{3}}(5x^2+3)^{\frac{-2}{3}} = (5x^2+3)^{\left(\frac{8}{3}\right)\left(\frac{2}{3}\right)}$
$= (5x^2+3)^{\frac{6}{3}}$
$= (5x^2+3)^2$

Since the bases are the same, we only have to apply the property $a^n a^m = a^{n+m}$.

d. $\sqrt[5]{\sqrt[3]{x^2}} = \sqrt[15]{x^2}$

Use the property stating $\sqrt[m]{\sqrt[n]{a}} = \sqrt[mn]{a}$ to simplify this expression.

e. $\dfrac{5x-y}{(5x-y)^{\frac{-1}{3}}} = (5x-y)^{1-\left(\frac{-1}{3}\right)}$
$= (5x-y)^{\frac{4}{3}}$

We use the property $\dfrac{a^n}{a^m} = a^{n-m}$ to simplify this expression.

section 1.4   Properties of Radicals

The following examples require a bit more work and/or caution.

example 7

a. Simplify the expression $\sqrt[4]{x^2}$.

b. Write $\sqrt[3]{2} \cdot \sqrt{3}$ as a single radical.

Solutions:

a. $\sqrt[4]{x^2} = (x^2)^{\frac{1}{4}}$
$= |x|^{\frac{1}{2}}$
$= \sqrt{|x|}$

We might be tempted to write $\sqrt[4]{x^2}$ as simply $\sqrt{x}$, but note that $\sqrt[4]{x^2}$ is defined for *all* real numbers x, while $\sqrt{x}$ is defined only for non-negative real numbers. We can rectify this disparity by first making sure the radicand is non-negative.

b. $\sqrt[3]{2} \cdot \sqrt{3} = 2^{\frac{1}{3}} \cdot 3^{\frac{1}{2}}$
$= 2^{\frac{2}{6}} \cdot 3^{\frac{3}{6}}$
$= (2^2)^{\frac{1}{6}} (3^3)^{\frac{1}{6}}$
$= 4^{\frac{1}{6}} \cdot 27^{\frac{1}{6}}$
$= 108^{\frac{1}{6}}$
$= \sqrt[6]{108}$

We can make use of the property $a^n b^n = (ab)^n$ if we can first make the exponents equal. We do so by finding the least common denominator of $\frac{1}{3}$ and $\frac{1}{2}$, writing both fractions with this common denominator, and then making use of the property $a^{nm} = (a^n)^m$.

example 8

A *regular hexagon* is a six-sided plane figure whose sides are all the same length and whose interior angles are all the same. We can use Heron's formula (see the inside front cover of the text) to derive a formula for the area of a regular hexagon with side-length d as follows.

Solution:

Begin by drawing three line segments joining each vertex to the vertex diagonally opposite it. The sum of the six angles meeting in the middle must be 360 degrees and all of the angles are equal, so each one must be a 60-degree angle. By symmetry, each of the six triangles making up the hexagon is isosceles, so each of the remaining two angles in any one triangle must measure 60 degrees as well. Thus each triangle is equilateral (equal sides and equal angles).

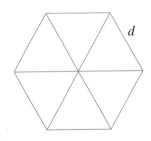

continued on next page...

chapter one Number Systems and Fundamental Concepts of Algebra

Heron's formula tells us that the area of an equilateral triangle of side-length d is $\sqrt{s(s-d)^3}$ where $s = \dfrac{3d}{2}$. Expressing this in terms of d alone, each triangle has area $\sqrt{\left(\dfrac{3d}{2}\right)\left(\dfrac{d}{2}\right)^3}$ or $\sqrt{\dfrac{3d^4}{16}}$. Simplifying this radical, we obtain $\dfrac{d^2\sqrt{3}}{4}$. Since the hexagon is made up of six of these triangles, the total area A of the hexagon is $A = \dfrac{3d^2\sqrt{3}}{2}$.

exercises

Evaluate the following radical expressions. See Example 1.

1. $-\sqrt{9}$
2. $\sqrt[3]{-27}$
3. $\sqrt{-25}$
4. $\sqrt[6]{-64}$
5. $-\sqrt[6]{64}$

6. $-\sqrt{169}$
7. $\sqrt[3]{-125}$
8. $\sqrt{-49}$
9. $\sqrt[4]{-256}$
10. $-\sqrt[3]{-64}$

11. $\sqrt[3]{-\dfrac{27}{125}}$
12. $\sqrt{\dfrac{25}{121}}$
13. $\sqrt[3]{\dfrac{-8}{64}}$
14. $\sqrt{\dfrac{1}{4}}$
15. $-\sqrt[3]{-8}$

16. $\sqrt[4]{\sqrt{16} - \sqrt[3]{-27} + \sqrt{81}}$
17. $\sqrt{\dfrac{\sqrt[3]{-64}}{-\sqrt{144} - \sqrt{169}}}$

18. $\sqrt{\sqrt[3]{64} + \sqrt[4]{81} + \sqrt[5]{32}}$

Simplify the following radical expressions. See Example 2.

19. $\sqrt{9x^2}$
20. $\sqrt[3]{-8x^6y^9}$
21. $\sqrt[4]{\dfrac{x^8z^4}{16}}$

22. $\sqrt{2x^6y}$
23. $\sqrt[7]{x^{14}y^{49}z^{21}}$
24. $\sqrt{\dfrac{x^2}{4x^4y^6}}$

25. $\sqrt[3]{\dfrac{a^3b^{12}}{27c^6}}$
26. $\sqrt[3]{-125x^{12}y^9}$
27. $\sqrt[4]{\dfrac{x^{12}y^8}{16}}$

28. $\sqrt[3]{81m^4n^7}$
29. $\sqrt[5]{\dfrac{y^{30}z^{25}}{32x^{35}}}$
30. $\sqrt[5]{32x^7y^{10}}$

section 1.4 Properties of Radicals

Simplify the following radicals by rationalizing the denominators. See Example 3.

31. $\sqrt[3]{\dfrac{4x^2}{3y^4}}$

32. $\dfrac{-\sqrt{3a^3}}{\sqrt{6a}}$

33. $\dfrac{3}{\sqrt{2}-\sqrt{5}}$

34. $\dfrac{10}{\sqrt{7}-\sqrt{2}}$

35. $\dfrac{3}{\sqrt{6}-\sqrt{3}}$

36. $\dfrac{5}{6-\sqrt{5}}$

37. $\dfrac{\sqrt{x}}{\sqrt{x}-\sqrt{2}}$

38. $\dfrac{x-y}{\sqrt{x}+\sqrt{y}}$

39. $\dfrac{\sqrt{x}+\sqrt{y}}{\sqrt{x}-\sqrt{y}}$

40. $\dfrac{1}{2-\sqrt{x}}$

41. $\dfrac{\sqrt{y}}{\sqrt{y}+2}$

42. $\dfrac{-\sqrt{6y^7}}{\sqrt{5y}}$

Rationalize the numerator of the following expressions (that is, rewrite each expression as an equivalent expression that has no radicals in the numerator). See Example 4.

43. $\dfrac{\sqrt{5}-3}{-4}$

44. $\dfrac{\sqrt{7}-6}{7}$

45. $\dfrac{3+\sqrt{y}}{6}$

46. $\dfrac{\sqrt{x}+\sqrt{y}}{\sqrt{x}}$

47. $\dfrac{\sqrt{13}+\sqrt{t}}{13-t}$

48. $\dfrac{2\sqrt{x}+\sqrt{y}}{\sqrt{x}-\sqrt{y}}$

49. $\dfrac{\sqrt{6}+\sqrt{y}}{\sqrt{6}-\sqrt{y}}$

50. $\dfrac{4\sqrt{xy}+y}{x-y}$

Combine the radical expressions, if possible. See Example 5.

51. $\sqrt[3]{-16x^4}+5x\sqrt[3]{2x}$

52. $\sqrt{27xy^2}-4\sqrt{3xy^2}$

53. $\sqrt{7x}-\sqrt[3]{7x}$

54. $|x|\sqrt{8xy^2z^3}-|yz|\sqrt{18x^3z}$

55. $-x^2\sqrt[3]{54x}+3\sqrt[3]{2x^7}$

56. $\sqrt[5]{32x^{13}}+3x\sqrt[5]{x^8}$

57. $\sqrt[3]{-16z^4}+6z\sqrt[3]{2z}$

58. $\sqrt[3]{7y}-\sqrt[4]{7y}$

59. $-x^2\sqrt[3]{16x}+2\sqrt[3]{2x^7}$

chapter one — Number Systems and Fundamental Concepts of Algebra

Simplify the following expressions, writing your answer using the same notation as the original expression. See Example 6.

60. $\sqrt[3]{\sqrt[4]{x^{36}}}$

61. $(3x^2-4)^{\frac{1}{3}}(3x^2-4)^{\frac{5}{3}}$

62. $32^{-\frac{3}{5}}$

63. $81^{\frac{3}{4}}$

64. $\dfrac{(x-z)^y}{(x-z)^4}$

65. $\sqrt[7]{n^9}\sqrt[7]{n^5}$

66. $(-8)^{\frac{2}{3}}$

67. $\dfrac{x^{\frac{1}{5}}y^{-\frac{2}{3}}}{x^{-\frac{3}{5}}y}$

68. $(1024)^{-\frac{2}{5}}$

69. $(625)^{-\frac{3}{4}}$

70. $\sqrt[8]{49a^2}$

71. $\sqrt[3]{\sqrt[5]{y^{25}}}$

72. $\dfrac{(a-b)^{-\frac{2}{3}}}{(a-b)^{-2}}$

73. $\dfrac{\sqrt[3]{a^2}}{\sqrt[3]{a^5}}$

74. $(ax^2+by)^{\frac{3}{4}}(by+ax^2)^{-\frac{2}{3}}$

Convert the following expressions from radical notation to exponential notation, or vice-versa. Simplify each expression in the process, if possible.

75. $\sqrt[4]{a^3}\cdot\sqrt[3]{a^9}$

76. $(256)^{-\frac{3}{4}}$

77. $\sqrt[12]{x^3}$

78. $(9y^2)^{\frac{3}{2}}(y^6)^{\frac{5}{3}}$

79. $\sqrt[6]{\dfrac{2}{72}}$

80. $(36n^4)^{\frac{5}{6}}$

Simplify the following expressions. See Example 7.

81. $\sqrt{5}\sqrt[4]{5}$

82. $\sqrt[4]{25}$

83. $\sqrt[16]{y^4}$

84. $\sqrt[4]{36}$

85. $\sqrt[3]{x^7}\sqrt[9]{x^6}$

86. $\sqrt[5]{y^{16}}\sqrt[25]{y^{20}}$

87. $\sqrt[4]{7}\sqrt[16]{7}$

88. $\sqrt{y^4}\sqrt[6]{y^3}$

Apply the definition of rational exponents to demonstrate the following properties.

89. $\sqrt[n]{ab}=\sqrt[n]{a}\cdot\sqrt[n]{b}$

90. $\sqrt[n]{\dfrac{a}{b}}=\dfrac{\sqrt[n]{a}}{\sqrt[n]{b}}$

91. $\sqrt[m]{\sqrt[n]{a}}=\sqrt[mn]{a}$

92. The prism shown below is a triangular right cylinder, where the triangular base is a right triangle. Find the surface area of the prism in terms of $b, h,$ and l.

93. A jeweler decides to construct a pendant for a necklace by simply attaching equilateral triangles to each edge of a regular hexagon. The edge length of one of the points of the resulting star is $d = 0.8$ cm. Find the formula for the area of the star in terms of d and then evaluate for $d = 0.8$ cm (rounding to the nearest thousandth). Remember that the area of an equilateral triangle of side length d is $A = \dfrac{d^2\sqrt{3}}{4}$. See *Example 8*.

$d = 0.8$ cm

94. The pyramids in Egypt each consist of a square base and four triangular sides. For a class project, Karim constructs a model pyramid with equilateral triangles as sides. The side length is $s = 43$ cm. Find the total surface area of the pyramid (rounding to the nearest square centimeter). See *Example 8*.

$s = 43$ cm

95. Ilyana has made a home for her pet guinea pig (Ralph) in the shape of a right triangular cylinder. Before she can put the new home in Ralph's cage, she must paint it with a non-toxic outer coat. If the front of the home has a base of 17.5 cm and a height of 15 cm and the length of the home is 25 cm, what is the surface area of Ralph's home, rounded to the nearest square centimeter? The small bottle of non-toxic coating will cover up to 1500 cm^2. Will the small bottle contain enough non-toxic coating to cover Ralph's home?

96. Conic Candles plans to design a new candle. The designers have determined that the diameter across the top should be 20 cm and the height of the candle should be 15 cm. Find the volume of the cone shaped candle (see the inside front cover of the text). Round to the nearest whole cubic centimeter.

97. Einstein's Theory of Special Relativity tells us that $E = mc^2$, where E is energy (in joules, J), m is mass (in kilograms, kg), and c is the speed of light (in meters per second, m/s). This equation may also be written as $\sqrt{\dfrac{E}{m}} = c$. Assume you know $E = 418{,}400$ J and $m = 4.655 \times 10^{-12}$ kg. Use this information to estimate the speed of light. Round your answer, as necessary.

98. Explain, in your own words, why the square root of a negative number is not a real number.

99. Explain, in your own words, why exponents and roots are evaluated at the same time in the order of operations.

1.5 Polynomials and Factoring

TOPICS

1. The terminology of polynomial expressions
2. The algebra of polynomials
3. Common factoring methods:
   - Greatest Common Factor
   - Factoring by Grouping
   - Factoring Special Binomials
   - Factoring Trinomials
   - Factoring Expressions Containing Fractional Exponents

Topic 1: The Terminology of Polynomial Expressions

Polynomials are algebraic expressions that meet further criteria. These criteria are that each term in a polynomial consists only of a number multiplied by variables (if it is multiplied by anything at all) raised to positive integer exponents. The number in any such term is called the **coefficient** of the term, and the sum of the exponents of the variables is the **degree of the term**. If a given term consists only of a non-zero number a (known as a **constant term**), its degree is defined to be 0, while the one specific term 0 is not assigned any degree at all. The polynomial as a whole is also given a degree: the **degree of a polynomial** is the largest of the degrees of the individual terms.

Every polynomial consists of some finite number of terms. Polynomials consisting of a single term are called **monomials**, those consisting of two terms are **binomials**, and those consisting of three terms are **trinomials**. The following example illustrates the use of the above terminology as applied to some specific polynomials.

chapter one   Number Systems and Fundamental Concepts of Algebra

example 1

a. The polynomial $-3x^4y^2 + 5.4x^3y^4$ is a binomial in the two variables x and y. The degree of the first term is 6, and the degree of the second term is 7, so the degree of the polynomial as a whole is 7. The coefficient of the first term is -3 and the coefficient of the second term is 5.4.

b. The single number 5 can be considered a polynomial. In particular, it is a monomial of degree 0. The rationale for assigning degree 0 to non-zero constants such as 5 is that 5 could be thought of as $5x^0$ (or 5 times *any* variable raised to the 0 power). The coefficient of this monomial is itself: 5.

c. The polynomial $\frac{2}{3}x^3y^5 - z + y^{10} + 3$ has four terms and is a polynomial in three varibles. If one of the terms of a polynomial consists only of a number, it is referred to as the constant term. The degree of this polynomial is 10, and the degrees of the individual terms are, from left to right, 8, 1, 10 and 0. The coefficients are $\frac{2}{3}$, -1, 1, and 3.

The majority of the polynomials that occur in this book will be polynomials of a single variable and can be described generically as follows.

Polynomials of a Single Variable

A polynomial in the variable x of degree n can be written in the form

$$a_n x^n + a_{n-1} x^{n-1} + \ldots + a_1 x + a_0,$$

where $a_n, a_{n-1}, \ldots, a_1, a_0$ are numbers, $a_n \neq 0$ and n is a positive integer. This form is called **descending order**, because the powers descend from left to right. The **leading coefficient** of this polynomial is a_n.

Topic 2: The Algebra of Polynomials

The variables in polynomials, as in all algebraic expressions, represent numbers (albeit possibly unknown), and consequently polynomials can be added, subtracted, multiplied and divided according to the field properties discussed in Section 1.2.

section 1.5 Polynomials and Factoring

Addition (and thus subtraction) and multiplication of polynomials will be discussed here, while division of polynomials will be taken up in Chapters 2 and 5.

To add two or more polynomials, we use the field properties to combine **like**, or **similar**, terms. These are the terms among all the polynomials being added that have the same variables raised to the same powers. Subtraction of a polynomial from another is accomplished by distributing the minus sign over the terms of the polynomial being subtracted and then adding.

example 2

Add or subtract the polynomials as indicated.

a. $(2x^3y - 3y + z^2x) + (3y + z^2 - 3xz^2 + 4)$ **b.** $(4ab^3 - b^3c) - (4 - b^3c)$

Solutions:

a.
$(2x^3y - 3y + z^2x) + (3y + z^2 - 3xz^2 + 4)$
$= 2x^3y + (-3y + 3y) + (xz^2 - 3xz^2) + z^2 + 4$
$= 2x^3y + y(-3 + 3) + xz^2(1 - 3) + z^2 + 4$
$= 2x^3y - 2xz^2 + z^2 + 4$

The first step is to identify like terms and group these together. Note that the terms z^2x and $-3xz^2$ are similar, as multiplication is commutative. The like terms are then combined by using the distributive property.

b.
$(4ab^3 - b^3c) - (4 - b^3c)$
$= 4ab^3 - b^3c - 4 + b^3c$
$= 4ab^3 - 4 + (-b^3c + b^3c)$
$= 4ab^3 - 4$

Again, like terms are identified and grouped together after distributing the minus sign over all the terms in the second polynomial.

Polynomials are multiplied using the same properties. The distributive property is of particular importance, and its use is the key to multiplying polynomials correctly. In general, the product of two polynomials, one with n terms and one with m terms, will consist initially of nm terms. If any of the resulting terms are similar, they are then combined in the last step.

example 3

Multiply the polynomials, as indicated.

a. $(2x^2y - z^3)(4 + 3z - 3xy)$ **b.** $(3ab - a^2)(ab + a^2)$

Solutions:

a.
$$(2x^2y - z^3)(4 + 3z - 3xy)$$
$$= 2x^2y(4 + 3z - 3xy) - z^3(4 + 3z - 3xy)$$
$$= 8x^2y + 6x^2yz - 6x^3y^2 - 4z^3 - 3z^4 + 3xyz^3$$

We use the distributive property first, and multiply each term of the first polynomial by each term of the second. None of the resulting terms are similar, so the final answer is a polynomial of 6 terms.

b.
$$(3ab - a^2)(ab + a^2)$$
$$= 3ab(ab + a^2) - a^2(ab + a^2)$$
$$= 3a^2b^2 + 3a^3b - a^3b - a^4$$
$$= 3a^2b^2 + 2a^3b - a^4$$

Again, we begin by distributing. In this case, however, the four terms that result from the initial multiplication contain two similar terms. We combine these to obtain the final trinomial.

When a binomial is multiplied by a binomial, as in *Example 3b* above, the acronym FOIL is commonly used as a reminder of the four necessary products. For instance, in the product $(3ab - a^2)(ab + a^2)$, the product of the First terms is $3a^2b^2$, the product of the Outer terms is $3a^3b$, the product of the Inner terms is $-a^3b$, and the product of the Last terms is $-a^4$. This acronym is, however, nothing more than a specialized application of the distributive property.

Topic 3: Common Factoring Methods

Factoring, in general, means reversing the process of multiplication in order to find two or more expressions whose product is the original expression. For instance, the factored form of the polynomial $x^2 - 3x + 2$ is $(x - 2)(x - 1)$, since if the product $(x - 2)(x - 1)$ is expanded we obtain $x^2 - 3x + 2$ (as you should verify). Factoring is thus a logical topic to discuss immediately after multiplication, but the process of factoring is sometimes more complicated to carry out than multiplication.

section 1.5 Polynomials and Factoring

Consequently, various methods have been devised to help. In some cases, more than one method may be used to factor a given algebraic expression, but all the methods will give the same result. In other cases, it may not be immediately obvious which method should be used, and the factoring process may require much trial and error.

The first four of the five methods presented here apply to polynomials. Unless otherwise indicated, we say that a polynomial with integer coefficients is **factorable** if it can be written as a product of two or more polynomials, all of which also have integer coefficients. If this cannot be done, we say the polynomial is **irreducible** (over the integers), or **prime**. The goal of factoring a polynomial is to **completely factor** it: to write it as a product of prime polynomials.

Method 1: Greatest Common Factor. Factoring out those factors common to all the terms in an expression is the least complex factoring method to apply and should be done first if possible. The **Greatest Common Factor** (GCF) among all the terms is the product of all the factors common to each. For instance, $2x$ is a factor common to all the terms in the polynomial $12x^5 - 4x^2 + 8x^3z^3$, but $4x^2$ is the greatest common factor. The Greatest Common Factor method is a matter of applying the distributive property to "undistribute" the greatest common factor.

example 4

Use the Greatest Common Factor method to factor the following polynomials.

a. $12x^5 - 4x^2 + 8x^3z^3$ b. $-24ax^2 + 60a$ c. $(a^2 - b) - 3(a^2 - b)$

Solutions:

a. $12x^5 - 4x^2 + 8x^3z^3$
$= (4x^2)(3x^3) + (4x^2)(-1) + (4x^2)(2xz^3)$
$= (4x^2)(3x^3 - 1 + 2xz^3)$

As we noted above, $4x^2$ is the greatest common factor. Applying the distributive property in reverse leads to the factored form of this degree 6 trinomial.

b. $-24ax^2 + 60a$
$= (-12a)(2x^2) + (-12a)(-5)$
$= -12a(2x^2 - 5)$

An alternative form of the final answer is $12a(-2x^2 + 5)$. We would have obtained this answer if we had factored out $12a$ initially. These two answers are equivalent.

continued on next page...

c. $(a^2 - b) - 3(a^2 - b)$
$= (a^2 - b)(1) + (a^2 - b)(-3)$
$= (a^2 - b)(1 - 3)$
$= -2(a^2 - b)$

In factoring out the greatest common factor $a^2 - b$, remember that it is being multiplied first by 1 and then by -3. One common source of error in factoring is to forget factors of 1.

Now that a few examples have been presented, one more word on factoring is in order. As mentioned above, the goal in factoring a polynomial is to write it as a product of prime polynomials, and we continue the factoring process until *all* the factors are prime. The exception to this is that we generally do not factor monomials further. For instance, in *Example 4a* we do not write the factor $4x^2$ as $(2)(2)(x)(x)$.

Method 2: Factoring by Grouping. Many polynomials have a GCF of 1, and the first factoring method is therefore not directly applicable. But if the terms of the polynomial are grouped in a suitable way, the GCF method may apply to each group, and a common factor might subsequently be found among the groups. **Factoring by Grouping** is the name given to this process, and it is important to realize that this is a trial and error process. Your first attempt at grouping and factoring may not succeed, and you may have to try several different ways of grouping the terms.

example 5

Use the Factor by Grouping method to factoring the following polynomials.

a. $6x^2 - y + 2x - 3xy$

b. $ax - ay - bx + by$

Solutions:

a. $6x^2 - y + 2x - 3xy$
$= (6x^2 + 2x) + (-y - 3xy)$
$= 2x(3x + 1) + y(-1 - 3x)$
$= 2x(3x + 1) - y(3x + 1)$
$= (3x + 1)(2x - y)$

The GCF of the four terms in the polynomial is 1, so the Greatest Common Factor method doesn't directly apply. The first and third terms have a GCF of $2x$, while the second and fourth have a GCF of y, so we group accordingly. After factoring the two groups, we notice that $3x + 1$ and $-1 - 3x$ differ only by a minus sign (and the order). This means $3x + 1$ can be factored out.

b. $ax - ay - bx + by$
$= a(x - y) - b(x - y)$
$= (x - y)(a - b)$

The first two terms have a common factor, as do the last two, so we proceed accordingly. In this problem, we could also have grouped the first and third terms, and the second and fourth terms, and obtained the same result.

section 1.5   **Polynomials and Factoring**

> **caution**
>
> One common error in factoring is to stop after groups within the original polynomial have been factored. For instance, while we have done some factoring to achieve the expression $2x(3x+1)+y(-1-3x)$ in *Example 5a*, this is *not* in factored form. An expression is only factored if it is written as a *product* of two or more factors. The expression $2x(3x+1)+y(-1-3x)$ is a sum of two smaller expressions.

Method 3: Factoring Special Binomials. Three types of binomials can always be factored by following the patterns outlined below. You should verify the truth of the patterns by multiplying out the products on the right-hand side of each one.

Factoring Special Binomials

In the following, A and B represent algebraic expressions.

Difference of Two Squares: $A^2 - B^2 = (A-B)(A+B)$

Difference of Two Cubes: $A^3 - B^3 = (A-B)(A^2 + AB + B^2)$

Sum of Two Cubes: $A^3 + B^3 = (A+B)(A^2 - AB + B^2)$

chapter one   Number Systems and Fundamental Concepts of Algebra

example 6

Factor the following binomials.

a. $49x^2 - 9y^6$ b. $27a^6b^{12} + c^3$ c. $64 - (x+y)^3$

Solutions:

a. $49x^2 - 9y^6$
$= (7x)^2 - (3y^3)^2$
$= (7x - 3y^3)(7x + 3y^3)$

The first step is to realize that the binomial is a difference of two squares, and to identify the two expressions that are being squared. Then follow the pattern to factor the binomial.

b. $27a^6b^{12} + c^3$
$= (3a^2b^4)^3 + (c)^3$
$= \left(\underbrace{3a^2b^4}_{A} + \underbrace{c}_{B}\right)\left(\underbrace{(3a^2b^4)^2}_{A^2} - \underbrace{(3a^2b^4)(c)}_{AB} + \underbrace{(c)^2}_{B^2}\right)$
$= (3a^2b^4 + c)(9a^4b^8 - 3a^2b^4c + c^2)$

First recognize the binomial as a sum of two cubes, and then identify the two expressions being cubed. Remember that all of A and B must be squared when applying the pattern. Also, remember the minus sign in front of the product AB in the second factor.

c. $64 - (x+y)^3$
$= 4^3 - (x+y)^3$
$= (4 - (x+y))(4^2 + 4(x+y) + (x+y)^2)$
$= (4 - x - y)(16 + 4x + 4y + x^2 + 2xy + y^2)$

In this difference of two cubes, the second cube is itself a binomial. But the factoring pattern still applies, leading to the final factored form of the original binomial.

Method 4: Factoring Trinomials. In factoring a trinomial of the form $ax^2 + bx + c$ the goal is to find two binomials $px + q$ and $rx + s$ such that

$$ax^2 + bx + c = (px + q)(rx + s).$$

Since $(px+q)(rx+s) = prx^2 + (ps+qr)x + qs$, we seek p, q, r, and s such that $a = pr$, $b = ps + qr$ and $c = qs$:

$$ax^2 + bx + c = \underbrace{pr}_{a}x^2 + \underbrace{(ps+qr)}_{b}x + \underbrace{qs}_{c}$$

section 1.5   Polynomials and Factoring

In general, this may require much trial and error, but the following guidelines will help.

Case 1: Leading Coefficient is 1. In this case, p and r will both be 1, so we only need q and s such that $x^2 + bx + c = x^2 + (q+s)x + qs$. That is, we need two integers whose sum is b, the coefficient of x, and whose product is c, the constant term.

example 7

To factor $x^2 + x - 12$ we can begin by writing $x^2 + x - 12 = (x + \boxed{?})(x + \boxed{?})$ and then try to find two integers to replace the question marks. The two integers we seek must have a product of -12, and the fact that the product is negative means that one integer must be positive and one negative. The only possibilities are $\{1,-12\}$, $\{-1,12\}$, $\{2,-6\}$, $\{-2,6\}$, $\{3,-4\}$, and $\{-3,4\}$, and when we add the requirement that the sum must be 1, we are left with $\{-3,4\}$. Thus $x^2 + x - 12 = (x-3)(x+4)$.

Case 2: Leading Coefficient is not 1. In this case, trial and error may still be an effective way to factor the trinomial $ax^2 + bx + c$ especially if a, b, and c are relatively small in magnitude and if you have lots of practice. If, however, trial and error seems to be taking too long, the following steps use factoring by grouping to minimize the amount of guessing required.

Factoring Trinomials by Grouping

For the trinomial $ax^2 + bx + c$:

Step 1: Multiply a and c.

Step 2: Factor ac into two integers whose sum is b. If no such factors exist, the trinomial is irreducible over the integers.

Step 3: Rewrite b in the trinomial with the sum found in step 2, and distribute. The resulting polynomial of four terms may now be factored by grouping.

chapter one — Number Systems and Fundamental Concepts of Algebra

example 8

To factor the trinomial $6x^2 - x - 12$ by trial and error, we would begin by noting that if it can be factored, the factors must be of the form $(x+\boxed{?})(6x+\boxed{?})$ or $(2x+\boxed{?})(3x+\boxed{?})$. If we use the method outlined above, we form the product $(6)(-12) = -72$ and then factor -72 into two integers whose sum is -1. The two numbers -9 and 8 work, so we write $6x^2 - x - 12 = 6x^2 + (-9+8)x - 12 = 6x^2 - 9x + 8x - 12$. Now proceed by grouping:

$$6x^2 - 9x + 8x - 12$$
$$= 3x(2x-3) + 4(2x-3)$$
$$= (2x-3)(3x+4)$$

Some trinomial expressions are known as "perfect square trinomials" because their factored form is the square of a binomial expression. As an example, $x^2 - 6x + 9 = (x-3)^2$. (Note that either the trial and error method or factoring by grouping suffices to give us this answer.) In general, such trinomials will have one of the following two forms.

Perfect Square Trinomials

In the following, A and B represent algebraic expressions.

$$A^2 + 2AB + B^2 = (A+B)^2$$
$$A^2 - 2AB + B^2 = (A-B)^2$$

Method 5: Factoring Expressions Containing Fractional Exponents. This last method does not apply to polynomials, as polynomials cannot have fractional exponents. It will, however, be very useful in solving certain problems later in this book and possibly in other math classes. The method applies to negative fractional exponents as well as positive.

To factor an algebraic expression (in one variable) that has fractional exponents, identify the least exponent among the various terms, and factor the variable raised to that least exponent from each of the terms. Factor out any other common factors as well, and simplify if possible.

section 1.5 Polynomials and Factoring

example 9

Factor each of the following algebraic expressions.

a. $3x^{-\frac{2}{3}} - 6x^{\frac{1}{3}} + 3x^{\frac{4}{3}}$

b. $(x-1)^{\frac{1}{2}} - (x-1)^{-\frac{1}{2}}$

Solutions:

a. $3x^{-\frac{2}{3}} - 6x^{\frac{1}{3}} + 3x^{\frac{4}{3}}$
$= 3x^{-\frac{2}{3}}(1 - 2x + x^2)$
$= 3x^{-\frac{2}{3}}(x^2 - 2x + 1)$
$= 3x^{-\frac{2}{3}}(x-1)(x-1)$
$= 3x^{-\frac{2}{3}}(x-1)^2$

Under the guidelines above, we factor out $3x^{-\frac{2}{3}}$. Note that we use the properties of exponents to obtain the terms in the second factor. After factoring out $3x^{-\frac{2}{3}}$, we notice that the second factor is a second-degree trinomial, and is itself factorable. In fact, it is an example of a perfect square trinomial (one that can be written as the square of a binomial).

b. $(x-1)^{\frac{1}{2}} - (x-1)^{-\frac{1}{2}}$
$= (x-1)^{-\frac{1}{2}}((x-1) - 1)$
$= (x-1)^{-\frac{1}{2}}(x-2)$

In this example we factor out $(x-1)^{-\frac{1}{2}}$ again using the properties of exponents to obtain the terms in the second factor. The second factor in this example can then be simplified.

exercises

Classify each of the following expressions as either a polynomial or a non-polynomial. For those that are polynomials, identify **(a)** the degree of the polynomial, and **(b)** the number of terms (use the words monomial, binomial, and trinomial if applicable). See Example 1.

1. $3x^{\frac{3}{2}} - 2x$

2. $17x^2y^5 + 2z^3 - 4$

3. $5x^{10} + 3x^3 - 2y^3z^8 + 9$

4. πx^3

5. 8

6. 0

7. $7^3xy^2 + 4y^4$

8. abc^2d^3

9. $4x^2 + 7xy + 5y^2$

10. $3n^4m^{-3} + n^2m$

11. $\frac{y^2z}{4} + 2yz^4$

12. $6x^4y + 3x^2y^2 + xy^5$

chapter one   Number Systems and Fundamental Concepts of Algebra

*Write each of the following polynomials in descending order, and identify **(a)** the degree of the polynomial, and **(b)** the leading coefficient.*

13. $8z^2 + \pi z^5 - 2z + 1$

14. $9x^8 - 9x^{10}$

15. $4s^3 - 10s^5 + 2s^6$

16. $4 - 2x^5 + x^2$

17. $9y^6 - 2 + y - 3y^5$

18. $4n + 6n^2 - 3$

19. $-4x^{10} - x^{13} + 9 + 7x^{11}$

20. $-6y^5 - 3y^7 + 12y^6$

Add or subtract the polynomials, as indicated. See Example 2.

21. $(-4x^3y + 2xz - 3y) - (2xz + 3y + x^2z)$

22. $(4x^3 - 9x^2 + 1) + (-2x^3 - 8)$

23. $(x^2y - xy - 6y) + (xy^2 + xy + 6x)$

24. $(5x^2 - 6x + 2) - (4 - 6x - 3x^2)$

25. $(a^2b + 2ab + ab^2) - (ab^2 + 5ab + a^2b)$

26. $(x^4 + 2x^3 - x + 5) - (x^3 - x - x^4)$

27. $(xy - 4y + xy^2) + (3y - x^2y - xy)$

28. $(-8x^4 + 13 - 9x^2) - (8 - 2x^4)$

Multiply the polynomials, as indicated. See Example 3.

29. $(3a^2b + 2a - 3b)(ab^2 + 7ab)$

30. $(x^2 - 2y)(x^2 + y)$

31. $(3a + 4b)(a - 2b)$

32. $(x + xy + y)(x - y)$

33. $(6x - 3y)(x + 6y)$

34. $(5y + x)(4y - 2x)$

35. $(7y^2 + x)(y^2 - 5x)$

36. $(y^2 + x)(3y^2 - 7x)$

37. $(6xy^2 - 3x + 4y)(x^2y + 6xy)$

38. $(2xy^2 + 4y - 6x)(x^2y - 5xy)$

section 1.5 Polynomials and Factoring

Factor each polynomial by factoring out the greatest common factor. See Example 4.

39. $3a^2b + 3a^3b - 9a^2b^2$

40. $(x^3 - y)^2 - (x^3 - y)$

41. $2x^6 - 14x^3 + 8x$

42. $27x^7y + 9x^6y - 9x^4yz$

43. $3x^3y - 9x^4y + 12x^3y^2$

44. $6xy^3 + 9y^3 - 12xy^4$

45. $12y^6 - 8y^2 - 16y^5$

46. $(2x + y^2)^4 - (2x + y^2)^6$

Factor each polynomial by grouping. See Example 5.

47. $a^3 + ab - a^2b - b^2$

48. $ax - 2bx - 2ay + 4by$

49. $z + z^2 + z^3 + z^4$

50. $x^2 + 3xy + 3y + x$

51. $nx^2 - 2y - 2x^2 + ny$

52. $2ac - 3bd + bc - 6ad$

53. $ax - 5bx + 5ay - 25by$

54. $3ac - 5bd + bc - 15ad$

Use the special factoring methods to factor the following binomials. See Example 6.

55. $25x^4y^2 - 9$

56. $27a^9 + 8b^{12}$

57. $x^3 - 1000y^3$

58. $64x^6 - 125y^3z^9$

59. $m^6 + 125n^9$

60. $49a^6 - 9b^2c^4$

61. $27x^6 - 8y^{12}z^3$

62. $(3x - 6)^2 - (y - 2x)^2$

63. $16z^2y^4 - 9x^8$

64. $512x^6 + 729y^3$

65. $343y^9 - 27x^3z^6$

66. $(2x + y^2)^2 - (y^2 - 3)^2$

chapter one   Number Systems and Fundamental Concepts of Algebra

Factor the following trinomials. See Examples 7 and 8.

67. $x^2 + 2x - 15$
68. $x^2 + 6x + 9$
69. $x^2 - 2x + 1$
70. $x^2 - 5x + 6$
71. $x^2 - 4x + 4$
72. $x^2 + 5x + 4$
73. $y^2 + 14y + 49$
74. $x^2 - 3x - 18$
75. $x^2 + 13x + 22$
76. $y^2 + y - 42$
77. $y^2 - 9y + 8$
78. $6x^2 + 5x - 6$
79. $5a^2 - 37a - 24$
80. $25y^2 + 10y + 1$
81. $5x^2 + 27x - 18$
82. $6y^2 - 13y - 8$
83. $16y^2 - 25y + 9$
84. $10m^2 + 29m + 10$
85. $8a^2 - 2a - 3$
86. $20y^2 + 21y - 5$
87. $12y^2 - 19y + 5$
88. $10y^2 - 11y - 6$

Factor the following algebraic expressions. See Example 9.

89. $(2x-1)^{-\frac{3}{2}} + (2x-1)^{-\frac{1}{2}}$
90. $2x^{-2} + 3x^{-1}$
91. $7a^{-1} - 2a^{-3}b$
92. $(3z+2)^{\frac{5}{3}} - (3z+2)^{\frac{2}{3}}$
93. $10y^{-2} - 2y^{-5}x$
94. $4y^{-3} + 12y^{-4}$
95. $(5x+7)^{\frac{7}{3}} - (5x+7)^{\frac{4}{3}}$
96. $(8x+6)^{-\frac{7}{2}} - (8x+6)^{-\frac{1}{2}}$
97. $7y^{-1} + 5y^{-4}$
98. $5x^{-4} - 4x^{-5}y$

section 1.5   Polynomials and Factoring

Use your knowledge of polynomials and factoring to answer the following questions.

99. Pneumothorax is a disease in which air or gas collects between the lung and the chest wall, causing the lung to collapse. When this disease is evident, the following formula is used to determine the degree of collapse of the lungs, represented as a percent:

$$Degree = 100\left(1 - \frac{L^3}{H^3}\right)$$

In this formula, L is the diameter of one lung and H is the diameter of one hemithorax (or half the chest cavity). Is this formula a polynomial? If so, find its degree and the number of terms. If not, explain.

100. You are trying to find a formula for the area of a certain trapezoid. You know the height of the trapezoid is x^2, the bottom base is $2x^2 + 4$, and the top base is $6x + 2$. Insert these values into the formula for the area of a trapezoid. Is the result a polynomial? If so, find the degree of the polynomial, the leading coefficient, and the number of terms in the polynomial. If not, explain.

101. a. Given a rectangular picture frame with sides of $2x + 1$ and $x^3 + 4$, find the area of the picture frame. Is the result a polynomial? If so, find the degree of the polynomial, the leading coefficient, and the number of terms in the polynomial. If not, explain.

b. Now find the perimeter of the picture frame. Is this a polynomial? If so, find the degree of the polynomial, the leading coefficient, and the number of terms in the polynomial. If not, explain.

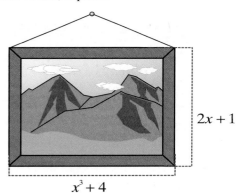

technology exercises

Computer algebra systems can be used to multiply polynomials, and are useful when the polynomials to be multiplied contain many terms or when many polynomials are to be multiplied. In *Mathematica*, the command is `Expand`, and its use is illustrated below.

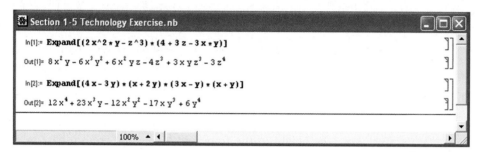

If available, use a computer algebra system to perform the polynomial multiplications that follow. (See Appendix A for guidance on using computer algebra systems like Mathematica.)

102. $(5x^2 - 2x + 3)(-x^2 + x - 3)(5x + 4)$

103. $(a + 2b - c)(3a - b + 4c)$

104. $(x^2 - 2xy + yz)(x - 3xy^2z + xz)$

105. $(pq - rs + t)(p^2 + r^2 + t^2)$

Computer algebra systems are also capable of reversing the result of polynomial multiplication, and such technology is useful when factoring by hand would be prohibitively tedious. In *Mathematica*, the command used to factor polynomials is, logically enough, `Factor`. The use of this command is illustrated below.

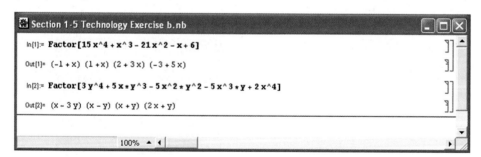

If available, use a computer algebra system to factor the following polynomials.

106. $3x^4 - 3x^2y + x^3y - xy^2 - 3x^3y^2 + 3xy^3 - x^2y^3 + y^4$

107. $16x^4 - 96x^3y + 216x^2y^2 - 216xy^3 + 81y^4$

108. $2p^4 - p^3q - 14p^2q^2 - 5pq^3 + 6q^4$

109. $ad + bd + cd + ae + be + ce + af + bf + cf$

1.6 The Complex Number System

TOPICS

1. The imaginary unit i and its properties
2. The algebra of complex numbers
3. Roots and complex numbers

Topic 1: The Imaginary Unit i and its Properties

Anything more than an elementary introduction to algebra soon reveals a troubling flaw with the real number system. The flaw manifests itself in a number of ways, one of which we have already alluded to: the lack of symmetry in the definition of roots of real numbers. Recall that we have thus far defined even roots only for non-negative numbers, but we have defined odd roots for both positive and negative numbers (as well as 0).

This asymmetry is a reflection of the fact that the real number system is not *algebraically complete*. The question of whether a field is algebraically complete is one that is dealt with in more advanced math classes (typically one called Abstract Algebra), but for the moment consider the following question:

For a given non-zero real number a, how many solutions does the equation $x^2 = a$ have?

As we have noted previously, the equation has two solutions $\left(x = \sqrt{a} \text{ and } x = -\sqrt{a}\right)$ if a is positive, but no (real) solutions if a is negative. We improve this situation with the following definition.

chapter one   Number Systems and Fundamental Concepts of Algebra

The Imaginary Unit *i*

> The imaginary unit i is defined as $i = \sqrt{-1}$. In other words, i has the property that its square is -1: $i^2 = -1$.

This allows us to immediately define square roots of negative numbers in general, as follows.

Square Roots of Negative Numbers

> If a is a positive real number, $\sqrt{-a} = i\sqrt{a}$. Note that by this definition, and by a logical extension of exponentiation, $\left(i\sqrt{a}\right)^2 = i^2 \left(\sqrt{a}\right)^2 = -a.$

example 1

a. $\sqrt{-16} = i\sqrt{16} = i(4) = 4i$. As is customary, we write a constant such as 4 before letters in algebraic expressions, even if, as in this case, the letter is not a variable. Remember that i has a fixed meaning: i is the square root of -1.

b. $\sqrt{-8} = i\sqrt{8} = i\left(2\sqrt{2}\right) = 2i\sqrt{2}$. As is customary, again, we write the radical factor last. You should verify that $\left(2i\sqrt{2}\right)^2$ is indeed -8.

c. $i^3 = i^2 i = (-1)(i) = -i$, and $i^4 = i^2 i^2 = (-1)(-1) = 1$. The simple fact that $i^2 = -1$ allows us, by our extension of exponentiation, to determine i^n for any natural number n.

d. $(-i)^2 = (-1)^2 i^2 = i^2 = -1$. This observation shows that $-i$ also has the property that its square is -1.

The definition of the imaginary unit i leads to the following definition of complex numbers.

section 1.6 The Complex Number System

Complex Numbers

> For any two real numbers a and b, the sum $a + bi$ is a **complex number**. The collection $\mathbb{C} = \{a + bi \,|\, a \text{ and } b \text{ are both real}\}$ is called the set of complex numbers and is another example of a field. The number a is called the **real part** of $a + bi$, and the number b is called the **imaginary part**. If the imaginary part of a given complex number is 0, the number is simply a real number. If the real part of a given complex number is 0, the number is a **pure imaginary number**.

Note that the set of real numbers is thus a subset of the complex numbers: every real number *is* a complex number with 0 as the imaginary part. The set of complex numbers is the largest set of numbers that will appear in this text.

Finally, a word on the nomenclature of complex numbers is in order. Don't be misled by the labels into thinking that complex numbers, with their possible imaginary parts, are somehow unimportant or physically meaningless. In many applications, complex numbers, even pure imaginary numbers, arise naturally and have important implications. As just one concrete example, the field of electrical engineering relies extensively on complex number arithmetic.

Topic 2:   The Algebra of Complex Numbers

The set of complex numbers is a field, so the field properties discussed in Section 1.2 apply. In particular, every complex number has an additive inverse (its negative), and every non-zero complex number has a multiplicative inverse (its reciprocal). Further, sums and products (and hence differences and quotients) of complex numbers are complex numbers, and so can be written in the standard form $a + bi$. Given several complex numbers combined by the operations of addition, subtraction, multiplication, or division, the goal is to *simplify* the expression into the form $a + bi$.

Sums, differences, and products of complex numbers are easily simplified by remembering the definition of i and by thinking of every complex number $a + bi$ as a binomial. The simplification process is outlined on the following page.

chapter one   Number Systems and Fundamental Concepts of Algebra

Simplifying Complex Expressions

> **Step 1:** Add, subtract, or multiply the complex numbers, as required, by treating every complex number $a + bi$ as a polynomial expression. Remember, though, that i is not actually a variable. Treating $a + bi$ as a binomial in i is just a handy device.
>
> **Step 2:** Complete the simplification by using the fact that $i^2 = -1$.

example 2

Simplify the following complex number expressions.

a. $(4+3i)+(-5+7i)$

b. $(-2+3i)-(-3+3i)$

c. $(3+2i)(-2+3i)$

d. $(2-3i)^2$

Solutions:

To enter complex numbers, you must first change the **MODE** from Real to a+bi by highlighting it with the arrow keys and pressing **ENTER**.

Remember to change it back when you are finished.

a. $(4+3i)+(-5+7i) = (4-5)+(3+7)i$
$= -1+10i$

Treating the two complex numbers as polynomials in i, we combine the real parts and then the imaginary parts.

b. $(-2+3i)-(-3+3i) = (-2+3i)+(-(-3)-3i)$
$= (-2+3)+(3-3)i$
$= 1+0i$
$= 1$

We begin by distributing the minus sign over the two terms of the second complex number, and then combine as in part a.

c. $(3+2i)(-2+3i) = -6+9i-4i+6i^2$
$= -6+(9-4)i+6(-1)$
$= -6+5i-6$
$= -12+5i$

The product of two complex numbers leads to four products via the distributive property, as illustrated here. After multiplying, we combine the two terms containing i, and rewrite i^2 as -1.

d. $(2-3i)^2 = (2-3i)(2-3i)$
$= 4-6i-6i+9i^2$
$= 4-12i+9(-1)$
$= -5-12i$

Squaring this complex number also leads to four products, which we simplify as in part c. Remember that a complex number is not simplified until it has the form $a + bi$.

section 1.6 The Complex Number System

Division of one complex number by another is slightly more complicated. In order to rewrite a quotient in the standard form $a + bi$, we make use of the following observation:

$$(a+bi)(a-bi) = a^2 - abi + abi - b^2i^2 = a^2 + b^2$$

Given any complex number $a + bi$, the complex number $a - bi$ is called its **complex conjugate**. We simplify the quotient of two complex numbers by multiplying the numerator and denominator of the fraction by the complex conjugate of the denominator.

example 3

Simplify the quotients.

a. $\dfrac{2+3i}{3-i}$ **b.** $(4-3i)^{-1}$ **c.** $\dfrac{1}{i}$

Solutions:

a.
$$\dfrac{2+3i}{3-i} = \left(\dfrac{2+3i}{3-i}\right)\left(\dfrac{3+i}{3+i}\right)$$
$$= \dfrac{(2+3i)(3+i)}{(3-i)(3+i)}$$
$$= \dfrac{6+2i+9i+3i^2}{9+3i-3i-i^2}$$
$$= \dfrac{3+11i}{10} = \dfrac{3}{10} + \dfrac{11}{10}i$$

We multiply the top and bottom of the fraction by $3+i$, which is the complex conjugate of the denominator. The rest of the simplification involves multiplying complex numbers as in the last example.

We would often leave the answer in the form $\dfrac{3+11i}{10}$ unless it is necessary to identify the real and imaginary parts.

To force your answer to be in fraction form, go to the **MATH** menu and select **Frac**. Enter i by selecting **2ND**, then **.**

b.
$$(4-3i)^{-1} = \dfrac{1}{4-3i}$$
$$= \left(\dfrac{1}{4-3i}\right)\left(\dfrac{4+3i}{4+3i}\right)$$
$$= \dfrac{4+3i}{(4-3i)(4+3i)}$$
$$= \dfrac{4+3i}{16-9i^2}$$
$$= \dfrac{4+3i}{25} = \dfrac{4}{25} + \dfrac{3}{25}i$$

In this example, we simplify the reciprocal of the complex number $4 - 3i$. After writing the original expression as a fraction, we multiply the top and bottom by the complex conjugate of the denominator and proceed as in part *a*.

continued on next page...

c. $\dfrac{1}{i} = \left(\dfrac{1}{i}\right)\left(\dfrac{-i}{-i}\right)$

$= \dfrac{-i}{-i^2}$

$= \dfrac{-i}{1} = -i$

This problem illustrates the process of writing the reciprocal of the imaginary unit as a complex number. Note that with this as a starting point, we could now calculate $i^{-2}, i^{-3}, \ldots$

Topic 3: Roots and Complex Numbers

We have now defined $\sqrt{a}$ for any real number a. We have done so in such a way that there is no ambiguity in the meaning of $\sqrt{a}$: given a positive real number a, $\sqrt{a}$ is defined to be the positive real number whose square is a, and $\sqrt{-a}$ is defined to be $i\sqrt{a}$. These are sometimes called the **principal square roots**, to distinguish them from $-\sqrt{a}$ and $-i\sqrt{a}$ respectively. (Remember, both $\sqrt{a}$ and $-\sqrt{a}$ are square roots of a.)

One more comment about roots should be made before we close this section. In simplifying radical expressions, we have made frequent use of the properties that if $\sqrt{a}$ and $\sqrt{b}$ are real numbers, then

$$\sqrt{a}\sqrt{b} = \sqrt{ab} \text{ and } \dfrac{\sqrt{a}}{\sqrt{b}} = \sqrt{\dfrac{a}{b}}.$$

There is a subtle but important condition in the above statement: $\sqrt{a}$ and $\sqrt{b}$ must both be real numbers. If this condition is not met, these properties of radicals do not necessarily hold. For instance,

$$\sqrt{(-9)(-4)} = \sqrt{36} = 6,$$

but

$$\sqrt{-9}\sqrt{-4} = (3i)(2i) = 6i^2 = -6.$$

In order to apply either of these two properties, then, first simplify any square roots of negative numbers by rewriting them as pure imaginary numbers.

example 4

Simplify the following expressions.

a. $\left(2 - \sqrt{-3}\right)^2$

b. $\dfrac{\sqrt{4}}{\sqrt{-4}}$

section 1.6 The Complex Number System

a. $(2-\sqrt{-3})^2 = (2-\sqrt{-3})(2-\sqrt{-3})$
$= 4 - 4\sqrt{-3} + \sqrt{-3}\sqrt{-3}$
$= 4 - 4i\sqrt{3} + (i\sqrt{3})^2$
$= 4 - 4i\sqrt{3} - 3$
$= 1 - 4i\sqrt{3}$

Each $\sqrt{-3}$ is converted to $i\sqrt{3}$ before carrying out the associated multiplications. Note that incorrect use of one of the properties of radicals would have led to adding 3 instead of subtracting 3.

b. $\dfrac{\sqrt{4}}{\sqrt{-4}} = \dfrac{2}{2i}$
$= \dfrac{1}{i}$
$= -i$

We have already simplified $\dfrac{1}{i}$ in Example 3c, so we quickly obtain the correct answer of $-i$. If we had incorrectly rewritten the original fraction as $\sqrt{\dfrac{4}{-4}}$, we would have obtained $\sqrt{-1}$ or i as the final answer.

exercises

Evaluate the following square root expressions. See Example 1.

1. $\sqrt{-25}$
2. $\sqrt{-12}$
3. $-\sqrt{-27}$
4. $-\sqrt{-100}$
5. $\sqrt{-32x}$
6. $\sqrt{-x^2}$
7. $\sqrt{-29}$
8. $(-i)^2\sqrt{-64}$

Simplify the following complex expressions by adding, subtracting, or multiplying as indicated. See Examples 1 and 2.

9. $(4-2i)-(3+i)$
10. $(1+i)+i$
11. $(4-i)(2+i)$
12. $i(5-i)$
13. $(3-i)^2$
14. i^7
15. $(3i)^2$
16. $(-5i)^3$
17. $(7i-2)+(3i^2-i)$
18. $(5-3i)^2$
19. $(3+i)(3-i)$
20. $(5+i)(2-9i)$
21. $(9-4i)(9+4i)$
22. i^{13}
23. $i^{11}\left(\dfrac{6}{i^3}\right)$
24. $11i^{314}$
25. i^{132}
26. $(7-3i)^2$
27. $(10i^2-9i)+(9+5i)$
28. $(4-3i)(7+i)$
29. $i^7\left(\dfrac{49}{7i^2}\right)$

Chapter One: Number Systems and Fundamental Concepts of Algebra

Simplify the following quotients. See Example 3.

30. $\dfrac{1+2i}{1-2i}$
31. $\dfrac{10}{3-i}$
32. $\dfrac{i}{2+i}$
33. $\dfrac{1}{i^9}$

34. $(2+5i)^{-1}$
35. i^{-25}
36. $\dfrac{1}{i^{27}}$
37. $\dfrac{52}{5+i}$

38. $(2-3i)^{-1}$
39. $\dfrac{4i}{5+7i}$
40. i^{-4}
41. $\dfrac{5+i}{4+i}$

Simplify the following square root expressions. See Example 4.

42. $(\sqrt{-9})(\sqrt{-2})$
43. $(1+\sqrt{-6})^2$
44. $\dfrac{\sqrt{18}}{\sqrt{-2}}$

45. $(\sqrt{-32})(-\sqrt{-2})$
46. $(3+\sqrt{-2})^2$
47. $\dfrac{\sqrt{-98}}{3i\sqrt{-2}}$

48. $(\sqrt{-8})(\sqrt{-2})$
49. $(5+\sqrt{-3})^2$
50. $\dfrac{\sqrt{-72}}{5i\sqrt{-2}}$

Use your knowledge of the complex number system to answer the following questions.

51. Electrical engineers often use j, rather than i, to represent imaginary numbers. This is to prevent confusion with their use of i, which often represents current. Under this convention, assume the impedance of a particular part of a series circuit is $4-3j$ ohms and the impedance of another part of the circuit is $2+6j$ ohms. Find the total impedance of the circuit. (Impedances in series are simply added.)

52. Consider the formula $V = IZ$, where V is voltage (in volts), I is current (in amps), and Z is impedance (in ohms). If you know the current of a circuit is $5-4j$ amps and the impedance is $8+2j$ ohms, find the voltage.

53. If you know the voltage of a circuit is $35+5j$ volts and the current is $3+j$ amps, find the impedance.

54. Explain why it may be useful to be able to use imaginary numbers in real world math.

section 1.6 The Complex Number System

technology exercises

The advice given in this section for simplifying complex number expressions is to initially treat complex numbers as polynomials in the "variable" i, and the command used in *Mathematica* to simplify complex expressions reflects this approach. The *Mathematica* command for multiplying out polynomials is **Expand**, and this same command can be used to reduce complex number expressions to the form $a + bi$.

If available, use a computer algebra system to simplify the expressions that follow the illustration below. Note that the imaginary root i appears as the capital letter I in Mathematica. (See Appendix A for guidance on using computer algebra systems like Mathematica.)

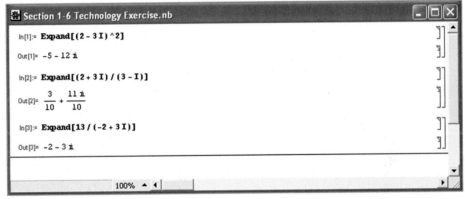

55. $\dfrac{3-2i}{1+i}$ **56.** $(3-2i)^4$ **57.** $\dfrac{2500}{(3+i)^4}$

58. $(2-5i)(3+7i)(1-4i)$ **59.** $(1+i)^5(3-i)^2$ **60.** $\dfrac{3+7i}{(2-5i)(1+3i)}$

61. $\dfrac{6+3i}{2-4i}$ **62.** $(5-3i)^5$ **63.** $\dfrac{400}{(6+2i)^3}$

64. $(6-3i)(8+i)(7-4i)$ **65.** $(5-3i)^4(7+2i)^3$ **66.** $\dfrac{4+3i}{(7-2i)(5+4i)}$

chapter one — Number Systems and Fundamental Concepts of Algebra

chapter one project

Polynomials

A chemistry professor calculates final grades for her class using the polynomial

$$A = 0.3f + 0.15h + 0.4t + 0.15p,$$

where A is the final grade, f is the final exam, h is the homework average, t is the chapter test average, and p is the semester project.

The following is a table containing the grades for various students in the class:

Name	Final Exam	Homework Avg.	Test Avg.	Project
Alex	77	95	79	85
Ashley	91	95	88	90
Barron	82	85	81	75
Elizabeth	75	100	84	80
Gabe	94	90	90	85
Lynn	88	85	80	75

1. Find the course average for each student, rounded to the nearest tenth.
2. Who has the highest total score?
3. Why is the total grade raised more with a grade of 100 on the final exam than with a grade of 100 on the semester project?
4. Assume you are a student in this class. With 1 week until the final exam, you have a homework average of 85, a test average of 85, and a 95 on the semester project. What score must you make on the final exam to achieve at least a 90.0 overall? (Round to the nearest tenth.)

chapter one summary

CHAPTER SUMMARY

A summary of concepts and skills follows each chapter. Refer to these summaries to make sure you feel comfortable with the material in the chapter. The concepts and skills are organized according to the section title and topic title in which the material is first discussed.

1.1 The Real Number System

topics	pages
Common Subsets of Real Numbers • The sets $\mathbb{N}, \mathbb{Z}, \mathbb{Q},$ and $\mathbb{R},$ as well as *whole* numbers and *irrational* numbers • Identifying numbers as elements of one or more of the common sets	p. 3 – 4
The Real Number Line • Plotting numbers on the real number line • The *origin* of the real line, and its relation to negative and positive numbers	p. 5 – 6
Order on the Real Number Line • The four inequality symbols $<, \leq, >,$ and $\geq$ • Distinguishing between strict and non-strict inequalities	p. 6 – 7
Set-Builder Notation and Interval Notation • Using set-builder notation to define sets • The empty set and its two common symbols • Interval notation and its relation to inequality statements	p. 7 – 8
Absolute Value and Distance on the Real Number Line • The definition of absolute value on the real line • The relationship between absolute value and distance • Distance between two real numbers • The triangle inequality	p. 9 – 11
Interlude: Working with Repeating Digits • Writing rational numbers that appear as decimal numbers with repeating digits as fractions	p. 11

chapter one — Number Systems and Fundamental Concepts of Algebra

1.2 The Arithmetic of Algebraic Expressions

topics	pages
Components and Terminology of Algebraic Expressions • Identifying *terms* of an expression, and identifying the *coefficient* of a term • Distinguishing between *constants* and *variables* in an expression • Identifying *factors* of a term	p. 16 – 17
The Field Properties and Their Use in Algebra • The application of the *closure, commutative, associative, identity, inverse,* and *distributive* properties of the real numbers • Using the *cancellation* and *zero-factor* properties	p. 17 – 19
The Order of Mathematical Operations • Applying the conventional order of operations when simplifying expressions and calculating numbers	p. 20 – 21
Basic Set Operations and Venn Diagrams • The meaning and use of Venn diagrams • The definition of the set operations *union* and *intersection* and their application to intervals	p. 21 – 22

1.3 Properties of Exponents

topics	pages
Natural Number Exponents • The meaning of exponential notation, and the distinction between *base* and *exponent*	p. 28 – 29
Integer Exponents • The extension of natural number exponents to integer exponents • The definition of exponentiation by 0 • The equivalence of a^{-n} and $\frac{1}{a^n}$ for $a \neq 0$	p. 30 – 31

chapter one summary

topics (continued)	pages
Properties of Exponents and Their Use • The properties of exponents and their use in simplifying expressions • Recognition of common errors made in trying to apply properties of exponents	p. 32 – 34
Scientific Notation • The definition of the scientific notation form of a number • Using scientific notation in expressing numerical values	p. 35 – 36
Interlude: Working with Geometric Formulas • Using geometric formulas with exponents	p. 37 – 38

1.4 Properties of Radicals

topics	pages
Roots and Radical Notation • The definition of nth roots, and the meaning of *index* and *radicand*	p. 45 – 47
Simplifying Radical Expressions • The meaning of *simplification* as applied to radical expressions • The properties of radicals and their use in simplifying expressions • The use of *conjugate radical expressions* in rationalizing denominators (and numerators)	p. 47 – 51
Combining Radical Expressions • The meaning of *like radicals*	p. 52
Rational Number Exponents • The extension of integer exponents to rational exponents • Applying properties of exponents to simplify expressions containing rational exponents	p. 53 – 56

chapter one — Number Systems and Fundamental Concepts of Algebra

1.5 Polynomials and Factoring

topics	pages
The Terminology of Polynomial Expressions • The definition of *polynomial* • The meaning of *coefficient, leading coefficient, descending order, degree of a term,* and *degree of a polynomial* as applied to polynomials • The meaning of *monomial, binomial,* and *trinomial*	p. 61 – 62
The Algebra of Polynomials • The meaning of *like,* or *similar,* terms • Polynomial addition, subtraction, and multiplication	p. 62 – 64
Common Factoring Methods • The meaning of *factorable* and *irreducible* (or *prime*) as applied to polynomials • The mechanics of using the *greatest common factor, factoring by grouping, special binomials,* and *factoring trinomials* methods • The extension of the methods to expressions containing fractional exponents	p. 64 – 71

1.6 The Complex Number System

topics	pages
The Imaginary Unit *i* and its Properties • The definition of *i* • The definition of *complex numbers,* and the identification of their *real* and *imaginary* parts	p. 77 – 79
The Algebra of Complex Numbers • Simplifying complex number expressions • The use of *complex conjugates*	p. 79 – 82
Roots and Complex Numbers • The meaning of *principal square root* • Understanding when properties of radicals apply and when they do not	p. 82 – 83

chapter one review

chapter review

Section 1.1

Write the following rational numbers as ratios of integers.

1. $7.\overline{6}$
2. $-2.0\overline{42}$

Evaluate the absolute value expressions.

3. $-|11-2|$
4. $-|-4-3|$
5. $\left|\sqrt{5}-\sqrt{11}\right|$

Section 1.2

Evaluate each expression for the given values of the variables.

6. $x^2z^3 + 5\sqrt{3x-2y}$ for $x=2, y=1,$ and $z=-1$

7. $7y^2 - \dfrac{1}{3}\pi xy + 8x^3$ for $x=-2$ and $y=2$

Identify the property that justifies each of the following statements. If one of the cancellation properties is being used to transform an equation, identify the quantity that is being added to both sides or the quantity that is multiplied by both sides.

8. $12a^2 = 8b \Leftrightarrow 3a^2 = 2b$
9. $(x-3)(z-2) = 0 \Rightarrow x-3 = 0$ or $z-2 = 0$

Section 1.3

Use the properties of exponents to simplify each of the following expressions, writing your answer with only positive exponents.

10. $\dfrac{-4t^0\left(s^2t^{-2}\right)^{-3}}{2^3 st^{-3}}$

11. $\left[\left(3y^{-2}z\right)^{-1}\right]^{-3}$

12. $\left(2^3 a^{-2} b^4\right)^{-1} c^{-3}$

13. $\dfrac{3^2 x^{-4}\left(y^2 z\right)^{-2}}{\left(2z^{-3}\right)^{-1} y^{-6}}$

Convert each number from scientific notation to standard notation, or vice versa, as indicated.

14. -2.004×10^{-4}; convert to standard
15. $52{,}240{,}000$; convert to scientific

chapter one Number Systems and Fundamental Concepts of Algebra

Section 1.4

Simplify the following radical expressions. Rationalize all denominators and use only positive exponents.

16. $\sqrt[3]{\dfrac{8x^2}{3y^{-4}}}$

17. $\dfrac{\sqrt{3a^3}}{\sqrt{12a}}$

18. $\dfrac{4}{\sqrt{2}-\sqrt{6}}$

19. $\sqrt{16x^2}$

20. $\sqrt[3]{-64x^{-9}y^3}$

21. $\sqrt[4]{\dfrac{a^9 b^{-4}}{81}}$

Section 1.5

Factor each of the following polynomials.

22. $nx + 3mx - 2ny - 6my$

23. $36x^6 - y^2$

24. $x^2 - x - 12$

25. $6a^2 - 7a - 5$

26. $2x^2 + x - 15$

27. $4a^2 - 9b^4$

Section 1.6

Simplify the following quotients.

28. $(7-2i)+(9i-5)$

29. $(5-3i)-(-12i)$

30. $(3-i)(6i^2 - 4)$

31. $\dfrac{3+4i}{3-4i}$

32. $\dfrac{17}{4-i}$

33. $\dfrac{2i}{3-i}$

chapter one test

chapter test

1. Determine if the following statement is true or false and explain your answer: If $|x| = -x$, then $x - 4 < 0$ for all x.

2. If $|a| < 6$, what is the smallest possible integer value of a?

Evaluate each of the following expressions.

3. $(-3^2 - 2^3)(1 - 2^4) + \left[-(-4^3) \cdot (-2)^2\right]$

4. $\left[2^3 - 3^2 \cdot (-3)^3 - (2 - 5^0)\right]\left[-4^2 + (-2)^3\right]$

5. $\dfrac{2 - 2^{-1}}{2 + 2^{-1}} \div \dfrac{1 + 2^{-1}}{1 - 2^{-1}}$

6. $3 + 4^2 \left(4^{-2}\right)^{-1} - 259$

7. $(-1)^{-10} \cdot (-1)^5 - (-2)^2 \cdot (-2^2)(-2)^3$

8. $\dfrac{3^{-3} \cdot (-3)^{-2} \left(-3^2\right)}{(1 - 2^2)(-3)(2^2 - 1)}$

Simplify each of the following expressions, writing your answer with only positive exponents.

9. $-2^3 \left(-2^2\right) \cdot \left(-2^{-2}\right)$

10. $\dfrac{2^{-3} \left(b^2\right)^{-2}}{\left(2^{-2}\right)^{-3} \left(b^{-1}\right)^5}$

11. $\left(2x^3 y^{-1}\right)^2 \cdot \left(-4xy^3\right)^2$

Evaluate each expression using the properties of exponents and express your answer using scientific notation.

12. $(3.46 \times 10^8)(1.2 \times 10^4)$

13. $(-3 \times 10^{-7})(1.2 \times 10^4)$

14. $\dfrac{2.4 \times 10^{-12}}{(1.2) \times 10^{-4}}$

15. $\dfrac{1.6 \times 10^{-4}}{8.0 \times 10^8}$

16. $\dfrac{2,000,000}{5,000,000,000}$

17. Arrange the following numbers from largest to smallest: $\sqrt{3}, \sqrt[3]{4}, \sqrt[6]{15}$

Evaluate each of the following radical expressions.

18. $-\sqrt{121}$

19. $\sqrt[3]{-125}$

20. $\sqrt{3^2 + 4^2}$

21. $\sqrt{-81}$

22. $\sqrt{8\sqrt{4}}$

23. $\sqrt{\sqrt{81}}$

chapter one — Number Systems and Fundamental Concepts of Algebra

24. $\dfrac{\sqrt[3]{\sqrt{15}}}{\sqrt{\sqrt[3]{5}}}$

25. $\sqrt[5]{-32} \cdot \sqrt[6]{64} \cdot \sqrt[4]{(-4)^4}$

26. $2^{\tfrac{3}{8}} \cdot \sqrt[8]{\dfrac{1}{2}} \cdot \sqrt[4]{\dfrac{1}{2}}$

27. $(2a-b)(a-2b)-(a+b)(a-b)$

28. $3x-(x-2y)+5y-(y-3x)$

29. $(a-2)(a+1)(a+2)-(a-1)$

30. $x^2 - y + (y-x)(y+x)$

31. $a - \left[-(1-3a)-(2+a)\right]-(a+3)$

Simplify the following radicals by rationalizing the denominators.

32. $\dfrac{\sqrt{2}}{\sqrt[3]{2}}$

33. $\dfrac{3}{\sqrt{5}+\sqrt{2}}$

34. $\dfrac{2}{\sqrt{a}-1}$

35. $\dfrac{\sqrt{3}-1}{\sqrt{3}+1} - \dfrac{\sqrt{3}+1}{\sqrt{3}-1}$

Simplify the following expressions.

36. $\sqrt{18} - 3\sqrt{50} + \sqrt{200}$

37. $(\sqrt{3}-\sqrt{2})(\sqrt{3}+\sqrt{2})$

38. $(2\sqrt{3}-5\sqrt{2})^2$

39. $\sqrt[3]{2} \cdot \sqrt{3} \cdot \sqrt[4]{2}$

40. $(2x-1)^{\tfrac{1}{3}} \cdot (2x-1)^{\tfrac{5}{3}}$

41. $\sqrt{5+2\sqrt{6}} \cdot \sqrt{5-2\sqrt{6}}$

42. If $A = 2x^3 - 4x^2 - x + 1$ and $B = -x^3 + 2x - 3$, find $3A + 6B$.

43. If $P(x) = -x^3 - 2x^2 - x + 1$, find $P(-2)$.

44. If $P(x) = x^2 - x$, find $P(2x - 1)$.

Factor each of the following polynomials.

45. $3x^2y - 9xy^2 + 6xy$

46. $(a^2 - b)^2 - (b - a^2)$

47. $2y^6 - 8y^3 - 10y^2$

48. $x^6 + x^2y^2 - x^4y^2 - y^4$

49. $m^3 + m^2 - m - 1$

50. $6ab - 2 + 4a - 3b$

51. $16a^2b^4 - 9c^2$

52. $(a-b+c)^2 - (a+b-c)^2$

53. $y^3 - 4z^2y$

54. $a^3 - 125b^3$

55. $64x^6 + 27y^9$

chapter one test

Factor the following trinomials.

56. $x^2 - x - 6$

57. $x^2 + 5x - 6$

58. $x^2 + 6x + 8$

59. $x^2 + 10x + 25$

60. $x^2 - 5x - 36$

61. $-x^2 + 14x - 49$

62. $x^4 + x^2 - 12$

63. $x^6 - 2x^3 + 1$

64. $9b^4 - 12ab^2 + 4a^2$

65. $4x^2 - 13x + 3$

66. $2x^2 + xy - y^2$

67. $2x^3 - x^2y - xy^2$

Factor the following expressions.

68. $a^{\frac{-1}{2}} - a^{\frac{-3}{2}}$

69. $9(a-1)^{-2} - 6(a-1)^{-3}$

70. $(3x-2y)^{\frac{4}{3}} - (3x-2y)^{\frac{2}{3}}$

71. $(4x-6y)^{-2} - (2x-3y)^{-4}$

Simplify the following expressions.

72. $i^2 \cdot i^4 \cdot i^{10}$

73. $i^{125} - i^{60} + (-i)^3$

74. $\sqrt{-50} + \sqrt{-8} - 3\sqrt{-18}$

75. $\sqrt{-48x} - 2\sqrt{-75x}$

76. $\dfrac{9}{\sqrt{-9}}$

77. $(-3i)^3 \cdot (2i)^2$

78. $(4-3i)(2+i)$

79. $(1-2i)(1+2i)$

80. $\dfrac{1-\sqrt{-9}}{1+\sqrt{-9}}$

81. $(3+2i)^{-1}$

chapter TWO

EQUATIONS AND INEQUALITIES OF ONE VARIABLE

OVERVIEW

2.1 Linear Equations in One Variable

2.2 Linear Inequalities in One Variable

2.3 Quadratic Equations in One Variable

2.4 Higher Degree Polynomial Equations

2.5 Rational Expressions and Equations

2.6 Radical Equations

By the end of this chapter you should be able to restate radical equations in terms of variables which appear inaccessible — they are "trapped" within a radical expression. When solving this type of equation, be sure to check your potential solutions in the original expression to see if they are true solutions. On page 163 you will find a problem in which the formula regarding impedance, resistance, capacitance, inductance, and frequency of current in a circuit with an AC power source is stated in terms of one variable and solved in terms of another. You will master this type of problem using tools such as the *Method of Solving Radical Equations* on page 159.

chapter two — Equations and Inequalities of One Variable

Introduction

This chapter reviews the basic algebraic methods we use to solve equations and inequalities. All of the problems that we encounter in this chapter will involve only one variable, but the methods we discuss here will still be useful when we consider problems containing more than one variable.

The mastery of methods that can be used to solve certain types of problems (mathematical and otherwise) has constituted much of the educational process for thousands of years, and the development of such methods has in turn made up a large part of scholarly activity. Archaeologists have discovered stone tablets showing that Babylonian and Egyptian mathematicians had devised methods for solving problems arising in commerce, agriculture, and engineering well before 2000 BC. One huge advantage that we enjoy today is that we can draw upon the achievements of all the people and cultures that have come before us as we seek mastery of various methods ourselves.

Babylonian numbers

One specific example of an advantage we possess is the subtle one of notation. Although we often don't realize it, much of our ability to solve a given problem lies in the value of the notation that we use to accurately and precisely state the problem. In fact, if the notation is sufficiently advanced, the mere act of using the notation to state a problem does much of the work of actually solving it.

For instance, early Babylonian, Hindu, Chinese, and Greek mathematicians, among others, long ago devised methods that could be used to solve certain types of quadratic equations (a class of equations discussed in this chapter). However, none of these cultures had yet devised an entirely satisfactory way to describe their procedures, and the methods all relied to a certain extent on interpreting quantities in the equations geometrically, as mathematicians of the time were still heavily dependent on pictures to convey ideas. As later mathematicians built on their work, our present algebraic notation evolved to the point that today we can describe a method for solving all quadratic equations with just a few symbols.

Similar progress has been made throughout mathematics, and the work is continuing in the present. In this chapter, you will learn methods that are guaranteed to find all the solutions of certain types of equations. For other types of equations, you will learn methods that are helpful but are not necessarily guaranteed to work. As mathematics continues to evolve, new approaches to various problems will be devised, with the result that some work that is considered difficult today will be considered trivial in the future.

2.1 Linear Equations in One Variable

TOPICS

1. Equivalent equations and the meaning of solutions
2. Solving linear equations
3. Solving absolute value equations
4. Solving linear equations for one variable
5. Interlude: distance and interest problems

Topic 1: Equivalent Equations and the Meaning of Solutions

In mathematics, an **equation** is a statement that two expressions are equal. If the statement is always true for any allowable value (or values) of the variable (respectively, variables), then the equation is an **identity**. (Unless otherwise stated, it is assumed that the context for a given equation is the set of real numbers, and variables are to be replaced by values that result in real numbers.) If the statement is never true, it is a **contradiction**. The third (and last) possibility is that the equation is true for some values of the variable(s) and false for others. These equations are called **conditional**, and the goal in solving such equations is to discover the set of values that the variable(s) can be replaced by to make the equation true. This set is called the **solution set** of the equation. Any one element of the solution set is called a **solution** of the equation.

example 1

a. The equations $x^{\frac{1}{2}}(x+1) = x^{\frac{3}{2}} + x^{\frac{1}{2}}$ and $5+3=8$ are identities. Note that in the first equation the only allowable replacements for x are non-negative real numbers, but for all such numbers the equation is true. There are no variables in the second equation, so any value for any variable automatically satisfies this true statement.

b. The equation $t+3=t$ is an example of a contradiction. The solution set of this equation is the empty set, $\emptyset$, since no value for t satisfies the equation.

c. The equation $x^2 = 9$ is conditional. The solution set of the equation is $\{-3, 3\}$, as any other value for x results in a false statement.

chapter two Equations and Inequalities of One Variable

The basic method for finding the solution set of an equation is to transform it into a simpler equation whose solution set is more clear. Two equations that have the same solution set are called **equivalent equations**, and transforming an equation into an equivalent but simpler equation constitutes much of the mechanical work of algebra. The field properties and the cancellation properties, discussed in Section 1.2, are the means by which equations are transformed into equivalent equations. Bear in mind, however, the strong possibility that more than one step may be required to transform the equation into an equivalent equation whose solution is apparent. Persistence and care are the key to successfully solving equations!

Topic 2:   Solving Linear Equations

Linear equations in one variable are arguably the least complicated type of equation to solve.

Linear Equations In One Variable

A **linear equation in one variable**, say the variable x, is an equation that can be transformed into the form $ax + b = 0$, where a and b are real numbers and $a \neq 0$. Such equations are also called **first-degree** equations, as x appears to the first power.

A linear equation is an example of a polynomial equation, and the general method of solving polynomial equations will be discussed in Section 2.4.

Note that linear equations in one variable take very few steps to solve. Once the equation has been written in the form $ax + b = 0$, the first cancellation property implies

$$ax + b = 0 \Leftrightarrow ax = -b, \qquad \text{(add } -b \text{ to both sides)}$$

and the second cancellation property implies

$$ax = -b \Leftrightarrow x = -\frac{b}{a}. \qquad \text{(multiply both sides by } \tfrac{1}{a}\text{)}$$

Of course, a given linear equation does not have to be written in the form $ax + b = 0$ initially, and it is likely that some of the field properties (such as the distributive and commutative properties) must be applied first.

example 2

Solve the following equations.

a. $3(x-2)+7x = 1-2\left(x+\dfrac{1}{2}\right)$

b. $3x-7 = 3(x-2)$

c. $\dfrac{y}{6}+\dfrac{2y-1}{2} = \dfrac{y+1}{3}$

d. $0.25(x-3)+0.08 = 0.15x$

Solutions:

a.
$$3(x-2)+7x = 1-2\left(x+\dfrac{1}{2}\right)$$
$$3x-6+7x = 1-2x-1$$
$$10x-6 = -2x$$
$$12x-6 = 0$$
$$12x = 6$$
$$x = \dfrac{1}{2}$$

As is typical, we must apply some of the field properties discussed in Section 1.2 in order to solve the equation. The distributive property leads to the second equation. Combining like terms leads to the third.

The cancellation properties then allow us to complete the process and solve the equation.

b.
$$3x-7 = 3(x-2)$$
$$3x-7 = 3x-6$$
$$3x-3x = -6+7$$
$$0 = 1 \quad \text{No Solution}$$

We first distribute, then add $-3x$ and 7 to both sides in order to combine like terms. In this problem, however, the variable disappears and we are left with a false statement. The solution set is $\varnothing$.

To get to equation solver, select **MATH** and **0: Solver...**. To solve, select **ALPHA** and **ENTER**. Press the up arrow to enter a new equation.

c.
$$\dfrac{y}{6}+\dfrac{2y-1}{2} = \dfrac{y+1}{3}$$
$$6\left(\dfrac{y}{6}+\dfrac{2y-1}{2}\right) = 6\left(\dfrac{y+1}{3}\right)$$
$$6\cdot\dfrac{y}{6}+6\cdot\dfrac{2y-1}{2} = 6\cdot\dfrac{y+1}{3}$$
$$y+3(2y-1) = 2(y+1)$$
$$y+6y-3 = 2y+2$$
$$y+6y-2y = 2+3$$
$$5y = 5$$
$$y = 1$$

Although it is not a necessary step, many people prefer to get rid of any fractions that might appear by multiplying both sides of the equation by the least common denominator (LCD). Remember to multiply every term by the LCD.

Note the cancellation that has occurred. We have replaced $\dfrac{6}{6}$ with 1, $\dfrac{6}{2}$ with 3, and $\dfrac{6}{3}$ with 2. Like terms are then combined, and multiplication by $\dfrac{1}{5}$ leads to the final answer.

d.
$$0.25(x-3)+0.08 = 0.15x$$
$$25(x-3)+8 = 15x$$
$$25x-75+8 = 15x$$
$$10x = 67$$
$$x = 6.7$$

One approach to solving an equation with decimals is to multiply both sides by the appropriate power of 10 to eliminate the decimals. In this problem, multiplying both sides by 100 results in a simpler linear equation. Another approach is to retain the decimals throughout the solution process.

chapter two — Equations and Inequalities of One Variable

Topic 3: Solving Absolute Value Equations

Linear absolute value equations in one variable are closely related to linear equations, and are solved in a similar fashion. The difference is that a linear absolute value equation contains at least one term inside the absolute value symbols; if these symbols were removed, the equation would be linear.

Such equations are solved by recognizing that the absolute value of any quantity is either the original quantity or its negative, depending on whether the quantity is positive or negative to start with. For instance, $|ax+b|$ means either $ax+b$ or $-(ax+b)$, depending on the sign of $ax+b$. Of course, until the equation is solved, we don't know the sign of $ax+b$, and so we must consider both cases. This means that, in general, every occurrence of an absolute value term in an equation leads to *two* equations with the absolute value signs removed. As an example,

$$|ax+b| = c \text{ means } ax+b = c \text{ or } -(ax+b) = c.$$

caution

A word of warning: the apparent solutions obtained by the above method may not solve the original absolute value equation! Absolute value equations are one class of equations (there are others, as we shall see) in which it is very important to check your final answer in the original equation. An apparent solution that does not solve the original problem is called an **extraneous solution**.

example 3

Solve the absolute value equations.

a. $|3x-2| = 1$ **b.** $|x-4| = |2x+1|$ **c.** $|6x-7| + 5 = 3$

Solutions:

a.
$$|3x-2| = 1$$
$$3x-2 = 1 \quad \text{or} \quad -(3x-2) = 1$$
$$3x = 3 \qquad\qquad 3x-2 = -1$$
$$x = 1 \qquad\qquad 3x = 1$$
$$\qquad\qquad\qquad x = \frac{1}{3}$$

The first step is to rewrite the absolute value equation without absolute values. The result is two linear equations that can be solved using the methods illustrated in this section. Note that this equation has two solutions, and they both solve the original equation.

section 2.1   Linear Equations in One Variable

b.
$$|x-4|=|2x+1|$$
$$x-4=\pm(2x+1) \quad \text{or} \quad -(x-4)=\pm(2x+1)$$
$$x-4=2x+1 \qquad\qquad -(x-4)=2x+1$$
$$-x=5 \qquad\qquad\qquad -3x=-3$$
$$x=-5 \qquad\qquad\qquad x=1$$

Because of the two absolute value quantities, the one original equation could potentially lead to four linear equations. But two of them are equivalent to the other two (as you should verify). We are thus left with two equations to solve, and the solution set is $\{-5, 1\}$. Note that both solutions satisfy the original equation.

c.
$$|6x-7|+5=3$$
$$6x-7=-2 \quad \text{or} \quad -(6x-7)=-2$$
$$6x=5 \qquad\qquad -6x=-9$$
$$x=\frac{5}{6} \qquad\qquad x=\frac{3}{2}$$
solution is $\varnothing$

If we proceed to solve this equation as the previous two, everything appears to work until the end. When we check the solutions, we find $\left|6\left(\frac{5}{6}\right)-7\right|+5 \neq 3$ and $\left|6\left(\frac{3}{2}\right)-7\right|+5 \neq 3$.

Example 3c illustrates the importance of checking the final answers in the original equation. If we had thought about the equation $|6x-7|+5=3$ before attempting to solve it, especially in the form $|6x-7|=-2$, we wouldn't have been surprised that there is no solution: any absolute value expression is automatically non-negative, so there is no value of x for which $|6x-7|$ could be -2.

Before leaving the subject of absolute value equations, we will use a different technique to solve one last equation. In Section 1.1 we saw that $|x-a|$ means, geometrically, the distance between the real numbers x and a. The solution set of the equation $|x-5|=|x+3|$ thus consists of all real numbers x which are an equal distance from 5 and -3 (note that $|x+3|=|x-(-3)|$). A moment's thought tells us that the only solution is the number 1, the number halfway between -3 and 5:

Figure 1: Solving $|x-5|=|x+3|$ Geometrically

Solve the equation $|x-5|=|x+3|$ algebraically, as in *Example 3b*, and verify that the solution set does indeed consist of only one number.

chapter two   Equations and Inequalities of One Variable

Topic 4:   Solving Linear Equations for One Variable

One common task in applied mathematics is to solve a given equation in two or more variables for one of the variables. **Solving for a variable** means to transform the equation into an equivalent one in which the specified variable is isolated on one side of the equation. For linear equations we accomplish this by the same methods we have used in the previous examples.

example 4

Solve the following equations for the specified variable. All of the equations are formulas that arise in various applications, and they are linear in the specified variable.

a. $P = 2l + 2w$. Solve for w.

b. $A = P\left(1 + \dfrac{r}{m}\right)^{mt}$. Solve for P.

c. $S = 2\pi r^2 + 2\pi rh$. Solve for h.

Solutions:

a.
$$P = 2l + 2w$$
$$P - 2l = 2w$$
$$\frac{P - 2l}{2} = w$$
$$w = \frac{P - 2l}{2}$$

This is the formula for the perimeter P of a rectangle of length l and width w. We first add $-2l$ to both sides, and then multiply by $\dfrac{1}{2}$.

The last equation is no different from the preceding one, but it is conventional to put the specified variable on the left side of the equation.

b.
$$A = P\left(1 + \frac{r}{m}\right)^{mt}$$
$$\frac{A}{\left(1 + \dfrac{r}{m}\right)^{mt}} = P$$
$$P = A\left(1 + \frac{r}{m}\right)^{-mt}$$

This is the equation for compound interest. If principal P is invested at an annual rate r for t years, compounded m times a year, the value of the investment at time t is A. This formula is linear in the variables P and A, though not in m, t, or r.

One step is all that is required to solve this equation for P, but the last equation uses one of the properties of exponents to make the result neater.

c.
$$S = 2\pi r^2 + 2\pi rh$$
$$S - 2\pi r^2 = 2\pi rh$$
$$\frac{S - 2\pi r^2}{2\pi r} = h$$
$$h = \frac{S - 2\pi r^2}{2\pi r}$$

This is the formula for the surface area of a right circular cylinder of radius r and height h. It is linear in the variables S and h, but not in r.

Two steps are all that are necessary to solve this formula for h. Solving this formula for r requires a technique that will be discussed in Section 2.3.

section 2.1   Linear Equations in One Variable

Topic 5:   Interlude: Distance and Interest Problems

Many applications lead to equations more complicated than those that we have studied so far, but good examples of linear equations arise from certain distance and simple interest problems. This is because the basic distance and simple interest formulas are linear in all of their variables:

$$d = rt, \text{ where } d \text{ is the distance traveled at rate } r \text{ for time } t$$
and
$$I = Prt, \text{ where } I \text{ is the interest earned on principal } P \text{ invested at rate } r \text{ for time } t$$

example 5

The distance from Shreveport, LA to Austin, TX by one route is 325 miles. If Kevin made the trip in five and a half hours, what was his average speed?

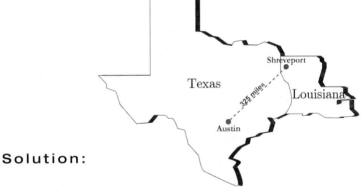

Solution:

We know d and t, and need to solve the linear equation $325 = \frac{11}{2}r$ for r (note that we have written five and a half as $\frac{11}{2}$). This is accomplished by multiplying both sides by $\frac{2}{11}$:

$$325 = \frac{11}{2}r$$

$$\frac{650}{11} = r$$

$$r = 59.1 \text{ miles/hour (rounded to nearest tenth)}$$

Alternatively, the time can be expressed in decimal form:

$$325 = 5.5r$$

$$\frac{325}{5.5} = r$$

$$r = 59.1 \text{ miles/hour (rounded to nearest tenth)}$$

chapter two Equations and Inequalities of One Variable

example 6

Julie invested $1500 in a risky high-tech stock on January 1st. On July 1st, her stock is worth $2100. She knows that her investment is very volatile and that it does not earn interest at a constant rate, but she wants to determine her average annual rate of return at this point in the year. What effective annual rate of return has she earned so far?

Solution:

The interest that Julie has earned in half a year is $600 (or $2100 − $1500). Replacing P with 1500, t with $\frac{1}{2}$, and I with 600 in the formula $I = Prt$, we have:

$$600 = (1500)\left(\frac{1}{2}\right)r$$

$$\frac{1200}{1500} = r$$

$$r = 0.8$$

$r = 80\%$ average rate of return per year

exercises

Solve the following linear equations. See Examples 1 and 2.

1. $-3(2t-4) = 7(1-t)$
2. $5(2x-1) = 3(1-x) + 5x$
3. $\dfrac{y+5}{4} = \dfrac{1-5y}{6}$
4. $3x + 5 = 3(x+3) - 4$
5. $3w + 5 = 2(w+3) - 4$
6. $3x + 5 = 3(x+3) - 5$
7. $\dfrac{4s-3}{2} + \dfrac{7}{4} = \dfrac{8s+1}{4}$
8. $\dfrac{4x-3}{2} + \dfrac{3}{8} = \dfrac{7x+3}{4}$
9. $\dfrac{4z-3}{2} + \dfrac{3}{8} = \dfrac{8z+3}{4}$
10. $3(2w+13) = 5w + w\left(7 - \dfrac{3}{w}\right)$
11. $\dfrac{6}{7}(m-4) - \dfrac{11}{7} = 1$
12. $0.08p + 0.09 = 0.65$
13. $0.6x + 0.08 = 2.3$
14. $0.9x + 0.5 = 1.3x$
15. $0.73x + 0.42(x-2) = 0.35x$

section 2.1 Linear Equations in One Variable

16. $\dfrac{8y-2}{4} + \dfrac{6}{8} = \dfrac{16y+2}{8}$

17. $\dfrac{3}{7}(y-2) - \dfrac{14}{7} = -5$

18. $6(5w-5) = -31(3-w)$

19. $\dfrac{7x-5}{4} + \dfrac{14}{8} = \dfrac{14x+4}{8}$

20. $\dfrac{3}{11}(y-2) - \dfrac{33}{11} = -6$

21. $3z + 3 = 3(z+4) - 9$

22. $4y + 9 = 4(y+4) - 10$

23. $2.8x + 1.2 = 3.2x$

24. $0.73z + 0.34 = 9.1$

25. $0.24x + 0.58(x-6) = 0.82x - 3.67$

Solve the following absolute value equations. See Example 3.

26. $|3x-2| = 5$

27. $-|3y+5| + 6 = 2$

28. $|4x+3| + 2 = 0$

29. $|6x-2| = 0$

30. $|-8x+2| = 14$

31. $|2x-109| = 731$

32. $|4x-4| - 40 = 0$

33. $|5x-3| = 7$

34. $|4x+15| = 3$

35. $-|6x+1| = 11$

36. $|-14y+3| + 3 = 2$

37. $|3x-2| - 1 = |5-x|$

Solve the following absolute value equations geometrically and algebraically. See Figure 1.

38. $|x+3| = |x-7|$

39. $|x-3| - |x-7| = 0$

40. $|2-x| = |2+x|$

41. $|x| = |x+1|$

42. $|x+97| = |x+101|$

43. $\left|x + \dfrac{1}{4}\right| = \left|x - \dfrac{3}{4}\right|$

44. $|z-51| - |z-5| = 0$

45. $\left|x - \dfrac{5}{7}\right| = \left|x + \dfrac{3}{7}\right|$

46. $|6y-3| = |5y+5|$

chapter two   Equations and Inequalities of One Variable

Solve the following equations for the indicated variable. See Example 4.

47. **Circumference of a Circle:** $C = 2\pi r$; solve for r

48. **Ideal Gas Law:** $PV = nRT$; solve for T

49. **Velocity:** $v^2 = v_0^2 + 2ax$; solve for a

50. **Area of a Trapezoid:** $A = \frac{1}{2}(B+b)h$; solve for h

51. **Temperature Conversions:** $C = \frac{5}{9}(F-32)$; solve for F

52. **Volume of a Right Circular Cone:** $V = \frac{1}{3}\pi r^2 h$; solve for h

53. **Surface Area of a Cube:** $A = 2lw + 2wh + 2hl$; solve for h

54. **Distance:** $d = rt_1 + rt_2$; solve for r

55. **Kinetic Energy of Protons:** $K = \frac{1}{2}mv^2$; solve for m

56. **Finance:** $A = P(1+rt)$; solve for t

Use a calculator with an equation solver to solve the following problems. Round your answers to the nearest hundredth when necessary.

57. $453x = 95(34x + 291)$

58. $-0.23 = 0.79x - 0.47(x + 0.98)$

59. $254 + 0.98(x - 1024) = 0$

60. $323x - 1745 = 531(68x - 887)$

Solve the following application problems. See Example 5 and 6.

61. A riverboat leaves port and proceeds to travel downstream at an average speed of 15 miles per hour. How long will it take for the boat to arrive at the next port, 95 miles downstream?

62. Two trucks leave a warehouse at the same time. One travels due east at an average speed of 45 miles per hour, and the other travels due west at an average speed of 55 miles per hour. After how many hours will they be 450 miles apart?

63. Two cars leave a rest stop at the same time and proceed to travel down the highway in the same direction. One travels at an average rate of 62 miles per hour, and the other at an average rate of 59 miles per hour. How far apart are the two cars after four and a half hours?

64. Two trains are 630 miles apart, heading directly toward each other. The first train is traveling at 95 mph, and the second train is traveling at 85 mph. How long will it be before the trains pass each other?

65. Two brothers, Rick and Tom, each inherit $10,000. Rick invests his inheritance in a savings account with an annual return of 2.25%, while Tom invests his in a CD paying 6.15% annually. How much more money does Tom have than Rick after 1 year?

66. Sarah, sister to Rick and Tom in the previous problem, also inherits $10,000, but she invests her inheritance in a global technology mutual fund. At the end of 1 year, her investment is worth $12,800. What has her effective annual rate of return been?

67. Bob buys a large screen digital TV priced at $9,500, but pays $10,212.50 with tax. What is the rate of tax where Bob lives?

68. Will and Matt are brothers. Will is 6 feet, 4 inches tall, and Matt is 6 feet, 7 inches tall. How tall is Will as a percentage of Matt's height? How tall is Matt as a percentage of Will's height?

69. A farmer wants to fence in three square garden plots situated along a road, as shown, and he decides not to install fencing along the edge of the road. If he has 182 feet of fencing material total, what dimensions should he make each square plot?

70. Find three consecutive integers whose sum is 288. (Hint: if n represents the smallest of the three, then $n + 1$ and $n + 2$ represent the other two numbers.)

71. Find three consecutive odd integers whose sum is 165. (Hint: if n represents the smallest of the three, then $n + 2$ and $n + 4$ represent the other two numbers.)

72. Kathy buys last year's best selling novel, in hardcover, for $15.05. This is a 30% discount from the original price. What was the original price?

73. The highest point on Earth is the peak of Mount Everest. If you climbed to the top, you would be approximately 29,035 feet above sea level. Remembering that a mile is 5280 feet, what percentage of the height of the mountain would you have to climb to reach a point two miles above sea level?

section 2.1   Linear Equations in One Variable

technology exercises

Computer algebra systems are capable of solving simple equations for specific variables. In *Mathematica*, the command to do this is **Solve.** For instance, the equation relating the area A of a trapezoid to the lengths of the two bases B and b and the height h can be solved first for B and then for h as shown below. Notice the form that *Mathematica* uses to present the answer, and note particularly that *Mathematica*'s answer may not be in the simplest possible form.

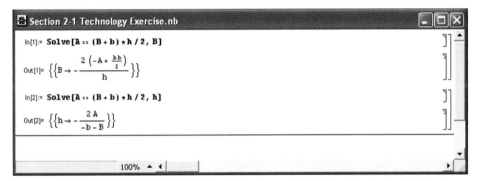

Use a computer algebra system, if available, to solve the following equations for the specified variable. (See Appendix A for guidance on using computer algebra systems like Mathematica.)

74. Surface Area of a Cylinder: $S = 2\pi r^2 + 2\pi rh$; solve for h

75. Force on an Object in Circular Motion: $F = \dfrac{mv^2}{r}$; solve for r

76. Actual Cash Value of Property: $V = p - \left(\dfrac{ap}{e}\right)$; solve for p

77. Sum of a Geometric Series: $S = \dfrac{a}{1-r}$; solve for r

chapter two: Equations and Inequalities of One Variable

2.2 Linear Inequalities in One Variable

TOPICS

1. Solving linear inequalities
2. Solving compound linear inequalities
3. Solving absolute value inequalities
4. Interlude: translating inequality phrases

Topic 1: Solving Linear Inequalities

If the equality symbol in a linear equation is replaced with $<$, $\leq$, $>$, or $\geq$, the result is a **linear inequality**. One difference between linear equations and linear inequalities is the way in which the solutions are described. Typically, the solution of a linear inequality consists of some interval of real numbers; such solutions can be described graphically or with interval notation. The process of obtaining the solution, however, is much the same as the process for solving linear equations, with the one important difference discussed below.

In solving linear inequalities, the field properties outlined in Section 1.2 are all still applicable, and it is often necessary to use, for example, the distributive and commutative properties in order to simplify one or both sides of an inequality. The additive version of the two cancellation laws is also used in the same way as in solving equations. The one difference lies in applying the multiplicative version of cancellation. Note that if both sides of the true statement $-3 < 2$ are multiplied by -1, and if no other changes are made, we obtain the false statement $3 < -2$. This is a basic illustration of the fact that the multiplicative cancellation law is slightly different for inequalities.

section 2.2 Linear Inequalities in One Variable

Cancellation Properties for Inequalities

Throughout this table, A, B, and C represent algebraic expressions. Each of the two properties is stated for the inequality symbol <, but they are also true for the other three symbols with the appropriate changes made.

Property	Description
$A < B \Leftrightarrow A + C < B + C$	Adding the same quantity to both sides of an inequality results in an equivalent inequality.
If $C > 0$, $A < B \Leftrightarrow A \cdot C < B \cdot C$ If $C < 0$, $A < B \Leftrightarrow A \cdot C > B \cdot C$	If both sides of an inequality are multiplied by a positive quantity, the sense of the inequality is unchanged. But if both sides are multiplied by a negative quantity, the sense of the inequality is reversed.

Keep in mind that multiplying (or dividing) both sides of an inequality by a negative quantity necessitates reversing, or "flipping," the inequality symbol. This will be illustrated several times in the examples to follow.

example 1

Solve the following inequalities, using interval notation to describe the solution set.

    a. $5 - 2(x - 3) \leq -(1 - x)$ **b.** $\dfrac{3(a-2)}{2} < \dfrac{5a}{4}$

Solutions:

a. $5 - 2(x - 3) \leq -(1 - x)$ We begin by using the distributive property, and then proceed to combine like terms.

$5 - 2x + 6 \leq -1 + x$

$-2x + 11 \leq -1 + x$

$-3x \leq -12$ At this point, division by -3 is all that is required to solve the inequality. *Note the reversal of the inequality symbol.*

$x \geq 4$

Solution is $[4, \infty)$

continued on next page...

chapter two   Equations and Inequalities of One Variable

b. $\dfrac{3(a-2)}{2} < \dfrac{5a}{4}$

$4\left(\dfrac{3(a-2)}{2}\right) < 4\left(\dfrac{5a}{4}\right)$

$6(a-2) < 5a$

$6a - 12 < 5a$

$a < 12$

Solution is $(-\infty, 12)$

Just as with equations, fractions in inequalities can be eliminated by multiplying both sides by the least common denominator.

The solutions in *Example 1* were described using interval notation. The solutions can also be described by graphing. Graphing a solution to an inequality can lead to a better understanding of which real numbers solve the inequality.

The symbols used in this text for graphing intervals are the same as the symbols in interval notation. Parentheses are used to indicate excluded endpoints of intervals and brackets are used when the endpoints are included in the interval. The portion of the number line that constitutes the interval is then shaded. (Other commonly used symbols in graphing are open circles for parentheses and filled-in circles for brackets.)

For example, the two solutions above are graphed as follows:

Topic 2: Solving Compound Linear Inequalities

A **compound inequality** is a statement containing two inequality symbols and can be interpreted as two distinct inequalities joined by the word "and." Compound inequalities often arise in applications as a result of constraints placed upon some quantity. As an example, suppose the course grade in a given class depends solely on the grades of 5 exams, all equally weighted and each worth a maximum of 100 points. Suppose Janice's scores on the first four tests are 67, 82, 73, and 85, and she is wondering what score she needs to make on the fifth test to get a B in the class. If the B range corresponds to an average greater than or equal to 80 and less than 90, and if x represents the fifth test score, she needs to solve the following compound inequality:

section 2.2 Linear Inequalities in One Variable

$$80 \le \frac{67 + 82 + 73 + 85 + x}{5} < 90$$

This could be solved by breaking it into the two inequalities

$$80 \le \frac{67 + 82 + 73 + 85 + x}{5} \quad \text{and} \quad \frac{67 + 82 + 73 + 85 + x}{5} < 90,$$

but a more efficient approach is to solve both at the same time. This is illustrated in the following example.

example 2

Solve the compound inequalities.

a. $80 \le \dfrac{67 + 82 + 73 + 85 + x}{5} < 90$ b. $-1 < 3 - 2x \le 5$

Solutions:

a. $80 \le \dfrac{67 + 82 + 73 + 85 + x}{5} < 90$

$400 \le 307 + x < 450$

$93 \le x < 143$

Solution is $[93, 143)$

Given either of the two inequalities alone, multiplying through by 5 would probably be the first step. We do this with the compound inequality as well, and then subtract 307 from all three expressions. Of course, if this compound inequality relates to test scores, as indicated in the paragraph above, the solution set is actually $[93, 100]$.

b. $\quad -1 < 3 - 2x \le 5$

$-4 < -2x \le 2$

$2 > x \ge -1$

$-1 \le x < 2$

Solution is $[-1, 2)$

We begin by subtracting 3 from all three expressions, and then divide each expression by -2. Don't forget to reverse each inequality, since we are dividing by a negative number. The final compound inequality is identical to the one before it, but has been written so that the smaller number appears first.

Topic 3: Solving Absolute Value Inequalities

An **absolute value inequality** is an inequality in which some variable expression appears inside absolute value symbols. In the problems that we will study, the inequality would be linear if the absolute value symbols were not there.

The geometric meaning of absolute value provides the method by which absolute value inequalities are solved. Recall that $|x|$ can be interpreted as the distance between x and 0 on the real number line. Assuming for the moment that a is a positive real number, the inequality $|x| < a$ means that x is less than a units from 0, and the inequality $|x| > a$ means that x is greater than a units from 0. (Similar interpretations hold for the symbols $\leq$ and $\geq$.) This means that absolute value inequalities can be written without absolute values as follows:

$$|x| < a \Leftrightarrow -a < x < a$$

and

$$|x| > a \Leftrightarrow x < -a \text{ or } x > a$$

While most absolute value inequalities will be more complicated than the two above, they will serve as paradigms for the rewriting of such inequalities without the absolute values.

example 3

Solve the absolute value inequalities.

a. $|3y - 2| + 2 \leq 6$ **b.** $|4 - 2x| > 6$ **c.** $|5 + 2s| \leq -3$

Solutions:

a.
$$|3y - 2| + 2 \leq 6$$
$$|3y - 2| \leq 4$$
$$-4 \leq 3y - 2 \leq 4$$
$$-2 \leq 3y \leq 6$$
$$-\frac{2}{3} \leq y \leq 2$$

Solution is $\left[-\frac{2}{3}, 2\right]$

In order to remove the absolute value symbols, we first subtract 2 from both sides.

After rewriting the inequality as described above, we are faced with a compound inequality to solve.

The graph of the solution is:

section 2.2   Linear Inequalities in One Variable

b. $|4-2x| > 6$

$4 - 2x < -6 \qquad 4 - 2x > 6$

$-2x < -10 \quad \text{or} \quad -2x > 2$

$x > 5 \qquad\qquad x < -1$

Solution is $(-\infty, -1) \cup (5, \infty)$

We can immediately rewrite the inequality without absolute values, and begin solving the two independent inequalities.

The graph of the solution is:

c. $|5+2s| \le -3$

solution is $\varnothing$

Just as in *Example 3c* of Section 2.1, we conclude that the solution set is the empty set, as it is impossible for the absolute value of *any* expression to be negative.

Topic 4: Interlude: Translating Inequality Phrases

Many real-world applications leading to inequalities involve notions such as "is no greater than", "at least as large as", "does not exceed", and so on. Phrases such as these all have precise mathematical translations that use one of the four inequality symbols. The following example will illustrate how some sample phrases are translated.

example 4

Express each of the following problems as an inequality, and then solve the inequality.

a. The average daily high temperature in Santa Fe, NM over the course of three days exceeded 75. Given that the high on the first day was 72 and the high on the third day was 77, what was the minimum high temperature on the second day?

b. As a test for quality at a plant manufacturing silicon wafers for computer chips, a random sample of 10 batches of 1000 wafers each must not detect more than 5 defective wafers per batch on average. In the first 9 batches tested, the average number of defective wafers per batch is found to be 4.78 (to the nearest hundredth). What is the maximum number of defective wafers that can be found in the 10th batch for the plant to pass the quality test?

continued on next page...

chapter two   Equations and Inequalities of One Variable

Solutions:

a.
$$\frac{72 + x + 77}{3} > 75$$
$$149 + x > 225$$
$$x > 76$$

Let x denote the high temperature on the second day. We obtain our inequality by setting the average of the three high temperatures greater than 75. We now know that the high temperature on the second day was greater than 76.

b.
$$\frac{(9)(4.78) + x}{10} \leq 5$$
$$43.02 + x \leq 50$$
$$x \leq 6.98$$

Let x denote the maximum allowable number of defective wafers in the 10th batch. The number of defective wafers found in the first 9 batches is $(9)(4.78) = (43.02)$. When this number is added to x and divided by 10, the result must be less than or equal to 5. Of course, it is not possible to have a fractional number of wafers, so there must have actually been 43 defective wafers in the first 9 batches, and the maximum allowable number of defective wafers in the last batch is 7.

exercises

Determine which elements of $S = \{12, -9, 3.14, -2.83, 1, 5.24, 8, -3, 4\}$ satisfy each inequality below.

1. $7y - 33.6 < -8.6 + 2y$

2. $-2.2y - 18.8 \geq 5.2(1 - y)$

3. $-40 < 4y - 8 \leq 4$

4. $-4 < -2(z - 2) \leq 2$

Solve the following linear inequalities. Describe the solution set using interval notation and by graphing. See Example 1.

5. $4 + 3t \leq t - 2$

6. $x - 7 \geq 5 + 3x$

7. $5y - 24 < -9.6 + 2y$

8. $-\dfrac{v+2}{3} > \dfrac{5-v}{2}$

9. $4.2x - 5.6 < 1.6 + x$

10. $8.5y - 3.5 \geq 2.5(3 - y)$

section 2.2   Linear Inequalities in One Variable

11. $-2(3-x) < -2x$

12. $\dfrac{1-x}{5} > \dfrac{-x}{10}$

13. $4w + 7 \leq -7w + 4$

14. $-5(p-3) > 19.8 - p$

15. $\dfrac{6f-2}{5} < \dfrac{5f-3}{4}$

16. $\dfrac{u-6}{7} \geq \dfrac{2u-1}{3}$

17. $0.04n + 1.7 < 0.13n - 1.45$

18. $2k + \dfrac{3}{2} < 5k - \dfrac{7}{3}$

19. $\dfrac{4x+4}{5} > \dfrac{3x+2.6}{4}$

20. $-1.4z - 19.6 \geq 4.4(1-z)$

21. $6m + \dfrac{7}{4} > \dfrac{4m+5.8}{5}$

22. $-3.9n - 5.4 \geq 6.2(2-3n)$

Solve the following compound inequalities. Describe the solution set using interval notation and by graphing. See Example 2.

23. $-4 < 3x - 7 \leq 8$

24. $5 \leq 2m - 3 \leq 13$

25. $-36 < 3x - 6 \leq 12$

26. $-10 < -2(4+y) \leq 9$

27. $-8 \leq \dfrac{z}{2} - 4 < -5$

28. $3 < \dfrac{w+3}{8} \leq 9$

29. $4 \leq \dfrac{p+7}{-2} < 9$

30. $\dfrac{1}{3} < \dfrac{7}{6}(l-3) < \dfrac{2}{3}$

31. $\dfrac{1}{4} \leq \dfrac{g}{2} - 3 < 5$

32. $-1.2 \leq \dfrac{x+3}{-5} \leq 0.2$

33. $0.08 < 0.03c + 0.13 \leq 0.16$

Solve the following absolute value inequalities. Describe the solution set using interval notation and by graphing. See Example 3.

34. $|x - 2| \geq 5$

35. $|4 - 2x| > 11$

36. $4 + |3 - 2y| \leq 6$

37. $4 + |3 - 2y| > 6$

38. $2|z + 5| < 12$

39. $7 - \left|\dfrac{q}{2} + 3\right| \geq 12$

chapter two   Equations and Inequalities of One Variable

40. $4|z+3| \leq 28$
41. $-3|4-t| < -6$
42. $-3|4-t| > -6$

43. $3|4-t| < -6$
44. $7 - |4-2y| \leq -5$
45. $11 - \left|\dfrac{w}{4}+1\right| \geq 12$

46. $5.5 + |x - 7.2| \leq 3.5$
47. $6 - 5|x+2| \geq -4$
48. $|2x-1| < x+4$

The words "and" and "or" can appear explicitly between two inequalities, and their meaning in such cases is the same as in absolute value inequalities. If two inequalities are joined by the word "and", the solution set consists of all those real numbers that satisfy both inequalities; that is, the solution set overall is the intersection of the two individual solution sets. If the word "or" appears between two inequalities, the solution set consists of all those real numbers that satisfy at least one of the two inequalities; in other words, the solution set overall is the union of the two individual solution sets.

Guided by the above paragraph, solve the following inequality problems. Describe the solution set using interval notation and by graphing.

49. $t < 2t - 3$ and $-3(t+4) > -57$
50. $7 - \dfrac{3x}{5} < \dfrac{2}{5}$ or $2 - 3x \geq 5$

51. $-2(a-1) < 4$ and $6 + a \leq 9$
52. $-2(a-1) < 4$ and $6 - a \leq 9$

53. $-2(a-1) < 4$ or $6 + a \leq 9$
54. $\dfrac{5n+6}{3} < -10$ and $-3(n-1) < -6$

55. $\dfrac{23x-3}{-7} \leq 7$ and $-x < -(4x-9)$
56. $7 - \dfrac{x}{3} \leq 14 + \dfrac{x}{2}$ or $-3x < 15$

section 2.2   Linear Inequalities in One Variable

Solve the following application problems. See Example 4.

57. In a class in which the final course grade depends entirely on the average of four equally weighted 100-point tests, Cindy has scored 96, 94, and 97 on the first three. The professor has announced that there will be a fifteen-point bonus problem on the fourth test, and that anyone who finishes the semester with an average of more than 100 will receive an A+. What range of scores on the fourth test will give Cindy an A for the semester (an average between 90 and 100, inclusive), and what range will give Cindy an A+?

58. In a series of 30 racquetball games played to date, Larry has won 10, giving him a winning average so far of 33.3% (to the nearest tenth). If he continues to play, what interval describes the number of games he must now win in a row to have an overall winning average greater than 50%?

59. Assume that the national average SAT score for high school seniors is 1020 out of 1600. A group of 7 students receive their scores in the mail, and 6 of them look at their scores. Two students scored 1090, one got an 1120, two others each got a 910, and the sixth student received an 880. What range of scores can the seventh student receive to pull the group's average above the national average?

60. The United States government tries to keep the inflation rate below 5.0% on an annual basis. Assume that inflation rates for the first 3 quarters of a given year are as follows: 5.2%, 4.3%, and 4.7%. What range of inflation rates for the final quarter would satisfy the government's goal?

chapter two — Equations and Inequalities of One Variable

2.3 Quadratic Equations in One Variable

TOPICS

1. Solving quadratic equations by factoring
2. Solving "perfect square" quadratic equations
3. Solving quadratic equations by completing the square
4. The quadratic formula
5. Interlude: gravity problems

Topic 1: Solving Quadratic Equations by Factoring

In Section 2.1, we studied first-degree polynomial equations in one variable. We will now expand the class of one-variable polynomial equations that we can solve to include **quadratic** (or **second-degree polynomial**) equations.

Recall that the method of solving linear equations was particularly straightforward, and that the method always works for *any* such equation. We will, by the end of this section, develop a method for solving one-variable second-degree equations that is also guaranteed to work. This is in contrast to polynomial equations in general. In fact, it can be shown that for polynomial equations of degree five and higher there is *no* method that always works.

Our development will begin with a formal definition of quadratic equations, and we will then proceed to study those quadratic equations that can be solved by applying the factoring skills learned in Section 1.5.

section 2.3 Quadratic Equations in One Variable

Quadratic Equations

A **quadratic equation in one variable**, say the variable x, is an equation that can be transformed into the form
$$ax^2 + bx + c = 0,$$
where $a, b,$ and c are real numbers and $a \neq 0$. Such equations are also called **second-degree** equations, as x appears to the second power. The name quadratic comes from the Latin word *quadrus*, meaning "square."

The key to using factoring to solve a quadratic equation, or indeed any polynomial equation, is to rewrite the equation so that 0 appears by itself on one side. This then allows us to use the Zero-Factor Property discussed in Section 1.2.

Zero-Factor Property

Let A and B represent algebraic expressions. If the product of A and B is 0, then at least one of A and B is itself 0. That is,
$$AB = 0 \Rightarrow A = 0 \text{ or } B = 0.$$

If the trinomial $ax^2 + bx + c$ can be factored, it can be written as a product of two linear factors A and B. The Zero-Factor Property then implies that the only way for $ax^2 + bx + c$ to be 0 is if one (or both) of A and B is 0. This is all we need to solve the equation.

chapter two   Equations and Inequalities of One Variable

example 1

Solve the quadratic equations by factoring.

a. $x^2 + \dfrac{5x}{2} = \dfrac{3}{2}$ **b.** $s^2 + 9 = 6s$ **c.** $5x^2 + 10x = 0$

Solutions:

a.
$$x^2 + \frac{5x}{2} = \frac{3}{2}$$
$$2x^2 + 5x = 3$$
$$2x^2 + 5x - 3 = 0$$
$$(2x-1)(x+3) = 0$$
$$2x - 1 = 0 \quad \text{or} \quad x + 3 = 0$$
$$x = \frac{1}{2} \quad \text{or} \quad x = -3$$

To make the polynomial easier to factor, we multiply both sides by the LCD.

Although we could factor $2x^2 + 5x$, this would not do us any good. We must have 0 on one side in order to apply the Zero-Factor Property.

After factoring, we have two linear equations to solve. The solution set is $\left\{\dfrac{1}{2}, -3\right\}$.

b.
$$s^2 + 9 = 6s$$
$$s^2 - 6s + 9 = 0$$
$$(s-3)^2 = 0$$
$$s - 3 = 0 \quad \text{or} \quad s - 3 = 0$$
$$s = 3$$

Again, we rewrite the equation with 0 on one side, and then factor the quadratic.

In this example, the two linear factors are the same. In such cases, the single solution is called a *double solution* or a *double root*.

c.
$$5x^2 + 10x = 0$$
$$5x(x+2) = 0$$
$$5x = 0 \quad \text{or} \quad x + 2 = 0$$
$$x = 0 \quad \text{or} \quad x = -2$$

An alternative approach in this example would be to divide both sides by 5 at the very beginning. This would lead to the equation $x(x+2) = 0$, which gives us the same solution set of $\{0, -2\}$.

section 2.3   **Quadratic Equations in One Variable**

Topic 2:   ## Solving "Perfect Square" Quadratic Equations

The factoring method is fine when it works, but there are two potential problems with the method: (1) the second degree polynomial in question might not factor over the integers, and (2) even if the polynomial does factor, the factored form may not be obvious.

In some cases where the factoring method is unsuitable, the solution can be obtained by using our knowledge of square roots. If A is an algebraic expression and if c is a constant, the equation $A^2 = c$ means $A = \sqrt{c}$ or $A = -\sqrt{c}$. We will find it convenient to summarize this as:

$$A^2 = c \text{ implies } A = \pm\sqrt{c}$$

If a given quadratic equation can be written in the form $A^2 = c$, we can use the above observation to obtain two linear equations that can be easily solved.

example 2

Solve the quadratic equations by taking square roots.

a. $(2x+3)^2 = 8$ **b.** $(x-5)^2 + 4 = 0$

Solutions:

a. $(2x+3)^2 = 8$

$2x+3 = \pm\sqrt{8}$

$2x+3 = \pm 2\sqrt{2}$

$2x = -3 \pm 2\sqrt{2}$

$x = \dfrac{-3 \pm 2\sqrt{2}}{2}$

We begin by taking the square root of each side, keeping in mind that there are two numbers whose square is 8.
We solve the two linear equations at once by subtracting 3 from both sides and then dividing both sides by 2. The solution set is $\left\{\dfrac{-3+2\sqrt{2}}{2}, \dfrac{-3-2\sqrt{2}}{2}\right\}$.

b. $(x-5)^2 + 4 = 0$

$(x-5)^2 = -4$

$x-5 = \pm\sqrt{-4}$

$x-5 = \pm 2i$

$x = 5 \pm 2i$

Before taking square roots, we isolate the perfect square algebraic expression on one side, and put the constant on the other.
In this example, taking square roots leads to two complex number solutions. (See Section 1.6 for a review of complex numbers.) The solution set is $\{5+2i, 5-2i\}$.

chapter two Equations and Inequalities of One Variable

Topic 3: Solving Quadratic Equations by Completing the Square

There are potential pitfalls, once again, with the method just developed. If the quadratic equation under consideration appears in the form $A^2 = c$, the method works well (even if the ultimate solutions wind up being complex, as in *Example 2b*). But what if the equation doesn't have the form $A^2 = c$?

The method of **completing the square** allows us to write an arbitrary quadratic equation $ax^2 + bx + c = 0$ in the desired form. This method is outlined below.

Completing the Square

Step 1: Write the equation $ax^2 + bx + c = 0$ in the form $ax^2 + bx = -c$.

Step 2: Divide by a, if $a \neq 1$, so that the coefficient of x^2 is 1: $x^2 + \dfrac{b}{a}x = -\dfrac{c}{a}$.

Step 3: Divide the coefficient of x by 2, square the result, and add this to both sides.

Step 4: The trinomial on the left side will now be a perfect square. That is, it can be written as the square of an algebraic expression.

At this point, the equation will have the form $A^2 = c$ and can be solved by taking the square root of both sides.

example 3

Solve the quadratic equations by completing the square.

a. $x^2 - 2x - 6 = 0$
b. $9x^2 + 3x = 2$

section 2.3 — Quadratic Equations in One Variable

Solutions:

a.
$$x^2 - 2x - 6 = 0$$
$$x^2 - 2x = 6$$
$$x^2 - 2x + 1 = 6 + 1$$
$$(x-1)^2 = 7$$
$$x - 1 = \pm\sqrt{7}$$
$$x = 1 \pm \sqrt{7}$$

After moving the constant term to the right-hand side, we divide -2 (the coefficient of x) by 2 to get -1, and add $(-1)^2$ to both sides. The trinomial on the left can now be factored.

Taking square roots leads to two readily solved linear equations.

b.
$$9x^2 + 3x = 2$$
$$x^2 + \frac{1}{3}x = \frac{2}{9}$$
$$x^2 + \frac{1}{3}x + \frac{1}{36} = \frac{2}{9} + \frac{1}{36}$$
$$\left(x + \frac{1}{6}\right)^2 = \frac{1}{4}$$
$$x + \frac{1}{6} = \pm\frac{1}{2}$$
$$x = -\frac{1}{6} \pm \frac{1}{2}$$
$$x = \frac{1}{3}, -\frac{2}{3}$$

The constant term is already isolated on the right-hand side, so our first step is to divide by 9 (and simplify the resulting fractions, if possible).

Half of the coefficient of x is $\frac{1}{6}$, and the square of this is $\frac{1}{36}$.

After simplifying the sum of fractions on the right, we take the square root of each side.

Since the answer of $-\frac{1}{6} \pm \frac{1}{2}$ can be simplified, we do so to obtain the final answer.

It is time to mention an incidental benefit of the method we have devised. The polynomial in *Example 3a*, $x^2 - 2x - 6$, does not factor over the integers. That is, it cannot be written as a product of two first-degree polynomials with integer coefficients. Nevertheless, it can be factored as a product of two first-degree polynomials:

$$x^2 - 2x - 6 = \left(x - 1 - \sqrt{7}\right)\left(x - 1 + \sqrt{7}\right)$$

We know this because we know that the two solutions of the equation $x^2 - 2x - 6 = 0$ are $1 + \sqrt{7}$ and $1 - \sqrt{7}$, and we know that there is a close relationship between factors of a quadratic and the solutions of the equation in which that quadratic is equal to 0.

Specifically, if a quadratic can be factored as $(x-p)(x-q)$, then p and q solve the equation $(x-p)(x-q)=0$.

How does the quadratic $9x^2+3x-2$ factor? This quadratic comes from *Example 3b*, so we might guess factors of $x-\frac{1}{3}$ and $x+\frac{2}{3}$. But,

$$\left(x-\frac{1}{3}\right)\left(x+\frac{2}{3}\right) = x^2 + \frac{1}{3}x - \frac{2}{9}.$$

It shouldn't be surprising that the product of these two factors has a leading coefficient of 1, since each of them individually has a leading coefficient of 1. To get the correct leading coefficient, we need to multiply by 9:

$$9\left(x-\frac{1}{3}\right)\left(x+\frac{2}{3}\right) = 9\left(x^2+\frac{1}{3}x-\frac{2}{9}\right) = 9x^2+3x-2$$

To make the factors look better, we can factor 9 into two factors of 3 and rearrange the products:

$$9\left(x-\frac{1}{3}\right)\left(x+\frac{2}{3}\right) = 3\left(x-\frac{1}{3}\right) \cdot 3\left(x+\frac{2}{3}\right) = (3x-1)(3x+2)$$

Topic 4: The Quadratic Formula

The method of completing the square will always serve to solve any equation of the form $ax^2+bx+c=0$. But this begs the question: why not just solve $ax^2+bx+c=0$ once and for all? Since a, b, and c represent arbitrary constants, the ideal situation would be to find a formula for the solutions of $ax^2+bx+c=0$ based on a, b, and c. That is exactly what the quadratic formula is—a formula that gives the solution to *any* equation of the form $ax^2+bx+c=0$. We will derive the formula now, using what we have learned.

section 2.3 Quadratic Equations in One Variable

$$ax^2 + bx + c = 0$$
$$x^2 + \frac{b}{a}x = -\frac{c}{a}$$
$$x^2 + \frac{b}{a}x + \frac{b^2}{4a^2} = -\frac{c}{a} + \frac{b^2}{4a^2}$$
$$\left(x + \frac{b}{2a}\right)^2 = -\frac{4ac}{4a^2} + \frac{b^2}{4a^2}$$
$$\left(x + \frac{b}{2a}\right)^2 = \frac{b^2 - 4ac}{4a^2}$$
$$x + \frac{b}{2a} = \pm\frac{\sqrt{b^2 - 4ac}}{2a}$$
$$x = \frac{-b}{2a} \pm \frac{\sqrt{b^2 - 4ac}}{2a}$$
$$x = \frac{-b \pm \sqrt{b^2 - 4ac}}{2a}$$

We begin, as always in completing the square, by moving the constant to the right-hand side and dividing by a.

We next divide $\frac{b}{a}$ by 2 to get $\frac{b}{2a}$, and add $\left(\frac{b}{2a}\right)^2 = \frac{b^2}{4a^2}$ to both sides. Note that to add the fractions on the right, we need a common denominator of $4a^2$.

Taking square roots leads to two linear equations, which we then solve for x.

Since the fractions have the same denominator, they are easily added to obtain the final formula.

The Quadratic Formula

The solutions of the equation $ax^2 + bx + c = 0$ are $x = \frac{-b \pm \sqrt{b^2 - 4ac}}{2a}$.

Note:

- The equation has real solutions if $b^2 - 4ac > 0$.
- The equation has a double solution if $b^2 - 4ac = 0$.
- The equation has complex solutions (which are conjugates of one another) if $b^2 - 4ac < 0$.

(It should be mentioned, for completeness' sake, that the quadratic formula also applies if a, b, and c are complex constants, not merely real constants.)

chapter two   Equations and Inequalities of One Variable

example 4

Solve the quadratic equations using the quadratic formula.

a. $8y^2 - 4y = 1$ **b.** $t^2 + 6t + 13 = 0$

Solutions:

a.
$$8y^2 - 4y = 1$$
$$8y^2 - 4y - 1 = 0$$
$$y = \frac{-(-4) \pm \sqrt{(-4)^2 - (4)(8)(-1)}}{(2)(8)}$$
$$y = \frac{4 \pm \sqrt{16 + 32}}{16}$$
$$y = \frac{4 \pm \sqrt{48}}{16}$$
$$y = \frac{4 \pm 4\sqrt{3}}{16}$$
$$y = \frac{1 \pm \sqrt{3}}{4}$$

Before applying the quadratic formula, move all the terms to one side so that $a, b,$ and c can be identified correctly.

Apply the quadratic formula by making the appropriate replacements for $a, b,$ and c.

Note that there is a common factor of 4 that can be cancelled from the numerator and denominator in this case.

The two solutions are $\frac{1+\sqrt{3}}{4}$ and $\frac{1-\sqrt{3}}{4}$.

To use the quadratic formula, first define the variables A, B, and C: enter the desired value, select **STO**, then **ALPHA MATH** (A). Select **ALPHA** and **.** to enter a colon, and continue storing variables. Then enter the quadratic formula using **ALPHA** and the defined variables (remember to use parentheses!) To obtain both answers, enter the formula with both the positive and the negative square root.

b. $t^2 + 6t + 13 = 0$
$$t = \frac{-6 \pm \sqrt{36 - 52}}{2}$$
$$t = \frac{-6 \pm \sqrt{-16}}{2}$$
$$t = \frac{-6 \pm 4i}{2}$$
$$t = -3 \pm 2i$$

The equation is already in the proper form to apply the quadratic formula.

Note that at this point we know the solutions will be complex numbers.

The two solutions are $-3 + 2i$ and $-3 - 2i$.

Topic 5:   Interlude: Gravity Problems

When an object near the surface of the Earth is moving under the influence of gravity alone, its height above the surface is described by a quadratic polynomial in the variable t, where t stands for time and is usually measured in seconds.

section 2.3   Quadratic Equations in One Variable

The phrase "moving under the influence of gravity alone" means that all other forces that could potentially affect the object's motion, such as air resistance or mechanical lifting forces, are either negligible or absent. The phrase "near the surface of the earth" means that we are considering objects that travel short vertical distances relative to the earth's radius; the following formula doesn't apply, for instance, to rockets shot into orbit. As an example, think of someone throwing a baseball into the air on a windless day. After the ball is released, gravity is the only force acting on it.

If we let h represent the height at time t of such an object,

$$h = -\frac{1}{2}gt^2 + v_0 t + h_0,$$

where g, v_0, and h_0 are all constants: g is the force due to gravity, v_0 is the initial velocity which the object has when $t=0$, and h_0 is the height of the object when $t=0$ (we normally say that ground level corresponds to a height of 0). If t is measured in seconds and h in feet, g is 32 ft/s^2. If t is measured in seconds and h in meters, g is 9.8 m/s^2.

Many applications involving the above formula will result in a quadratic equation that must be solved for t. In some cases, one of the two solutions must be discarded as meaningless in the given problem.

example 5

Robert stands on the topmost tier of seats in a baseball stadium, and throws a ball out onto the field with a vertical upward velocity of 60 ft/s. The ball is 50 feet above the ground at the moment he releases the ball. When does the ball land?

Solution:

First, note that although the thrown ball has a horizontal velocity as well as a vertical velocity (otherwise it would go straight up and come straight back down), the horizontal velocity is irrelevant in such questions. All we are interested in is when the ball lands on the ground ($h = 0$). If we wanted to determine where in the field the ball lands, we would have to know the horizontal velocity as well.

We are given the following: $h_0 = 50$ ft and $v_0 = 60$ ft/s. Since the units in the problem are feet and seconds, we know to use $g = 32$ ft/s^2. What we are interested in determining is the time, t, when the height, h, of the ball is 0. That is, we need to solve the equation $0 = -16t^2 + 60t + 50$ for t.

continued on next page...

chapter two Equations and Inequalities of One Variable

$0 = -16t^2 + 60t + 50$

$0 = 8t^2 - 30t - 25$

$t = \dfrac{30 \pm \sqrt{900 + 800}}{16}$

$t = \dfrac{30 \pm 10\sqrt{17}}{16}$

$t = \dfrac{15 \pm 5\sqrt{17}}{8}$

$t \approx \cancel{-0.70}, 4.45$

The quadratic polynomial in the equation doesn't factor over the integers, so the quadratic formula is a good method to use. To simplify the calculations, we can begin by dividing both sides of the equation by -2.

We then proceed to reduce the radical and simplify the resulting fraction (which actually describes two solutions).

These numbers are most meaningful in decimal form. The negative solution is immaterial in this problem, as this represents a time before the ball is thrown, so it is discarded. The ball lands 4.45 seconds after being thrown.

exercises

Solve the following quadratic equations by factoring. See Example 1.

1. $2x^2 - x = 3$
2. $3x^2 - 7x = 0$
3. $x^2 - 14x + 49 = 0$
4. $9x - 5x^2 = -2$
5. $y(2y + 9) = -9$
6. $2x^2 - 3x = x^2 + 18$
7. $(3x + 2)(x - 1) = 7 - 7x$
8. $3x^2 + 33 = 2x^2 + 14x$
9. $5x^2 + 2x + 3 = 4x^2 + 6x - 1$
10. $15x^2 + x = 2$
11. $(x - 7)^2 = 16$
12. $4x^2 - 9 = 0$

132

section 2.3   Quadratic Equations in One Variable

Solve the following quadratic equations by the square root method. See Example 2.

13. $(x-3)^2 = 9$
14. $(a-2)^2 = -5$
15. $(8t-3)^2 = 0$
16. $(2x+1)^2 - 7 = 0$
17. $(y-18)^2 - 1 = 0$
18. $9 = (3s+2)^2$
19. $(2x-1)^2 = 8$
20. $x^2 - 6x + 9 = -16$
21. $x^2 - 4x + 4 = 49$
22. $-3(n+7)^2 = -27$
23. $(3x-6)^2 = 4x^2$
24. $(2x+3)^2 + 9 = 0$

Solve the following quadratic equations by completing the square. See Example 3.

25. $x^2 + 8x + 7 = -8$
26. $2x^2 + 6x - 10 = 10$
27. $2x^2 + 7x - 15 = 0$
28. $4x^2 - 4x - 63 = 0$
29. $u^2 + 10u + 9 = 0$
30. $4x^2 - 56x + 195 = 0$
31. $4x^2 + 32x - 260 = 0$
32. $z^2 + 26z + 2 = -23$
33. $y^2 + 22y + 96 = 0$

Solve the following quadratic equations by using the quadratic formula. See Example 4.

34. $4x^2 - 3x = -1$
35. $3x^2 - 4 = -x$
36. $2.1y^2 - 3.5y = 4$
37. $2.6z^2 - 0.9z + 2 = 0$
38. $a(a+2) = -1$
39. $3x^2 - 2x = 0$
40. $6x^2 + 5x - 4 = 3x - 2$
41. $7x^2 - 4x = 51$
42. $4x^2 - 14x - 27 = 3$

Solve the following quadratic equations using any appropriate method.

43. $y^2 + 9y = -40.50$
44. $(z-11)^2 = 9$
45. $x^2 + 20x + 36 = -48$
46. $256t^2 - 324 = 0$
47. $(y-8)^2 = 36$
48. $(9y-6)^2 = 121y^2$
49. $2x^2 + 8x - 3 = 6x$
50. $4z^2 + 14z = 10z - 3$
51. $x^2 - 6x = 27$
52. $y^2 - 2y + 1 = -289$
53. $3a^2 + 12a - 576 = 0$
54. $-3(b+5)^2 = -768$
55. $y^2 + 13y + 42 = 0$
56. $3x^2 - 6x = 0$
57. $7x^2 - 42x = 0$
58. $y^2 + 24y + 23 = 0$
59. $5x^2 - 5x - 10 = 0$
60. $4w^2 + 10w + 5 = 3w^2 + 18w - 10$

chapter two   Equations and Inequalities of One Variable

Solve the following application problems. See Example 5.

61. How long would it take for a ball dropped from the top of a 144-foot building to hit the ground?

62. Suppose that instead of being dropped, as in problem 61, a ball is thrown upward with a velocity of 40 feet per second from the top of a 144-foot building. Assuming it misses the building on the way back down, how long after being thrown will it hit the ground?

63. A slingshot is used to shoot a BB at a velocity of 96 feet per second straight up from ground level. When will the BB reach its maximum height of 144 feet?

64. A rock is thrown upward with a velocity of 20 meters per second from the top of a 24 meter high cliff, and it misses the cliff on the way back down. When will the rock be 7 meters from ground level? (Round your answer to the nearest tenth.)

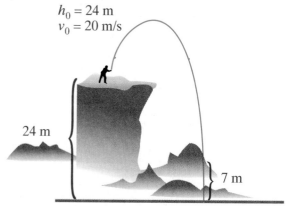

section 2.3   Quadratic Equations in One Variable

65. Luke, an experienced bungee jumper, leaps from a tall bridge and falls toward the river below. The bridge is 170 feet above the water and Luke's bungee cord is 110 feet long unstretched. When will Luke's cord begin to stretch? (Round your answer to the nearest tenth.)

Use the connection between solutions of quadratic equations and polynomial factoring to answer the following questions. See the discussion after Example 3.

66. Factor the quadratic $x^2 - 6x + 13$.

67. Factor the quadratic $9x^2 - 6x - 4$.

68. Factor the quadratic $4x^2 + 12x + 1$.

69. Factor the quadratic $25x^2 - 10x + 2$.

70. Determine b and c so that the equation $x^2 + bx + c = 0$ has the solution set $\{-3, 8\}$.

technology exercises

The Technology Exercises in Section 2.1 introduced *Mathematica*'s `Solve` command as used to solve a multi-variable formula for a specific variable. Not surprisingly, considering the name of the command, `Solve` can also be used to solve many simple single-variable equations. Care must be taken, however, when using a computer algebra system to solve equations. The user must understand the math well enough to know the limitations of the computer algorithms used, to spot errors introduced by the user or the computer, and to correctly interpret the computer's output. For instance, when *Mathematica* is used to solve the problem in *Example 5* of this section (as shown below), it is up to the user to realize that only the second answer makes physical sense.

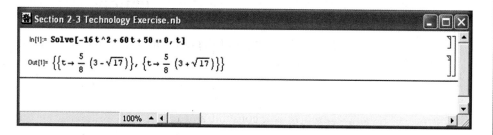

Use a computer algebra system, if available, to solve the following single-variable equations. (See Appendix A for guidance on using computer algebra systems like *Mathematica*.)

71. $5x^2 - 3x = 17$

72. $5x^2 - 3x = -17$

73. $(a+4)(4a-3) = 5$

74. $10\pi r + \pi r^2 = 107$

75. $4.8x^2 + 3.5x - 9.2 = 0$

76. $(3x-1)(3-x) = 2x+5$

2.4 Higher Degree Polynomial Equations

TOPICS

1. Solving quadratic-like equations
2. Solving general polynomial equations by factoring
3. Solving polynomial-like equations by factoring

Topic 1: Solving Quadratic-Like Equations

A polynomial equation of degree n in one variable, say x, is an equation that can be written in the form

$$a_n x^n + a_{n-1} x^{n-1} + \ldots + a_1 x + a_0 = 0,$$

where each a_i is a constant and $a_n \neq 0$. As we have seen, such equations can always be solved if $n = 1$ or $n = 2$, but in general there is no method for solving polynomial equations that is guaranteed to find all solutions. (There are formulas, called the *cubic* and *quartic* formulas, that solve third and fourth degree polynomial equations, but we will not use them in this text.) There are, moreover, many other types of equations for which no method is guaranteed to work. We are not entirely helpless, however. The Zero-Factor Property applies whenever a product of any finite number of factors is equal to 0; if $A_1 \cdot A_2 \cdot \ldots \cdot A_n = 0$, then at least one of the A_i's must equal 0.

We can use the Zero-Factor Property and our knowledge of quadratic equations to solve quadratic-like equations.

Quadratic-Like Equations

An equation is **quadratic-like**, or **quadratic in form**, if it can be written in the form
$$aA^2 + bA + c = 0,$$
where a, b, and c are constants, $a \neq 0$, and A is an algebraic expression. Such equations can be solved by first solving for A and then solving for the variable in the expression A.

chapter two — Equations and Inequalities of One Variable

example 1

Solve the quadratic-like equations.

a. $(x^2+2x)^2 - 7(x^2+2x) - 8 = 0$ **b.** $y^{\frac{2}{3}} + 4y^{\frac{1}{3}} - 5 = 0$

Solutions:

a.
$$(x^2+2x)^2 - 7(x^2+2x) - 8 = 0$$
$$A^2 - 7A - 8 = 0$$
$$(A-8)(A+1) = 0$$
$$A = 8 \quad \text{or} \quad A = -1$$
$$x^2 + 2x = 8 \qquad x^2 + 2x = -1$$
$$x^2 + 2x - 8 = 0 \qquad x^2 + 2x + 1 = 0$$
$$(x+4)(x-2) = 0 \qquad (x+1)^2 = 0$$
$$x = -4 \text{ or } x = 2 \text{ or } x = -1$$

By letting $A = x^2 + 2x$, the quadratic-like equation becomes a quadratic equation that can be solved by factoring.

Once the solutions for A have been obtained, we replace A with $x^2 + 2x$ and solve for x. Note that -1 is a double root, while -4 and 2 are single roots.

b.
$$y^{\frac{2}{3}} + 4y^{\frac{1}{3}} - 5 = 0$$
$$\left(y^{\frac{1}{3}}\right)^2 + 4\left(y^{\frac{1}{3}}\right) - 5 = 0$$
$$\left(\left(y^{\frac{1}{3}}\right) + 5\right)\left(\left(y^{\frac{1}{3}}\right) - 1\right) = 0$$
$$y^{\frac{1}{3}} = -5 \quad \text{or} \quad y^{\frac{1}{3}} = 1$$
$$y = (-5)^3 \qquad y = 1^3$$
$$y = -125 \quad \text{or} \quad y = 1$$

This equation is not even a polynomial equation, but it is a polynomial equation in the expression $y^{\frac{1}{3}}$. In fact, it is another quadratic that can be solved for $y^{\frac{1}{3}}$, and then for y.

Verify that these do indeed solve the equation $y^{\frac{2}{3}} + 4y^{\frac{1}{3}} - 5 = 0$.

These screens are performed using the quadratic formula as described on (page 130).

Topic 2: Solving General Polynomial Equations by Factoring

As we have noted above, if an equation consists of a polynomial on one side and 0 on the other, and if the polynomial can be factored completely, then the equation can be solved by using the Zero-Factor Property. If the coefficients in the polynomial are all real, the polynomial can, in principle, be factored into a product of first-degree and second-degree factors. In practice, however, this may be difficult to accomplish unless the degree of the polynomial is small or the polynomial is easily recognizable as a special product. (Several more factoring aids will be discussed in Chapter 5.)

section 2.4   Higher Degree Polynomial Equations

example 2

Solve the equations by factoring.

a. $x^4 = 9$ **b.** $y^3 + y^2 - 4y - 4 = 0$ **c.** $8t^3 - 27 = 0$

Solutions:

a.
$$x^4 = 9$$
$$x^4 - 9 = 0$$
$$(x^2 - 3)(x^2 + 3) = 0$$
$$x^2 = 3 \text{ or } x^2 = -3$$
$$x = \sqrt{3} \text{ or } x = -\sqrt{3} \text{ or } x^2 = -3$$
$$x = \sqrt{3} \text{ or } x = -\sqrt{3} \text{ or } x = i\sqrt{3} \text{ or } x = -i\sqrt{3}$$

After isolating 0 on one side, the polynomial is seen to be a difference of two squares, which can always be factored (see Section 1.5).

The Zero-Factor Property gives us two equations, both of which can be solved by taking square roots.

b.
$$y^3 + y^2 - 4y - 4 = 0$$
$$y^2(y+1) - 4(y+1) = 0$$
$$(y+1)(y^2 - 4) = 0$$
$$(y+1)(y-2)(y+2) = 0$$
$$y = -1 \text{ or } y = 2 \text{ or } y = -2$$

The polynomial on the left can be factored by grouping. After doing so, and after further factoring the difference of two squares, we obtain three solutions by the Zero-Factor Property.

A word of caution – only one answer is generated here using equation solver, but there are three solutions. Additional work must be done to find multiple solutions.

c.
$$8t^3 - 27 = 0$$
$$(2t)^3 - 3^3 = 0$$
$$(2t - 3)(4t^2 + 6t + 9) = 0$$
$$2t - 3 = 0 \text{ or } 4t^2 + 6t + 9 = 0$$
$$2t = 3 \qquad t = \frac{-6 \pm \sqrt{36 - 144}}{8}$$
$$t = \frac{3}{2} \qquad t = \frac{-6 \pm \sqrt{-108}}{8}$$
$$t = \frac{-6 \pm 6i\sqrt{3}}{8}$$
$$\text{or } t = \frac{-3 \pm 3i\sqrt{3}}{4}$$

The polynomial in this case is a difference of two cubes, which can always be factored. The Zero-Factor Property gives us two equations, one of which is linear and straightforward to solve. The second equation is a quadratic equation, and we can use the quadratic formula to solve it.

After simplifying the radical, and after further simplifying the fraction, we have three solutions of the original equation.

chapter two   Equations and Inequalities of One Variable

Topic 3: Solving Polynomial-Like Equations by Factoring

The last equations that we will consider in this section are equations that are not polynomial, but which can be solved using the methods we have developed above. We have already seen one such equation in *Example 1b*: the equation $y^{\frac{2}{3}} + 4y^{\frac{1}{3}} - 5 = 0$ is quadratic-like, and can be solved using polynomial methods. The general idea will be to rewrite the equation so that 0 appears on one side, and then to apply the Zero-Factor Property. You may want to review some of the factoring techniques in Section 1.5 as you work through these problems.

example 3

Solve the following equations by factoring.

a. $x^{\frac{7}{3}} + x^{\frac{4}{3}} - 2x^{\frac{1}{3}} = 0$

b. $(x-1)^{\frac{1}{2}} - (x-1)^{-\frac{1}{2}} = 0$

Solutions:

a.
$$x^{\frac{7}{3}} + x^{\frac{4}{3}} - 2x^{\frac{1}{3}} = 0$$
$$x^{\frac{1}{3}}(x^2 + x - 2) = 0$$
$$x^{\frac{1}{3}}(x+2)(x-1) = 0$$
$$x^{\frac{1}{3}} = 0 \quad \text{or} \quad x+2 = 0 \quad \text{or} \quad x-1 = 0$$
$$x = 0 \quad \text{or} \quad x = -2 \quad \text{or} \quad x = 1$$

Recall that in cases like this, we factor out x raised to the lowest exponent. In this case, the remaining factor is a factorable trinomial.

The Zero-Factor Property leads to three simple equations.

b.
$$(x-1)^{\frac{1}{2}} - (x-1)^{-\frac{1}{2}} = 0$$
$$(x-1)^{-\frac{1}{2}}((x-1) - 1) = 0$$
$$(x-1)^{-\frac{1}{2}}(x-2) = 0$$
$$(x-1)^{-\frac{1}{2}} = 0 \quad \text{or} \quad x-2 = 0$$
$$\frac{1}{(x-1)^{\frac{1}{2}}} = 0 \quad \text{or} \quad x = 2$$

undefined, therefore no solution

Again, we factor out the common algebraic expression raised to the lowest exponent.

This equation leads to two equations, only one of which has a solution. (Note that there is no value for x which would solve the first of the two equations.) The original equation has only one solution.

This screen was performed using the equation solver.

section 2.4 Higher Degree Polynomial Equations

exercises

Solve the following quadratic-like equations. See Example 1.

1. $(x-1)^2 + (x-1) - 12 = 0$
2. $(z-8)^2 - 7(z-8) + 12 = 0$
3. $(y-5)^2 - 11(y-5) + 24 = 0$
4. $(x^2-1)^2 + (x^2-1) - 12 = 0$
5. $(x^2+1)^2 + (x^2+1) - 12 = 0$
6. $(x^2-13)^2 + (x^2-13) - 12 = 0$
7. $(x^2-2x+1)^2 + (x^2-2x+1) - 12 = 0$
8. $2y^{\frac{2}{3}} + y^{\frac{1}{3}} - 1 = 0$
9. $2x^{\frac{2}{3}} - 7x^{\frac{1}{3}} + 3 = 0$
10. $(x^2-6x)^2 + 4(x^2-6x) - 5 = 0$
11. $(y^2-5)^2 + 5(y^2-5) - 36 = 0$
12. $(x^2+7)^2 + 8(x^2+7) + 12 = 0$
13. $(t^2-t)^2 - 8(t^2-t) + 12 = 0$
14. $2x^{\frac{1}{2}} - 5x^{\frac{1}{4}} + 2 = 0$
15. $3x^{\frac{2}{3}} - x^{\frac{1}{3}} - 2 = 0$
16. $y^{\frac{1}{2}} - 5y^{\frac{1}{4}} + 6 = 0$
17. $(z^2+4z)^2 + 7(z^2+4z) + 12 = 0$
18. $5y^{\frac{2}{3}} + 33y^{\frac{1}{3}} + 18 = 0$

Solve the following polynomial equations by factoring. See Example 2.

19. $a^3 - 3a^2 = a - 3$
20. $2x^3 + x^2 + 2x + 1 = 0$
21. $2x^3 - x^2 = 15x$
22. $x^4 + 5x^2 - 36 = 0$
23. $y^4 + 21y^2 - 100 = 0$
24. $y^3 + 8 = 0$
25. $5s^3 + 6s^2 - 20s = 24$
26. $8a^3 - 27 = 0$
27. $16a^4 = 81$
28. $6x^3 + 8x^2 = 14x$
29. $14x^3 + 27x^2 - 20x = 0$
30. $5z^3 + 28z^2 = 49z$
31. $27x^3 + 64 = 0$
32. $x^3 - 4x^2 + x = 4$
33. $x^3 + 27 = 0$

chapter two   Equations and Inequalities of One Variable

Solve the following equations by factoring. See Example 3.

34. $3x^{\frac{11}{3}} + 2x^{\frac{8}{3}} - 5x^{\frac{5}{3}} = 0$

35. $(x-3)^{\frac{-1}{2}} + 2(x-3)^{\frac{1}{2}} = 0$

36. $(y-6)^{\frac{-5}{2}} + 7(y-6)^{\frac{-3}{2}} = 0$

37. $2x^{\frac{13}{5}} - 5x^{\frac{8}{5}} + 2x^{\frac{3}{5}} = 0$

38. $(2x-5)^{\frac{1}{3}} - 3(2x-5)^{\frac{-2}{3}} = 0$

39. $y^{\frac{7}{2}} - 5y^{\frac{5}{2}} + 6y^{\frac{3}{2}} = 0$

40. $(t+4)^{\frac{2}{3}} + 2(t+4)^{\frac{8}{3}} = 0$

41. $x^{\frac{11}{2}} - 6x^{\frac{9}{2}} + 9x^{\frac{7}{2}} = 0$

42. $5y^{\frac{11}{3}} + 3y^{\frac{8}{3}} - 2y^{\frac{5}{3}} = 0$

43. $5y^{\frac{12}{5}} - 43y^{\frac{7}{5}} + 24y^{\frac{2}{5}} = 0$

44. $(3x-3)^{\frac{-1}{3}} - 5(3x-3)^{\frac{-4}{3}} = 0$

45. $(y+3)^{\frac{2}{5}} + 4(y+3)^{\frac{7}{5}} = 0$

Use the connection between solutions of polynomial equations and polynomial factoring to answer the following questions.

46. Find $b, c,$ and d so the equation $x^3 + bx^2 + cx + d = 0$ has solutions of $-3, -1,$ and 5.

47. Find $b, c,$ and d so the equation $x^3 + bx^2 + cx + d = 0$ has solutions of $-2, 0,$ and 6.

48. Find b and c so the equation $x^3 + bx^2 + cx = 0$ has solutions of $0, 1,$ and -7.

49. Find $a, c,$ and d so the equation $ax^3 + 4x^2 + cx + d = 0$ has solutions of $-4, 6,$ and -6.

50. Find $a, b,$ and d so the equation $ax^3 + bx^2 + 3x + d = 0$ has solutions of $-3, -\frac{1}{2},$ and 0.

51. Find $a, b,$ and c so the equation $ax^3 + bx^2 + cx + 6 = 0$ has solutions of $-\frac{3}{5}, \frac{2}{3},$ and 1.

2.5 Rational Expressions and Equations

TOPICS

1. Simplifying rational expressions
2. Combining rational expressions
3. Simplifying complex rational expressions
4. Solving rational equations
5. Interlude: work-rate problems

Topic 1: Simplifying Rational Expressions

Many equations contain fractions in which the variable appears in the denominator, and the presence of such fractions can make the solution process challenging. In this section, we will learn how to work with a class of fractions called *rational expressions* and develop, at the end, a general method for solving equations that contain such expressions.

Rational Expressions

A **rational expression** is an expression that can be written as a *ratio* of two polynomials $\frac{P}{Q}$. Of course, such a fraction is undefined for any value(s) of the variable(s) for which $Q = 0$. A given rational expression is *simplified* or *reduced* when P and Q contain no common factors (other than 1 and -1).

To simplify rational expressions, we factor the polynomials in the numerator and denominator completely and then cancel any common factors. It is important to remember, however, that the simplified rational expression may be defined for values of the variable (or variables) that the original (unsimplified) expression is not, and the two versions are equal only where they are both defined. That is, if A, B, and C represent algebraic expressions,

$$\frac{AC}{BC} = \frac{A}{B} \text{ only where } B \neq 0 \text{ and } C \neq 0.$$

example 1

Simplify the following rational expressions, and indicate values of the variable that must be excluded.

a. $\dfrac{x^3 - 8}{x^2 - 2x}$ \qquad **b.** $\dfrac{x^2 - x - 6}{3 - x}$

Solutions:

Notice that x-coordinates of 0 and 2 result in an error. To arrive at this screen, select **Y=** and enter the equation. Next, select **2ND** and **GRAPH** to view the table.

a. $\dfrac{x^3 - 8}{x^2 - 2x} = \dfrac{\cancel{(x-2)}(x^2 + 2x + 4)}{x\cancel{(x-2)}}$

$= \dfrac{x^2 + 2x + 4}{x}, \quad x \neq 0, 2$

After factoring both polynomials, we cancel the common factor of $x - 2$. Note that even though the final expression is defined when $x = 2$, the first and last expressions are equal only where both are defined.

b. $\dfrac{x^2 - x - 6}{3 - x} = \dfrac{(x+2)\cancel{(x-3)}}{-\cancel{(x-3)}}$

$= \dfrac{x + 2}{-1}$

$= -x - 2, \quad x \neq 3$

In this rational expression, the denominator is already factored, but we bring out a factor of -1 from the denominator in order to cancel a common factor of $x - 3$. Note that the original and simplified versions are only equal for values of x not equal to 3.

section 2.5   **Rational Expressions and Equations**

> **caution**
>
> Remember that only common *factors* can be cancelled! A very common error is to think that common terms from the numerator and denominator can be cancelled. For instance, the statement $\frac{x+4}{x^2} = \frac{4}{x}$ is **incorrect**. It is not possible to factor $x + 4$ at all, and the x that appears in the numerator is not a factor that can be cancelled with one of the x's in the denominator. The expression $\frac{x+4}{x^2}$ is already simplified as far as possible.

Topic 2: Combining Rational Expressions

Rational expressions are combined by the operations of addition, subtraction, multiplication, and division the same way that numerical fractions are. That is, in order to add or subtract two rational expressions, a common denominator must first be found. In order to multiply two rational expressions, the two numerators are multiplied and the two denominators are multiplied. Finally, in order to divide one rational expression by another, the first is multiplied by the reciprocal of the second. Remember to check to see if the resulting rational expression can be simplified in each case.

No matter which operation is being considered, it is generally best to factor all the numerators and denominators before combining rational expressions. This is the first step in finding the Least Common Denominator (LCD) of two rational expressions, and a common denominator is essential before adding or subtracting. Factoring first is also an efficient way to identify any common factors that can be cancelled if the operation is multiplication or division.

chapter two Equations and Inequalities of One Variable

example 2

Add or subtract the rational expressions, as indicated.

a. $\dfrac{2x-1}{x^2+x-2} - \dfrac{2x}{x^2-4}$

b. $\dfrac{x+1}{x+3} + \dfrac{x^2+x-2}{x^2-x-6} - \dfrac{x^2-2x+9}{x^2-9}$

Solutions:

a.
$$\dfrac{2x-1}{x^2+x-2} - \dfrac{2x}{x^2-4}$$
$$= \dfrac{2x-1}{(x+2)(x-1)} - \dfrac{2x}{(x+2)(x-2)}$$
$$= \dfrac{x-2}{x-2} \cdot \dfrac{2x-1}{(x+2)(x-1)} - \dfrac{x-1}{x-1} \cdot \dfrac{2x}{(x+2)(x-2)}$$
$$= \dfrac{2x^2-5x+2}{(x-2)(x+2)(x-1)} - \dfrac{2x^2-2x}{(x-1)(x+2)(x-2)}$$
$$= \dfrac{-3x+2}{(x-2)(x+2)(x-1)}$$

We begin by factoring both denominators, and note that the LCD is $(x-2)(x+2)(x-1)$. In order to obtain this denominator in the first fraction, we multiply the top and bottom by $x-2$. In order to obtain the LCD in the second fraction, we multiply the top and bottom by $x-1$.

After subtracting the second numerator from the first, we are done. Note that there are no common factors to cancel.

b.
$$\dfrac{x+1}{x+3} + \dfrac{x^2+x-2}{x^2-x-6} - \dfrac{x^2-2x+9}{x^2-9}$$
$$= \dfrac{x+1}{x+3} + \dfrac{\cancel{(x+2)}(x-1)}{(x-3)\cancel{(x+2)}} - \dfrac{x^2-2x+9}{(x-3)(x+3)}$$
$$= \dfrac{x-3}{x-3} \cdot \dfrac{x+1}{x+3} + \dfrac{x+3}{x+3} \cdot \dfrac{x-1}{x-3} - \dfrac{x^2-2x+9}{(x-3)(x+3)}$$
$$= \dfrac{x^2-2x-3+x^2+2x-3-x^2+2x-9}{(x-3)(x+3)}$$
$$= \dfrac{x^2+2x-15}{(x-3)(x+3)}$$
$$= \dfrac{(x+5)\cancel{(x-3)}}{\cancel{(x-3)}(x+3)}$$
$$= \dfrac{x+5}{x+3}$$

We again factor all the polynomials, and note that the second rational expression can be reduced. We do this before determining that the LCD is $(x-3)(x+3)$.

After multiplying each numerator and denominator by the required factors in order to obtain a common denominator, we combine the numerators and simplify.

After factoring the resulting numerator, there is a common factor that can be cancelled.

section 2.5 **Rational Expressions and Equations**

example 3

Multiply or divide the rational expressions, as indicated.

a. $\dfrac{x^2+3x-10}{x+3} \cdot \dfrac{x-3}{x^2-x-2}$

b. $\dfrac{x^2+5x-14}{3x} \div \dfrac{x^2-4x+4}{9x^3}$

Solutions:

a.
$$\dfrac{x^2+3x-10}{x+3} \cdot \dfrac{x-3}{x^2-x-2}$$
$$= \dfrac{(x+5)(x-2)}{x+3} \cdot \dfrac{x-3}{(x-2)(x+1)}$$
$$= \dfrac{(x+5)\cancel{(x-2)}(x-3)}{(x+3)\cancel{(x-2)}(x+1)}$$
$$= \dfrac{(x+5)(x-3)}{(x+3)(x+1)}$$

We begin by factoring both numerators and denominators, and then write the product of the two rational expressions as a single fraction.

Since we have already factored the polynomials completely, any common factors that can be cancelled are easily identified.

b.
$$\dfrac{x^2+5x-14}{3x} \div \dfrac{x^2-4x+4}{9x^3}$$
$$= \dfrac{(x+7)(x-2)}{3x} \cdot \dfrac{9x^3}{(x-2)^2}$$
$$= \dfrac{\cancel{9}^3 x^{\cancel{3}^2}(x+7)\cancel{(x-2)}}{\cancel{3}x \cancel{(x-2)^2}}$$
$$= \dfrac{3x^2(x+7)}{x-2}$$

We divide the first rational expression by the second by inverting the second fraction and multiplying. Note that we have factored all the polynomials and inverted the second fraction in one step.

Now we proceed to cancel common factors (including any common factors in the purely numerical factors) to obtain the final answer.

Topic 3: **Simplifying Complex Rational Expressions**

A **complex rational expression** is a fraction in which the numerator or denominator (or both) contains at least one rational expression. Complex rational expressions can always be rewritten as simple rational expressions. One way to do this is to simplify the numerator and denominator individually and then divide the numerator by the denominator as in *Example 3b*. Another way, which is frequently faster, is to multiply the numerator and denominator by the LCD of all the fractions that make up the complex rational expression. This method will be illustrated in the next two examples.

chapter two Equations and Inequalities of One Variable

example 4

Simplify the complex rational expressions.

a. $\dfrac{\dfrac{1}{x+h} - \dfrac{1}{x}}{h}$

b. $\dfrac{x^{-1} - y^{-1}}{x^{-2} - y^{-2}}$

Solutions:

a. $\dfrac{\dfrac{1}{x+h} - \dfrac{1}{x}}{h} = \dfrac{\dfrac{1}{x+h} - \dfrac{1}{x}}{\dfrac{h}{1}} \cdot \dfrac{(x+h)(x)}{(x+h)(x)}$

$= \dfrac{x - (x+h)}{(h)(x+h)(x)}$

$= \dfrac{-\cancel{h}}{(\cancel{h})(x+h)(x)}$

$= \dfrac{-1}{x(x+h)}$

This complex rational expression contains two rational expressions in the numerator. It may be helpful to write the denominator as a fraction, as we have done here, in order to determine that the LCD of all the fractions making up the overall expression is $(x+h)(x)$. We multiply the numerator and denominator by the LCD (so we are multiplying the overall expression by 1), then cancel the common factor of h to get the final answer.

b. $\dfrac{x^{-1} - y^{-1}}{x^{-2} - y^{-2}} = \dfrac{\dfrac{1}{x} - \dfrac{1}{y}}{\dfrac{1}{x^2} - \dfrac{1}{y^2}}$

$= \dfrac{\dfrac{1}{x} - \dfrac{1}{y}}{\dfrac{1}{x^2} - \dfrac{1}{y^2}} \cdot \dfrac{x^2 y^2}{x^2 y^2}$

$= \dfrac{xy^2 - x^2 y}{y^2 - x^2}$

$= \dfrac{xy(\cancel{y-x})}{(\cancel{y-x})(y+x)}$

$= \dfrac{xy}{y+x}$

This expression is a complex rational expression, a fact that is more clear once we rewrite the terms that have negative exponents as fractions.

The LCD in this case is $x^2 y^2$, so we multiply top and bottom by this and factor the resulting polynomials.

There is a common factor that can be cancelled, so we do so to obtain the final simplified expression.

section 2.5 Rational Expressions and Equations

> **technology note**
>
> *Mathematica*, and other computer algebra systems, are capable of simplifying combinations of rational expressions. The command to do this in *Mathematica* is `Together`; when invoked, the command attempts to combine fractional expressions into one, eliminating common factors in the process. Of course, it is up to the user to realize that the resulting single fraction might not be defined for some values of the variable(s).
>
> Output [1] below illustrates how *Mathematica* can be used to combine three rational expressions. Output [2] simply echoes Input [2], one of the complex rational expressions seen in *Example 4*. Output [3] then illustrates that `Together` can also be used to simplify this complex rational expression. (*See Appendix A for guidance on using computer algebra systems like Mathematica.*)
>
>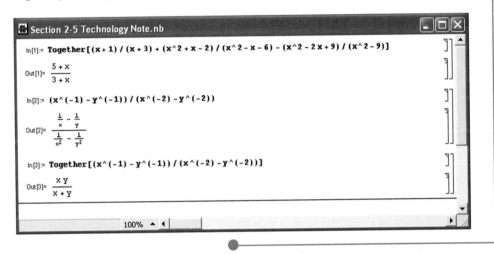

Topic 4: Solving Rational Equations

A **rational equation** is an equation that contains at least one rational expression, while any non-rational expressions are polynomials. Our general approach to solving such equations is to multiply each term in the equation by the LCD of all the rational expressions; this has the effect of converting rational equations into polynomial equations, which we have already learned how to solve. There is, however, one important difference between rational equations and polynomial equations: it is quite possible that one or more rational expressions in a rational equation are not defined for some values of the variable. Of course, these values cannot possibly be solutions of the equation, and must be excluded from the solution set. These excluded values may appear, however, as solutions of the polynomial equation that is derived from the original rational equation. Potential error can be avoided by keeping track of what values of the variable are disallowed and/or checking all solutions in the original equation.

example 5

Solve the following rational equations.

a. $\dfrac{x^3+3x^2}{x^2-2x-15} = \dfrac{4x+5}{x-5}$

b. $\dfrac{3x^2}{5x-1} - 1 = 0$

Solutions:

a.
$$\dfrac{x^3+3x^2}{x^2-2x-15} = \dfrac{4x+5}{x-5}$$
$$\dfrac{x^2\cancel{(x+3)}}{(x-5)\cancel{(x+3)}} = \dfrac{4x+5}{x-5}$$
$$(x-5)\cdot\dfrac{x^2}{(x-5)} = (x-5)\cdot\dfrac{4x+5}{x-5}$$
$$x^2 = 4x+5$$
$$x^2 - 4x - 5 = 0$$
$$(x-5)(x+1) = 0$$
$$x = \cancel{5}, -1$$
$$x = -1$$

In order to cancel any factors that might be cancelable, and in order to determine the LCD, we begin by factoring all the numerators and denominators. This also tells us the very important fact that 5 and −3 cannot be solutions of the equation.

After multiplying both sides of the equation by the LCD, we have a second-degree polynomial equation that can be solved by factoring. Note, however, that we have already determined that 5 cannot be a solution. It must be discarded.

b.
$$\dfrac{3x^2}{5x-1} - 1 = 0$$
$$(5x-1)\cdot\dfrac{3x^2}{5x-1} - (5x-1) = 0$$
$$3x^2 - 5x + 1 = 0$$
$$x = \dfrac{5\pm\sqrt{25-12}}{6}$$
$$x = \dfrac{5\pm\sqrt{13}}{6}$$

There is no factoring possible in this problem, so we merely note that $\dfrac{1}{5}$ is the one value for x that must be excluded. After multiplying through by the LCD, we have a second-degree polynomial equation that can be solved by the quadratic formula.

Since neither of the roots is $\dfrac{1}{5}$, both will solve the original equation.

Topic 5: Interlude: Work-Rate Problems

Many seemingly different applications fall into a class of problems known as work-rate problems. What these applications have in common is two or more "workers" acting in unison to complete a task. The workers can be, for example, employees on a job, machines manufacturing a part, or inlet and outlet pipes filling or draining a tank. Typically, each worker is capable of doing the task alone, but each worker works at his, her, or its own individual rate regardless of whether others are involved.

section 2.5 Rational Expressions and Equations

The goal in a work-rate problem is usually to determine how fast the task at hand can be completed, either by the workers together or by one of the workers individually. There are two keys to solving a work-rate problem:

(1) **The rate of work is the reciprocal of the time needed to complete the task.** If a given job can be done by a worker in x units of time, the worker works at a rate of $\frac{1}{x}$ jobs per unit of time. For instance, if Jane can overhaul an engine in 2 hours, her rate of work is $\frac{1}{2}$ of the job per hour. If a faucet can fill a sink in 5 minutes, its rate is $\frac{1}{5}$ of the job per minute. Of course, this also means that the time needed to complete a task is the reciprocal of the rate of work: if a faucet fills a sink at the rate of $\frac{1}{5}$ of the job per minute, it takes 5 minutes to fill the sink.

(2) **Rates of work are "additive."** This means that, in the ideal situation, two workers working together on the same task will have a combined rate of work that is the sum of their individual rates. For instance, if Jane's rate in overhauling an engine is $\frac{1}{2}$ and Ted's rate is $\frac{1}{3}$, their combined rate is $\frac{1}{2}+\frac{1}{3}$, or $\frac{5}{6}$. That is, together they can overhaul an engine in $\frac{6}{5}$ of an hour, or one hour and twelve minutes.

example 6

One hose can fill a swimming pool in 12 hours. The owner buys a second hose that can fill the pool at twice the rate of the first one. If both hoses are used together, how long does it take to fill the pool?

Solution:
The rate of work of the first hose is $\frac{1}{12}$, so the rate of the second hose is $\frac{1}{6}$. If we let x denote the time needed to fill the pool when both hoses are used together, the sum of the two individual rates must equal $\frac{1}{x}$. So we need to solve the equation $\frac{1}{12}+\frac{1}{6}=\frac{1}{x}$.

$$\frac{1}{12}+\frac{1}{6}=\frac{1}{x}$$
$$x+2x=12$$
$$3x=12$$
$$x=4$$

As is typical, this work-rate problem leads to a rational equation, and we can solve it as we solve any rational equation: multiply both sides by the LCD (which is $12x$ in this case) and proceed to solve the resulting polynomial equation. Together, the two hoses can fill the pool in 4 hours.

chapter two — Equations and Inequalities of One Variable

example 7

The pool owner in *Example 6* is a bit clumsy, and one day proceeds to fill his empty pool with the two hoses but accidentally turns on the pump that drains the pool also. Fortunately, the pump rate is slower than the combined rate of the two hoses, and the pool fills anyway, but it takes 10 hours to do so. At what rate can the pump empty the pool?

Solution:

If we let x denote the time it takes the pump to empty the pool, we can say that the pump has a *filling* rate of $-\frac{1}{x}$ (since emptying is the opposite of filling). The combined filling rate of the two hoses is $\frac{1}{4}$, and the total rate at which the pool is filled on this unfortunate day is $\frac{1}{10}$.

$$\frac{1}{4} - \frac{1}{x} = \frac{1}{10}$$
$$5x - 20 = 2x$$
$$3x = 20$$
$$x = \frac{20}{3}$$

We again use the fact that the sum of the individual rates is equal to the combined rate, but in this case one of the individual rates is negative. We solve this rational equation as always, by multiplying through by $20x$, the LCD.

The pump can empty the pool in $\frac{20}{3}$ hours, or 6 hours and 40 minutes.

exercises

Simplify the following rational expressions, indicating which real values of the variable must be excluded. See Example 1.

1. $\dfrac{2x^2 + 7x + 3}{x^2 - 2x - 15}$
2. $\dfrac{x^2 + 5x - 6}{x^3 + 2x^2 - 3x}$
3. $\dfrac{x^3 + 2x^2 - 3x}{x + 3}$
4. $\dfrac{x^2 - 4x + 4}{x^2 - 4}$

5. $\dfrac{x^2 + 5x - 6}{x^2 + 4x - 5}$
6. $\dfrac{2x^2 + 7x - 15}{x^2 + 3x - 10}$
7. $\dfrac{x + 1}{x^3 + 1}$
8. $\dfrac{x^3 + x}{3x^2 + 3}$

9. $\dfrac{2x^2 + 11x + 5}{x + 5}$
10. $\dfrac{x^4 - x^3}{x^2 - 3x + 2}$
11. $\dfrac{2x^2 + 11x - 21}{x + 7}$
12. $\dfrac{8x^3 - 27}{2x - 3}$

Add or subtract the rational expressions, as indicated, and simplify your answer. See Example 2.

13. $\dfrac{x - 3}{x + 5} + \dfrac{x^2 + 3x + 2}{x - 3}$
14. $\dfrac{x^2 - 1}{x - 2} - \dfrac{x - 1}{x + 1}$

section 2.5 — **Rational Expressions and Equations**

15. $\dfrac{x+2}{x-3} - \dfrac{x-3}{x+5} - \dfrac{1}{x^2+2x-15}$

16. $\dfrac{x+1}{x-3} + \dfrac{x^2+3x+2}{x^2-x-6} - \dfrac{x^2-2x-3}{x^2-6x+9}$

17. $\dfrac{x^2+1}{x-3} + \dfrac{x-5}{x+3}$

18. $\dfrac{x-37}{(x+3)(x-7)} + \dfrac{3x+6}{(x-7)(x+2)} - \dfrac{3}{x+3}$

19. $\dfrac{x^2+2x-35}{x-5} + \dfrac{x-4}{x+3}$

20. $\dfrac{y+2}{y-2} + \dfrac{y-6}{y+4} + \dfrac{4}{y^2+2y-8}$

21. $\dfrac{x+2}{x-6} + \dfrac{x^2+5x+6}{x^2-3x-18} - \dfrac{x^2-4x-12}{x^2-12x+36}$

22. $\dfrac{y^2+2}{y+3} - \dfrac{y-4}{y-3}$

Multiply or divide the rational expressions, as indicated, and simplify your answer. See Example 3.

23. $\dfrac{y-2}{y+1} \cdot \dfrac{y^2-1}{y-2}$

24. $\dfrac{a^2-3a-4}{a-2} \div \dfrac{a^2-2a-8}{a-2}$

25. $\dfrac{2x^2-5x-12}{x-3} \cdot \dfrac{x^2-x-6}{x-4}$

26. $\dfrac{z^2+2z+1}{2z^2+3z+1} \cdot \dfrac{2z^2-5z-3}{z+1}$

27. $\dfrac{y^2-11y+24}{y+6} \div \dfrac{y^2+5y-24}{y+6}$

28. $\dfrac{y^2+8y+16}{5y^2+22y+8} \cdot \dfrac{5y^2-13y-6}{y+4}$

29. $\dfrac{5y^2-27y-18}{y-5} \cdot \dfrac{y^2-6y+5}{y-6}$

30. $\dfrac{4z^2+20z-56}{z^2-8z+12} \div \dfrac{5z^2+43z+56}{15z^2-66z-144}$

31. $\dfrac{3b^2+9b-84}{b^2-5b+4} \div \dfrac{5b^2+37b+14}{-10b^2+6b+4}$

32. $\dfrac{3x^2-x-10}{x-1} \cdot \dfrac{x^2-1}{6x^2+x-15} \div \dfrac{x^2-x-2}{2x^2+5x-12}$

chapter two   Equations and Inequalities of One Variable

Simplify the complex rational expressions. See Example 4.

33. $\dfrac{\dfrac{3}{x}+\dfrac{x}{3}}{2-\dfrac{1}{x}}$

34. $\dfrac{\dfrac{1}{x}-\dfrac{1}{y}}{\dfrac{1}{x}+\dfrac{1}{y}}$

35. $\dfrac{6x-6}{3-\dfrac{3}{x^2}}$

36. $\dfrac{x^{-2}-y^{-2}}{y-x}$

37. $\dfrac{\dfrac{1}{r}-\dfrac{1}{s}}{r+\dfrac{1}{r}}$

38. $\dfrac{\dfrac{1}{x^2}-\dfrac{1}{y^2}}{\dfrac{1}{y^3}-\dfrac{1}{xy^2}}$

39. $\dfrac{\dfrac{m}{n}-\dfrac{n}{m}}{m-n}$

40. $\dfrac{\dfrac{1}{y}-\dfrac{1}{x+3}}{\dfrac{1}{x}-\dfrac{y}{x^2+3x}}$

41. $\dfrac{x+y^{-1}}{x^{-1}+y}$

42. $\dfrac{1+xy}{x^{-2}-y^2}$

43. $\dfrac{x^2-y^2}{y^{-2}-x^{-2}}$

44. $\dfrac{xy^{-1}+\left(\dfrac{x}{y}\right)^{-1}}{x^{-2}+y^{-2}}$

45. $\dfrac{\dfrac{1}{7y}+\dfrac{1}{x-2}}{\dfrac{1}{11x}+\dfrac{7y}{11x^2-22x}}$

46. $\dfrac{8z+8}{2-\dfrac{2}{z^2}}$

47. $\dfrac{25x^{-2}-9z^{-2}}{\dfrac{5z+3x}{x^2}}$

48. $\dfrac{\dfrac{3y}{5}-\dfrac{5}{3y}}{3-\dfrac{5}{y}}$

Solve the following rational equations. See Example 5.

49. $\dfrac{3}{x-2}+\dfrac{2}{x+1}=1$

50. $\dfrac{x}{x-1}+\dfrac{2}{x-3}=-\dfrac{2}{x^2-4x+3}$

51. $\dfrac{1}{t-3}+\dfrac{1}{t+2}=\dfrac{t}{t-3}$

52. $\dfrac{z}{6+z}+\dfrac{z-1}{6-z}=\dfrac{z}{6-z}$

53. $\dfrac{1}{x-3}+\dfrac{1}{x+3}=\dfrac{2x}{x^2-9}$

54. $\dfrac{y}{y-1}+\dfrac{2}{y-3}=\dfrac{y^2}{y^2-4y+3}$

55. $\dfrac{2}{2x+1}-\dfrac{x}{x-4}=\dfrac{-3x^2+x-4}{2x^2-7x-4}$

56. $\dfrac{2}{2b+1}+\dfrac{2b^2-b+4}{2b^2-7b-4}=\dfrac{b}{b-4}$

57. $\dfrac{3}{x-1}-\dfrac{3}{x+2}=\dfrac{9}{x^2+x-2}$

58. $\dfrac{2}{n+3}+\dfrac{3}{n+2}=\dfrac{6}{n}$

section 2.5   Rational Expressions and Equations

Perform the indicated operations on the following rational expressions, and simplify your answer.

59. $\left(\dfrac{x^2-3x}{x^2+6x-27}-\dfrac{2}{x+9}\right)\cdot\dfrac{x+9}{x+2}$

60. $\dfrac{2y(y-1)}{y^2+6y-16}\div\dfrac{2}{y+8}-\dfrac{2}{y-2}$

61. $\left(\dfrac{z^2-17z+30}{z^2+2z-8}+\dfrac{6}{z-2}\right)\div\dfrac{1}{z^2-5z-36}$

62. $\dfrac{y+3}{2y+18}+\dfrac{y^2+2y+4}{y^2+3y-54}\cdot\dfrac{y-6}{y+3}$

63. $\dfrac{y^2+2y-15}{y+1}\cdot\left(\dfrac{y^2+3y+4}{y^2+3y-10}+\dfrac{y+4}{y+5}\right)\div\dfrac{y-3}{y-2}$

64. $\dfrac{y+6}{y-3}\left(\dfrac{y+5}{y-3}+\dfrac{y-3}{y+6}-\dfrac{y^2+4}{y^2+3y-18}\right)$

Solve the following work-rate problems. See Examples 6 and 7.

65. If Joanne were to paint her living room alone, it would take 5 hours. Her sister Lisa could do the job in 7 hours. How long would it take them working together?

66. The hot water tap can fill a given sink in 4 minutes. If the cold water tap is turned on as well, the sink fills in 1 minute. How long would it take for the cold water tap to fill the sink alone?

67. The hull of Jack's yacht needs to be cleaned. He can clean it by himself in 5 hours, but he asks his friend Thomas to help him. If it takes 3 hours for the two men to clean the hull of the boat, how long would it have taken Thomas alone?

68. Two hoses, one of which has a flow-rate three times the other, can together fill a tank in 3 hours. How long does it take each of the hoses individually to fill the tank?

69. Officials begin to release water from a full man-made lake at a rate that would empty the lake in 12 weeks, but a river that can fill the lake in 30 weeks is replenishing the lake at the same time. How long does it take to empty the lake?

70. In order to flush deposits from a radiator, a drain that can empty the entire radiator in 45 minutes is left open at the same time it is being filled at a rate that would fill it in 30 minutes. How long does it take for the radiator to fill?

71. Jimmy and Janice are picking strawberries. Janice can fill a bucket in a half hour, but Jimmy continues to eat the strawberries that Janice has picked at a rate of one bucket per 1.5 hours. How long does it take Janice to fill her bucket?

72. A farmer can plow a given field in 2 hours less time than it takes his son. If they acquire two tractors and work together, they can plow the field in 5 hours. How long does it take the father alone? Round your answer to the nearest tenth of an hour.

section 2.5 Rational Expressions and Equations

technology exercises

Use a computer algebra system to combine the rational expressions or simplify the complex rational expressions, as indicated. See the Technology Note after Example 4 for guidance.

73. $\dfrac{3}{2x-10} - \dfrac{1}{2x+10}$

74. $\dfrac{5}{8x-40} + \dfrac{3}{8x+24}$

75. $\dfrac{1}{5} + \dfrac{6}{25x-5}$

76. $\dfrac{-3}{7x+14} + \dfrac{9}{21x-7} + \dfrac{2}{x+2}$

77. $\dfrac{x^{-3} - y^{-3}}{x^{-1} - y^{-1}}$

78. $\dfrac{\dfrac{1}{x} - \dfrac{y}{x^2}}{\dfrac{y}{x^2} + \dfrac{x}{y}}$

chapter two — Equations and Inequalities of One Variable

2.6 Radical Equations

TOPICS

1. Solving radical equations
2. Solving equations with positive rational exponents

Topic 1: Solving Radical Equations

The last one-variable equations that we will discuss are those that contain radical expressions. A **radical equation** is an equation that has at least one radical expression containing a variable, while any non-radical expressions are polynomial terms. As with the rational equations discussed in the previous section, we will develop a general method of solution that converts a given radical equation into a polynomial equation. We will see, however, that just as with rational equations, we must check our potential solutions carefully to see if they actually solve the original radical equation.

Our method of solving rational equations involved multiplying both sides of the equation by an algebraic expression (the LCD of all the rational expressions), and in some cases potential solutions had to be discarded because they led to division by 0 in one or more of the rational expressions. Something similar can happen with radical equations. Since our goal is to convert a given radical equation into a polynomial equation, one reasonable approach is to raise both sides of the equation to whatever power is necessary to "undo" the radical (or radicals). The problem is that this does *not* result in an equivalent equation; the only means we have of transforming an equation into an equivalent equation is to add the same quantity to both sides or to multiply both sides by a non-zero quantity. We won't *lose* any solutions by raising both sides of an equation to the same power, but we may *gain* some extraneous solutions. We identify these and discard them by simply checking all of our eventual solutions in the original equation.

A simple example will make this clear. Consider the equation

$$x = -3.$$

This equation is so basic that it is its own solution. But if, solely for the purposes of demonstration, we square both sides, we obtain the equation

$$x^2 = 9.$$

This second-degree equation can be solved by factoring the polynomial $x^2 - 9$ or by taking the square root of both sides, and in either case we obtain the solution set $\{-3, 3\}$. That is, by squaring both sides of the original equation, we gained a second (and false) solution.

The specifics of our method are as follows.

Method of Solving Radical Equations

Step 1: Begin by isolating the radical expression on one side of the equation. If there is more than one radical expression, choose one to isolate on one side.

Step 2: Raise both sides of the equation by the power necessary to "undo" the isolated radical. That is, if the radical is an nth root, raise both sides to the nth power.

Step 3: If any radical expressions remain, simplify the equation if possible and then repeat steps 1 and 2 until the result is a polynomial equation. When a polynomial equation has been obtained, solve the equation using polynomial methods.

Step 4: Check your solutions in the original equation! Any extraneous solutions must be discarded.

(If the equation contains many radical expressions, and especially if they are of differing indices, eliminating all the radicals may be a long process! The equations that we will solve will not require more than a few repetitions of steps 1 and 2.)

chapter two   Equations and Inequalities of One Variable

example 1

Solve the radical equations.

a. $\sqrt{1-x} - 1 = x$ **b.** $\sqrt{x+1} + \sqrt{x+2} = 1$ **c.** $\sqrt[4]{x^2 + 8x + 7} - 2 = 0$

Solutions:

a.
$$\sqrt{1-x} - 1 = x$$
$$\sqrt{1-x} = x+1$$
$$\left(\sqrt{1-x}\right)^2 = (x+1)^2$$
$$1-x = x^2 + 2x + 1$$
$$0 = x^2 + 3x$$
$$0 = x(x+3)$$
$$x = 0, \cancel{-3}$$
$$x = 0$$

There is only one radical expression, so we isolate it and proceed to square both sides. The result is a second-degree polynomial equation that can be solved by factoring.

Note, though, that $\sqrt{1-(-3)} \neq -3+1$, so -3 must be discarded. The solution is the single number 0.

b.
$$\sqrt{x+1} + \sqrt{x+2} = 1$$
$$\sqrt{x+1} = 1 - \sqrt{x+2}$$
$$\left(\sqrt{x+1}\right)^2 = \left(1 - \sqrt{x+2}\right)^2$$
$$x+1 = 1 - 2\sqrt{x+2} + x + 2$$
$$2\sqrt{x+2} = 2$$
$$\sqrt{x+2} = 1$$
$$x+2 = 1$$
$$x = -1$$

This equation has two radical expressions, so we isolate one of them initially.

We square both sides to eliminate the isolated radical, and then proceed to simplify and isolate the one remaining radical.

At this point, we square both sides again and solve the polynomial equation. Our one solution solves the original equation in this problem.

c.
$$\sqrt[4]{x^2 + 8x + 7} - 2 = 0$$
$$\sqrt[4]{x^2 + 8x + 7} = 2$$
$$\left(\sqrt[4]{x^2 + 8x + 7}\right)^4 = 2^4$$
$$x^2 + 8x + 7 = 16$$
$$x^2 + 8x - 9 = 0$$
$$(x+9)(x-1) = 0$$
$$x = -9, 1$$

We first isolate the radical (a fourth root in this case) and then raise both sides to the fourth power.

The resulting second-degree equation can again be solved by factoring.

Both roots satisfy the original equation.

section 2.6 **Radical Equations**

Topic 2: **Solving Equations with Positive Rational Exponents**

We have already encountered several equations with rational exponents in Section 2.4, in which we solved such equations either by factoring or by quadratic methods. In the problems that we will consider here, equations containing terms with positive rational exponents can be viewed as radical equations. Rewriting each term that has a positive rational exponent as a radical will allow us to use the method developed previously to solve rational equations.

example 2

Solve the following equations with rational exponents.

a. $x^{\frac{2}{3}} - 9 = 0$

b. $(32x^2 - 32x + 17)^{\frac{1}{4}} = 3$

Solutions:

To use the equation solver for example **a** as shown above:
1. Press the **MATH** button.
2. Select "0: Solver" by highlighting it and pressing enter.
3. Press the up arrow to enter a new equation. Enter the equation $x^{\frac{2}{3}} - 9$ and press the enter key.
4. Now press the **ALPHA** button and then press **ENTER**.

a. $x^{\frac{2}{3}} - 9 = 0$

$x^{\frac{2}{3}} = 9$

$\sqrt[3]{x^2} = 9$

$x^2 = 9^3$

$x = \pm 9^{\frac{3}{2}}$

$x = \left(\pm 9^{\frac{1}{2}}\right)^3$

$x = (\pm 3)^3$

$x = \pm 27$

Since the term containing the rational exponent can be rewritten as a radical expression, we will begin by isolating that term.

Cubing both sides eliminates the cube root.

Raising both sides to the $\frac{1}{2}$ power solves the equation for x, but we can evaluate the expression on the right-hand side. Note that both +3 and −3 must be considered. Verify that both numbers satisfy the original equation.

b. $(32x^2 - 32x + 17)^{\frac{1}{4}} = 3$

$\sqrt[4]{32x^2 - 32x + 17} = 3$

$32x^2 - 32x + 17 = 3^4$

$32x^2 - 32x + 17 = 81$

$32x^2 - 32x - 64 = 0$

$32(x^2 - x - 2) = 0$

$x^2 - x - 2 = 0$

$(x-2)(x+1) = 0$

$x = 2, -1$

The exponent of $\frac{1}{4}$ indicates we should raise both sides to the fourth power in order to eliminate the radical.

We are left with a second-degree polynomial equation that can be solved by factoring. Note that both solutions again satisfy the original equation.

161

chapter two   Equations and Inequalities of One Variable

exercises

Solve the following radical equations. See Example 1.

1. $\sqrt{4-x} - x = 2$
2. $\sqrt{3y+4} + \sqrt{5y+6} = 2$
3. $\sqrt{3-3x} - 3 = \sqrt{3x+2}$
4. $\sqrt{x^2 - 4x + 5} - x + 2 = 0$
5. $\sqrt{x^2 - 4x + 4} + 2 = 3x$
6. $\sqrt{50 + 7s} - s = 8$
7. $\sqrt[3]{3-2x} - \sqrt[3]{x+1} = 0$
8. $\sqrt[4]{x^2 - x} = \sqrt[4]{x-1}$
9. $\sqrt[4]{2x+3} = -1$
10. $\sqrt{11x+3} + 4x = 18$
11. $\sqrt{2b-1} + 3 = \sqrt{10b-6}$
12. $\sqrt{5x+5} = \sqrt{4x-7} + 2$
13. $\sqrt{x+10} + 1 = x - 1$
14. $\sqrt{x+1} + 10 = x - 1$
15. $\sqrt{x^2 - 10} - 1 = x + 1$
16. $\sqrt[3]{5x^2 - 14x} = -2$
17. $\sqrt[5]{7t^2 + 2t} = \sqrt[5]{5t^2 + 4}$
18. $\sqrt[3]{y^3 - 7y + 2} = \sqrt[3]{2 - 3y}$
19. $\sqrt{14y^2 - 18y + 4} + 2 = 2y$
20. $\sqrt{9x+4} = \sqrt{7x+1} + 1$
21. $\sqrt{4z+41} + 3 = z + 2$

Solve the following equations. See Example 2.

22. $(x+3)^{\frac{1}{4}} + 2 = 0$
23. $(2x-5)^{\frac{1}{4}} = (x-1)^{\frac{1}{4}}$
24. $(2x-1)^{\frac{2}{3}} = x^{\frac{1}{3}}$
25. $(3y^2 + 9y - 5)^{\frac{1}{2}} = y + 3$
26. $(3x-5)^{\frac{1}{5}} = (x+1)^{\frac{1}{5}}$
27. $w^{\frac{3}{5}} + 8 = 0$
28. $z^{\frac{4}{3}} - \frac{16}{81} = 0$
29. $x^{\frac{2}{3}} - \frac{25}{49} = 0$
30. $(x^2 + 21)^{\frac{-3}{2}} = \frac{1}{125}$
31. $(x-2)^{\frac{2}{3}} = (14-x)^{\frac{1}{3}}$
32. $(x^2 + 7)^{\frac{-3}{2}} = \frac{1}{64}$
33. $(y-2)^{\frac{2}{3}} = (13y - 66)^{\frac{1}{3}}$

Solve the following formulas for the indicated variable.

34. $T = 2\pi\sqrt{\dfrac{l}{g}}$. This is the formula for the period T of a pendulum of length l. Solve this formula for l.

35. $c = \sqrt{a^2 + b^2}$. This is the formula for the length of the hypotenuse c of a right triangle. Solve this formula for a.

section 2.6   Radical Equations

36. Einstein's Theory of Relativity states that $E = mc^2$. Solve this equation for c.

37. $\omega = \sqrt{\dfrac{k}{m}}$. This is the formula for the angular frequency ω of a mass m suspended from a spring of spring constant k. Solve this formula for m.

38. $V = \dfrac{4}{3}\pi r^3$. This is the formula for the volume of a sphere with radius r. Solve the equation for r.

39. $F = \dfrac{mv^2}{r}$. This is the formula for the force on an object in circular motion. Solve the equation for v.

40. The formula for lateral acceleration, used in automotives, is $a = \dfrac{1.227r}{t^2}$. Solve this equation for t.

41. The ideal body weight for a male may be found using the formula $w = 23h^2$. Solve this equation for h.

42. Kepler's Third Law is $T^2 = \dfrac{4\pi^2 r^3}{GM}$. It relates the period T of a planet to the radius r of its orbit and the sun's mass M. Solve this formula for r.

43. $r = \dfrac{2gm}{c^2}$. This is the Schwarzschild Radius Formula used to find the radius of a black hole in space. Solve the equation for c.

44. The total mechanical energy of an object with mass m at height h in a closed system can be written as $ME = \dfrac{1}{2}mv^2 + mgh$. Solve for v, the velocity of the object, in terms of the given quantities.

45. Recall, the Pythagorean Theorem states that $a^2 + b^2 = c^2$. Solve the Pythagorean Theorem for b.

46. In a circuit with an AC power source, the total impedance Z depends on the resistance R, the capacitance C, the inductance L, and the frequency of the current ω according to: $Z = \sqrt{R^2 + \left(\omega L - \dfrac{1}{\omega C}\right)^2}$. Solve this equation for the inductance L.

47. The formula used to find the orbital period for circular Keplerian orbits is $P = \dfrac{2\pi}{\sqrt{\dfrac{u}{a^3}}}$. Solve this equation for a.

chapter two — Equations and Inequalities of One Variable

chapter two project

Purchasing a New Car

There are many financing options for new car buyers, and sometimes comparing offers between dealerships can be confusing. Newspaper and television ads often seem much more complicated once the fine print is read. If you decide to purchase a new car, be sure to get all the details and remember that the dealerships might be negotiating on different variables. To do a thorough comparison, you must take all the variables into consideration.

Assume you have decided to purchase a new car with a manufacturer's suggested retail price (MSRP) of $22,000, including all the options you have selected. There are two local dealerships that carry this car and you have collected offers from both of them. You plan to use the trade-in value of your old car as a down payment. The dealership offers and the assessed values for your car are listed in the table below.

Dealership	Factory MSRP	Dealer Incentive	Trade-in Value	Financed Amount	Term of Loan	Annual Rate of Interest
City Motors	$22,000	$1200	$2500	$18,300	48 months	11%
City Motors	$22,000	$1000	$2500	$18,500	36 months	4.5%
City Motors	$22,000	$1000	$2500	$18,500	48 months	7.9%
Arrow Imports	$22,000	$900	$3000	$18,100	48 months	9.9%
Arrow Imports	$22,000	$500	$3000	$18,500	24 months	3.9%

1. Your monthly payment can be calculated using the formula

$$P = A \left[\frac{1-(1+i)^{-n}}{i} \right]^{-1},$$

where P represents your payment, A is the financed amount, i is the monthly interest rate in decimal form (i = annual rate/12), and n is the duration of the loan in months. Compute the monthly payment for each of the scenarios above.
2. What is the total cost of the car for each of the scenarios?
3. How much interest is paid in total for each of the scenarios?
4. Which of these scenarios is the best for you? What is best for you may not be what is best for everyone, so explain the reasons for your selection.

chapter two summary

CHAPTER SUMMARY

A summary of concepts and skills follows each chapter. Refer to these summaries to make sure you feel comfortable with the material in the chapter. The concepts and skills are organized according to the section title and topic title in which the material is first discussed.

2.1 Linear Equations in One Variable

topics	pages
Equivalent Equations and the Meaning of Solutions • The three categories of equations: *identities*, *contradictions*, and *conditionals* • The *solution set* of an equation, and *equivalent* equations	p. 99 – 100
Solving Linear Equations • The definition of a *linear*, or *first-degree*, *equation in one variable* • General method of solving linear equations	p. 100 – 101
Solving Absolute Value Equations • Algebraic and geometric meaning of absolute value expressions in equations • *Extraneous solutions* to equations	p. 102 – 103
Solving Linear Equations for One Variable • Solving an equation (or a formula) for a specified variable	p. 104
Interlude: Distance and Interest Problems • Basic distance and interest problems leading to linear equations	p. 105 – 106

2.2 Linear Inequalities in One Variable

topics	pages
Solving Linear Inequalities • The definition of a *linear inequality* • The use of *Cancellation Properties* in solving inequalities	p. 112 – 114

chapter two   Equations and Inequalities of One Variable

topics (continued)	pages
Solving Compound Linear Inequalities • The definition of a *compound inequality*	p. 114 – 115
Solving Absolute Value Inequalities • The definition of an *absolute value inequality* • The geometric meaning of inequalities containing an absolute value expression	p. 116 – 117
Interlude: Translating Inequality Phrases • The meaning of commonly encountered inequality phrases	p. 117 – 118

2.3 Quadratic Equations in One Variable

topics	pages
Solving Quadratic Equations by Factoring • The definition of a *quadratic*, or *second-degree*, *equation in one variable* • Using the Zero-Factor Property to solve quadratic equations	p. 122 – 124
Solving "Perfect Square" Quadratic Equations • Solving quadratic equations of the general form $A^2 = c$	p. 125
Solving Quadratic Equations by Completing the Square • Using the method of *completing the square* to solve arbitrary quadratic equations	p. 126 – 128
The Quadratic Formula • Applying the method of completing the square to a generic quadratic equation • The *quadratic formula*	p. 128 – 130
Interlude: Gravity Problems • The general equation of the motion of an object under the influence of gravity	p. 130 – 132

chapter two summary

2.4 Higher Degree Polynomial Equations

topics	pages
Solving Quadratic-Like Equations • The definition of a *degree n polynomial equation in one variable* • *Quadratic-like equations*	p. 137 – 138
Solving General Polynomial Equations by Factoring • Extending the use of the Zero-Factor Property to higher-degree polynomial equations	p. 138 – 139
Solving Polynomial-Like Equations by Factoring • Extending the use of the Zero-Factor Property to selected non-polynomial equations	p. 140

2.5 Rational Expressions and Equations

topics	pages
Simplifying Rational Expressions • The definition of a *rational expression* • Simplifying rational expressions	p. 143 – 145
Combining Rational Expressions • Addition, subtraction, multiplication, and division of rational expressions	p. 145 – 147
Simplifying Complex Rational Expressions • The definition of a *complex rational expression* • Two methods for simplifying complex rational expressions	p. 147 – 149
Solving Rational Equations • The meaning of the phrase *rational equation* • A general method of solving rational equations	p. 149 – 150

chapter two — Equations and Inequalities of One Variable

topics (continued)	pages
Interlude: Work-Rate Problems • Types of problems that fall into the classification of *work-rate problems* • The concept of rate of work as the reciprocal of time needed to complete the work • The concept of the *additivity* of rates of work	p. 150 – 152

2.6 Radical Equations

topics	pages
Solving Radical Equations • The definition of a *radical equation* • The importance of checking for extraneous solutions to radical equations • A general method of solving radical equations	p. 158 – 160
Solving Equations with Positive Rational Exponents • Using the method of solving radical equations to solve equations with positive rational exponents	p. 161

chapter two review

chapter review

Section 2.1

Solve the following absolute value equations.

1. $|2x-7|=1$
2. $|2y-5|-1=|3-y|$
3. $|7z+5|+3=8$
4. $|w-5|=|3w+1|$

Solve the following application problems.

5. Two trains leave the station at the same time in opposite directions. One travels at an average rate of 90 miles per hour, and the other at an average rate of 95 miles per hour. How far apart are the two trains after an hour and twenty minutes? Round your answer to the nearest tenth of a mile.

6. Two firefighters, Jake and Rose, each have $5000 to invest. Jake invests his money in a money market account with an annual return of 3.25%, while Rose invests hers in a CD paying 4.95% annually. How much more money does Rose have than Jake after 1 year?

Section 2.2

Solve the following inequalities. Describe the solution set using interval notation and by graphing.

7. $-8 < 3x - 5 \leq 16$
8. $-14 < -2(3+y) \leq 8$
9. $-5|3+t| > -10$
10. $-3(x-1) < 12$ and $x - 4 \leq 9$

Section 2.3

Solve the following quadratic equations.

11. $2x^2 + 3x - 10 = 10$
12. $x^2 - 7x - 2 = -12$
13. $1.7z^2 - 3.8z - 2 = 0$
14. $2x^2 + 7x = x^2 + 2x - 6$

Section 2.4

Solve the following quadratic-like equations.

15. $(x^2+2)^2 - 7(x^2+2) + 12 = 0$

16. $y^{\frac{2}{3}} + y^{\frac{1}{3}} - 6 = 0$

Solve the following polynomial equations by factoring.

17. $x^3 - 4x^2 - 2x + 8 = 0$

18. $2x^3 + 2x = 5x^2$

Section 2.5

Simplify the complex rational expressions.

19. $\dfrac{\dfrac{x}{3} - \dfrac{3}{x}}{-\dfrac{3}{x} + 1}$

20. $\dfrac{\dfrac{1}{2a} - \dfrac{1}{2b}}{\dfrac{2}{a} + \dfrac{2}{b}}$

Solve the following rational equations.

21. $\dfrac{y}{y-1} + \dfrac{1}{y-4} = \dfrac{y^2}{y^2 - 5y + 4}$

22. $\dfrac{2}{x+1} - \dfrac{x}{x-3} = \dfrac{3x-21}{x^2 - 2x - 3}$

Section 2.6

Solve the following equations.

23. $\sqrt{2x^2 + 8x + 1} - x - 3 = 0$

24. $\sqrt{10x^2 - 14x + 16} + 1 = 3x$

25. $(2x-5)^{\frac{1}{6}} = (x-2)^{\frac{1}{6}}$

26. $(x^2 + x - 16)^{\frac{1}{3}} = 2(x-1)^{\frac{1}{3}}$

chapter test

Solve the following equations.

1. $-(x-3)+1=2(x-1)-1$
2. $2y-(1-y)=y+2(y-1)$
3. $\dfrac{x-1}{2}-\dfrac{1}{3}=1-\dfrac{x}{4}$
4. $\dfrac{x}{2}-\dfrac{1}{3}=x-\dfrac{1}{3}-\dfrac{x}{2}$
5. $-0.2x-0.5=-0.4x+0.75$

Solve the following absolute value equations.

6. $|-2x+1|=7$
7. $3+|5x-1|=0$
8. $|x+1|+|2x+2|-3=0$
9. $|x+4|-|x-1|=0$
10. If $-2<x<5,$ find $|x+2|+|x-5|.$

Solve the following inequalities. Describe the solution set using interval notation and by graphing.

11. $2(x-1)<-x+7$
12. $\dfrac{2x-1}{3}\leq\dfrac{x+2}{2}$
13. $-5<3m+1<13$
14. $-1<\dfrac{a-1}{3}\leq 5$
15. $1+2|x-3|\leq 11$
16. $3+|2x-1|<1$
17. $3|2x+3|-5>1$
18. $-2|x-1|+|3x-3|\geq 7$
19. $1+|-4x+3|>-3$

Solve the following quadratic equations.

20. $x^2-x=2$
21. $x^2-6x-7=0$
22. $x(x+2)=8$
23. $x^2-5x=0$
24. $-x^2+9=0$
25. $3x^2+75=0$
26. $(2x-1)^2=49$
27. $a^2-4a+4=196$
28. $(x-3)(x+2)=x+9$
29. $x^2+x+2=0$
30. $x^2+x-1=0$

31. If a rectangle has a length of $(x+3)$, a width of $(x-2)$, and an area of 14, what is x?

chapter two — Equations and Inequalities of One Variable

Solve the following quadratic-like equations.

32. $(t+2)^2 - 2(t+2) = 24$

33. $x^4 - 13x^2 + 36 = 0$

34. $(x^2 - x)^2 - 8(x^2 - x) + 12 = 0$

35. $2y^{\frac{1}{2}} - 9y^{\frac{1}{4}} + 4 = 0$

Solve the following equations by factoring.

36. $x^3 - 4x^2 = x - 4$

37. $x^3 - x^2 + 4x - 4 = 0$

38. $27x^3 - 125 = 0$

39. $(2x+1)^{\frac{-1}{3}} - 3(2x+1)^{\frac{-4}{3}} = 0$

40. $x^{\frac{7}{2}} - 3x^{\frac{5}{2}} - 4x^{\frac{3}{2}} = 0$

41. Find b, c, and d so the equation $x^3 + bx^2 + cx + d = 0$ has solutions of $-1, 2,$ and 3.

42. Given that the equation $x^2 - 6x + m - 1 = 0$ has only one root, find m.

43. If the sum of the roots of the equation $x^2 + mx - 6 = 0$ is 5, then what is m?

Simplify the following rational expressions.

44. $\dfrac{2x^2 + 5x - 3}{x^2 + x - 6}$

45. $\dfrac{x+2}{x^3 + 8}$

46. $\dfrac{3x^2 + 5x - 2}{9x^2 - 1}$

47. $\dfrac{x^3 + 6x^2 + 9x}{x^3 - 9x}$

Add or subtract the following rational expressions as indicated and simplify.

48. $\dfrac{1}{x-1} + \dfrac{2}{x+1}$

49. $\dfrac{1}{x} - \dfrac{3}{x+2} - \dfrac{6}{x^2 + 2x}$

50. $\dfrac{x+2}{x^2 - 4} - \dfrac{1}{x+2} - \dfrac{4}{x^2 - 4}$

51. $\dfrac{a^2 - 1}{a - 1} - \dfrac{a - 1}{a + 1}$

Multiply or divide the following rational expressions as indicated.

52. $\dfrac{x^2 + 9x}{x^2 - 81} \cdot \dfrac{x^2 - 8x - 9}{x^2 + x}$

53. $\dfrac{x^3 + 2x^2 + x}{x^3 + 1} \div \dfrac{x^2 - 1}{x^2 - x + 1}$

54. $\dfrac{4y^2 + 5y - 6}{2y^2 + 5y + 2} \cdot \dfrac{4y^2 - 1}{8y^2 - 10y + 3}$

55. $\dfrac{a^3 - 8}{a^2 - 4} \div \dfrac{a^3 + 2a^2 + 4a}{a^3 + 2a^2} \cdot \dfrac{1}{a^2 + a}$

56. $\dfrac{x^2 - 3x}{4x^2 - 4x + 1} \cdot \dfrac{2x^2 - x}{x^2 - 9} \div \dfrac{x+2}{x+3}$

chapter two test

Simplify the following complex rational expressions.

57. $\dfrac{1 - \dfrac{1}{x}}{1 + \dfrac{1}{x}}$

58. $\dfrac{x^{-1} - y^{-1}}{x^{-2} - y^{-2}}$

59. $\left(x - 1 + \dfrac{2}{x+1}\right) \div \left(1 + \dfrac{1}{x^2}\right)$

60. $\dfrac{x-3}{x+2} \cdot \left(\dfrac{x+2}{x-3} + \dfrac{x-3}{x-4} - \dfrac{17 - 4x}{x^2 - 7x + 12}\right)$

Solve the following rational equations.

61. $\dfrac{1}{x+2} + \dfrac{1}{x-3} - \dfrac{x}{x-3} = 0$

62. $\dfrac{1}{x-2} - \dfrac{x}{x+2} = \dfrac{2}{x^2 - 4}$

63. $\dfrac{1}{y+1} + \dfrac{y}{y+3} = \dfrac{6}{y^2 + 4y + 3}$

64. $\dfrac{1 - \dfrac{2}{x}}{1 + \dfrac{2}{x}} + \dfrac{2}{x+2} = 0$

65. $\dfrac{3}{x-2} - \dfrac{1}{x+2} = \dfrac{2x+8}{x^2 - 4}$

66. Number a is 3 more than the number b. If the product of these numbers is 28, then find the numbers.

67. Water tap A can fill a given sink in 2 hours, tap A and tap B can fill the same sink in one hour and 15 minutes. How long does it take tap B to fill the sink?

68. Jim cleans a house in 6 hours. John cleans the same house in 8 hours. How long does it take together?

69. The formula used to find the area of a trapezoid is $A = \dfrac{(b_1 + b_2)h}{2}$. Solve this equation for b_2.

70. Given $\dfrac{1}{x} + \dfrac{1}{y} = \dfrac{2}{z}$, solve for y.

Solve the following radical equations.

71. $\sqrt{x+5} - x = -1$

72. $\sqrt[3]{2x+19} - \sqrt[3]{5x+12} = 0$

chapter THREE

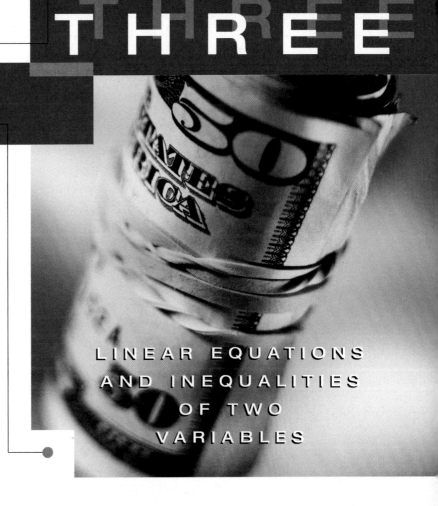

LINEAR EQUATIONS AND INEQUALITIES OF TWO VARIABLES

OVERVIEW

3.1 The Cartesian Coordinate System

3.2 Linear Equations in Two Variables

3.3 Forms of Linear Equations

3.4 Parallel and Perpendicular Lines

3.5 Linear Inequalities in Two Variables

3.6 Introduction to Circles

By the end of this chapter you should be able to use skills you have learned regarding linear inequalities in two variables to discover the possible range of activity for situations limited by maximum capacities for things such as price, weight, volume, time, materials, and more. On page 234 you will find a problem in which a sizeable amount of money is inherited, but comes with stipulations regarding how it must be invested. You will master this type of problem using tools such as the steps for *Solving Linear Inequalities in Two Variables* on page 224.

chapter three   Linear Equations and Inequalities of Two Variables

Introduction

This chapter introduces the Cartesian coordinate system, the two-dimensional framework that underlies most of the material throughout the remainder of this text.

Many problems that we seek to solve with mathematics are most naturally described with two variables. The existence of two variables in a problem leads naturally to the use of a two-dimensional system in which to work, especially when we attempt to depict the situation graphically, but it took many centuries for the ideas presented in this chapter to evolve. Some of the greatest accomplishments of the French mathematician and philosopher René Descartes (1596 - 1650) were his contributions to the then fledgling field of *analytic geometry*, the marriage of algebra and geometry. In *La géométrie*, an appendix to a volume of scientific philosophy, Descartes laid out the basic principles by which algebraic problems could be construed as geometric problems, and the methods by which solutions to the geometric problems could be interpreted algebraically. As many later mathematicians expanded upon Descartes' work, analytic geometry came to be an indispensable tool in understanding and solving problems of both an algebraic and geometric nature.

Descartes

In this chapter we will use the Cartesian coordinate system, named in honor of Descartes, primarily to study linear equations and inequalities in two variables. As we will see, a *graph* is one of the best ways to describe the solutions of such problems. Sections 3.2 through 3.5 will introduce the basic means by which we can construct graphs and consequently shift between the algebraic and geometric views of a given linear equation or linear inequality.

The Cartesian coordinate system will continue to play a prominent role as we proceed to other topics, such as relations and functions, in later chapters. Mastery of the foundational concepts in this chapter will be essential to understanding these related ideas.

3.1 The Cartesian Coordinate System

TOPICS

1. The components of the Cartesian coordinate system
2. The graph of an equation
3. The distance and midpoint formulas

Topic 1: The Components of the Cartesian Coordinate System

The last chapter dealt, essentially, with algebra of a single variable. Many of the methods learned in Chapter 2 revolve around solving equations and inequalities of one variable, and even in those equations with two or more variables we were interested in tasks such as solving the equation for one particular variable.

Our focus now shifts. While many important problems can be phrased, mathematically, as one-variable equations or inequalities, many more problems are most naturally expressed with two or more variables, and solving such a problem requires determining *all* of the values of *all* of the variables that make the equation or inequality true. This seemingly simple observation serves as the motivation for the material in this section. Consider, for example, an equation in the two variables x and y. A particular solution of the equation, if there is one, will consist of a value for x and a *corresponding* value for y; a solution of the equation cannot consist of a value for only one of the variables.

This leads us naturally to the concept of a two-dimensional coordinate system. Just as the real number line, a one-dimensional coordinate system, is the natural arena in which to depict solutions of one-variable equations and inequalities, a two-dimensional coordinate system is the natural place to graph solutions of two-variable equations and inequalities. The coordinate system we use is named after René Descartes, the 17th century French mathematician largely responsible for its development.

chapter three Linear Equations and Inequalities of Two Variables

The Cartesian Coordinate System

The **Cartesian coordinate system**, also referred to as the **Cartesian plane** or the **rectangular coordinate system**, consists of two perpendicular real number lines (each of which is an **axis**) intersecting at the point 0 of each line (see Figure 1). The point of intersection is called the **origin** of the system, and the four quarters defined by the two lines are called the **quadrants** of the plane, numbered as indicated in Figure 1. Because the Cartesian plane consists of two crossed real lines, it is often given the symbol $\mathbb{R} \times \mathbb{R}$, or $\mathbb{R}^2$. Each point P in the plane is identified by a unique pair of numbers (a, b), called an **ordered pair**. In a given ordered pair (a, b), the number a (the **first coordinate**) indicates the horizontal displacement of the point from the origin, and the number b (the **second coordinate**) indicates the vertical displacement. Figure 1 is an example of a Cartesian coordinate system, and illustrates how several ordered pairs are **graphed**, or **plotted**.

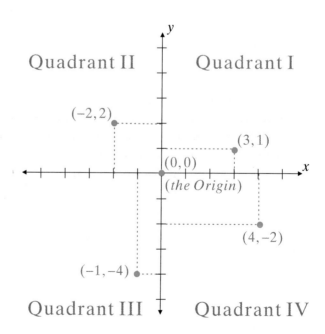

Figure 1: The Cartesian Plane $\mathbb{R}^2$

section 3.1   **The Cartesian Coordinate System**

> **caution**
>
> Unfortunately mathematics uses parentheses to denote ordered pairs as well as open intervals, which sometimes leads to confusion. You must rely on the context to determine the meaning of any parentheses you encounter. For instance, in the context of solving a one-variable inequality, the notation $(-2, 5)$ most likely refers to the open interval with endpoints at -2 and 5, whereas in the context of solving an equation in two variables, $(-2, 5)$ probably refers to a point in the Cartesian plane.

Topic 2: The Graph of an Equation

As previously mentioned, a solution of an equation in x and y must consist of a value a for x and a corresponding value b for y. It is natural to present such a solution as an ordered pair (a, b), and equally natural to depict the solution, graphically, as a point in the plane whose coordinates are a and b. In this context, the horizontal number line would be referred to as the ***x*-axis**, the vertical number line as the ***y*-axis**, and the two coordinates of the ordered pair (a, b) as the ***x*-coordinate** and the ***y*-coordinate**.

As we shall see, a given equation in x and y will usually consist of far more than one ordered pair (a, b). The **graph of an equation** consists of a depiction in the Cartesian plane of *all* of those ordered pairs that make up the solution set of the equation.

At this point, we can make rough sketches of the graphs of many equations just by plotting enough ordered pairs that solve the equation to give us a sense of the entire solution set. We can find individual ordered pair solutions of a given equation by selecting numbers that seem appropriate for one of the variables and then solving the equation for the other variable. Doing so changes the task of solving a two-variable equation into that of solving a one-variable equation, and we have all the methods of Chapter 2 at our disposal to accomplish this. The process is illustrated in *Example 1*.

chapter three Linear Equations and Inequalities of Two Variables

example 1

Sketch graphs of the following equations by plotting points.

a. $2x - 5y = 10$ **b.** $x^2 + y^2 - 6x = 0$ **c.** $y = x^2 - 2x$

Solutions:

You can view a table of specific points on the graph of an equation by selecting **Y=** and entering the equation, and then selecting **2ND** and **GRAPH**.

a.

x	y
−3	?
0	?
?	0
?	5
1	?

$2x - 5y = 10$ →

x	y
−3	−16/5
0	−2
5	0
35/2	5
1	−8/5

In each row in the first table, values have been chosen essentially at random for one of the variables. Once these values have been chosen, the equation $2x - 5y = 10$ can be solved for the other variable. This gives us a list of 5 ordered pairs that can be plotted, though the ordered pair $\left(\dfrac{35}{2}, 5\right)$ is off the coordinate system drawn below.

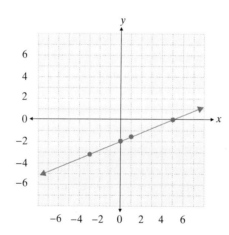

It appears that the four ordered pairs plotted lie on a straight line, and this is indeed the case. To gain more confidence in this fact, we could continue to plot more solutions of the equation, and we would find they all lie along the line that has been drawn through the four plotted ordered pairs. The infinite number of solutions of the equation are thus depicted by the blue line drawn as shown.

b.

x	y
0	?
1	?
2	?
3	?
4	?
5	?
6	?

$x^2 + y^2 - 6x = 0$ →

x	y
0	0
1	$\pm\sqrt{5} \approx \pm 2.2$
2	$\pm 2\sqrt{2} \approx \pm 2.8$
3	± 3
4	$\pm 2\sqrt{2} \approx \pm 2.8$
5	$\pm\sqrt{5} \approx \pm 2.2$
6	0

As in *Example 1a*, we determine enough ordered pairs that solve the equation $x^2 + y^2 - 6x = 0$ to feel confident in sketching the entire solution set. First, note that for any $x < 0$ and for any $x > 6$, the corresponding y would be imaginary, and thus irrelevant as far as graphing the equation is concerned. Similarly, for any $y < -3$ and for any $y > 3$, the corresponding x would be a complex number (use the quadratic formula to verify this fact).

section 3.1   The Cartesian Coordinate System

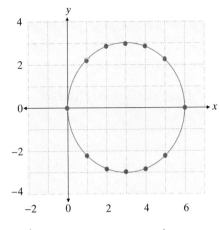

Once a sufficient number of ordered pairs have been calculated, the graph of the equation begins to take the shape of a circle. In fact, the graph is a circle, with center $(3, 0)$ and a radius of 3, but we will not be able to prove this claim until Section 3.6.

c.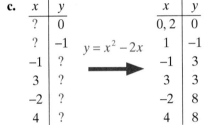

x	y
?	0
?	−1
−1	?
3	?
−2	?
4	?

$y = x^2 - 2x$ ⟶

x	y
0, 2	0
1	−1
−1	3
3	3
−2	8
4	8

We again start with a table of values, with a number chosen for one of the two variables in the first table. Note that when y is 0, there are two corresponding values for x (this is determined by solving the resulting quadratic equation in x, which can be done by factoring). Again, enough points should be plotted to give some idea of the nature of the entire solution set of the equation.

The graph of $y = x^2 - 2x$ is a shape known as a *parabola*. We will encounter these shapes again in Section 4.2, and we will be able, at that time, to prove that our rough sketch at the left is indeed the graph of the solution set of $y = x^2 - 2x$.

At this point, you may be uneasy about the graphs that were sketched in *Example 1*. You may be wondering if we have really plotted enough points to accurately "fill in" the gaps and sketch the entire solution set of each equation. Or you may be wondering if filling in the gaps is justified in the first place: what proof do we have that *all* the ordered pairs along our sketches actually solve the corresponding equation? Finally, you may be wondering if there isn't a faster and more sophisticated way to determine the graph of an equation. Plotting points is, for the most part, easily accomplished, but it is also a rather crude and tedious method.

chapter three Linear Equations and Inequalities of Two Variables

These concerns are not trivial, and the resolution of them will constitute much of Chapters 3 through 7. We will begin in the next section by undertaking a careful study of equations like the one in *Example 1a*. We will finish up this section, however, with two useful formulas regarding pairs of points in the plane.

Topic 3: The Distance and Midpoint Formulas

Throughout the rest of this text, we will have reasons for wanting to know, on occasion, the *distance* between two points in the Cartesian plane. We already have the tools necessary to answer this question, and we will now derive a formula that we can apply whenever necessary.

Let (x_1, y_1) and (x_2, y_2) be the coordinates of two arbitrary points in the plane. By drawing the dotted lines parallel to the coordinate axes as shown in Figure 2, we can form a right triangle. Note that we are able to determine the coordinates of the vertex at the right angle from the two vertices (x_1, y_1) and (x_2, y_2).

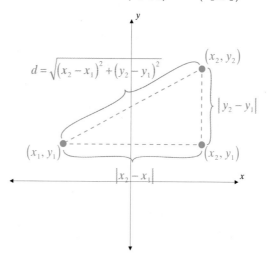

Figure 2: Distance Formula

The lengths of the two perpendicular sides of the triangle are easily calculated, as these lengths correspond to distances between numbers on real number lines. (The absolute value symbols are present as we don't necessarily know, in general, whether $x_2 - x_1$ or $y_2 - y_1$ is non-negative.) Recall the Pythagorean Theorem states that for a right triangle the sum of the squares of its sides is equal to the square of the hypotenuse ($a^2 + b^2 = c^2$). We can apply the Pythagorean Theorem to determine the distance, as labeled in Figure 2.

$$d^2 = (x_2 - x_1)^2 + (y_2 - y_1)^2, \text{ so}$$
$$d = \sqrt{(x_2 - x_1)^2 + (y_2 - y_1)^2}$$

(Notice that the absolute value symbols are not necessary in the formula, as any quantity squared is automatically non-negative.)

section 3.1   The Cartesian Coordinate System

Distance Formula

Letting (x_1, y_1) and (x_2, y_2) represent two points on the Cartesian plane, the distance between these two points may be found using the following formula, derived from the Pythagorean Theorem:

$$d = \sqrt{(x_2 - x_1)^2 + (y_2 - y_1)^2}$$

We will also, on occasion, want to be able to determine the midpoint of a line segment in the plane. That is, given two points (x_1, y_1) and (x_2, y_2), we will want to know the coordinates of the point exactly halfway between the two given points. Of course, such a formula will logically take the form of an ordered pair, as it must give us two numbers.

Consider the points as plotted in Figure 3. The x-coordinate of the midpoint should be the average of the two x-coordinates of the given points, and similarly for the y-coordinate.

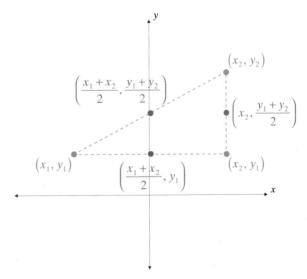

Figure 3: Midpoint Formula

Since x_1 and x_2 are numbers on a real number line, and y_1 and y_2 are numbers on a (different) real number line, determining the averages of these two pairs of numbers is straightforward: the average of the x-coordinates is $\frac{x_1 + x_2}{2}$ and the average of the y-coordinates is $\frac{y_1 + y_2}{2}$. Putting these two coordinates together gives us the desired formula:

chapter three   Linear Equations and Inequalities of Two Variables

Midpoint Formula

Letting (x_1, y_1) and (x_2, y_2) represent two points on the Cartesian plane, the midpoint between these two points may be found using the following formula, which finds the average of the two *x*-values and the average of the two *y*-values:

$$\left(\frac{x_1 + x_2}{2}, \frac{y_1 + y_2}{2}\right)$$

We will end with an example of sample calculations.

example 2

Determine **a.** the distance between $(5, 1)$ and $(-1, 3)$, and **b.** the midpoint of the line segment joining $(5, 1)$ and $(-1, 3)$.

Solutions:

When entering the distance formula, be sure to include all the necessary parentheses.

a. $d = \sqrt{(5-(-1))^2 + (1-3)^2}$

$= \sqrt{36 + 4}$

$= \sqrt{40}$

$= 2\sqrt{10}$

All we have to do is apply the distance formula that we have derived, keeping in mind that it is the *differences* of the respective coordinates that are squared and added.

The two points $(5, 1)$ and $(-1, 3)$ are $2\sqrt{10}$ units apart.

b. $\left(\dfrac{5+(-1)}{2}, \dfrac{1+3}{2}\right) = (2, 2)$

In determining the coordinates of the midpoint of the line segment, remember to add the respective coordinates (since we are finding an average), and then divide by 2.

184

exercises

Plot the following sets of points in the Cartesian plane.

1. $\{(-3, 2), (5, -1), (0, -2), (3, 0)\}$
2. $\{(-4, 0), (0, -4), (-3, -3), (3, -3)\}$
3. $\{(3, 4), (-2, -1), (-1, -3), (-3, 0)\}$
4. $\{(2, 2), (0, 3), (4, -5), (-1, 3)\}$
5. $\{(0, 5), (-3, 2), (2, 4), (1, 1)\}$
6. $\{(8, 3), (-3, 4), (-4, -6), (3, -4)\}$
7. $\{(-5, -4), (3, 2), (4, 5), (-2, -1), (-4, -4), (1, 1)\}$
8. $\{(-2, 5), (0, 1), (1, -1), (1, -3), (0, 0), (-1, 2), (0, -2)\}$

Identify the quadrant in which each point lies, if possible. If a point lies on an axis, specify which part (positive or negative) of which axis (x or y).

9. $(-2, -4)$
10. $(0, -12)$
11. $(4, -7)$
12. $(-2, 0)$
13. $(9, 0)$
14. $(3, 26)$
15. $(-4, -7)$
16. $(0, 1)$
17. $(17, -2)$
18. $(-\sqrt{2}, 4)$
19. $(-1, 1)$
20. $(-4, 0)$
21. $(3, -9)$
22. $(0, 0)$
23. $(4, 3)$
24. $(-3, -11)$
25. $(0, -97)$
26. $\left(\frac{1}{3}, 0\right)$

Determine appropriate settings on a graphing calculator so that each of the given points will lie within the viewing window. Answers may vary slightly.

27. $\{(-4, 1), (2, 8), (5, -7)\}$
28. $\{(12, 3), (5, -11), (-9, 6)\}$
29. $\{(3, 2), (-2, 4), (5, -3)\}$
30. $\{(30, 55), (40, 25), (-80, -10)\}$
31. $\{(3.75, -8.5), (-5.25, 6.0), (7.5, -2.25)\}$
32. $\{(63, 99), (-87, 34), (45, -22)\}$

chapter three Linear Equations and Inequalities of Two Variables

For each of the following equations, determine the value of the missing entries in the accompanying table of ordered pairs. Then plot the ordered pairs and sketch your guess of the complete graph of the equation. See Example 1.

33. $6x - 4y = 12$

x	y
0	?
?	0
3	?
?	3

34. $y = x^2 + 2x + 1$

x	y
?	0
1	?
?	1
2	?
-3	?

35. $x = y^2$

x	y
0	?
1	?
4	?
9	?
?	$-\sqrt{2}$

36. $5x - 2 = -y$

x	y
?	0
0	?
1	?
?	7
-2	?

37. $x^2 + y^2 = 9$

x	y
0	?
?	0
-1	?
1	?
?	2

38. $y = -x^2$

x	y
0	?
-1	?
1	?
-2	?
2	?

*Determine **a.** the distance between the following pairs of points, and **b.** the midpoint of the line segment joining each pair of points. See Example 2.*

39. $(-2, 3)$ and $(-5, -2)$ **40.** $(-1, -2)$ and $(2, 2)$ **41.** $(0, 7)$ and $(3, 0)$

42. $\left(-\dfrac{1}{2}, 5\right)$ and $\left(\dfrac{9}{2}, -7\right)$ **43.** $(-2, 0)$ and $(0, -2)$ **44.** $(5, 6)$ and $(-3, -2)$

45. $(13, -14)$ and $(-7, -2)$ **46.** $(-8, 3)$ and $(2, 11)$ **47.** $(-3, -3)$ and $(5, -9)$

48. $(7, -7)$ and $(-7, -6)$ **49.** $(5, -4)$ and $(-1, 5)$ **50.** $(4, 6)$ and $(2, -7)$

51. $(8, 8)$ and $(-2, -2)$

52. Given $(10, 4)$ and $(x, -2)$, find x such that the distance between these two points is 10.

53. Given $(1, y)$ and $(13, -3)$, find y such that the distance between these two points is 15.

54. Given $(x, 3)$ and $(-6, y)$, find x and y such that the midpoint between these two points is $(2, 2)$.

section 3.1   The Cartesian Coordinate System

Find the perimeter of the triangle whose vertices are the specified points in the plane.

55. $(-2, 3), (-2, 1),$ and $(-5, -2)$ **56.** $(-1, -2), (2, -2),$ and $(2, 2)$

57. $(6, -1), (-6, 4),$ and $(9, 3)$ **58.** $(3, -4), (-7, 0),$ and $(-2, -5)$

59. $(-3, 7), (5, 1),$ and $(-3, -14)$ **60.** $(-12, -3), (-7, 9),$ and $(9, -3)$

61. Use the distance formula to prove that the triangle with vertices at the points $(1, 1)$, $(-2, -5)$, and $(3, 0)$ is a right triangle. Then determine the area of the triangle.

62. Use the distance formula to prove that the triangle with vertices at the points $(-2, 2)$, $(1, -2)$, and $(2, 5)$ is isosceles. Then determine the area of the triangle. (**Hint:** Make use of the midpoint formula.)

63. Use the distance formula to prove that the triangle with vertices at the points $(5, 1)$, $(-3, 7)$, and $(8, 5)$ is a right triangle. Then determine the area of the triangle.

64. Use the distance formula to prove that the triangle with vertices at the points $(1, 2)$, $(-2, 0)$, and $(3, 5)$ is isosceles. Then determine the area of the triangle. (**Hint:** Make use of the midpoint formula.)

65. Use the distance formula to prove that the triangle with vertices at the points $(2, 2)$, $(6, 3)$, and $(4, 11)$ is a right triangle. Then determine the area of the triangle.

66. Use the distance formula to prove that the triangle with vertices at the points $(2, -1), (4, 3),$ and $(-2, -3)$ is isosceles. Then determine the area of the triangle. (**Hint:** Make use of the midpoint formula.)

67. Use the distance formula to prove that the polygon with vertices at the points $(-2, -1), (6, 5), (-2, 5),$ and $(6, -1)$ is a rectangle. Then determine the area of the rectangle. (**Hint:** It may help to plot the points before you begin.)

68. Plot the points $(-3, 3), (-5, -2), (3, -2),$ and $(1, 3)$ to demonstrate they are the vertices of a trapezoid. Then determine the area of the trapezoid.

69. Two college friends are taking a weekend road trip. Friday they leave home and drive 87 miles north for a night of dinner and dancing in the city. The next morning they drive 116 miles east to spend a day at the beach. If they drive straight home from the beach the next day, how far do they have to travel on Sunday?

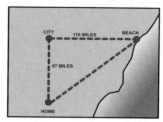

70. Your backpacker's guide contains a grid map of Paris, with each unit on the grid representing 0.25 kilometers. If the Eiffel Tower is located at (−8, −1) and the Arc de Triomphe is located at (−8, 4), what is the direct distance (not walking distance, which would have to account for bridges and roadways) between the two monuments in kilometers?

71. Your hotel, located at (−1, −2) on the map from Exercise 70, is advertised as exactly halfway between the Eiffel Tower and Notre Dame. What are the grid coordinates of Notre Dame on your map? Find the direct distance from the Eiffel Tower to Notre Dame, rounded to the nearest hundredth of a kilometer.

72. The navigator of a submarine plots the position of the submarine and surrounding objects using a rectangular coordinate system, where each block is one square meter.
   a. If his submarine is located at (50, 231) and the mobile base to which he is heading is located at (83, 478), how far is he from the mobile base?
   b. Suppose there is another submarine located halfway between the first submarine and the mobile base. What is the position of the second sub?

73. At the entrance to Paradise Island Theme Park you are given a map of the park that is in the form of a grid, with the park entrance located at (−5, −5). After walking past three rides and the restrooms, you arrive at the Tsunami Water Ride, which is located at (−3, −1) on the grid. If you have traveled halfway along a straight line to your favorite ride, Thundering Tower, where on the grid is your favorite ride located? How far is Thundering Tower from the park entrance on the map?

3.2 Linear Equations in Two Variables

TOPICS

1. Recognizing linear equations in two variables
2. Intercepts of the coordinate axes
3. Horizontal and vertical lines

Topic 1: Recognizing Linear Equations in Two Variables

We have encountered one linear equation in two variables already, in *Example 1a* of Section 3.1. Our first goal in this section is to be able to recognize when an equation in two variables is linear; that is, we wish to know when the solution set of an equation consists of a straight line in the Cartesian plane.

Linear Equations in Two Variables

A **linear equation in two variables**, say the variables x and y, is an equation that can be written in the form $ax + by = c$, where a, b, and c are constants and a and b are not both zero. This form of such an equation is called the **standard form**.

Of course, an equation may be linear but not appear in standard form. Some algebraic manipulation is often necessary in order to determine if a given equation is linear.

example 1

Determine if the following equations are linear equations.

a. $3x - (2 - 4y) = x - y + 1$

b. $3x + 2(x + 7) - 2y = 5x$

c. $\dfrac{x+2}{3} - y = \dfrac{y}{5}$

d. $7x - (4x - 2) + y = y + 3(x - 1)$

e. $4x^3 - 2y = 5x$

f. $x^2 - (x - 3)^2 = 3y$

continued on next page...

chapter three   Linear Equations and Inequalities of Two Variables

Solutions:

a. $3x - (2 - 4y) = x - y + 1$
$3x - 2 + 4y = x - y + 1$
$3x - x + 4y + y = 1 + 2$
$2x + 5y = 3$

As is typical, we need to use some algebra to rewrite the equation in an equivalent form which will be clearly either linear or not. After using the distributive and cancellation properties, we see that the equation is indeed linear.

b. $3x + 2(x + 7) - 2y = 5x$
$3x + 2x + 14 - 2y = 5x$
$5x - 5x - 2y = -14$
$-2y = -14$
$y = 7$

We proceed as in the problem above, using familiar properties to rewrite the equation as an equivalent one in standard form. In this problem, however, one of the variables, namely x, is absent from the result. But since the coefficient of y is non-zero, this equation is also linear.

c. $\dfrac{x+2}{3} - y = \dfrac{y}{5}$
$\dfrac{1}{3}x + \dfrac{2}{3} - y = \dfrac{y}{5}$
$\dfrac{1}{3}x - y - \dfrac{y}{5} = -\dfrac{2}{3}$
$\dfrac{1}{3}x - \dfrac{6}{5}y = -\dfrac{2}{3}$

Although this equation may appear at first to not fit the criteria to be linear, appearances are deceiving.

After splitting apart one of the fractions and moving all constant terms to the right-hand side and all variables to the left-hand side, we see that this is another example of a linear equation. Note that we could also have begun by clearing the equation of fractions first.

d. $7x - (4x - 2) + y = y + 3(x - 1)$
$7x - 4x + 2 + y = y + 3x - 3$
$3x - 3x + y - y = -3 - 2$
$0 = -5$

In this equation, both variables are absent in the final (and equivalent) form. This tells us immediately that the equation is not linear, as the coefficients of x and y are both 0. Moreover, this equation has no solution: no values for x and y result in a true statement.

e. $4x^3 - 2y = 5x$

The presence of the cubed term in this already simplified equation makes it clearly not linear.

f. $x^2 - (x - 3)^2 = 3y$
$x^2 - x^2 + 6x - 9 = 3y$
$6x - 3y = 9$

In contrast to the last equation, when we simplify this equation the result clearly is linear.

section 3.2   Linear Equations in Two Variables

Topic 2: Intercepts of the Coordinate Axes

Quite often, the goal in working with a given linear equation is to graph its solution set. If the straight line whose points constitute the solution set crosses the horizontal and vertical axes in two distinct points, knowing the coordinates of these two points is sufficient to graph the complete solution; this follows from the fact that two points are enough to completely determine a line.

If the equation under consideration is in the two variables x and y, it is natural to call the point where the graph crosses the x-axis the **x-intercept** and the point where it crosses the y-axis the **y-intercept**. If the line does indeed cross both axes, the two intercepts are easy to find: the y-coordinate of the x-intercept is 0, and the x-coordinate of the y-intercept is 0.

example 2

Find the x- and y-intercepts of the following equations, and then graph each equation.

a. $3x - 4y = 12$

b. $4x - (3 - x) + 2y = 7$

Solutions:

a.
$$3x - 4y = 12$$

$$3(0) - 4y = 12 \qquad 3x - 4(0) = 12$$
$$y = -3 \qquad\qquad x = 4$$
y-intercept: $(0, -3)$ $\quad$ x-intercept: $(4, 0)$

To find the two intercepts, we first set x equal to 0 and solve for y, then set y equal to 0 and solve for x. The coordinates of the two intercepts are shown to the left. Although it is slightly imprecise, it is common to say, for instance, that the y-intercept is -3 and the x-intercept is 4.

Once we have plotted the two intercepts, drawing a straight line through them gives us the graph of the equation.

Notice that the y-value at $x=4$ is 0. To get to this screen, select `Y=` and enter the equation. Then select `2ND` and `GRAPH` to view the table.

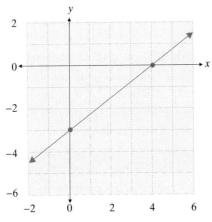

continued on next page...

chapter three Linear Equations and Inequalities of Two Variables

b.
$$4x - (3-x) + 2y = 7$$
$$5x + 2y = 10$$

$$5(0) + 2y = 10 \qquad 5x + 2(0) = 10$$
$$y = 5 \qquad\qquad x = 2$$

y-intercept: $(0, 5)$ x-intercept: $(2, 0)$

Again, find the two intercepts by setting the appropriate variables equal to 0. The y-intercept is 5 and the x-intercept is 2.

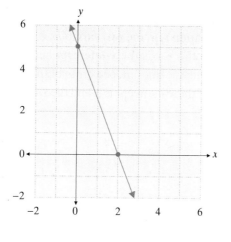

Note that in both graphs, the location of the origin has been chosen in order to conveniently plot the intercepts.

Topic 3: Horizontal and Vertical Lines

The procedure used in *Example 2* for graphing linear equations may be undesirable or inadequate for several reasons. One such reason is that the x-intercept and y-intercept of a given line may be the same point–the origin. In order to graph the equation in this case, a second point distinct from the origin must be found in order to have two points to connect with a line. Another reason is that a given linear equation may not have one of the two types of intercepts. A moment's thought will reveal that this can happen only when the graph of the equation is a horizontal or vertical line.

Equations whose graphs are horizontal or vertical lines are missing one of the two variables (though it may be necessary to simplify the equation before this is apparent). However, in the absence of any other information, it is impossible to know if the solution of such a linear equation consists of a point on the real number line or a line in the Cartesian plane (or, indeed, of a higher-dimensional set in a higher-dimensional coordinate system). You must rely on the context of the problem to know how many variables should be considered. Throughout this chapter, all equations are assumed to be in two variables unless otherwise stated, so an equation of the form $ax = c$ or $by = c$ should be thought of as representing a line in $\mathbb{R}^2$.

section 3.2 Linear Equations in Two Variables

Consider an equation of the form $ax = c$. The variable y is absent, so *any* value for this variable will suffice as long as we pair it with $x = \frac{c}{a}$. Thinking of the solution set as a set of ordered pairs, the solution consists of ordered pairs with a fixed first coordinate and arbitrary second coordinate. This describes, geometrically, a vertical line with an x-intercept of $\frac{c}{a}$. Similarly, the equation $by = c$ represents a horizontal line with y-intercept equal to $\frac{c}{b}$.

example 3

Graph the following equations.

a. $3x + 2(x + 7) - 2y = 5x$ **b.** $5x = 0$ **c.** $2x - 2 = 3$

Solutions:

a. $3x + 2(x + 7) - 2y = 5x$

$y = 7$

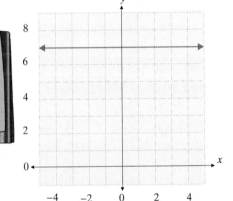

We encountered this equation in *Example 1b*, and have already written it in standard form as shown.

The graph of this equation is the horizontal line consisting of all those ordered pairs whose y-coordinate is 7.

b. $5x = 0$

$x = 0$

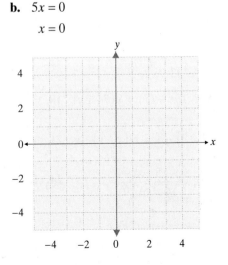

The first step in this example is to divide both sides by 5, leaving the simple equation $x = 0$.

Note that the graph of this equation is the y-axis, as all ordered pairs on the y-axis have an x-coordinate of 0. This equation is unique in that it has an infinite number of y-intercepts (each point on the graph is also on the y-axis) and one x-intercept (the origin). Similarly, the equation $y = 0$ has an infinite number of x-intercepts and one y-intercept.

continued on next page...

To graph an equation, select [Y=] and enter the equation in terms of y. Then select [GRAPH] to view the graph. Note that this is not possible for vertical lines since no y term is present.

chapter three Linear Equations and Inequalities of Two Variables

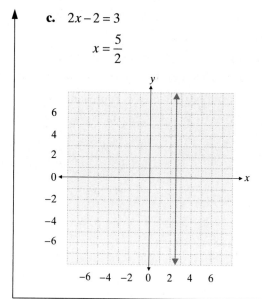

c. $2x - 2 = 3$

$x = \dfrac{5}{2}$

Upon simplifying, it is apparent that this equation also represents a vertical line, this time passing through $\dfrac{5}{2}$ on the x-axis.

exercises

Determine if the following equations are linear. See Example 1.

1. $3x + 2(x - 4y) = 2x - y$
2. $9x + 4(y - x) = 3$
3. $9x^2 - (x + 1)^2 = y - 3$
4. $3x + xy = 2y$
5. $8 - 4xy = x - 2y$
6. $\dfrac{x - y}{2} + \dfrac{7y}{3} = 5$
7. $\dfrac{6}{x} - \dfrac{5}{y} = 2$
8. $3x - 3(x - 2y) = y + 1$
9. $2y - (x + y) = y + 1$
10. $(3 - y)^2 - y^2 = x + 2$
11. $x^2 - (x - 1)^2 = y$
12. $(x + y)^2 - (x - y)^2 = 1$
13. $x(y + 1) = 16 - y(1 - x)$
14. $\dfrac{x - 3}{2} = \dfrac{4 + y}{5}$
15. $x - 2x^2 + 3 = \dfrac{x - 7}{2}$
16. $x - 3 = \dfrac{4x + 17}{5}$
17. $13x - 17y = y(7 - 2x)$
18. $y^2 - 3y = (1 + y)^2 - 2x$
19. $x - 1 = \dfrac{2y}{x} - x$
20. $3x - 4 = 89(x - y) - y$
21. $x - x(1 + x) = y - 3x$
22. $x^2 - 2x = 3 - x^2 + y$
23. $\dfrac{2y - 5}{14} = \dfrac{x - 3}{9}$
24. $16x = y(4 + (x - 3)) - xy$

section 3.2 Linear Equations in Two Variables

Determine the x- and y-intercepts of the following linear equations, if possible, and then graph the equations. See Examples 2 and 3.

25. $4x - 3y = 12$

26. $y - 3x = 9$

27. $5 - y = 10x$

28. $y - 2x = y - 4$

29. $3y = 9$

30. $2x - (x + y) = x + 1$

31. $x + 2y = 7$

32. $y - x = x - y$

33. $y = -x$

34. $2x - 3 = 1 - 4y$

35. $3y + 7x = 7(3 + x)$

36. $4 - 2y = -2 - 6x$

37. $x + y = 1 + 2y$

38. $3y + x = 2x + 3y + 4$

39. $3(x + y) + 1 = x - 5$

Match each equation to the correct graph.

40. $y = 2x + 3$

41. $2x + 3y = 4$

42. $2x - 1 = 5$

43. $y + 3 - x = 3$

44. $4y + 3 = 11$

45. $5y - x - 1 = 4y + 3x + 5$

a.

b.

c.

d.

e.

f.

chapter three Linear Equations and Inequalities of Two Variables

Solve each equation for the specified variable.

46. Standard Form of a Line: $ax + by = c$. Solve for y.

47. Perimeter of a Triangle: $P = a + b + c$. Solve for a.

48. Surface Area of a Rectangular Box: $S = 2lw + 2wh + 2lh$. Solve for w.

49. Population: In your history class, you were told that the current population of Jamaica is approximately 24,000 more than 9 times the population of the Bahamas. Using j to represent the population of Jamaica and b to represent the population of the Bahamas, write this in the form of an equation. Then solve your equation for b to find an equation representing the population of the Bahamas. Are these equations linear?

50. Geography: The lowest point in the ocean, the bottom of the Mariana Trench, is about 1100 feet deeper than 26 times the depth of the lowest point on land, the Dead Sea. Find an equation to express the depth of the Mariana Trench, m, in terms of the depth of the Dead Sea, d. Then solve your equation for d to find the depth of the Dead Sea in terms of the depth of the Mariana Trench. Are these equations linear?

3.3 Forms of Linear Equations

TOPICS

1. The slope of a line
2. Slope-intercept form of a line
3. Point-slope form of a line
4. Interlude: linear regression

Topic 1: The Slope of a Line

The methods we have used for graphing linear equations have been adequate for our purposes so far, but for many problems we will want more sophisticated ways to graph lines and to pick out the important characteristics of linear equations.

There are several ways to characterize a given line in the plane. We have already used one way repeatedly: two distinct points in the Cartesian plane determine a unique line. Another, often more useful, approach is to identify just one point on the line and to indicate how "steeply" the line is rising or falling as we scan the plane from left to right. It turns out that a single number is sufficient to convey this notion of "steepness."

Slope of a Line

Let L stand for a given line in the Cartesian plane, and let (x_1, y_1) and (x_2, y_2) be the coordinates of any two distinct points on L. The **slope** of the line L is the ratio $\dfrac{y_2 - y_1}{x_2 - x_1}$, which can be described in words as "change in y over change in x" or "rise over run."

The phrase "rise over run" is motivated by the diagram in Figure 1.

chapter three Linear Equations and Inequalities of Two Variables

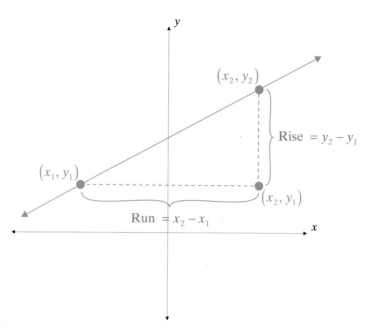

Figure 1: Rise and Run Between Two Points

As drawn in Figure 1, the ratio $\frac{y_2 - y_1}{x_2 - x_1}$ is positive, and we say that the line has a positive slope. If the rise and run have opposite signs, the slope of the line would be negative and the line under consideration would be falling from the upper left to the lower right.

caution

It doesn't matter how you assign the labels (x_1, y_1) and (x_2, y_2) to the two points you are using to calculate slope, but it *is* important that you are consistent as you apply the formula. That is, don't change the order in which you are subtracting as you determine the numerator and denominator in the formula $\frac{y_2 - y_1}{x_2 - x_1}$.

The formula $\frac{y_2 - y_1}{x_2 - x_1}$ is only valid if $x_2 - x_1 \neq 0$, as division by zero is never defined. For what sorts of lines, then, are we unable to find the slope? If two points on a line have the same first coordinate, that is if $x_1 = x_2$, the line must be perfectly vertical. The other extreme is that the numerator is 0 (and the denominator is non-zero). In this case, the slope as a whole is 0. For what sorts of lines will this happen? If two points on a line have the same second coordinate, that is if $y_1 = y_2$, the line must be horizontal.

section 3.3   **Forms of Linear Equations**

Slopes of Horizontal and Vertical Lines

Horizontal lines all have slopes of 0, and horizontal lines are the *only* lines with slope equal to 0. The equation of a horizontal line can be written in the form $y = c$, where c is a constant.

Vertical lines all have undefined slopes, and vertical lines are the *only* lines for which the slope is undefined. The equation of a vertical line can be written in the form $x = c$, where c is a constant.

example 1

Determine the slopes of the lines passing through the following pairs of points in $\mathbb{R}^2$.

a. $(-4, -3)$ and $(2, -5)$ **b.** $\left(\dfrac{3}{2}, 1\right)$ and $\left(1, -\dfrac{4}{3}\right)$ **c.** $(-2, 7)$ and $(1, 7)$

Solutions:

a. $\dfrac{-3-(-5)}{-4-2} = \dfrac{2}{-6} = -\dfrac{1}{3}$

In the calculation to the left, the coordinates of the second point have been subtracted from the coordinates of the first. The result would be the same if we reversed the order and subtracted the coordinates of the first from those of the second.

b. $\dfrac{1-\left(-\dfrac{4}{3}\right)}{\dfrac{3}{2}-1} = \dfrac{\dfrac{7}{3}}{\dfrac{1}{2}} = \dfrac{7}{3} \cdot \dfrac{2}{1} = \dfrac{14}{3}$

The final answer tells us that the line through the two points rises 14 units for every run of 3 units horizontally.

c. $\dfrac{7-7}{-2-1} = \dfrac{0}{-3} = 0$

The two points in part **c** have the same second coordinate and thus lie on a horizontal line.

We already know how to identify any number of ordered pairs that lie on a line, given the equation for the line. Identifying just two such ordered pairs allows us to calculate the slope of a line defined by an equation.

chapter three   Linear Equations and Inequalities of Two Variables

example 2

Determine the slopes of the lines defined by the following equations.

a. $4x - 3y = 12$ b. $2x + 7y = 9$ c. $x = -\dfrac{3}{4}$

Solutions:

a. x-intercept: $(3, 0)$
   y-intercept: $(0, -4)$
   slope $= \dfrac{-4 - 0}{0 - 3} = \dfrac{-4}{-3} = \dfrac{4}{3}$

Determining the x- and y-intercepts of this line is fairly straightforward. Recall that the x-intercept is found by setting y equal to 0 and solving for x, and vice versa for the y-intercept. Once we have two points, we apply the slope formula.

b. x-intercept: $\left(\dfrac{9}{2}, 0\right)$
   second point on the line: $(1, 1)$
   slope $= \dfrac{1 - 0}{1 - \dfrac{9}{2}} = \dfrac{1}{-\dfrac{7}{2}} = -\dfrac{2}{7}$

In this example, we have found the x-intercept, but have used an ordered pair that was more easily determined than the y-intercept for the second point.

c. first point: $\left(-\dfrac{3}{4}, 2\right)$
   second point: $\left(-\dfrac{3}{4}, 10\right)$
   slope is undefined

As soon as we realize that the line defined by the equation is vertical, we can state that the slope is undefined. Two arbitrarily chosen points have been identified as lying on the line, but the critical observation is that *any* two points in the slope formula result in division by 0.

Note that the line in *Example 2a* has a positive slope and the line in *Example 2b* has a negative slope. So even without graphing these lines, we know that the first line will rise from the lower left to the upper right part of the plane, while the second line will fall from the upper left to the lower right. You should practice your graphing skills and verify that these observations are indeed correct.

Topic 2:   Slope-Intercept Form of a Line

Example 2 illustrates the most elementary way of determining the slope of a line from an equation. With a small amount of work, we can develop a faster method for determining not only the slope of a line, but also the y-intercept (assuming the line *has* a y-intercept).

section 3.3 Forms of Linear Equations

Consider a fixed non-vertical line in the plane. The variable y must appear in the linear equation that describes the line (otherwise the line would be vertical), so the equation can be solved for y. The result will be an equation of the form $y = mx + b$, where m and b (the letters used by convention in this context) are constants particular to the given line (note that the constant m will be 0 if the line is horizontal). Now suppose that (x_1, y_1) and (x_2, y_2) are two points that lie on the line $y = mx + b$. That is, suppose that (x_1, y_1) and (x_2, y_2) both solve the equation $y = mx + b$. Then it must be the case that $y_1 = mx_1 + b$ and $y_2 = mx_2 + b$. If we use these two points to determine the slope of the line, we obtain:

$$\text{slope} = \frac{y_2 - y_1}{x_2 - x_1} = \frac{(mx_2 + b) - (mx_1 + b)}{x_2 - x_1} = \frac{m(x_2 - x_1)}{x_2 - x_1} = m.$$

What we have just shown is that the slope of the line is automatically the coefficient of the variable x when the line's equation is solved for y.

Let us continue investigating the line in the form $y = mx + b$. Since we are assuming the line is non-vertical, it must have a y-intercept. We know how to find the y-intercept by setting x equal to 0 and solving for y. This calculation can be performed directly, as we have already solved the equation for y. We find that the y-intercept is the point $(0, b)$, and conclude that the two constants in the equation $y = mx + b$ both have physical meaning. These observations are summarized below.

Slope-Intercept Form of a Line

If the equation of a non-vertical line in x and y is solved for y, the result is an equation of the form

$$y = mx + b.$$

The constant m is the slope of the line, and the line crosses the y-axis at b; that is, the y-intercept of the line is $(0, b)$. If the variable x does not appear in the equation, the slope is 0 and the equation is simply of the form $y = b$.

We can make use of the slope-intercept form of a line to graph the line, as illustrated in the following example.

chapter three Linear Equations and Inequalities of Two Variables

example 3

Use the slope-intercept form of the line to graph the equation $4x - 3y = 6$.

Solution:

$$4x - 3y = 6$$
$$-3y = -4x + 6$$
$$y = \frac{4}{3}x - 2$$

The first step is to solve the equation for y. Once we have done this, we know that the line has a slope of $\frac{4}{3}$ and crosses the y-axis at -2.

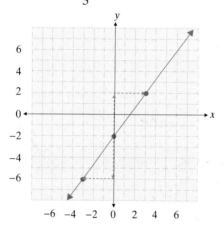

To use this information to sketch the graph of the line, we begin by plotting the y-intercept. A second point can now be found by using the fact that slope is "rise over run." This means a second point must lie, in this case, 3 units to the right and 4 units up, i.e. at $(3, 2)$. (Alternatively, we could locate a second point by moving to the left 3 units and moving down 4 units.)

In some cases, we can also make use of the slope-intercept form to find the equation of a line that has certain properties.

example 4

Find the equation of the line that passes through the point $(0, 3)$ and has a slope of $-\frac{3}{5}$. Then graph the line.

Solution:

$$y = -\frac{3}{5}x + 3$$
or
$$3x + 5y = 15$$

We can immediately write down the equation, as we have been given the y-intercept and the slope. If we wish, we can then write the equation in standard form.

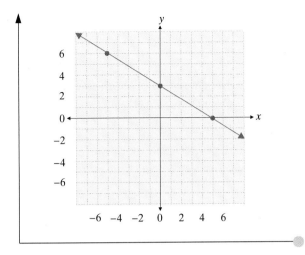

We can plot the line given only information about the slope and the y-intercept, just as we did in *Example 3*. We first plot the y-intercept, then move down 3 units and to the right 5 units to locate a second point. Or, we could have plotted a second point by moving up 3 units and to the left 5 units. Either method makes use of the fact that the slope is $-\dfrac{3}{5}$.

Topic 3: Point-Slope Form of a Line

A close examination of *Example 4* illustrates a point of weakness in the slope-intercept form. We were able to find the equation of a line having certain properties in that example, but only because the point given to us happened to be the y-intercept. In the most general case, we need to be able to construct an equation of a line given the slope of the line and *any* point on the line (as opposed to the y-intercept). Or, we might need to be able to find the equation of a line given two points on the line. In order to meet these needs, we introduce here the point-slope form of a line, the final form of linear equations in two variables.

Suppose we know that a given line L has slope m and passes through the point (x_1, y_1). Given these three numbers (m, x_1, and y_1), we want to come up with an equation in the variables x and y whose solution set is the line L. We can achieve this goal by assuming that (x, y) represents an arbitrary point on the line. Then by the definition of slope, it must be the case that

$$\frac{y - y_1}{x - x_1} = m.$$

This is true because for any point (x, y) on the line, the slope formula applied to the two points (x, y) and (x_1, y_1) must yield the number m.

If we now multiply the above equation by $x - x_1$, we obtain the desired form, as summarized on the next page.

chapter three Linear Equations and Inequalities of Two Variables

Point-Slope Form of a Line

Given an ordered pair (x_1, y_1) and a real number m, an equation for the line passing through the point (x_1, y_1) with slope m is
$$y - y_1 = m(x - x_1).$$
Note that m, x_1, and y_1 are all constants, and that x and y are variables. Note also that since the line, by definition, has slope m, vertical lines cannot be described in this form.

example 5

Find the equation, in slope-intercept form, of the line that passes through the point $(-2, 5)$ with slope 3.

Solution:

$$y - 5 = 3(x - (-2))$$
$$y - 5 = 3(x + 2)$$
$$y - 5 = 3x + 6$$
$$y = 3x + 11$$

A point and a slope are exactly the information we need to use the point-slope form. Replacing y_1 with 5 and x_1 with -2, we have an equation with the desired properties. The one remaining task is to write the equation in slope-intercept form, which we accomplish by solving for y.

We know that two distinct points in the plane are sufficient to determine a line. How do we use two such points to find the equation for a line?

example 6

Find the equation, in slope-intercept form, of the line that passes through the two points $(-3, -2)$ and $(1, 6)$.

Solution:

$$m = \frac{-2 - 6}{-3 - 1} = \frac{-8}{-4} = 2$$
$$y - 6 = 2(x - 1)$$
$$y - 6 = 2x - 2$$
$$y = 2x + 4$$

We begin by using the slope formula to determine the slope of the line from the two given points.

Once we have found m, we can use this number and either of the two points that we know are on the line to find the equation in point-slope form. If we had used the point $(-3, -2)$ instead of the point $(1, 6)$, we would have obtained the same equation once it is written in slope-intercept form.

Topic 4:   Interlude: Linear Regression

We have seen that, given a linear equation in the form $y = mx + b$, we can easily determine any number of points lying on the line. We have also seen how to determine an equation for a line passing through any two distinct points (x_1, y_1) and (x_2, y_2). In many practical applications, the goal is to do just the opposite to the greatest extent possible. That is, given a number of points $(x_1, y_1), (x_2, y_2), \ldots, (x_n, y_n)$, we would like to find the equation $y = mx + b$ whose graph comes closest to "fitting" the points.

An example will best illustrate this goal. Figure 2 below contains a plot of the growth of a new financial advising firm, with the vertical axis representing the number of clients acquired and the horizontal axis representing the number of weeks after the firm's opening. The actual data points are listed in the accompanying table.

Figure 2: Growth of Number of Clients

x (weeks after opening)	y (number of clients)
1	14
3	18
6	21
10	25
14	28
18	30
22	35

The owner of the business would like to be able to make a projection about the number of customers in the near future, based on the assumption that the growth can be approximately modeled by a straight line. A crude method would be to "eyeball" a straight line – that is, to literally draw a straight line on the above graph coming as close

continued on next page...

chapter three Linear Equations and Inequalities of Two Variables

as possible to the given data points. If the line were extended far enough to the right, a very approximate guess for the number of clients obtained by a particular week could be made.

This method's disadvantage is its imprecision. Different people are likely to draw different straight lines, and there is no resultant equation with which to work. Both of these flaws can be avoided by using **linear regression** to find the line that most closely fits given data.

The method of linear regression results in the slope-intercept form for the line whose graph minimizes the deviations between the line and the actual data points. The formulas in the method are easy enough to state, but their derivation is best left to a course in either calculus or linear algebra. We will illustrate the steps in the method by finding the best-fitting line for the seven data points in Figure 2 (with all calculations rounded to two decimal points).

Step 1: Calculate $\bar{x}$ and $\bar{y}$.

$\bar{x}$ and $\bar{y}$ (read "x bar" and "y bar") are the averages (or *means*) of the x-values and y-values, respectively, of all the given data points. For our example,

$$\bar{x} = \frac{1+3+6+10+14+18+22}{7} = 10.57$$

and

$$\bar{y} = \frac{14+18+21+25+28+30+35}{7} = 24.43.$$

Step 2: Calculate Δx and Δy values.

For each x-value there is a corresponding Δx, which represents the difference between the x-value and $\bar{x}$. Similarly, each y-value gives rise to a Δy. The Greek letter Δ ("delta") is often used to indicate difference, and these particular differences are read as "delta x" and "delta y". The tables below contain these calculations.

x	$\Delta x = x - \bar{x}$
1	−9.57
3	−7.57
6	−4.57
10	−0.57
14	3.43
18	7.43
22	11.43

y	$\Delta y = y - \bar{y}$
14	−10.43
18	−6.43
21	−3.43
25	0.57
28	3.57
30	5.57
35	10.57

Step 3: Calculate $\sum \Delta x \Delta y$ and $\sum \Delta x \Delta x$.

The Greek letter Σ ("sigma") is often used to indicate a sum of terms; you will encounter the symbol again and learn more about its use in Section 9.1. In this context, $\sum \Delta x \Delta y$ is just a convenient way to indicate that corresponding Δx's and Δy's from the above tables should be multiplied together, and the resulting products added. Similarly, $\sum \Delta x \Delta x$ represents the sum of the squares of all the Δx's. The following tables show the computations of each quantity.

Δx	Δy	$\Delta x \Delta y$
−9.57	−10.43	99.82
−7.57	−6.43	48.68
−4.57	−3.43	15.68
−0.57	0.57	−0.32
3.43	3.57	12.25
7.43	5.57	41.39
11.43	10.57	120.82
		$\Sigma \Delta x \Delta y = 338.32$

Δx	Δx	$\Delta x \Delta x$
−9.57	−9.57	91.58
−7.57	−7.57	57.30
−4.57	−4.57	20.88
−0.57	−0.57	0.32
3.43	3.43	11.76
7.43	7.43	55.20
11.43	11.43	130.64
		$\Sigma \Delta x \Delta x = 367.68$

Step 4: Calculate slope m and y-intercept b for the linear regression "best-fit" line.

The remaining tasks are easy. The slope for the linear regression line is

$$m = \frac{\sum \Delta x \Delta y}{\sum \Delta x \Delta x} = \frac{338.32}{367.68} = 0.92$$

and the y-intercept is

$$b = \bar{y} - m\bar{x} = 24.43 - (0.92)(10.57) = 14.71.$$

So the equation for the linear regression line fitting the data in this example is $y = 0.92x + 14.71$. Figure 3 contains the original data points and the graph of the line we have just found.

continued on next page...

chapter three Linear Equations and Inequalities of Two Variables

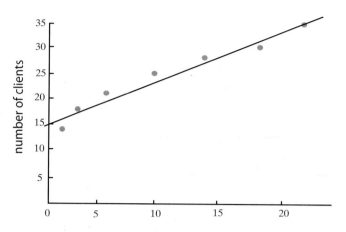

Figure 3: Linear Regression Line and Data

technology note

Computer software and graphing calculators can also be used to find a line of best fit, and can be used to check your understanding of the material in this chapter. Using a TI-84 Plus, you can find a line of best fit through the following steps.

First, press **STAT** and select **1:Edit**. To clear any preexisting lists, move your cursor over the title bar and press **CLEAR**, then **ENTER**. Enter your data vertically by variable, then press **2ND** and **Y=**.

Select **1:** to get the screen for **PLOT1**. Highlight the appropriate options, as shown to the right. Make sure the names for **Xlist** and **Ylist** match the names of the data lists you created. (To change the names shown, select **2ND** and **STAT**, then select the appropriate name from the list.)

To plot the points, press **ZOOM** and select **9: ZoomStat**.

section 3.3   Forms of Linear Equations

Find the equation for the line of best fit by pressing **STAT**, then scroll to the right to the **CALC** menu and select **4: LinReg(ax+b)**. This pastes the LinReg command into the home screen. Now, use **2ND** and **STAT** to paste your list names into the home screen, separated by a **,** . Place another **,** and then press **VARS**, scroll right to the **Y-VARS** menu, and select **1: Function**, then **1: Y1**. Press **ENTER** and you will see the variables to plug into $y=ax+b$ to get the line of best fit (shown to the left).

Now press **ZOOM** and select **9: ZoomStat** to see the data and regression line.

exercises

Determine the slopes of the lines passing through the specified points. See Example 1.

1. $(0, -3)$ and $(-2, 5)$
2. $(-3, 2)$ and $(7, -10)$
3. $(4, 5)$ and $(-1, 5)$

4. $(3, -1)$ and $(-7, -1)$
5. $(3, -5)$ and $(3, 2)$
6. $(0, 0)$ and $(-2, 5)$

7. $(-2, 1)$ and $(-5, -1)$
8. $(-2, 4)$ and $(6, 9)$
9. $(0, -21)$ and $(-3, 0)$

10. $(-3, -5)$ and $(-2, 8)$
11. $(29, -17)$ and $(31, -29)$
12. $(7, 4)$ and $(-6, 13)$

Determine the slopes of the lines defined by the following equations. See Example 2.

13. $8x - 2y = 11$
14. $2x + 8y = 11$
15. $12x - 4y = -9$

16. $4y = 13$
17. $\dfrac{x-y}{3} + 2 = 4$
18. $7x = 2$

19. $3y - 2 = \dfrac{x}{5}$
20. $3 - y = 2(5 - x)$
21. $3(2y - 1) = 5(2 - x)$

22. $\dfrac{x+2}{3} + 2(1 - y) = -2x$
23. $2y - 7x = 4y + 5x$
24. $x - 7 = \dfrac{2y - 1}{-5}$

chapter three Linear Equations and Inequalities of Two Variables

Use the slope-intercept form of each line to graph the equations. See Example 3.

25. $6x - 2y = 4$
26. $3y + 2x - 9 = 0$
27. $5y - 15 = 0$
28. $x + 4y = 20$
29. $\dfrac{x - y}{2} = -1$
30. $3x + 7y = 8y - x$
31. $-4x - 4y = 8$
32. $-5x + 3y + 16 = 0$
33. $3x = 3y - 21$

Find the equation, in slope-intercept form, of the line passing through the specified points with the specified slopes. See Example 4.

34. point $(0, -3)$; slope of $\dfrac{3}{4}$
35. point $(0, 5)$; slope of -3
36. point $(0, -7)$; slope of $-\dfrac{5}{2}$
37. point $(0, 6)$; slope of 4
38. point $(0, -9)$; slope of -5
39. point $(0, 2)$; slope of $\dfrac{1}{2}$

Find the equation, in standard form, of the line passing through the specified points with the specified slopes. See Example 5.

40. point $(-1, -3)$; slope of $\dfrac{3}{2}$
41. point $(3, -1)$; slope of 10
42. point $(-3, 5)$; slope of 0
43. point $(-2, -13)$; undefined slope
44. point $(6, 0)$; slope of $\dfrac{5}{4}$
45. point $(-1, 3)$; slope of $-\dfrac{2}{7}$
46. point $(5, 11)$; slope of -3
47. point $(5, -9)$; slope of $-\dfrac{1}{2}$

Find the equation, in standard form, of the line passing through the specified points. See Example 6.

48. $(-1, 3)$ and $(2, -1)$
49. $(1, 3)$ and $(-2, 3)$
50. $(2, -2)$ and $(2, 17)$
51. $(-9, 2)$ and $(1, 5)$
52. $(3, -1)$ and $(8, -1)$
53. $\left(\dfrac{4}{3}, 1\right)$ and $\left(\dfrac{2}{5}, \dfrac{3}{7}\right)$
54. $(-2, 8)$ and $(5, 6)$
55. $(8, -10)$ and $(8, 0)$
56. $(7, 5)$ and $(-9, 5)$
57. $(7, 7)$ and $(9, -8)$
58. $\left(\dfrac{2}{3}, \dfrac{5}{4}\right)$ and $\left(\dfrac{3}{5}, \dfrac{9}{8}\right)$
59. $(-5, -5)$ and $(10, -11)$

section 3.3 Forms of Linear Equations

Match each equation or description to the correct graph below.

60. $-3x - 2y = 17$ **61.** $-4y + 10 = -4x$ **62.** $-6y + 9 = \dfrac{x}{-2}$

63. point $(-9, 7)$; slope $\dfrac{4}{3}$ **64.** point $(-5, 10)$; slope -2 **65.** point $(0, -5)$; slope -9

a.

b.

c.

d.

e.

f.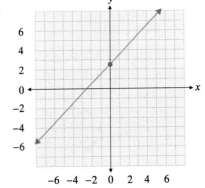

chapter three   Linear Equations and Inequalities of Two Variables

66. A bottle manufacturer has determined that the total cost (C) in dollars of producing x bottles is $C = 0.25x + 2100$.
    a. What is the cost of producing 500 bottles?
    b. What are the fixed costs (costs incurred even when 0 bottles are produced)?
    c. What is the increase in cost for each bottle produced?

67. Sales at Glover's Golf Emporium have been increasing linearly for the past couple of years. Last year's sales were $163,000. This year's sales were $215,000. If sales continue to increase at this rate, predict next year's sales.

68. Amy owns stock in Trimetric Technologies. If the stock had a value of $2500 in 2003 when she purchased it, what has been the average change in value per year if in 2005 the stock was worth $3150?

69. For tax and accounting purposes, businesses often have to depreciate equipment values over time. One method of depreciation is the straight-line method. Three years ago Hilde Construction purchased a bulldozer for $51,500. Using the straight-line method, the bulldozer has now depreciated to a value of $43,200. If V equals the value at the end of year t, write a linear equation expressing the value of the bulldozer over time. How many years from the purchase date will the value equal $0? (Round your answer to the nearest hundredth.)

In exercises 70 – 73, **a.** *find the equation of the regression line for the given points, and* **b.** *graph the data and the regression line on the same set of axes.*

70. $(0, 3), (1, 5), (2, 7), (3, 8), (5, 9), (6, 9)$

71. $(1, 10), (2, 8), (3, 7), (4, 6), (5, 5), (6, 5), (7, 4)$

72. $(1, 9.6), (2, 8.7), (3, 7.7), (4, 6.1), (5, 5.0)$

73. $(1, 5.2), (2, 6.4), (3, 8.1), (4, 9.2), (5, 10.6)$

74. An automobile company is running a new television commercial in five cities with approximately the same population. The following table shows the number of times the commercial is run on TV in each city and the number of car sales (in hundreds).

   a. Find the linear regression line for the data given in the table.
   b. Graph the data and the regression line on the same set of axes.

Number of TV commercials x	4	5	12	16	18
Car sales y (in hundreds)	3	5	5	8	7

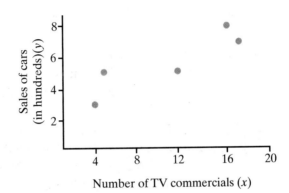

75. The following table shows the high school grade-point averages (HS-GPA) and the college grade-point averages (C-GPA) after 1 year of college for 10 students.
   a. Find the regression line for the data given in the table.
   b. Graph the data and the regression line on the same set of axes.

HS-GPA (x)	2.0	2.0	2.2	2.2	2.7	3.2	3.2	3.3	3.5	3.7
C-GPA (y)	1.5	1.8	2.0	1.5	2.0	2.8	3.0	3.5	3.5	3.4

chapter three   Linear Equations and Inequalities of Two Variables

technology exercises

The `Solve` command in *Mathematica* (and its equivalent in other computer algebra systems) is useful as an aid in a variety of contexts. When given a linear equation in two variables, the command can be used to quickly determine the slope and y-intercept of the line. In the first of the three commands illustrated below, `Solve` is used to write the equation in slope-intercept form (the result is an equation if we replace the arrow with the equality symbol; the `Simplify` command has been used to simplify the output of the `Solve` command so that the slope can be determined more easily.) The result tells us that the slope of the line is $\frac{9}{2}$ and that the line has a y-intercept of $\left(0, \frac{7}{2}\right)$. In the second use of the command the output is the empty set, meaning the variable y disappears upon simplifying the equation in question. As we know, this indicates the line is vertical, so if we `Solve` for x (as illustrated by the third command) we can determine where it crosses the x-axis.

If available, use a computer algebra system to determine the slope and y-intercept of each of the following linear equations. (See Appendix A for guidance on using computer algebra systems like Mathematica.)

76. $\dfrac{3x-7y}{4} + 5 = y - \dfrac{x}{9}$

77. $4(2x - 3y) + 17y = 3 + 5y$

78. $\dfrac{x-3y}{5} = \dfrac{1+2x}{7}$

79. $3 - \dfrac{x-y}{7} = x + \dfrac{2y}{3}$

80. $\dfrac{x+y}{3} - \dfrac{x-y}{2} = \dfrac{y-x}{6}$

81. $\dfrac{7x-3y}{2} = \dfrac{2y-5x}{9}$

3.4 Parallel and Perpendicular Lines

TOPICS

1. Slopes of parallel lines
2. Slopes of perpendicular lines

Topic 1: Slopes of Parallel Lines

In this short section, we will explore the concept of slope in greater detail by studying the relationship between slope and the geometric concepts of parallel and perpendicular lines. We begin with a study of parallel lines.

As the slope of a line is a precise indication of its "steepness", it should come as no surprise that two lines are parallel if and only if they have the same slope (though in the case of vertical lines this means they both have undefined slope). This fact is also clear from the formula for slope: two lines are parallel if and only if they both "rise" vertically the same amount relative to the same horizontal "run." We can use this observation to derive equations for lines that are described in terms of other lines, as illustrated below.

example 1

Find the equation, in slope-intercept form, for the line which is parallel to the line $3x + 5y = 23$ and which passes through the point $(-2, 1)$.

Solution:

We begin by writing the equation $3x + 5y = 23$ in slope-intercept form: $y = -\frac{3}{5}x + \frac{23}{5}$; this tells us that the slope of the line whose equation we seek is $-\frac{3}{5}$. With this slope and the knowledge that the line is to pass through $(-2, 1)$, we use the point-slope form to obtain the equation below.

$y - 1 = -\frac{3}{5}(x - (-2))$

$y - 1 = -\frac{3}{5}(x + 2)$

$y - 1 = -\frac{3}{5}x - \frac{6}{5}$

$y = -\frac{3}{5}x - \frac{1}{5}$

The first equation to the left is the point-slope form of the line that passes through $(-2, 1)$ with a slope of $-\frac{3}{5}$.

The instructions in this example asked for the equation in slope-intercept form, so we solve for y to obtain the final answer.

The screen above is the graphs of both $3x+5y=23$ and the parallel line passing through $(-2,1)$. To view multiple graphs, select **Y=** and enter the equations in terms of y. Then select **GRAPH**.

chapter three Linear Equations and Inequalities of Two Variables

We can also use the knowledge that parallel lines have the same slope to answer questions that are more geometric in nature.

example 2

Determine if the quadrilateral (four-sided figure) whose vertices are $(-4, 2), (1, 1), (2, 3)$ and $(-3, 4)$ is a parallelogram (a quadrilateral in which both pairs of opposite sides are parallel).

Solution:

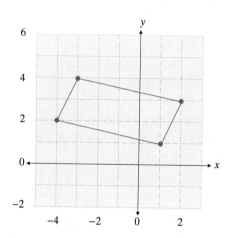

The four vertices are plotted in the picture to the left, and the sides of the quadrilateral drawn. The figure is a parallelogram if the left and right sides are parallel and the top and bottom sides are parallel. The slopes of the left and right sides are, respectively,

$$\frac{4-2}{-3-(-4)} = 2 \quad \text{and} \quad \frac{3-1}{2-1} = 2,$$

and the slopes of the top and bottom sides are, respectively,

$$\frac{4-3}{-3-2} = -\frac{1}{5} \quad \text{and} \quad \frac{2-1}{-4-1} = -\frac{1}{5}.$$

Thus the figure is indeed a parallelogram.

In summary, we have the following.

Slopes of Parallel Lines

Two non-vertical lines with slopes m_1 and m_2 are parallel if and only if $m_1 = m_2$. Also, two vertical lines (with undefined slopes) are always parallel to each other.

Topic 2: Slopes of Perpendicular Lines

The relationship between the slopes of perpendicular lines is a bit less obvious. Consider a non-vertical line L_1, and consider two points (x_1, y_1) and (x_2, y_2) on the line, as shown in Figure 1. These two points can be used, of course, to calculate the slope m_1 of L_1, with the result that $m_1 = \frac{a}{b}$, where $a = y_2 - y_1$ and $b = x_2 - x_1$.

section 3.4 Parallel and Perpendicular Lines

Figure 1: Definition of a and b

If we now draw a line L_2 perpendicular to L_1, we can use a and b to determine the slope m_2 of line L_2. There are an infinite number of lines that are perpendicular to L_1; one of them is drawn in Figure 2.

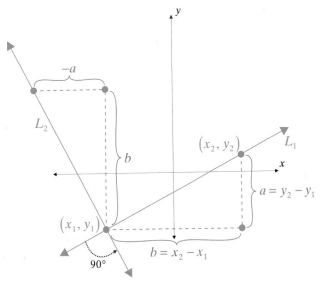

Figure 2: Perpendicular Lines

Note that in rotating the line L_1 by 90 degrees to obtain L_2, we have also rotated the right triangle drawn with dashed lines. But to travel along the line L_2 from the point (x_1, y_1) to the second point drawn requires a positive rise and a negative run, whereas the rise and run between (x_1, y_1) and (x_2, y_2) are both positive. In other words, $m_2 = -\dfrac{b}{a}$, the negative reciprocal of the slope m_1.

chapter three Linear Equations and Inequalities of Two Variables

This relationship always exists between the slopes of two perpendicular lines, assuming neither one is vertical. Of course, if one line is vertical, any line perpendicular to it will be horizontal with a slope of zero, while if one line is horizontal, any line perpendicular to it will be vertical with undefined slope. This is summarized below.

Slopes of Perpendicular Lines

Suppose m_1 and m_2 represent the slopes of two lines, neither of which is vertical. The two lines are perpendicular if and only if $m_1 = -\dfrac{1}{m_2}$ (equivalently, $m_2 = -\dfrac{1}{m_1}$). If one of two perpendicular lines is vertical, the other is horizontal, and the slopes are, respectively, undefined and zero.

The next two examples illustrate how we can use this knowledge of the relationship between slopes of perpendicular lines.

example 3

Find the equation, in standard form, of the line that passes through the point $(-3, 13)$ and that is perpendicular to the line $y = -7$.

Solution:

The line $y = -7$ is a horizontal line, and hence any line perpendicular to it must have undefined slope. In particular, the perpendicular line passing through $(-3, 13)$ must have the form $x = -3$, and this is the desired solution.

example 4

Prove that the two lines $3x - 7y = 12$ and $14x + 6y = -5$ are perpendicular to each other.

Solution:

This is most easily done by rewriting each equation in slope-intercept form. In this form, the first equation is $y = \dfrac{3}{7}x - \dfrac{12}{7}$ and the second equation is $y = -\dfrac{7}{3}x - \dfrac{5}{6}$. Hence the slope of the first line is $\dfrac{3}{7}$, which is the negative reciprocal of the slope of the second line, $-\dfrac{7}{3}$.

section 3.4 Parallel and Perpendicular Lines

exercises

Find the equation, in slope-intercept form, for the line parallel to the given line and passing through the indicated point. See Example 1.

1. Parallel to $y - 4x = 7$ and passing through $(-1, 5)$.

2. Parallel to $6x + 2y = 19$ and passing through $(-6, -13)$.

3. Parallel to $3x + 2y = 3y - 7$ and passing through $(3, -2)$.

4. Parallel to $2 - \dfrac{y - 3x}{3} = 5$ and passing through $(0, -2)$.

5. Parallel to $y - 4x = 7 - 4x$ and passing through $(23, -9)$.

6. Parallel to $2(y - 1) + \dfrac{x + 3}{5} = -7$ and passing through $(-5, 0)$.

7. Parallel to $6y - 4 = -3(1 - 2x)$ and passing through $(-2, -2)$.

8. Parallel to $5 - \dfrac{7y + 5x}{2} = 1$ and passing through $(4, 1)$.

9. Parallel to $2(y - 1) - \dfrac{7x + 1}{3} = -3$ and passing through $(1, 10)$.

10. Parallel to $8y - 6 = -3(4 - x)$ and passing through $(11, -5)$.

Each set of four ordered pairs below defines the vertices, in counterclockwise order, of a quadrilateral. Determine if the quadrilateral is a parallelogram. See Example 2.

11. $\{(-2, 2), (-5, -2), (2, -3), (5, 1)\}$

12. $\{(-1, 6), (-4, 7), (-2, 3), (1, 1)\}$

13. $\{(-3, 3), (-2, -2), (3, -1), (2, 4)\}$

14. $\{(-2, -3), (-3, -6), (1, -2), (2, 1)\}$

15. $\{(-6, -2), (-1, 0), (-3, 4), (-8, 2)\}$

16. $\{(-3, -2), (3, -3), (5, 2), (-1, 3)\}$

17. $\{(-1, -1), (5, 1), (3, 5), (-2, 3)\}$

18. $\{(0, 1), (6, 0), (7, 4), (1, 6)\}$

Determine if the two lines in each problem below are parallel.

19. $x - 5y = 2$ and $5x - y = 2$

20. $2x - 3y = (x - 1) - (y - x)$ and $-2y - x = 9$

21. $\dfrac{2x-3y}{3} = \dfrac{x-1}{6}$ and $2y - x = 3$

22. $3 - (2y + x) = 7(x - y)$ and $\dfrac{5y+1}{4} = 3 + 2x$

23. $\dfrac{x-y}{2} = \dfrac{x+y}{3}$ and $\dfrac{2x+3}{5} - 4y = 1 + 2y$

24. $6 = -12(x - y) + y$ and $13y = -12x + 3$

25. $5 - (4y + 3x) = 5(x - y)$ and $y + 4 = 5 + 8x$

26. $7x - 2(x + 3) = 5y - x$ and $-6x = 1 - 5y$

27. $\dfrac{2y + 11x}{3} = x + 1$ and $7x - 8y = 9x + 7$

28. $\dfrac{x-y}{5} = \dfrac{x+y}{3} - 1$ and $7 = -2(x - y) + 6y$

Find the equation, in slope-intercept form, for the line perpendicular to the given line and passing through the indicated point. See Example 3.

29. Perpendicular to $3x + 2y = 3y - 7$ and passing through $(3, -2)$.

30. Perpendicular to $6y + 2x = 1$ and passing through $(-4, -12)$.

31. Perpendicular to $-y + 3x = 5 - y$ and passing through $(-2, 7)$.

32. Perpendicular to $x + y = 5$ and passing through the origin.

33. Perpendicular to $x = \dfrac{1}{4}y - 3$ and passing through $(1, -1)$.

34. Perpendicular to $2(y + x) - 3(x - y) = -9$ and passing through $(2, 5)$.

35. Perpendicular to $4x + 8y = 4y - 3$ and passing through $(-2, 1)$.

36. Perpendicular to $\dfrac{3x - y}{4} = \dfrac{4x - 5}{2}$ and passing through $(8, 5)$.

37. Perpendicular to $4(y + x) - 8(x - y) = -1$ and passing through $(6, 10)$.

38. Perpendicular to $\dfrac{3x + 4}{3} - 3y = 1 - 4y$ and passing through $(2, -8)$.

section 3.4 Parallel and Perpendicular Lines

Determine if the two lines in each problem below are perpendicular. See Example 4.

39. $x - 5y = 2$ and $5x - y = 2$

40. $3x + y = 2$ and $x + 3y = 2$

41. $\dfrac{3x - y}{3} = x + 2$ and $x = 9$

42. $5x - 6(x + 1) = 2y - x$ and $2y - (x + y) = 4y + x$

43. $\dfrac{x - 1}{2} + \dfrac{3y + 2}{3} = -9$ and $3y - 5x = x + 5$

44. $-x = -\dfrac{2}{5}y + 2$ and $5y = 2x$

45. $\dfrac{7x - 5y}{4} = x + 2$ and $-3y - 3x = 2x + 4$

46. $3(4 - x) = 6y + 3$ and $-3y - 2x = 3 - 8x$

47. $-6y + 3x = 7$ and $8x - 3(x + 1) = 3y - x$

48. $1 - \dfrac{2y - 5x}{2} = 7x + 4$ and $9x - 2y = 11$

Each set of four ordered pairs below defines the vertices, in counterclockwise order, of a quadrilateral. Use the ideas in this section to determine if the quadrilateral is a rectangle.

49. $\{(-2, 2), (-5, -2), (2, -3), (5, 1)\}$ 50. $\{(2, -1), (-2, 1), (-3, -1), (1, -3)\}$

51. $\{(1, 2), (3, -3), (9, -1), (7, 4)\}$ 52. $\{(5, -7), (1, -13), (28, -31), (32, -25)\}$

53. $\{(-5, -1), (0, -6), (5, -1), (0, 4)\}$ 54. $\{(-3, -3), (3, -2), (1, 2), (-5, 1)\}$

55. A construction company is building a new suspension bridge that has support cables attached to a center tower at various heights. One cable is attached at a height of 30 feet and connects to the ground 50 feet from the base of the tower. If the support cables should run parallel to each other, how far from the base should the company attach a cable whose other end is connected to the tower at a height of 25 feet?

56. A light hits a mirror and is reflected off the mirror at a trajectory perpendicular to the original beam of light. If the line formed by the original beam of light can be described by the equation $y = -3.2x + b$, write a linear equation to describe the trajectory formed by the reflected beam.

technology exercises

Use a computer algebra system to determine the slopes of each of the following lines, as illustrated in the Technology Exercises of Section 3.3. Then use the results to classify the lines in each pair as parallel, perpendicular, or neither. As an example, we know the two equations in the illustration below correspond to perpendicular lines, as the slopes are negative reciprocals of one another ($\frac{6}{5}$ and $-\frac{5}{6}$, respectively). (See Appendix A for guidance on using computer algebra systems like Mathematica.)

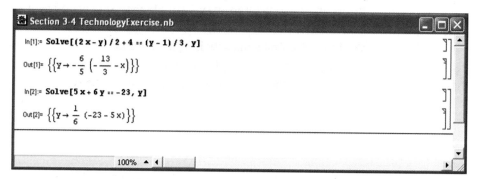

57. $3y = 2x - 7$ and $3x + 2y + 8 = 0$

58. $\dfrac{x-3y}{2} + \dfrac{y}{5} = 2x - \dfrac{y}{3}$ and $29y + 45x = 17$

59. $\dfrac{2x-y}{3} + 1 = \dfrac{4y-x}{5}$ and $13y + 17x = 0$

60. $\dfrac{y-4x}{3} = x + 2$ and $y - \dfrac{y-x}{5} = 1 - \dfrac{y}{2}$

61. $\dfrac{x-5}{3} + y = \dfrac{y-2x}{7}$ and $13y - 18x = 7$

62. $\dfrac{2y-x}{7} = \dfrac{x-2y}{3} + 5$ and $2y = x - 5$

63. $\dfrac{4y+6x}{3} = x + 1$ and $2x - 3y = 6x + 13$

64. $\dfrac{x-y}{3} = \dfrac{x+y}{2} - 1$ and $7 = -(x-y) + 4y$

3.5 Linear Inequalities in Two Variables

TOPICS

1. Solving linear inequalities in two variables
2. Solving linear inequalities joined by "and" or "or"
3. Interlude: regions of constraint

Topic 1: Solving Linear Inequalities in Two Variables

Just as linear inequalities in one variable are similar to linear equations in one variable, linear inequalities in two variables have much in common with linear equations in two variables. The first similarity is in the definition: if the equality symbol in a linear equation in two variables is replaced with $<$, $>$, $\leq$, or $\geq$, the result is a **linear inequality in two variables**. In other words, a linear inequality in the two variables x and y is an inequality that can be written in the form

$$ax + by < c, \ ax + by > c, \ ax + by \leq c, \ \text{or} \ ax + by \geq c,$$

where a, b, and c are constants and a and b are not both 0.

Another similarity lies in the solution process. The solution set of a linear inequality in two variables consists of all the ordered pairs in the Cartesian plane that lie on one side of a line in the plane, possibly including those points on the line. The first step in solving such an inequality then is to identify and graph this line. This is easily done, as the line is simply the graph of the equation that results from replacing the inequality symbol in the original problem with the equality symbol.

Any line in the Cartesian plane divides the plane into two **half-planes**, and in the context of linear inequalities all of the points in one of the two half-planes will solve the inequality. In addition, the points on the line (called the **boundary line** in this context) will also solve the inequality if the inequality symbol is $\leq$ or $\geq$, and this fact must somehow be denoted graphically. The entire process of solving linear inequalities in two variables, including the graphing conventions, is summarized on the following page.

chapter three Linear Equations and Inequalities of Two Variables

Solving Linear Inequalities in Two Variables

Step 1: Graph the line in $\mathbb{R}^2$ that results from replacing the inequality symbol with =. Make the line solid if the inequality symbol is ≤ or ≥ (non-strict) and dashed if the symbol is < or > (strict). A solid line indicates that points on the line will be included in the eventual solution set while a dashed line indicates that points on the line are to be excluded from the solution set.

Step 2: Determine which of the half-planes defined by the boundary line solve the inequality by substituting a **test point** from one of the two half-planes into the inequality. If the resulting numerical statement is true, all the points in the same half-plane as the test point solve the inequality. Otherwise, the points in the other half-plane solve the inequality. Shade in the half-plane that solves the inequality.

example 1

The line below was entered in the usual manner by selecting `Y=`. Before selecting `GRAPH`, choose the area to shade by pressing the left arrow until the cursor is to the left of Y. Then select `ENTER` until the correct area is shaded.

Solve the following linear inequalities by graphing their solution sets.

a. $3x + 2y < 12$ b. $x - y \leq 0$ c. $x > 3$

Solutions:

a. $3x + 2y < 12$
    x-intercept of boundary: $(4, 0)$
    y-intercept of boundary: $(0, 6)$

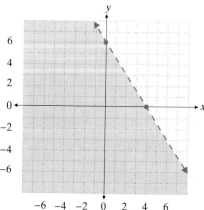

We begin by constructing the linear equation that defines the boundary line of the solution set. The equation $3x + 2y = 12$ is graphed by locating the x- and y-intercepts, as we have done to the left, or by determining the slope and y-intercept. The graph of the equation is then drawn with a dashed line since the inequality is strict.

A good test point in this problem is $(0, 0)$, as it clearly lies on one side of the boundary line. When both x and y in the inequality are replaced with 0, we obtain the statement $0 < 12$, which is true. This tells us to shade the half-plane that contains $(0, 0)$.

224

b. $x - y \leq 0$

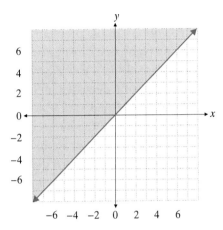

As in the first problem, we begin by graphing the boundary line. In slope-intercept form, the equation for the boundary is $y = x$, a line with a y-intercept of 0 and a slope of 1. The graph of the boundary has been drawn with a solid line since the inequality is not strict; that is, points on the boundary will solve the inequality.

This time, the origin cannot be used as a test point, as it lies directly on the boundary line. The point $(1, -1)$ clearly lies below the boundary, and if the values $x = 1$ and $y = -1$ are substituted into the inequality $x - y \leq 0$, we obtain the false statement $2 \leq 0$. Thus the half-plane that does not contain $(1, -1)$ is the one we shade.

c. $x > 3$

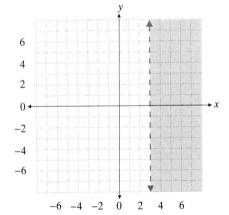

In this last example, the boundary line is a vertical line passing through 3 on the x-axis. It has been drawn with a dashed line since the inequality is strict.

The origin can be used as a convenient test point, and results in the false statement $0 > 3$, leading us to shade in the half-plane to the right of the boundary.

As an alternative to the test-point method demonstrated above, consider each of the inequalities in *Example 1* after solving for y (or for x if y is not present, as in *Example 1c*). The inequality in *Example 1a* is satisfied by all those ordered pairs for which y is less than $-\frac{3}{2}x + 6$. That is, if the line $y = -\frac{3}{2}x + 6$ is graphed (with a dashed line), the solution set of the inequality consists of all those ordered pairs whose y-coordinate is *less than* (lies below) $-\frac{3}{2}$ times the x-coordinate, plus 6. Similarly, in *Example 1b*, we shade in those ordered pairs for which the y-coordinate is *greater than or equal to* the x-coordinate. Finally, in *Example 1c*, we shade in the ordered pairs for which the x-coordinate is *greater than* (lies to the right of) 3.

chapter three Linear Equations and Inequalities of Two Variables

Topic 2: Solving Linear Inequalities Joined by "And" or "Or"

It is often necessary, in a variety of applications, to identify those points in the Cartesian plane that satisfy more than one condition. In the problems that we will examine, we will want to identify the portion of the plane that satisfies two or more linear inequalities joined by the word "and" or by the word "or." You may want to review the basic set operations of union and intersection that were discussed in Section 1.2. In that section, we defined the union of two sets A and B, denoted $A \cup B$, as the set containing all elements that are in set A **or** set B, and we defined the intersection of two sets A and B, denoted $A \cap B$, as the set containing all elements that are in both A **and** B. If we let A denote the portion of $\mathbb{R}^2$ that solves one inequality and B the portion of $\mathbb{R}^2$ that solves a second inequality, $A \cup B$ then represents the solution set of the two inequalities joined by the word "or" and $A \cap B$ represents the solution set of the two inequalities joined by the word "and." Keep in mind that $A \cup B$ *contains* both of the sets A and B (and so is at least as large as either one individually), while $A \cap B$ is *contained in* both A and B (and so is no larger than either individual set).

To find the solution sets in the following problems, we will solve each linear inequality individually and then form the union or the intersection of the individual solutions, as appropriate.

example 2

Choose the area to shade for each equation from the screen by pressing the left arrow until the cursor is to the left of Y. Then select ENTER until the correct area is shaded.

Graph the solution sets that satisfy the following inequalities.

a. $5x - 2y < 10$ and $y \leq x$

b. $x + y < 4$ or $x \geq 4$

Solutions:

a. $5x - 2y < 10$ and $y \leq x$

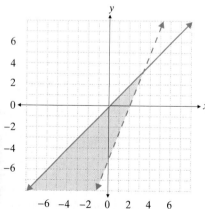

To solve the first inequality, we first graph the line $5x - 2y = 10$ with a dashed line and the line $y = x$ with a solid line. We then note that the half-plane lying above $5x - 2y = 10$ solves the first inequality (verify this by using the test point $(0, 0)$) and that the half-plane lying below $y = x$ solves the second inequality (this can be verified by using a test point such as $(1, -1)$).

Since the two inequalities are joined by the word "and", we shade in the *intersection* of the two half-planes, resulting in the picture at left.

b. $x + y < 4$ or $x \geq 4$

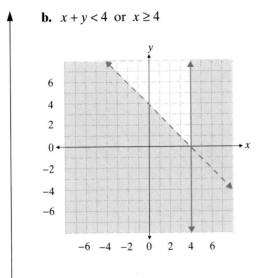

As always, we begin by solving each inequality separately. We graph the line $x + y = 4$ with a dashed line, and the line $x = 4$ with a solid line, and note that the half-plane below $x + y = 4$ solves the first inequality and that the half-plane to the right of $x = 4$ solves the second.

Since the two inequalities are joined by "or," we proceed to shade the *union* of the two half-planes. Note that this results in a larger solution set than either of the two individual solutions. Any ordered pair in the shaded region will satisfy one or the other (or both) of the inequalities.

Unions and intersections of regions of the plane can also arise when solving inequalities involving absolute values. In Section 2.2, we saw that an inequality of the form $|x| < a$ can be rewritten as the compound inequality $-a < x < a$. This in turn can be rewritten as the joint condition $x > -a$ and $x < a$, so an absolute value inequality of this form corresponds to the intersection of two sets. Similarly, an inequality of the form $|x| > a$ can be rewritten as $x < -a$ or $x > a$, so the solution of this form of absolute value inequality is a union of two sets.

If we are working in the Cartesian plane, solutions of such inequalities will be, respectively, intersections and unions of half-planes. The next example demonstrates how such solution sets can be found.

example 3

Graph the solution set in $\mathbb{R}^2$ that satisfies the joint conditions $|x - 3| > 1$ and $|y - 2| \leq 3$.

Solution:

We need to identify all ordered pairs (x, y) for which $x - 3 < -1$ or $x - 3 > 1$ while $-3 \leq y - 2 \leq 3$. That is, we need $x < 2$ or $x > 4$ while $-1 \leq y \leq 5$. The solution sets of the two conditions individually are:

continued on next page...

chapter three Linear Equations and Inequalities of Two Variables

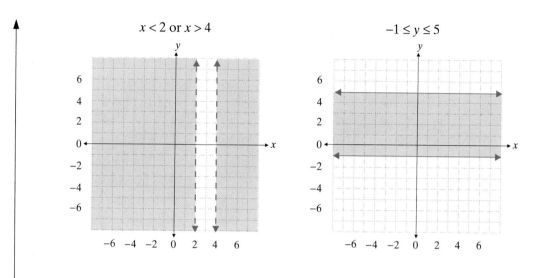

We now intersect the solution sets to obtain the final answer:

$$|x-3|>1 \text{ and } |y-2|\leq 3$$

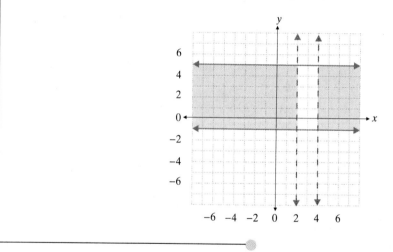

Topic 3: Interlude: Regions of Constraint

One noteworthy application of the ideas in this section is **linear programming**, an important mathematical tool in business and the social sciences. In brief, linear programming is a method used to maximize or minimize a variable expression, subject to certain linear constraints on the values of the variables. For instance, a given corporation might reasonably wish to determine the values of several variables that will maximize an expression representing profit, or it may want to determine the values of variables that will minimize an expression representing cost of operations.

section 3.5   Linear Inequalities in Two Variables

A more complete discussion of linear programming must wait until Chapter 8, but it is appropriate to introduce one aspect now. The first step in linear programming is to determine the **region of constraint**, a mathematical description of all the possible values that can be taken on by the variables (in some contexts, this is called the **feasible region**). If the variable expression to be maximized or minimized contains just two variables, the region of constraint will be a portion of the Cartesian plane, and it will usually be defined by the intersection of a number of half-planes. The following example illustrates one such region of constraint.

example 4

A family orchard is in the business of selling peaches and nectarines. The members of the family know that to prevent a certain pest infestation, the number of nectarine trees in their orchard cannot exceed the number of peach trees. Also, because of the space requirements of each type of tree, the number of nectarine trees plus twice the number of peach trees cannot exceed 100 trees. Graph the region of constraint for this situation.

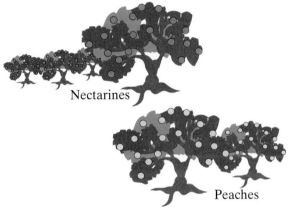

Solution:

If we let p represent the number of peach trees and n the number of nectarine trees, the paragraph above gives us various constraints on p and n. For instance, the second sentence translates into the inequality $n \leq p$, and the third sentence translates into the inequality $n + 2p \leq 100$. In addition, we can assume that $n \geq 0$ and that $p \geq 0$, as it is not possible to have a negative number of trees. Altogether, we need to solve four linear inequalities joined by the word "and," as we wish to satisfy all four conditions at once. The graph to the right is the intersection of the four half-planes that solve each individual inequality, with the horizontal axis representing p and the vertical axis n.

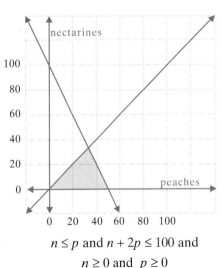

$n \leq p$ and $n + 2p \leq 100$ and
$n \geq 0$ and $p \geq 0$

chapter three Linear Equations and Inequalities of Two Variables

exercises

Graph the solution sets of the following linear inequalities. See Examples 1 through 3.

1. $x - 3y < 6$
2. $y < 2x - 1$
3. $x > \dfrac{3}{4} y$
4. $x - 3y \geq 6$
5. $3x - y \leq 2$
6. $\dfrac{2x - y}{4} > 1$
7. $y < -2$
8. $x + 1 \geq 0$
9. $x + y < 0$
10. $x + y > 0$
11. $-(y - x) > -\dfrac{5}{2} - y$
12. $-2y \leq -x + 4$
13. $5(y + 1) \geq -x$
14. $3x - 7y \geq 7(1 - y) + 2$
15. $x - y < 2y + 3$
16. $y > -3x - 6$ or $y \leq 2x - 7$
17. $y \geq -2$ and $y > 1$
18. $y \geq -2x - 5$ and $y \leq -6x - 9$
19. $y \leq 4x + 4$ and $y > 7x + 7$
20. $x - 3y \geq 6$ and $y > -4$
21. $x - 3y \geq 6$ or $y > -4$
22. $3x - y \leq 2$ and $x + y > 0$
23. $x > 1$ and $y > 2$
24. $x > 1$ or $y > 2$
25. $x + y > -2$ and $x + y < 2$
26. $y > -2$ and $2y > -3x - 4$
27. $3y > x + 2$ or $4y \leq -x - 2$
28. $y \leq -x$ and $2y + 3x > -4$
29. $5x + 6y < -30$ and $x \geq 2$
30. $6y - 2x > -6$ or $y > 6$
31. $x > -3$ or $y \geq 4$
32. $-2y < -3x - 6$ or $-3y \geq -6x - 18$
33. $x < 6$ and $x \geq -5$
34. $|x - 3| < 2$
35. $|x - 3| > 2$
36. $|3y - 1| \leq 2$
37. $|2x - 4| > 2$
38. $1 - |y + 3| < -1$
39. $|x + 1| < 2$ and $|y - 3| \leq 1$

section 3.5   Linear Inequalities in Two Variables

40. $|x-3|\geq 1$ or $|y-2|\leq 1$ **41.** $|x-y|<1$ **42.** $|x+y|\geq 1$

43. $|4x-2y-3|\leq 5$ **44.** $|2x-3|\geq 1$ or $|2y+3|\geq 1$ **45.** $|y-3x|\leq 2$ and $|y|<2$

Match the following inequalites to the appropriate graph.

46. $-8y+5x \geq -8y+5$

47. $x<-2$ and $x \geq -5$

48. $|-7x-4y+23|\leq 16$

49. $y \leq 3x-6$ and $y > 2x-4$

50. $|3y-2x|>17$ and $|y+6|\geq 1$

51. $4(y+2)<-x$

52. $-y<6x+3$ or $4y \geq 3x-6$

53. $|7x+4|\leq 5$ or $|7y+4|\leq 5$

a.

b.

c.

d.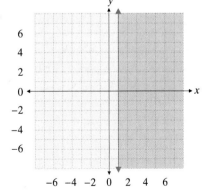

chapter three Linear Equations and Inequalities of Two Variables

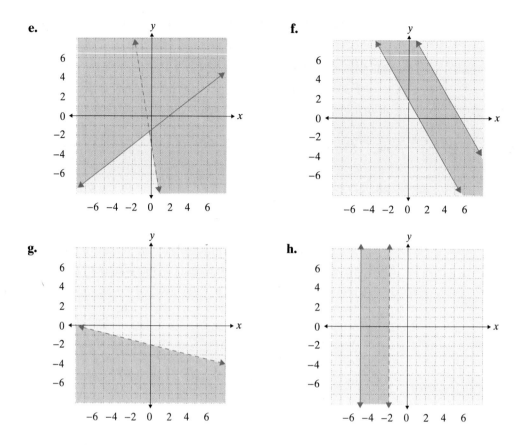

In each of the following problems, construct an appropriate linear inequality as directed.

54. It costs Happy Land Toys $5.50 in variable costs per doll produced. If total costs must remain less than $200, write a linear inequality describing the relationship between cost and dolls produced.

55. Trish is having a garden party where she wants to have several arrangements of lilies and orchids for decoration. The lily arrangements cost $12 each and the orchids cost $22 each. If Trish wants to spend less than $150 on flowers, write a linear inequality describing the number of each arrangement she can purchase. Graph the inequality.

56. Rob has 300 feet of fencing he can use to enclose a small rectangular area of his yard for a garden. Assuming Rob may or may not use all the fencing, write a linear inequality describing the possible dimensions of his garden. Graph the inequality.

section 3.5 Linear Inequalities in Two Variables

57. Flowertown Canoes produces two types of canoes. The two person model costs $73 to produce and the one person model costs $46 to produce. Write a linear inequality describing the number of each canoe the company can produce and keep costs under $1750. Graph the inequality.

Graph the regions of constraint defined by the following situations. See Example 4

58. A plane carrying relief food and water can carry a maximum of 50,000 pounds, and is limited in space to carrying no more than 6000 cubic feet. Each container of water weighs 60 pounds and takes up 1 cubic foot, and each container of food weighs 50 pounds and takes up 10 cubic feet. What is the region of constraint for the number of containers of food and water that the plane can carry?

59. A furniture company makes two kinds of sofas, the standard model and the deluxe model. The standard model requires 40 hours of labor to build, and the deluxe model requires 60 hours of labor to build. The finish of the deluxe model, however, uses both teak and fabric, while the standard uses only fabric, with the result that each deluxe sofa requires 5 square yards of fabric and each standard sofa requires 8 square yards of fabric. Given that the company can use 200 hours of labor and 25 square yards of fabric per week building sofas, what is the region of constraint for the number of deluxe and standard sofas the company can make per week?

Standard model

Deluxe model

chapter three Linear Equations and Inequalities of Two Variables

60. Sarah is looking through a clothing catalog, and she is willing to spend up to $80 on clothes and $10 for shipping. Shirts cost $12 each plus $2 shipping, and a pair of pants costs $32 plus $3 shipping. What is the region of constraint for the number of shirts and pairs of pants Sarah can buy?

SALE!
Tank top $12

Jeans $32

61. Suppose you inherit $75,000 from a previously unknown (and highly eccentric) uncle, and that the inheritance comes with certain stipulations regarding investments. First, the dollar amount invested in bonds must not exceed the dollar amount invested in stocks. Second, a minimum of $10,000 must be invested in stocks and a minimum of $5000 must be invested in bonds. Finally, a maximum of $40,000 can be invested in stocks. What is the region of constraint for the dollar amount that can be invested in the two categories of stocks and bonds?

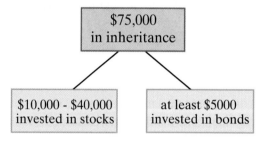

3.6 Introduction to Circles

TOPICS

1. Standard form of a circle
2. Graphing circles

Topic 1: Standard Form of a Circle

The focus of this chapter so far has been on equations that define lines in $\mathbb{R}^2$. But as you no doubt suspect, linear equations are just the beginning. Much of the rest of this text will deal, directly or indirectly, with more interesting graphs in the plane. As a preview of the discussion to come, we will close out this chapter with a brief introduction to a particular class of nonlinear equations.

Circles in $\mathbb{R}^2$ can be visualized, graphed, and described mathematically. To begin with, note that two pieces of information are all we need to completely characterize a particular circle: the circle's center and the circle's radius. To be specific, suppose (h, k) is the ordered pair corresponding to the circle's center, and suppose the radius is given by the positive real number r. Our goal is to develop an equation in the two variables x and y so that every solution (x, y) of the equation corresponds to a point on the circle.

The main tool that we need to achieve this goal is the distance formula derived in Section 3.1. Since every point (x, y) on the circle is a distance r from the circle's center (h, k), that formula tells us that

$$r = \sqrt{(x-h)^2 + (y-k)^2}.$$

This form of the equation for a circle is in many ways the most natural, but such equations are commonly presented in the radical-free form that results from squaring both sides:

$$r^2 = (x-h)^2 + (y-k)^2$$

chapter three Linear Equations and Inequalities of Two Variables

Standard Form of a Circle

The **standard form** of the equation for a circle of radius r and center (h, k) is

$$(x-h)^2 + (y-k)^2 = r^2.$$

example 1

Find the standard form of the equation for the circle with radius 3 and center $(-2, 7)$.

Solution:

We are given $h = -2$, $k = 7$, and $r = 3$, so the equation is determined to be

$$(x-(-2))^2 + (y-7)^2 = 3^2.$$

This is better written as

$$(x+2)^2 + (y-7)^2 = 9.$$

example 2

Find the standard form of the equation for each of the following circles:
  a. A circle with a diameter whose endpoints are $(-4, -1)$ and $(2, 5)$.
  b. A circle which is tangent to the line $x = -1$ and whose center is $(3, 5)$.

Solutions:

a. The midpoint of a diameter of a circle is the circle's center, so the first step is to use the midpoint formula as follows:

$$(h, k) = \left(\frac{-4+2}{2}, \frac{-1+5}{2}\right) = (-1, 2).$$

The distance from either diameter endpoint to the center defines the circle's radius. Since we ultimately will want r^2, we can use a slight variation of the distance formula to determine

$$r^2 = (-4-(-1))^2 + (-1-2)^2 = 9+9 = 18.$$

Thus the equation we seek is

$$(x+1)^2 + (y-2)^2 = 18.$$

Note that $(-4, -1)$ and $(2, 5)$, two points that by definition are on the circle, both satisfy this equation.

b. The word tangent in this context means that the circle just touches the line $x = -1$. In fact, by referring to a sketch, we can see that the described circle must touch the vertical line $x = -1$ at the point $(-1, 5)$. The distance between the center $(3, 5)$ and the point $(-1, 5)$ must then correspond to the radius of the circle, giving us $r = 4$. At this point, we are ready to construct the equation of the circle:
$$(x - 3)^2 + (y - 5)^2 = 16.$$

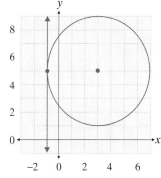

Topic 2: Graphing Circles

We will often need to reverse the process illustrated in the first two examples. That is, given an equation for a circle, we will need to determine the circle's center and radius and, possibly, graph the circle. If the equation is given in standard form, this is very easily accomplished.

example 3

Sketch the graph of the circle defined by $(x - 2)^2 + (y + 3)^2 = 4$.

Solution:

The only preliminary step required is to slightly rewrite the equation in the form
$$(x - 2)^2 + (y - (-3))^2 = 2^2.$$

From this, we see that $(h, k) = (2, -3)$ and that $r = 2$. Thus the graph of the equation is as follows:

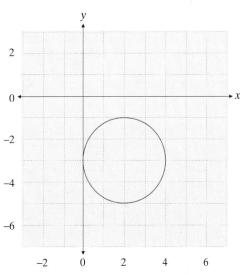

Typically, though, the equation for a circle will not be given to us in quite so neat a fashion. We may have to resort to a small amount of algebraic manipulation in order to determine that a given equation describes a circle and to determine the specifics of that circle. Fortunately, the algebraic technique of *completing the square* (Section 2.3) is usually all that is required.

example 4

Sketch the graph of the equation $x^2 + y^2 + 8x - 2y = -1$.

Solution:

We need to complete the square in the variable x and the variable y, and we do so as follows:

$$\left(x^2 + 8x + 16\right) + \left(y^2 - 2y + 1\right) = -1 + 16 + 1 \qquad \text{Add 16 and 1 to both sides.}$$

$$(x+4)^2 + (y-1)^2 = 16.$$

We now see that the equation does indeed describe a circle, and that the center of the circle is (−4, 1) and the radius is 4. The graph appears below.

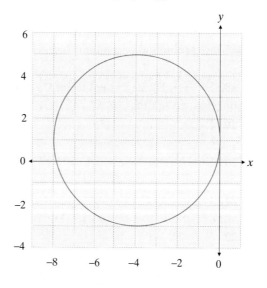

section 3.6 Introduction to Circles

exercises

Find the standard form of the equation for the circle. See Examples 1 and 2.

1. Center $(-4, -3)$; radius 5
2. Center $(0, 0)$; radius $\sqrt{6}$
3. Center $(7, -9)$; radius 3
4. Center $(-2, 2)$; radius 2
5. Center at origin; radius 3
6. Center $(6, 3)$; radius 8
7. Center $\left(\sqrt{5}, \sqrt{3}\right)$; radius 4
8. Center $\left(\dfrac{5}{3}, \dfrac{8}{5}\right)$; radius $\sqrt{8}$
9. Center $(7, 2)$; tangent to the x-axis
10. Center at $(3, 3)$; tangent to the x-axis
11. Center at $(-3, 8)$; passes through $(-4, 9)$
12. Center at $(0, 0)$; passes through $(2, 10)$
13. Center $(4, 8)$; passes through $(1, 9)$
14. Center $(12, -4)$; passes through $(-9, 5)$
15. Center at the origin; passes through $(6, -7)$
16. Center $(13, -2)$; passes through $(8, -3)$
17. Endpoints of a diameter are $(-8, 6)$ and $(1, 11)$
18. Endpoints of a diameter are $(5, 3)$ and $(8, -3)$
19. Endpoints of a diameter are $(-7, -4)$ and $(-5, 7)$
20. Endpoints of a diameter are $(2, 3)$ and $(7, 4)$
21. Endpoints of a diameter are $(0, 0)$ and $(-13, -14)$
22. Endpoints of a diameter are $(4, 10)$ and $(0, 3)$
23. Endpoints of a diameter are $(0, 6)$ and $(8, 0)$
24. Endpoints of a diameter are $(6, 9)$ and $(4, 9)$

25. **26.**

27. **28.**

29.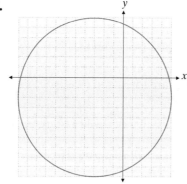

Sketch a graph of the equation and find the center and radius of each circle. Then use a graphing calculator or computer to check your work. See Examples 3 and 4.

30. $x^2 + y^2 = 25$ **31.** $x^2 + y^2 = 36$

32. $x^2 + (y - 3)^2 = 16$ **33.** $x^2 + (y - 8)^2 = 9$

34. $(x + 2)^2 + y^2 = 169$ **35.** $(x - 8)^2 + y^2 = 8$

36. $(x - 9)^2 + (y - 4)^2 = 49$ **37.** $(x + 5)^2 + (y + 4)^2 = 4$

section 3.6   Introduction to Circles

38. $(x+2)^2 + (y-7)^2 = 64$

39. $(x-5)^2 + (y+5)^2 = 5$

40. $x^2 + y^2 - 2x + 10y + 1 = 0$

41. $x^2 + y^2 - 4x + 4y - 8 = 0$

42. $x^2 + y^2 + 6x + 5 = 0$

43. $x^2 + y^2 + 10y + 9 = 0$

44. $x^2 + y^2 - x - y = 2$

45. $x^2 + y^2 + 6y - 2x = -2$

46. $(x-5)^2 + y^2 = 225$

47. $4x^2 + 4y^2 = 256$

48. $(x-3)^2 + (y+2)^2 = 81$

49. $x^2 + y^2 - 6x + 4y - 3 = 0$

50. $(x+2)^2 + (y-1)^2 = 16$

51. $(x-1)^2 + y^2 = 9$

52. $x^2 + (y+2)^2 = 49$

53. $x^2 + y^2 - 4x + 8y - 16 = 0$

54. $x^2 + y^2 + 8x = 9$

55. $4x^2 + 4y^2 - 24x + 24y = 28$

technology note

Graphing calculators and computer algebra systems, such as *Mathematica*, can be used to sketch accurate graphs of circles. To use *Mathematica* to do so, the command `<<Graphics`ImplicitPlot`` must be executed. This command allows *Mathematica* to graph equations, and needs to be executed just once per *Mathematica* session. The command used to then graph an equation is **ImplicitPlot**, and the details of its use are shown below.

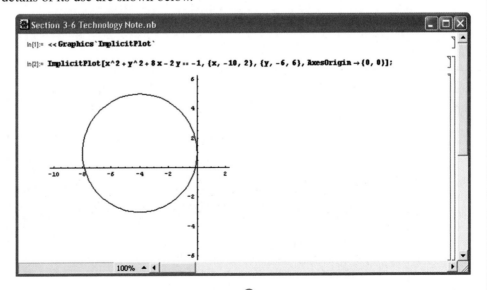

chapter three Linear Equations and Inequalities of Two Variables

chapter three project

Using the Pythagorean Theorem

Assume that a company that builds radio towers has hired you to supervise the installation of steel support cables for several newly built structures. Your task is to find the point at which the cables should be secured to the ground. Assume that the cables reach from ground level to the top of each tower. The cables have been pre-cut by a subcontractor and have been labeled for each tower. Finding the correct distance from the base is necessary because each cable must be grounded before being attached to a tower to avoid damaging the equipment by electric shock.

The following is your work list for this week.

Tower Name	Tower Height	Cable Length	Distance from Base
Shelbyville Tower	58 ft	75 ft	_____
Brockton Tower	100 ft	125 ft	_____
Springfield Tower	77 ft	98 ft	_____
Ogdenville Tower	130 ft	170 ft	_____

1. Use the Pythagorean formula $(a^2 + b^2 = c^2)$ to determine how far from the base of the towers to attach the cables to the ground.

2. Could you determine the length of cable needed if you knew the height of a particular tower and the distance to the grounding point? If so, what length of cable would you need for a 150 foot tower if the grounding point should be 100 feet from the base?

chapter three summary

CHAPTER SUMMARY

A summary of concepts and skills follows each chapter. Refer to these summaries to make sure you feel comfortable with the material in the chapter. The concepts and skills are organized according to the section title and topic title in which the material is first discussed.

3.1 The Cartesian Coordinate System

topics	pages
The Components of the Cartesian Coordinate System • The meaning of the *Cartesian plane*, as well as the terms *axes*, *origin*, *quadrant*, and *ordered pair* • The relation between ordered pairs and points in the plane	p. 177 – 179
The Graph of an Equation • The correspondence between an equation in two variables and its *graph* in the Cartesian plane	p. 179 – 182
The Distance and Midpoint Formulas • The meaning of *distance* in the plane, and the use of the *distance formula* in determining distance • The use of the *midpoint formula* in finding the point midway between two given points in the plane	p. 182 – 184

3.2 Linear Equations in Two Variables

topics	pages
Recognizing Linear Equations in Two Variables • The definition of a *linear equation in two variables* • The *standard form* of a linear equation	p. 189 – 190
Intercepts of the Coordinate Axes • The geometric meaning of the *x-intercept* and the *y-intercept* of a line, and how to determine them	p. 191 – 192

chapter three Linear Equations and Inequalities of Two Variables

topics (continued)	pages
Horizontal and Vertical Lines • The forms of equations that correspond to horizontal and vertical lines	p. 192 – 194

3.3 Forms of Linear Equations

topics	pages
The Slope of a Line • The meaning of *slope*, and how to determine the slope of the line passing through two given points • The meaning of a slope of zero, and the meaning of undefined slope	p. 197 – 200
Slope-Intercept Form of a Line • The definition of the *slope-intercept form* of a line • Obtaining the slope-intercept form of a given line	p. 200 – 203
Point-Slope Form of a Line • The definition of *point-slope form* of a line • Using the point-slope form to obtain equations of lines with prescribed properties	p. 203 – 204
Interlude: Linear Regression • To find the equation, in standard form, of the line close to the specified points	p. 205 – 209

3.4 Parallel and Perpendicular Lines

topics	pages
Slopes of Parallel Lines • The relation between the slopes of parallel lines in the plane	p. 215 – 216
Slopes of Perpendicular Lines • The relation between the slopes of perpendicular lines in the plane • The relation between slopes of perpendicular lines when one of the lines is vertical	p. 216 – 218

chapter three summary

3.5 Linear Inequalities in Two Variables

topics	pages
Solving Linear Inequalities in Two Variables • The meaning of a *linear inequality in two variables*, and its solution set in the plane • The difference between strict and non-strict inequalities, and how the difference is indicated graphically • The use of *test points* in determining the solution set of an inequality	p. 223 – 225
Solving Linear Inequalities Joined by "And" or "Or" • The meaning of "and" and "or" when used to join two or more linear inequalities • The solution set of two or more joined inequalities	p. 226 – 228
Interlude: Regions of Constraint • The meaning of *region of constraint* and how to determine it in a given problem	p. 228 – 229

3.6 Introduction to Circles

topics	pages
Standard Form of a Circle • The definition of the standard form of the equation of a circle with radius r and center (h, k): $(x-h)^2 + (y-k)^2 = r^2$.	p. 235 – 237
Graphing Circles • Obtaining the center and radius of a circle given only the equation • Using the method of *completing the square* to convert the equation of a circle to standard form	p. 237 – 238

chapter three Linear Equations and Inequalities of Two Variables

chapter review

Section 3.1

*Determine **a.** the distance between the following pairs of points, and **b.** the midpoint of the line segment joining each pair of points.*

1. $(2, -6)$ and $(3, -7)$
2. $(-4, -3)$ and $(4, -9)$
3. $(-3, 6)$ and $(-7, 0)$

Find the perimeter of the triangle whose vertices are the specified points in the plane.

4. $(-3, 2), (-3, 0),$ and $(-6, -3)$
5. $(8, -3), (2, -3),$ and $(2, 5)$

Section 3.2

Determine if the following equations are linear.

6. $3x + y(4 - 2x) = 8$
7. $y - 3(y - x) = 8x$
8. $9x^2 - (3x + 1)^2 = y - 3$

Section 3.3

Determine the slopes of the lines passing through the specified points.

9. $(-2, 5)$ and $(-3, -7)$
10. $(3, 6)$ and $(7, -10)$
11. $(3, 5)$ and $(3, -7)$

Use the slope-intercept form of each line to graph the equations.

12. $6x - 3y = 9$
13. $2y + 5x + 9 = 0$
14. $15y - 5x = 0$

Find the equation, in standard form, of the line passing through the specified points with the specified slopes.

15. point $(4, -1)$; slope of 1
16. point $(-2, 3)$; slope of $\dfrac{3}{2}$

Find the equation, in slope-intercept form, of the line passing through the specified points with the specified slopes.

17. point $(0, -2)$; slope of $\dfrac{5}{9}$
18. point $(0, 9)$; slope of $-\dfrac{7}{3}$

Find the equation, in standard form, of the line passing through the specified points.

19. $(5, 7)$ and $(3, -2)$
20. $\left(\dfrac{3}{2}, 1\right)$ and $\left(-3, \dfrac{5}{2}\right)$

chapter three review

Section 3.4

Determine if the two lines in each problem below are perpendicular, parallel, or neither.

21. $x - 4y = 3$ and $4x - y = 2$

22. $3x + y = 2$ and $x - 3y = 25$

23. $\dfrac{3x - y}{3} = x + 2$ and $\dfrac{y}{3} + x = 9$

Each set of four ordered pairs below defines the vertices, in counterclockwise order, of a quadrilateral. Determine if the quadrilateral is a rectangle.

24. $\{(-2, 1), (-1, -1), (3, 1), (2, 3)\}$

25. $\{(-2, 2), (-3, -1), (2, -3), (2, 1)\}$

26. Find the equation, in slope-intercept form, for the line parallel to $3y - 2 = -5(2x - 1)$ and passing through the point $(2, -5)$.

27. Find the equation, in slope-intercept form, for the line perpendicular to $5x + 7y - 2 = 10$ and passing through the point $\left(\dfrac{2}{7}, -1\right)$.

Section 3.5

Graph the solution sets of the following linear inequalities.

28. $7x - 2y \geq 8$ and $y < 5$ 29. $x - 4y \geq 6$ or $y > -2$ 30. $|2x + 5| < 3$

Section 3.6

31. Find the standard form of the equation for the circle with a diameter whose endpoints are $(1, 2)$ and $(-5, 8)$.

Sketch a graph of the circle defined by the given equation. Then state the radius and center of the circle.

32. $(x + 5)^2 + (y - 2)^2 = 16$ 33. $x^2 + y^2 + 6x - 10y = -5$

chapter three   Linear Equations and Inequalities of Two Variables

chapter test

Identify the quadrant in which each point lies, if possible. If a point lies on an axis, specify which part (positive or negative) of which axis (x or y).

1. $(-3, -1)$ **2.** $(0, -2)$ **3.** $(0, 0)$

4. $(1, 0)$ **5.** $(3, -2)$

For each of the following equations, determine the value of the missing entries in the accompanying table of ordered pairs. Then plot the ordered pairs and sketch your guess of the complete graph of the equation.

6. $x^2 + y^2 = 4$

x	y
0	?
?	0
-1	?
1	?
?	2

7. $3x - 2y = 6$

x	y
?	0
0	?
-1	?
?	-2
-2	?

8. Given $A(-4, 2)$, $B(x, y)$, and $C(1, -1)$ find $x + y$ if C is the midpoint of the line segment $\overline{AB}$.

9. Given $(3, 2)$ and $(4, y)$, find y such that the distance between these two points is $\sqrt{10}$.

10. Find the perimeter of the triangle whose vertices are $(2, 3)$, $(1, -2)$, and $(-1, 2)$.

11. Use the distance formula to prove that the triangle with vertices at the points $(-2, 2)$, $(0, 3)$, and $(4, -5)$ is a right triangle and determine the area of the triangle.

12. The vertices of $\triangle ABC$ are $A(-4, 1)$, $B(2, 4)$, and $C(4, -2)$. If $\overline{AD}$ is the median of the triangle find the measure of $\overline{AD}$.

13. $\triangle KLM$ is isosceles and $KL = KM$. If $L(-2, 0)$, $M(4, 0)$, and $K(x, 6)$, find x.

Determine if the following equations are linear.

14. $6x - 5y + xy = 1$ **15.** $\dfrac{x}{2} - \dfrac{y^2 - 1}{y + 1} - 3 = 0$

16. Find the x- and y-intercepts of $2x - 6y = 12$, then graph the equation.

chapter three test

Determine the slopes of the lines passing through the specified points.

17. $(1, -2)$ and $(-3, 1)$ **18.** $(-2, 3)$ and $(-2, 1)$ **19.** $(2, 2)$ and $(-1, 2)$

20. Given the points $A(-2, 4)$, and $B(x, -1)$, find x if they have a slope of $\dfrac{3}{2}$.

21. Find the slope of the line $\dfrac{x-2}{3} - y = \dfrac{1-2y}{2}$.

22. Find the equation, in standard form, of the line with a slope of $\dfrac{1}{2}$ passing through point $(2, -3)$.

23. Given that point $A(a-1, -a)$ is on the line $2x - 3y = 4$, find a.

24. Find the equation, in standard form, of the line passing through the points $(1, -1)$ and $(-2, 3)$.

25. Graph the line that passes through point $(4, -3)$ and has a slope of -2.

26. Assume your salary was $30,000 in 1998 and $40,000 in 2000. If your salary follows a linear growth pattern, what will your salary be in 2003?

27. A sales person receives a monthly salary of $2800 plus a commission of 8% of sales. Write a linear equation for the sales person's monthly wage W, in terms of monthly sales, s.

28. Find the area of the region that is bound by the x-axis, the y-axis and the line $\dfrac{x}{4} + \dfrac{y}{3} = 1$.

29. Find the equation, in slope-intercept form, of the line parallel to $1 - \dfrac{y-3x}{2} = 4$ and passing through $(2, -5)$.

30. If points $A(-2, 3)$, $B(1, -2)$, and $C(a-1, 4)$ are collinear, then find a.

31. Given that the line $4x + (a-2)y = 6$ and the line $ax + 2y = 9$ are parallel to each other. Find a.

32. Find the equation, in slope-intercept form, of the line perpendicular to $3x - y = 2(x-1) - 3y$ and passing through $(-4, 3)$.

33. Assume that the line $y = (2a-1)x - 4$ and the line $\dfrac{x}{3} - \dfrac{y}{2} = 1$ are perpendicular to each other. Find a.

Determine if each of the following sets of lines is parallel, perpendicular, or neither.

34. $y = 2x - 1$ and $y = 2x + 1$ **35.** $y = \dfrac{2}{3}x - 3$ and $y = \dfrac{-3}{2}x + 1$

36. $2(x-3) - \dfrac{y-1}{2} = 0$ and $-x + 3y = 6$

37. $\dfrac{x}{2} - \dfrac{2y-3}{4} = \dfrac{1}{8}$ and $2(x-y) = 4(x-4)$

38. Is it possible for two lines with negative slopes to be perpendicular? Explain.

Given points P(0, –4), Q(8, –3), R(4, 4), and S(–4, 3), show each of the following.

39. Quadrilateral *PQRS* is a parallelogram.

40. Parallelogram *PQRS* is a rhombus.

41. Parallelogram *PQRS* is not a rectangle.

42. Given that three consecutive vertices of a square are $(4, 0), (0, 3),$ and $(-3, -1),$ find the fourth.

Determine the slopes and y-intercepts of the following linear equations, if possible, and then graph the equations.

43. $4x - y + 1 = 0$

44. $3x + 2 = 0$

45. $2y - 4 = 0$

46. $x - y - 4 = 0$

For each linear equation below, find the equation in standard form of a line **a.** *parallel to the given line at the specified point and* **b.** *perpendicular to the given line at the specified point.*

47. $4x - 2y = 1$ at $(2, 1)$

48. $x + y = 3$ at $\left(\dfrac{1}{2}, \dfrac{-3}{2}\right)$

49. $y = -3$ at $(-1, 0)$

50. $x = 4$ at $(2, 5)$

Graph the solution sets of the following linear inequalities.

51. $x - 2y < 4$

52. $\dfrac{x}{2} \geq 1 - y$

53. $x < -2$ and $y \geq -1$

54. $x > 1$ or $y < -2$

55. $y - x > 0$ and $x < 2$

56. $-2 < x < 3$

57. $1 \leq y \leq 4$

58. $2 < x + y \leq 4$

59. $|y+2| < 3$

60. $|2x-1| < 5$

61. $|x+y| \geq 2$

62. $|x-y| < 3$

Find the standard form of the equation for each circle described below.

63. Center $(-3, -2)$; radius $\sqrt{5}$

64. Radius 4; center $\left(\sqrt{5}, -\sqrt{2}\right)$

65. End points of a diameter are $(1, -3)$ and $(-5, 3)$

66. Center at $(2, -1)$; passes through $(4, 3)$

67. Center at the origin; passes through $(5, 12)$

68. Given that point $A(a, 4)$ is on the circle $x^2 + y^2 = 25$, find a.

69. What is the radius and center of the circle $(x+3)^2 + (y-1)^2 = 8$?

70. Find the radius and center of the circle $x^2 + y^2 + 2x - 6y - 6 = 0$.

71. Sketch a graph of the following equation: $x^2 + y^2 + 2x + 4y = 4$

72. Given the diameter of a circle is $\overline{AB}$ and the center is at $(2, 2)$, find the coordinates of point B if point A lies at $(-2, 1)$. Find the equation of this circle.

chapter FOUR

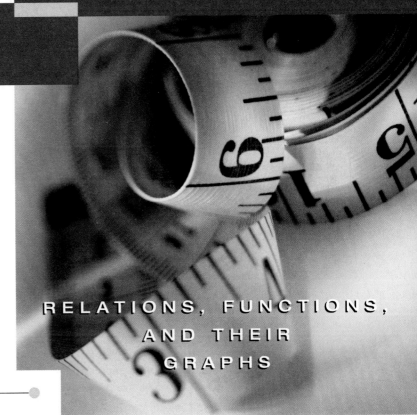

RELATIONS, FUNCTIONS, AND THEIR GRAPHS

OVERVIEW

4.1 Relations and Functions

4.2 Linear and Quadratic Functions

4.3 Other Common Functions

4.4 Transformations of Functions

4.5 Combining Functions

4.6 Inverses of Functions

Many examples of variation exist in the natural and social realms. Distance fallen varies based on time traveled, demand varies based on supply, and the list goes on. By the end of this chapter, you should be able to find the constant of proportionality given certain information about a variation, and formulate unknown information using this constant. On page 301 you will find a problem concerning a person's Body Mass Index (BMI) — a general indicator of acceptable weight based on height. Given the height, weight, and BMI of one person, you will be able to find the BMI of another, using tools such as the definition of *Direct and Inverse Variation* on page 297.

chapter four Relations, Functions, and Their Graphs

Introduction

This chapter begins with a study of *relations*, which are generalizations of the equations in two variables discussed in Chapter 3, and then moves on to the more specialized topic of *functions*. As concepts, relations and functions are more abstract, but at the same time far more powerful and useful than the equations studied thus far in this text. Functions, in particular, lie at the heart of a great deal of the mathematics that you will encounter from this point on.

Leibniz

The history of the function concept serves as a good illustration of how mathematics develops. One of the first people to use the idea in a mathematical context was the German mathematician and philosopher Gottfried Leibniz (1646 - 1716), one of two people (along with Isaac Newton) usually credited with the development of calculus. Initially, Leibniz and other mathematicians tended to use the term to indicate that one quantity could be defined in terms of another by some sort of algebraic expression, and this (incomplete) definition of function is often encountered even today in elementary mathematics. As the problems that mathematicians were trying to solve increased in complexity, however, it became apparent that functional relations between quantities existed in situations where no algebraic expression defining the function was possible. One example came from the study of heat flow in materials, in which a description of the temperature at a given point at a given time was often given in terms of an infinite sum, not an algebraic expression.

The result of numerous refinements and revisions of the function concept is the definition that you will encounter in this chapter, and is essentially due to the German mathematician Lejeune Dirichlet (1805 - 1859). Dirichlet also refined our notion of what is meant by a *variable*, and gave us our modern understanding of *dependent* and *independent* variables, all of which you will soon encounter.

The proof of the power of functions lies in the multitude and diversity of their applications. The subtle and easily overlooked advantage of functional notation deserves special mention; as noted at the start of Chapter 2, innovations in notation often go a long way toward solving difficult problems. As you work through Chapter 4, pay special attention to how functional notation works. A solid understanding of what functional notation means is essential to using functions.

4.1 Relations and Functions

TOPICS

1. Relations, domains, and ranges
2. Functions and the vertical line test
3. Functional notation and function evaluation
4. Implied domain of a function

Topic 1: Relations, Domains, and Ranges

In Chapter 3 we saw many examples of equations in two variables. Any such equation automatically defines a relation between the two variables present, in the sense that each ordered pair on the graph of the equation relates a value for one variable (namely, the first coordinate of the ordered pair) to a value for the second variable (the second coordinate). Many applications of mathematics involve relating one variable to another, and this notion merits much further study.

We begin by generalizing the above observation and define a relation to be simply a collection of ordered pairs; any collection of ordered pairs constitutes a relation, whether the ordered pairs correspond to the graph of an equation or not. This definition is repeated below, along with two related ideas.

Relations, Domains, and Ranges

A **relation** is a set of ordered pairs. Any set of ordered pairs automatically relates the set of first coordinates to the set of second coordinates, and these sets have special names. The **domain** of a relation is the set of all the first coordinates, and the **range** of a relation is the set of all second coordinates.

It is important to understand that relations can be described in many different ways. We have already noted that an equation in two variables describes a relation, as the solution set of the equation is a collection of ordered pairs. Relations can also be described with a simple list of ordered pairs (if the list is not too long), with a picture in the Cartesian plane, and by many other means.

chapter four   Relations, Functions, and Their Graphs

The following example demonstrates some of the common ways of describing relations and identifies the domain and range of each relation.

example 1

a. The set $R = \{(-4, 2), (6, -1), (0, 0), (-4, 0), (\pi, \pi^2)\}$ is a relation consisting of five ordered pairs. The domain of R is the set $\{-4, 6, 0, \pi\}$, as these four numbers appear as first coordinates in the relation. Note that it is not necessary to list the number -4 twice in the domain, even though it appears twice as a first coordinate in the relation. The range of R is the set $\{2, -1, 0, \pi^2\}$, as these are the numbers that appear as second coordinates. Again, it is not necessary to list 0 twice in the range, even though it is used twice as a second coordinate in the relation. The *graph* of this relation is simply a picture of the five ordered pairs plotted in the Cartesian plane, as shown below.

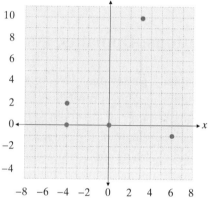

b. The equation $-3x + 7y = 13$ describes a relation. In contrast to the last example, this relation consists of an infinite number of ordered pairs, so it is not possible to list them all as a set. As an example, one of the ordered pairs in the relation is $(-2, 1)$, since $-3(-2) + 7(1) = 13$. Although it is not possible to list all the elements of this relation, it is possible to graph it; it is the graph of the solution set of the equation, as drawn below. The domain and range of this relation are both the set of real numbers, as every real number appears as both a first coordinate and a second coordinate in the relation.

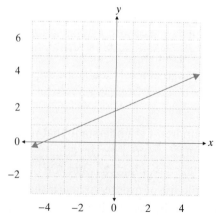

c. The picture below describes a relation. Some of the elements of the relation are (−1, 1), (−1, −2), (−0.3, 2), (0, −2), and (1, −0.758), but this is another example of a relation with an infinite number of elements so we cannot list all of them. Using the interval notation defined in Section 1.1, the domain of this relation is the closed interval [−1, 1] and the range is the closed interval [−2, 2].

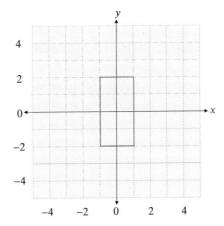

d. The picture below describes another relation, similar to the last but still different. The shading in the picture indicates that all ordered pairs lying inside the rectangle drawn in blue, as well as those actually *on* the rectangle, are elements of the relation. The domain is again the closed interval [−1, 1] and the range is again the closed interval [−2, 2], but this relation is not identical to the last example. For instance, the ordered pairs (0, 0) and (0.2, 1.5) are elements of this relation but are not elements of the relation in *Example 1c*.

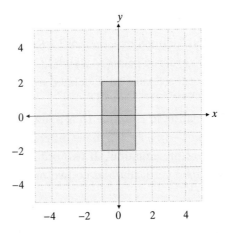

e. Although we will almost never encounter relations in this text that do not consist of ordered pairs of real numbers, there is nothing in our definition to prevent us from considering more exotic relations.

continued on next page...

chapter four   Relations, Functions, and Their Graphs

For example, the set $S = \{(x, y) | x \text{ is the mother of } y\}$ is a relation among people. The domain of S is the set of all mothers, and the range of S is the set of all people. (Although advances in cloning are occurring rapidly, as of the writing of this text, no one has yet been born without a mother!)

Topic 2: Functions and the Vertical Line Test

As important as relations are in mathematics, a subclass of relations, called functions, has proven to be of even greater utility.

Functions

> A **function** is a relation in which every element of the domain is paired with *exactly one* element of the range. Equivalently, a function is a relation in which no two distinct ordered pairs have the same first coordinate.

Note that there is an asymmetry in the way domains and ranges are treated in the definition of function: the definition allows for the possibility that two distinct ordered pairs may have the same second coordinate, as long as their first coordinates differ.

To gain some intuition into which relations qualify as functions, consider the five relations defined in *Example 1*. *Example 1a* is not a function, for the sole reason that the two ordered pairs $(-4, 2)$ and $(-4, 0)$ have the same first coordinate. If either one of these ordered pairs were deleted from the relation, the relation would be a function.

The relation in *Example 1b* is a function. Any two distinct ordered pairs that solve the equation $-3x + 7y = 13$ have different first coordinates. This can also be seen from the graph of the equation. If two ordered pairs have the same first coordinate, they must be aligned vertically, and no two ordered pairs on the graph of $-3x + 7y = 13$ have this property.

The relations in *Examples 1c* and *1d* are also not functions. To prove that a relation is not a function, it is only necessary to find two distinct ordered pairs with the same first coordinate, and the pair $(0, 2)$ and $(0, -2)$ works to show that both of these relations fail to be functions. To again illustrate the difference between these relations, note that $(0, 1)$, $(0, 0.005)$, and $(0, -1.9)$ are all elements of the relation in *Example 1d*, but not of the relation in *Example 1c*.

Finally, the relation in *Example 1e* also fails to be a function. It is only necessary to think of two people who have the same mother to convince yourself of this fact.

In determining that the relation in *Example 1b* is a function, we noted that two distinct ordered pairs in the plane have the same first coordinate only if they are aligned vertically, and that no two points on the graph of the equation $-3x + 7y = 13$ have this property. We could also have used this criterion to determine that the relation in *Example 1a* is not a function, since in the graph of the relation the two ordered pairs $(-4, 2)$ and $(-4, 0)$ clearly lie on the same vertical line. This visual method of determining whether a relation is a function is quick and easily applied whenever an accurate graph of the relation is available to study. It is restated and given a name below.

The Vertical Line Test

If a relation can be graphed in the Cartesian plane, the relation is a **function** if and only if no vertical line passes through the graph more than once. If even *one* vertical line intersects the graph of the relation two or more times, the relation fails to be a function.

Note that vertical lines that miss the graph of a relation entirely don't prevent the relation from being a function; it is only the presence of a vertical line that hits the graph two or more times that indicates the relation isn't a function. The next example illustrates some more applications of the vertical line test.

example 2

a. The relation $R = \{(-3, 2), (-1, 0), (0, 2), (2, -4), (4, 0)\}$, graphed below, is a function. Any given vertical line in the plane either intersects the graph once or not at all.

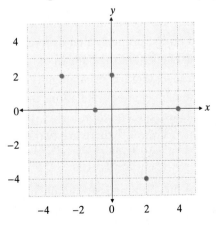

continued on next page...

b. The relation graphed below is not a function, as there are many vertical lines that intersect the graph more than once. The dashed line is one such vertical line.

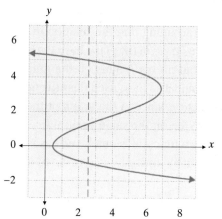

c. The relation graphed below is a function. In this case, every vertical line in the plane intersects the graph exactly once.

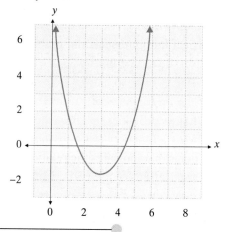

Topic 3: Functional Notation and Function Evaluation

Functions are useful in mathematics for a variety of reasons, all of which ultimately depend on the property that differentiates functions from other relations. Specifically, given an element of the domain of a function, the function relates it to a *unique* element of the range; there is no ambiguity in which element of the range to associate with a given domain element. Relations that are not functions fail this criterion of uniqueness. For instance, the equation $x = y^2$ describes a perfectly good relation, but for a given value of x, say $x = 9$, there are two elements of the range associated with it: both $(9, 3)$ and $(9, -3)$ are on the graph of $x = y^2$.

Because functions assign a unique element of the range to each element of the domain, functions are often defined by means of a notation not yet seen in this text, called functional notation.

Functional notation accomplishes a number of things at once: it gives a name to the function being defined, it provides a formula to work with when using the function, and it makes the solving of many, many mathematical problems far easier.

As an example of functional notation, consider the function defined by the graph of $10x - 5y = 15$. We know this represents a function because if we graph $10x - 5y = 15$ we obtain a non-vertical line, and such a line always passes the vertical line test. If we need to know which specific element of the range corresponds to a given element of the domain, say $x = 4$, we can substitute 4 for x in the equation and solve for y. But if we have to do this for more than one domain element, it is foolish to repeat the process over and over. Why not just solve for y once and for all? If we do this, we obtain the equivalent equation $y = 2x - 3$. Now, given any element of the domain (that is, given any value for x), we can determine the corresponding range element (the value for y). In this context, x is called the **independent variable** and y is called the **dependent variable**, as the value for y depends on the value for x.

In functional notation, the function defined by $10x - 5y = 15$ is written as $f(x) = 2x - 3$, where we have arbitrarily chosen to name the function f. The key point here is that we have used a mathematical formula to describe what the function named f does to whatever element of the domain it is given. The notation $f(x) = 2x - 3$ is read "f of x equals two times x, minus three," and indicates that when given a specific value for x, the function f returns two times that value, minus three. For instance, $f(4) = 2(4) - 3$, or $f(4) = 5$ (this last line is read "f of 4 equals 5"). Notice that the only difference between the equation $y = 2x - 3$ and the function $f(x) = 2x - 3$ is one of notation: we have replaced y with $f(x)$.

caution

By far the most common error made when encountering functions for the first time is to think that $f(x)$ stands for the product of f and x. This is entirely wrong! While it is true that parentheses are often used to indicate multiplication, they are also used in defining functions. This is another example of the unfortunate reuse of symbols.

A word on the use of the variable x is in order here. In defining the function f as $f(x) = 2x - 3$, the critical idea being conveyed is the formula. We can use absolutely any symbol at all as a placeholder in defining the formula that we have named f. For instance, $f(n) = 2n - 3$, $f(z) = 2z - 3$, $f(\square) = 2\square - 3$ and $f(\_) = 2(\_) - 3$ all define exactly the same function: the function that multiplies what it is given by two and then subtracts three. The variable (or symbol) that is used in defining a given function is called its **argument**, and serves as nothing more than a placeholder.

chapter four   Relations, Functions, and Their Graphs

example 3

Each of the following equations in x and y represents a function. Rewrite each one using functional notation, and then evaluate each function at $x = -3$.

a. $7x + 3 = 2y - 1$ **b.** $y - 5 = x^2$ **c.** $\sqrt{1-x} - 2y = 6$

Solutions:

This screen was performed using equation solver.

a. $7x + 3 = 2y - 1$
$7x - 2y = -4$
$-2y = -7x - 4$
$y = \frac{7}{2}x + 2$
$f(x) = \frac{7}{2}x + 2$
$f(-3) = \frac{7}{2}(-3) + 2 = -\frac{17}{2}$

The first step is to solve the equation for the dependent variable y.

We can name the function anything at all. Typical names of functions are f, g, h, etc.

We now evaluate f at -3 to obtain the desired answer.

b. $y - 5 = x^2$
$y = x^2 + 5$
$g(x) = x^2 + 5$
$g(-3) = (-3)^2 + 5 = 14$

Again, we begin by solving for y.

To distinguish this function from the last, we have used a different name.

Replace the argument with -3 to finish.

c. $\sqrt{1-x} - 2y = 6$
$-2y = 6 - \sqrt{1-x}$
$y = -3 + \frac{\sqrt{1-x}}{2}$
$h(x) = -3 + \frac{\sqrt{1-x}}{2}$
$h(-3) = -3 + \frac{\sqrt{1-(-3)}}{2}$
$= -3 + \frac{2}{2} = -2$

We have solved for y and named the function h.

The resulting evaluation tells us that h of -3 is equal to -2.

section 4.1   **Relations and Functions**

technology note

Functions and functional notation are immensely important not just as mathematical concepts but also in the way we use technology as mathematical aids. Graphing calculators and computer algebra systems all have methods for defining functions and then making use of them. The illustration below shows how the function $f(x) = x^3 + 7x^{\frac{3}{2}} - 3x$ is defined in *Mathematica*, and how the function is then evaluated at 4 and 10. Graphing calculators and other computer algebra systems have their own particular syntax for defining and using functions (*consult your user's manual for the specific steps for your calculator or software, or see Appendix A for guidance on using Mathematica.*)

This screen also used equation solver.

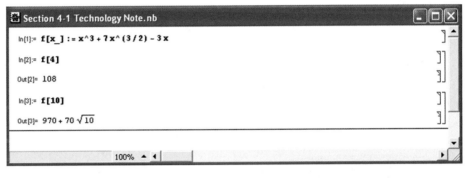

We will not always be replacing the arguments of functions with numbers. In many instances, we will have reason to replace the argument of a function with another variable or possibly a more complicated algebraic expression. Keep in mind that this just involves substituting something for the placeholder used in defining the function.

example 4

Given the function $f(x) = 3x^2 - 2$, evaluate:

a. $f(a)$ **b.** $f(x+h)$ **c.** $\dfrac{f(x+h) - f(x)}{h}$

continued on next page...

chapter four   Relations, Functions, and Their Graphs

Solutions:

a. $f(a) = 3a^2 - 2$

This is just a matter of replacing x by a.

b. $f(x+h) = 3(x+h)^2 - 2$
$= 3(x^2 + 2xh + h^2) - 2$
$= 3x^2 + 6xh + 3h^2 - 2$

In this problem, we replace x by $x + h$ and simplify the result.

c. $\dfrac{f(x+h) - f(x)}{h} = \dfrac{(3x^2 + 6xh + 3h^2 - 2) - (3x^2 - 2)}{h}$
$= \dfrac{6xh + 3h^2}{h}$
$= \dfrac{h(6x + 3h)}{h}$
$= 6x + 3h$

We can use the result from above in simplifying this expression. In the original expression, the h in the denominator is rewritten until, eventually, it cancels a factor of h in the numerator.

technology note

The illustration below shows how *Mathematica* can be used to carry out tasks like those seen in *Example 4*. The first command defines the function $f(x) = 2x^4 - 5x^2$, and the second command then simplifies the expression $\dfrac{f(x+h) - f(x)}{h}$. Other computer algebra systems and some graphing calculators can be used to perform the same task. (*Consult your user's manual for the specific steps for your calculator or software, or see Appendix A for guidance on using Mathematica.*)

Topic 4: Implied Domain of a Function

Sometimes, especially in applications, the domain of a function under consideration is defined at the same time as the function itself. Often, however, no mention is made explicitly of the values that may be "plugged into" the function, and it is up to us to determine what the domain is. In these cases, the domain of the function is implied by the formula used in defining the function. It is assumed that the domain of the function consists of all those real numbers at which the function can be evaluated to obtain a real number. This means, for instance, that any values for the argument that result in division by zero or an even root of a negative number must be excluded from the domain.

example 5

Determine the domain of the following functions.

a. $f(x) = 5x - \sqrt{3-x}$ **b.** $g(x) = \dfrac{x-3}{x^2-1}$

Solutions:

a. The $5x$ in the formula $5x - \sqrt{3-x}$ can't cause any problems, as any real number can be safely multiplied by 5. However, for $\sqrt{3-x}$ to be a real number, we need $3 - x \geq 0$. Solving this inequality for x results in the statement $x \leq 3$. Using interval notation, the domain of the function f is the interval $(-\infty, 3]$.

b. The formula in this function is a rational expression or a ratio of polynomials. A polynomial in x can be evaluated for any value of x, so the only potential problem we must guard against is division by zero. The denominator of this formula is zero if $x^2 - 1 = 0$, so we need to solve this quadratic equation and exclude the solutions from the domain of g. This equation can be solved by factoring: $(x+1)(x-1) = 0$ tells us that we must exclude $x = -1$ and $x = 1$ from the domain. In interval notation, the domain of g is $(-\infty, -1) \cup (-1, 1) \cup (1, \infty)$.

chapter four   Relations, Functions, and Their Graphs

exercises

For each relation below, describe the domain and range. See Example 1.

1. $R = \{(-2, 5), (-2, 3), (-2, 0), (-2, -9)\}$
2. $S = \{(0, 0), (-5, 2), (3, 3), (5, 3)\}$
3. $A = \{(\pi, 2), (-2\pi, 4), (3, 0), (1, 7)\}$
4. $B = \{(3, 3), (-4, 3), (3, 8), (3, -2)\}$
5. $T = \{(x, y) \mid x \in \mathbb{Z} \text{ and } y = 2x\}$
6. $U = \{(\pi, y) \mid y \in \mathbb{Q}\}$
7. $C = \{(x, 3x + 4) \mid x \in \mathbb{Z}\}$
8. $D = \{(5x, 3y) \mid x \in \mathbb{Z} \text{ and } y \in \mathbb{Z}\}$
9. $3x - 4y = 17$
10. $x + y = 0$
11. $x = |y|$
12. $y = x^2$
13. $y = -1$
14. $x = 3$
15. $x = 4x$
16. $y = 7\pi^2$

17.
18.

19.
20.

21.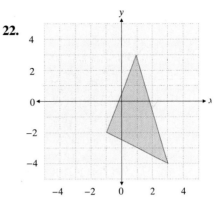

22.

23. $V = \{(x, y) | x \text{ is the brother of } y\}$

24. $W = \{(x, y) | y \text{ is the daughter of } x\}$

Determine which of the relations below is a function. For those that are not, identify two ordered pairs with the same first coordinate. See Example 2.

25. $R = \{(-2, 5), (2, 4), (-2, 3), (3, -9)\}$

26. $S = \{(3, -2), (4, -2)\}$

27. $T = \{(-1, 2), (1, 1), (2, -1), (-3, 1)\}$

28. $U = \{(4, 5), (2, -3), (-2, 1), (4, -1)\}$

29.

30.

31.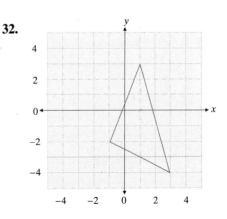

32.

33. $V = \{(6,-1), (3,2), (6,4), (-1,5)\}$ 34. $W = \{(2,-3), (-2,4), (-3,2), (4,-2)\}$

35.

36.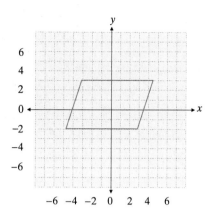

Rewrite each of the relations below as a function of x. Then evaluate the function at $x = -1$. See Example 3.

37. $6x^2 - x + 3y = x + 2y$

38. $2y - \sqrt[3]{x} = x - (x-1)^2$

39. $\dfrac{x+3y}{5} = 2$

40. $x^2 + y = 3 - 4x^2 + 2y$

41. $y - 2x^2 = -2(x + x^2 + 5)$

42. $\dfrac{9y+2}{6} = \dfrac{3x-1}{2}$

*For each function below, determine **a.** $f(x-1)$, **b.** $f(x+a) - f(x)$, and **c.** $f(x^2)$. See Example 4.*

43. $f(x) = x^2 + 3x$

44. $f(x) = \sqrt{x}$

45. $f(x) = 3x + 2$

46. $f(x) = -x^2 - 7$

47. $f(x) = 2(5 - 3x)$

48. $f(x) = 2x^2 + \sqrt[4]{x}$

49. $f(x) = \sqrt{1-x} - 3$

50. $f(x) = \dfrac{-\sqrt{1-x} + 5}{2}$

section 4.1   **Relations and Functions**

Given a function f(x), the fraction $\dfrac{f(x+h)-f(x)}{h}$ is the difference quotient associated with f. The determination of a function's difference quotient is the first step in determining the function's derivative and is an important topic in Calculus. For each of the functions below, determine its difference quotient. See Example 4c.

51. $f(x) = x^2 - 5x$ **52.** $t(x) = x^3 + 2$

53. $h(x) = \dfrac{1}{x+2}$ **54.** $g(x) = 6x^2 - 7x + 3$

55. $f(x) = 5x^2$ **56.** $f(x) = (x+3)^2$

57. $f(x) = 2x - 7$ **58.** $f(x) = \sqrt{x}$

59. $f(x) = x^{\frac{1}{2}} - 4$ **60.** $f(x) = \dfrac{3}{x}$

Determine the implied domain of each of the following functions. See Example 5.

61. $f(x) = \sqrt{x-1}$ **62.** $g(x) = \sqrt[5]{x+3} - 2$

63. $h(x) = \dfrac{3x}{x^2 - x - 6}$ **64.** $f(x) = (2x+6)^{\frac{1}{2}}$

65. $g(x) = \sqrt[4]{2x^2 + 3}$ **66.** $h(x) = \dfrac{3x^2 - 6x}{x^2 - 6x + 9}$

67. $s(x) = \dfrac{2x}{1 - 3x}$ **68.** $f(x) = (x^2 - 5x + 6)^3$

69. $c(x) = \sqrt{\dfrac{x-1}{2-x}}$ omit   **70.** $g(x) = \dfrac{5}{\sqrt{3-x^2}}$

71. $f(x) = \sqrt{x+6} + 1$ **72.** $g(x) = -5x^2 - 4x$

73. $h(x) = \dfrac{-3(-5 + 5x)}{x}$ **74.** $h(x) = \sqrt{3-x}$

chapter four   Relations, Functions, and Their Graphs

technology exercises

Use a graphing calculator or a computer algebra system such as Mathematica to first define each of the following functions and then perform the indicated function evaluations. See the Technology Note after Example 3 for guidance.

75. $f(x) = \dfrac{7x^{\frac{5}{3}} - 2x^{\frac{1}{3}}}{x^{\frac{1}{2}}}$; find $f(8)$ and $f(12)$

76. $g(x) = \sqrt{x^3 - 4x^2 + 2x + 31}$; find $g(2)$ and $g(3)$

77. $f(x) = \dfrac{2x^5 - 9x^3 + 12}{4x^3 - 7x + 6}$; find $f(-3)$ and $f(2)$

78. $g(x) = (5x^2 - 7x + 1)^3$; find $g(-19)$ and $g(12)$

79. $f(x) = \dfrac{\sqrt{x^4 + 6x^3 - 4x + 13}}{4x^3 + 2x^2 - 12}$; find $f(-4)$ and $f(6)$

80. $g(x) = \dfrac{(3x^3 - 2x + 9)^4}{(7x^2 - 5x)^2}$; find $g(-5)$ and $g(4)$

Use a graphing calculator, if possible, or a computer algebra system such as Mathematica to first define each of the following functions and then simplify the expression $\dfrac{f(x+h) - f(x)}{h}$. See the Technology Note after Example 4 for guidance.

81. $f(x) = x^3$

82. $f(x) = (x^2 + 1)^3$

83. $f(x) = 3x^2 - x$

84. $f(x) = \dfrac{1}{x-1}$

4.2 Linear and Quadratic Functions

TOPICS

1. Linear functions and their graphs
2. Quadratic functions and their graphs
3. Interlude: maximization / minimization problems

Topic 1: Linear Functions and Their Graphs

Much of the next several sections of this chapter will be devoted to gaining familiarity with some of the types of functions that commonly arise in mathematics. We will discuss two classes of functions in this section, beginning with the class of linear functions.

Recall that a linear equation in two variables is an equation whose graph consists of a straight line in the Cartesian plane. Similarly, a linear function (of one variable) is a function whose graph is a straight line. We can define such functions algebraically as follows.

Linear Functions

A **linear function**, say f, of one variable, say the variable x, is any function that can be written in the form $f(x) = ax + b$, where a and b are real numbers. If $a \neq 0$, a function $f(x) = ax + b$ is also called a **first-degree function**.

In the last section, we learned that a function defined by an equation in x and y can be written in functional form by solving the equation for y and then replacing y with $f(x)$. Of course, the process can be reversed, so the function $f(x) = ax + b$ appears in equation form as $y = ax + b$, and we recognize this as a linear equation written in slope-intercept form. This means the graph of the function $f(x) = ax + b$ is a straight line with a y-intercept of b (the constant term) and a slope of a (the coefficient of the variable x). For this reason, some texts define a linear function as any function that can be written in the form $f(x) = mx + b$, since m is the letter traditionally used to denote slope and b is commonly used to refer to the y-intercept.

chapter four   Relations, Functions, and Their Graphs

The observations in the last paragraph give us a quickly applied method of graphing linear functions, but before we look at some examples, a few words on graphing functions in general are in order. As we noted in Section 4.1, the graph of a function is a plot of all the ordered pairs that make up the function; that is, the graph of a function f is the plot of all the ordered pairs in the set $\{(x, y) \mid (x, y) \text{ is an element of } f\}$. We have a great deal of experience in plotting such sets if the ordered pairs are defined by an equation in x and y, but we have only plotted a few functions that have been defined with functional notation. Any function of x defined with functional notation can be written as an equation in x and y by replacing $f(x)$ with y, so the graph of a function f consists of a plot of the ordered pairs in the set $\{(x, f(x)) \mid x \in \text{ domain of } f\}$.

Consider, for example, the function $f(x) = -3x + 5$. Figure 1 contains a table of four ordered pairs defined by the function, followed by a graph of the entire function with the four specific ordered pairs noted.

x	$f(x)$
-1	8
0	5
1	2
2	-1

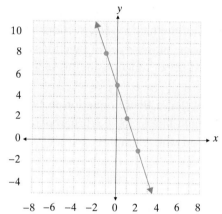

Figure 1: Graph of $f(x) = -3x + 5$

Again, note that every point on the graph of the function in Figure 1 is an ordered pair of the form $(x, f(x))$; we have simply highlighted four of them with teal dots.

We could have graphed the function $f(x) = -3x + 5$ by noting that it is a straight line with a slope of -3 and a y-intercept of 5. We will use this approach in the following examples.

example 1

Graph the following linear functions.

a. $f(x) = \dfrac{4 + 6x}{2}$

b. $g(x) = 3$

section 4.2 Linear and Quadratic Functions

Solutions:

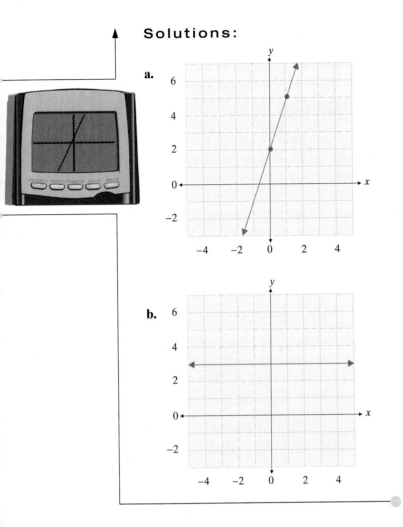

a. The function f can be rewritten as $f(x) = 3x + 2$, and in this form we recognize it as a line with a slope of 3 and a y-intercept of 2.

To graph the function, we can plot the ordered pair $(0, 2)$ and locate another point on the line by moving up 3 units and over to the right 1 unit, giving us the ordered pair $(1, 5)$. Once these two points have been plotted, connecting them with a straight line completes the process.

b. The graph of the function g is a straight line with a slope of 0 and a y-intercept of 3. A linear function with a slope of 0 is also called a **constant** function, as it turns any input into one fixed constant–in this case the number 3. The graph of a constant function is always a horizontal line.

Example 1b is an example of a function whose graph is a horizontal line. Of course, no function can have a vertical line as its graph, as a vertical line clearly can't pass the vertical line test. Thus vertical lines can represent the graphs of equations, but not functions.

Topic 2: Quadratic Functions and Their Graphs

In Section 2.3, we learned how to solve quadratic equations in one variable. We will now study quadratic *functions* of one variable and learn how to relate this new material to what we already know.

Quadratic Functions

A **quadratic function**, or **second-degree function**, say f, of one variable, say the variable x, is any function that can be written in the form $f(x) = ax^2 + bx + c$, where a, b, and c are real numbers and $a \neq 0$.

The graph of any quadratic function is a roughly U-shaped curve known as a **parabola**. We will study parabolas in more generality in Chapter 6, but in this section we will learn how to graph parabolas as they arise in the context of quadratic functions.

The graph in Figure 2 is an example of a parabola. Specifically, it is the graph of the quadratic function $f(x) = x^2$, and the table that precedes the graph contains a few of the ordered pairs on the graph.

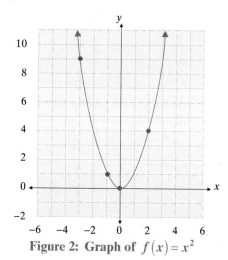

Figure 2: Graph of $f(x) = x^2$

As Figure 2 demonstrates, one key characteristic of a parabola is that it has one (and only one) point at which the graph "changes direction." This point is known as the **vertex**; scanning the plane from left to right, it is the point on a parabola where the graph stops going down and begins to go up (if the parabola opens upward) or stops going up and begins to go down (if the parabola opens downward). Another characteristic of a parabola is that it is symmetric with respect to its **axis**, a straight line passing through the vertex and dividing the parabola into two halves that are mirror images of each other. Every parabola that represents the graph of a quadratic function has a vertical axis, but we will see parabolas later in the text that have non-vertical axes. Finally, parabolas can be relatively skinny or relatively broad, meaning that the curve of the parabola at the vertex can range from very sharp to very flat.

Our knowledge of linear equations gives us a very satisfactory method for sketching the graphs of linear functions. Our goal now is to develop a method that allows us to sketch the graphs of quadratic functions. We *could*, given a quadratic function, simply plot a number of points and then connect them with a U-shaped curve, but this method has several drawbacks, chief among them being that it is slow and not very accurate. Ideally, we will develop a method that is fast and that pinpoints some of the important characteristics of a parabola (such as the location of its vertex) exactly.

We will develop our graphing method by working from the answer backward. We will first see what effects various mathematical operations have on the graphs of parabolas, and then see how this knowledge lets us graph a generic quadratic function.

To begin, let us say that the graph of the function $f(x) = x^2$, graphed in Figure 2, is the prototypical parabola. We already know its characteristics: its vertex is at the origin, its axis is the y-axis, it opens upward, and the sharpness of the curve at its vertex will serve as a convenient reference when discussing other parabolas. Now consider the function $g(x) = (x-3)^2$, obtained by replacing x in the formula for f with $x - 3$. Both of the formulas x^2 and $(x-3)^2$ are non-negative for any value of x, and x^2 is equal to 0 when $x = 0$. What value of x results in $(x-3)^2$ equaling 0? The answer is $x = 3$. In other words, the point $(0,0)$ on the graph of f corresponds to the point $(3,0)$ on the graph of g. With this in mind, the table and graph in Figure 3 should not be too surprising.

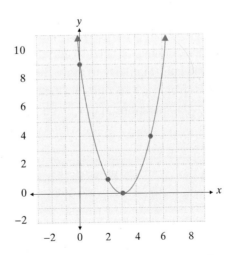

x	$g(x)$
0	9
2	1
3	0
5	4

Figure 3: Graph of $g(x) = (x-3)^2$

Notice that the shape of the graph of g is identical to that of f, but it has been shifted over to the right by 3 units. This is our first example of how we can manipulate graphs of functions.

Consider now the function h obtained by replacing the x in x^2 by $x + 7$. As with the functions f and g, $h(x) = (x+7)^2$ is non-negative for all values of x, and only one value for x will return a value of 0: $h(-7) = 0$. Compare the table and graph in Figure 4 with those in Figures 2 and 3.

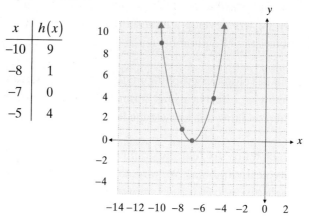

Figure 4: Graph of $h(x) = (x+7)^2$

We have thus demonstrated the ability to shift the prototypical parabola to the left and right. To sum up, the graph of $g(x) = (x-h)^2$ has the same shape as the graph of $f(x) = x^2$, but shifted h units to the right if h is positive and h units to the left if h is negative.

How do we shift a parabola up and down? To move the graph of $f(x) = x^2$ up by a fixed number of units, we need to add that number of units to the second coordinate of each ordered pair. Similarly, to move the graph down we need to subtract the desired number of units from each second coordinate. To demonstrate this, consider the table and graphs for the two functions $i(x) = x^2 + 5$ and $j(x) = x^2 - 2$ in Figure 5.

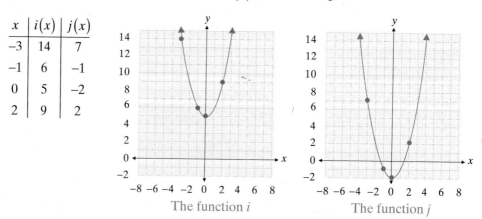

Figure 5: Graph of $i(x) = x^2 + 5$ and $j(x) = x^2 - 2$

section 4.2   Linear and Quadratic Functions

Now, how do we make a parabola skinnier or broader? To make the prototypical parabola skinnier (or to make the curve at the vertex sharper), we need to stretch the graph vertically. That is, we want to leave the vertex untouched but increase the second coordinates of all other points on the graph. We can do this by multiplying the formula x^2 by a constant a greater than 1 to obtain the formula ax^2. Multiplying the formula x^2 by a constant a that lies between 0 and 1 makes the parabola broader (that is, it makes the curve at the vertex flatter). Finally, multiplying x^2 by a negative constant a turns all of the non-negative outputs of f into non-positive outputs, resulting in a parabola that opens downward instead of upward.

Compare the graphs of $k(x) = 6x^2$ and $l(x) = -\frac{1}{2}x^2$ in Figure 6 to the prototypical quadratic $f(x) = x^2$, which is shown as a teal dotted line.

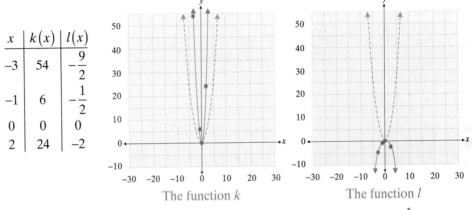

x	$k(x)$	$l(x)$
-3	54	$-\frac{9}{2}$
-1	6	$-\frac{1}{2}$
0	0	0
2	24	-2

The function k The function l

Figure 6: Graph of $k(x) = 6x^2$ and $l(x) = -\frac{1}{2}x^2$

The following box summarizes all of the above ways of altering the prototypical parabola $f(x) = x^2$.

Vertex Form of a Quadratic Function

The graph of the function $g(x) = a(x-h)^2 + k$, where a, h, and k are real numbers and $a \neq 0$, is a parabola whose vertex is at (h, k). The parabola is narrower than $f(x) = x^2$ if $|a| > 1$ and is broader than $f(x) = x^2$ if $0 < |a| < 1$. The parabola opens upward if a is positive and downward if a is negative.

chapter four Relations, Functions, and Their Graphs

The question now is: given a quadratic function $f(x) = ax^2 + bx + c$, how do we determine the location of its vertex, whether it opens upward or downward, and whether it is skinnier or broader than the prototypical parabola? The answer is to use the method of completing the square.

example 2

Sketch the graph of the function $f(x) = -x^2 - 2x + 3$. Locate the vertex and the x-intercepts exactly.

Solution:

We identify the vertex of the function by completing the square on the expression $-x^2 - 2x + 3$, as shown below.

$$f(x) = -x^2 - 2x + 3$$
$$= -(x^2 + 2x) + 3$$
$$= -(x^2 + 2x + 1) + 1 + 3$$
$$= -(x+1)^2 + 4$$

The first step is to factor out the leading coefficient from the first two terms. In this problem, the leading coefficient is -1. We then complete the square on the x^2 and $2x$ terms, as shown. Note, though, that this amounts to adding -1 to the function, so we compensate by adding 1 as well.

Completing the square thus tells us that the vertex of the parabola lies at the point $(-1, 4)$. Note that we have to interpret the expression $-(x+1)^2 + 4$ as $-(x-(-1))^2 + 4$, so h is -1 and k is 4.

The instructions also ask us to identify the x-intercepts. The graph of any quadratic function will intersect the x-axis in 0, 1, or 2 places, depending on the orientation of the corresponding parabola. An x-intercept of the function f will be a point on the x-axis where $f(x) = 0$, so we need to solve the equation $-x^2 - 2x + 3 = 0$. This can be done by factoring:

$$-x^2 - 2x + 3 = 0$$
$$x^2 + 2x - 3 = 0$$
$$(x+3)(x-1) = 0$$
$$x = -3, 1$$

section 4.2   Linear and Quadratic Functions

The vertex form of the function, $f(x) = -(x+1)^2 + 4$, tells us that this quadratic opens downward, has its vertex at $(-1, 4)$, and is neither skinnier nor broader than the prototypical parabola. We now also know that it crosses the *x*-axis at -3 and 1. Putting this all together, we obtain the following graph.

After graphing this equation in the usual manner by selecting **Y=** and then **GRAPH**, individual coordinates can be seen by selecting **TRACE** and using the left and right arrows to move along the graph.

Topic 3: Interlude: Maximization/Minimization Problems

Many applications of mathematics involve determining the value (or values) of the variable *x* that return either the maximum possible value or the minimum possible value of some function $f(x)$. Such problems are called Max/Min problems for short. Examples from business include minimizing a cost function and maximizing a profit function. Examples from physics include maximizing a function that measures the height of a rocket as a function of time and minimizing a function that measures the energy required by a particle accelerator.

We are now in a position to solve Max/Min problems if the function involved is a quadratic. Specifically, if the parabola corresponding to a given quadratic function opens upward, we know that the vertex is the lowest (minimum) point on the graph. Similarly, if the parabola opens downward, the vertex is the highest (maximum) point on the graph. In either case, we locate the vertex by the process of completing the square.

We can even shorten the process of locating the vertex of a parabola by completing the square on the generic quadratic $f(x) = ax^2 + bx + c$ once and for all. The work on the following page should remind you of the derivation of the quadratic formula.

chapter four   Relations, Functions, and Their Graphs

$$f(x) = ax^2 + bx + c$$
$$= a\left(x^2 + \frac{b}{a}x\right) + c$$
$$= a\left(x^2 + \frac{b}{a}x + \frac{b^2}{4a^2}\right) - \frac{b^2}{4a} + c$$
$$= a\left(x + \frac{b}{2a}\right)^2 + \frac{4ac - b^2}{4a}$$

As always, we begin by factoring the leading coefficient a from the first two terms.

To complete the square, we add the square of half of $\frac{b}{a}$ inside the parentheses. This means we also have to subtract a times this quantity outside the parentheses. The vertex is at $\left(-\frac{b}{2a}, \frac{4ac - b^2}{4a}\right)$.

An easier way to remember this formula is to realize that *every* point on the graph of a function f has the form $(x, f(x))$, so the vertex must lie at $\left(-\frac{b}{2a}, f\left(-\frac{b}{2a}\right)\right)$. That is, if $f(x) = ax^2 + bx + c$, then $f\left(-\frac{b}{2a}\right) = \frac{4ac - b^2}{4a}$. You should verify this claim.

example 3

A farmer plans to use 100 feet of spare fencing material to form a rectangular garden plot against the side of a long barn, using the barn as one side of the plot. How should he split up the fencing among the other three sides in order to maximize the area of the garden plot?

Solution:

If we let x represent the length of one side of the plot, as shown in the diagram below, then the dimensions of the plot are x feet by $100 - 2x$ feet. The letter A, for area, is a reasonable name for the function that we wish to maximize in this problem, and so we want to find the maximum possible value of $A(x) = x(100 - 2x)$.

If we multiply out the formula for A, we recognize it as the quadratic function $A(x) = -2x^2 + 100x$. This is a parabola opening downward, so the vertex will be the maximum point on the graph of A.

section 4.2 Linear and Quadratic Functions

Using our work from the previous page, we know that the vertex of A is the ordered pair $\left(-\dfrac{100}{2(-2)}, A\left(-\dfrac{100}{2(-2)}\right)\right)$, or $(25, A(25))$. That is, to maximize area, we should let $x = 25$, and so $100 - 2x = 50$. The resulting maximum possible area, 25×50, or 1250 square feet, is also the value $A(25)$.

exercises

Graph the following linear functions. See Example 1.

1. $f(x) = -5x + 2$
2. $g(x) = \dfrac{3x-2}{4}$
3. $h(x) = -x + 2$
4. $p(x) = -2$
5. $g(x) = 3 - 2x$
6. $r(x) = 2 - \dfrac{x}{5}$
7. $f(x) = -2(1-x)$
8. $a(x) = 3\left(1 - \dfrac{1}{3}x\right) + x$
9. $f(x) = 2 - 4x$
10. $g(x) = \dfrac{2x-8}{4}$
11. $h(x) = 5x - 10$
12. $k(x) = 3x - \dfrac{2+6x}{2}$
13. $m(x) = \dfrac{-x+25}{10}$
14. $q(x) = 1.5x - 1$
15. $w(x) = (x-2) - (2+x)$

Graph the following quadratic functions, locating the vertices and x-intercepts (if any) accurately. See Example 2.

16. $f(x) = (x-2)^2 + 3$
17. $g(x) = -(x+2)^2 - 1$
18. $h(x) = x^2 + 6x + 7$
19. $F(x) = 3x^2 + 2$
20. $G(x) = x^2 - x - 6$
21. $p(x) = -2x^2 + 2x + 12$
22. $q(x) = 2x^2 + 4x + 3$
23. $r(x) = -3x^2 - 1$
24. $s(x) = \dfrac{(x-1)^2}{4}$
25. $m(x) = x^2 + 2x + 4$
26. $n(x) = (x+2)(2-x)$
27. $p(x) = -x^2 + 2x - 5$
28. $f(x) = 4x^2 - 6$
29. $k(x) = 2x^2 - 4x$
30. $i(x) = (x+10)(x-2) + 36$

chapter four   Relations, Functions, and Their Graphs

Match the following functions with their graphs.

31. $f(x) = (8x - 14) - (-17 + 2x)$

32. $f(x) = -x^2 + 2x$

33. $f(x) = x^2 + 7x + 6$

34. $f(x) = 3x - \dfrac{7 + 8x}{3}$

35. $f(x) = \dfrac{6}{2} - \dfrac{2}{8}x$

36. $f(x) = 2\left(2 - \dfrac{8}{5}x\right) + x$

37. $f(x) = \dfrac{x^2 - 8x + 16}{2}$

38. $f(x) = (x - 5)(x + 3) + 16$

a.
b.

c.

e.
f.

g.
h.

section 4.2   Linear and Quadratic Functions

Solve the following maximization/minimization problems by analyzing the appropriate quadratic function. See Example 3.

39. Cindy wants to construct three rectangular dog-training arenas side-by-side, as shown, using a total of 400 feet of fencing. What should the overall length and width be in order to maximize the area of the three combined arenas? (Suggestion: let x represent the width, as shown, and find an expression for the overall length in terms of x.)

40. Among all the pairs of numbers with a sum of 10, find the pair whose product is maximum.

41. Among all rectangles that have a perimeter of 20, find the dimensions of the one whose area is largest.

42. Find the point on the line $2x + y = 5$ that is closest to the origin. (Hint: instead of trying to minimize the distance between the origin and points on the line, minimize the square of the distance.)

43. Among all the pairs of numbers (x, y) such that $2x + y = 20$, find the pair for which the sum of the squares is minimum.

44. A rancher has a rectangular piece of sheet metal that is 20 inches wide by 10 feet long. He plans to fold the metal into a three-sided channel and weld two other sheets of metal to the ends to form a watering trough 10 feet long, as shown. How should he fold the metal in order to maximize the volume of the resulting trough?

10 feet

45. Find a pair of numbers whose product is maximum if the pair must have a sum of 16.

chapter four   Relations, Functions, and Their Graphs

46. Search the Seas cruise ship has a conference room offering unlimited internet access that can hold up to 60 people. Companies can reserve the room for groups of 38 or more. If the group contains 38 people, the company pays $60 per person. The cost per person is reduced by $1 for each person in excess of 38. Find the size of the group that maximizes the income for the owners of the ship and find this income.

47. The back of George's property is a creek. George would like to enclose a rectangular area, using the creek as one side and fencing for the other three sides, to create a pasture for his two horses. If he has 300 feet of material, what is the maximum possible area of the pasture?

48. Find a pair of numbers whose product is maximum if two times the first number plus the second number is 48.

49. The total revenue for Thompson's Studio Apartments is given as the function
$$R(x) = 100x - 0.1x^2,$$
where x is the number of rooms rented. What number of rooms rented produces the maximum revenue?

50. The total revenue of Tran's Machinery Rental is given as the function
$$R(x) = 300x - 0.4x^2,$$
where x is the number of units rented. What number of units rented produces the maximum revenue?

51. The total cost of producing a type of small car is given by
$$C(x) = 9000 - 135x + 0.045x^2,$$
where x is the number of cars produced. How many cars should be produced to incur minimum cost?

52. The total cost of manufacturing a set of golf clubs is given by
$$C(x) = 800 - 10x + 0.20x^2,$$
where x is the number of sets of golf clubs produced. How many sets of golf clubs should be manufactured to incur minimum cost?

53. The owner of a parking lot is going to enclose a rectangular area with fencing, using an existing fence as one of the sides. The owner has 220 feet of new fencing material (which is much less than the length of the existing fence). What is the maximum possible area that the owner can enclose?

For each of the following three problems, use the formula $h(t) = -16t^2 + v_0 t + h_0$ for the height at time t of an object thrown vertically with velocity v_0 (in feet per second) from an initial height of h_0 (in feet).

54. Sitting in a tree, 48 feet above ground level, Sue shoots a pebble straight up with a velocity of 64 feet per second. What is the maximum height attained by the pebble?

55. A ball is thrown upward with a velocity of 48 feet per second from the top of a 144-foot building. What is the maximum height of the ball?

56. A rock is thrown upward with a velocity of 80 feet per second from the top of a 64-foot high cliff. What is the maximum height of the rock?

Use a graphing calculator to graph the following quadratic functions. Then determine the vertex and x-intercepts.

57. $f(x) = 2x^2 - 16x + 31$ **58.** $f(x) = -x^2 - 2x + 3$ **59.** $f(x) = x^2 - 8x - 20$

60. $f(x) = x^2 - 4x$ **61.** $f(x) = 25 - x^2$ **62.** $f(x) = 3x^2 + 18x$

63. $f(x) = x^2 + 2x + 1$ **64.** $f(x) = 3x^2 - 8x + 2$ **65.** $f(x) = -x^2 + 10x - 4$

66. $f(x) = \dfrac{1}{2}x^2 + x - 1$

4.3 Other Common Functions

TOPICS

1. Commonly occurring functions
2. Interlude: variation problems

Topic 1: Commonly Occurring Functions

In Section 4.2, we investigated the behavior of linear and quadratic functions at great length. These are just two types of commonly occurring functions; there are many other functions that arise naturally in the course of solving various problems. In this section, we will briefly explore a variety of other classes of functions, with the aim of building up a useful portfolio of functions with which you are familiar.

Functions of the Form ax^n

We already know what the graph of any function of the form $f(x) = ax$ or $f(x) = ax^2$ looks like, as these are, respectively, simple linear and quadratic functions. What happens to the graphs as we increase the exponent, and consider functions of the form $f(x) = ax^3$, $f(x) = ax^4$, etc.?

Roughly speaking, the behavior of a function of the form $f(x) = ax^n$, where a is a real number and n is a natural number, falls into one of two categories. Consider the graphs in Figure 1:

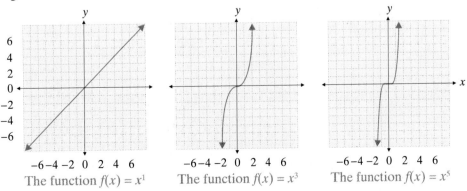

The function $f(x) = x^1$ The function $f(x) = x^3$ The function $f(x) = x^5$

Figure 1: Odd Exponents

chapter four   Relations, Functions, and Their Graphs

The three graphs in Figure 1 show the behavior of $f(x) = x^n$ for the first three odd exponents. Note that in each case, the domain and the range of the function are both the entire set of real numbers; the same is true for higher odd exponents as well. Now, consider the graphs in Figure 2:

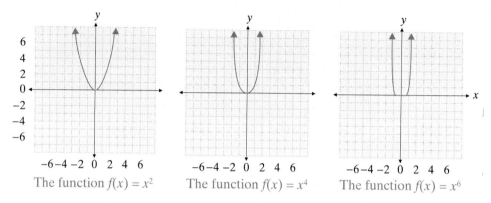

The function $f(x) = x^2$ The function $f(x) = x^4$ The function $f(x) = x^6$

Figure 2: Even Exponents

These three functions are also similar to one another. The first one is the prototypical parabola that we studied in Section 4.2. The other two bear some similarity to parabolas, but are flatter near the origin and rise more steeply for $|x| > 1$. For any function of the form $f(x) = x^n$ where n is an even natural number, the domain is the entire set of real numbers and the range is the interval $[0, \infty)$.

Multiplying a function of the form x^n by a constant a has the effect that we noticed in Section 4.2. If $|a| > 1$, the graph of the function is stretched vertically; if $0 < |a| < 1$, the graph is compressed vertically; and if $a < 0$, the graph is reflected with respect to the x-axis. We can use this knowledge, along with a few specific points, to quickly sketch reasonably accurate graphs of any function of the form $f(x) = ax^n$.

example 1

Sketch the graphs of the following functions.

a. $f(x) = \dfrac{x^4}{5}$ **b.** $g(x) = -x^3$

Solutions:

a.

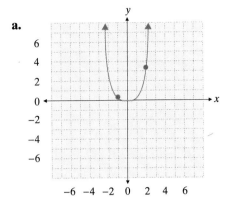

The graph of the function f will have the same basic shape as the function x^4, but compressed a bit vertically because of the factor of $\frac{1}{5}$. To make our sketch reasonably accurate, we can calculate the coordinates of a few points on the graph. The graph to the left illustrates that $f(-1) = \frac{1}{5}$ and that $f(2) = \frac{16}{5}$.

b.

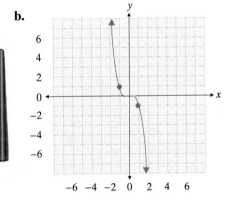

We know that the function g will have the same shape as the function x^3, but reflected with respect to the x-axis because of the factor of -1. The graph at left illustrates this point.

We have also plotted a few points on the graph of g, namely $(-1, 1)$ and $(1, -1)$, just to verify our analysis.

Functions of the Form $\frac{a}{x^n}$

We could also describe the following functions as having the form ax^{-n}, where a is a real number and n is a natural number. Again, the graphs of these functions fall roughly into two categories, as illustrated in Figures 3 and 4.

chapter four   Relations, Functions, and Their Graphs

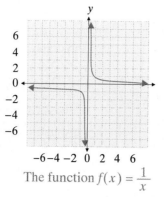
The function $f(x) = \dfrac{1}{x}$

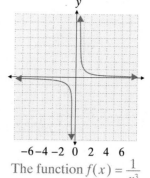
The function $f(x) = \dfrac{1}{x^3}$

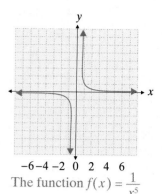
The function $f(x) = \dfrac{1}{x^5}$

Figure 3: Odd Exponents

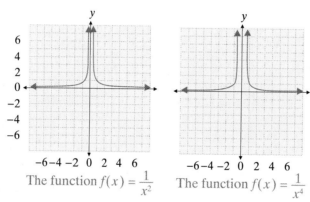
The function $f(x) = \dfrac{1}{x^2}$

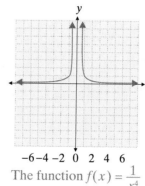
The function $f(x) = \dfrac{1}{x^4}$

The function $f(x) = \dfrac{1}{x^6}$

Figure 4: Even Exponents

As with functions of the form ax^n, increasing the exponent on functions of the form $\dfrac{a}{x^n}$ sharpens the curve of the graph near the origin. Note that the domain of any function of the form $f(x) = \dfrac{a}{x^n}$ is $(-\infty, 0) \cup (0, \infty)$, but that the range depends on whether n is even or odd. When n is odd, the range is also $(-\infty, 0) \cup (0, \infty)$, and when n is even the range is $(0, \infty)$.

example 2

Sketch the graph of the function $f(x) = -\dfrac{1}{4x}$.

Solution:

The graph of the function f is similar to that of the function $\dfrac{1}{x}$, with two differences. We obtain the formula $-\dfrac{1}{4x}$ by multiplying $\dfrac{1}{x}$ by $-\dfrac{1}{4}$, a negative number between -1 and 1. So one difference is that the graph of f is the reflection of $\dfrac{1}{x}$ with respect to the x-axis. The other difference is that the graph of f is compressed vertically, compared to $\dfrac{1}{x}$.

section 4.3 **Other Common Functions**

With the above in mind, we can calculate the coordinates of a few points (such as $\left(-\frac{1}{4}, 1\right)$ and $\left(1, -\frac{1}{4}\right)$) and sketch the graph of f as shown below.

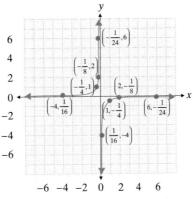

Functions of the Form $ax^{\frac{1}{n}}$

Using radical notation, these are functions of the form $a\sqrt[n]{x}$, where a is again a real number and n is a natural number. Square root and cube root functions, in particular, are commonly seen in mathematics.

Functions of this form again fall into one of two categories, depending on whether n is odd or even. To begin with, note that the domain and range are both the entire set of real numbers when n is odd, and that both are the interval $[0, \infty)$ when n is even. Figures 5 and 6 illustrate the two basic shapes of functions of this form.

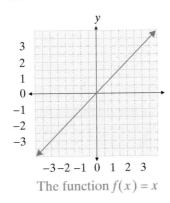
The function $f(x) = x$

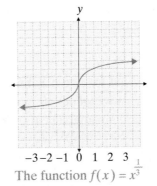
The function $f(x) = x^{\frac{1}{3}}$

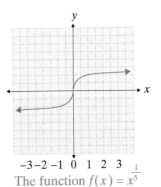
The function $f(x) = x^{\frac{1}{5}}$

Figure 5: Odd Roots

chapter four   Relations, Functions, and Their Graphs

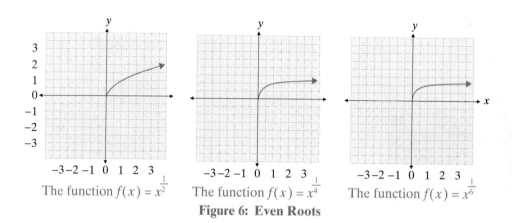

Figure 6: Even Roots

At this point, you may be thinking that the graphs in Figures 5 and 6 appear familiar. The shapes in Figure 5 are the same as those seen in Figure 1, but rotated by 90 degrees and reflected with respect to the x-axis. Similarly, the shapes in Figure 6 bear some resemblance to those in Figure 2, except that half of the graphs appear to have been erased. This resemblance is no accident, and is perhaps not surprising given that n^{th} roots undo n^{th} powers. We will explore this observation in much more detail in Section 4.6.

The Absolute Value Function

The basic absolute value function is $f(x) = |x|$. Note that for any value of x, $f(x)$ is non-negative, so the graph of f should lie on or above the x-axis. One way to determine its exact shape is to review the definition of absolute value:

$$|x| = \begin{cases} x & \text{if } x \geq 0 \\ -x & \text{if } x < 0 \end{cases}$$

This means that for non-negative values of x, $f(x)$ is a linear function with a slope of 1, and for negative values of x, $f(x)$ is a linear function with a slope of -1. Both linear functions have a y-intercept of 0, so the complete graph of f is as shown in Figure 7.

section 4.3   Other Common Functions

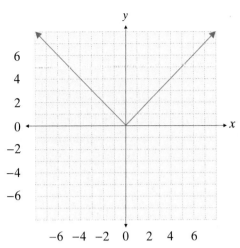

Figure 7: The Absolute Value Function

The effect of multiplying $|x|$ by a real number a is what we have come to expect: if $|a| > 1$, the graph is stretched vertically; if $0 < |a| < 1$, the graph is compressed vertically; and if a is negative, the graph is reflected with respect to the x-axis.

example 3

Sketch the graph of the function $f(x) = -2|x|$.

Solution:

The graph of f will be a stretched version of $|x|$, reflected with respect to the x-axis. As always, we can plot a few points to verify that our reasoning is correct. In the graph below, we have plotted the values of $f(-4)$ and $f(2)$.

To graph an absolute value, select **Y=** and enter the equation. To enter the absolute value, select **MATH** and then use the right arrow to select **Num**. Press **1**, and you will see **abs** in your equation.

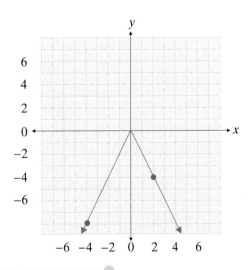

chapter four Relations, Functions, and Their Graphs

The Greatest Integer Function

The greatest integer function, $f(x) = [\![x]\!]$, is a function commonly encountered in computer science applications. It is defined as follows: the **greatest integer of x** is the largest integer less than or equal to x. For instance, $[\![4.3]\!] = 4$ and $[\![-2.9]\!] = -3$ (note that -3 is the largest integer to the left of -2.9 on the real number line).

Careful study of the greatest integer function reveals that its graph must consist of intervals where the function is constant, and that these portions of the graph must be separated by discrete "jumps," or breaks, in the graph. For instance, any value for x chosen from the interval $[1, 2)$ results in $f(x) = 1$, but $f(2) = 2$. Similarly, any value for x chosen from the interval $[-3, -2)$ results in $f(x) = -3$, but $f(-2) = -2$.

Our graph of the greatest integer function must somehow indicate this repeated pattern of jumps. In cases like this, it is conventional to use an open circle on the graph to indicate that the function is either undefined at that point or is defined to be another value. Closed circles are used to emphasize that a certain point really does lie on the graph of the function. With these conventions in mind, the graph of the greatest integer function appears in Figure 8.

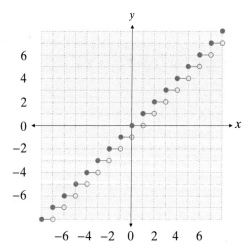

Figure 8: The Greatest Integer Function

Piecewise-Defined Function

The last category of functions that we will discuss in this section is that of piecewise-defined functions. A **piecewise-defined** function is a function defined in terms of two or more formulas, each valid for its own unique portion of the real number line. In evaluating a piecewise-defined function f at a certain value for x, it is important to correctly identify which formula is valid for that particular value.

We have worked with one piecewise-defined function already; in evaluating the absolute value of x, we use one formula if x is greater than or equal to 0 and a different formula if x is less than 0. Keep the absolute value function in mind as you study the piecewise-defined function in *Example 4*.

example 4

Sketch the graph of the function $f(x) = \begin{cases} -2x - 2 & \text{if } x \leq -1 \\ x^2 & \text{if } x > -1 \end{cases}$.

Solution:

The function f is a linear function on the interval $(-\infty, -1]$ and a quadratic function on the interval $(-1, \infty)$. To graph f, we graph each portion separately, making sure that each formula is applied only on the appropriate interval. The complete graph appears below, with the points $f(-4) = 6$ and $f(2) = 4$ noted in particular. Also note the use of a closed circle at $(-1, 0)$ to emphasize that this point is part of the graph, and the use of an open circle at $(-1, 1)$ to indicate that this point is not part of the graph. That is, the value of $f(-1)$ is 0, not 1.

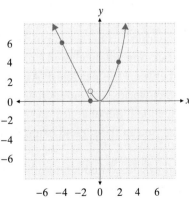

chapter four   Relations, Functions, and Their Graphs

technology note

As their name implies, graphing calculators are very useful for sketching graphs of functions, though care must be used in interpreting the results. Computer algebra systems like *Mathematica* are also very good at graphing, and generally provide better quality sketches. The command in *Mathematica* for graphing a function is `Plot`, and its use in sketching the graph of $f(x) = x^3 - 2x^2 + 1$ is illustrated below.

Consult your user's manual for instructions on how to use your graphing calculator or software package to graph functions, or see Appendix A for more guidance on using Mathematica.

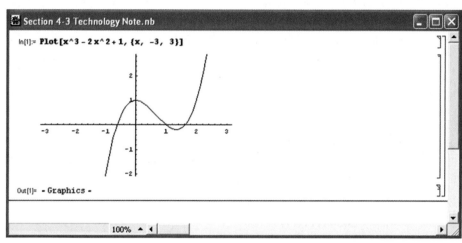

Topic 2:   Interlude: Variation Problems

A number of natural phenomena exhibit the mathematical property of variation: one quantity varies (or changes) as a result of a change in another quantity. One example is the electrostatic force of attraction between two oppositely charged particles, which varies in response to the distance between the particles. Another example is the distance traveled by a falling object, which varies as time increases. Of course, the principle underlying variation is that of functional dependence; in the first example, the force of attraction is a function of distance, and in the second example the distance traveled is a function of time.

We have now gained enough familiarity with functions that we can define the two most common forms of variation: direct and inverse.

Direct and Inverse Variation

We say that **y varies directly as the nth power of x** (or that **y is directly proportional to the nth power of x**) if there is a non-zero constant k (called the **constant of proportionality**) such that
$$y = kx^n.$$

We say that **y varies inversely as the nth power of x** (or that **y is inversely proportional to the nth power of x**) if there is a non-zero constant k such that
$$y = \frac{k}{x^n}.$$

Many variation problems involve determining exactly what the constant of proportionality is in a given situation. This can be easily done if enough information is given about how the various quantities in the problem vary with respect to one another, and once k is determined many other questions can be answered. The following examples illustrate the solution of typical variation problems.

example 5

The weight of a person, relative to the Earth, is inversely proportional to the square of the person's distance from the center of the Earth. Using a radius for the Earth of 6370 kilometers, how much does a 180 pound man weigh when flying in a jet 9 kilometers above the Earth's surface?

Solution:

If we let W stand for the weight of a person and let d stand for the distance between the person and the Earth's center, the first sentence tells us that
$$W = \frac{k}{d^2}.$$

continued on next page...

chapter four Relations, Functions, and Their Graphs

The second sentence gives us enough information to determine k. Namely, we know that $W = 180$ (pounds) when $d = 6370$ (kilometers). Solving the equation for k and substituting in the values that we know, we obtain

$$k = Wd^2 = (180)(6370)^2 \approx 7.3 \times 10^9.$$

When the man is 9 kilometers above the Earth's surface, $d = 6379$, so the man's weight while flying is

$$W = \frac{(180)(6370)^2}{(6379)^2}$$

$$\approx 179.49 \text{ pounds.}$$

Flying is not, therefore, a terribly effective way to lose weight.

example 6

Hooke's Law says that the force exerted by the spring in a spring scale varies directly with the distance that the spring is stretched. If a 5 pound mass suspended on a spring scale stretches the spring 2 inches, how far will a 13 pound mass stretch it?

Solution:

The first sentence tells us that $F = kx$, where F represents the force exerted by the spring and x represents the distance that the spring is stretched. When a mass is suspended on a spring scale (and is stationary), the force exerted upward by the spring must equal the force downward due to gravity, so the spring exerts a force of 5 pounds when a 5 pound mass is suspended from it. So the second sentence tells us that

$$5 = 2k,$$

or $k = \dfrac{5}{2}$. We can now answer the question:

$$13 = \left(\frac{5}{2}\right)x$$

$$\frac{26}{5} = x$$

So the spring stretches 5.2 inches when a 13 pound mass is suspended from it.

exercises

Sketch the graphs of the following functions. Pay particular attention to intercepts, if any, and locate these accurately. See Examples 1 through 4.

1. $f(x) = -x^3$
2. $g(x) = 2x^2$
3. $F(x) = \sqrt{x}$
4. $h(x) = \dfrac{1}{x}$
5. $p(x) = -\dfrac{2}{x}$
6. $q(x) = -\sqrt[3]{x}$
7. $G(x) = -|x|$
8. $k(x) = \dfrac{1}{x^3}$
9. $f(x) = 2[\![x]\!]$
10. $P(x) = -[\![x]\!]$
11. $G(x) = \dfrac{\sqrt{x}}{2}$
12. $H(x) = 0.5\sqrt[3]{x}$
13. $r(x) = 3|x|$
14. $p(x) = \dfrac{-1}{x^2}$
15. $W(x) = \dfrac{x^4}{16}$
16. $k(x) = \dfrac{x^3}{9}$
17. $h(x) = 2\sqrt[3]{x}$
18. $S(x) = \dfrac{4}{x^2}$
19. $d(x) = 2x^5$
20. $f(x) = -x^2$
21. $m(x) = \left[\!\!\left[\dfrac{x}{2}\right]\!\!\right]$
22. $r(x) = \dfrac{\sqrt[3]{x}}{3}$
23. $s(x) = -2|x|$
24. $t(x) = \dfrac{x^6}{4}$

25. $f(x) = \begin{cases} 3-x & \text{if } x < -2 \\ \sqrt[3]{x} & \text{if } x \geq -2 \end{cases}$

26. $g(x) = \begin{cases} -x^2 & \text{if } x \leq 1 \\ x^2 & \text{if } x > 1 \end{cases}$

27. $h(x) = \begin{cases} -|x| & \text{if } x < 2 \\ [\![x]\!] & \text{if } x \geq 2 \end{cases}$

28. $r(x) = \begin{cases} \dfrac{1}{x} & \text{if } x < 1 \\ -x & \text{if } x > 1 \end{cases}$

29. $p(x) = \begin{cases} x+1 & \text{if } x < -2 \\ x^3 & \text{if } -2 \leq x < 3 \\ -1-x & \text{if } x \geq 3 \end{cases}$

30. $q(x) = \begin{cases} -1 & \text{if } x \in \mathbb{Z} \\ 1 & \text{if } x \notin \mathbb{Z} \end{cases}$

31. $s(x) = \begin{cases} \dfrac{x^2}{3} & \text{if } x < 0 \\ -\dfrac{x^2}{3} & \text{if } x \geq 0 \end{cases}$

32. $t(x) = \begin{cases} x^4 & \text{if } x \leq 1 \\ [\![x]\!] & \text{if } x > 1 \end{cases}$

33. $u(x) = \begin{cases} [\![x]\!] & \text{if } x \leq 1 \\ 2x-2 & \text{if } x > 1 \end{cases}$

34. $v(x) = \begin{cases} x^2 & \text{if } -1 \leq x \leq 1 \\ |x| & \text{if } x < -1 \text{ or } x > 1 \end{cases}$

chapter four Relations, Functions, and Their Graphs

35. $M(x) = \begin{cases} x & \text{if } x \in \mathbb{Z} \\ -x & \text{if } x \notin \mathbb{Z} \end{cases}$

36. $N(x) = \begin{cases} x^2 & \text{if } x \in \mathbb{Z} \\ [\![x]\!] & \text{if } x \notin \mathbb{Z} \end{cases}$

Match the following functions to their graphs.

37. $f(x) = -2x^4$

38. $f(x) = -\dfrac{7}{9x^4}$

39. $f(x) = -\dfrac{7\sqrt[3]{x}}{3}$

40. $f(x) = -\dfrac{8}{9}|x|$

41. $f(x) = -4\left[\!\left[\dfrac{x}{4}\right]\!\right]$

42. $f(x) = \dfrac{3}{7}|x|$

43. $f(x) = \begin{cases} -4x - 12 & \text{if } x \leq -3 \\ \dfrac{5}{10}x^2 & \text{if } x > -3 \end{cases}$

44. $f(x) = \begin{cases} \dfrac{-1}{3}|x| & \text{if } x < 2 \\ \left[\!\left[\dfrac{x}{2}\right]\!\right] & \text{if } x \geq 2 \end{cases}$

a.

b.

c.

d.

e.

f.

g.

h.

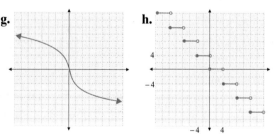

section 4.3   Other Common Functions

Solve the following variation problems. See Examples 5 and 6.

45. The distance that an object falls from rest, when air resistance is negligible, varies directly as the square of the time. A stone dropped from rest travels 144 feet in the first 3 seconds. How far does it travel in the first 4 seconds?

46. A record store manager observes that the number of CD's sold seems to vary inversely as the price per CD. If the store sells 840 CD's per week when the price per CD is $15.99, how many does he expect to sell if he lowers the price to $14.99? Round your answer to the nearest CD.

47. Suppose that y varies directly as the square root of x, and that $y = 36$ when $x = 16$. What is y when $x = 20$?

48. Suppose that y varies inversely as the cube of x, and that $y = 0.005$ when $x = 10$. What is y when $x = 5$?

49. Suppose that y varies inversely as the square of x, and that $y = 8$ when $x = 6$. What is y when $x = 20$?

It is also possible for one variable to depend on several other variables. In some situations, one variable varies directly with respect to some variables and inversely with respect to others. For instance, the force F between two bodies of mass m_1 and mass m_2 varies directly as the product of the masses and inversely as the square of the distance between the masses: $F = \dfrac{km_1m_2}{d^2}$.

50. A person's Body Mass Index (BMI) is used by physicians to determine if a patient's weight falls within reasonable guidelines relative to the patient's height. The BMI varies directly as a person's weight in pounds and inversely as the square of a person's height in inches. Rounded to the nearest whole number, a 6-foot tall man weighing 180 pounds has a BMI of 24. What is the BMI of a woman weighing 120 pounds whose height is 5 feet 4 inches?

51. The force necessary to keep a car from skidding as it travels along a circular arc varies directly as the product of the weight of the car and the square of the car's speed, and inversely as the radius of the arc. If it takes 241 pounds of force to keep a 2200 pound car moving 35 miles per hour on an arc whose radius is 750 feet, how many pounds of force would be required if the car were to travel 40 miles per hour?

52. If a beam of width w, height h, and length l is supported at both ends, the maximum load that the beam can hold varies directly as the product of the width and the square of the height, and inversely as the length. A given beam 10 meters long with a width of 10 centimeters and a height of 5 centimeters can hold a load of 200 kilograms when the beam is supported at both ends. If the supports are moved inward so that the effective length of the beam is shorter, the beam can support more load. What should the distance between the supports be if the beam has to hold a load of 300 kilograms?

53. In a simple electric circuit connecting a battery and a light bulb, the current I varies directly with the voltage V but inversely with the resistance R. When a 1.5 volt battery is connected to a light bulb with resistance 0.3 ohms (Ω), the current that travels through the circuit is 5 amps. Find the current if the same light bulb is connected to a 6 volt battery.

54. The amount of time it takes for water to flow down a drainage pipe is inversely proportional to the square of the radius of the pipe. If a pipe of radius 1 inch can empty a pool in 25 hours, find the radius of a pipe that would allow the pool to drain completely in 16 hours.

55. Hooke's Law says that the force exerted by the spring in a spring scale varies directly with the distance that the spring is stretched. If a 32 pound mass suspended on a spring scale stretches the spring 17 inches, how far will a 37 pound mass stretch the spring?

56. The gravitational force, F, between an object and the Earth is inversely proportional to the square of the distance from the object and the center of the Earth. If an astronaut weighs 193 pounds on the surface of the Earth, what will this astronaut weigh 1000 miles above the Earth? Assume that the radius of the Earth is 4000 miles.

57. The volume of a gas in a container varies inversely as the pressure on the gas. If a gas has a volume of 252 cubic inches under a pressure of 5 pounds per square inch, what will its volume be if the pressure is increased to 6 pounds per square inch?

technology exercises

Use a graphing calculator or a computer algebra system such as Mathematica to graph the following functions. Experiment by instructing your calculator or software to draw different portions of the plane until you obtain a sketch that seems to capture the meaningful parts of the graph. See the Technology Note after Example 4 for guidance.

58. $f(x) = 10x^5 - x^3$

59. $g(x) = x^5 + x^2$

60. $f(x) = x^3 - 5x^2 + x$

61. $g(x) = \sqrt{x} - x^2$

62. $f(x) = \sqrt{x} + 3x - 1$

63. $g(x) = x^4 - 3x^3 + 2$

chapter four   Relations, Functions, and Their Graphs

4.4 Transformations of Functions

TOPICS

1. Shifting, stretching, and reflecting graphs
2. Symmetry of functions and equations
3. Intervals of monotonicity

Topic 1: Shifting, Stretching, and Reflecting Graphs

Much of the material in this section was introduced in Section 4.2, in our discussion of quadratic functions. You may want to review the ways in which the prototypical quadratic function $f(x) = x^2$ can be shifted, stretched, and reflected as you work through the more general ideas here.

Horizontal Shifting

Let $f(x)$ be a function whose graph is known, and let h be a fixed real number. If we replace x in the definition of f by $x - h$, we obtain a new function $g(x) = f(x - h)$. The graph of g is the same shape as the graph of f, but shifted to the right by h units if $h > 0$ and shifted to the left by h units if $h < 0$.

caution

It is easy to forget that the minus sign in the expression $x - h$ is critical. It may help to remember a few specific examples: replacing x with $x - 5$ shifts the graph 5 units to the *right*, since 5 is positive. Replacing x with $x + 5$ shifts the graph 5 units to the *left*, since we have actually replaced x with $x - (-5)$. With practice, knowing the effect that replacing x with $x - h$ has on the graph of a function will become natural.

section 4.4   Transformations of Functions

example 1

Sketch the graphs of the following functions.

a. $f(x)=(x+2)^3$ **b.** $g(x)=|x-4|$

Solutions:

a.

The screen above is the graph of both $y = x^3$ and $y = (x+2)^3$.

The shape of $(x+2)^3$ is the same as the shape of x^3, since one expression is obtained from the other by replacing x by $x + 2$. We simply draw the basic cubic shape (the shape of $y = x^3$, shown in teal) shifted to the left by 2 units. Note, for example, that $(-2, 0)$ is one point on the graph.

b.

The graph of $g(x)=|x-4|$ has the same shape as the graph of the standard absolute value function (shown in teal), but shifted to the right by 4 units. Note, for example, that $(4, 0)$ lies on the graph of g.

Vertical Shifting

Let $f(x)$ be a function whose graph is known, and let k be a fixed real number. The graph of the function $g(x)=f(x)+k$ is the same shape as the graph of f, but shifted upward if $k > 0$ and downward if $k < 0$.

305

chapter four   Relations, Functions, and Their Graphs

The effect of adding a constant k to a function can be most easily remembered by recalling that every point on the graph of a function f has the form $(x, f(x))$, so adding k shifts every point $(x, f(x))$ to $(x, f(x)+k)$. These new points are above the original points if k is positive and below the original points if k is negative.

example 2

Sketch the graphs of the following functions.

a. $f(x) = \dfrac{1}{x} + 3$

b. $g(x) = \sqrt[3]{x} - 2$

Solutions:

a.

The screen above is the graph of both
$y = \dfrac{1}{x}$ and
$y = \dfrac{1}{x} + 3$.

The graph of $f(x) = \dfrac{1}{x} + 3$ is the graph of $y = \dfrac{1}{x}$ (shown in teal) shifted up 3 units. Note that this doesn't affect the domain: the domain of f is $(-\infty, 0) \cup (0, \infty)$, the same as the domain of $y = \dfrac{1}{x}$. However, the range is affected. The range of f is $(-\infty, 3) \cup (3, \infty)$.

b.

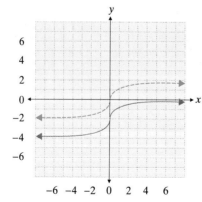

To graph $g(x) = \sqrt[3]{x} - 2$, we shift the graph of $y = \sqrt[3]{x}$ (shown in teal) down by 2 units.

306

section 4.4 Transformations of Functions

Reflecting With Respect to the Axes

Let $f(x)$ be a function whose graph is known.
1. The graph of the function $g(x) = -f(x)$ is the reflection of the graph of f with respect to the x-axis.
2. The graph of the function $g(x) = f(-x)$ is the reflection of the graph of f with respect to the y-axis.

In other words, a function is reflected with respect to the x-axis by multiplying the entire function by -1, and reflected with respect to the y-axis by replacing x with $-x$.

example 3

Sketch the graphs of the following functions.

a. $f(x) = -x^2$ **b.** $g(x) = \sqrt{-x}$

Solutions:

a.

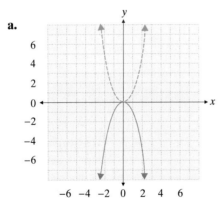

To graph $f(x) = -x^2$, we turn the graph of the prototypical parabola $y = x^2$ upside down. Note that the domain is still the entire real line, but the range of f is the interval $(-\infty, 0]$.

b.

To graph $g(x) = \sqrt{-x}$, we reflect the graph of $y = \sqrt{x}$ (shown in teal) with respect to the y-axis. Note that this changes the domain, but not the range. The domain of g is the interval $(-\infty, 0]$ and the range is $[0, \infty)$.

chapter four   Relations, Functions, and Their Graphs

Stretching and Compressing

Let $f(x)$ be a function whose graph is known, and let a be a positive real number.
1. The graph of the function $g(x) = a f(x)$ is stretched vertically compared to the graph of f if $a > 1$.
2. The graph of the function $g(x) = a f(x)$ is compressed vertically compared to the graph of f if $0 < a < 1$.

If the function g is obtained from the function f by multiplying f by a negative real number, think of the number as the product of -1 and a positive real number (namely, its absolute value). The box above tells us what multiplication by a positive constant does to a graph, and we already know that multiplying a function by -1 reflects the graph with respect to the x-axis.

example 4

Sketch the graphs of the following functions.

a. $f(x) = \dfrac{\sqrt{x}}{10}$　　　　**b.** $g(x) = 5|x|$

Solutions:

a.

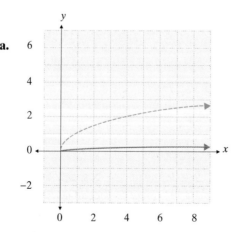

The graph of the function f is compressed considerably compared to the graph of $\sqrt{x}$ (shown in teal) because of the factor of $\dfrac{1}{10}$. The shape is similar to the shape of $\sqrt{x}$ but all of the second coordinates have been multiplied by the factor of $\dfrac{1}{10}$, and are consequently smaller.

308

b.

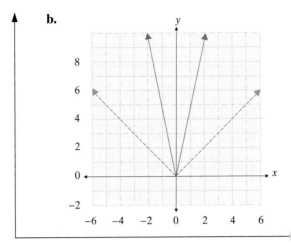

In contrast to the last example, the graph of $g(x) = 5|x|$ is stretched compared to the standard absolute value function. Every second coordinate has been multiplied by a factor of 5 and is consequently larger.

We can now put all of the above together, and consider functions that have been derived through a sequence of transformations from simpler functions. To do so, consider the following:

Order of Transformations

If a function g has been obtained from a simpler function f through a number of transformations, g can usually be understood by looking for the transformations in this order:

1. horizontal shifts
2. stretching and compressing
3. reflections
4. vertical shifts

Consider, for example, the function $g(x) = -2\sqrt{x+1} + 3$. The function g has been "built up" from the basic square root function through a variety of transformations. First, $\sqrt{x}$ has been transformed into $\sqrt{x+1}$ by replacing x with $x + 1$, and we know that this corresponds graphically to a shift leftward by 1 unit. Next, the function $\sqrt{x+1}$ has been multiplied by 2 to get the function $2\sqrt{x+1}$, and we know that this has the effect of stretching the graph of $\sqrt{x+1}$ vertically. The function $2\sqrt{x+1}$ has then been multiplied by -1, giving us $-2\sqrt{x+1}$, and the graph of this is the reflection of $2\sqrt{x+1}$ with respect to the x-axis. Finally, the constant 3 has been added to $-2\sqrt{x+1}$, shifting the entire graph upward by 3 units. These transformations are illustrated in order in Figure 1, culminating in the graph of $g(x) = -2\sqrt{x+1} + 3$.

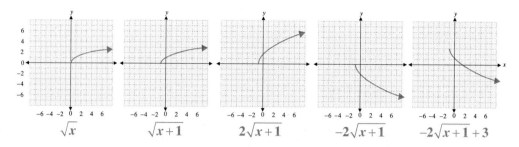

Figure 1: Building the Graph of $g(x) = -2\sqrt{x+1} + 3$

example 5

Sketch the graph of the function $f(x) = \dfrac{1}{2-x}$.

Solution:

A less complex function that f is similar to is $\dfrac{1}{x}$. If we replace x by $x + 2$ (shifting the graph 2 units to the left), we obtain the function $\dfrac{1}{x+2}$, which is closer to what we want. If we now replace x by $-x$, we have $\dfrac{1}{-x+2}$, which is the same as f. This last transformation reflects the graph of $\dfrac{1}{x+2}$ with respect to the y-axis. The entire sequence of transformations is shown below, ending with the graph of f.

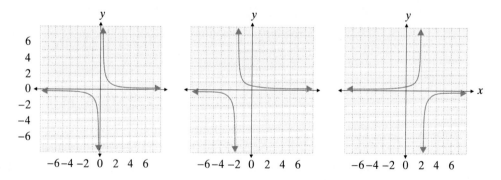

Note: An alternative approach to graphing $f(x) = \dfrac{1}{2-x}$ is to rewrite the function in the form $f(x) = -\dfrac{1}{x-2}$. In this form, the graph of f is the graph of $\dfrac{1}{x}$ shifted two units to the right, and then reflected with respect to the x-axis. The result is the same, as you should verify.

Topic 2: Symmetry of Functions and Equations

For some functions, replacing x with $-x$ has no effect, in the sense that the resulting function is identical to the original. We know that replacing x with $-x$ reflects the graph of a function with respect to the y-axis, so if f is one of these functions, the graph of f must be symmetric with respect to the y-axis. Algebraically, this means that $f(-x) = f(x)$ for all x in the domain of f.

y-axis Symmetry

> The graph of a function f has **y-axis symmetry**, or is **symmetric with respect to the y-axis**, if $f(-x) = f(x)$ for all x in the domain of f. Such functions are called **even functions**.

Functions whose graphs have y-axis symmetry are called even functions because polynomial functions with only even exponents form one large class of functions with this property. For instance, consider the function $f(x) = 7x^8 - 5x^4 + 2x^2 - 3$. This function is a polynomial of four terms, all of which have even degree (8, 4, 2, and 0, respectively). If we replace x with $-x$ and simplify the result, we obtain the function f again:

$$f(-x) = 7(-x)^8 - 5(-x)^4 + 2(-x)^2 - 3$$
$$= 7x^8 - 5x^4 + 2x^2 - 3$$
$$= f(x)$$

Be aware, however, that such polynomial functions are not the only even functions. We will see many more examples as we proceed.

There is another class of functions for which replacing x with $-x$ results in the exact opposite of the original function. That is, $f(-x) = -f(x)$ for all x in the domain. What does this mean geometrically? Suppose f is such a function, and that $(x, f(x))$ is a point on the graph of f. If we change the sign of both coordinates, we obtain a new point that is the reflection through the origin of the original. For instance, if $(x, f(x))$ lies in the first quadrant, $(-x, -f(x))$ lies in the third, and if $(x, f(x))$ lies in the second quadrant, $(-x, -f(x))$ lies in the fourth. But since $f(-x) = -f(x)$, the point $(-x, -f(x))$ can be rewritten as $(-x, f(-x))$. Written in this form, we know that $(-x, f(-x))$ is a point on the graph of f, since *any* point of the form $(?, f(?))$ lies on the graph of f. So a function with the property $f(-x) = -f(x)$ has a graph that is symmetric with respect to the origin.

Origin Symmetry

> The graph of a function f has **origin symmetry**, or is **symmetric with respect to the origin**, if $f(-x) = -f(x)$ for all x in the domain of f. Such functions are called **odd functions**.

As you might guess, such functions are called odd because polynomial functions with only odd exponents serve as simple examples. For instance, the function $f(x) = -2x^3 + 8x$ is odd:

$$\begin{aligned} f(-x) &= -2(-x)^3 + 8(-x) \\ &= -2(-x^3) + 8(-x) \\ &= 2x^3 - 8x \\ &= -f(x) \end{aligned}$$

As far as functions are concerned, y-axis and origin symmetry are the two principal types of symmetry. You may have a sense, though, that something is missing. What about x-axis symmetry? It is certainly possible to draw a graph that displays x-axis symmetry; but unless the graph lies entirely on the x-axis, such a graph cannot represent a function. (Why not? Draw a variety of graphs that are symmetric with respect to the x-axis and convince yourself that they cannot be functions.)

This brings us back to relations. Recall that any equation in x and y defines a relation between the two variables. There are three principal types of symmetry that equations can possess.

Symmetry of Equations

> We say that an equation in x and y is symmetric with respect to:
>
> 1. the **y-axis** if replacing x with $-x$ results in an equivalent equation.
> 2. the **x-axis** if replacing y with $-y$ results in an equivalent equation.
> 3. the **origin** if replacing x with $-x$ and y with $-y$ results in an equivalent equation.

section 4.4 Transformations of Functions

Noting the symmetry of a function or an equation can serve as a useful aid in graphing. For instance, in graphing an even function it is only necessary to graph the part to the right of the y-axis, as the left half of the graph is the reflection of the right half with respect to the y-axis. Similarly, if a function is odd, the left half of its graph is the reflection of the right half through the origin.

example 6

Sketch the graphs of the following relations.

a. $f(x) = \dfrac{1}{x^2}$ **b.** $g(x) = x^3 - x$ **c.** $x = y^2$

Solutions:

a.

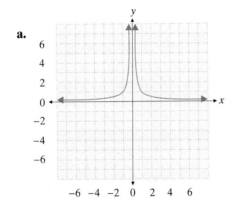

This relation is actually a function, one that we have already graphed in Section 4.3. Note that it is indeed an even function:

$$f(-x) = \dfrac{1}{(-x)^2}$$
$$= \dfrac{1}{x^2}$$
$$= f(x)$$

b.

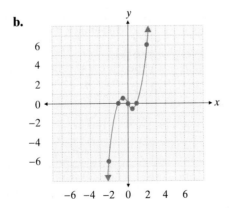

We do not quite have the tools yet to graph general polynomial functions, but $g(x) = x^3 - x$ can be done. For one thing, g is odd: $g(-x) = -g(x)$ (as you should verify). If we now calculate a few values, such as $g(0) = 0$, $g\left(\dfrac{1}{2}\right) = -\dfrac{3}{8}$, $g(1) = 0$, and $g(2) = 6$, and reflect these through the origin, we begin to get a good idea of the shape of g.

continued on next page...

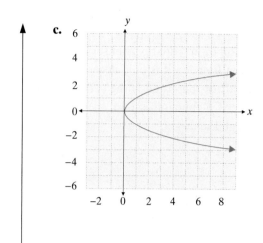

c.

The equation $x = y^2$ does not represent a function, but it is a relation in x and y that has x-axis symmetry. If we replace y with $-y$ and simplify the result, we obtain the original equation:

$$x = (-y)^2$$
$$x = y^2$$

The upper half of the graph is the function $y = \sqrt{x}$, so drawing this and its reflection gives us the complete graph of $x = y^2$.

Summary of Symmetry

To tie together the types of symmetry that have been described, it's helpful to consider a graph of a relation between x and y without regard, initially, to whether the graph comes from an equation or a function (or neither). The first column in the table below summarizes the behavior of a graph in the Cartesian plane if it possesses any of the three types of symmetry defined so far. If the graph is of an equation in x and y, the algebraic method in the second column can be used to identify the symmetry. The third column contains the algebraic method used to identify the type of symmetry if the graph is that of a function $f(x)$. Finally, the fourth column contains an example of each type of symmetry.

A graph is symmetric with respect to:	If the graph is of an equation in x and y, the equation is symmetric with respect to:	If the graph is of a function $f(x)$, the function is symmetric with respect to:	Example:
The y-axis if whenever the point (x, y) is on the graph, the point $(-x, y)$ is also on the graph.	The y-axis if replacing x with $-x$ results in an equivalent equation.	The y-axis if $f(-x) = f(x)$. We say the function is even.	$(-x, y)$ (x, y)
The x-axis if whenever the point (x, y) is on the graph, the point $(x, -y)$ is also on the graph.	The x-axis if replacing y with $-y$ results in an equivalent equation.	Not applicable (unless the graph consists only of points on the x-axis.)	(x, y) $(x, -y)$

A graph is symmetric with respect to:	If the graph is of an equation in x and y, the equation is symmetric with respect to:	If the graph is of a function $f(x)$, the function is symmetric with respect to:	Example:
The origin if whenever the point (x, y) is on the graph, the point $(-x, -y)$ is also on the graph.	The origin if replacing x with $-x$ and y with $-y$ results in an equivalent equation.	The origin if $f(-x) = -f(x)$. We say the function is odd.	*(graph showing points (x, y) and $(-x, -y)$ symmetric about origin)*

Topic 3: Intervals of Monotonicity

In many applications, it is useful to identify the intervals of the x-axis for which a function f is increasing in value, decreasing in value, or remaining constant. In this context, we say that these are intervals on which f is **monotone**. In practice, we are usually interested in identifying the largest open intervals of monotonicity (recall that open intervals do not include their endpoints).

Increasing, Decreasing, and Constant

We say that a function f is:

1. **Increasing on an interval** if, for any x_1 and x_2 in the interval with $x_1 < x_2$, it is the case that $f(x_1) < f(x_2)$.
2. **Decreasing on an interval** if, for any x_1 and x_2 in the interval with $x_1 < x_2$, it is the case that $f(x_1) > f(x_2)$.
3. **Constant on an interval** if, for any x_1 and x_2 in the interval, it is the case that $f(x_1) = f(x_2)$.

Determining the intervals of monotonicity of a function is a task that can be quite demanding, and in many cases is best tackled with the tools of calculus. We will look at some problems now in which algebra and our intuition will be sufficient.

example 7

Determine the open intervals of monotonicity of the function $f(x) = (x-2)^2 - 1$.

Solution:

We know that the graph of f is the prototypical parabola shifted 2 units to the right and down 1 unit, as shown on the following page.

continued on next page...

chapter four   Relations, Functions, and Their Graphs

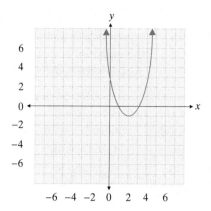

From the graph, we can see that f is decreasing on the interval $(-\infty, 2)$ and increasing on the interval $(2, \infty)$. Remember that these are intervals of the x-axis: if x_1 and x_2 are any two points in the interval $(-\infty, 2)$, with $x_1 < x_2$, then $f(x_1) > f(x_2)$. In other words, f is falling on this interval as we scan the graph from left to right. On the other hand, f is rising on the interval $(2, \infty)$ as we scan the graph from left to right.

example 8

The water level of a certain river varied over the course of a year as follows. In January, the level was 13 feet. From that level, the water increased linearly to a level of 18 feet in May. The water remained constant at that level until July, at which point it began to decrease linearly to a final level of 11 feet in December. Graph the water level as a function of time and determine the intervals of monotonicity.

Solution:

If we let 1 to 12 represent January through December, the intervals of monotonicity are as follows: increasing on $(1, 5)$, constant on $(5, 7)$, and decreasing on $(7, 12)$. The graph of the water level as a function of the month is shown below.

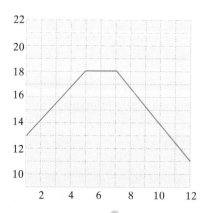

section 4.4 Transformations of Functions

exercises

For each function or graph below, determine the basic function that has been shifted, reflected, stretched, or compressed.

1. $f(x) = -(1-x)^2 + 2$

2. $f(x) = \dfrac{1}{x-4} + 5$

3. $f(x) = \sqrt[3]{x+6} - 2$

4. $f(x) = -2 + 2|x-3|$

5. $f(x) = [\![-2-x]\!]$

6. $f(x) = \dfrac{1}{(x+2)^2} + 1$

7. $f(x) = \dfrac{\sqrt{-x}}{2} + 4$

8. $f(x) = (x+6)^3$

9.

10.

11.

12.

chapter four Relations, Functions, and Their Graphs

Sketch the graphs of the following functions by first identifying the more basic functions that have been shifted, reflected, stretched or compressed. Then determine the domain and range of each function. See Examples 1 through 5.

13. $f(x) = (x+2)^3$
14. $G(x) = |x-4|$
15. $p(x) = -(x+1)^2 + 2$
16. $g(x) = \sqrt{x+3} - 1$
17. $q(x) = (1-x)^2$
18. $r(x) = -\sqrt[3]{x}$
19. $s(x) = \sqrt{2-x}$
20. $F(x) = \dfrac{|x+2|}{3} + 3$
21. $w(x) = \dfrac{1}{(x-3)^2}$
22. $v(x) = \dfrac{1}{3x} - 2$
23. $f(x) = \dfrac{1}{2-x}$
24. $k(x) = \sqrt{-x} + 2$
25. $b(x) = [\![x-4]\!] + 4$
26. $R(x) = 4 - |2x|$
27. $S(x) = (3-x)^3$
28. $g(x) = -\dfrac{1}{x+1}$
29. $h(x) = \dfrac{x^2}{2} - 3$
30. $W(x) = 1 - |4-x|$
31. $g(x) = x^2 - 6x + 9$ (Hint: find a better way to write the function.)
32. $h(x) = \dfrac{|x|}{x}$ (Hint: evaluate h at a few points to understand its behavior.)
33. $W(x) = \dfrac{x-1}{|x-1|}$
34. $S(x) = [\![x-2]\!]$
35. $V(x) = -3\sqrt{x-1} + 2$

Write a formula for each of the functions described below.

36. Use the function $f(x) = x^2$. Move the function 3 units to the left and 4 units down.

37. Use the function $f(x) = x^2$. Move the function 4 units to the right and 2 units up.

38. Use the function $f(x) = x^2$. Reflect the function across the x-axis and move it 6 units up.

39. Use the function $f(x) = x^2$. Move the function 2 units to the right and reflect across the y-axis.

40. Use the function $f(x) = x^3$. Move the function 1 unit to the left and reflect across the y-axis.

41. Use the function $f(x) = x^3$. Move the function 10 units to the right and 4 units up.

42. Use the function $f(x) = \sqrt{x}$. Move the function 5 units to the left and reflect across the x-axis.

43. Use the function $f(x) = \sqrt{x}$. Reflect the function across the y-axis and move it down 3 units.

44. Use the function $f(x) = |x|$. Move the function 7 units to the left, reflect across the x-axis and reflect across the y-axis.

45. Use the function $f(x) = |x|$. Move the function 8 units to the right, 2 units up and reflect across the x-axis.

Determine if each of the following relations is a function. If so, determine whether it is even, odd, or neither. Also determine if it has y-axis symmetry, x-axis symmetry, origin symmetry, or none of the above, and then sketch the graph of the relation. See Example 6.

46. $f(x) = |x| + 3$ **47.** $g(x) = x^3$ **48.** $h(x) = x^3 - 1$

49. $w(x) = \sqrt[3]{x}$ **50.** $x = -y^2$ **51.** $3y - 2x = 1$

52. $x + y = 1$ **53.** $F(x) = (x-1)^2$ **54.** $x = y^2 + 1$

55. $x = |2y|$ **56.** $g(x) = \dfrac{x^2}{5} - 5$ **57.** $s(x) = \left| x + \dfrac{1}{2} \right|$

58. $m(x) = \sqrt[3]{x} - 1$ **59.** $xy = 2$ **60.** $x + y^2 = 3$

For each function below, find the open intervals where the function is increasing, decreasing, or constant. See Examples 7 and 8.

61. $f(x) = (x+3)^2$ **62.** $g(x) = -|x-2|$ **63.** $h(x) = \dfrac{1}{x-1}$

64. $H(x) = \dfrac{1}{(x+3)^2}$ **65.** $G(x) = \sqrt{x+1}$ **66.** $F(x) = -2$

67. $p(x) = -30|x-1|$ **68.** $q(x) = (4-x)^2 + 1$ **69.** $r(x) = \dfrac{(x-7)^4}{-2} + 4$

70. $P(x) = \begin{cases} (x+3)^2 & \text{if } x < -1 \\ 1 & \text{if } x \geq -1 \end{cases}$ **71.** $Q(x) = \begin{cases} |x-1| & \text{if } x \leq 3 \\ 5-x & \text{if } x > 3 \end{cases}$

72. During the summer months, the water level of a garden pool varies as water is added and as it evaporates. On May 1st the pool was 3.4 feet deep. After a steady and linear increase due to rain, the depth had increased to 4.9 feet on June 1st. By July 1st the water level had decreased linearly to 4.2 feet. Knowing that the pool would be covered for the winter, the owner filled the pool (in an essentially linear fashion) until it reached 5 feet on August 1st. Graph the function as a function of time and determine the intervals of monotonicity.

73. The cost of a calling card is $2 for the first minute and $0.23 for each additional minute (even if the full minute is not used). A model of the function is given by $C(t) = 2 - 0.23[\![-t+1]\!]$ where $t > 0$ is the number of minutes the call lasted. Graph this cost function for the time interval $(0, 10]$ and determine the intervals of monotonicity.

74. The profit made by a hot dog vendor is given by the function
$$P(x) = \begin{cases} 2x - 3, & x \geq 0 \text{ and } x < 7 \\ \dfrac{1}{4}x^2, & x \geq 7 \end{cases}$$
where x is the number of hot dogs sold. Graph the function as a function of time and determine the intervals of monotonicity.

75. The cost incurred by a newspaper stand is given by the function
$$P(x) = \begin{cases} -2\sqrt{x} + 8, & x \geq 0 \text{ and } x < 3 \\ -x + 8, & x \geq 3 \end{cases}$$
where x is the number of newspapers sold. Graph the function as a function of time and determine the intervals of monotonicity.

section 4.4 Transformations of Functions

technology exercises

Mentally sketch the graph of each of the following functions by identifying the basic shape that has been shifted, reflected, stretched, or compressed. Then use a graphing calculator or a computer algebra system (if available) to graph each function and check your reasoning. (See Appendix A for guidance on using Mathematica.)

76. $f(x) = -2(3-x)^3 + 5$ **77.** $f(x) = \dfrac{3}{x+5} - 1$ **78.** $f(x) = \dfrac{-1}{(x-2)^2} - 3$

79. $f(x) = -3|x+2| - 4$ **80.** $f(x) = -\sqrt{1-x} + 2$ **81.** $f(x) = \sqrt[3]{2+x} - 1$

Write a possible equation for each of the functions depicted on the graphing calculator screens.

82.

83.

84.

85.

86.

87.

chapter four Relations, Functions, and Their Graphs

4.5 Combining Functions

TOPICS

1. Combining functions arithmetically
2. Composing functions
3. Decomposing functions
4. Interlude: recursive graphics

Topic 1: Combining Functions Arithmetically

In Section 4.4, we gained considerable experience in building new functions from old by shifting, reflecting and stretching the old functions. In this section, we will explore other ways of building functions.

We begin with four arithmetic ways of combining two or more functions to obtain new functions. The four arithmetic operations are very familiar to you: addition, subtraction, multiplication and division. The only possibly new idea is that we are applying these operations to functions, not numbers. But as we will see, the arithmetic combination of functions is based entirely on the arithmetic combination of numbers.

Addition, Subtraction, Multiplication and Division of Functions

Let f and g be two functions. The **sum** $f + g$, **difference** $f - g$, **product** fg and **quotient** $\dfrac{f}{g}$ are four new functions defined as follows:

1. $(f+g)(x) = f(x) + g(x)$
2. $(f-g)(x) = f(x) - g(x)$
3. $(fg)(x) = f(x)g(x)$
4. $\left(\dfrac{f}{g}\right)(x) = \dfrac{f(x)}{g(x)}$, provided that $g(x) \neq 0$

The domain of each of these new functions consists of the common elements (or the intersection of elements) of the domains of f and g individually, with the added condition that in the quotient function we have to omit those elements for which $g(x) = 0$.

section 4.5 Combining Functions

We have actually combined functions arithmetically already, though we didn't use the terminology of functions at the time, when we studied addition, subtraction, multiplication and division of polynomials in Section 1.5.

With the above definition, we can determine the sum, difference, product or quotient of two functions at one particular value for x, or find a formula for these new functions based on the formulas for f and g, if they are available.

example 1

Given that $f(-2) = 5$ and $g(-2) = -3$, find $(f-g)(-2)$ and $\left(\dfrac{f}{g}\right)(-2)$.

Solution:

By the definition of the difference and quotient of functions,

$$(f-g)(-2) = f(-2) - g(-2)$$
$$= 5 - (-3)$$
$$= 8,$$

and

$$\left(\dfrac{f}{g}\right)(-2) = \dfrac{f(-2)}{g(-2)}$$
$$= \dfrac{5}{-3}$$
$$= -\dfrac{5}{3}.$$

example 2

Given the two functions $f(x) = 4x^2 - 1$ and $g(x) = \sqrt{x}$, find $(f+g)(x)$ and $(fg)(x)$.

Solution:

By the definition of the sum and product of functions,

$$(f+g)(x) = f(x) + g(x)$$
$$= 4x^2 - 1 + \sqrt{x},$$

continued on next page...

chapter four Relations, Functions, and Their Graphs

and

$$(fg)(x) = (4x^2 - 1)(\sqrt{x})$$
$$= 4x^{\frac{5}{2}} - x^{\frac{1}{2}}.$$

The domain of both $f + g$ and fg is the interval $[0, \infty)$, although the domain of f alone is the entire real line.

example 3

Based on the graphs of f and g below, determine the domain of $\dfrac{f}{g}$ and evaluate $\left(\dfrac{f}{g}\right)(1)$.

Solution:

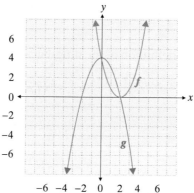

Based on the graphs, it appears that g is 0 when $x = -2$ and $x = 2$. The domain of both f and g individually is the entire set of real numbers, so the domain of $\dfrac{f}{g}$ is $(-\infty, -2) \cup (-2, 2) \cup (2, \infty)$. Also based on the graphs, it appears that $f(1) = 1$ and $g(1) = 3$, so $\left(\dfrac{f}{g}\right)(1) = \dfrac{1}{3}$.

Topic 2: Composing Functions

A fifth way of combining functions is to form the *composition* of one function with another. Informally speaking, this means to apply one function, say f, to the output of another function, say g. The symbol for composition is an open circle.

Composing Functions

Let f and g be two functions. The **composition** of f and g, denoted $f \circ g$, is the function defined by $(f \circ g)(x) = f(g(x))$. The domain of $f \circ g$ consists of all x in the domain of g for which $g(x)$ is in turn in the domain of f. The function $f \circ g$ is read "f composed with g", or "f of g."

caution

Note that the order of f and g is important. In general, we can expect the function $f \circ g$ to be different from the function $g \circ f$. In formal terms, the composition of two functions, unlike the sum and product of two functions, is not commutative.

The diagram in Figure 1 is a sort of schematic of the composition of two functions. The ovals represent sets, with the leftmost oval being the domain of the function g. The arrows indicate the element that x is associated with by the various functions.

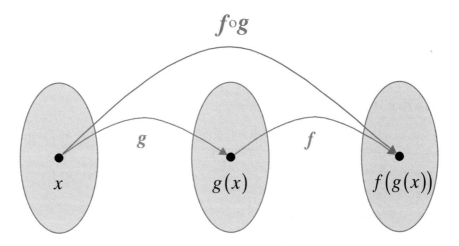

Figure 1: Composition of f and g

As with the four arithmetic ways of combining functions, we can evaluate the composition of two functions at a single point, or find a formula for the composition if we have been given formulas for the individual functions.

example 4

Given $f(x) = x^2$ and $g(x) = x - 3$, find:

a. $(f \circ g)(6)$ **b.** $(f \circ g)(x)$

Solutions:

a. $g(6) = 6 - 3 = 3$ and so $(f \circ g)(6) = f(g(6)) = f(3) = 3^2 = 9$.

b. To find the formula for $f \circ g$, we simply apply the definition of composition and then simplify:

$$(f \circ g)(x) = f(g(x))$$
$$= f(x - 3)$$
$$= (x - 3)^2$$
$$= x^2 - 6x + 9$$

Enter the functions using **Y=**. To return to the main screen, select **2ND** and **MODE** (Quit). To enter Y_1 and Y_2, select **VARS**, use the right arrow to select Y-Vars and press **ENTER**. Then **ENTER** on the desired variable.

Note that once we have found a formula for $f \circ g$, we have an alternative way of answering the first question: $(f \circ g)(6) = (6)^2 - (6)(6) + 9 = 9$

example 5

Let $f(x) = x^2 - 4$ and $g(x) = \sqrt{x}$. Find formulas and state the domains for:

a. $f \circ g$ **b.** $g \circ f$

Solutions:

a. $(f \circ g)(x) = f(\sqrt{x})$
$$= (\sqrt{x})^2 - 4$$
$$= x - 4$$

The answer at left may seem to indicate that the domain of $f \circ g$ is all real numbers, but this is incorrect. The domain of $f \circ g$ is actually the interval $[0, \infty)$, because only non-negative numbers can be plugged into g.

b. $(g \circ f)(x) = g(x^2 - 4)$
$$= \sqrt{x^2 - 4}$$

The domain of $g \circ f$ consists of all x for which $x^2 - 4 \geq 0$, or $x^2 \geq 4$. In interval form, the domain is the set $(-\infty, -2] \cup [2, \infty)$.

Topic 3: Decomposing Functions

Often, functions can be best understood by recognizing them as a composition of two or more simpler functions. We have already seen an instance of this: shifting, reflecting, stretching, and compressing can all be thought of as a composition of two or more functions. For example, the function $h(x) = (x-2)^3$ can be thought of as the composition of the functions $f(x) = x^3$ and $g(x) = x - 2$. Note that

$$f(g(x)) = f(x-2)$$
$$= (x-2)^3$$
$$= h(x).$$

To "decompose" a function into a composition of simpler functions, it is usually best to identify what the function does to its argument from the inside out. That is, identify the first thing that is done to the argument, then the second, and so on. Each action describes a less complex function, and can be identified as such. The composition of these functions, with the innermost function corresponding to the first action, the next innermost corresponding to the second action, and so on, is then equivalent to the original function.

Decomposition can often be done in several different ways. Consider, for example, the function $f(x) = \sqrt[3]{5x^2 - 1}$. The following table illustrates just some of the ways f can be written as a composition of functions. Be sure you understand how each of the different compositions is equivalent to f.

a. $g(x) = \sqrt[3]{x}$
  $h(x) = 5x^2 - 1$

  $g(h(x)) = g(5x^2 - 1)$
  $= \sqrt[3]{5x^2 - 1}$
  $= f(x)$

b. $g(x) = \sqrt[3]{x - 1}$
  $h(x) = 5x^2$

  $g(h(x)) = g(5x^2)$
  $= \sqrt[3]{5x^2 - 1}$
  $= f(x)$

c. $g(x) = \sqrt[3]{x}$
  $h(x) = 5x - 1$
  $i(x) = x^2$

  $g(h(i(x))) = g(h(x^2))$
  $= g(5x^2 - 1)$
  $= \sqrt[3]{5x^2 - 1}$
  $= f(x)$

example 6

Decompose the function $f(x) = |x^2 - 3| + 2$ into:

a. a composition of two functions **b.** a composition of three functions

Solutions:

a. $g(x) = |x| + 2$
$h(x) = x^2 - 3$

$$g(h(x)) = g(x^2 - 3)$$
$$= |x^2 - 3| + 2$$
$$= f(x)$$

b. $g(x) = x + 2$
$h(x) = |x - 3|$
$i(x) = x^2$

$$g(h(i(x))) = g(h(x^2))$$
$$= g(||x^2 - 3||)$$
$$= |x^2 - 3| + 2$$
$$= f(x)$$

Note: These are **not** the only possible solutions for the decompositions of $f(x)$.

Topic 4: Interlude: Recursive Graphics

Recursion, in general, refers to using the output of a function as its input, and repeating the process a certain number of times. In other words, recursion refers to the composition of a function with itself, possibly many times. Recursion has many varied uses, one of which is a branch of mathematical art.

Some special nomenclature and notation have evolved to describe recursion. If f is a function, $f^2(x)$ is used in this context to stand for $f(f(x))$, or $(f \circ f)(x)$ (not $(f(x))^2$!) Similarly, $f^3(x)$ stands for $f(f(f(x)))$, or $(f \circ f \circ f)(x)$, and so on. The functions f^2, f^3, ... are called **iterates** of f, with f^n being the nth **iterate** of f.

Some of the most famous recursively generated mathematical art is based on functions whose inputs and outputs are complex numbers. Recall from Section 1.6 that every complex number can be expressed in the form $a + bi$, where a and b are real numbers and i is the imaginary unit. A one-dimensional coordinate system, such as the real number line, is insufficient to graph complex numbers, but complex numbers are easily graphed in a two-dimensional coordinate system.

> So naturalists observe, a flea
> Hath smaller fleas that on him prey
> And these have smaller still to bite 'em
> And so proceed *ad infinitum*.
>
> - Jonathan Swift

To graph the number $a + bi$, we treat it as the ordered pair (a, b) and plot the point (a, b) in the Cartesian plane, where the horizontal axis represents pure real numbers and the vertical axis represents pure imaginary numbers.

Benoit Mandelbrot used the function $f(z) = z^2 + c$, where both z and c are variables representing complex numbers, to generate the image known as the Mandelbrot set in the 1970s. The basic idea is to evaluate the sequence of iterates $f(0) = 0^2 + c = c$, $f^2(0) = f(c) = c^2 + c$, $f^3(0) = f(c^2 + c) = (c^2 + c)^2 + c$, ... for various complex numbers c and determine if the sequence of complex numbers stays close to the origin or not. Those complex numbers c that result in so-called "bounded" sequences are colored black, while those that lead to unbounded sequences are colored white. The author has used similar ideas to generate his own recursive art, as described below.

The image "i of the storm" reproduced here is based on the function $f(z) = \dfrac{(1-i)z^4 + (7+i)z}{2z^5 + 6}$, where again z is a variable that will be replaced with complex

numbers. The image is actually a picture of the complex plane, with the origin in the very center of the golden ring. The golden ring consists of those complex numbers that lie between 0.9 and 1.1 in distance from the origin. The rules for coloring other complex numbers in the plane are as follows: given an initial complex number z not on the gold ring, $f(z)$ is calculated. If the complex number $f(z)$ lies somewhere on the gold ring, the original number z is colored the deepest shade of green. If not, the iterate $f^2(z)$ is calculated.

If this result lies in the gold ring, the original z is colored a bluish shade of green. If not, the process continues up to the 12th iterate $f^{12}(z)$, using a different color each time. If $f^{12}(z)$ lies in the gold ring, z is colored red, and if not the process halts and z is colored black.

The idea of recursion can be used to generate any number of similar images, with the end result usually striking and often surprising even to the creator.

chapter four   Relations, Functions, and Their Graphs

exercises

In each of the following problems, use the information given to determine **(a)** $(f+g)(-1)$, **(b)** $(f-g)(-1)$, **(c)** $(fg)(-1)$, *and* **(d)** $\left(\dfrac{f}{g}\right)(-1)$. *See Examples 1, 2, and 3.*

1. $f(-1) = -3$ and $g(-1) = 5$

2. $f(-1) = 0$ and $g(-1) = -1$

3. $f(x) = x^2 - 3$ and $g(x) = x$

4. $f(x) = \sqrt[3]{x}$ and $g(x) = x - 1$

5. $f(-1) = 15$ and $g(-1) = -3$

6. $f(x) = \dfrac{x+5}{2}$ and $g(x) = 6x$

7. $f(x) = x^4 + 1$ and $g(x) = x^{11} + 2$

8. $f(x) = \dfrac{6-x}{2}$ and $g(x) = \sqrt{\dfrac{x}{-4}}$

9. $f = \{(5, 2), (0, -1), (-1, 3), (-2, 4)\}$ and $g = \{(-1, 3), (0, 5)\}$

10. $f = \{(3, 15), (2, -1), (-1, 1)\}$ and $g(x) = -2$

11.

12.

13.

14.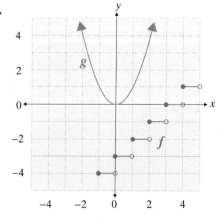

In each of the following problems, find (a) the formula and domain for f + g and (b) the formula and domain for $\frac{f}{g}$. See Examples 2 and 3.

15. $f(x) = |x|$ and $g(x) = \sqrt{x}$

16. $f(x) = x^2 - 1$ and $g(x) = \sqrt[3]{x}$

17. $f(x) = x - 1$ and $g(x) = x^2 - 1$

18. $f(x) = x^{\frac{3}{2}}$ and $g(x) = x - 3$

19. $f(x) = 3x$ and $g(x) = x^3 - 8$

20. $f(x) = x^3 + 4$ and $g(x) = \sqrt{x - 2}$

21. $f(x) = -2x^2$ and $g(x) = [\![x + 4]\!]$

22. $f(x) = 6x - 1$ and $g(x) = x^{\frac{2}{3}}$

In each of the following problems, use the information given to determine $(f \circ g)(3)$. See Example 4.

23. $f(-5) = 2$ and $g(3) = -5$

24. $f(\pi) = \pi^2$ and $g(3) = \pi$

25. $f(x) = x^2 - 3$ and $g(x) = \sqrt{x}$

26. $f(x) = \sqrt{x^2 - 9}$ and $g(x) = 1 - 2x$

27. $f(x) = 2 + \sqrt{x}$ and $g(x) = x^3 + x^2$

28. $f(x) = x^{\frac{3}{2}} - 3$ and $g(x) = \left[\!\left[\frac{3x}{2}\right]\!\right]$

29. $f(x) = \sqrt{x + 6}$ and $g(x) = \sqrt{4x - 3}$

30. $f(x) = \sqrt{\frac{3x}{14}}$ and $g(x) = x^4 - x^3 - x^2 - x$

31.

32.

33.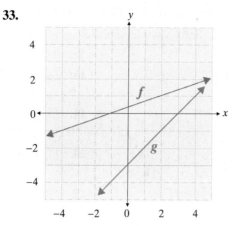

In each of the following problems, find **(a)** the formula and domain for $f \circ g$ and **(b)** the formula and domain for $g \circ f$. See Example 5.

34. $f(x) = \sqrt{x-1}$ and $g(x) = x^2$

35. $f(x) = \dfrac{1}{x}$ and $g(x) = x - 1$

36. $f(x) = \dfrac{4x-2}{3}$ and $g(x) = \dfrac{1}{x}$

37. $f(x) = 1 - x$ and $g(x) = \sqrt{x}$

38. $f(x) = [\![x-3]\!]$ and $g(x) = x^3 + 1$

39. $f(x) = x^2 + 2x$ and $g(x) = x - 3$

40. $f(x) = \sqrt{x-1}$ and $g(x) = \dfrac{x+1}{2}$

41. $f(x) = x^3 + 4x^2$ and $g(x) = |x| - 1$

42. $f(x) = -3x + 2$ and $g(x) = x^2 + 2$

43. $f(x) = [\![x+2]\!]$ and $g(x) = \dfrac{x^2+3}{2}$

Write the following functions as a composition of two functions. Answers will vary. See Example 6.

44. $f(x) = \sqrt[3]{3x^2 - 1}$

45. $f(x) = \dfrac{2}{5x-1}$

46. $f(x) = |x - 2| + 3$

47. $f(x) = x + \sqrt{x+2} - 5$

48. $f(x) = |x^3 - 5x| + 7$

49. $f(x) = \dfrac{\sqrt{x-3}}{x^2 - 6x + 9}$

50. $f(x) = \sqrt{2x^3 - 3} - 4$

51. $f(x) = |x^2 + 3x| - 3$

52. $f(x) = \dfrac{3}{4x-2}$

section 4.5   Combining Functions

In each of the following problems, use the information given to find g(x).

53. $f(x) = [\![x+3]\!]$ and $(f+g)(x) = [\![x+3]\!] + \sqrt{x+5}$

54. $f(x) = x$ and $(f \circ g)(x) = \dfrac{x+12}{-3}$

55. $f(x) = x^2 - 3$ and $(f-g)(x) = x^3 + x^2 + 4$

56. $f(x) = x^2$ and $(g \circ f)(x) = \sqrt{-x^2 + 5} + 4$

Solve the following application problems.

57. The volume of a right circular cylinder is given by the formula $V = \pi r^2 h$. If the height h is three times the radius r, show the volume V as a function of r.

58. The surface area S of a windsock is given by the formula $S = \pi r \sqrt{r^2 + h^2}$, where r is the radius of the base of the windsock and h is the height of the windsock. As the windsock is being knitted by an automated knitter, the height h increases with time t according to the formula $h(t) = \dfrac{1}{4} t^2$. Find the surface area S of the windsock as a function of time t.

59. The volume V of the windsock described in the previous question is given by the formula $V = \dfrac{1}{3} \pi r^2 h$ where r is the radius of the windsock and h is the height of the windsock. If the height h increases with time t according to the formula $h(t) = \dfrac{1}{4} t^2$, find the volume V of the windsock as a function of time t.

60. A widget factory produces n widgets in t hours of a single day. The number of widgets the factory produces is given by the formula $n(t) = 10,000t - 25t^2, 0 \le t \le 9$ The cost c in dollars of producing n widgets is given by the formula $c(n) = 2040 + 1.74n$. Find the cost c as a function of time t.

61. Given two odd functions f and g, show that $f \circ g$ is also odd. Verify this fact with the particular functions $f(x) = \sqrt[3]{x}$ and $g(x) = \dfrac{-x^3}{3x^2 - 9}$. Recall that a function is odd if $f(-x) = -f(x)$ for all x in the domain of f.

62. Given two even functions f and g, show that the product is also even. Verify this fact with the particular functions $f(x) = 2x^4 - x^2$ and $g(x) = \dfrac{1}{x^2}$. Recall that a function is even if $f(-x) = f(x)$ for all x in the domain of f.

As mentioned in the Interlude, a given complex number c is said to be in the Mandelbrot set if, for the function $f(z) = z^2 + c$, the sequence of iterates $f(0)$, $f^2(0)$, $f^3(0)$, ... stays close to the origin (which is the complex number $0 + 0i$). It can be shown that if any single iterate falls more than 2 units in distance (magnitude) from the origin, then the remaining iterates will grow larger and larger in magnitude. In practice, computer programs that generate the Mandelbrot set calculate the iterates up to a pre-decided point in the sequence, such as $f^{50}(0)$, and if no iterate up to this point exceeds 2 in magnitude, the number c is admitted to the set. The magnitude of a complex number $a + bi$ is the distance between the point (a, b) and the origin, so the formula for the magnitude of $a + bi$ is $\sqrt{a^2 + b^2}$.

Use the above criterion to determine, without a calculator or computer, if the following complex numbers are in the Mandelbrot set or not.

63. $c = 0$ **64.** $c = 1$ **65.** $c = i$ **66.** $c = -1$ **67.** $c = 1 + i$

68. $c = -i$ **69.** $c = 1 - i$ **70.** $c = -1 - i$ **71.** $c = 2$ **72.** $c = -2$

section 4.5   Combining Functions

technology exercises

A computer algebra system like *Mathematica* can be very useful in combining functions. The illustration below shows how the two functions $f(x) = x^2 - 3x - 5$ and $g(x) = (x-2)^3$ are defined in *Mathematica*, and then shows how to ask *Mathematica* for $(f+g)(7)$, $(fg)(x)$, $(f \circ g)(x)$, $(g \circ f)(x)$, and $(f \circ g)(-3)$. (Note how the composition of two functions is expressed in the *Mathematica* command).

Use a computer algebra system, if available, to evaluate $(f+g)(x)$, $(fg)(x)$, $(f \circ g)(x)$ and $(g \circ f)(x)$ for each of the following pairs of functions. (See Appendix A for guidance on using Mathematica.)

To enter Y_1 and Y_2, select **VARS** as described on page 326.

73. $f(x) = (3x+2)^2$ and $g(x) = \sqrt{x^2+5}$

74. $f(x) = \dfrac{1}{3x-5}$ and $g(x) = (x+2)^3$

75. $f(x) = \dfrac{x+1}{x-1}$ and $g(x) = \dfrac{x-1}{x}$

76. $f(x) = [\![x^2]\!] + 3$ and $g(x) = \sqrt{2x+6}$

77. $f(x) = \sqrt{-x^2+5} + 4$ and $g(x) = |x^3 - 5x^2| - 1$

chapter four: Relations, Functions, and Their Graphs

4.6 Inverses of Functions

TOPICS

1. Inverses of relations
2. Inverse functions and the horizontal line test
3. Finding inverse function formulas

Topic 1: Inverses of Relations

In many problems, "undoing" one or more mathematical operations plays a critical role in the solution process. For instance, to solve the equation $3x + 2 = 8$, the first step is to "undo" the addition of 2 on the left-hand side (by subtracting 2 from both sides) and the second step is to "undo" the multiplication by 3 (by dividing both sides by 3). In the context of more complex problems, the "undoing" process is often a matter of finding and applying the inverse of a function.

We begin our discussion with the more general idea of the inverse of a relation. Recall that a relation is just a set of ordered pairs; the inverse of a given relation is the set of these ordered pairs with the first and second coordinates of each exchanged.

Inverse of a Relation

Let R be a relation. The **inverse of R**, denoted R^{-1}, is the set

$$R^{-1} = \{(b, a) | (a, b) \in R\}.$$

example 1

Determine the inverse of each of the following relations. Then graph each relation and its inverse, and determine the domain and range of both.

a. $R = \{(4, -1), (-3, 2), (0, 5)\}$ **b.** $y = x^2$

Solutions:

a. $R = \{(4, -1), (-3, 2), (0, 5)\}$ $\text{Dom}(R) = \{4, -3, 0\}$ and $\text{Ran}(R) = \{-1, 2, 5\}$

$R^{-1} = \{(-1, 4), (2, -3), (5, 0)\}$ $\text{Dom}(R^{-1}) = \{-1, 2, 5\}$ and $\text{Ran}(R^{-1}) = \{4, -3, 0\}$

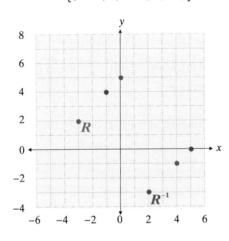

In the graph to the left, R is in blue and its inverse is in red. The relation R consists of three ordered pairs, and its inverse is simply these three ordered pairs with the coordinates exchanged. Note that the domain of R is the range of R^{-1} and vice versa.

b. $R = \{(x, y) \mid y = x^2\}$ $\text{Dom}(R) = \mathbb{R}$ and $\text{Ran}(R) = [0, \infty)$

$R^{-1} = \{(x, y) \mid x = y^2\}$ $\text{Dom}(R^{-1}) = [0, \infty)$ and $\text{Ran}(R^{-1}) = \mathbb{R}$

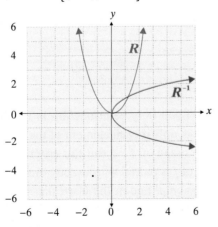

In this problem, R is described by the given equation in x and y. The inverse relation is the set of ordered pairs in R with the coordinates exchanged, so we can describe the inverse relation by just exchanging x and y in the equation, as shown at left.

Consider the graphs of the two relations and their respective inverses in *Example 1*. By definition, an ordered pair (b, a) lies on the graph of a relation R^{-1} if and only if (a, b) lies on the graph of R, so it shouldn't be surprising that the graphs of a relation and its inverse bear some resemblance to one another. Specifically, they are mirror images of one another with respect to the line $y = x$. If you were to fold the Cartesian plane in half along the line $y = x$ in the two examples above, you would see that the points in R and R^{-1} coincide with one another.

The two relations in *Example 1* illustrate another important point. Note that in both cases, R is a function, as its graph passes the vertical line test. By the same criterion, R^{-1} in *Example 1a* is also a function, but R^{-1} in *Example 1b* is not. The conclusion to be drawn is that even if a relation is a function, its inverse may or may not be a function.

Topic 2: Inverse Functions and the Horizontal Line Test

With a bit more thought, we can draw a stronger conclusion from *Example 1* about when the inverse of a relation is a function.

In practice, we will only be concerned with the question of when the inverse of a function f, denoted f^{-1}, is itself a function. Note that f^{-1} has already been defined: f^{-1} stands for the inverse of f, where we are making use of the fact that a function is also a relation.

caution

We are faced with another example of reuse of notation. f^{-1} does *not* stand for $\frac{1}{f}$ when f is a function! We use an exponent of -1 to indicate the reciprocal of a number or an algebraic expression, but when applied to a function or a relation it stands for the inverse relation.

Assume that f is a function. f^{-1} will only be a function in its own right if its graph passes the vertical line test; that is, only if each element of the domain of f^{-1} is paired with exactly one element of the range of f^{-1}. But this criterion is identical to saying that each element of the range of f is paired with exactly one element of the domain of f. In other words, every *horizontal* line in the Cartesian plane must intersect the graph of f no more than once. We say that functions meeting this condition pass the horizontal line test.

The Horizontal Line Test

> Let f be a function. We say that the graph of f passes the **horizontal line test** if every horizontal line in the plane intersects the graph no more than once.

Of course, the horizontal line test is only useful if the graph of f is available to study. We can also phrase the above condition in a non-graphical manner as follows. The inverse of f will only be a function if for every pair of distinct elements x_1 and x_2 in the domain of f, we have $f(x_1) \neq f(x_2)$. This criterion is important enough to merit a name.

One-to-One Functions

> A function f is **one-to-one** if for every pair of distinct elements x_1 and x_2 in the domain of f, we have $f(x_1) \neq f(x_2)$. This means that every element of the range of f is paired with exactly one element of the domain of f.

To sum up: the inverse f^{-1} of a function f is also a function if and only if f is one-to-one. If the graph of f is available for examination, f is one-to-one if and only if its graph passes the horizontal line test.

If we now examine *Example 1* again, we see that the function R in *Example 1a* is one-to-one, and so we know that its inverse is also a function. On the other hand, the function R in *Example 1b* is not one-to-one (plenty of horizontal lines pass through the graph twice), so its inverse is not a function.

example 2

Determine if the following functions have inverse functions.

a. $f(x) = |x|$ **b.** $g(x) = (x+2)^3$

Solutions:

a. The function f does not have an inverse function, a fact easily demonstrated by showing that its graph does not pass the horizontal line test (recall that the graph of f is an upward-opening V shape, as shown to the left). An algebraic proof that f does not have an inverse function is the following: although $-3 \neq 3$, we have $f(-3) = f(3)$. Of course, there are an infinite number of pairs of numbers that show f is not one-to-one, but one such pair is all it takes to show that f does not have an inverse function.

b. The graph of g is the standard cubic shape shifted horizontally two units to the left, and the standard cubic shape passes the horizontal line test, so g has an inverse function. But again, it is good practice to prove this algebraically. Note how each line in the following argument implies the next line:

$$x_1 \neq x_2$$
$$\Rightarrow x_1 + 2 \neq x_2 + 2$$
$$\Rightarrow (x_1 + 2)^3 \neq (x_2 + 2)^3$$
$$\Rightarrow g(x_1) \neq g(x_2)$$

This argument shows that any two distinct elements of the domain of g lead to different values when plugged into g, so g is one-to-one and hence has an inverse function.

Consider the function in *Example 2a* again. As we noted, the function $f(x) = |x|$ is not one-to-one on its implied domain, and so cannot have an inverse function. However, if we *restrict* the domain of f explicitly by specifying that the domain is the interval $[0, \infty)$, the new function, with its restricted domain, is one-to-one and has an inverse function. Of course, this **restriction of domain** changes the function; in this case the graph of the new function is the right-hand half of the graph of the absolute value function.

Topic 3: Finding Inverse Function Formulas

In applying the notion of the inverse of a function, we will most often begin with a formula for f and want to find a formula for f^{-1}. This will allow us, for instance, to transform equations of the form
$$f(x) = y$$
into the form
$$x = f^{-1}(y).$$

Before we discuss the general algorithm for finding a formula for f^{-1}, consider the problem with which we began this section. If we define $f(x) = 3x+2$, the equation $3x+2 = 8$ can be written as $f(x) = 8$. Note that f is one-to-one, so f^{-1} does exist (as a function). If we can find a formula for f^{-1}, we can transform the equation into $x = f^{-1}(8)$. This is an overly complicated way to solve this equation, but it illustrates the point well.

What should the formula for f^{-1} be? Consider what f does to its argument. The first action is to multiply x by 3, and the second is to add 2. To "undo" f, we need to negate these two actions in reverse order: subtract 2 and then divide the result by 3. So,
$$f^{-1}(x) = \frac{x-2}{3}.$$
Applying this to the problem at hand, we obtain
$$x = f^{-1}(8) = \frac{8-2}{3} = 2.$$
This method of analyzing a function f and then finding a formula for f^{-1} by undoing the actions of f in reverse order is conceptually important and works for simple functions. For other functions, however, the following algorithm may be necessary.

To Find a Formula for f^{-1}

Let f be a one-to-one function, and assume that f is defined by a formula. To find a formula for f^{-1}, perform the following steps:

1. Replace $f(x)$ in the definition of f with the variable y. The result is an equation in x and y that is solved for y at this point.
2. Interchange x and y in the equation.
3. Solve the new equation for y.
4. Replace the y in the resulting equation with $f^{-1}(x)$.

chapter four Relations, Functions, and Their Graphs

example 3

Find the inverse of each of the following functions.

a. $f(x) = (x-1)^3 + 2$ **b.** $g(x) = \dfrac{x-3}{2x+1}$

Solutions:

a. $f(x) = (x-1)^3 + 2$

$f^{-1}(x) = (x-2)^{\frac{1}{3}} + 1$

We can find the inverse of this function either by the algorithm, or by undoing the actions of f in reverse order. The function f subtracts 1 from x, cubes the result, and adds 2; its inverse will first subtract 2, take the cube root of the result, and then add 1.

b. $g(x) = \dfrac{x-3}{2x+1}$

$y = \dfrac{x-3}{2x+1}$

$x = \dfrac{y-3}{2y+1}$

$x(2y+1) = y - 3$

$2xy + x = y - 3$

$2xy - y = -x - 3$

$y(2x-1) = -x - 3$

$y = \dfrac{-x-3}{2x-1}$

$g^{-1}(x) = \dfrac{-x-3}{2x-1}$

The inverse of the function g is most easily found by the algorithm.

The first step is to replace $g(x)$ with y.

The second step is to interchange x and y in the equation.

We now have to solve the equation for y. We begin by clearing the equation of fractions, and then proceed to collect all the terms that contain y on one side.

Factoring out the y on the left-hand side and dividing by $2x - 1$ completes the process.

The last step is to rename the formula g^{-1}.

Remember that the graphs of a relation and its inverse are mirror images of one another with respect to the line $y = x$; this is still true if the relations are functions. We can demonstrate this fact by graphing the function and its inverse from *Example 3a* above, as shown in Figure 1.

section 4.6 Inverses of Functions

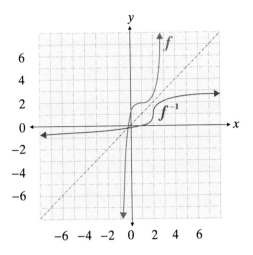

Figure 1: Graph of a Function and its Inverse

We can use the functions and their inverses from *Example 3* to illustrate one last important point. The key characteristic of the inverse of a function is that it undoes the function. This means that if a function and its inverse are composed together, in either order, the resulting function has no effect on any allowable input; specifically,

$$f(f^{-1}(x)) = x \text{ for all } x \in \text{Dom}(f^{-1}), \text{ and}$$
$$f^{-1}(f(x)) = x \text{ for all } x \in \text{Dom}(f).$$

For example, given $f(x) = (x-1)^3 + 2$ and $f^{-1}(x) = (x-2)^{\frac{1}{3}} + 1$:

$$\begin{aligned} f(f^{-1}(x)) &= f\left((x-2)^{\frac{1}{3}} + 1\right) \\ &= \left((x-2)^{\frac{1}{3}} + 1 - 1\right)^3 + 2 \\ &= \left((x-2)^{\frac{1}{3}}\right)^3 + 2 \\ &= x - 2 + 2 \\ &= x. \end{aligned}$$

A similar calculation shows that $f^{-1}(f(x)) = x$, as you should verify.

chapter four Relations, Functions, and Their Graphs

As another example, consider $g(x) = \dfrac{x-3}{2x+1}$ and $g^{-1}(x) = \dfrac{-x-3}{2x-1}$:

$$g^{-1}(g(x)) = g^{-1}\left(\dfrac{x-3}{2x+1}\right)$$

$$= \dfrac{-\dfrac{x-3}{2x+1} - 3}{2\left(\dfrac{x-3}{2x+1}\right) - 1}$$

$$= \left(\dfrac{-\dfrac{x-3}{2x+1} - 3}{2\left(\dfrac{x-3}{2x+1}\right) - 1}\right)\left(\dfrac{2x+1}{2x+1}\right)$$

$$= \dfrac{-x+3-6x-3}{2x-6-2x-1}$$

$$= \dfrac{-7x}{-7}$$

$$= x$$

Similarly, $g(g^{-1}(x)) = x$, as you should verify.

exercises

Graph the inverse of each of the following relations, and state its domain and range. See Example 1.

1. $R = \{(-4, 2), (3, 2), (0, -1), (3, -2)\}$
2. $S = \{(-3, -3), (-1, -1), (0, 1), (4, 4)\}$
3. $y = x^3$
4. $y = |x| + 2$
5. $x = |y|$
6. $x = -\sqrt{y}$
7. $y = \dfrac{1}{2}x - 3$
8. $y = -x + 1$
9. $y = [\![x]\!]$
10. $T = \{(4, 2), (3, -1), (-2, -1), (2, 4)\}$
11. $x = y^2 - 2$
12. $y = 2\sqrt{x}$

section 4.6 Inverses of Functions

Determine if each of the following functions is a one-to-one function. If so, graph the inverse of the function and state its domain and range.

13.

$y = 2x + 3$

14.

$y = x^2 + 4x$

15.

$y = \dfrac{1}{x^2}$

16.
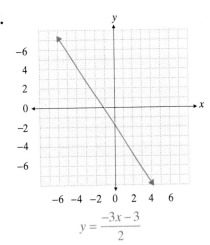
$y = \dfrac{-3x - 3}{2}$

Determine if the following functions have inverse functions. If not, suggest a domain to restrict the function to so that it would have an inverse function (answers may vary). See Example 2.

17. $f(x) = x^2 + 1$
18. $g(x) = (x - 2)^3 - 1$
19. $h(x) = \sqrt{x + 3}$
20. $s(x) = \dfrac{1}{x^2}$
21. $G(x) = 3x - 5$
22. $F(x) = -x^2 + 5$
23. $r(x) = -\sqrt{x^3}$
24. $b(x) = [\![x]\!]$
25. $f(x) = x^2 - 4x$
26. $m(x) = \dfrac{13x - 2}{4}$
27. $H(x) = |x - 12|$
28. $p(x) = 10 - x^2$

chapter four Relations, Functions, and Their Graphs

Find a formula for the inverse of each of the following functions. See Example 3.

29. $f(x) = x^{\frac{1}{3}} - 2$
30. $g(x) = 4x - 3$
31. $r(x) = \dfrac{x-1}{3x+2}$

32. $s(x) = \dfrac{1-x}{1+x}$
33. $F(x) = (x-5)^3 + 2$
34. $G(x) = \sqrt[3]{3x-1}$

35. $V(x) = \dfrac{x+5}{2}$
36. $W(x) = \dfrac{1}{x}$
37. $h(x) = x^{\frac{3}{5}} - 2$

38. $A(x) = (x^3 + 1)^{\frac{1}{5}}$
39. $J(x) = \dfrac{2}{1-3x}$
40. $k(x) = \dfrac{x+4}{3-x}$

41. $h(x) = x^7 + 6$
42. $F(x) = \dfrac{3-x^5}{-9}$
43. $r(x) = \sqrt[5]{2x}$

44. $P(x) = (2+3x)^3$
45. $f(x) = 3(2x)^{\frac{1}{3}}$
46. $q(x) = (x-2)^2 + 2,\ x \geq 2$

In each of the following problems, verify that $f(f^{-1}(x)) = x$ and that $f^{-1}(f(x)) = x$.

47. $f(x) = \dfrac{3x-1}{5}$ and $f^{-1}(x) = \dfrac{5x+1}{3}$

48. $f(x) = \sqrt[3]{x+2} - 1$ and $f^{-1}(x) = (x+1)^3 - 2$

49. $f(x) = \dfrac{2x+7}{x-1}$ and $f^{-1}(x) = \dfrac{x+7}{x-2}$
50. $f(x) = x^2,\ x \geq 0$ and $f^{-1}(x) = \sqrt{x}$

51. $f(x) = 2x - 3$ and $f^{-1}(x) = \dfrac{x+3}{2}$
52. $f(x) = \sqrt[3]{x+1}$ and $f^{-1}(x) = x^3 - 1$

53. $f(x) = \dfrac{1}{x}$ and $f^{-1}(x) = \dfrac{1}{x}$
54. $f(x) = \dfrac{x-5}{2x+3}$ and $f^{-1}(x) = \dfrac{3x+5}{1-2x}$

55. $f(x) = (x-2)^2,\ x \geq 2$ and $f^{-1}(x) = \sqrt{x} + 2,\ x \geq 0$

56. $f(x) = \dfrac{1}{1+x}$ and $f^{-1}(x) = \dfrac{1-x}{x}$

*Match the following functions with the graphs of the inverses of the functions. The graphs are labeled **a.** through **f.***

57. $f(x) = x^3$

58. $f(x) = x - 5$

59. $f(x) = \sqrt{x-4}$

60. $f(x) = x^2, x \geq 0$

61. $f(x) = \dfrac{x}{4}$

62. $f(x) = \sqrt[3]{x+1}$

a.

b.

c.

d.

e.

f.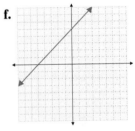

An inverse function can be used to encode and decode words and sentences by assigning each letter of the alphabet a numerical value (A = 1, B = 2, C = 3, ..., Z = 26). Example: Use the function $f(x) = x^2$ to encode the word PRECALCULUS. The encoded message would be 256 324 25 9 1 144 9 441 144 441 361. The word can then be decoded by using the inverse function $f^{-1}(x) = \sqrt{x}$. The inverse values are 16 18 5 3 1 12 3 21 12 21 19 which translates back to the word PRECALCULUS. Encode or decode the following words using the numerical values (A = 1, B = 2, C = 3, ..., Z = 26).

63. Encode the message SANDY SHOES using the function $f(x) = 4x - 3$.

64. Encode the message WILL IT RAIN TODAY using the function $f(x) = x^2 - 8$.

65. The following message was encoded using the function $f(x) = 8x - 7$. Decode the message.
   41 137 65 145 9 33 33 169 113 89 89 33 193 9 1 89 89 1 105 25 57 113 137 145 33 145 57 113 33 145

66. The following message was encoded using the function $f(x) = 5x + 1$. Decode the message.
   91 26 66 26 66 11 26 91 126 76 106 91 96 106 71 11 61 76 16 56

67. The following message was encoded using the function $f(x) = x^3$. Decode the message.

27 1 8000 27 512 1 12167 1 10648 125

68. The following message was encoded using the function $f(x) = -3 - 5x$. Decode the message.

−13 −28 −8 −18 −43 −33 −108 −73 −48 −73 −103 −43 −28 −98 −108 −73

A graphing calculator can be used to verify the inverse of a function. Use a graphing calculator to graph the following functions. Determine the domain and range of the inverse.

69. $f(x) = \sqrt{x+5}$ and $f^{-1}(x) = x^2 - 5$ **70.** $f(x) = x^3 - 1$ and $f^{-1}(x) = \sqrt[3]{x+1}$

71. $f(x) = \dfrac{2x+1}{x-1}$ and $f^{-1}(x) = \dfrac{x+1}{x-2}$ **72.** $f(x) = \dfrac{4}{\sqrt{x}}$ and $f^{-1}(x) = \dfrac{16}{x^2}$

73. $f(x) = -\sqrt{x^2 - 16},\ x \geq 4$ and $f^{-1}(x) = \sqrt{(x)^2 + 16}$

74. $f(x) = x^2 + 3,\ x \geq 0$ and $f^{-1}(x) = \sqrt{x-3}$

chapter four project

The Ozone Layer

As time goes on, there is continually increasing awareness, controversy, and legislation regarding the ozone layer and other environmental issues. The hole in the ozone layer over the south pole disappears and reappears annually, and one model for its growth assumes the hole is circular and that its radius grows at a constant rate of 2.6 kilometers per hour.

1. Write the area of the circle as a function of the radius, r.
2. Assuming that t is measured in hours, that $t=0$ corresponds to the start of the annual growth of the hole, and that the radius of the hole is initially 0, write the radius as a function of time, t.
3. Write the area of the circle as a function of time, t.
4. What is the radius after 3 hours?
5. What is the radius after 5.5 hours?
6. What is the area of the circle after 3 hours?
7. What is the area of the circle after 5.5 hours?
8. What is the average rate of change of the area from 3 hours to 5.5 hours?
9. What is the average rate of change of the area from 5.5 hours to 8 hours?
10. Is the average rate of change of the area increasing or decreasing as time passes?

chapter four Relations, Functions, and Their Graphs

CHAPTER SUMMARY

A summary of concepts and skills follows each chapter. Refer to these summaries to make sure you feel comfortable with the material in the chapter. The concepts and skills are organized according to the section title and topic title in which the material is first discussed.

4.1 Relations and Functions

topics	pages
Relations, Domains, and Ranges • The definition of *relation* as a set of ordered pairs • The definition of *domain* and *range* as, respectively, the set of first coordinates and the set of second coordinates for a given relation • The correspondence between a relation and its graph in the Cartesian plane	p. 255 – 258
Functions and the Vertical Line Test • The definition of a *function* as a special type of relation • The meaning of the *vertical line test* as applied to the graph of a relation and in identifying functions	p. 258 – 260
Functional Notation and Function Evaluation • The meaning of *functional notation* • Evaluation of a function for a given *argument* • The role of the argument as a placeholder in defining a function	p. 260 – 264
Implied Domain of a Function • Determining the domain of a function when it is not stated explicitly	p. 265

4.2 Linear and Quadratic Functions

topics	pages
Linear Functions and Their Graphs • The definition of a *linear*, or *first-degree*, *function* • The graph of a linear function	p. 271 – 273
Quadratic Functions and Their Graphs • The definition of a *quadratic*, or *second-degree*, *function* • The graph of a quadratic function, including the location of the vertex and the *x*- and *y*-intercepts • Finding the vertex form of a quadratic function	p. 273 – 279

chapter four summary

topics (continued)	pages
Interlude: Maximization/Minimization Problems • The role of *completing the square* in locating the maximum or minimum value of a quadratic function	p. 279 – 281

4.3 Other Common Functions

topics	pages
Commonly Occurring Functions • Functions of the form ax^n, ax^{-n}, and $ax^{\frac{1}{n}}$ • The absolute value function • The greatest integer function • Piecewise-defined functions	p. 287 – 296
Interlude: Variation Problems • Direct variation • Inverse variation • Solving variation application problems	p. 296 – 298

4.4 Transformations of Functions

topics	pages
Shifting, Stretching, and Reflecting Graphs • Replacing the argument x with $x - h$ to shift a graph h units horizontally • Adding k to a function to shift its graph k units vertically • Multiplying a function by -1 to reflect its graph with respect to the x–axis • Replacing the argument x with $-x$ to reflect a graph with respect to the y–axis • Multiplying a function by an appropriate constant to stretch or compress its graph • Determining the order in which to evaluate transformations	p. 304 – 310
Symmetry of Functions and Equations • The meaning of *y–axis symmetry* • The meaning of *x–axis symmetry* • The meaning of *origin symmetry* • The meaning of *even* and *odd* functions	p. 315 – 316
Intervals of Monotonicity • The meaning of *increasing*, *decreasing*, and *constant* as applied to the graph of a function	p. 315 – 316

chapter four   Relations, Functions, and Their Graphs

4.5 Combining Functions

topics	pages
Combining Functions Arithmetically • *Sums*, *differences*, *products*, and *quotients* of functions, and how to evaluate such combinations • Identifying the domain of an arithmetic combination of functions	p. 322 – 324
Composing Functions • The meaning of *composition* of functions • Determining a formula for the composition of two functions, and evaluating a composition of functions for a given argument • Identifying the domain of a composition of functions	p. 324 – 326
Decomposing Functions • The decomposition of complicated functions into simpler functions	p. 327 – 328
Interlude: Recursive Graphics • *Recursion* as an application of composition, and the meaning of the *iterates* of a function	p. 328 – 329

4.6 Inverses of Functions

topics	pages
Inverses of Relations • The definition of the *inverse* of a relation, and the correspondence between the graph of a relation and the graph of its inverse	p. 336 – 338
Inverse Functions and the Horizontal Line Test • The concept of the *inverse* of a function, and the notation used • Identifying when a function has an inverse function • The correspondence between the *horizontal line test* and whether a function is *one-to-one* • Restriction of domain	p. 338 – 340
Finding Inverse Function Formulas • Constructing the formula for the inverse of a function by "undoing" the function step by step • The algorithmic approach to finding the inverse of a function	p. 341 – 344

chapter review

Section 4.1

For each relation below, describe the domain and range and determine whether or not the relation is a function.

1. $R = \{(-2, 9), (-3, -3), (-2, 2), (-2, -9)\}$
2. $3x - 4y = 17$
3. $x = y^2 - 6$
4. $x = \sqrt{y - 4}$

Section 4.2

Graph the following linear functions.

5. $f(x) = 7x - 2$
6. $g(x) = \dfrac{2x - 6}{3}$

Graph the following quadratic functions, locating the vertices and x-intercepts (if any) accurately.

7. $f(x) = (x - 1)^2 - 1$
8. $g(x) = -(x + 3)^2 - 2$

Section 4.3

Sketch the graphs of the following functions. Pay particular attention to intercepts, if any, and locate these accurately.

9. $f(x) = \begin{cases} x^2 & \text{if } x < 1 \\ \dfrac{1}{x} & \text{if } x \geq 1 \end{cases}$

10. $g(x) = \begin{cases} (x+1)^2 - 1 & \text{if } x \leq 0 \\ \sqrt[3]{x} & \text{if } x > 0 \end{cases}$

Solve the following variation problems.

11. The distance that an object falls from rest, when air resistance is negligible, varies directly as the square of the time. A stone dropped from rest travels 400 feet in the first 5 seconds. How far did it travel in the first 2 seconds?

12. A video store manager observes that the number of videos rented seems to vary inversely as the price of a rental. If the store's customers rent 1050 videos per month when the price per rental is $3.49, how many videos per month does he expect to rent if he lowers the price to $2.99?

Section 4.4

Sketch the graphs of the following functions by first identifying the more basic functions that have been shifted, reflected, stretched, or compressed. Then determine the domain and range of each function.

13. $f(x) = (x-1)^3 + 2$

14. $G(x) = 4|x+3|$

For each function below, find the intervals where the function is increasing, decreasing, or constant.

15. $f(x) = (x-2)^4 - 6$

16. $R(x) = \begin{cases} (x+2)^2 & \text{if } x < -1 \\ -x & \text{if } x \geq -1 \end{cases}$

Section 4.5

In each of the following problems, find (a) the formula and domain for $f + g$ and (b) the formula and domain for $\dfrac{f}{g}$.

17. $f(x) = x^2$ and $g(x) = \sqrt{x}$

18. $f(x) = \dfrac{1}{x-2}$ and $g(x) = \sqrt[3]{x}$

In each of the following problems, use the information given to determine $(f \circ g)(3)$.

19. $f(x) = -x + 1$ and $g(x) = -x - 1$

20. $f(x) = \dfrac{x^{-1}}{18} - 3$ and $g(x) = \dfrac{x-4}{x^3}$

In each of the following problems, find the formula for $f \circ g$.

21. $f(x) = 4x - 1$ and $g(x) = x^3 + 2$

22. $f(x) = \dfrac{1}{\sqrt{x^2 - 4}}$ and $g(x) = x + 2$

Section 4.6

Find a formula for the inverse of each of the following functions.

23. $r(x) = \dfrac{2}{7x-1}$

24. $g(x) = \dfrac{4x-3}{x}$

25. $f(x) = x^{\frac{1}{5}} - 6$

In the following problem, verify that $f(f^{-1}(x)) = x$ and that $f^{-1}(f(x)) = x$.

26. $f(x) = \dfrac{6x-7}{2-x}$ and $f^{-1}(x) = \dfrac{2x+7}{6+x}$

chapter four — Relations, Functions, and Their Graphs

chapter test

Given $f(x) = -x^2 + 2x$, evaluate each of the following.

1. $f(x-h)$
2. $\dfrac{f(x+h) - f(x)}{h}$

Rewrite each of the relations below as a function of x. Then evaluate the function at $x = -2$.

3. $y + 3x - x^2 = 2y - x^2$
4. $xy + 2 = x - 3y$
5. $\dfrac{1}{x} - \dfrac{2}{y} = \dfrac{3}{4}$

Determine the implied domain of each of the following functions.

6. $f(x) = \sqrt{2 - 4x}$
7. $f(x) = \sqrt[3]{1 - x^2} - 1$
8. $f(x) = \dfrac{x+1}{x^2 - 4}$
9. $f(x) = \sqrt{\dfrac{x+3}{4-x}}$
10. $f(x) = (1 - x + 2x^2)^3$

Graph the following quadratic functions, locating the vertices, x-intercepts (if any), and y-intercepts.

11. $y = x^2 - 2$
12. $y = -x^2 + 4x$
13. $y = -(x-1)^2 + 2$
14. $y = x^2 + 2x - 8$

15. Given $f(x) = \begin{cases} x - 1 & \text{if } x < 2 \\ -3 & \text{if } x = 2 \\ 2 - x^2 & \text{if } x > 2 \end{cases}$, find $f(-3) - f(2) + 2f(3)$.

16. Given $f(x) = \begin{cases} -|2x| & \text{if } x > 1 \\ [\![x+1]\!] & \text{if } x \leq 1 \end{cases}$, find $f(3) - f\left(-\dfrac{5}{2}\right)$.

Sketch the graphs of following functions.

17. $y = \begin{cases} -x^2 & \text{if } |x| \leq 1 \\ x & \text{if } x < -1 \text{ or } x > 1 \end{cases}$
18. $y = \begin{cases} -|x| & \text{if } x > 3 \\ [\![x]\!] & \text{if } x \leq 3 \end{cases}$

Write a formula for each of the functions described below.

19. Use the function $f(x) = x^2$. Move the function 2 units left and 4 units up.

20. Use the function $f(x) = x^2$. Move the function 1 unit right and 2 units down.

chapter four test

21. Use the function $f(x) = x^2$. Move the functions 3 units right and reflect across the y-axis.

In each of the following problems, use the information given to determine **(a)** $(f+g)(-2)$ **(b)** $(f-g)(-2)$ **(c)** $(f \cdot g)(-2)$ **(d)** $\left(\dfrac{f}{g}\right)(-2)$.

22. $f(-2) = -4$ and $g(-2) = 2$

23. $f(x) = -x^2 + x$ and $g(x) = \dfrac{1}{x}$

In each of the following problems, use the information given to determine $(f \circ g)(2)$.

24. $f(-3) = 4$ and $g(2) = -3$

25. $f(x) = 1 - \sqrt[3]{x}$ and $g(x) = -x^2$

26. $f(x) = \left[\dfrac{x}{3}\right]$ and $g(x) = -\sqrt{x}$

27. Given $f(x) = 2x - a + 1$, $g(x) = -x^2 - 2$ and $(f \circ g)(-3) = 4$, find a.

28. Given $f(x) = -x^2 - 1$, find $(f \circ f)(x)$.

29. Given $f(x) = x^2 - 1$, $g(x) = x + 1$, and $(f \circ g)(x) = 0$, find x.

30. Give an example that satisfies $(f \circ g)(x) \neq (g \circ f)(x)$.

31. Given $f(x) = \dfrac{2x-1}{x-2}$, find $(f \circ f)(x)$.

Write the following functions as a composition of two functions. Answers may vary.

32. $f(x) = \sqrt{2x^2 + 1}$

33. $f(x) = |x^2 - x| - 2$

34. $f(x) = \dfrac{3}{3x^2 + 1}$

35. $f(x) = \dfrac{\sqrt{x+2}}{x^2 + 4x + 4}$

36. Given $f(x) = 6x - 1$, and $(f \circ g)(x) = x + 3$, find $g(x)$.

Find a formula for the inverse of each of the following functions.

37. $f(x) = x^{\frac{1}{4}} + 3$

38. $f(x) = -x^2 + 4$

39. $f(x) = \dfrac{2}{x-1}$

40. $f(x) = \sqrt[7]{3x}$

41. $f(x) = \dfrac{2x-3}{x+1}$

42. $2f(x) \cdot x - 3x = f(x) + 1$

43. $f(x) = \sqrt[3]{x+2} - 1$

44. Given $f(x) = 2x - a + 4$, if $f^{-1}(3) = -2$, then find a.

45. Given $f^{-1}(2x-1) = 4x + 3$, find $f(x)$.

chapter four   Relations, Functions, and Their Graphs

46. Given $f(x) = \dfrac{x-a}{2x}$, $g(x) = 2x^2$, and $(f^{-1} \circ g)(-2) = 6$, find a.

47. Given $f(x) = x + a$ and $(f \circ f)^{-1}(-2) = 4$, find a.

48. If $f(x)$ is linear function such that $f(1) = 1$ and $f(2) = 3$, find $f(4)$.

49. Find the domain of the function: $f = \{(-3, 0), (-1, 4), (0, 3), (3, 3), (4, -1)\}$

50. Find the domain of the volume of a sphere, $V = \dfrac{4}{3}\pi r^3$.

For each function below, determine if it has y-axis symmetry, x-axis symmetry, origin symmetry, or none of these.

51. $y = x^4 - 2x^2 + 3$

52. $y = \dfrac{x^3 - x}{2}$

53. $y = |2x - 1|$

54. $y = \dfrac{x}{x^2 + 1}$

55. What is the minimum value of the function $f(x) = x^2 - 2x - 3$?

56. Given that the maximum value of the function $f(x) = -x^2 + 4x - a + 2$ is 2, find a.

57. If the graph of the function $y = x^2 + 4x + a$ is tangent to the x-axis, find a.

58. If $(f \circ f)(x) = 4x - 12$, find $f(x)$.

59. Find the inverse of $f(x) = \dfrac{ax + b}{cx + d}$.

60. If $f(x) = \dfrac{x + 4}{3}$ and $f(x) + f^{-1}(x) = 0$, find x.

In each of the following problems, use the information given to determine:
(a) $(f^{-1} \circ g)(-2)$ (b) $(f \circ g^{-1})(-1)$.

61. $f(x) = \sqrt[3]{x}$ and $g(x) = x - 2$

62. $f(x) = x^2$ and $g(x) = \dfrac{1}{x}$

Given $f(x) = \sqrt{x}$ and $g(x) = \sqrt[3]{x^2}$, evaluate:

63. $f(x + h)$

64. $\dfrac{f(x + h) - f(x)}{x + h}$

65. $g(x + h)$

66. $\dfrac{g(x + h) - g(x)}{h}$

chapter four test

Determine the implied domain of each of the following functions.

67. $f(x) = \sqrt{|-x-3|}$

68. $f(x) = \dfrac{x}{[\![x]\!]}$

69. $f(x) = \dfrac{2x-1}{x^2+x+1}$

70. $f(x) = \sqrt[4]{1-x^2}$

chapter FIVE

POLYNOMIAL FUNCTIONS

OVERVIEW

5.1 Introduction to Polynomial Equations and Graphs

5.2 Polynomial Division and the Division Algorithm

5.3 Locating Real Zeros of Polynomials

5.4 The Fundamental Theorem of Algebra

The information in this chapter will guide you in your work with polynomials — graphing, dividing, solving equations and inequalities, etc. Business applications may often make use of polynomial inequalities to find values for which various situations are profitable. On page 375 you will find a problem describing a skateboard company; given information about their revenue and expenses, you will be asked to find the number of skateboards they should manufacture in order to make a profit. You will master this type of problem, using techniques for *Solving Polynomial Inequalities* as shown in *Example 3* on page 370.

chapter five   Polynomial Functions

Introduction

In Chapter 4, we studied properties of functions in general and learned the nomenclature and notation commonly used when working with functions and their graphs. In this chapter, we narrow our focus and concentrate on polynomial functions.

We have, of course, already seen many examples of polynomial functions but have barely scratched the surface as far as obtaining a deep understanding of polynomials is concerned. We will soon see many mathematical methods that are peculiar to polynomials and make many observations that are relevant only to polynomials. Some of these methods include polynomial division and the Rational Zero Theorem, and many of the observations point out the strong connection between factors of polynomials, solutions of polynomial equations, and (when appropriate) graphs of polynomial functions. Our deeper understanding of polynomial functions will then make solving polynomial inequalities a far easier task than it would be otherwise.

The chapter ends with the Fundamental Theorem of Algebra, a theorem that makes a deceptively simple claim about polynomials, but one that nevertheless manages to tie together nearly all of the methods and observations of the chapter. The German mathematician Carl Friedrich Gauss (1777 - 1855), one of the towering figures in the history of mathematics, first proved this theorem in 1799. In doing so, he accomplished something that many brilliant people (among them Isaac Newton and Leonhard Euler) had attempted, and it is all the more remarkable that Gauss did so in his doctoral dissertation at the age of twenty-two. The Fundamental Theorem of Algebra operates on many levels: philosophically, it can be seen as one of the most elegant and fundamental arguments for the necessity of complex numbers, while pragmatically it is of great importance in solving polynomial equations. It thus ties together observations about polynomials that originated with Italian mathematicians of the 16th century, and points the way toward later work on polynomials by such people as Niels Abel and Evariste Galois.

Gauss

5.1 Introduction to Polynomial Equations and Graphs

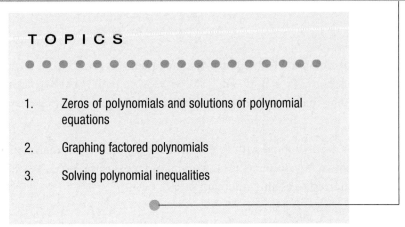

TOPICS

1. Zeros of polynomials and solutions of polynomial equations
2. Graphing factored polynomials
3. Solving polynomial inequalities

Topic 1: Zeros of Polynomials and Solutions of Polynomial Equations

At this point, we have a firm grasp of how linear and quadratic polynomial functions behave, and we have tools guaranteed to solve all linear and quadratic equations. We have also studied some elementary higher-degree polynomial functions (those of the basic form ax^n for some natural number n). In this section, we begin a more complete exploration of higher-degree polynomials.

Ideally, by the end of this chapter our mastery of linear and quadratic polynomials would extend to all polynomials. Unfortunately, this is too much to expect: not surprisingly, the complexity of polynomial functions increases with the degree. This means, for instance, that higher-degree polynomials are generally more difficult to graph accurately, and that we cannot necessarily expect to find exact solutions to higher-degree polynomial equations. Nevertheless, there is much that we can say about a given polynomial function.

In order to make our discussion as general as possible, we will often refer to a generic n^{th} degree polynomial function $f(x) = a_n x^n + a_{n-1} x^{n-1} + \ldots + a_1 x + a_0$ where $a_n, a_{n-1}, \ldots, a_1, a_0$ all represent constants (possibly non-real complex number constants) and $a_n \neq 0$. We will begin by identifying the values of the variable that make a polynomial function equal to zero.

chapter five — Polynomial Functions

Zeros of a Polynomial

> The number k (k may be a complex number) is said to be a **zero** of the polynomial function $f(x) = a_n x^n + a_{n-1} x^{n-1} + \ldots + a_1 x + a_0$ if $f(k) = 0$. This is also expressed by saying that k is a **root** or a **solution** of the equation $f(x) = 0$.

If f is a polynomial with real coefficients, and if k is a real number zero of f, then the statement $f(k) = 0$ means the graph of f crosses the x-axis at $(k, 0)$. In this case, k may be referred to as an x-intercept of f.

The task of determining the zeros of a polynomial arises in many contexts, two of which are solving polynomial equations and graphing polynomials. We will discuss polynomial equations first.

Polynomial Equations

> A **polynomial equation in one variable**, say the variable x, is an equation that can be written in the form $a_n x^n + a_{n-1} x^{n-1} + \ldots + a_1 x + a_0 = 0$, where a_n, a_{n-1}, ..., a_1, a_0 are constants. Assuming $a_n \neq 0$, we say such an equation is of degree n.

Just as with linear and quadratic equations (which are polynomial equations of degree 1 and 2, respectively), a given polynomial equation may not initially appear as in the above definition. The first task, then, may be to add or subtract terms as necessary so that one side of the equation is zero. Once this is done, the zeros of the polynomial constitute the solutions of the equation.

section 5.1 Introduction to Polynomial Equations and Graphs

example 1

Verify that the given values of x solve the corresponding polynomial equations.

a. $6x^2 - x^3 = 12 + 5x$; $x = 4$

b. $x^2 = 2x - 5$; $x = 1 + 2i$

c. $\dfrac{x}{1-i} = 3x^2$; $x = 0$

Solutions:

This screen was performed using equation solver.

a.
$$6x^2 - x^3 = 12 + 5x$$
$$6(4)^2 - (4)^3 \stackrel{?}{=} 12 + 5(4)$$
$$96 - 64 \stackrel{?}{=} 12 + 20$$
$$32 = 32$$

Verifying that 4 is a solution of the equation is just a matter of substituting 4 for x and simplifying both sides. This has also verified that 4 is a zero of the polynomial function $f(x) = -x^3 + 6x^2 - 5x - 12$. That is, $f(4) = 0$.

b.
$$x^2 = 2x - 5$$
$$x^2 - 2x + 5 = 0$$
$$x = \dfrac{2 \pm \sqrt{4 - 20}}{2}$$
$$x = \dfrac{2 \pm 4i}{2}$$
$$x = 1 \pm 2i$$

We could verify that $1 + 2i$ is a solution of the equation $x^2 = 2x - 5$ by substituting $1 + 2i$ for x and simplifying both sides, but since this is a quadratic equation we have the tools necessary to solve the equation for ourselves.

The quadratic formula (see Section 2.3) tells us that $1 + 2i$ and $1 - 2i$ are the two solutions of the equation.

c.
$$\dfrac{x}{1-i} = 3x^2$$
$$\dfrac{0}{1-i} \stackrel{?}{=} 3(0)^2$$
$$0 = 0$$

Don't be misled by the complex number in the denominator; this is another polynomial equation, and verification that 0 is a solution is shown at left. For practice, verify that the polynomial $f(x) = 3x^2 - \dfrac{x}{1-i}$ could be rewritten as $f(x) = 3x^2 - \left(\dfrac{1}{2} + \dfrac{i}{2}\right)x$ (see Section 1.6 for review). Of course, 0 is a zero of f.

chapter five   **Polynomial Functions**

Topic 2:   # Graphing Factored Polynomials

Consider again the generic polynomial function $f(x) = a_n x^n + a_{n-1} x^{n-1} + \ldots + a_1 x + a_0$, where $a_n \neq 0$ and, for the moment, all of the coefficients $a_n, a_{n-1}, \ldots, a_1, a_0$ are assumed to be real. Our goal is to be able to sketch the graph of such a function, paying particular attention to the behavior of f as $x \to -\infty$ and as $x \to \infty$, the y-intercept of f, and the x-intercepts of f (if there are any).

The behavior of a polynomial function as $x \to \pm\infty$ can be determined as follows. In brief, the graph of $f(x) = a_n x^n + a_{n-1} x^{n-1} + \ldots + a_1 x + a_0$ is similar to the graph of $a_n x^n$ for values of x that are very large in magnitude; that is, as $x \to -\infty$ and as $x \to \infty$, the leading term of $f(x) = a_n x^n + a_{n-1} x^{n-1} + \ldots + a_1 x + a_0$ dominates the behavior. Since we have already studied functions of the form $a_n x^n$, we know roughly how the sketch of an nth degree polynomial function should appear far away from the origin. Specifically, we know from Section 4.3 that if n is even, $x^n \to \infty$ as $x \to -\infty$ and as $x \to \infty$, and if n is odd, then $x^n \to -\infty$ as $x \to -\infty$ and $x^n \to \infty$ as $x \to \infty$. We also know that if a_n is positive, multiplying x^n by a_n merely compresses (if $0 < a_n < 1$) or stretches (if $a_n > 1$) the graph of x^n, while if a_n is negative, the graph of $a_n x^n$ is the reflection with respect to the x-axis of the graph of $|a_n| x^n$. Figure 1 below contains a few examples of these observations.

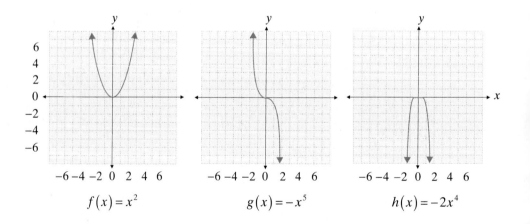

Figure 1: Examples of Behavior at $\pm\infty$

section 5.1   Introduction to Polynomial Equations and Graphs

Near the origin, however, the graph of $f(x) = a_n x^n + a_{n-1} x^{n-1} + \ldots + a_1 x + a_0$ is likely to be quite different from the graph of $a_n x^n$. For one thing, the y-intercept of f is $(0, a_0)$ and a_0 is possibly non-zero, whereas $a_n x^n$ always crosses the y-axis at $(0, 0)$. (Note that to determine the y-intercept of f, we simply evaluate f at 0; for our generic polynomial, $f(0) = a_0$.) Also, an nth degree polynomial may have, in general, a very different number of x-intercepts than the function $a_n x^n$ (which has exactly one x-intercept, namely the origin). The y-intercept of a polynomial function will never be difficult to determine, but finding the x-intercepts (if any exist) may require more effort.

At this point, the value of writing a polynomial function in factored form will be very apparent. If we are able to factor a given polynomial f into a product of linear factors, every linear factor with real coefficients will correspond to an x-intercept of the graph of f. To see why this is true, consider the polynomial function

$$f(x) = (3x - 5)(x + 2)(2x - 6).$$

To determine the x-intercepts of this polynomial, we need to find all real numbers k for which $f(k) = 0$; that is, we need to solve the equation

$$(3x - 5)(x + 2)(2x - 6) = 0.$$

Since the polynomial in this polynomial equation appears in factored form, the Zero-Factor Property (see Section 2.3) tells us immediately that the only solutions are those values of x for which $3x - 5 = 0$, $x + 2 = 0$, or $2x - 6 = 0$. Solving these three linear equations gives us the x-coordinates of the three x-intercepts of f: $\left\{\frac{5}{3}, -2, 3\right\}$.

What else can we say about the graph of $f(x) = (3x - 5)(x + 2)(2x - 6)$? If we were to multiply out the three linear factors, we would obtain a third degree polynomial. In order to know how the graph of f behaves as $x \to -\infty$ and as $x \to \infty$, we also need to know the leading coefficient. Instead of actually multiplying out f completely, we can quickly determine the leading coefficient by considering how the x^3 term arises. The third degree term comes from multiplying together the $3x$ from the first factor, the x from the second factor, and the $2x$ from the third factor. That is, the third degree term of the expanded polynomial is $6x^3$. Since the leading coefficient, 6, is positive, we know that $f(x) \to -\infty$ as $x \to -\infty$ and $f(x) \to \infty$ as $x \to \infty$.

chapter five   Polynomial Functions

The last observation we will make about the graph of f is that $f(0) = (0-5)(0+2)(0-6)$, or $f(0) = 60$, so the y-intercept is $(0, 60)$. Putting all of the above together, along with a few more computed values of f, we obtain the sketch in Figure 2 below (note the difference in the horizontal and vertical scales).

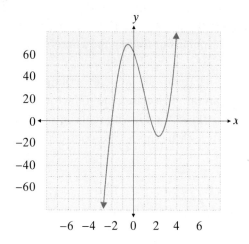

Figure 2: Graph of $f(x) = (3x-5)(x+2)(2x-6)$

The above method of graphing factored polynomials leaves unanswered one very important question: What do we do if we are given a polynomial in non-factored form? Since the only potentially difficult step in the method is determining the x-intercepts, this question is equivalent to: How do we solve the generic n^{th} degree polynomial equation $a_n x^n + a_{n-1} x^{n-1} + \ldots + a_1 x + a_0 = 0$? The answers to these questions will constitute much of the rest of this chapter, so for the remainder of this section all polynomials will either be given in factored form or else will be readily factored with the tools we already possess.

example 2

Sketch the graphs of the following polynomial functions, paying particular attention to the x-intercept(s), the y-intercept, and the behavior as $x \to \pm\infty$.

    a. $f(x) = -x(2x+1)(x-2)$ **b.** $g(x) = x^4 - 1$ **c.** $h(x) = x^2 + 2x - 3$

section 5.1 Introduction to Polynomial Equations and Graphs

Solutions:

a.

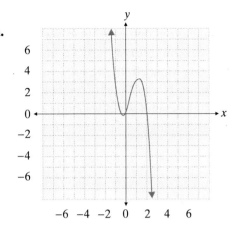

The solutions of $-x(2x+1)(x-2) = 0$ are $-\frac{1}{2}$, 0, and 2, so these are the x-coordinates of the three x-intercepts.

The graph of f crosses the y-axis at $f(0) = 0$.

Finally, if we were to multiply out the three linear factors of f, the highest degree term would be $(-x)(2x)(x) = -2x^3$. The degree of f and the fact that the leading coefficient is negative indicates how f behaves far away from the origin.

b.

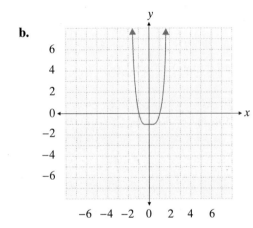

We know, from the discussion in Chapter 4, that the graph of $g(x) = x^4 - 1$ is the graph of x^4 shifted down one unit. But we can also determine the x-intercepts by solving the equation $x^4 - 1 = 0$. Since the polynomial is a difference of two squares, we can factor it to obtain the equation $(x^2 - 1)(x^2 + 1) = 0$, and then factor again to obtain $(x-1)(x+1)(x^2+1) = 0$. Only the two linear factors correspond to x-intercepts (the factor $x^2 + 1$ leads to zeros of i and $-i$, which are irrelevant for graphing purposes).

c.

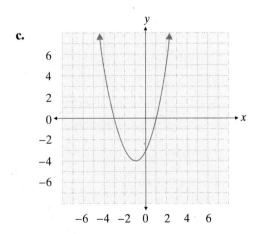

We can solve the quadratic equation $x^2 + 2x - 3 = 0$ by factoring, which gives us $(x+3)(x-1) = 0$, and hence x-intercepts for the function h of -3 and 1. We also know that the shape of h is an upward-opening parabola, and that the graph crosses the y-axis at $h(0) = -3$. We also know that the axis of symmetry must be a vertical line halfway between the x-intercepts, so the x-coordinate of the vertex must be -1. Note that $h(-1) = -4$.

chapter five   Polynomial Functions

Topic 3: Solving Polynomial Inequalities

We first encountered inequalities in one variable in Section 2.2. As in that section, the inequalities we study here contain just one variable, but the variable may now appear with exponents larger than one. Specifically, we will learn how to solve inequalities that would be polynomial equations if the inequality symbol were replaced by equality. The techniques learned here will be useful when we study rational functions in Chapter 6, and will also aid your understanding of just what, exactly, the graph of a function tells us.

Every polynomial inequality can be rewritten in the form $f(x)<0$, $f(x)\leq 0$, $f(x)>0$, or $f(x)\geq 0$, where f is a polynomial function, and this will be the key to solving the inequality. By graphing the polynomial f, we will be able to easily pick out the intervals that solve the inequality. Example 3 illustrates this process.

example 3

Before selecting the `Y=` screen, change the `MODE` from Sequential to Simul. To enter Y_1, select `VARS` (as described on pg. 326), and to enter >, select `2ND` `MATH` (Test) and enter the desired symbol.

Solve the following polynomial inequalities.

a. $2x^2 - 3x > 9$ b. $x^4 - 2x^2 < -x^3$ c. $(x+3)(x+1)(x-2) \geq 0$

Solutions:

a.

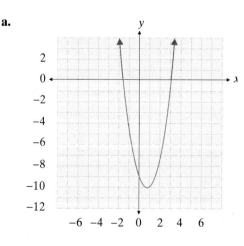

We need to identify the intervals for which $2x^2 - 3x - 9 > 0$; that is, we need to locate the intervals on which the graph of the polynomial $f(x) = 2x^2 - 3x - 9$ is strictly above the x-axis. Factoring is the first step: $f(x) = (2x+3)(x-3)$. This tells us that $-\frac{3}{2}$ and 3 are the x-intercepts of this upward-opening parabola. After graphing, we see the solution of the inequality is $\left(-\infty, -\frac{3}{2}\right) \cup (3, \infty)$.

b.

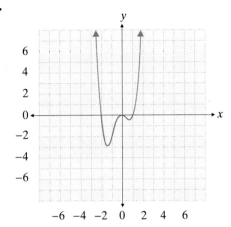

Again, we begin by writing the inequality in the form $x^4 + x^3 - 2x^2 < 0$. The function $f(x) = x^4 + x^3 - 2x^2$ can be factored more easily after first factoring out x^2, and we obtain $f(x) = x^2(x+2)(x-1)$. This means that the x-intercepts are -2, 0, and 1, but this function behaves a bit differently near 0 than others we have graphed so far. By plotting a few points near 0, we see that f touches the x-axis at this x-intercept, but does not cross.

The solution of the inequality consists of those intervals where f is strictly below the x-axis: $(-2, 0) \cup (0, 1)$.

c.

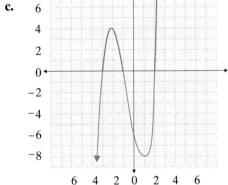

The polynomial in $(x+3)(x+1)(x-2) \geq 0$ is already factored, and the graph of the function $f(x) = (x+3)(x+1)(x-2)$ is as shown. The solution of the inequality consists of those intervals where the graph of f is above or on the x-axis. Hence, the solution is $[-3, -1] \cup [2, \infty)$. Note that -3, -1, and 2 are included in the solution set.

chapter five — Polynomial Functions

exercises

Verify that the given values of x solve the corresponding polynomial equations. See Example 1.

1. $9x^2 - 4x = 2x^3 + 15$; $x = -1$
2. $x^2 - 4x = -13$; $x = 2 - 3i$
3. $x^2 + 13 = 4x$; $x = 2 + 3i$
4. $3x^3 + (5 - 3i)x^2 = (2 + 5i)x - 2i$; $x = i$
5. $9x^2 - 4x = 2x^3 + 15$; $x = 3$
6. $9x^2 - 4x = 2x^3 + 15$; $x = \dfrac{5}{2}$
7. $3x^3 + (5 - 3i)x^2 = (2 + 5i)x - 2i$; $x = -2$
8. $x^5 - 10x^4 - 80x^2 = 32 - 80x - 40x^3$; $x = 2$
9. $4x^5 - 8x^4 - 12x^3 = 16x^2 - 25x - 69$; $x = 3$
10. $x^2 - 4x - 12 = 0$; $x = 6$
11. $23x^7 - 12x^5 = 63x^4 - 3x^2$; $x = 0$
12. $x^2 + 74 = 10x$; $x = 5 + 7i$
13. $4x^2 + 32x + (8 + i)x^3 = -8$; $x = 2i$
14. $8x - 17 = x^2$; $x = 4 - i$
15. $(5 - 3i)x - 3x = 4 - 6i$; $x = 2$
16. $x^6 - x^5 + 7x^4 + x^3 - 9x = -1$; $x = 1$
17. $6x^7 - 3x^5 = 3x^4 - 6x^2$; $x = -1$

Determine if the given values of x are zeros of the corresponding polynomial equation. See Example 1.

18. $16x = x^3 + x^2 + 20$; $x = -5$
19. $x^4 - 13x^2 + 12 = -x^3 + x$; $x = -1$
20. $x^4 - 3x^3 - 10x^2 = 0$; $x = 2$
21. $4x^5 - 216x^2 = 36x^3 - 24x^4$; $x = -6$
22. $x^3 - 8ix + 30 = 15x + 2x^2 + 16i$; $x = -i$
23. $x^3 - 7x^2 + 4x - 28 = 0$; $x = 2i$

Solve the following polynomial equations by factoring and/or using the quadratic formula, making sure to identify all the solutions. See Section 2.3 for review, if necessary.

24. $x^3 - x^2 - 6x = 0$
25. $x^2 - 2x + 5 = 0$
26. $x^4 + x^2 - 2 = 0$
27. $2x^2 + 5x = 3$
28. $9x^2 = 6x - 1$
29. $x^4 - 8x^2 + 15 = 0$

section 5.1 Introduction to Polynomial Equations and Graphs

30. $x^3 - x^2 = 72x$
31. $x^2 + 5x = -\dfrac{25}{4}$
32. $2x^2 + 5 = 11x$
33. $x^4 - 8x^3 + 25x^2 = 0$
34. $x^4 - 13x^2 + 36 = 0$
35. $x^4 + 7x^2 = 8$

For each of the following polynomials, determine the degree and the leading coefficient; then describe the behavior of the graph as $x \to \pm\infty$.

36. $p(x) = 2x^4 - 3x^3 - 6x^2 - x - 23$
37. $i(x) = 4x^7 + 5x^5 + 12$
38. $r(x) = (3x+5)(x-2)(2x-1)(4x-7)$
39. $h(x) = -6x^5 + 2x^3 - 7x$
40. $g(x) = (x-5)^3(2x+1)(-x-1)$
41. $f(x) = -2(x+4)(x-4)(x^2)$

For each of the following polynomial functions, describe the behavior of its graph as $x \to \pm\infty$ and identify the x- and y-intercepts. Use this information to then sketch the graph of each polynomial. See Example 2.

42. $f(x) = (x-3)(x+2)(x+4)$
43. $g(x) = (3-x)(x+2)(x+4)$
44. $f(x) = (x-2)^2(x+5)$
45. $h(x) = -(x+2)^3$
46. $r(x) = x^2 - 2x - 3$
47. $s(x) = x^3 + 3x^2 + 2x$
48. $f(x) = -(x-2)(x+1)^2(x+3)$
49. $g(x) = (x-3)^5$

In Exercises 50 through 55, use the behavior as $x \to \pm\infty$ and the intercepts to match each polynomial with one of the graphs labeled a. through f.

50. $g(x) = (x+1)^2(x-3)^2$

51. $h(x) = 1 - (x+2)^2$

52. $f(x) = (x-1)(x+2)(3-x)$

53. $r(x) = x^2 - x - 6$

54. $s(x) = (x-1)^3 - 2$

55. $j(x) = (x-2)^2(4x+1)(x+2)(x-2)$

a.
b.
c.
d.
e.
f.

chapter five   Polynomial Functions

Match each of the following functions to the appropriate description.

56. $z(x) = (x-1)(x+2)(4-x)$

57. $r(x) = x^2 - 6x - 7$

58. $s(x) = x^3 + 3x^2 + 2x$

59. $g(x) = (x-1)(x+4)(3-x)$

60. $s(x) = x^3 + 5x^2 + 4x$

61. $s(x) = x^2 - 5x - 6$

a. x-cubed curve increasing to infinity, has x-intercepts of 0, −1, and −2, and crosses the y-axis at 0.

b. parabola that opens up, has x-intercepts at 6 and −1, crosses the y-axis at −6.

c. x-cubed curve increasing to infinity, has x-intercepts of 0, −1, and −4, and crosses the y-axis at 0.

d. parabola that opens up, has x-intercepts at 7 and −1, crosses the y-axis at −7.

e. x-cubed curve decreasing to infinity, has x-intercepts at 1, 4, and −2, crosses the y-axis at −8.

f. x-cubed curve decreasing to infinity, has x-intercepts of 1, 3, and −4, and crosses the y-axis at −12.

Solve the following polynomial inequalities. See Example 3.

62. $x^2 - x - 6 \leq 0$

63. $x^2 > x + 6$

64. $(x+2)^2 (x-1)^2 > 0$

65. $x^3 + 3x^2 + 2x < 0$

66. $(x-2)(x+1)(x+3) \geq 0$

67. $(x-1)(x+2)(3-x) \leq 0$

68. $-x^3 - x^2 + 30x > 0$

69. $(x^2 - 1)(x-4)(x+5) \leq 0$

70. $x^4 + x^2 > 0$

71. $4x^2 < 6x + 4$

72. $x^2(x+4)(x-3) > 0$

73. $(x-3)(x+4)(2-x) > 0$

section 5.1   Introduction to Polynomial Equations and Graphs

Solve the following application problems.

74. A small start-up skateboard company projects that the cost per month of manufacturing x skateboards will be $C(x) = 10x + 300$, and the revenue per month from selling x skateboards will be $r(x) = -x^2 + 50x$. Given that profit is revenue minus cost, what value(s) for x will give the company a non-negative profit?

75. A manufacturer has determined that the revenue from the sale of x cordless telephones is given by $r(x) = -x^2 + 15x$. The cost of producing x telephones is $C(x) = 135 - 17x$. Given that profit is revenue minus cost, what value(s) for x will give the company a non-negative profit?

76. The revenue from the sale of x fire extinguishers is estimated to be $r(x) = 9 - x^2$. The total cost of producing x fire extinguishers is $C(x) = 209 - 33x$. Given that profit is revenue minus cost, what value(s) for x will give the company a non-negative profit?

77. A manufacturer has determined that the cost and revenue of producing and selling x telescopes are $C(x) = 253 - 7x$ and $r(x) = 27x - x^2$, respectively. Given that profit is revenue minus cost, what value(s) for x will give the company a non-negative profit?

78. A company that produces and sells compact refrigerators has found that the revenue from the sale of x compact refrigerators is $r(x) = -x^2 + 30x - 370$. The cost function is given by $C(x) = 6 - 25x$. Given that profit is revenue minus cost, what value(s) for x will give the company a non-negative profit?

chapter five   **Polynomial Functions**

79. The population of sea lions on an island is represented by the function $L(m) = 110m^2 - 0.35m^4 + 750$, where m is the number of months the sea lions have been observed on the island. Given this information, how many more months will there be sea lions on the island?

80. An electronics company is deciding whether or not to begin producing MP3 players. The company must determine if doing so is profitable enough to start producing MP3 players. The profit function is modeled by the equation $P(x) = x + 0.27x^2 - 0.0015x^3 - 300$, where x is the number of MP3 players produced in hundreds. Given this equation, how many MP3 players must the company produce to make a profit?

81. The population of mosquitoes in a city in Florida is modeled by the function $M(w) = 200w^2 - 0.01w^4 + 1200$, where w is the number of weeks since the town began spraying for mosquitoes. How many weeks will it take for all the mosquitoes to have died off?

5.2 Polynomial Division and the Division Algorithm

TOPICS

1. The Division Algorithm and the Remainder Theorem
2. Polynomial long division and synthetic division
3. Constructing polynomials with given zeros

Topic 1: The Division Algorithm and the Remainder Theorem

In Section 5.1, we made use of the factored form of a polynomial in graphing polynomials and in solving polynomial equations and inequalities. In this section, we will formalize the observations we made and introduce some useful techniques for working with polynomials.

We begin with a statement about polynomials that is the analog of an elementary number fact: If a natural number is divided by a smaller one, the result can be expressed as a whole number (called the quotient) plus a fraction less than 1 (the fraction may in fact be 0). The polynomial version of this is called the *Division Algorithm*.

The Division Algorithm

Let $p(x)$ and $d(x)$ be polynomials such that $d(x) \neq 0$ and with the degree of d less than or equal to the degree of p. Then there are unique polynomials $q(x)$ and $r(x)$, called the quotient and the remainder, respectively, such that

$$\underbrace{p(x)}_{\text{dividend}} = \underbrace{q(x)}_{\text{quotient}} \underbrace{d(x)}_{\text{divisor}} + \underbrace{r(x)}_{\text{remainder}}.$$

The degree of the remainder, r, is less than the degree of the divisor d, or else the remainder is 0, in which case we say d divides evenly into the polynomial p. If the remainder is 0, the two polynomials q and d are factors of p.

If we divide every term in the equation $p(x) = q(x)d(x) + r(x)$ by the polynomial d, we obtain a form that is very similar to the number fact mentioned above:

$$\frac{p(x)}{d(x)} = q(x) + \frac{r(x)}{d(x)}$$

This fact may be stated in words as something like "If one polynomial is divided by another of smaller degree, the result is a polynomial plus, possibly, a ratio of two polynomials, the numerator of which has a smaller degree than the denominator." This version of the division algorithm will be very useful when we study rational functions in Chapter 6.

In many cases, we will be interested in dividing a given polynomial p by a divisor of the form $d(x) = x - k$. The division algorithm tells us that the remainder is guaranteed to be either 0 (which is a polynomial of undefined degree), or else a polynomial of degree 0 (since the degree of d is 1). In either case, the remainder polynomial is guaranteed to be simply a number, so $p(x) = q(x)(x - k) + r$ where r is a constant.

What does this mean if k happens to be a zero of the polynomial p? If this is the case, $p(k) = 0$, so

$$0 = p(k)$$
$$= q(k)(k - k) + r$$
$$= r.$$

In other words, if k is a zero of the polynomial p, $p(x) = q(x)(x - k)$. On the other hand, if $x - k$ divides a given polynomial p evenly, then we know $p(x) = q(x)(x - k)$ and hence $p(k) = q(k)(k - k) = 0$, so k is a zero of p. Together, these two observations constitute a major tool that we use in graphing polynomials and in solving polynomial equations, summarized below for reference.

Zeros and Linear Factors

The number k is a zero of a polynomial $p(x)$ if and only if the linear polynomial $x - k$ is a factor of p. In this case, $p(x) = q(x)(x - k)$ for some quotient polynomial q. This also means that k is a solution of the polynomial equation $p(x) = 0$, and if p is a polynomial with real coefficients and if k is a real number, then k is an x-intercept of p.

The above reasoning also leads to the more general conclusion called the *remainder theorem*.

The Remainder Theorem

If the polynomial $p(x)$ is divided by $x - k$, the remainder is $p(k)$. That is,
$$p(x) = q(x)(x-k) + p(k).$$

Topic 2: Polynomial Long Division and Synthetic Division

To make use of the division algorithm and the remainder theorem, we need to be able to actually divide one polynomial by another. Polynomial long division is the analog of numerical long division, and provides the means for dividing any polynomial by another of equal or smaller degree. Synthetic division is a shortcut that allows us to quickly divide a polynomial by a polynomial of the form $x - k$.

Polynomial long division is best explained with an example. As you study *Example 1*, notice the similarities between polynomial division and numerical division.

example 1

Divide the polynomial $6x^5 - 5x^4 + 10x^3 - 15x^2 + 9x - 19$ by the polynomial $2x^2 - x + 3$.

Solution:

$$\begin{array}{r} 3x^3 \\ 2x^2 - x + 3 \overline{\smash{)}6x^5 - 5x^4 + 10x^3 - 15x^2 + 9x - 19} \\ -(6x^5 - 3x^4 + 9x^3) \\ \hline -2x^4 + x^3 - 15x^2 + 9x - 19 \end{array}$$

The first step is to arrange the dividend and the divisor in descending order. The first term of the quotient is then the first term of the dividend divided by the first term of the divisor, giving us $3x^3$ in this case. Multiply $3x^3$ by $2x^2 - x + 3$ to obtain the polynomial $6x^5 - 3x^4 + 9x^3$, which we subtract from the dividend.

continued on next page...

chapter five   Polynomial Functions

$$\begin{array}{r} 3x^3 - x^2 \\ 2x^2 - x + 3 \overline{) 6x^5 - 5x^4 + 10x^3 - 15x^2 + 9x - 19} \\ -(6x^5 - 3x^4 + 9x^3) \\ \overline{-2x^4 + x^3 - 15x^2 + 9x - 19} \\ -(-2x^4 + x^3 - 3x^2) \\ \overline{-12x^2 + 9x - 19} \end{array}$$

To determine the second term of the quotient, we repeat the above process, this time dividing the first term of $-2x^4 + x^3 - 15x^2 + 9x - 19$ by $2x^2$ to get $-x^2$ in the quotient. Multiply this by $2x^2 - x + 3$ to obtain the next polynomial we subtract, $-2x^4 + x^3 - 3x^2$.

$$\begin{array}{r} 3x^3 - x^2 - 6 \\ 2x^2 - x + 3 \overline{) 6x^5 - 5x^4 + 10x^3 - 15x^2 + 9x - 19} \\ -(6x^5 - 3x^4 + 9x^3) \\ \overline{-2x^4 + x^3 - 15x^2 + 9x - 19} \\ -(-2x^4 + x^3 - 3x^2) \\ \overline{-12x^2 + 9x - 19} \\ -(-12x^2 + 6x - 18) \\ \overline{3x - 1} \end{array}$$

To complete the division, we repeat the procedure one more time. At this point, the process halts, as the degree of $3x - 1$ is smaller than the degree of the divisor.

We can summarize the result as:

$$\frac{6x^5 - 5x^4 + 10x^3 - 15x^2 + 9x - 19}{2x^2 - x + 3} = 3x^3 - x^2 - 6 + \frac{3x - 1}{2x^2 - x + 3}$$

caution

Although polynomial long division is a straightforward process, one common error is to forget to distribute the minus sign in each step as one polynomial is subtracted from the one above it. A good way to avoid this error is to put parentheses around the polynomial being subtracted, as in *Example 1* above.

Synthetic division is a shortened version of polynomial long division, and can be used when the divisor is of the form $x - k$ for some constant k. Synthetic division does not do anything that long division can't do (and in fact is only applicable in certain circumstances), but the speed of synthetic division is often convenient.

section 5.2   Polynomial Division and the Division Algorithm

Compare the division of $-2x^3 + 8x^2 - 9x + 7$ by $x - 2$ below, using long division on the left and synthetic division on the right.

$$\begin{array}{r} -2x^2 + 4x - 1 \\ x-2 \overline{\smash{\big)}\, -2x^3 + 8x^2 - 9x + 7} \\ \underline{-(-2x^3 + 4x^2)} \\ 4x^2 - 9x + 7 \\ \underline{-(4x^2 - 8x)} \\ -x + 7 \\ \underline{-(-x + 2)} \\ 5 \end{array}$$

$$\begin{array}{r|rrrr} 2 & -2 & 8 & -9 & 7 \\ & & -4 & 8 & -2 \\ \hline & -2 & 4 & -1 & 5 \end{array}$$

The increase in efficiency in synthetic division comes from omitting the variables in the division process. Instead of various powers of the variable, synthetic division uses a tabular arrangement to keep track of the coefficients of the dividend and, ultimately, the coefficients of the quotient and the remainder. In the example above, note that the numbers in red are the coefficients of the dividend and the numbers in blue are the coefficients of the quotient and remainder. More importantly, note that the number 2, which corresponds to k in the form $x - k$, appears without the minus sign. This is a very important and easily overlooked fact: When dividing by $x - k$ using synthetic division, the number that appears in the upper left spot is k.

Synthetic division is an iterative process based on simple numerical multiplication and addition. Once we have written k and the coefficients of the dividend as above, the synthetic division process begins by copying the leading coefficient of the dividend (-2 in the above example) in the first slot below the horizontal line. This number is then multiplied by the number k (which is 2 above) and the result is written directly below the second coefficient of the dividend (8 in the example). The two numbers in that column are then added and the result written in the second slot below the horizontal line. The process is repeated until the last column is completed, and the last number written down is the remainder. The other numbers in the bottom row constitute the coefficients of the quotient, which will be a polynomial of one degree less than the dividend. For instance, in our example above and to the right, the remainder is 5 and the quotient is $-2x^2 + 4x - 1$, read from the bottom row.

chapter five   Polynomial Functions

Figure 1 is an illustration of the calculations described on the previous page, for the same division problem.

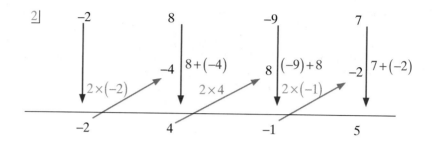

Figure 1: Synthetic Division

Because synthetic division is much faster than long division, it is very useful in determining if $x - k$ is a factor of a given polynomial. By the remainder theorem, synthetic division also provides a quick means of determining $p(k)$ for a given polynomial p, since $p(k)$ is the remainder when $p(x)$ is divided by $x - k$. This fact is used in *Example 2*.

example 2

For each polynomial p below, determine if the given k is a zero. If not, determine $p(k)$.

a. $p(x) = 3x^8 + 9x^7 - x^3 - 3x^2 + x - 1; \quad k = -3$

b. $p(x) = -2x^4 + 11x^3 - 5x^2 - 3x + 15; \quad k = 5$

Solutions:

a.
```
-3│  3   9   0   0   0  -1  -3   1  -1
       -9   0   0   0   0   3   0  -3
   ─────────────────────────────────────
       3   0   0   0   0  -1   0   1  -4
```

First of all, note that it is essential that we place a number of 0's in certain slots in the first row, as these serve as placeholders for the missing terms of the dividend (namely, x^6, x^5, and x^4). The fact that the last number is non-zero means -3 is not a zero of p. What we *can* conclude is that $p(-3) = -4$.

section 5.2 Polynomial Division and the Division Algorithm

$$\frac{3x^8 + 9x^7 - x^3 - 3x^2 + x - 1}{x + 3}$$

$$= 3x^7 - x^2 + 1 - \frac{4}{x+3}$$

Note also that we can write the result of the division as a polynomial plus a rational function. And since $k = -3$, the divisor is $x - (-3)$, or $x + 3$.

b.
$$\begin{array}{r|rrrrr} 5 & -2 & 11 & -5 & -3 & 15 \\ & & -10 & 5 & 0 & -15 \\ \hline & -2 & 1 & 0 & -3 & 0 \end{array}$$

$$-2x^4 + 11x^3 - 5x^2 - 3x + 15$$
$$= (-2x^3 + x^2 - 3)(x - 5)$$

In this case, the remainder is 0, and hence 5 is a zero of the polynomial p.

Since the remainder is 0, we now know of two factors of p, as illustrated at left.

To convince yourself of the utility of synthetic division in evaluating polynomials, examine *Example 2a* again. The direct way to determine $p(-3)$ is, of course, to simplify the following: $p(-3) = 3(-3)^8 + 9(-3)^7 - (-3)^3 - 3(-3)^2 + (-3) - 1$. This is a tedious process, but the result, $3(6561) + 9(-2187) - (-27) - 3(9) + (-3) - 1$, does indeed equal -4. Compare these calculations with the far simpler synthetic division used in the example.

Before leaving the subject of polynomial division, let us consider polynomials with complex coefficients and/or complex zeros. When graphing polynomials, we will be concerned with those that have only real coefficients, but complex zeros and coefficients may still arise in intermediate stages of the graphing process. In solving polynomial equations, we have already seen (in the case of quadratic equations) that complex numbers may be the *only* solutions. For these reasons, it is important for us to be able to handle complex numbers as they arise.

chapter five   Polynomial Functions

example 3

a. Divide $p(x) = x^4 + 1$ by $d(x) = x^2 + i$.

b. Perform the indicated division: $\dfrac{-3x^3 + (5-2i)x^2 + (-4+i)x + (1-i)}{x - 1 + i}$

Solutions:

a.
$$
\begin{array}{r}
x^2 - i \\
x^2 + i \overline{\smash{\big)} x^4 + 1} \\
\underline{-(x^4 + ix^2)} \\
-ix^2 + 1 \\
\underline{-(-ix^2 + 1)} \\
0
\end{array}
$$

Note that the divisor is not of the form $x - k$, so long division is called for.

The procedure is the same as always, though complex number arithmetic is called for (for instance, we need to know $-i^2 = 1$). The remainder of 0 tells us that $x^4 + 1 = (x^2 + i)(x^2 - i)$.

b.
$$
\begin{array}{r|rrrr}
1-i & -3 & 5-2i & -4+i & 1-i \\
 &   & -3+3i & 3-i & -1+i \\
\hline
 & -3 & 2+i & -1 & 0
\end{array}
$$

In this problem, the divisor can be written as $x - (1-i)$, so we can use synthetic division. Note that we have proved that $1-i$ is a zero of the numerator of the original fraction.

Topic 3:   Constructing Polynomials with Given Zeros

The last topic in this section concerns reversing the division process. We now know well the connection between zeros and factors: k is a zero of the polynomial $p(x)$ if and only if $x - k$ is a factor of $p(x)$. We can make use of this fact to construct polynomials that have certain desired properties, as illustrated in *Example 4*.

example 4

Construct a polynomial that has the given properties.

    a. Third degree, zeros of $-3, 2,$ and $5,$ and goes to $-\infty$ as $x \to \infty$.
    b. Fourth degree, zeros of $-5, -2, 1,$ and $3,$ and a y-intercept of 15.

Solutions:

a. $p(x) = -(x+3)(x-2)(x-5)$
$ = -x^3 + 4x^2 + 11x - 30$

We begin by noting that $x+3$, $x-2$, and $x-5$ must be factors of the polynomial p we are to construct, since these factors give rise to the desired zeros. Further, there can be no other linear (or higher degree) factors, since the polynomial is to be third degree. But a cubic with a positive leading coefficient goes to $-\infty$ as $x \to \infty$, so we must multiply the three linear factors by a negative constant to achieve the desired behavior. If we multiply by -1, the result is as shown at left.

b. $p(x) = a(x+5)(x+2)(x-1)(x-3)$
$p(0) = a(5)(2)(-1)(-3) = 30a$
$30a = 15 \Rightarrow a = \dfrac{1}{2}$

$p(x) = \dfrac{1}{2}(x+5)(x+2)(x-1)(x-3)$
$ = \dfrac{1}{2}x^4 + \dfrac{3}{2}x^3 - \dfrac{15}{2}x^2 - \dfrac{19}{2}x + 15$

The given zeros and the desired degree of the polynomial lead to the four linear factors shown at left. And as in the first problem, the linear factors can be multiplied by any non-zero constant a without affecting the zeros. Since we want the y-intercept to be 15, it must be the case that $30a = 15$, allowing us to solve for a.

If we multiply out the four linear factors and the constant, the result is as shown.

chapter five   Polynomial Functions

exercises

Use polynomial long division to rewrite each of the following fractions in the form $q(x) + \dfrac{r(x)}{d(x)}$, where $d(x)$ is the denominator of the original fraction, $q(x)$ is the quotient, and $r(x)$ is the remainder. See Examples 1 and 3.

1. $\dfrac{6x^4 - 2x^3 + 8x^2 + 3x + 1}{2x^2 + 2}$

2. $\dfrac{5x^2 + 9x - 6}{x + 2}$

3. $\dfrac{x^3 - 6x^2 + 12x - 10}{x^2 - 4x + 4}$

4. $\dfrac{7x^5 - x^4 + 2x^3 - x^2}{x^2 + 1}$

5. $\dfrac{4x^3 - 6x^2 + x - 7}{x + 2}$

6. $\dfrac{x^3 + 2x^2 - 4x - 8}{x - 3}$

7. $\dfrac{3x^5 + 18x^4 - 7x^3 + 9x^2 + 4x}{3x^2 - 1}$

8. $\dfrac{9x^5 - 10x^4 + 18x^3 - 28x^2 + x + 3}{9x^2 - x - 1}$

9. $\dfrac{2x^5 - 5x^4 + 7x^3 - 10x^2 + 7x - 5}{x^2 - x + 1}$

10. $\dfrac{14x^5 - 2x^4 + 27x^3 - 3x^2 + 9x}{2x^3 + 3x}$

11. $\dfrac{x^4 + x^2 - 20x - 8}{x - 3}$

12. $\dfrac{2x^5 - 3x^2 + 1}{x^2 + 1}$

13. $\dfrac{9x^3 + 2x}{3x - 5}$

14. $\dfrac{-4x^5 + 8x^3 - 2}{2x^3 + x}$

15. $\dfrac{2x^2 + x - 8}{x + 3}$

16. $\dfrac{5x^5 + x^4 - 13x^3 - 2x^2 + 6x}{x^3 - 2x}$

17. $\dfrac{2x^3 - 3ix^2 + 11x + (1 - 5i)}{2x - i}$

18. $\dfrac{9x^3 - (18 + 9i)x^2 + x + (-2 - i)}{x - 2 - i}$

19. $\dfrac{3x^3 + ix^2 + 9x + 3i}{3x + i}$

20. $\dfrac{35x^4 + (14 - 10i)x^3 - (7 + 4i)x^2 + 2ix}{7x - 2i}$

Use synthetic division to determine if the given value for k is a zero of the corresponding polynomial. If not, determine $p(k)$. See Examples 2 and 3.

21. $p(x) = 32x^5 - 80x^4 + 80x^3 - 40x^2 + 10x + 2;\ k = 1$

22. $p(x) = 32x^5 - 80x^4 + 80x^3 - 40x^2 + 10x + 2;\ k = \dfrac{1}{2}$

section 5.2   Polynomial Division and the Division Algorithm

23. $p(x) = 12x^4 - 7x^3 - 32x^2 - 7x + 6; \ k = 2$

24. $p(x) = 12x^4 - 7x^3 - 32x^2 - 7x + 6; \ k = -1$

25. $p(x) = 12x^4 - 7x^3 - 32x^2 - 7x + 6; \ k = \dfrac{1}{3}$

26. $p(x) = 2x^2 - (3-5i)x + (3-9i); \ k = -2$

27. $p(x) = 8x^4 - 2x + 6; \ k = 1$

28. $p(x) = x^4 - 1; \ k = 1$

29. $p(x) = x^5 + 32; \ k = -2$

30. $p(x) = 3x^5 + 9x^4 + 2x^2 + 5x - 3; \ k = -3$

31. $p(x) = 2x^2 - (3-5i)x + (3-9i); \ k = -3i$

32. $p(x) = x^2 - 6x + 13; \ k = 2$

33. $p(x) = x^2 - 6x + 13; \ k = 3 - 2i$

34. $p(x) = 3x^3 - 13x^2 - 28x - 12; \ k = -2$

35. $p(x) = 3x^3 - 13x^2 - 28x - 12; \ k = 6$

36. $p(x) = 2x^3 - 8x^2 - 23x + 63; \ k = 2$

37. $p(x) = 2x^3 - 8x^2 - 23x + 63; \ k = 5$

38. $p(x) = x^4 - 3x^3 - 3x^2 + 11x - 6; \ k = 1$

39. $p(x) = x^4 - 3x^3 - 3x^2 + 11x - 6; \ k = -2$

40. $p(x) = x^4 - 3x^3 - 3x^2 + 11x - 6; \ k = 3$

chapter five — Polynomial Functions

Use synthetic division to rewrite each of the following fractions in the form $q(x) + \dfrac{r(x)}{d(x)}$, where $d(x)$ is the denominator of the original fraction, $q(x)$ is the quotient, and $r(x)$ is the remainder. See Examples 2 and 3.

41. $\dfrac{x^3 + x^2 - 18x + 9}{x + 5}$

42. $\dfrac{-2x^5 + 4x^4 + 3x^3 - 7x^2 + 3x - 2}{x - 2}$

43. $\dfrac{x^8 + x^7 - 3x^3 - 3x^2 + 3}{x + 1}$

44. $\dfrac{x^8 - 5x^7 - 3x^3 + 15x^2 - 2}{x - 5}$

45. $\dfrac{4x^3 - (16 + 4i)x^2 + (14 + 4i)x + (-6 - 2i)}{x - 3 - i}$

46. $\dfrac{x^6 - 2x^5 + 2x^4 + 4x^2 - 8x + 8}{x - 1 + i}$

47. $\dfrac{x^5 - 3x^4 + x^3 - 5x^2 + 18}{x - 2}$

48. $\dfrac{x^5 - 3x^4 + x^3 - 5x^2 + 18}{x - 3}$

49. $\dfrac{x^4 + (i - 1)x^3 + (1 - i)x^2 + ix}{x + i}$

50. $\dfrac{x^6 + 8x^5 + x^3 + 8x^2 - 14x - 112}{x + 8}$

51. $\dfrac{2x^3 - 10ix^2 + 5x + (8 - 3i)}{x - 3i}$

52. $\dfrac{4x^5 - 6x^4 + 10x^3 - 4x^2 - 4x}{x - 1}$

Construct polynomial functions with the stated properties (answers may not be unique). See Example 4.

53. Second degree, zeros of -4 and 3, and goes to $-\infty$ as $x \to -\infty$.

54. Third degree, zeros of -2, 1, and 3, and a y-intercept of -12.

55. Second degree, zeros of $2 - 3i$ and $2 + 3i$, and a y-intercept of -13.

56. Third degree, zeros of $1 - i$, $2 + i$, and -1, and a leading coefficient of -2.

57. Fourth degree and a single x-intercept of 3.

58. Second degree, zeros of $-\dfrac{3}{4}$ and 2, and a y-intercept of 6.

59. Fourth degree, zeros of -3, -2, and 1, and a y-intercept of 18.

60. Third degree, zeros of $1, 2,$ and 3, and passes through the point $(4, 12)$.

61. A box company makes a variety of boxes, all with volume given by the formula $x^3 + 10x^2 + 31x + 30$. If the height is given by $x + 3$, what is the formula for the surface area of the base?

section 5.2 Polynomial Division and the Division Algorithm

technology exercises

Wise use of a computer algebra system can allow us to easily solve problems in cases where hand computation would be depressingly tedious. Suppose, for instance, we are asked to construct a fifth degree polynomial function that passes through the point $(2,10)$ and has zeros of $-4, -1, 3, -3+i,$ and $-3-i$. The illustration below shows how *Mathematica* can be used to define a function with the appropriate zeros (note that the zeros are unaffected by any non-zero choice for the constant a) and then further refined so as to pass through the required point.

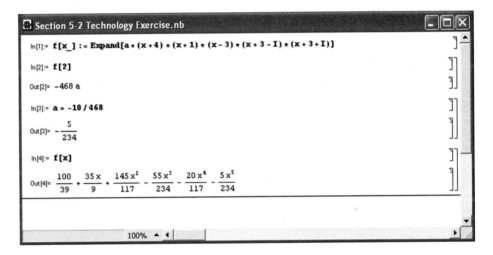

If possible, make similar use of technology to construct polynomial functions with the stated properties.

62. Fifth degree, zeros of $1+i, 1-i, -3+2i, -3-2i,$ and 7, and a y-intercept of 2.

63. Third degree, zeros of $2-5i, -1+3i,$ and $3-i,$ and with the property that $f(1) = -78 - 26i$.

64. Third degree, zeros of $2-5i, -1+3i,$ and $3-i,$ and with the property that $f(1+i) = 24 + 4i$.

chapter five   Polynomial Functions

5.3 Locating Real Zeros of Polynomials

TOPICS

1. The Rational Zero Theorem
2. Descartes' Rule of Signs
3. Bounds of real zeros
4. The Intermediate Value Theorem

Topic 1: The Rational Zero Theorem

For a given polynomial function $f(x) = a_n x^n + a_{n-1} x^{n-1} + \ldots + a_1 x + a_0$, we know that the number k is a zero if and only if $x - k$ is a factor of f. Furthermore, if $x - k$ is a factor of f, we can use either polynomial long division or synthetic division to actually divide f by $x - k$ and find the quotient polynomial q, allowing us to write $f(x) = (x - k)q(x)$. What we are lacking, at this point, is a method for determining the zeros of f. This means, in effect, that the skills we have acquired in the last two sections cannot be put to use until we have some way of locating a zero k as a starting point.

If there were a method guaranteed to find all the zeros of a given polynomial, we would be able to solve any polynomial equation and (if the coefficients of the polynomial were all real) we would be able to sketch with fair accuracy the graph of any polynomial. As hinted at in Section 5.1, however, no such method exists. In fact, it can be proven (using ideas from a branch of mathematics called *abstract algebra*) that there is no formula based on elementary mathematical operations that identifies all the zeros of an arbitrary polynomial of degree five or higher. We do, however, have some tools that give us hints about where to look for zeros of a given polynomial. One such tool is the Rational Zero Theorem.

section 5.3   Locating Real Zeros of Polynomials

The Rational Zero Theorem

If $f(x) = a_n x^n + a_{n-1} x^{n-1} + \ldots + a_1 x + a_0$ is a polynomial with integer coefficients, then any rational zero of f must be of the form $\dfrac{p}{q}$, where p is a factor of the constant term a_0 and q is a factor of the leading coefficient a_n.

The proof of this statement is outlined in exercise 63.

Before demonstrating how the Rational Zero Theorem is used, some cautions on what it doesn't do are in order. First, it doesn't necessarily find even a single zero of a polynomial; instead, it identifies a list of rational numbers that could *potentially* be zeros. In order to determine if any of the potential zeros actually *are* zeros, we have to test them. Second, the theorem says nothing about irrational zeros or complex zeros. If a given polynomial has some zeros that are either irrational or else have a non-zero imaginary part, we must resort to other means to find them.

example 1

For each of the polynomials that follow, list all of the potential rational zeros. Then write the polynomial in factored form and identify the actual zeros.

a. $f(x) = 2x^3 + 5x^2 - 4x - 3$ **b.** $g(x) = 27x^4 - 9x^3 - 33x^2 - x - 4$

Solutions:

Note in the table below that the y value is zero for both $x=-3$ and $x=1$.

a. Factors of a_0: $\pm\{1, 3\}$

Factors of a_3: $\pm\{1, 2\}$

Possible rational zeros: $\pm\left\{1, 3, \dfrac{1}{2}, \dfrac{3}{2}\right\}$

$\underline{1 \rfloor}\ \ 2\quad 5\quad -4\quad -3$
$\ \ \ \ 2\quad\ \ 7\quad\ \ 3$
$\overline{\ \ 2\quad 7\quad\ \ 3\quad\ \ 0}$

We begin by listing the factors of the constant term, -3, and the leading coefficient, 2. Note that each number has two positive factors and two negative factors; this fact is denoted with the $\pm$ symbol in front of each set. By the Rational Zero Theorem, any rational zero of the given polynomial must be one of the eight numbers generated by dividing factors of a_0 by factors of a_3.

continued on next page...

$$f(x) = (x-1)(2x^2 + 7x + 3)$$
$$= (x-1)(2x+1)(x+3)$$

Actual zeros: $\left\{1, -\dfrac{1}{2}, -3\right\}$

We can then use synthetic division to identify which of the potential zeros actually *are* zeros. In the work on the previous page, we first discover that 1 is a zero and then finish by factoring the trinomial $2x^2 + 7x + 3$. You may have to try several potential zeros before finding an actual zero.

b. Factors of $a_0 : \pm\{1, 2, 4\}$

Factors of $a_4 : \pm\{1, 3, 9, 27\}$

Possible rational zeros:

$$\pm\left\{1, 2, 4, \dfrac{1}{3}, \dfrac{2}{3}, \dfrac{4}{3}, \dfrac{1}{9}, \dfrac{2}{9}, \dfrac{4}{9}, \dfrac{1}{27}, \dfrac{2}{27}, \dfrac{4}{27}\right\}$$

$$\begin{array}{r|rrrrr}
-1 & 27 & -9 & -33 & -1 & -4 \\
 & & -27 & 36 & -3 & 4 \\
\hline
\dfrac{4}{3} & 27 & -36 & 3 & -4 & 0 \\
 & & 36 & 0 & 4 & \\
\hline
 & 27 & 0 & 3 & 0 & \\
\end{array}$$

$$g(x) = (x+1)\left(x - \dfrac{4}{3}\right)(27x^2 + 3)$$

$$= 27(x+1)\left(x - \dfrac{4}{3}\right)\left(x - \dfrac{i}{3}\right)\left(x + \dfrac{i}{3}\right)$$

$$= (x+1)(3x-4)(3x-i)(3x+i)$$

Actual zeros: $\left\{-1, \dfrac{4}{3}, \dfrac{i}{3}, -\dfrac{i}{3}\right\}$

In this problem, our list of potential rational zeros contains 24 candidates (12 positive and 12 negative). Although the task of trying so many may seem daunting, the Rational Zero Theorem has at least eliminated all rational numbers except these 24. The work at left illustrates that two of the 24 numbers, -1 and $\dfrac{4}{3}$, are actual zeros of g. (Note how the result of the first synthetic division is used in the second synthetic division.)

Once we have found two zeros and factored out the corresponding linear factors, we are left with the task of factoring $27x^2 + 3$, which comes from the results of our synthetic division. By solving the equation $27x^2 + 3 = 0$, we find $27x^2 + 3 = 27\left(x - \dfrac{i}{3}\right)\left(x + \dfrac{i}{3}\right)$. If we wish, we can distribute the three factors of 3 in 27 among the last three linear factors to obtain factors without fractions, as shown. Of course, this doesn't affect the four zeros of g.

section 5.3 Locating Real Zeros of Polynomials

Topic 2: Descartes' Rule of Signs

Descartes' Rule of Signs is another tool to aid us in our search for zeros. Unlike the Rational Zeros Theorem, Descartes' Rule doesn't identify candidates; instead, it gives us guidelines on how many positive real zeros and how many negative real zeros we can expect a given polynomial to have. Again, this tool tells us nothing about zeros with non-zero imaginary parts.

Descartes' Rule of Signs

Let $f(x) = a_n x^n + a_{n-1} x^{n-1} + \ldots + a_1 x + a_0$ be a polynomial with real coefficients, and assume $a_0 \neq 0$. A *variation in sign* of f is a change in the sign of one coefficient of f to the next, either from positive to negative or vice versa.

1. The number of *positive real zeros* of f is either the number of variations in sign of $f(x)$ or is less than this number by a positive even integer.

2. The number of *negative real zeros* of f is either the number of variations in sign of $f(-x)$ or is less than this number by a positive even integer.

Note that in order to apply Descartes' Rule of Signs, it is critical to first write the terms of the polynomial in descending order. Note also that unless the number of variations in sign is 0 or 1, the rule does not give us a definitive answer for the number of zeros to expect. For instance, if the number of variations in sign of $f(x)$ is 4, we know only that there will be 4, 2, or 0 positive real zeros.

example 2

Use Descartes' Rule of Signs to determine the possible number of positive and negative real zeros of each of the following polynomials. Then use the Rational Zeros Theorem and other means to find the zeros, if possible.

a. $f(x) = 2x^3 + 3x^2 - 14x - 21$ **b.** $g(x) = 3x^3 - 10x^2 + \dfrac{51}{4}x - \dfrac{13}{4}$

continued on next page...

Polynomial Functions

Solutions:

a. $f(x) = 2x^3 + \underset{\text{sign change}}{3x^2 \quad -14x} \quad -21$

$f(-x) = 2(-x)^3 + 3(-x)^2 - 14(-x) - 21$

$ = \underset{\text{sign change}}{-2x^3 \quad +3x^2} \underset{\text{sign change}}{+14x \quad -21}$

Factors of $-21 : \pm\{1, 3, 7, 21\}$

Factors of $2 : \pm\{1, 2\}$

Possible rational zeros: $\pm\left\{1, 3, 7, 21, \dfrac{1}{2}, \dfrac{3}{2}, \dfrac{7}{2}, \dfrac{21}{2}\right\}$

$$\begin{array}{r|rrrr} -\dfrac{3}{2} & 2 & 3 & -14 & -21 \\ & & -3 & 0 & 21 \\ \hline & 2 & 0 & -14 & 0 \end{array}$$

$f(x) = \left(x + \dfrac{3}{2}\right)(2x^2 - 14)$

$ = 2\left(x + \dfrac{3}{2}\right)(x - \sqrt{7})(x + \sqrt{7})$

Zeros: $\left\{-\dfrac{3}{2}, \sqrt{7}, -\sqrt{7}\right\}$

There is only one variation in sign in the function $f(x)$, and so we know there is exactly one positive real zero. We next simplify the function $f(-x)$ and determine that it has two variations in sign. Thus f has either 2 or 0 negative real zeros.

We can now use the Rational Zero Theorem to try to actually find the zeros. The calculations at left show there are 16 rational numbers that are possible zeros of f, and trial and error eventually leads to the discovery that $-\dfrac{3}{2}$ is actually a zero. Now, since we have found 1 negative zero, we know there must be a second.

Synthetic division allows us to factor f as shown, and as always we know we can factor the quadratic quotient. In this case, we can solve the equation $2x^2 - 14 = 0$ fairly easily (though the quadratic formula could also be used). Note that the end result is indeed 1 positive zero and 2 negative zeros.

b. $g(x) = 3x^3 - 10x^2 + \dfrac{51}{4}x - \dfrac{13}{4}$

$g(-x) = 3(-x)^3 - 10(-x)^2 + \dfrac{51}{4}(-x) - \dfrac{13}{4}$

$ = -3x^3 - 10x^2 - \dfrac{51}{4}x - \dfrac{13}{4}$

There are 3 variations in sign in the function $g(x)$, so there are either 3 or 1 positive real zeros. And since $g(-x)$ has no variations in sign, $g(x)$ has no negative real zeros.

section 5.3 Locating Real Zeros of Polynomials

$g(x) = \frac{1}{4}(12x^3 - 40x^2 + 51x - 13)$

Factors of $-13: \pm\{1, 13\}$

Factors of $12: \pm\{1, 2, 3, 4, 6, 12\}$

Possible rational zeros:

$\pm\left\{1, 13, \frac{1}{2}, \frac{13}{2}, \frac{1}{3}, \frac{13}{3}, \frac{1}{4}, \frac{13}{4}, \frac{1}{6}, \frac{13}{6}, \frac{1}{12}, \frac{13}{12}\right\}$

$\begin{array}{r|rrrr} \frac{1}{3} & 12 & -40 & 51 & -13 \\ & & 4 & -12 & 13 \\ \hline & 12 & -36 & 39 & 0 \end{array}$

$g(x) = \frac{1}{4}\left(x - \frac{1}{3}\right)(12x^2 - 36x + 39)$

$= \frac{(12)}{4}\left(x - \frac{1}{3}\right)\left(x - \left(\frac{3-2i}{2}\right)\right)\left(x - \left(\frac{3+2i}{2}\right)\right)$

Zeros: $\left\{\frac{1}{3}, \frac{3-2i}{2}, \frac{3+2i}{2}\right\}$

Before applying the Rational Zero Theorem, we have to make one small adjustment. The theorem requires that all the coefficients of the polynomial be integers, and g does not meet this condition. However, we can factor out $\frac{1}{4}$ as shown and apply the theorem to $12x^3 - 40x^2 + 51x - 13$, since the zeros of this polynomial are the same as the zeros of g (multiplying a function by a non-zero constant has no effect on the zeros.)

After a trial and error search, synthetic division shows that $\frac{1}{3}$ is an actual zero of g. At left, g is shown in completely factored form (don't forget to include the factor of $\frac{1}{4}$.) We can use the quadratic formula to factor $12x^2 - 36x + 39$, keeping in mind that we must multiply the product of the two linear factors that result by 12 to obtain the correct leading coefficient.

Topic 3: **Bounds of Real Zeros**

Although the Rational Zero Theorem and Descartes' Rule of Signs are useful for determining the zeros of a polynomial, *Examples 1* and *2* illustrate that more guidance would certainly be welcome, especially guidance that reduces the number of potential zeros that must be tested by trial and error. The following theorem does just that.

chapter five Polynomial Functions

Upper and Lower Bounds of Zeros

Assume $f(x)$ is a polynomial with real coefficients, a positive leading coefficient, and degree ≥ 1. Let a and b be fixed numbers, with $a < 0 < b$. Then:

1. No real zero of f is larger than b (we say b is an upper bound of the zeros of f) if the last row in the synthetic division of $f(x)$ by $x-b$ contains no negative numbers. That is, b is an upper bound of the zeros if the quotient and remainder have no negative coefficients when $f(x)$ is divided by $x-b$.

2. No real zero of f is smaller than a (we say a is a lower bound of the zeros of f) if the last row in the synthetic division of $f(x)$ by $x-a$ has entries that alternate in sign (0 can be counted as either positive or negative, as necessary).

Example 3 revisits the polynomial $f(x) = 2x^3 + 3x^2 - 14x - 21$ that we studied in *Example 2a* and illustrates the use of the above theorem.

example 3

Use synthetic division to identify upper and lower bounds of the real zeros of the polynomial $f(x) = 2x^3 + 3x^2 - 14x - 21$.

Solution:

$$\begin{array}{r|rrrr} 2 & 2 & 3 & -14 & -21 \\ & & 4 & 14 & 0 \\ \hline & 2 & 7 & 0 & -21 \end{array}$$

We can begin by trying out any positive number whatsoever as an upper bound. The work at left shows that 2 is not necessarily an upper bound, as the last row contains a negative number.

section 5.3 Locating Real Zeros of Polynomials

```
3 | 2   3   -14   -21
  |     6    27    39
  ——————————————————
    2   9    13    18
```

The number 3, however, is an upper bound according to the theorem, as all of the coefficients in the last row are non-negative. In other words, all real zeros of f (including any irrational zeros) are less than or equal to 3.

```
-3 | 2   3   -14   -21
   |    -6     9    —
   ——————————————————
     2  -3    -5    —
```

The work to the left tests whether -3 is a lower bound, according to the theorem. The synthetic division has not been completed, because as soon as the signs in the last row cease to alternate, the theorem doesn't apply.

```
-4 | 2   3   -14   -21
   |    -8    20   -24
   ——————————————————
     2  -5     6   -45
```

When we try a lower number, we find that -4 is a lower bound, as the signs in the last row alternate. Remember that if a 0 appears, it can be counted as either positive or negative, whichever leads to a sequence of alternating signs.

To summarize, we now know that all real zeros of f lie in the interval $[-4, 3]$, meaning there is no need to test the numbers $\left\{7, 21, \frac{7}{2}, \frac{21}{2}, -7, -21, -\frac{21}{2}\right\}$ out of the list of potential rational zeros $\pm\left\{1, 3, 7, 21, \frac{1}{2}, \frac{3}{2}, \frac{7}{2}, \frac{21}{2}\right\}$. We have already determined in *Example 2* that the zeros of f are $\left\{-\frac{3}{2}, \sqrt{7}, -\sqrt{7}\right\}$, all of which do indeed lie in the interval $[-4, 3]$.

chapter five   Polynomial Functions

caution

Don't read more into the Upper and Lower Bounds Theorem than is actually there. For instance, -3 actually *is* a lower bound of the zeros of $f(x) = 2x^3 + 3x^2 - 14x - 21$, but the theorem is not powerful enough to indicate this. The work in *Example 3* shows that -4 is a lower bound, but the theorem fails to spot the fact that -3 is a better lower bound. The tradeoff for this weakness in the theorem is that it is quickly and easily applied.

Topic 4: The Intermediate Value Theorem

The last technique for locating zeros that we study makes use of a property of polynomials called *continuity*. Although continuity of functions will not be discussed in this text, one consequence of continuity is that the graph of a continuous function has no "breaks" in it. That is, assuming that the function can be graphed at all, it can be drawn without lifting the drawing tool.

Intermediate Value Theorem

Assume that $f(x)$ is a polynomial with real coefficients, and that a and b are real numbers with $a < b$. If $f(a)$ and $f(b)$ differ in sign, then there is at least one point c such that $a < c < b$ and $f(c) = 0$. That is, at least one zero of f lies between a and b.

We can use the Intermediate Value Theorem to prove that a zero of a given polynomial must lie in a particular interval, and then through repeated application to "home in" on the zero. *Example 4* illustrates this process.

section 5.3 Locating Real Zeros of Polynomials

example 4

a. Show that $f(x) = x^3 + 3x - 7$ has a zero between 1 and 2.
b. Find an approximation of the zero to the nearest tenth.

Solutions:

a.
```
1│ 1   0   3   -7
       1   1    4
   ─────────────────
   1   1   4   -3
```

```
2│ 1   0   3   -7
       2   4   14
   ─────────────────
   1   2   7    7
```

For the function $f(x) = x^3 + 3x - 7$, we could determine $f(1)$ and $f(2)$ by direct computation, but for many polynomials it is quicker to use synthetic division as illustrated at left. Remember that the remainder upon dividing $f(x)$ by $x - k$ is $f(k)$; thus $f(1) = -3$ and $f(2) = 7$. The critical point is that these two values are opposite in sign, so a zero of f must lie between 1 and 2.

This table was reached by entering the equation in **Y=** and selecting **2ND** **GRAPH** (Table). Notice that the signs of the y values change between *x*=1 and *x*=2, indicating that there is a zero there.

b.
```
1.5│ 1   0     3     -7
         1.5  2.25  7.875
     ──────────────────────
     1  1.5  5.25  0.875
```

Since we know f has a zero between 1 and 2, we might guess 1.5 as a first approximation for the zero. If we continue to use synthetic division as shown, we find $f(1.5) = 0.875$, which tells us that the actual zero of f must lie between 1 and 1.5. We might further guess that the zero is closer to 1.5 than it is to 1, since $f(1.5)$ is closer to 0 than is $f(1)$.

$$f(1.4) = (1.4)^3 + 3(1.4) - 7$$
$$= -0.056$$

With this in mind, we calculate $f(1.4)$. Again, we can use synthetic division or direct computation as shown at left. The result is negative, so we now know the actual zero must lie between 1.4 and 1.5.

continued on next page...

399

$$f(1.45) = (1.45)^3 + 3(1.45) - 7$$
$$= 0.398625$$

The value of f midway between 1.4 and 1.5 is positive, so the zero must lie between 1.4 and 1.45 (and in fact is probably closer to 1.4). In any event, we have shown that the zero, to the nearest tenth, is 1.4. (Continuing the process would lead to the better approximation 1.41).

technology note

Graphing calculators, while not as powerful as computer algebra systems such as *Mathematica*, are nonetheless useful aids in finding real zeros of polynomials. Their graphing capability can be used, for example, to quickly eliminate some candidates for zeros given by the Rational Zero Theorem. And the "Zoom" feature available on most graphing calculators is a convenient tool when applying the Intermediate Value Theorem (see your owner's manual for help with your specific calculator). The illustration below shows successive screen shots of the graph of $f(x) = x^3 + 3x - 7$ as we "zoom in" on the approximate zero identified in *Example 4*.

y-axis: $[-20, 60]$; x-axis: $[-1, 4]$
y scale $= 20$; x scale $= 1$

y-axis: $[-10, 25]$; x-axis: $[0, 3]$
y scale $= 5$; x scale $= 0.5$

y-axis: $[-4, 6]$; x-axis: $[0, 2]$
y scale $= 2$; x scale $= 0.2$

section 5.3 Locating Real Zeros of Polynomials

exercises

Throughout these exercises, a graphing calculator or a computer algebra system may be used as an additional tool in identifying zeros, if permitted by your instructor.

List all of the potential rational zeros of the following polynomials. Then use polynomial division and the quadratic formula, if necessary, to identify the actual zeros. See Example 1.

1. $f(x) = 3x^3 + 5x^2 - 26x + 8$
2. $g(x) = -2x^3 + 11x^2 + x - 30$
3. $p(x) = x^4 - 5x^3 + 10x^2 - 20x + 24$
4. $h(x) = x^3 - 3x^2 + 9x + 13$
5. $q(x) = x^3 - 10x^2 + 23x - 14$
6. $r(x) = x^4 + x^3 + 23x^2 + 25x - 50$
7. $s(x) = 2x^3 - 9x^2 + 4x + 15$
8. $t(x) = x^3 - 6x^2 + 13x - 20$
9. $j(x) = 3x^4 - 3$
10. $k(x) = x^4 - 10x^2 + 24$
11. $m(x) = x^3 + 11x^2 - x - 11$
12. $g(x) = x^3 - 6x^2 - 5x + 30$

Using the Rational Zero Theorem or your answers to the preceding problems, solve the following polynomial equations.

13. $x^4 + x - 2 = -2x^4 + x + 1$
14. $x^4 + 10 = 10x^2 - 14$
15. $x^3 - 3x^2 + 9x + 13 = 0$
16. $3x^3 + 5x^2 = 26x - 8$
17. $x^4 + 10x^2 - 20x = 5x^3 - 24$
18. $-2x^3 + 11x^2 + x = 30$
19. $2x^3 - 12x^2 + 26x = 40$
20. $2x^3 + 9x^2 + 4x = 15$
21. $x^4 + x^3 + 23x^2 = 50 - 25x$
22. $x^3 + 23x = 10x^2 + 14$
23. $x^3 + 11x^2 = 11 + x$
24. $-6x^2 + x^3 = 5x - 30$

chapter five — Polynomial Functions

Use Descartes' Rule of Signs to determine the possible number of positive and negative real zeros of each of the following polynomials. See Example 2.

25. $f(x) = x^3 + 8x^2 + 17x + 10$

26. $g(x) = x^3 + 2x^2 - 5x - 6$

27. $f(x) = x^3 - 6x^2 + 3x + 10$

28. $g(x) = x^3 + 6x^2 + 11x + 6$

29. $f(x) = x^4 - 5x^3 - 2x^2 + 40x - 48$

30. $g(x) = x^3 + 3x^2 + 3x + 9$

31. $f(x) = x^4 - 25$

32. $g(x) = x^4 - 7x^3 + 5x^2 + 31x - 30$

33. $f(x) = 5x^5 - x^4 + 2x^3 + x - 9$

34. $g(x) = -6x^7 - x^5 - 7x^3 - 2x$

35. $f(x) = -5x^{11} - 14x^9 - 10x^7 - 15x^5$

36. $g(x) = 2x^4 + 7x^3 + 28x^2 + 112x - 64$

Use synthetic division to identify upper and lower bounds of the real zeros of the following polynomials (answers may vary). See Example 3.

37. $f(x) = x^3 + 4x^2 - x - 4$

38. $f(x) = 2x^3 - 3x^2 - 8x - 3$

39. $f(x) = x^3 - 6x^2 + 3x + 10$

40. $g(x) = x^3 + 6x^2 + 11x + 6$

41. $f(x) = x^4 - 5x^3 - 2x^2 + 40x - 48$

42. $g(x) = x^3 + 3x^2 + 3x + 9$

43. $f(x) = x^4 - 25$

44. $g(x) = x^4 - 7x^3 + 5x^2 + 31x - 30$

45. $f(x) = 2x^3 - 7x^2 - 28x - 12$

46. $g(x) = x^5 + x^4 - 9x^3 - x^2 + 20x - 12$

Using your answers to the preceding problems, polynomial division, and the quadratic formula, if necessary, find all of the zeros of the following polynomials.

47. $f(x) = x^3 + 4x^2 - x - 4$

48. $f(x) = 2x^3 - 3x^2 - 8x - 3$

49. $f(x) = x^3 - 6x^2 + 3x + 10$

50. $g(x) = x^3 + 6x^2 + 11x + 6$

51. $f(x) = x^4 - 5x^3 - 2x^2 + 40x - 48$

52. $g(x) = x^3 + 3x^2 + 3x + 9$

53. $f(x) = x^4 - 25$

54. $g(x) = x^4 - 7x^3 + 5x^2 + 31x - 30$

55. $f(x) = 2x^3 - 7x^2 - 28x - 12$

56. $g(x) = x^5 + x^4 - 9x^3 - x^2 + 20x - 12$

section 5.3 Locating Real Zeros of Polynomials

Use the Intermediate Value Theorem to show that each of the following polynomials has a real zero between the indicated values. See Example 4.

57. $f(x) = 5x^3 - 4x^2 - 31x - 6$; -3 and -1 **58.** $f(x) = x^4 - 9x^2 - 14$; 1 and 4

59. $f(x) = x^4 + 2x^3 - 10x^2 - 14x + 21$; 2 and 3

60. $f(x) = -x^3 + 2x^2 + 13x - 26$; -4 and -3

Show that each of the following equations must have a solution between the indicated real numbers.

61. $14x + 10x^2 = x^4 + 2x^3 + 21$; 2 and 3

62. $x^3 - 2x^2 = 13(x-2)$; -4 and -3

63. Construct a proof of the Rational Zero Theorem by following the suggested steps.

  a. Assuming that $\dfrac{p}{q}$ is a zero of the polynomial $f(x) = a_n x^n + a_{n-1} x^{n-1} + \ldots + a_1 x + a_0$, show that the equation $a_n \left(\dfrac{p}{q}\right)^n + a_{n-1}\left(\dfrac{p}{q}\right)^{n-1} + \ldots + a_1 \left(\dfrac{p}{q}\right) + a_0 = 0$ can be written in the form $a_n p^n + a_{n-1} p^{n-1} q + \ldots + a_1 p q^{n-1} = -a_0 q^n$.

  b. It can be assumed that $\dfrac{p}{q}$ is written in lowest terms (that is, the greatest common divisor of p and q is 1). By examining the left-hand side of the last equation above, show that p must be a divisor of the right-hand side, and hence a factor of a_0.

  c. By rearranging the equation so that all terms with a factor of q are on one side, use a similar argument to show that q must be a factor of a_n.

Using any of the methods discussed in this section as guides, find all of the real zeros of the following functions.

64. $f(x) = 3x^3 - 18x^2 + 9x + 30$ **65.** $f(x) = -4x^3 - 19x^2 + 29x - 6$

66. $f(x) = 3x^5 + 7x^4 + 12x^3 + 28x^2 - 15x - 35$ **67.** $f(x) = 2x^4 + 5x^3 - 9x^2 - 15x + 9$

68. $f(x) = -15x^4 + 44x^3 + 15x^2 - 72x - 28$ **69.** $f(x) = 2x^4 + 13x^3 - 23x^2 - 32x + 20$

chapter five   Polynomial Functions

70. $f(x) = 3x^4 + 7x^3 - 25x^2 - 63x - 18$ **71.** $f(x) = x^5 + 7x^4 + 5x^3 - 43x^2 - 42x + 72$

72. $f(x) = 2x^5 - 3x^4 - 47x^3 + 103x^2 + 45x - 100$ **73.** $f(x) = x^6 - 125x^4 + 4804x^2 - 57{,}600$

Using any of the methods discussed in this section as guides, solve the following equations.

74. $x^3 + 6x^2 + 11x = -6$ **75.** $x^3 - 7x = 6(x^2 - 10)$

76. $x^3 + 9x^2 = 2x + 18$ **77.** $6x^3 + 14 = 41x^2 + 9x$

78. $4x^3 = 18x^2 + 106x + 48$ **79.** $3x^3 + 15x^2 - 6x = 72$

80. $8x^4 + 24 + 8x = 2x^3 + 38x^2$ **81.** $x^4 + 7x^2 = 3x^3 + 21x$

82. $6x^6 - 10x^5 - 9x^4 + 27x^3 = 20x^2 + 18x - 30$ **83.** $4x^5 - 5x^4 + 20x^2 = 6x^3 + 25x + 30$

5.4 The Fundamental Theorem of Algebra

TOPICS

1. The Fundamental Theorem of Algebra
2. Multiple zeros and their geometric meaning
3. Conjugate pairs of zeros
4. Summary of polynomial methods

Topic 1: The Fundamental Theorem of Algebra

We are now ready to tie together all that we have learned about polynomials, and we begin with a powerful but deceptively simple-looking statement called the Fundamental Theorem of Algebra.

The Fundamental Theorem of Algebra

If f is a polynomial of degree n, with $n \geq 1$, then f has at least one zero. That is, the equation $f(x) = 0$ has at least one root, or solution. It is important to note that the zero of f, and consequently the solution of $f(x) = 0$, may be a non-real complex number.

Mathematicians began to suspect the truth of this statement in the first half of the 17th century, but a convincing proof did not appear until the 22-year-old German mathematician Carl Friedrich Gauss (1777 - 1855) provided one in his doctoral dissertation in 1799.

Although the proof of the Fundamental Theorem of Algebra is beyond the scope of this book, we can use it to prove a consequence that summarizes much of the previous three sections. The following theorem has great implications in solving polynomial equations and in graphing real-coefficient polynomial functions.

chapter five Polynomial Functions

The Linear Factors Theorem

> Given the polynomial $f(x) = a_n x^n + a_{n-1} x^{n-1} + \ldots + a_1 x + a_0$, where $n \geq 1$ and $a_n \neq 0$, f can be factored as $f(x) = a_n (x - c_1)(x - c_2) \cdots (x - c_n)$, where $c_1, c_2, \ldots, c_n$ are constants (possibly non-real complex constants and not necessarily distinct). In other words, an nth degree polynomial can be factored as a product of n linear factors.

Proof: The Fundamental Theorem of Algebra tells us that $f(x)$ has at least one zero; call it c_1. From our work in Section 5.2, then, we know that $x - c_1$ is a factor of f, and we can write

$$f(x) = (x - c_1) q_1(x),$$

where $q_1(x)$ is a polynomial of degree $n - 1$. Note that the leading coefficient of q_1 must be a_n, since we divided f by $x - c_1$, a polynomial with a leading coefficient of 1.

If the degree of q_1 is 0 (that is, if $n = 1$), then $q_1(x) = a_n$ and $f(x) = a_n(x - c_1)$. If this is not the case, then q_1 is of degree 1 or larger, and by the Fundamental Theorem of Algebra, q_1 itself has at least one zero; call it c_2. By the same reasoning, then, we can write

$$f(x) = (x - c_1)(x - c_2) q_2(x),$$

where q_2 is a polynomial of degree $n - 2$, also with leading coefficient a_n. We can perform this process a total of n times (and no more), at which point we have the desired result:

$$f(x) = a_n (x - c_1)(x - c_2) \cdots (x - c_n)$$

caution

Keep in mind that some of the linear factors obtained may be identical, and that some of the constants $c_1, c_2, \ldots, c_n$ may be non-real complex numbers. The meaning of these possible outcomes will be explored shortly.

section 5.4 The Fundamental Theorem of Algebra

The Linear Factors Theorem tells us that, in theory, any polynomial can be written as a product of linear factors; what it does *not* do is tell us how to determine the linear factors. To accomplish that task, we must still rely on the techniques developed in Section 5.3.

In the case where all of the coefficients of $f(x) = a_n x^n + a_{n-1} x^{n-1} + \ldots + a_1 x + a_0$ are real, the Linear Factors Theorem tells us that the graph of f has *at most n* x-intercepts (and can only have *exactly n* x-intercepts if all n zeros are real and distinct). Indirectly, the theorem tells us something more: the graph of f can have at most $n-1$ *turning points*. A **turning point** of a graph is a point where the graph changes behavior from decreasing to increasing or vice versa. These facts are summarized in the box below.

Interpreting the Linear Factors Theorem

> The graph of an nth degree polynomial function has *at most n* x-intercepts and *at most* $n-1$ turning points.

Figure 1 illustrates that the degree of a polynomial merely gives us an upper bound on the number of x-intercepts and turning points. Note that the graph of the 4th degree polynomial f has just two x-intercepts and only one turning point, while the graph of the 3rd degree polynomial g has three x-intercepts and two turning points.

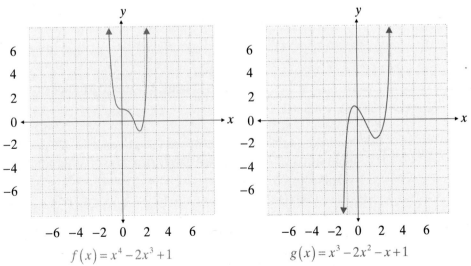

$f(x) = x^4 - 2x^3 + 1$ $g(x) = x^3 - 2x^2 - x + 1$

Figure 1: *x*-Intercepts and Turning Points

Topic 2: Multiple Zeros and Their Geometric Meaning

If the linear factor $x-c$ appears k times in the factorization of a polynomial, we say the number c is a *zero of multiplicity* k. If we are graphing a polynomial f for which c is a real zero of multiplicity k, then c is certainly an x-intercept of the graph of f, but the behavior of the graph near c depends on whether k is even or odd. We have already seen numerous examples of this. Consider, for instance, the function $f(x) = (x-1)^3$: clearly, 1 is the only zero of f, and it is evidently a zero of multiplicity 3. (Note that f is a 3rd degree polynomial, so if it had been given to us in non-factored form, we would still know it could be written as a product of three linear factors). From our work in Chapter 4, we know that the graph of f is the prototypical cubic shape shifted to the right by 1 unit, as shown in Figure 2.

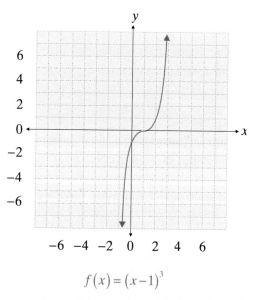

$f(x) = (x-1)^3$

Figure 2: Zero of Multiplicity 3

Compare the behavior of $f(x) = (x-1)^3$ near its zero to the behavior of the function $g(x) = (x+2)^4$ near its own zero of -2, a zero of multiplicity 4. Figure 3 contains the graph of g.

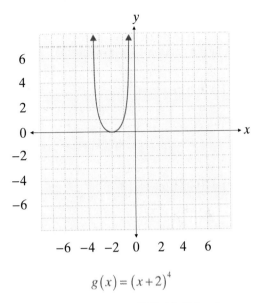

$g(x) = (x+2)^4$

Figure 3: Zero of Multiplicity 4

Geometric Meaning of Multiplicity

In general, if c is a real zero of multiplicity k of a polynomial f (alternatively, if $(x-c)^k$ is a factor of f), the graph of f will touch the x-axis at $(c, 0)$ and:
- cross through the x-axis if k is odd
- stay on the same side of the x-axis if k is even

Further, if $k \geq 2$, the graph of f will "flatten out" near $(c, 0)$.

With an understanding of how a zero's multiplicity affects a polynomial, constructing a reasonably accurate sketch of the graph becomes easier. *Example 1* illustrates this point.

chapter five — Polynomial Functions

example 1

Sketch the graph of the polynomial $f(x) = (x+2)(x+1)^2(x-3)^3$.

Solution:

$f(x) \to \infty$ as $x \to \pm\infty$

The polynomial f has degree 6, and so we know the behavior of the graph as $x \to -\infty$ and $x \to \infty$.

$f(0) = (0+2)(0+1)^2(0-3)^3$
$= (2)(1)(-27)$
$= -54$

The y-intercept is easy to determine, as always.

We can also determine at which of the x-intercepts the graph actually crosses the axis.

f crosses at -2 and 3, but not at -1, and flattens out at -1 and 3

The sketch at left takes into account all of the above observations. Note that the graph flattens out at -1 and at 3, but not at -2. Note also the extreme difference in the scales of the two axes.

To fill in portions of the graph between zeros accurately, we must still compute a few values of the function. For instance, $f(1) = -96$ and $f(2) = -36$. This also gives us a way to double-check our analysis of the behavior on either side of a zero.

Topic 3: Conjugate Pairs of Zeros

If all the coefficients of a given polynomial $f(x) = a_n x^n + a_{n-1} x^{n-1} + \ldots + a_1 x + a_0$ are real, then f is a function that transforms real numbers into other real numbers, and consequently f can be graphed in the Cartesian plane. Nonetheless, it is very possible that some of the constants $c_1, c_2, \ldots, c_n$ in the factored form of f, $f(x) = a_n(x - c_1)(x - c_2) \cdots (x - c_n)$, might be non-real complex numbers. If this happens, such complex numbers must occur in pairs, as described on the next page.

section 5.4 The Fundamental Theorem of Algebra

The Conjugate Roots Theorem

> Let $f(x) = a_n x^n + a_{n-1} x^{n-1} + \ldots + a_1 x + a_0$ be a polynomial with only real coefficients. If the complex number $a + bi$ is a zero of f, then so is the complex number $a - bi$. In terms of the linear factors of f, this means that if $x - (a + bi)$ is a factor of f, then so is $x - (a - bi)$.

We can make use of this fact in several ways. For instance, if we are given one non-real zero of a real-coefficient polynomial, we automatically know a second zero. We may also find the theorem useful when constructing polynomials with certain specified properties.

example 2

Given that $4 - 3i$ is a zero of the polynomial $f(x) = x^4 - 8x^3 + 200x - 625$, factor f completely.

Solution:

Since $4 - 3i$ is a zero of f, we know that $4 + 3i$ is a zero as well. This gives us two ways to proceed: we could divide f by $x - (4 - 3i)$ and then divide the result by $x - (4 + 3i)$ (this would be most efficiently done with synthetic division), or we could multiply $x - (4 - 3i)$ and $x - (4 + 3i)$ and divide f by their product (using polynomial long division). In either case, we will be left with a quadratic polynomial that we know we can factor.

If we take the second approach, the first step is as follows:

$$(x - (4 - 3i))(x - (4 + 3i)) = (x - 4 + 3i)(x - 4 - 3i)$$
$$= x^2 - 4x - \cancel{3ix} - 4x + 16 + \cancel{12i} + \cancel{3ix} - \cancel{12i} - 9i^2$$
$$= x^2 - 8x + 25$$

continued on next page...

chapter five   Polynomial Functions

Now, we divide f by this product:

$$x^2-8x+25 \overline{\smash{\big)}\, x^4 - 8x^3 + 200x - 625} \quad \text{quotient: } x^2 - 25$$

$$\underline{-(x^4 - 8x^3 + 25x^2)}$$
$$-25x^2 + 200x - 625$$
$$\underline{-(-25x^2 + 200x - 625)}$$
$$0$$

The quotient, $x^2 - 25$, is a difference of two squares and is easily factored, giving us our final result:

$$f(x) = (x - 4 + 3i)(x - 4 - 3i)(x - 5)(x + 5)$$

example 3

Construct a 4th degree real-coefficient polynomial function f with zeros of 2, −5, and $1+i$ such that $f(1) = 12$.

Solution:

As $1+i$ is to be one of the zeros and f is to have only real coefficients, $1-i$ must be a zero as well. Based on this, f must be of the form

$$f(x) = a(x - (1+i))(x - (1-i))(x - 2)(x + 5)$$

for some real constant a. Of course, a must be chosen so that $f(1) = 12$. In order to do this, we might begin by multiplying out $(x - (1+i))(x - (1-i))$:

$$(x - (1+i))(x - (1-i)) = (x - 1 - i)(x - 1 + i)$$
$$= x^2 - 2x + 2$$

Thus

$$f(1) = a(1^2 - 2(1) + 2)(1 - 2)(1 + 5)$$
$$= a(1)(-1)(6)$$
$$= -6a.$$

section 5.4   **The Fundamental Theorem of Algebra**

Since we want $f(1) = 12$, we need $12 = -6a$ or $a = -2$. In factored form, then, the polynomial is

$$f(x) = -2(x-1-i)(x-1+i)(x-2)(x+5),$$

which, multiplied out, is $f(x) = -2x^4 - 2x^3 + 28x^2 - 52x + 40$.

Topic 4: Summary of Polynomial Methods

All of the methods that you have learned in this chapter may be useful in solving a given polynomial problem, whether the problem focuses on graphing a polynomial function, solving a polynomial equation, or solving a polynomial inequality. Now that all of the methods have been introduced, it makes sense to summarize them and see how they contribute to the big picture.

Recall that, in general, an nth degree polynomial function has the form

$$f(x) = a_n x^n + a_{n-1} x^{n-1} + \ldots + a_1 x + a_0,$$

where $a_n \neq 0$ and any (or all) of the coefficients may be non-real complex numbers. Keep in mind that it only makes sense to talk about graphing f in the Cartesian plane if all of the coefficients are real. Similarly, a polynomial inequality in which f appears on one side only makes sense if all the coefficients are real. For this reason, most of the polynomials in this text have only real coefficients.

Nevertheless, complex numbers arise in a very natural way when working with polynomials, as some of the numbers $c_1, c_2, \ldots, c_n$ in the factored form of f,

$$f(x) = a_n (x - c_1)(x - c_2) \cdots (x - c_n),$$

may be non-real even if all of $a_1, a_2, \ldots, a_n$ are real. The fact that f can, in principle, be factored is a direct consequence of the Fundamental Theorem of Algebra.

Factoring f into a product of linear factors as shown is the central point in solving a polynomial equation and (when the coefficients of f are real) in graphing a polynomial and solving a polynomial inequality. Specifically:

chapter five   Polynomial Functions

- The solutions of the polynomial equation $f(x) = 0$ are the numbers $c_1, c_2, \ldots, c_n$.
- When $a_1, a_2, \ldots, a_n$ are all real, the x-intercepts of the graph of f are the real numbers in the list $c_1, c_2, \ldots, c_n$. Keep in mind that if a given c_i appears in the list k times, it is a *zero of multiplicity k*. If an x-intercept of f is of multiplicity k, the behavior of f near that x-intercept depends on whether k is even or odd. Any non-real zeros in the list $c_1, c_2, \ldots, c_n$ will appear in conjugate pairs.
- When $a_1, a_2, \ldots, a_n$ are all real, the solution of the polynomial inequality $f(x) > 0$ consists of all the open intervals on the x-axis where the graph of f lies strictly above the x-axis. The solution of $f(x) < 0$ consists of all the open intervals where the graph of f lies strictly below the x-axis. The solutions of $f(x) \geq 0$ and $f(x) \leq 0$ consist of closed intervals.

The remaining topics discussed in this chapter are observations and techniques that aid us in filling in the details of the big picture.

- The observation that the degree of a polynomial and the sign of its leading coefficient tell us how the graph of the polynomial behaves as $x \to -\infty$ and as $x \to \infty$.
- The observation that the graph of f crosses the y-axis at the easily computed point $(0, f(0))$.
- The technique of polynomial long division, useful in dividing one polynomial by another of the same or smaller degree.
- The technique of synthetic division, a shortcut that applies when dividing a polynomial by a polynomial of the form $x - k$. Recall that the remainder of this division is the value $f(k)$.
- The Rational Zero Theorem, which provides a list of potential rational zeros for polynomials with integer coefficients.
- Descartes' Rule of Signs, which provides guidance on the number of positive and negative real zeros that a real-coefficient polynomial might have.
- The Upper and Lower Bounds rule, which indicates an interval in which to search for all the zeros of a real-coefficient polynomial.
- The Intermediate Value Theorem, which can be used to "home in" on a real zero of a given polynomial.

section 5.4 The Fundamental Theorem of Algebra

As you solve various polynomial problems, try to keep the big picture in mind. Often, it is useful to literally keep a picture, namely the graph of the polynomial, in mind even if the problem does not specifically involve graphing.

exercises

Throughout these exercises, a graphing calculator or a computer algebra system may be used as an additional tool in identifying zeros and in checking your graphing, if permitted by your instructor.

Sketch the graph of each factored polynomial. See Example 1.

1. $f(x) = (x+1)^4 (x-2)^3 (x-1)$
2. $g(x) = -x^3 (x-1)(x+2)^2$
3. $f(x) = -x(x+2)(x-1)^2$
4. $g(x) = (x+2)(x-1)^3$
5. $f(x) = (x-1)^4 (x-2)(x-3)$
6. $g(x) = (x+1)^2 (x-2)^3$
7. $f(x) = (x-4)(x+2)^2 (x-3)^3$
8. $g(x) = (x+3)(x-1)^5$

Use all available methods (e.g. the Rational Zero Theorem, Descartes' Rule of Signs, polynomial division, etc.) to factor each of the following polynomials completely, and then sketch the graph of each one. See Example 1.

9. $f(x) = x^5 + 4x^4 + x^3 - 10x^2 - 4x + 8$
10. $p(x) = 2x^3 - x^2 - 8x - 5$
11. $s(x) = -x^4 + 2x^3 + 8x^2 - 10x - 15$
12. $f(x) = -x^3 + 6x^2 - 12x + 8$
13. $H(x) = x^4 - x^3 - 5x^2 + 3x + 6$
14. $h(x) = x^5 - 11x^4 + 46x^3 - 90x^2 + 81x - 27$
15. $f(x) = 2x^3 + 11x^2 + 20x + 12$
16. $g(x) = x^4 + 3x^3 - 5x^2 - 21x - 14$

chapter five   Polynomial Functions

Use all available methods (e.g. the Rational Zero Theorem, Descartes' Rule of Signs, polynomial division, etc.) to solve each polynomial equation. Use the Linear Factors Theorem to make sure you find the appropriate number of solutions, counting multiplicity.

17. $x^5 + 4x^4 + x^3 = 10x^2 + 4x - 8$

18. $x^4 + 15 = 2x^3 + 8x^2 - 10x$

19. $x^4 + x^3 + 3x^2 + 5x - 10 = 0$

20. $x^3 - 9x^2 = 30 - 28x$

21. $x^5 + x^4 - x^3 + 7x^2 - 20x + 12 = 0$

22. $2x^4 - 5x^3 - 2x^2 + 15x = 0$

23. $x^5 + 15x^3 + 16 = x^4 + 15x^2 + 16x$

24. $x^3 - 5 = 5x^2 - 9x$

Use all available methods (in particular, the Conjugate Roots Theorem, if applicable) to factor each of the following polynomials completely, making use of the given zero if one is given. See Example 2.

25. $f(x) = x^4 - 9x^3 + 27x^2 - 15x - 52$; $3 - 2i$ is a zero.

26. $g(x) = x^3 - (1-i)x^2 - (8-i)x + (12-6i)$; $2-i$ is a zero.

27. $f(x) = x^3 - (2+3i)x^2 - (1-3i)x + (2+6i)$; 2 is a zero.

28. $p(x) = x^4 - 2x^3 + 14x^2 - 8x + 40$; $2i$ is a zero.

29. $n(x) = x^4 - 4x^3 + 6x^2 + 28x - 91$; $2+3i$ is a zero.

30. $G(x) = x^4 - 14x^3 + 98x^2 - 686x + 2401$; $7i$ is a zero.

31. $f(x) = x^4 - 3x^3 + 5x^2 - x - 10$

32. $g(x) = x^6 - 8x^5 + 25x^4 - 40x^3 + 40x^2 - 32x + 16$

33. $r(x) = x^4 + 7x^3 - 41x^2 + 33x$

34. $d(x) = x^5 - x^4 - 18x^3 + 18x^2 + 81x - 81$

35. $P(x) = x^3 - 6x^2 + 28x - 40$

36. $g(x) = x^6 - x^4 - 16x^2 + 16$

Construct polynomial functions with the stated properties. See Example 3.

37. Third degree, only real coefficients, -1 and $5+i$ are two of the zeros, y-intercept is -52.

38. Fourth degree, only real coefficients, $\sqrt{7}$ and $i\sqrt{5}$ are two of the zeros, y-intercept is -35.

39. Fifth degree, 1 is a zero of multiplicity 3, -2 is the only other zero, leading coefficient is 2.

40. Fifth degree, only real coefficients, 0 is the only real zero, $1+i$ is a zero of multiplicity 1, leading coefficient is 1.

41. Fourth degree, only real coefficients, x-intercepts are 0 and 6, $-2i$ is a zero, leading coefficient is 3.

42. Fifth degree, -2 is a zero of multiplicity 2, another integer is a zero of multiplicity 3, y-intercept is -108, leading coefficient is 1.

43. Third degree, only real coefficients, -4 and $3+i$ are two of the zeros, y-intercept is -40.

44. Fifth degree, 1 is a zero of multiplicity 4, -2 is the only other zero, leading coefficient is 4.

45. Third degree, only real coefficients, -4 and $4+i$ are two of the zeros, y-intercept is -68.

chapter five   Polynomial Functions

Solve the following application problems.

46. An open-top box is to be constructed from a 10" by 18" sheet of tin by cutting out squares from each corner as shown and then folding up the sides. Let $V(x)$ denote the volume of the resulting box.

    a. Write $V(x)$ as a product of linear factors.
    b. For which values of x is $V(x) = 0$?
    c. Which of your answers from part **b** are physically possible?

47. An open-top box is to be constructed from a 10" by 15" sheet of tin by cutting out squares from each corner as shown and then folding up the sides. Let $V(x)$ denote the volume of the resulting box.

    a. Write $V(x)$ as a product of linear factors.
    b. For which values of x is $V(x) = 0$?
    c. Which of your answers from part **b** are physically possible?

48. An open-top box is to be constructed from a 9" by 17" sheet of tin by cutting out squares from each corner as shown and then folding up the sides. Let $V(x)$ denote the volume of the resulting box.

    a. Write $V(x)$ as a product of linear factors.
    b. For which values of x is $V(x) = 0$?
    c. Which of your answers from part **b** are physically possible?

49. Assume $f(x)$ is an nth degree polynomial with real coefficients. Explain why the following statement is true: If n is even, the number of turning points is odd and if n is odd, the number of turning points is even.

chapter five project

Polynomial Functions

Ace Automobiles is an international auto manufacturer that has just finished the design for a new sports car. In order to produce this car, they must lease a new assembly plant, purchase new robotic equipment, and hire new staff.

Ace has projected the monthly costs of these expenditures for budgeting purposes. According to their estimates, the plant lease and all utilities will cost $72,000; the depreciation on the new equipment will be $130,000; salaries will cost $480,000; and all other combined overhead will be $47,000. Thus the monthly total for all expenses will be $729,000.

The cost of raw materials to produce each car will be $4500.

The cost function per month for manufacturing x new cars will be
$$C(x) = 4500x + 729000.$$
The revenue per month from selling x cars will be
$$r(x) = 13500x.$$

1. How many cars per month must the new plant make in order to show a profit, if profit is revenue minus cost?
2. If the plant manager decides to increase staff and create two shifts, resulting in an additional $300,000 in salaries with all other expenses remaining unchanged, how many cars would the plant then have to produce in a month to be profitable?
3. Assume now that the plant at full capacity (including the second shift) can produce a maximum of 75 cars per month. What minimum price per car must Ace Automobiles set in order to maintain a profit?

chapter five   Polynomial Functions

CHAPTER SUMMARY

A summary of concepts and skills follows each chapter. Refer to these summaries to make sure you feel comfortable with the material in the chapter. The concepts and skills are organized according to the section title and topic title in which the material is first discussed.

5.1 Introduction to Polynomial Equations and Graphs

topics	pages
Zeros of Polynomials and Solutions of Polynomial Equations • The connection between *zeros* of polynomials and *roots* or *solutions* of polynomial equations • The connection between *x*-intercepts and real zeros of polynomials with real coefficients	p. 363 – 365
Graphing Factored Polynomials • The geometric meaning of linear factors of a polynomial • Behavior of a polynomial as $x \to \pm \infty$ based on its degree and the sign of its leading coefficient	p. 366 – 369
Solving Polynomial Inequalities • The use of the graph of a polynomial in solving a polynomial inequality	p. 370 – 371

5.2 Polynomial Division and the Division Algorithm

topics	pages
The Division Algorithm and the Remainder Theorem • The meaning of the terms *quotient*, *divisor*, *dividend*, and *remainder* as applied to polynomial division • The *Division Algorithm* and what it implies about the degree of the remainder • The connection between zeros of a polynomial and its linear factors • The *Remainder Theorem*	p. 377 – 379
Polynomial Long Division and Synthetic Division • The method of *polynomial long division* • The method of *synthetic division* • Using synthetic division to evaluate polynomials for given values	p. 379 – 384

chapter five summary

topics (continued)	pages
Constructing Polynomials with Given Zeros • Using the theorems to construct polynomials with desired properties	p. 384 – 385

5.3 Locating Real Zeros of Polynomials

topics	pages
The Rational Zero Theorem • Using the *Rational Zero Theorem* to identify possible zeros	p. 390 – 392
Descartes' Rule of Signs • Using *Descartes' Rule of Signs* to identify the possible number of positive and negative real zeros	p. 393 – 395
Bounds of Real Zeros • Determining *upper* and *lower bounds* for the real zeros of a polynomial	p. 395 – 398
The Intermediate Value Theorem • Using the *Intermediate Value Theorem* to approximate real zeros of a polynomial	p. 398 – 400

5.4 The Fundamental Theorem of Algebra

topics	pages
The Fundamental Theorem of Algebra • The *Fundamental Theorem of Algebra* and what it says about solutions of polynomial equations • The *Linear Factors Theorem* and what it says about the number of zeros of a polynomial based on its degree • How to determine the maximum number of *x*-intercepts and *turning points* of the graph of a polynomial	p. 405 – 407
Multiple Zeros and Their Geometric Meaning • The meaning of *multiplicity* as applied to zeros of a polynomial • The geometric implications of zeros of a given multiplicity	p. 408 – 410

chapter five — Polynomial Functions

topics (continued)	pages
Conjugate Pairs of Zeros	p. 410 – 413
• The *Conjugate Roots Theorem* and what it says about polynomials with real coefficients	
• Using the conjugate roots theorem to construct polynomials with desired properties	
Summary of Polynomial Methods	p. 413 – 415
• Using all the tools of the chapter to graph polynomials, factor polynomials, and solve polynomial equations and inequalities	

chapter review

Section 5.1

Verify that the given values of x solve the corresponding polynomial equations.

1. $4x^3 - 5x^2 = -3x + 18;\ x = 2$

2. $x^2 - 6x = -13;\ x = 3 + 2i$

3. $x^3 + x = 6x^2 - 164;\ x = 5 - 4i$

4. $x^3 + (1 + 4i)x = (7 - 2i)x^2 - 2i + 36;\ x = -2i$

Solve the following polynomial inequalities.

5. $2x^2 + 15 \leq 11x$

6. $(x-3)^2(x+1)^2 > 0$

7. $(x-4)(x+2)(x^2-1) \leq 0$

Section 5.2

Use polynomial long division to rewrite each of the following fractions in the form $q(x) + \dfrac{r(x)}{d(x)}$, where $d(x)$ is the denominator of the original fraction, $q(x)$ is the quotient, and $r(x)$ is the remainder.

8. $\dfrac{8x^4 - 6x^3 + 2x^2 + 3x + 4}{2x^2 - 1}$

9. $\dfrac{11x^2 + 2x - 5}{x - 3}$

10. $\dfrac{x^4 - 3x^2 + x - 8}{x^2 + 3x + 2}$

Use synthetic division to determine if the given value for k is a zero of the corresponding polynomial. If not, determine $p(k)$.

11. $p(x) = 6x^5 - 23x^4 - 95x^3 + 70x^2 + 204x - 72;\ k = 1$

12. $p(x) = 48x^4 + 10x^3 - 51x^2 - 10x + 3;\ k = \dfrac{1}{6}$

13. $p(x) = 18x^5 - 87x^4 + 110x^3 - 28x^2 - 16x + 3;\ k = \dfrac{2}{3}$

chapter five   Polynomial Functions

Section 5.3

List all of the potential rational zeros of the following polynomials. Then use polynomial division and the quadratic formula, if necessary, to identify the actual zeros.

14. $f(x) = x^4 + 3x^3 - 3x^2 - 11x - 6$ **15.** $g(x) = 2x^3 - 11x^2 + 18x - 9$

Use Descartes' Rule of Signs to determine the possible number of positive and negative real zeros of each of the following polynomials.

16. $f(x) = 2x^4 - 3x^3 - x^2 + 3x + 10$ **17.** $g(x) = x^6 - 4x^5 - 2x^4 + x^3 - 6x^2 - 11x + 6$

Use synthetic division to identify integer upper and lower bounds of the real zeros of the following polynomials.

18. $f(x) = 2x^3 - 11x^2 + 3x + 36$ **19.** $g(x) = 4x^3 - 16x^2 - 79x - 35$

Section 5.4

Use all available methods (e.g. the Rational Zero Theorem, Descartes' Rule of Signs, polynomial division, etc.) to solve each polynomial equation. Use the Linear Factors Theorem to make sure you find the appropriate number of solutions, counting multiplicity.

20. $3x^5 + x^4 + 5x^3 = x^2 + 28x + 20$ **21.** $8x^5 + 12x^4 - 18x^3 - 35x^2 = 18x + 3$

22. $x^5 + 3x^4 + 3x^3 + 9x^2 = 4(x+3)$

Use all available methods (in particular, the Conjugate Roots Theorem, if applicable) to factor each of the following polynomials completely, making use of the given zero.

23. $f(x) = 14x^4 - 109x^3 + 296x^2 - 321x + 70$; $2+i$ is a zero

24. $f(x) = x^4 - 5x^3 + 19x^2 - 125x - 150$; $-5i$ is a zero

25. $f(x) = 2x^4 + 3x^3 - 7x^2 + 8x + 6$; $1+i$ is a zero

chapter five review

Construct polynomial functions with the stated properties.

26. Fourth degree, only real coefficients, $\dfrac{1}{2}$ and $1+2i$ are two of the zeros, y-intercept is -30, leading coefficient is 2.

27. Fifth degree, only real coefficients, -1 is a zero of multiplicity 3, $\sqrt{6}$ is a zero, y-intercept is -6, leading coefficient is 1.

28. Fifth degree, only real coefficients, 1 is a zero of multiplicity 3, $\sqrt{3}$ is a zero, y-intercept is 3, leading coefficient is 1.

chapter five   Polynomial Functions

chapter test

Verify that the given values of x solve the corresponding polynomial equations.

1. $x^2 - 2x = -3$; $x = 1 + i\sqrt{2}$
2. $x^3 - 2ix^2 = 2i - x$; $x = i$

Solve the following polynomial equations by factoring and/or using the quadratic formula, making sure to identify all the solutions.

3. $x^3 - x^2 = 20x$
4. $x^2 - 3x + 3 = 0$
5. $x^4 - 7x^2 + 10 = 0$
6. $x^5 - x^3 - 2x = 0$

For each of the following polynomial functions, describe the behavior of its graph as $x \to \pm\infty$ and identify the x- and y-intercepts. Use this information to sketch the graph of each polynomial equation.

7. $f(x) = (1-x)(x^2 - 4)$
8. $f(x) = (x-2)^3$
9. $f(x) = (x+2)(x-1)(x-3)$
10. $f(x) = (x-2)^2(x+1)^2$

Solve the following polynomial inequalities.

11. $x^2 + 5x < 6$
12. $x^3 - 2x^2 - 8x \geq 0$
13. $(x+1)^2(x-3)^2 > 0$
14. $x^2(x-2)(1-x) < 0$
15. $x^2 + x \geq -1$
16. $-3x^2 + 7x - 2 > 0$

17. A manufacturer has determined that the revenue from the sale of x shoes is given by $S(x) = -x^2 + 60x$. The cost of producing x shoes is $C(x) = 2000 - 30x$. Given that profit is revenue minus cost, what value(s) for x will give the company non-negative profit?

Use polynomial long division to rewrite each of the following fractions in the form $q(x) + \dfrac{r(x)}{d(x)}$, where $d(x)$ is the denominator of the original fraction, $q(x)$ is the quotient and r(x) is remainder.

18. $\dfrac{x^4 - 1}{x - 1}$
19. $\dfrac{2x^5 - 4x^3 - x^2 + x - 2}{x^2 - x}$
20. $\dfrac{2x^3 + ix^2 - 12x - 4 + i}{2x + i}$
21. $\dfrac{6x^3 i - 3x^2 + (20i - 6)x + 3 + 9i}{2x - 3i}$

22. $\dfrac{x^4 - x^3 + 2x^2 - 4}{x + 2}$

Use synthetic division to determine if the given value of k is a zero of the corresponding polynomial. If not, determine p(k).

23. $p(x) = 10x^4 - 20x^3 + 8x^2 - x + 3, \ k = 1$

24. $p(x) = 16x^4 - 20x^3 + 8x^2 - 2x + 3, \ k = \dfrac{1}{2}$

25. $p(x) = 4x^3 - x^2 + x - 3, \ k = -2$

26. $p(x) = x^4 - 2x^3 - x^2 + x - 21, \ k = 3$

27. $p(x) = -x^4 - x^3 - x^2 + 2x + 69, \ k = -3$

Construct polynomial functions with the stated properties (answers may not be unique).

28. Second degree, zeros of −2 and 1 and goes to −∞ as $x \to -\infty$.

29. Third degree, zeros of −3, 0, 2 and y-intercept 0.

30. Fourth degree and a single x-intercept of −4 and y-intercept 128.

31. Third degree, zeros of ±2 and 3 and passing through the point (4, 24).

Use methods from this chapter to solve the following polynomial equations.

32. $x^3 - 2x^2 - 5x + 6 = 0$

33. $x^4 + 2x^3 + 4x^2 + 8x = 0$

34. $x^4 - 5x^3 + 5x^2 + 5x - 6 = 0$

Use Descartes' Rule of Signs to determine the possible number of positive and negative real zeros of each of the following polynomials.

35. $f(x) = 2x^3 + 3x^2 + 5x + 2$

36. $f(x) = x^3 - 7x^2 + x + 10$

37. $f(x) = 4x^5 - 2x^4 + 3x^3 + x - 9$

38. $f(x) = 3x^4 - 5x^3 + 4x^2 + 10x - 9$

39. $p(x) = x^3 - x^2 + x + a - 1$ is given. If $(x + 2)$ is a factor of the given polynomial, then find a.

40. Given that $p(x) = 2x^2 + x + a + 2$ divided by $(x - 3)$ has a remainder of 4, find a.

41. If $p(x - 2) = x^2 - 4x + 3$, what is the remainder when $p(x)$ is divided by $(x - 3)$?

chapter five   Polynomial Functions

Use all available methods to factor each of the following polynomials completely and then sketch the graph of each one.

42. $f(x) = 2x^3 - 3x^2 - 3x + 2$

43. $f(x) = x^3 - 3x^2 + x - 3$

44. $f(x) = x^4 - 3x^3 - 6x^2 + 28x - 24$

45. $f(x) = x^5 - x^4 - 2x^3 - x^2 + x + 2$

Construct polynomial functions with the stated properties.

46. Second degree, only real coefficients, $1 + i\sqrt{2}$ is one zero, y-intercept is –6.

47. Third degree, real coefficients, zeros 2, 1 + i, and 1 – i, leading coefficient is 1.

48. Fourth degree, real coefficients, x-intercepts are 0 and 3, and 3 + i is a zero, leading coefficient is 1.

49. $p(x)$ is a first degree polynomial and $x \cdot p(x) - 2 = x^2 + 3x - 2$. Find $p(-2)$.

50. Use synthetic division to determine if $(x + 3)$ is a factor of
$f(x) = 3x^3 + 4x^2 - 18x - 3$.

51. Find the constant c such that the denominator will divide evenly into the numerator of $\dfrac{x^5 - 2x^3 + x + c}{x + 2}$.

Express each of the following functions in the form of $f(x) = (x - k)q(x) + r$ for the given value of k, and demonstrate that f(k) = r.

52. $f(x) = 15x^4 + 10x^3 - 6x^2 + 14, \quad k = -\dfrac{2}{3}$

53. $f(x) = x^3 + 3x^2 - 2x - 14, \quad k = \sqrt{2}$

Determine (if possible) the zeros of the function g if the function f has zeros $x = r_1, x = r_2,$ and $x = r_3$.

54. $g(x) = -f(x)$

55. $g(x) = 3f(x)$

56. $g(x) = f(x - 5)$

57. $g(x) = f(2x)$

58. $g(x) = 3 + f(x)$

59. $g(x) = f(-x)$

chapter SIX

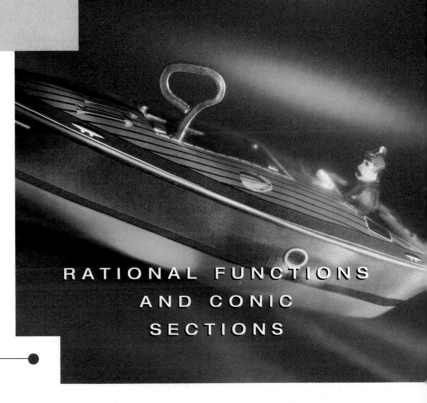

RATIONAL FUNCTIONS AND CONIC SECTIONS

OVERVIEW

6.1 Rational Functions and Rational Inequalities

6.2 The Ellipse

6.3 The Parabola

6.4 The Hyperbola

By the end of this chapter you should be able to graph hyperbolas, find their vertices, and even describe them with equations. One interesting application which makes use of hyperbolas and their characteristics is the "LORAN" radio-communications system. On page 491 you will find a problem regarding the position of a ship at sea. Given information about the distances between transmitters and the ship, you will use hyperbolas to discover an equation describing its possible locations. You will master this type of problem, using tools such as the *Standard Form of a Hyperbola* on page 480.

chapter six — Rational Functions and Conic Sections

Introduction

In this chapter, we will have an opportunity to put to use many of the skills acquired in previous chapters. An understanding of the subject of the first section, rational functions, depends not only on a knowledge of polynomials (of which rational functions are ratios), but also of x- and y-intercepts, factoring, and transformations of functions. Similarly, our study of conic sections, which takes up the remaining three sections of the chapter, is made easier by our general familiarity with the Cartesian plane and our experience in graphing equations.

The study of conic sections, the curves obtained by intersecting a plane with a cone, has a history extending back to Greek mathematics of the third century BC and continuing very much to the present. In fact, few branches of mathematics display such a long-lived vitality. Such famous early mathematical figures as Euclid and Archimedes studied conics and discovered many of their properties, while later mathematicians like Isaac Newton, René Descartes, and Carl Friedrich Gauss continued the tradition and made significant contributions of their own. But perhaps the most interesting aspect of this history is the long span of time that lies between the early (almost purely intellectual) formulation of the theory and its modern (thoroughly pragmatic) applications.

The Greek mathematician Apollonius (c. 262-190 BC) is largely remembered today for his eight volume book entitled *Conic Sections*. Apollonius improved upon the slightly earlier work of Euclid and Archimedes, and proceeded to develop almost all of the theory of conic sections you will encounter in this chapter. Apparently, he is also responsible for the names of the three varieties of conic sections: ellipses, parabolas, and hyperbolas. The names adhere to the Pythagorean tradition of identifying mathematical objects by their geometric properties, and the Greek words refer to the behavior of certain projections of the curves. (Incidentally, the three figures of speech known as *ellipsis*, *parabole*, and *hyperbole* have a similar root). And while the mathematicians of Apollonius' time did have a few worldly uses for their knowledge of conics, it must be stressed that their driving force was the joy of intellectual accomplishment.

Contrast this with the applications of the theory that were discovered several millennia later. The German astronomer and mathematician Johann Kepler (1571-1630) made the observation that planets orbit the sun in elliptical paths, and proceeded to quantify those paths. The unique focusing property of parabolas leads to their use in modern lighting, in the design of satellite dishes, and in telescopes. And an analysis of the problems of long-range navigation leads naturally to a use of hyperbolas in such navigational systems as LORAN. All of these applications, along with many others, make use of geometric properties identified by people with no foreknowledge of their eventual use.

Kepler

6.1 Rational Functions and Rational Inequalities

TOPICS

1. Definitions and useful notation
2. Vertical asymptotes
3. Horizontal and oblique asymptotes
4. Graphing rational functions
5. Solving rational inequalities

Topic 1: Definitions and Useful Notation

Our study of polynomials in the last chapter leads naturally to a study of rational functions, as rational functions are ratios of polynomials. But the behavior of rational functions can be significantly more complex than that of polynomials, and it makes sense to begin by introducing some ideas and notation that will aid us.

Rational Functions

A **rational function** is a function that can be written in the form

$$f(x) = \frac{p(x)}{q(x)},$$

where $p(x)$ and $q(x)$ are both polynomial functions and $q(x) \neq 0$. Of course, even though q is not allowed to be identically zero, there will often be values of x for which $q(x)$ is zero, and at these values the fraction is undefined. Consequently, the **domain of** f is the set $\{x \in \mathbb{R} \mid q(x) \neq 0\}$.

chapter six   Rational Functions and Conic Sections

We have already seen a few rational functions and are familiar with their graphs. For instance, in Chapter 4 we studied the basic reciprocal function – one of the simplest rational functions – and modifications of it. You already know how to construct the graphs, shown in Figure 1, of the following three rational functions:

$$f(x) = \frac{1}{x},$$

$$g(x) = \frac{1}{x-2},$$

$$h(x) = \frac{1}{x+1} + 2 = \frac{1}{x+1} + \frac{2x+2}{x+1} = \frac{2x+3}{x+1}$$

(Note that h can be written as a ratio of polynomials, so h is a rational function)

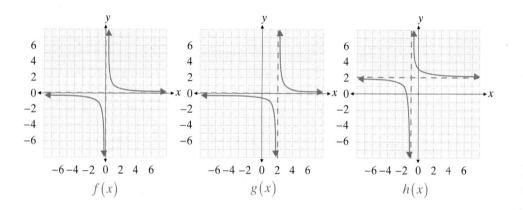

Figure 1: Graphs of Three Rational Functions

Each of the three graphs in Figure 1 is augmented by two dashed lines, one vertical and one horizontal. It is important to realize that in each graph the two dashed lines are *not* part of the function: they are examples of *asymptotes*, and they serve as guides to understanding the function. Roughly speaking, an asymptote is an auxiliary line that the graph of a function approaches. Three kinds of asymptotes will appear in our study of rational functions–vertical, horizontal, and oblique.

Vertical Asymptotes

The vertical line $x = c$ is a **vertical asymptote** of a function f if $f(x)$ increases in magnitude without bound as x approaches c. Examples of vertical asymptotes appear in Figure 2. The graph of a rational function cannot intersect a vertical asymptote.

Figure 2: Vertical Asymptotes

Horizontal Asymptotes

The horizontal line $y = c$ is a **horizontal asymptote** of a function f if $f(x)$ approaches the value c as $x \to -\infty$ or as $x \to \infty$. Examples of horizontal asymptotes appear in Figure 3. The graph of a rational function may intersect a horizontal asymptote near the origin, but will eventually approach the asymptote from one side only as x increases in magnitude.

chapter six   Rational Functions and Conic Sections

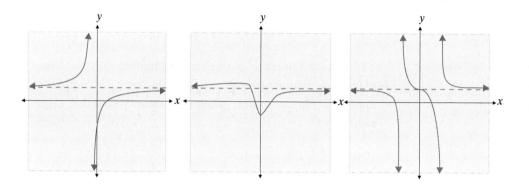

Figure 3: Horizontal Asymptotes

Oblique Asymptotes

A non-vertical, non-horizontal line may also be an asymptote of a function f. Examples of **oblique** (or **slant**) **asymptotes** appear in Figure 4. Again, the graph of a rational function may intersect an oblique asymptote, but will eventually approach the asymptote from one side only as $x \to \infty$ or $x \to -\infty$.

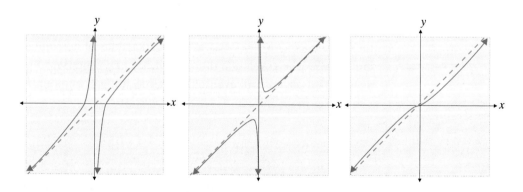

Figure 4: Oblique Asymptotes

section 6.1   Rational Functions and Rational Inequalities

As Figures 2, 3, and 4 illustrate, the behavior of rational functions with respect to asymptotes can vary considerably. In order to describe the behavior of a given rational function more easily, we introduce the following notation.

Asymptote Notation

The notation $x \to c^-$ is used in describing the behavior of a graph as x approaches the value c from the left (the negative side). The notation $x \to c^+$ is used in describing behavior as x approaches c from the right (the positive side). The notation $x \to c$ is used in describing behavior that is the same on both sides of c.

Figure 5 illustrates how the above notation can be used to describe the behavior of functions.

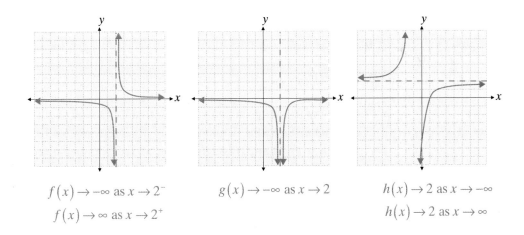

$f(x) \to -\infty$ as $x \to 2^-$
$f(x) \to \infty$ as $x \to 2^+$

$g(x) \to -\infty$ as $x \to 2$

$h(x) \to 2$ as $x \to -\infty$
$h(x) \to 2$ as $x \to \infty$

Figure 5: Asymptote Notation

Topic 2:   Vertical Asymptotes

With the above notation and examples as background, we are ready to delve into the details of identifying asymptotes for rational functions.

Equations for Vertical Asymptotes

If the rational function $f(x) = \dfrac{p(x)}{q(x)}$ has been written in reduced form (so that p and q have no common factors), the vertical line $x = c$ is a vertical asymptote for f if and only if c is a real zero of the polynomial q. In other words, f has vertical asymptotes at the x-intercepts of q.

Note that the numerator of a rational function is irrelevant in locating the vertical asymptotes, assuming that all common factors in the fraction have been canceled.

example 1

Find the equations for the vertical asymptotes of the following functions.

a. $f(x) = \dfrac{x^2 - x}{x - 1}$ **b.** $g(x) = \dfrac{x^2 + 1}{x^2 + 2x - 15}$

Solutions:

a. One of the first steps in analyzing a rational function should always be determining the domain. Since the denominator of f is zero when $x = 1$, the domain of f is the set $\{x \in \mathbb{R} \mid x \neq 1\}$; that is, f is defined for all real numbers except 1. Once the domain has been noted, we can proceed to look for common factors to cancel. In this case, a factor of $x - 1$ can be canceled from the numerator and denominator, leaving $f(x) = x$ (but only for $x \neq 1$). Since the reduced form of f has no denominator, f has no vertical asymptotes.

b. As usual, in order to determine the domain of the rational function we have to factor the denominator. And after making note of the domain, we will look for common factors to cancel, so we may as well factor the numerator, if possible:

$$g(x) = \frac{x^2 + 1}{x^2 + 2x - 15} = \frac{x^2 + 1}{(x + 5)(x - 3)}$$

What we see in this example is that the domain is $\{x \in \mathbb{R} \mid x \neq -5 \text{ and } x \neq 3\}$ and, since the numerator is a sum of two squares and cannot be factored over the real numbers, the fraction is automatically in reduced form. Thus the equations of the two vertical asymptotes are $x = -5$ and $x = 3$.

section 6.1   Rational Functions and Rational Inequalities

Topic 3:   **Horizontal and Oblique Asymptotes**

To determine horizontal and oblique asymptotes, we are interested in the behavior of a function $f(x)$ as $x \to -\infty$ and as $x \to \infty$. If f is a rational function, f is a ratio of two polynomials p and q, so we can begin by considering the effect p and q have on one another. There are many possibilities, but in order to develop some intuition about rational functions, consider the following examples:

1) $f(x) = \dfrac{6x+1}{5x^2 - 2x + 3}$. Any polynomial of degree greater than or equal to 1 gets larger and larger in magnitude (without bound) as $x \to \pm\infty$, and that is true of the numerator and denominator of this function. The critical point, though, is that the degree of the denominator is larger than the degree of the numerator, and hence the behavior of the denominator dominates. As $x \to -\infty$ and as $x \to \infty$, the increasingly large values of the denominator overpower the large values of the numerator, and the fraction as a whole approaches 0. (Think about the result of dividing a fixed number, no matter how large, by a much larger number.) In symbols, $f(x) \to 0$ as $x \to \pm\infty$.

2) $f(x) = \dfrac{6x^2 - 3x + 2}{3x^2 + 5x - 17}$. In this case, the degrees of the numerator and denominator are the same, and neither dominates the other as $x \to \pm\infty$. We know, however, that in any given polynomial the highest power term dominates the rest far away from the origin. In this case, the numerator "looks" more and more like $6x^2$ and the denominator "looks" more and more like $3x^2$ as $x \to -\infty$ and as $x \to \infty$. As a whole, then, the fraction approaches $\dfrac{6x^2}{3x^2} = \dfrac{6}{3} = 2$ far away from the origin. In symbols, $f(x) \to 2$ as $x \to \pm\infty$.

3) $f(x) = \dfrac{7x^3 + 2x - 1}{x^2 + 4x}$. A comparison of the degrees indicates that the numerator of this fraction dominates the denominator far away from the origin. From our work in Chapter 5, we know that dividing a 3rd degree polynomial by a 2nd degree polynomial results in a 1st degree quotient with a possible remainder. As we will see, the quotient (whose graph is a straight line since its degree is 1) is the oblique asymptote of f.

4) $f(x) = \dfrac{3x^5 - 2x^3 + 7x^2 - 1}{4x^3 + 19x^2 - 3x + 5}$. If the degree of the numerator is larger than the degree of the denominator by more than 1 (as in this case), the rational function increases in magnitude without bound as $x \to \pm\infty$, but does not approach any straight line asymptote.

chapter six — Rational Functions and Conic Sections

The following table summarizes the situation as far as horizontal and oblique asymptotes are concerned.

Equations for Horizontal and Oblique Asymptotes

Let $f(x) = \dfrac{p(x)}{q(x)}$ be a rational function, where p is an nth degree polynomial with leading coefficient a_n and q is an mth degree polynomial with leading coefficient b_m. Then:

1. If $n < m$, the horizontal line $y = 0$ (the x-axis) is the horizontal asymptote for f.
2. If $n = m$, the horizontal line $y = \dfrac{a_n}{b_m}$ is the horizontal asymptote for f.
3. If $n = m + 1$, the line $y = g(x)$ is an oblique asymptote for f, where g is the quotient polynomial obtained by dividing p by q. (The remainder polynomial is irrelevant.)
4. If $n > m + 1$, there is no straight line horizontal or oblique asymptote for f.

example 2

Find the equation for the horizontal or oblique asymptote of the following functions.

a. $g(x) = \dfrac{x^2 + 1}{x^2 + 2x - 15}$
b. $h(x) = \dfrac{x^3 + x^2 + 2x + 2}{x^2 + 9}$

Solutions:

a. Because the numerator and denominator of g have the same degree, the asymptote is the horizontal line $y = 1$ (since 1 is the ratio of the leading coefficients).

b. Because the degree of the numerator is one more than the degree of the denominator, we know h has an oblique asymptote. To get the equation for it, we need to do the polynomial division:

section 6.1   Rational Functions and Rational Inequalities

When entering this equation in [Y=], pay close attention to the placement of parentheses. Then select [GRAPH].

$$\begin{array}{r} x+1 \\ x^2+9 \overline{\smash{\big)}\, x^3+x^2+2x+2} \\ -\left(x^3+0x^2+9x\right) \\ \hline x^2-7x+2 \\ -\left(x^2+0x+9\right) \\ \hline -7x-7 \end{array}$$

This tells us that

$$h(x) = x+1 + \frac{-7x-7}{x^2+9},$$

but we only need the quotient, $x+1$, for the task at hand. The equation for the oblique asymptote is $y = x+1$.

Topic 4: Graphing Rational Functions

Much of our experience in graph-sketching will be useful as we graph rational functions. But in addition to the standard steps of identifying the x-intercepts (if any), the y-intercept (if there is one), useful symmetry, and so on, we will make use of asymptotes when graphing rational functions. The following is a list of suggested steps in such work.

Graphing Steps

Given a rational function f,

1. Factor the denominator in order to determine the domain of f;
2. Factor the numerator as well and cancel any common factors;
3. Examine the remaining factors in the denominator to determine the equations for any vertical asymptotes;
4. Compare the degrees of the numerator and denominator to determine if there is a horizontal or oblique asymptote, and proceed to find the equation for such;
5. Determine the y-intercept, assuming 0 is in the domain of f;
6. Determine the x-intercepts, if there are any, by setting the numerator of the reduced fraction equal to 0;
7. Determine if there is any symmetry that can be used as an aid;
8. Plot enough points to determine the behavior of f between x-intercepts and between vertical asymptotes.

chapter six — Rational Functions and Conic Sections

The work in *Example 3* will illustrate the above steps, using three functions that we have already studied in part.

example 3

Sketch the graphs of the following rational functions.

a. $f(x) = \dfrac{x^2 - x}{x - 1}$ **b.** $g(x) = \dfrac{x^2 + 1}{x^2 + 2x - 15}$ **c.** $h(x) = \dfrac{x^3 + x^2 + 2x + 2}{x^2 + 9}$

Solutions:

a.

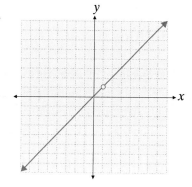

We have already noted, in *Example 1a*, that the domain of f is all real numbers except for 1, and that after canceling common factors the function reduces to $f(x) = x$. Graphing f then is simply a matter of graphing the straight line with a slope of 1 and a y-intercept of 0, and then noting that f is not defined when $x = 1$. We denote this with an open circle at the point of the graph where $x = 1$. (Note that the resolution on a graphing calculator is inadequate to show an open circle on the graph.)

b.

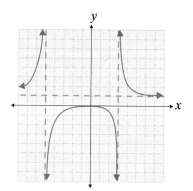

In *Example 1b*, we determined that the function g has the vertical asymptotes $x = -5$ and $x = 3$, and in *Example 2a* we determined it has the horizontal asymptote $y = 1$. These have been drawn at left.

The y-intercept can be calculated by finding the value of the function at 0: $g(0) = -\dfrac{1}{15}$. Note that g has no x-intercepts as there is no real number solution to $x^2 + 1 = 0$. The remainder of the work consists in plotting enough points to allow us to sketch the function as shown.

section 6.1   **Rational Functions and Rational Inequalities**

c.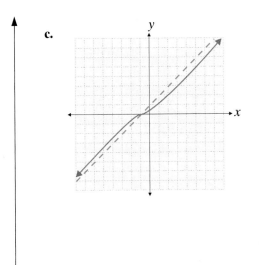

In *Example 2b*, we determined that h has an oblique asymptote of $y = x + 1$. Note that h has no vertical asymptotes, as the denominator of h does not factor over the real numbers.

The function crosses the y-axis at $\dfrac{2}{9}$, but to determine the x-intercept(s) we have to work a bit harder. One of the potential rational zeros of the numerator, -1, actually is a zero, and the numerator factors as $(x+1)(x^2+2)$, so -1 is the only x-intercept.

Topic 5: Solving Rational Inequalities

A rational inequality is an inequality that can be written with a rational function on one side and 0 on the other, and this definition is the key to their solution. For instance, if we were asked to solve the inequality

$$\frac{x^2+1}{x^2+2x-15} > 0,$$

we would only have to examine our graph in *Example 3b* and note that the only intervals on which the function $g(x) = \dfrac{x^2+1}{x^2+2x-15}$ is positive are $(-\infty, -5)$ and $(3, \infty)$, and so the solution of the inequality is the union of these two intervals.

Forgetting this observation can lead to disasters. Suppose, for example, we tried to solve the inequality

$$\frac{x}{x+2} < 3$$

by multiplying through by $x + 2$ and thus clearing the inequality of fractions. The result would be

$$x < 3x + 6,$$

a simple linear inequality. We can solve this to obtain $x > -3$, or the interval $(-3, \infty)$. But does this really work? The number $-\dfrac{5}{2}$ is in this interval, but

$$\frac{-\dfrac{5}{2}}{-\dfrac{5}{2}+2} = 5,$$

which is not less than 3. Also, note that -4 solves the inequality, but -4 is not in the interval $(-3, \infty)$.

What went wrong? The answer is subtle but very important. By multiplying through by $x + 2$, we implicitly made the assumption that $x + 2$ was positive, since we did not worry about possibly reversing the inequality symbol. Of course we don't know beforehand if $x + 2$ is positive or not, since we are trying to solve for the variable x.

To solve this inequality correctly, we have to write it in the standard form of a rational inequality, as shown in *Example 4*.

example 4

Solve the rational inequality $\dfrac{x}{x+2} < 3$.

Solution:

We begin by subtracting 3 from both sides and writing the left-hand side as a single rational function, giving us:

$$\frac{-2x-6}{x+2} < 0$$

We can factor -2 from the numerator of the fraction and then divide both sides by -2 (remembering to reverse the inequality symbol) to obtain the simpler inequality

$$\frac{x+3}{x+2} > 0.$$

section 6.1 Rational Functions and Rational Inequalities

Thus, the solution of the original inequality will consist of those intervals for which the rational function $f(x) = \dfrac{x+3}{x+2}$ is positive. (Alternatively, we can recognize that the only way for the fraction to be positive is if the factor $x+3$ in the numerator and the factor $x+2$ in the denominator have the same sign. We could check the intervals $(-\infty, -3), (-3, -2)$, and $(-2, \infty)$ one by one to see when the factors agree in sign).

The graph of $f(x) = \dfrac{x+3}{x+2}$ is as shown to the right, and from the graph we can see that the solution to the inequality is $(-\infty, -3) \cup (-2, \infty)$.

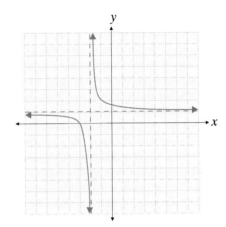

exercises

Find equations for the vertical asymptotes, if any, for each of the following rational functions. See Example 1.

1. $f(x) = \dfrac{5}{x-1}$

2. $f(x) = \dfrac{x^2+3}{x+3}$

3. $f(x) = \dfrac{x^2-4}{x+2}$

4. $f(x) = \dfrac{-3x+5}{x-2}$

5. $f(x) = \dfrac{3x^2+1}{x-2}$

6. $f(x) = \dfrac{x^2+2x}{x+1}$

7. $f(x) = \dfrac{x^2-4}{2x-x^2}$

8. $f(x) = \dfrac{x+2}{x^2-9}$

9. $f(x) = \dfrac{x^2-2x-3}{2x^2-5x-3}$

10. $f(x) = \dfrac{2x^2+2x-4}{x^2+2x+1}$

11. $f(x) = \dfrac{x^3-27}{x^2+5}$

12. $f(x) = \dfrac{x^2+5}{x^3-27}$

13. $f(x) = \dfrac{x^2-1}{x^2-8x+7}$

14. $f(x) = \dfrac{2x^2+7x-14}{2x^2+7x-15}$

15. $f(x) = \dfrac{x^3-6x^2+11x-6}{x^3+8}$

16. $f(x) = \dfrac{x^2-2x-15}{x-5}$

17. $f(x) = \dfrac{x^2-16}{x^2-4}$

18. $f(x) = \dfrac{x^2+4x+4}{x^2+x-2}$

chapter six   Rational Functions and Conic Sections

Find equations for the horizontal or oblique asymptotes, if any, for each of the following rational functions. See Example 2.

19. $f(x) = \dfrac{5}{x-1}$ **20.** $f(x) = \dfrac{x^2+3}{x+3}$ **21.** $f(x) = \dfrac{x^2-4}{x+2}$

22. $f(x) = \dfrac{x^2-4}{2x-x^2}$ **23.** $f(x) = \dfrac{x+2}{x^2-9}$ **24.** $f(x) = \dfrac{x^2-2x-3}{2x^2-5x-3}$

25. $f(x) = \dfrac{2x^2+2x-4}{x^2+2x+1}$ **26.** $f(x) = \dfrac{-3x+5}{x-2}$ **27.** $f(x) = \dfrac{3x^2+1}{x-2}$

28. $f(x) = \dfrac{x^3-27}{x^2+5}$ **29.** $f(x) = \dfrac{x^2+5}{x^3-27}$ **30.** $f(x) = \dfrac{x^2+2x}{x+1}$

31. $f(x) = \dfrac{x^2-81}{x^3+7x-12}$ **32.** $f(x) = \dfrac{x^3-3x^2+2x}{x-7}$ **33.** $f(x) = \dfrac{x^2-9x+4}{x+2}$

34. $f(x) = \dfrac{-x^5+2x^2}{5x^5+3x^3-7}$ **35.** $f(x) = \dfrac{5x^2-x+12}{x-1}$ **36.** $f(x) = \dfrac{2x^2-5x+6}{x-3}$

Sketch the graphs of the following rational functions, making use of your work in the problems above and additional information about intercepts and any other points that may be useful. See Example 3.

37. $f(x) = \dfrac{5}{x-1}$ **38.** $f(x) = \dfrac{x^2+3}{x+3}$ **39.** $f(x) = \dfrac{x^2-4}{x+2}$

40. $f(x) = \dfrac{x^2-4}{2x-x^2}$ **41.** $f(x) = \dfrac{x+2}{x^2-9}$ **42.** $f(x) = \dfrac{x^2-2x-3}{2x^2-5x-3}$

43. $f(x) = \dfrac{2x^2+2x-4}{x^2+2x+1}$ **44.** $f(x) = \dfrac{-3x+5}{x-2}$ **45.** $f(x) = \dfrac{3x^2+1}{x-2}$

46. $f(x) = \dfrac{x^3-27}{x^2+5}$ **47.** $f(x) = \dfrac{x^2+5}{x^3-27}$ **48.** $f(x) = \dfrac{x^2+2x}{x+1}$

section 6.1   Rational Functions and Rational Inequalities

For each graph below, find any **a.** *vertical asymptotes,* **b.** *horizontal asymptotes,* **c.** *oblique asymptotes,* **d.** *visible x-intercepts, or* **e.** *visible y-intercepts.*

49.

50.

51.

52.

53.

54.

55.

56.

Solve the following rational inequalities. See Example 4.

57. $2x < \dfrac{4}{x+1}$

58. $\dfrac{5}{x-2} \geq \dfrac{3x}{x-2}$

59. $\dfrac{5}{x-2} > \dfrac{3}{x+2}$

60. $\dfrac{x}{x^2-x-6} \leq \dfrac{-1}{x^2-x-6}$

61. $\dfrac{x}{x^2-x-6} \leq \dfrac{-2}{x^2-x-6}$

62. $x > \dfrac{1}{x}$

63. $\dfrac{4}{x-3} \leq \dfrac{4}{x}$

64. $\dfrac{x-7}{x-3} \geq \dfrac{x}{x-1}$

65. $\dfrac{x}{x^2+3x+2} > \dfrac{1}{x^2+3x+2}$

66. $\dfrac{1}{x-4} \geq \dfrac{1}{x+1}$

67. $\dfrac{x}{x+1} \geq \dfrac{x+1}{x}$

68. $\dfrac{x}{x^2-2x-3} > \dfrac{3}{x^2-2x-3}$

The following application problems involve rational functions.

69. Joan raises rabbits, and the population of her rabbit colony follows the formula
$$p(t) = \dfrac{200t}{t+1}$$
where $t \geq 0$ represents the number of months since she began operation.

   a. Sketch the graph of $p(t)$ for $t \geq 0$.
   b. What happens to Joan's rabbit population in the long run?

70. If an object is placed a distance x from a lens with a focal length of F, the image of the object will appear a distance y on the opposite side of the lens, where x, F, and y are related by the equation $\dfrac{1}{x} + \dfrac{1}{y} = \dfrac{1}{F}$.

   a. Express y as a function of x and F.
   b. Graph your function for a lens with a focal length of 30 mm ($F = 30$). What happens to y as the distance x increases?

71. At t minutes after injection, the concentration (in mg/L) of a given drug in the bloodstream of a patient is given by the formula

$$c(t) = \frac{20t}{t^2 + 1}.$$

   a. Sketch the graph of $c(t)$ for $t \geq 0$.
   b. What happens to the concentration of the drug in the long run?

chapter six — Rational Functions and Conic Sections

technology exercises

As you know by now, graphing calculators and computer algebra systems can be very useful tools for analyzing functions. In particular, several of the *Mathematica* commands you have seen previously are useful in understanding rational functions. The simple **Plot** command can be used to obtain a graph of a rational function, though always with the caution that such mechanically derived plots can be misleading. The **Factor** command can be used to factor the numerator and denominator of a rational function, and is thus helpful in identifying the domain, the vertical asymptotes, and the x-intercepts. The illustration below shows the use of these two commands in understanding one particular rational function (comments in red are the conclusions that can be drawn from *Mathematica*'s output).

If available, make similar use of technology to plot, identify the domain, and identify the vertical asymptotes of the rational functions that follow.

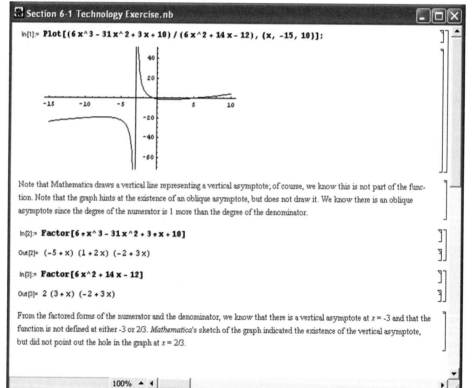

72. $f(x) = \dfrac{8x^3 + 30x^2 - 47x + 15}{4x^3 - 8x^2 - 15x + 9}$

73. $f(x) = \dfrac{8x^3 + 30x^2 - 47x + 15}{4x^2 + 5x - 6}$

74. $f(x) = \dfrac{15x^3 - 61x^2 + 2x + 8}{5x^2 + 8x - 4}$

75. $f(x) = \dfrac{5x^2 + 8x - 4}{15x^3 - 61x^2 + 2x + 8}$

76. $f(x) = \dfrac{6x^3 - 11x^2 - 26x + 40}{6x^2 + 5}$

77. $f(x) = \dfrac{6x^3 - 11x^2 - 26x + 40}{6x^3 - 5x^2 - 49x + 60}$

6.2 The Ellipse

TOPICS

1. Geometric and algebraic meaning of the conic sections
2. Derivation of the standard form of an ellipse
3. Using the standard form of an ellipse
4. Interlude: planetary orbits

Topic 1: Geometric and Algebraic Meaning of the Conic Sections

The three types of conic sections – ellipses, parabolas, and hyperbolas – are so named because all three types of curves arise from intersecting a plane with a circular cone. As shown in Figure 1, an **ellipse** is a closed curve resulting from the intersection of a cone with a plane that intersects only one *nappe* of the cone (that is, the part of the cone on one side of the vertex). A **parabola** results from intersecting a cone with a plane that is parallel to a line on the surface of the cone passing through the vertex; note that a parabola also intersects only one nappe of the cone. Finally, a **hyperbola** is the intersection of a cone with a plane that intersects both nappes and doesn't pass through the vertex.

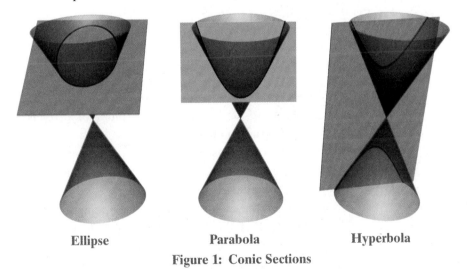

Ellipse   Parabola   Hyperbola

Figure 1: Conic Sections

chapter six Rational Functions and Conic Sections

The three varieties of conic sections share similarities in their algebraic definitions as well. As curves in the Cartesian plane, each conic section we will study is the graph of an equation that can be written in the form $Ax^2 + Cy^2 + Dx + Ey + F = 0$, where A, C, D, E and F are real constants.

Algebraic Definition of Conics

A conic section described by an equation of the form $Ax^2 + Cy^2 + Dx + Ey + F = 0$, where at least one of the two coefficients A and C is not equal to 0, is:

1. an **ellipse** if the product AC is positive
2. a **parabola** if the product AC is 0
3. a **hyperbola** if the product AC is negative

(You may be wondering at the choice of the letters above; although we will not study them in this text, equations of the form $Ax^2 + Bxy + Cy^2 + Dx + Ey + F = 0$ describe more general conics, including those that have been rotated by some angle.)

Finally, the three conic sections are characterized by certain geometric properties that are similar in nature. We will make use of these properties in the work to follow.

Geometric Properties of Conics

1. An ellipse consists of the set of points in the plane for which the sum of the distances d_1 and d_2 to two *foci* (plural of *focus*) is a fixed constant. See the first diagram in Figure 2.
2. A parabola consists of the set of points that are the same distance d from a line (called the *directrix*), and a point not on the line (called the *focus*). See the second diagram in Figure 2.
3. A hyperbola consists of the set of points for which the magnitude of the difference of the distances d_1 and d_2 to two *foci* is fixed. See the third diagram in Figure 2.

section 6.2   The Ellipse

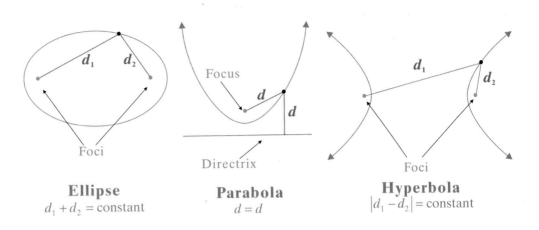

Figure 2: Properties of Conics

Topic 2: Derivation of the Standard Form of an Ellipse

Our goal is to use the properties of the conic sections to derive useful forms of their equations. Specifically, we want to be able to easily construct the graph of a conic section from its equation and, reversing the process, be able to find the equation for a conic section with known properties. We begin with the ellipse.

In order to simplify our work, we will initially assume we are working with an ellipse centered at the origin with foci at $(-c, 0)$ and $(c, 0)$; after deriving a useful form of the equation for this ellipse, it will be an easy matter to generalize it for an arbitrary ellipse. What we seek is an equation in x and y that identifies those points (x, y) on the ellipse, and the only property we need is that the sum $d_1 + d_2$ from Figure 3 is fixed. To make our algebra easier, we will denote this sum as $2a$; that is, $d_1 + d_2 = 2a$.

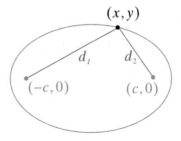

Figure 3

chapter six   Rational Functions and Conic Sections

The rest of the derivation consists of applying the distance formula from Section 3.1 followed by algebraic simplification. Using the distance formula to find d_1 and d_2, we know that

$$\sqrt{(x+c)^2 + y^2} + \sqrt{(x-c)^2 + y^2} = 2a.$$

In one sense, we have achieved our goal, as this is indeed an equation in x and y that describes all the points on the ellipse. What we desire, though, is to make the equation easier to use and interpret. We can do this by eliminating the radicals in the equation (as you learned to do in Section 2.6) and by ultimately renaming some of the constants that appear in the equation. We begin by isolating one radical, squaring both sides, and eliminating some terms:

$$\left(\sqrt{(x+c)^2 + y^2}\right)^2 = \left(2a - \sqrt{(x-c)^2 + y^2}\right)^2$$

$$(x+c)^2 + y^2 = 4a^2 - 4a\sqrt{(x-c)^2 + y^2} + (x-c)^2 + y^2$$

$$\cancel{x^2} + 2cx + \cancel{c^2} + \cancel{y^2} = 4a^2 - 4a\sqrt{(x-c)^2 + y^2} + \cancel{x^2} - 2cx + \cancel{c^2} + \cancel{y^2}$$

There is still one radical left, so we isolate it and repeat the process:

$$4cx - 4a^2 = -4a\sqrt{(x-c)^2 + y^2}$$

$$a^2 - cx = a\sqrt{(x-c)^2 + y^2}$$

$$(a^2 - cx)^2 = a^2\left((x-c)^2 + y^2\right)$$

$$a^4 - \cancel{2a^2cx} + c^2x^2 = a^2x^2 - \cancel{2a^2cx} + a^2c^2 + a^2y^2$$

$$a^4 - a^2c^2 = a^2x^2 - c^2x^2 + a^2y^2$$

This form of the equation has the virtue of no square root terms, but we can make a superficial change to improve the appearance even more. From the diagram in Figure 3, it's clear that $2a$ is larger than $2c$; that is, even in the extreme case of very skinny ellipses, the sum $d_1 + d_2$ (which we have called $2a$) will approach $2c$ but will always be larger than $2c$. We can conclude that $a > c$, so $a^2 - c^2$ is a positive number. All of this is simply to justify renaming $a^2 - c^2$ as b^2. In other words, we replace the constant $a^2 - c^2$ with the simpler symbol b^2, and we are safe in replacing $a^2 - c^2$ with what is clearly a non-negative number because we have shown that $a^2 - c^2$ is positive.

Thus

$$a^4 - a^2c^2 = a^2x^2 - c^2x^2 + a^2y^2$$
$$a^2(a^2 - c^2) = (a^2 - c^2)x^2 + a^2y^2$$
$$a^2b^2 = b^2x^2 + a^2y^2.$$

If we write this as $b^2x^2 + a^2y^2 - a^2b^2 = 0$, we have a simpler equation that does indeed describe an ellipse by the algebraic definition (since the product of the coefficients of x^2 and y^2 is positive). For our purposes, we will find it useful to make one more cosmetic adjustment and divide the equation by a^2b^2:

$$b^2x^2 + a^2y^2 = a^2b^2$$
$$\frac{x^2}{a^2} + \frac{y^2}{b^2} = 1$$

Before generalizing this equation to arbitrary ellipses, let's examine its utility with an example.

example 1

Graph the equation $\frac{x^2}{16} + \frac{y^2}{9} = 1$, and determine the foci of the resulting ellipse.

Solution:

This equation is in the form derived above, so we can see by inspection that $a = 4$ and $b = 3$. We can calculate the x and y intercepts of such an equation: when $y = 0$ it must be the case that $x^2 = 16$, so $x = \pm 4$, and when $x = 0$ the equation reduces to $y^2 = 9$, so $y = \pm 3$. So $(-4, 0)$ and $(4, 0)$ are the two x-intercepts and $(0, -3)$ and $(0, 3)$ are the two y-intercepts. Further, these four points are extremes of the graph, as any x-value larger than 4 in magnitude leads to imaginary values for y, and any y-value larger than 3 in magnitude leads to imaginary values for x.

Before plotting these intercepts and filling in the rest of the ellipse, we will use the relation $a^2 - c^2 = b^2$ from the derivation above to determine c, and hence the location of the two foci. Rearranging the terms of the equation, we obtain $c^2 = a^2 - b^2$, and so

$$c^2 = 16 - 9 = 7$$
$$c = \pm\sqrt{7}.$$

continued on next page...

chapter six — Rational Functions and Conic Sections

Hence the two foci are at $\left(-\sqrt{7},0\right)$ and $\left(\sqrt{7},0\right)$.

Putting these pieces together, and connecting the four extreme points with an elliptically shaped curve, we obtain the picture below.

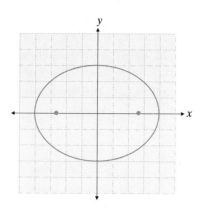

To display the graph of an ellipse, select the APPS button. Select Conics, then 2:Ellipse. Now select **1** to enter a horizontal ellipse, and plug in the values from the equation. To view the graph, select

Topic 3: Using the Standard Form of an Ellipse

(You may need to visit the Texas Instruments website to download the "Conic Graphing" application.)

We have nearly, but not quite, derived the general form of an ellipse. At this point, we know that

$$\frac{x^2}{a^2}+\frac{y^2}{b^2}=1$$

describes an ellipse centered at the origin that is elongated horizontally. If the foci of an origin-centered ellipse are instead at $(0,c)$ and $(0,-c)$, so that the ellipse is elongated vertically, then the equation is

$$\frac{x^2}{b^2}+\frac{y^2}{a^2}=1,$$

where we are again assuming $a > b$ and $c^2 = a^2 - b^2$. The last task is to come up with the equation for an ellipse that is centered at the point (h, k). But we know how to do this: replacing x with $x - h$ in an equation shifts the graph of the equation h units horizontally, and replacing y with $y - k$ shifts the graph k units vertically.

To simplify our discussion, we will say the **major axis** of an ellipse is the line segment extending from one extreme point of the ellipse to the other and passing through the two foci and the center. The two ends of the major axis are the **vertices** of the ellipse. The **minor axis** is the line segment that also passes through the center and is perpendicular to the major axis; it spans the distance between the other two extremes of the ellipse. Thus $2a$ and $2b$ are the lengths of the major and minor axes, respectively. The following box uses these terms to summarize our work.

Standard Form of an Ellipse

Assume a and b are fixed positive numbers with $a > b$. The standard form of the equation for the ellipse centered at (h, k) with major axis of length $2a$ and minor axis of length $2b$ is:

- $\dfrac{(x-h)^2}{a^2} + \dfrac{(y-k)^2}{b^2} = 1$ if the major axis is horizontal
- $\dfrac{(x-h)^2}{b^2} + \dfrac{(y-k)^2}{a^2} = 1$ if the major axis is vertical

In both cases, the two foci of the ellipse are located on the major axis c units away from the center of the ellipse, where $c^2 = a^2 - b^2$.

The next two examples will demonstrate how we can use the standard form to perform two different tasks.

example 2

Graph the equation $25x^2 + 4y^2 + 100x - 24y + 36 = 0$, and determine the foci of the resulting ellipse.

Solution:

The equation $25x^2 + 4y^2 + 100x - 24y + 36 = 0$ definitely represents an ellipse, as the product of the coefficients of x^2 and y^2 is positive, but the equation in this form tells us almost nothing about the specifics of the ellipse. We have encountered similar situations previously, and a familiar technique will assist us here just as it has in the past. That technique is the method of completing the square, as shown below.

$$25x^2 + 4y^2 + 100x - 24y + 36 = 0$$
$$25(x^2 + 4x) + 4(y^2 - 6y) = -36$$
$$25(x^2 + 4x + 4) + 4(y^2 - 6y + 9) = -36 + 100 + 36$$
$$25(x+2)^2 + 4(y-3)^2 = 100$$
$$\frac{(x+2)^2}{4} + \frac{(y-3)^2}{25} = 1$$

We begin by rearranging the terms as shown. Recall that the method involves completing the square (for both x and y in this case) in order to obtain perfect square trinomials. This requires adding 4 to the expression $x^2 + 4x$ and 9 to $y^2 - 6y$, and compensating by adding numbers to the right-hand side as well.

continued on next page...

chapter six
Rational Functions and Conic Sections

Note that the last step in the work above is to divide both sides by 100 in order to obtain the number 1 on the right-hand side. At this point, we have an equivalent equation that is in the standard form for an ellipse.

From the standard form, we can tell that the center of the ellipse is $(-2, 3)$, that the ellipse is elongated vertically (since the denominator under the y term is larger than that under the x term), and that $a = 5$ and $b = 2$. So the length of the major (vertical) axis is 10 and the length of the minor axis is 4. This gives us enough information to sketch the graph at right.

This ellipse is graphed in the same manner as the previous one, except the second equation form was selected for a vertical ellipse. Note that A and B are now switched. Also, remember that the value for H in this case is negative.

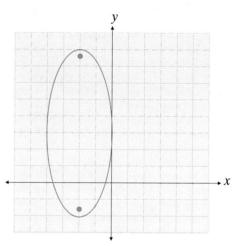

To determine the locations of the two foci, we solve the equation $c^2 = a^2 - b^2 = 21$ for c, obtaining $c = \pm\sqrt{21} \approx \pm 4.6$. So the two foci are approximately 4.6 units above and below the center of the ellipse, at $(-2, 7.6)$ and $(-2, -1.6)$.

technology note

A computer algebra system, such as *Mathematica*, can sketch the graphs of equations as well as functions, and therefore can be used to construct graphs of ellipses. Such graphs can't be relied on to identify features of an ellipse such as the lengths of the major and minor axes or the locations of the foci, but they can be useful in checking your work. In *Mathematica*, the command used to graph an equation is `ImplicitPlot`. To use the command, you must first execute the command `<<Graphics`ImplicitPlot`` (this extends the basic capabilities of *Mathematica* and must be executed only one time in a *Mathematica* session and only if you intend to use `ImplicitPlot`). The use of these commands is illustrated in sketching the graph of $x^2 - 6x + 2y^2 + 8y = 3$.

section 6.2   The Ellipse

Again, select APPS, Conics, and 2:Ellipse. This graph uses equation 1 for a horizontal ellipse.

The next example reverses the process, and is faster.

example 3

Construct the equation, in standard form, for the ellipse whose graph is given below.

Solution:

Careful examination of the graph shows that the center of the ellipse is at $(1, -2)$ and that the major axis has length 12 and the minor axis has length 2. This means that $a = 6$ and $b = 1$, and since the ellipse is elongated horizontally, the equation must be $\dfrac{(x-1)^2}{36} + (y+2)^2 = 1.$

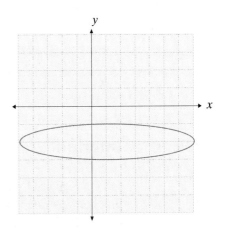

chapter six   Rational Functions and Conic Sections

Example 2 demonstrated that the technique of graphing ellipses bears much similarity to that of graphing circles, and this similarity warrants further investigation. To begin, we might wonder about the result if the numbers a and b in the standard form of an ellipse happen to be the same. If $a = b = r$, then the standard form is

$$\frac{(x-h)^2}{r^2} + \frac{(y-k)^2}{r^2} = 1,$$

which, if we multiply through by r^2, gives us $(x-h)^2 + (y-k)^2 = r^2$, the standard form of a circle. So a circle, not surprisingly, is an ellipse whose major and minor axes are the same length.

A circle, viewed as a type of ellipse, has only one focus, and that focus coincides with the center of the circle. Ellipses that are nearly circular have two foci, but they are relatively close to the center, while narrow ellipses have foci far away from the center. This gives us a convenient way to characterize the relative "skinniness" of an ellipse.

Eccentricity of an Ellipse

> Given an ellipse with major and minor axes of lengths $2a$ and $2b$, respectively, the **eccentricity** of the ellipse, often given the symbol e, is defined by
>
> $$e = \frac{c}{a} = \frac{\sqrt{a^2 - b^2}}{a}.$$
>
> If $e = 0$, then $c = 0$ and the ellipse is actually a circle. At the other extreme, c may be close to (but cannot equal) a, in which case e is close to 1. An eccentricity close to 1 indicates a relatively skinny ellipse.

Topic 4:   Interlude: Planetary Orbits

Johann Kepler (1571-1630) was the German-born astronomer and mathematician who first demonstrated that the planets in our solar system follow elliptical orbits. He did this by laboring, over a period of twenty-one years, to mathematically model the astronomical observations of his predecessor Tycho Brahe. The ultimate result was Kepler's Three Laws of Planetary Motion:

section 6.2 The Ellipse

> I. The planets orbit the sun in elliptical paths, with the sun at one focus of each orbit.
>
> II. A line segment between the sun and a given planet sweeps over equal areas of space in equal intervals of time.
>
> III. The square of the time needed for a planet's complete revolution about the sun is proportional to the cube of the orbit's semimajor (half of the major) axis.

You are probably already familiar with the fact that the shapes of the various planetary orbits in our solar system vary widely, and that these differences greatly influence seasons on the planets. For instance, Pluto has a much more eccentric orbit than Earth, in both the informal and formal meaning of the term. The eccentricity of Pluto's orbit is approximately 0.248, while the eccentricity of Earth's orbit is approximately 0.017. The eccentricities can be easily calculated from astronomical data, but we can also use the eccentricity of a planet's orbit to answer particular questions.

example 4

Given that the furthest Earth gets from the sun is approximately 94.56 million miles, and that the eccentricity of Earth's orbit is approximately 0.017, estimate the closest approach of the Earth to the sun.

continued on next page...

chapter six — Rational Functions and Conic Sections

Solution:

Kepler's first law states that each planet follows an elliptical orbit, and that the sun is positioned at one focus of the ellipse. Thus the Earth is furthest from the sun when Earth is at the end of the ellipse's major axis on the other side of the center from the sun. Using our standard terminology, this means that

$$a + c = 94.56 \text{ (million)}.$$

By the definition of eccentricity, we also know that

$$0.017 = \frac{c}{a},$$

or $c = 0.017a$. Combining these pieces of information, we get $1.017a = 94.56$, leading to $a = 92.98$ million miles. From this, we know $c = (0.017)(92.98) = 1.58$ million miles. The closest approach to the sun must be $a - c$, which is thus approximately 91.4 million miles.

exercises

Find the center, foci, and vertices of each ellipse that the equation describes. See Examples 1 and 2.

1. $\dfrac{(x-5)^2}{4} + \dfrac{(y-2)^2}{25} = 1$

2. $\dfrac{(x+3)^2}{9} + \dfrac{(y+1)^2}{16} = 1$

3. $(x+2)^2 + 3(y+5)^2 = 9$

4. $4(x-4)^2 + (y-2)^2 = 8$

5. $x^2 + 6x + 2y^2 - 8y + 13 = 0$

6. $2x^2 + y^2 - 4x + 4y - 10 = 0$

7. $4x^2 + y^2 + 40x - 2y + 85 = 0$

8. $x^2 + 2y^2 - 6x + 16y + 37 = 0$

9. $x^2 + 3y^2 + 8x - 12y + 1 = 0$

10. $4x^2 + 3y^2 - 8x + 18y + 19 = 0$

11. $x^2 - 4x + 5y^2 - 1 = 0$

12. $x^2 + 4y^2 + 24y + 28 = 0$

Match the following equations to their graphs.

13. $\dfrac{(x-1)^2}{4} + \dfrac{y^2}{81} = 1$

14. $\dfrac{(x-3)^2}{49} + \dfrac{(y-2)^2}{25} = 1$

15. $\dfrac{(x+4)^2}{9} + \dfrac{(y-3)^2}{4} = 1$

16. $\dfrac{(x+3)^2}{36} + \dfrac{(y-1)^2}{64} = 1$

17. $\dfrac{(x+3)^2}{4} + \dfrac{(y+3)^2}{9} = 1$

18. $\dfrac{x^2}{16} + \dfrac{(y+5)^2}{4} = 1$

19. $\dfrac{(x-4)^2}{9} + \dfrac{(y+3)^2}{49} = 1$

20. $\dfrac{(x-5)^2}{16} + \dfrac{(y+1)^2}{9} = 1$

a.

b.

c.

d.

e.

f.

g.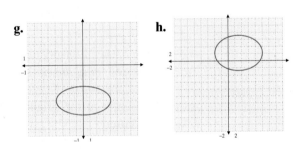

h.

Sketch the graphs of the following ellipses, and determine the coordinates of the foci. See Examples 1 and 2.

21. $\dfrac{(x-3)^2}{9} + \dfrac{(y+1)^2}{1} = 1$

22. $\dfrac{(x+5)^2}{4} + \dfrac{(y+2)^2}{16} = 1$

23. $\dfrac{(x-3)^2}{9} + \dfrac{(y-4)^2}{4} = 1$

24. $\dfrac{x^2}{25} + \dfrac{(y-3)^2}{16} = 1$

25. $(x-1)^2 + \dfrac{(y-4)^2}{4} = 1$

26. $\dfrac{(x-4)^2}{16} + \dfrac{(y-4)^2}{4} = 1$

27. $\dfrac{(x+1)^2}{25} + \dfrac{(y+5)^2}{4} = 1$

28. $\dfrac{(x-2)^2}{9} + \dfrac{(y+1)^2}{9} = 1$

29. $\dfrac{(x+2)^2}{16} + \dfrac{(y+1)^2}{9} = 1$

30. $\dfrac{x^2}{25} + (y+2)^2 = 1$

31. $9x^2 + 16y^2 + 18x - 64y = 71$

32. $9x^2 + 4y^2 - 36x - 24y + 36 = 0$

33. $16x^2 + y^2 + 160x - 6y = -393$

34. $25x^2 + 4y^2 - 100x + 8y + 4 = 0$

35. $4x^2 + 9y^2 + 40x + 90y + 289 = 0$

36. $16x^2 + y^2 - 64x + 6y + 57 = 0$

37. $4x^2 + y^2 + 4y = 0$

38. $9x^2 + 4y^2 + 108x - 32y = -352$

In each of the following problems, an ellipse is described either by picture or by properties it possesses. Find the equation, in standard form, for each ellipse. See Example 3.

39. Center at the origin, major axis of length 10 on the y-axis, foci 3 units from the center.

40. Center at $(-2, 3)$, major axis of length 8 oriented horizontally, minor axis of length 4.

41. Vertices at $(1, 4)$ and $(1, -2)$, foci $2\sqrt{2}$ units from the center.

42. Vertices at $(5, -1)$ and $(1, -1)$, minor axis of length 2.

43. Foci at $(0, 0)$ and $(6, 0)$, $e = \dfrac{1}{2}$.

44. Vertices at $(-1, 4)$ and $(-1, 0)$, $e = 0$.

45. Vertices at $(-2, -1)$ and $(-2, -5)$, minor axis of length 2.

46. Vertices at $(-4, 6)$ and $(-14, 6)$, $e = \dfrac{2}{5}$.

47. Vertices at $(1, 3)$ and $(9, 3)$, one of the foci at $(6, 3)$.

48. Foci at $(2, -4)$ and $(2, -8)$, minor axis of length 6.

49.

50.

51.

52.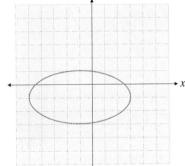

chapter six — Rational Functions and Conic Sections

Use your knowledge of ellipses to answer the following questions.

53. The orbit of Halley's Comet is an ellipse with an eccentricity of 0.967. Its closest approach to the sun is approximately 54,591,000 miles. What is the farthest Halley's Comet ever gets from the sun?

54. Pluto's closest approach to the sun is approximately 4.43×10^9 kilometers, and its maximum distance from the sun is approximately 7.37×10^9 kilometers. What is the eccentricity of Pluto's orbit?

55. Use the information given in *Example 4* to determine the length of the minor axis of the ellipse formed by Earth's orbit around the sun.

56. The archway supporting a bridge over a river is in the shape of half an ellipse. The archway is 60 feet wide and is 15 feet tall at the middle. A boat is 10 feet wide and 14 feet, 9 inches tall. Is the boat capable of passing under the archway?

57. Since the sum of the distances from each of the two foci to any point on an ellipse is constant, we can draw an ellipse using the following method. Tack the ends of a length of string at two points (the foci) and, keeping the string taut by pulling outward with the tip of a pencil, trace around the foci to form an ellipse (the total length of the string remains constant). If you want to create an ellipse with a major axis of length 5 cm and a minor axis of length 3 cm, how long should your string be and how far apart should you place the tacks? Use the relationships of distances and formulas that you have learned in this section.

58. Using the method described in exercise 57, describe the change in your ellipse when you move the two foci closer together. What happens when you move them further apart?

59. *The Whispering Gallery* in Chicago's Museum of Science and Industry is a giant ellipsoid that transmits the slightest whisper from one focus to the other focus. This giant ellipse is known to have a length of about 568 inches and a width of about 162 inches. Find the eccentricity of *The Whispering Gallery*. About how far apart are two whisperers when communicating in this gallery? Round your answers to four decimal places.

chapter six — Rational Functions and Conic Sections

technology exercises

Use a computer algebra system such as Mathematica (or a suitably equipped graphing calculator) to graph the following equations. See the Technology Note after Example 2 for guidance.

60. $15x^2 + 9y^2 + 150x - 36y = -276$

61. $5x^2 + 12y^2 - 20x + 144y + 392 = 0$

62. $3x^2 + 2y^2 = 3 - 18x$

63. $2x^2 + 5y^2 = 70y - 205$

6.3 The Parabola

TOPICS

1. Derivation of the standard form of a parabola (as a conic section)
2. Using the standard form of a parabola
3. Interlude: parabolic mirrors

Topic 1: Derivation of the Standard Form of a Parabola (as a Conic Section)

In Section 4.2, we studied parabolas as they arise in the context of quadratic functions. Specifically, we know that any function of the form

$$f(x) = a(x-h)^2 + k$$

describes a vertically-oriented parabola in the plane (opening upward or downward, depending on the sign of a) whose vertex is at (h, k). (We also know how to use the method of completing the square to transform a function $f(x) = ax^2 + bx + c$ into the more useful form above; we will have occasion to use this skill again in this section.) The material in this section does not replace what we have learned. Instead, viewing parabolas as one of the three varieties of conic sections broadens our understanding of them, and allows us to work with parabolic curves that are not defined by functions.

Recall from Section 6.2 that a principal geometric property of parabolas is that each point on a given parabola is equidistant from a fixed point called the *focus* and a fixed line called the *directrix*. Just as with ellipses, we will derive a useful form of the equation for a parabola from one geometric property. The form that we seek will allow us to quickly sketch fairly accurate graphs of parabolas if we are given the equation, and in turn construct equations for a parabola if we are given some geometric information about it.

Following convention, we will let p denote the distance between the vertex of the parabola and the focus, and hence also the distance between the vertex and the directrix, as shown in Figure 1. As in Section 6.2, we will begin our derivation by assuming the parabola is oriented vertically and that the vertex is at the origin. After our initial work, it will be a relatively simple matter to generalize to horizontally-oriented parabolas and parabolas whose vertices are not at the origin.

Given the information above, we know the focus of our parabola is at $(0, p)$ and the equation for the directrix is $y = -p$. The equation below then characterizes all those points that are of equal distance d from the focus and the directrix:

$$\sqrt{x^2 + (y-p)^2} = \sqrt{(x-x)^2 + (y+p)^2}$$

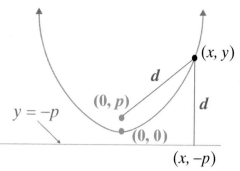

Figure 1: Parts of a Parabola

One way to make the equation look better is to square both sides. We can then proceed to eliminate some terms and combine others to arrive at a useful form:

$$x^2 + (y-p)^2 = (y+p)^2$$
$$x^2 + y^2 - 2py + p^2 = y^2 + 2py + p^2$$
$$x^2 = 4py$$

The conclusion is that any equation that can be written in the above form describes a parabola whose vertex is at the origin and whose focus is at $(0, p)$. This is true even if p is negative; if this is the case then the parabola opens downward.

As we have seen in other contexts, replacing x with y and vice versa reflects a graph about the line $y = x$, so the equation $y^2 = 4px$ describes a horizontally-oriented parabola with vertex at $(0, 0)$. Swapping x and y reflects the focus and directrix of a parabola as well, so that the focus of $y^2 = 4px$ is at $(p, 0)$ and the directrix is the vertical line $x = -p$. And, as always, replacing x with $x - h$ and y with $y - k$ shifts a graph h units horizontally and k units vertically.

Standard Form of a Parabola

Assume p is a fixed non-zero real number. The **standard form** of the equation for the parabola with vertex at (h, k) is:

- $(x-h)^2 = 4p(y-k)$ if the parabola is vertically oriented (the focus is at $(h, k+p)$ and the equation of the directrix is $y = k - p$)
- $(y-k)^2 = 4p(x-h)$ if the parabola is horizontally oriented (the focus is at $(h+p, k)$ and the equation of the directrix is $x = h - p$)

As you know, parabolas can be relatively flat or relatively skinny. The constant p in the standard form is the parameter that determines the degree of flatness of the parabola, a fact that will be visually clear to you if you draw several parabolas whose foci are of varying distances from their vertices.

Topic 2: Using the Standard Form of a Parabola

Putting the standard form of a parabola to use usually requires the use of some familiar algebraic techniques, as the next several examples demonstrate.

---

example 1

Graph the equation $-y^2 + 2x + 2y + 5 = 0$, and determine the focus and directrix of the resulting parabola.

Solution:

We know the equation $-y^2 + 2x + 2y + 5 = 0$, represents a parabola, as the product of the coefficients of x^2 and y^2 is 0 (that is, only one of the variables appears to the second power), but the equation in this form is difficult to graph. We can put the equation into the standard form (as a conic) by completing the square with respect to y (the squared variable) and then identifying p.

continued on next page...

$$-y^2 + 2x + 2y + 5 = 0$$
$$y^2 - 2y = 2x + 5$$
$$y^2 - 2y + 1 = 2x + 5 + 1$$
$$(y-1)^2 = 2(x+3)$$
$$(y-1)^2 = 4\left(\frac{1}{2}\right)(x+3)$$

We begin by rearranging the terms as shown. In order to complete the square, we have to add 1 to both sides.

The last step is to figure out what value the constant p has in the equation. To do this, we need to put the right-hand side into the form $4p(x-h)$, as shown.

Now that the equation is in standard form, by inspection we can tell that the vertex is at $(-3, 1)$ and that $p = \frac{1}{2}$. Since the focus is p units to the right of the vertex and the directrix is a vertical line p units to the left of the vertex, we obtain:

$$\text{focus: } \left(-\frac{5}{2}, 1\right)$$

$$\text{directrix: } x = -\frac{7}{2}$$

The only remaining task is to sketch the graph. We can begin by plotting the vertex at $(-3, 1)$. We can plot two other points by noting that the two points whose x-coordinates are $-\frac{5}{2}$ will be 1 unit away from the directrix; therefore they must lie 1 unit away from the focus as well. That is, $\left(-\frac{5}{2}, 0\right)$ and $\left(-\frac{5}{2}, 2\right)$ must also lie on the parabola. This gives us some idea of the "flatness" of the parabola.

To duplicate the graph below, select APPS, then 6: Conics and 4: Parabola. Select equation 1 for a horizontal parabola, and plug in the values from the equation. To view the graph, select GRAPH.

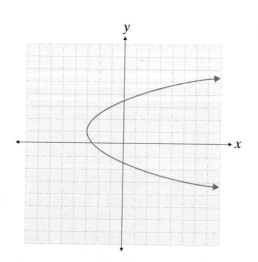

The last step in *Example 1* warrants further discussion. Because every point on a given parabola is equidistant from the parabola's focus and its directrix, the two points that are each 2p units away from the directrix and from the focus can be plotted as shown in Figure 2.

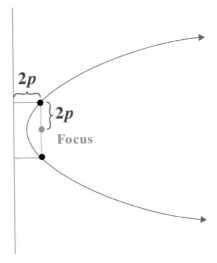

Figure 2: Two Easily Plotted Points

Example 2 asks for an equation based on some geometric information.

example 2

Given that the directrix of a parabola is the line $y = 2$, that $p = -1$, and that the parabola is symmetric with respect to the line $x = 2$, find the equation that defines the parabola.

Solution:

From the given information, we can immediately draw the picture at right to check our work. We know the parabola is oriented vertically because the directrix is horizontal, and from the equation for the line of symmetry we know the *x*-coordinate of the vertex (and the focus) must be 2. Moving down 1 unit from the directrix (as *p* is negative) puts the vertex at (2, 1). So the equation must be:

$$(x-2)^2 = -4(y-1)$$

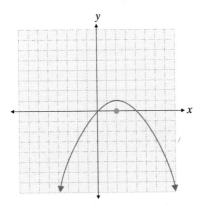

chapter six — Rational Functions and Conic Sections

Topic 3: Interlude: Parabolic Mirrors

Parabolas arise naturally in many diverse contexts, and knowledge of parabolas is a key to solving many practical applications. We have already seen one very important application: the path described by a thrown object (in the absence of friction and other forces) is a parabola. More exactly, we know that the height $h(t)$ of an object with initial velocity v_0 and initial height h_0 is

$$h(t) = -\frac{1}{2}gt^2 + v_0 t + h_0,$$

where g, a constant, is the acceleration due to gravity (see Section 2.3 for details). This is a classic example of a quadratic function, and we experience the fact that a parabola is the shape of its graph whenever we see, for example, a thrown baseball.

But parabolas also arise in contexts where a quadratic function may not be the most natural way to describe them or may simply be irrelevant. In some applications, the geometric properties of parabolas are the key issue. For instance, the fact that every point on a parabola is equidistant from its focus and its directrix has the consequence that the opposite angles shown in Figure 3 are equal to one another at each point. That is, if we imagine rays of light emanating from the focus of the parabola, each ray reflected from the inner surface of the parabola is parallel to the axis of symmetry.

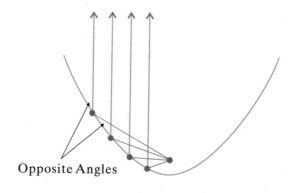

Figure 3: Equal Angles of Reflection in a Parabola

This property can be used in a variety of concrete applications. If a parabola is rotated about its axis of symmetry, a three-dimensional shape called a *paraboloid* is the result. A *parabolic mirror* is then made by coating the inner surface of a paraboloid with a reflecting material. Parabolic mirrors are the basis of searchlights and vehicle headlights, with a light source placed at the focus of the paraboloid, and are also the basis of one design of telescope, in which incoming (parallel) starlight is reflected to an eyepiece at the focus.

example 3

The Hale Telescope at the Mount Palomar observatory in California is a very large reflecting telescope. The paraboloid is the top surface of a large cylinder of Pyrex glass 200 inches in diameter. Along the outer rim, the cylinder of Pyrex is approximately 26.8 inches thick, while at the center the Pyrex is 23 inches thick. Given this, where is the focus of the parabolic mirror located?

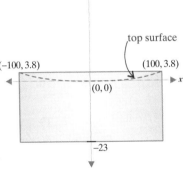

Solution:

In order to make the math as easy as possible, we can locate the origin of our coordinate system at the vertex of a parabolic cross-section of the Hale mirror, and we can assume the parabola opens upward. What we know, then, is that the equation $x^2 = 4py$ describes the shape of the cross-section for some p. We will be able to locate the focus of the parabola if we can determine p.

$(0,0)$ is one point (namely the vertex) on the graph of $x^2 = 4py$. What we have been told will allow us to find two other points on the graph. Since the difference in thickness of the mirror between the center and the outer rim is 3.8 inches, and since the mirror has a diameter of 200 inches, the two points $(-100, 3.8)$ and $(100, 3.8)$ must also lie on the graph. Using either one of these points in the equation $x^2 = 4py$, we can solve for p:

$$(100)^2 = 4p(3.8)$$
$$10000 = 15.2p$$
$$p \approx 657.9 \text{ inches}$$
$$p \approx 54.8 \text{ feet}$$

We know that the focus of a parabola is p units from the vertex, so the focus of the Hale Telescope is nearly 55 feet from the mirror.

chapter six — Rational Functions and Conic Sections

exercises

Graph the following parabolas and determine the focus and directrix of each. See Example 1.

1. $(x+1)^2 = 4(y-3)$
2. $y^2 - 4y = 8x + 4$
3. $(y-4)^2 = -2(x-1)$
4. $(y-1)^2 = 8(x+3)$
5. $(x-2)^2 = 4(y+1)$
6. $(y+1)^2 = -12(x+1)$
7. $y^2 = 6x$
8. $x^2 = 2y$
9. $x^2 = 7y$
10. $x^2 = -5y$
11. $y = -12x^2$
12. $x = -4y^2$
13. $x = \dfrac{1}{6}y^2$
14. $\dfrac{1}{5}x = -y^2$
15. $y^2 + 16x = 0$
16. $-6x - 2y^2 = 0$
17. $4y + 2x^2 = 4$
18. $2y^2 - 10x = 10$
19. $y^2 + 2y + 12x + 37 = 0$
20. $x^2 - 8y = 6x - 1$
21. $x^2 + 6x + 8y = -17$
22. $x^2 + 2x + 8y = 31$
23. $y^2 + 6y - 2x + 13 = 0$
24. $y^2 - 2y - 4x + 13 = 0$

Match the following equations to the appropriate graph on the following page.

25. $(x+2)^2 = 3(y-1)$
26. $(y-1)^2 = 2(x+2)$
27. $y^2 = 4(x+1)$
28. $x^2 = 2(y+1)$
29. $(x-1)^2 = -(y-2)$
30. $(y+2)^2 = 3x$
31. $(x-2)^2 = 4y$
32. $y^2 = -2(x+1)$

section 6.3   The Parabola

a.

b.

c.

d.

e.

f.

g.

h.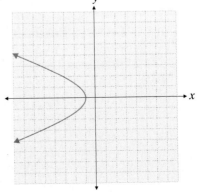

Find the equation, in standard form, for the parabola with the given properties or with the given graph. See Example 2.

33. Focus at $(-2, 1)$, directrix is the y-axis.

34. Focus at $(-2, 1)$, directrix is the x-axis.

35. Vertex at $(3, -1)$, focus at $(3, 1)$.

36. Symmetric with respect to the line $y = 1$, directrix is the line $x = 2$, and $p = -3$.

37. Vertex at $(3, -2)$, directrix is the line $x = -3$.

38. Vertex at $(7, 8)$, directrix is the line $x = \dfrac{27}{4}$.

39. Focus at $\left(-3, -\dfrac{3}{2}\right)$, directrix is the line $y = -\dfrac{1}{2}$.

40. Vertex at $(3, 16)$, focus at $(3, 11)$.

41. Vertex at $(-4, 3)$, focus at $\left(-\dfrac{3}{2}, 3\right)$.

42. Symmetric with respect to the x-axis, focus at $(-3, 0)$, and $p = 2$.

43.

44.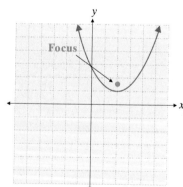

Use your knowledge of parabolas to answer the following questions.

45. One design for a solar furnace is based on the paraboloid formed by rotating the parabola $x^2 = 8y$ around its axis of symmetry. The object to be heated in the furnace is then placed at the focus of the paraboloid (assume that x and y are in units of feet). How far from the vertex of the paraboloid is the hottest part of the furnace?

46. A certain brand of satellite dish antenna is a paraboloid with a diameter of 6 feet and a depth of 1 foot, as shown. How far from the vertex of the dish should the receiver of the antenna be placed?

$d = 6$ ft
depth $= 1$ ft

47. A spotlight is made by placing a strong light bulb inside a reflective paraboloid formed by rotating the parabola $x^2 = 6y$ around its axis of symmetry (assume that x and y are in units of inches). In order to have the brightest, most concentrated light beam, how far from the vertex should the bulb be placed?

technology exercises

Use a computer algebra system such as Mathematica (or a suitably equipped graphing calculator) to graph the following equations. See the Technology Note after Example 2 in Section 6.2 for guidance.

48. $3x^2 - 4y + 24x = -56$

49. $y^2 + 2y = 8x - 41$

50. $y^2 - 6y + 4x = -17$

51. $x^2 - 6x + 12y + 21 = 0$

6.4 The Hyperbola

TOPICS

1. Derivation of the standard form of a hyperbola
2. Using the standard form of a hyperbola
3. Interlude: guidance systems

Topic 1: Derivation of the Standard Form of a Hyperbola

The pattern for exploring hyperbolas, the last of the three varieties of conic sections, will be familiar to you by now. We begin with a characteristic geometric property of hyperbolas, and use this to derive a useful form of the equation for a hyperbola. This we will call the standard form.

The characteristic geometric property of hyperbolas is similar to the one for ellipses. Recall that the points of an ellipse are those for which the sum of the distances to two foci is a fixed constant. For the points on a hyperbola, the magnitude of the *difference* of the distances to two foci is a fixed constant. This results in two disjoint pieces, called *branches*, of a hyperbola. The point halfway between the two branches is the *center* of the hyperbola, and the two points on the hyperbola closest to the center are the two *vertices*. Figure 1 illustrates how these parts of a hyperbola relate to one another.

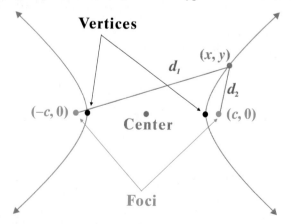

Figure 1: Parts of a Hyperbola

To begin, we will work with a hyperbola oriented as in Figure 1, with the center at the origin and the two foci at $(-c, 0)$ and $(c, 0)$. We know that for every point (x, y) on the hyperbola, the quantity $|d_1 - d_2|$ is fixed; in order to simplify the algebra, we will let $2a$ denote this fixed quantity. Using the distance formula for d_1 and d_2, then, we have:

$$\left| \sqrt{(x+c)^2 + (y-0)^2} - \sqrt{(x-c)^2 + (y-0)^2} \right| = 2a$$

As in our work with ellipses, we wish to manipulate this equation into a more useful form. Because of the absolute value symbols, we know that the one equation above actually represents two equations, as the left-hand side represents either $d_1 - d_2$ or $d_2 - d_1$ depending on which quantity is non-negative. Fortunately, the two cases quickly converge after we isolate either one of the radicals and square both sides. In what follows, d_1 has been isolated and both sides squared:

$$\left(\sqrt{(x+c)^2 + (y-0)^2} \right)^2 = \left(2a + \sqrt{(x-c)^2 + (y-0)^2} \right)^2$$

$$(x+c)^2 + y^2 = 4a^2 + 4a\sqrt{(x-c)^2 + (y-0)^2} + (x-c)^2 + y^2$$

This should look very familiar; our work so far closely parallels the derivation of the standard form for an ellipse. In fact, if we follow the same steps we will arrive at the form

$$\frac{x^2}{a^2} - \frac{y^2}{c^2 - a^2} = 1,$$

and as with ellipses we will make a change of variables to improve the appearance. Note that a (the distance of either vertex from the center) is less than c (the distance of either focus from the center), so $c^2 - a^2$ is positive and we are safe in renaming $c^2 - a^2$ as b^2. This changes the appearance of the above equation to

$$\frac{x^2}{a^2} - \frac{y^2}{b^2} = 1,$$

with the understanding that $b^2 = c^2 - a^2$ (we will find it more useful to write this as $c^2 = a^2 + b^2$ later, so we can solve for c).

To generalize this equation, we make the usual observations that swapping x and y reflects the graph in Figure 1 with respect to the line $y = x$ (giving us a hyperbola with foci on the y-axis) and replacing x with $x - h$ and y with $y - k$ moves the center to (h, k).

The following box summarizes our work.

Standard Form of a Hyperbola

The **standard form** of the equation for the hyperbola with center at (h, k) is:

- $\dfrac{(x-h)^2}{a^2} - \dfrac{(y-k)^2}{b^2} = 1$ if the foci are aligned horizontally (where the y-values are equal)

- $\dfrac{(y-k)^2}{a^2} - \dfrac{(x-h)^2}{b^2} = 1$ if the foci are aligned vertically (where the x-values are equal)

In either case, the foci are located c units away from the center, where $c^2 = a^2 + b^2$, and the vertices are located a units away from the center.

The standard form for a hyperbola is undeniably useful, as it tells us by inspection where the center is and hence where the two vertices are (as well as the foci if we wish). But this knowledge alone leaves a lot of uncertainty about the shape of the hyperbola. We always have the option of calculating and then plotting additional points to help us fill in the graph, but fortunately there is a more efficient aid. The analysis that follows shows that the branches of a hyperbola approach two *oblique asymptotes* far away from the center.

Consider again a hyperbola centered at the origin with foci aligned horizontally, as in Figure 1. Then for some pair of constants a and b, the hyperbola is described by

$$\frac{x^2}{a^2} - \frac{y^2}{b^2} = 1.$$

To understand how this hyperbola behaves far away from the center, we can solve the equation for y:

$$y^2 = b^2 \left(\frac{x^2}{a^2} - 1 \right)$$

We are wondering how y behaves when x is large in magnitude, but the answer is not clear from the equation in this form. As is often the case, a little algebraic manipulation sheds light on the issue. Factoring out the fraction from the parentheses gives us the equation

$$y^2 = \frac{b^2 x^2}{a^2} \left(1 - \frac{a^2}{x^2} \right),$$

section 6.4 The Hyperbola

and taking the square root of both sides leads to

$$y = \pm \frac{b}{a} x \sqrt{1 - \frac{a^2}{x^2}}.$$

The advantage of this last form is that as x goes to ∞ or $-\infty$, it is clear that the radicand approaches 1. Thus y gets closer and closer to the value

$$\frac{b}{a} x \quad \text{or} \quad -\frac{b}{a} x$$

for values of x that are large in magnitude. In other words, the two straight lines

$$y = \frac{b}{a} x \quad \text{and} \quad y = -\frac{b}{a} x$$

(which intersect at the center) are the asymptotes of the hyperbola.

For hyperbolas whose foci are aligned vertically, the equations for the asymptotes are

$$x = \frac{b}{a} y \quad \text{and} \quad x = -\frac{b}{a} y;$$

that is, x and y exchange places. If we solve these last two equations for y, and also consider hyperbolas not centered at the origin, we obtain the general situation described below.

Asymptotes of Hyperbolas

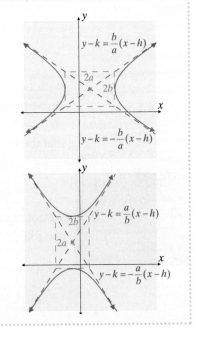

- The asymptotes of the hyperbola
$$\frac{(x-h)^2}{a^2} - \frac{(y-k)^2}{b^2} = 1$$
are the two lines
$$y - k = \frac{b}{a}(x - h) \quad \text{and} \quad y - k = -\frac{b}{a}(x - h).$$

- The asymptotes of the hyperbola
$$\frac{(y-k)^2}{a^2} - \frac{(x-h)^2}{b^2} = 1$$
are the two lines
$$y - k = \frac{a}{b}(x - h) \quad \text{and} \quad y - k = -\frac{a}{b}(x - h).$$

chapter six Rational Functions and Conic Sections

Topic 2: Using the Standard Form of a Hyperbola

It's time to see how the above analysis can be put to use.

example 1

To graph a hyperbola, select **APPS**, then 6:Conics and 3:Hyperbola. Select equation 2 for a vertical hyperbola, and plug in the values from the equation. To view the graph as shown below, select **GRAPH**. (Recall you may need to visit the Texas Instruments website to download the "Conic Graphing" application.)

Graph the equation $-16x^2 + 9y^2 + 96x + 18y = 279$, indicating the asymptotes of the resulting hyperbola.

Solution:

The fact that $-16x^2 + 9y^2 + 96x + 18y = 279$ represents a hyperbola follows from the fact that the product of the coefficients of x^2 and y^2 is negative. As with many ellipse and parabola problems, though, we will need to use the method of completing the square to write the equation in standard form and thus identify details of the hyperbola.

$$-16x^2 + 9y^2 + 96x + 18y = 279$$
$$-16(x^2 - 6x) + 9(y^2 + 2y) = 279$$
$$-16(x^2 - 6x + 9) + 9(y^2 + 2y + 1) = 279 - 144 + 9$$
$$-16(x-3)^2 + 9(y+1)^2 = 144$$
$$\frac{(y+1)^2}{16} - \frac{(x-3)^2}{9} = 1$$

We begin by factoring out the coefficients of x^2 and y^2 from the terms containing x and y, respectively. We then add the appropriate constant to complete each perfect square trinomial. Remember to compensate by adding the appropriate numbers to the right-hand side as well. The last step is to divide by 144.

With the equation in standard form, we now know that the foci of the hyperbola are aligned vertically (since the positive fraction on the left is in the variable y), that $a = 4$ and $b = 3$, and that the center of the hyperbola is at $(3, -1)$. This tells us immediately that the two vertices of the hyperbola must lie at

$$(3, -1 - 4) = (3, -5) \text{ and } (3, -1 + 4) = (3, 3).$$

We could also, if we wish, determine the foci of the hyperbola: $c^2 = a^2 + b^2 = 25$ so the two foci are at

$$(3, -1 - 5) = (3, -6) \text{ and } (3, -1 + 5) = (3, 4).$$

But the best aid in graphing the hyperbola will be the two asymptotes. The diagram at right illustrates a way to sketch the asymptotes of a hyperbola. We know that both asymptotes pass through the center of the hyperbola, and we know that one has a slope of $\frac{4}{3}$ and the other a slope of $-\frac{4}{3}$. The dashed-line rectangle drawn in the figure is centered at $(3, -1)$, with the bottom and top edges 4 units away from the center and the left and right edges 3 units away.

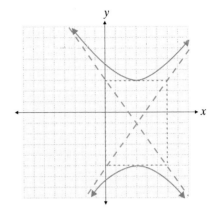

In other words, the rectangle has a height of $2a$ and a width of $2b$ (if the hyperbola was oriented horizontally, the width of the corresponding rectangle would be $2a$ and the height would be $2b$). Two dashed lines drawn through diagonally opposite corners of this rectangle then correspond to the two asymptotes, and the hyperbola can be drawn accordingly.

example 2

Given that the asymptotes of a certain hyperbola have slopes of $\frac{1}{2}$ and $-\frac{1}{2}$ and that the vertices of the hyperbola are at $(-1, 0)$ and $(7, 0)$, find the equation (in standard form) for the hyperbola.

Solution:

The fact that the vertices (and hence the foci) are aligned horizontally tells us immediately that we are going to be constructing an equation of the form

$$\frac{(x-h)^2}{a^2} - \frac{(y-k)^2}{b^2} = 1.$$

Our only remaining task is to determine h, k, a, and b. Since the vertices are 8 units apart, we know that $2a = 8$, or $a = 4$. We also know that the center lies halfway between the vertices, and that halfway point is $(3, 0)$. All that remains is to determine b.

As the foci are aligned horizontally, we know the asymptotes are of the form

$$y - k = \frac{b}{a}(x - h) \text{ and } y - k = -\frac{b}{a}(x - h),$$

continued on next page...

chapter six — Rational Functions and Conic Sections

and we already know a, h, and k. Specifically, we know that the two given slopes must correspond to the fractions $\dfrac{b}{a}$ and $-\dfrac{b}{a}$. That is,

$$\frac{1}{2} = \frac{b}{a}, \text{ so } \frac{1}{2} = \frac{b}{4}, \text{ and so } b = 2.$$

The equation we seek is thus

$$\frac{(x-3)^2}{16} - \frac{y^2}{4} = 1.$$

A sketch of the hyperbola, along with its asymptotes, appears at right.

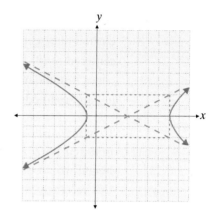

Topic 3: Interlude: Guidance Systems

Hyperbolas, like ellipses and parabolas, arise in a wide variety of contexts. For instance, hyperbolas are seen in architecture and structural engineering (think of the shape of a nuclear power plant's cooling towers) and in astronomy (comets that make a single pass through our solar system don't have elliptical orbits, but instead trace one branch of a hyperbola).

One of the most natural applications of hyperbolas concerns guidance systems, such as LORAN (LOng RAnge Navigation). LORAN is a radio-communication system that can be used to determine the location of a ship at sea, and the basis of LORAN is an understanding of hyperbolic curves.

Consider a situation in which two land-based radio transmitters, located at sites A and B in Figure 2, send out a pulse signal simultaneously. A receiver on a ship, located at C, would receive the two signals at slightly different times due to the difference in the distances the signal must travel. Since the times for signal travel are proportional to the respective distances d_1 and d_2, the difference in time between receipt of the two signals is proportional to $|d_1 - d_2|$. In other words, a person on the ship could determine $|d_1 - d_2|$ by noting the time-delay in receiving the two signals.

section 6.4   The Hyperbola

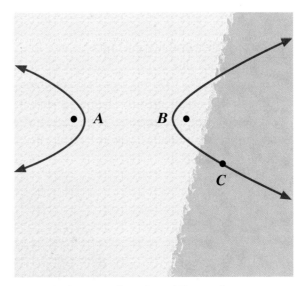

Figure 2: One Set of Transmitters

But as Figure 2 indicates, any point along the hyperbola (only one branch of which is relevant as far as the ship is concerned) could correspond to the position of the ship, as $|d_1 - d_2|$ is the same for all the points on the hyperbola. One way to actually determine the location of the ship is to perform the same computations for another pair of simultaneous signals sent out from two additional transmitters, located at A' and B' in Figure 3, and note the point of intersection of the two hyperbolas.

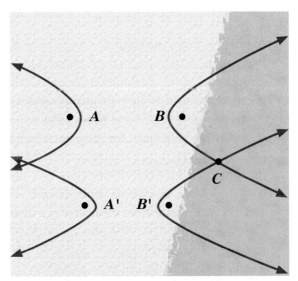

Figure 3: Two Sets of Transmitters

example 3

Three high-sensitivity microphones are located in a forest preserve, with microphone A two miles due north of microphone B and microphone C two miles due east of microphone B. During an early morning thunderstorm, microphone A detects a thunderclap (and possible lightning strike) at 3:28:15 AM. The same thunderclap is detected by microphone B at 3:28:19 AM, and by microphone C at 3:28:25 AM. Assuming that sound travels at 1100 feet per second, graphically approximate the source of the thunderclap.

Solution:

This situation is similar to the LORAN navigational system just described, but in this case there is one transmitter (the thunderclap), three receivers (the microphones), and the waves are sound instead of radio. But an understanding of how hyperbolas relate to the problem will resolve the issue.

In order to make the math as easy as possible, we will set up a coordinate system with microphone B at the origin, microphone A located 10,560 units (two miles = 10,560 feet) up on the y-axis, and microphone C located 10,560 units to the right on the x-axis. Now consider the thunderclap in relation to microphones A and B. Clearly, the thunder originated at a point closer to A than B, but all we know is that A detected the thunder 4 seconds earlier than B. If we let T stand for the point where the thunder originated, then T is 4400 feet (4 seconds × 1100 feet/second) closer to A than to B; if we let d_1 and d_2 represent the distances between T and the two microphones A and B, respectively, then $|d_1 - d_2| = 4400$. Recall that in our derivation of the standard form, we let $|d_1 - d_2| = 2a$, and a is the distance from the center of a hyperbola to either one of its vertices. Hence, $2a = 4400$, or $a = 2200$. We now know that T lies somewhere along a hyperbola with foci at the two microphones, center at (0, 5280) (the halfway point between A and B), and distance from the center to either vertex equal to 2200 feet. The equation for such a hyperbola is

$$\frac{(y-5280)^2}{(2200)^2} - \frac{x^2}{b^2} = 1,$$

with the constant b still undetermined. But since c is the distance from the center to either focus, and since $b^2 = c^2 - a^2$, we know $b^2 = (5280)^2 - (2200)^2 = 23{,}038{,}400$ and so the equation is

$$\frac{(y-5280)^2}{(2200)^2} - \frac{x^2}{23{,}038{,}400} = 1.$$

We can now do a similar analysis for microphones B and C, knowing that the difference in time is 6 seconds and hence $|d_1 - d_2| = 6600$. This means that for the hyperbola with foci at B and C, $a = 3300$ and $b^2 = c^2 - a^2 = 16{,}988{,}400$. Note that the center of this hyperbola is $(5280, 0)$, so the equation for the possible location of T relative to these two microphones is

$$\frac{(x-5280)^2}{(3300)^2} - \frac{y^2}{16{,}988{,}400} = 1.$$

We are now ready to graph these two hyperbolas and approximate the source of the thunder. Figure 4 contains the graph, marked off in square-mile grids, with the three microphones marked in teal.

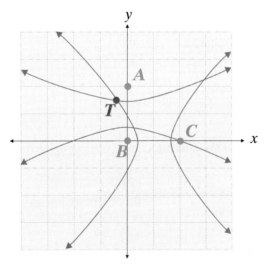

Figure 4: Three Microphones and a Possible Lightning Strike

Figure 4 indicates that there are four points of intersection of the two hyperbolas. How do we know which one corresponds to T? The point we are looking for should be closest to A and next closest to B, and this is the point marked in red. The coordinates of this point are approximately $(-1646, 7606)$, and this is the place to look for lightning damage.

chapter six — Rational Functions and Conic Sections

exercises

Sketch the graphs of the following hyperbolas, using asymptotes as guides. Determine the coordinates of the foci in each case. See Example 1.

1. $\dfrac{(x+3)^2}{4} - \dfrac{(y+1)^2}{9} = 1$

2. $\dfrac{(y-2)^2}{25} - \dfrac{(x+2)^2}{9} = 1$

3. $4y^2 - x^2 - 24y + 2x = -19$

4. $x^2 - 9y^2 + 4x + 18y - 14 = 0$

5. $9x^2 - 25y^2 = 18x - 50y + 241$

6. $9x^2 - 16y^2 + 116 = 36x + 64y$

7. $\dfrac{x^2}{16} - \dfrac{(y-2)^2}{4} = 1$

8. $\dfrac{(y-1)^2}{9} - (x+3)^2 = 1$

9. $9y^2 - 25x^2 - 36y - 100x = 289$

10. $9x^2 + 18x = 4y^2 + 27$

11. $9x^2 - 16y^2 - 36x + 32y - 124 = 0$

12. $x^2 - y^2 + 6x - 6y = 4$

13. $\dfrac{(y-2)^2}{64} - \dfrac{(x+7)^2}{49} = 1$

14. $\dfrac{(y-4)^2}{49} - \dfrac{(x+2)^2}{16} = 1$

15. $\dfrac{(x+1)^2}{64} - \dfrac{(y+7)^2}{4} = 1$

16. $\dfrac{(x+10)^2}{16} - \dfrac{(y+8)^2}{25} = 1$

Find the center, foci, and vertices of each hyperbola that the equation describes.

17. $\dfrac{(x+3)^2}{4} - \dfrac{(y-2)^2}{9} = 1$

18. $\dfrac{(y-2)^2}{16} - \dfrac{(x+1)^2}{9} = 1$

19. $3(x-1)^2 - (y+4)^2 = 9$

20. $(y-2)^2 - 2(x-4)^2 = 4$

21. $(x+2)^2 - 5(y-1)^2 = 25$

22. $6(y+2)^2 - (x+1)^2 = 12$

23. $2x^2 + 12x - y^2 - 2y + 9 = 0$

24. $y^2 - 9x^2 + 6y + 72x - 144 = 0$

25. $x^2 - 4y^2 - 2x = 0$

26. $4y^2 - x^2 + 32y + 2x + 47 = 0$

27. $4x^2 - y^2 - 64x + 10y + 167 = 0$

28. $4x^2 - 9y^2 - 36y - 72 = 0$

Match the corresponding equation to the appropriate graph.

29. $\dfrac{x^2}{9} - y^2 = 1$ **30.** $y^2 - \dfrac{x^2}{4} = 1$ **31.** $x^2 - \dfrac{(y-3)^2}{4} = 1$

32. $\dfrac{(x-3)^2}{4} - \dfrac{(y+1)^2}{9} = 1$ **33.** $(y+2)^2 - \dfrac{(x-2)^2}{4} = 1$ **34.** $\dfrac{x^2}{9} - \dfrac{(y+2)^2}{4} = 1$

35. $\dfrac{y^2}{4} - (x-1)^2 = 1$ **36.** $x^2 - y^2 = 1$

a.

b.

c.

d.

e.

f.

g.

h.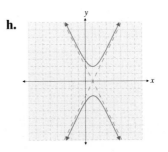

Find the equation, in standard form, for the hyperbola with the given properties or with the given graph. See Example 2.

37. Foci at $(-3, 0)$ and $(3, 0)$ and vertices at $(-2, 0)$ and $(2, 0)$.

38. Foci at $(1, 5)$ and $(1, -1)$ and vertices at $(1, 3)$ and $(1, 1)$.

39. Asymptotes of $y = \pm 2x$ and vertices at $(0, -1)$ and $(0, 1)$.

40. Asymptotes of $y = \pm(x - 2) + 1$ and vertices at $(-1, 1)$ and $(5, 1)$.

41. Foci at $(2, 4)$ and $(-2, 4)$ and asymptotes of $y = \pm 3x + 4$.

42. Foci at $(-1, 3)$ and $(-1, -1)$ and asymptotes of $y = \pm(x + 1) + 1$.

43. Foci at $(2, 5)$ and $(10, 5)$ and vertices at $(3, 5)$ and $(9, 5)$.

44. Foci at $(7, 4)$ and $(7, -4)$ and vertices at $(7, 1)$ and $(7, -1)$.

45. Asymptotes of $y = \pm(2x + 8) + 3$ and vertices at $(-6, 3)$ and $(-2, 3)$.

46. Asymptotes of $y = \pm \dfrac{4}{3}x - 3$ and vertices at $(0, -7)$ and $(0, 1)$.

47.

48.

49.

50.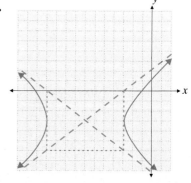

Using a graphing calculator, graph the following.

51. $x^2 - 6y^2 = 15$

52. $4y^2 - 9x^2 = 18$

53. $3y^2 - 18x^2 = 36$

54. $x^2 - 6 = 3y^2$

55. $(y+2)^2 - 20 = 5x^2$

56. $(x+5)^2 = 3(y-2)^2 + 15$

Use your knowledge of hyperbolas to answer the following questions. See Example 3

57. As mentioned in this section, some comets trace one branch of a hyperbola through the solar system, with the sun at one focus. Suppose a comet is spotted that appears to be headed straight for Earth, as shown in the figure. As the comet gets closer, however, it becomes apparent that it will pass between the Earth, which lies at the center of the hyperbolic path of the comet, and the sun. In the end, the closest the comet comes to Earth is 60,000,000 miles. Using a figure of 94,000,000 miles for the distance from the Earth to the sun, and positioning the Earth at the origin of a coordinate system, find the equation for the path of the comet.

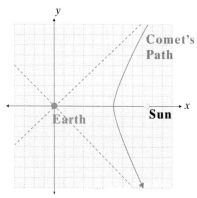

58. Suppose two LORAN radio transmitters are 26 miles apart. A ship at sea receives signals sent simultaneously from the two transmitters and is able to determine that the difference in the distances between the ship and each of the transmitters is 24 miles. By positioning the two transmitters on the y-axis, each 13 miles from the origin, find the equation for the hyperbola that describes the set of possible locations for the ship.

Rational Functions and Conic Sections

technology exercises

Use a computer algebra system such as Mathematica (or a suitably equipped graphing calculator) to graph the following equations. See the Technology Note after Example 2 in Section 6.2 for guidance.

59. $x^2 - 2y^2 = 4x + 12y + 26$

60. $2x^2 - y^2 + 12x + 2y = 3$

61. $5x^2 - y^2 + 20x = 10y + 25$

62. $x^2 - 5y^2 = 14x + 20y - 4$

chapter six project

Constructing a Bridge

Plans are in process to develop an uninhabited coastal island into a new resort. Before development can begin, a bridge must be constructed joining the island to the mainland.

Two possibilities are being considered for the support structure of the bridge. The archway could be built as a parabola, or in the shape of a semi-ellipse.

Assume all measurements that follow refer to dimensions at high tide. The county building inspector has deemed that in order to establish a solid foundation, the space between supports must be at least 300 feet and the height at the center of the arch should be 80 feet. There is a commercial fishing dock located on the mainland whose fishing vessels travel constantly along this intercoastal waterway. The tallest of these ships requires a 60 ft clearance to pass comfortably beneath the bridge. With these restrictions, the width of a channel with a minimum height of 60 ft has to be determined for both possible shapes of the bridge to confirm that it will be suitable for the water traffic beneath it.

1. Find the equation of a parabola that will fit these constraints.
2. How wide is the channel with a minimum of 60 ft vertical clearance for the parabola in question 1?

3. Find the equation of a semi-ellipse that will fit these constraints.
4. How wide is the channel with a minimum of 60 ft vertical clearance for the semi-ellipse in question 3?
5. Which of these bridge designs would you choose, and why?
6. Suppose the tallest fishing ship installs a new antenna which raises the center height by 12 ft. How far off of center (to the left or right) can the ship now travel and still pass under the bridge without damage to the antenna:
   a. For the parabola?
   b. For the semi-ellipse?

chapter six summary

CHAPTER SUMMARY

A summary of concepts and skills follows each chapter. Refer to these summaries to make sure you feel comfortable with the material in the chapter. The concepts and skills are organized according to the section title and topic title in which the material is first discussed.

6.1 Rational Functions and Rational Inequalities

topics	pages
Definitions and Useful Notation • The definition of a *rational function*, and how to determine domains of rational functions • The meaning of *vertical*, *horizontal*, and *oblique asymptotes*	p. 431 – 435
Vertical Asymptotes • How to find the vertical asymptotes for a given rational function	p. 435 – 436
Horizontal and Oblique Asymptotes • How to determine if a rational function has a horizontal or oblique asymptote, based on the degrees of the numerator and denominator • How to find the horizontal asymptote of a given rational function, if there is one • How to find the oblique asymptote of a given rational function, if there is one	p. 437 – 439
Graphing Rational Functions • How to use knowledge of a rational function's domain, intercepts, asymptotes, and symmetry to sketch its graph	p. 439 – 441
Solving Rational Inequalities • How to use graphing knowledge to solve *rational inequalities*	p. 441 – 443

chapter six   Rational Functions and Conic Sections

6.2 The Ellipse

topics	pages
Geometric and Algebraic Meaning of the Conic Sections • The geometric definitions of the three varieties of conic sections: *ellipses*, *parabolas*, and *hyperbolas* • How to identify, algebraically, equations whose graphs are ellipses, parabolas, or hyperbolas • The meaning of the *foci* (plural of *focus*) and *vertices* of an ellipse	p. 449 – 451
Derivation of the Standard Form of an Ellipse • How the standard form of the equation for an ellipse relates to its geometric properties	p. 451 – 454
Using the Standard Form of an Ellipse • The meaning of an ellipse's *major* and *minor axes,* and *vertices* • Using the standard form of an ellipse to sketch its graph • How to construct the equation for an ellipse with prescribed properties • The geometric and algebraic meaning of elliptical *eccentricity*	p. 454 – 458
Interlude: Planetary Orbits • Using knowledge of ellipses to answer questions about planetary orbit	p. 458 – 460

6.3 The Parabola

topics	pages
Derivation of the Standard Form of a Parabola (as a Conic Section) • How the standard form of the equation for a parabola relates to its geometric properties • The meaning of the *focus*, *vertex*, and *directrix* of a parabola	p. 467 – 469
Using the Standard Form of a Parabola • Using the standard form of a parabola to sketch its graph • How to construct the equation for a parabola with prescribed properties	p. 469 – 471

chapter six summary

topics (continued)	pages
Interlude: Parabolic Mirrors • The use of a parabola's focusing property, and how to answer questions related to this use	p. 472 – 473

6.4 The Hyperbola

topics	pages
Derivation of the Standard Form of a Hyperbola • How the standard form of the equation for a hyperbola relates to its geometric properties • The meaning of the *foci*, *vertices*, and *asymptotes* of a hyperbola	p. 478 – 481
Using the Standard Form of a Hyperbola • Using the standard form of a hyperbola to sketch its graph • How to construct the equation for a hyperbola with prescribed properties	p. 482 – 484
Interlude: Guidance Systems • How properties of hyperbolas are used in designing guidance systems	p. 484 – 487

chapter six
Rational Functions and Conic Sections

chapter review

Section 6.1

Find equations for the vertical asymptotes, if any, for each of the following rational functions.

1. $f(x) = \dfrac{4}{2x-5}$
2. $f(x) = \dfrac{x^2 - 3x + 2}{x - 1}$

Find equations for the horizontal or oblique asymptotes, if any, for each of the following rational functions.

3. $f(x) = \dfrac{2x^3 + 5x^2 - 1}{x^2 - 2x}$
4. $f(x) = \dfrac{x^2 - x + 8}{3x^2 - 7}$

Sketch the graphs of the following rational functions.

5. $\dfrac{2x}{x+1}$
6. $\dfrac{4x^2}{x^2 + 3x}$

Solve the following rational inequalities.

7. $\dfrac{7}{x+3} \geq \dfrac{2x}{x+3}$
8. $\dfrac{x}{x^2 - 5x + 6} \leq \dfrac{3}{x^2 - 5x + 6}$

Section 6.2

Sketch the graphs of the following ellipses, and determine the coordinates of the foci.

9. $\dfrac{(x+1)^2}{16} + \dfrac{(y-2)^2}{9} = 1$
10. $x^2 + 9y^2 - 6x + 18y = -9$

In each of the following problems, an ellipse is described by properties it possesses. Find the equation, in standard form, for each ellipse.

11. Center at $(-1, 4)$, major axis is vertical and of length 8, foci $\sqrt{7}$ units from the center.

12. Foci at $(1, 2)$ and $(7, 2)$, $e = \dfrac{1}{2}$.

13. Vertices at $\left(\dfrac{7}{2}, -1\right)$ and $\left(\dfrac{1}{2}, -1\right)$, $e = 0$.

14. Vertices at $(1, -8)$ and $(1, 2)$, minor axis of length 6.

Section 6.3

Graph the following parabolas and determine the focus and directrix of each.

15. $(y+1)^2 = -12(x+3)$

16. $y^2 - 8y + 2x + 14 = 0$

Find the equation, in standard form, for the parabola with the given properties.

17. Vertex at $(-2, 3)$, directrix is the line $y = 2$.

18. Vertex at $(5, -3)$, focus at $(5, 1)$.

19. Focus at $(3, -1)$, directrix is the line $x = 2$.

Use your knowledge of parabolas to answer the following question.

20. A motorcycle headlight is made by placing a strong light bulb inside a reflective paraboloid formed by rotating the parabola $x^2 = 5y$ around its axis of symmetry (assume that x and y are in units of inches). In order to have the brightest, most concentrated light beam, how far from the vertex should the bulb be placed?

Section 6.4

Sketch the graphs of the following hyperbolas, using asymptotes as guides. Determine the coordinates of the foci in each case.

21. $\dfrac{(y+2)^2}{9} - \dfrac{(x-2)^2}{16} = 1$

22. $9x^2 - 4y^2 + 54x - 8y + 41 = 0$

Find the equation, in standard form, for the hyperbola with the given properties.

23. Vertices at $(4, -1)$ and $(-2, -1)$ and foci at $(5, -1)$ and $(-3, -1)$.

24. Asymptotes of $y = \pm \dfrac{5}{2}(x+1) - 2$ and vertices at $(-3, -2)$ and $(1, -2)$.

25. Foci at $(-1, -2)$ and $(-1, 8)$ and asymptotes of $y = \pm \left(\dfrac{3}{4}x + \dfrac{3}{4}\right) + 3$.

26. Asymptotes of $y = \pm(3x - 6) + 2$ and vertices at $(2, -1)$ and $(2, 5)$.

chapter six
Rational Functions and Conic Sections

chapter test

Find equations for the vertical asymptotes, if any, for each of the following rational functions.

1. $y = \dfrac{-2}{x+1}$

2. $y = \dfrac{x^2+4}{x+2}$

3. $f(x) = \dfrac{x^2-9}{x+3}$

4. $f(x) = \dfrac{x^2-1}{x-x^2}$

5. $f(x) = \dfrac{x}{x^2+1}$

6. $f(x) = \dfrac{x-1}{x^2-4}$

7. $f(x) = \dfrac{x^2-x-6}{x^2-6x+9}$

8. $f(x) = \dfrac{x^3+2x^2-x-2}{x^3+8}$

Find the equations for the horizontal or oblique asymptotes, if any, for each of the following rational functions.

9. $f(x) = \dfrac{3}{x-3}$

10. $f(x) = \dfrac{x^2+2}{x+2}$

11. $f(x) = \dfrac{x^2-9}{x+3}$

12. $f(x) = \dfrac{x^2-1}{x-x^2}$

13. $f(x) = \dfrac{2x^3-x^2+1}{-x^3+3x}$

14. $f(x) = \dfrac{2x^3-x^2+x-2}{-x+4}$

15. $f(x) = \dfrac{-x^5+2x-1}{3x^5+2x^2-2}$

16. $f(x) = \dfrac{x^2+2x-3}{(x+1)^3}$

Sketch the graphs of the following rational functions.

17. $f(x) = \dfrac{2}{x+2}$

18. $f(x) = \dfrac{x^2+2}{x+2}$

19. $f(x) = \dfrac{x^2-1}{x-1}$

20. $f(x) = \dfrac{x^2-9}{3x-x^2}$

21. $f(x) = \dfrac{x+1}{x^2-4}$

22. $f(x) = \dfrac{2x^2-2x-4}{x^2-2x+1}$

23. $f(x) = \dfrac{-2x+1}{x-3}$

24. $f(x) = \dfrac{x^3-8}{x^2+3}$

25. $f(x) = \dfrac{2}{x^2-4}$

Solve the following rational inequalities.

26. $3x > \dfrac{-3}{x-2}$

27. $\dfrac{1}{x-1} \geq \dfrac{2x}{x-1}$

28. $\dfrac{3}{x+2} \leq \dfrac{1}{x-2}$

29. $\dfrac{x}{x^2+x-2} \leq \dfrac{-1}{x^2+x-2}$

30. $x > \dfrac{4}{x}$

31. $\dfrac{2}{1-x} \geq \dfrac{1}{x+4}$

32. $\dfrac{x-4}{x+3} < \dfrac{x}{x-2}$

33. $\dfrac{x}{x^2+4x+3} > \dfrac{-1}{x^2+4x+3}$

34. $\dfrac{x-2}{x+3} < 2$

35. $\dfrac{1}{x-2} + \dfrac{2}{x-1} \geq 0$

36. If the graph of a rational function f has a vertical asymptote at $x = -2$, is it possible to sketch the graph without lifting your pencil from the paper? Explain.

Find the equation of a rational function f with the given properties. Answers may vary.

37. Vertical asymptotes: $x = -3, x = 1$

38. Vertical asymptote: None
    Horizontal asymptote: $y = -2$

39. Vertical asymptotes: $x = 0, x = \dfrac{3}{2}$
    Horizontal asymptote: $y = -2$

40. Vertical asymptotes: $x = 1, x = -2$
    Horizontal asymptote: $y = 0$

Find the center, foci, and vertices of each ellipse that the equation describes.

41. $\dfrac{(x-2)^2}{16} + \dfrac{(y+1)^2}{81} = 1$

42. $(x-3)^2 + 4(y+1)^2 = 16$

43. $3x^2 + y^2 = 27$

44. $9x^2 + 4y^2 + 18x - 16y + 9 = 0$

45. $x^2 + 4y^2 + 6x - 8y + 4 = 0$

Find the equation, in standard form, for the ellipse with the specified properties.

46. Vertices at $(0, 2)$ and $(4, 2)$, minor axis of length 2.

47. Foci at $(0, 0)$ and $(4, 0)$, major axis of length 8.

48. Center at $(2, -1)$, vertex at $\left(2, \dfrac{1}{2}\right)$, minor axis of length 2.

49. Center at $(3, 2)$, $a = 3c$, and foci at $(1, 2)$ and $(5, 2)$.

50. Center at $(0, 4)$, $a = 2c$, and vertices at $(-4, 4)$ and $(4, 4)$.

51. Vertices at $(5, 0)$ and $(5, 12)$ and endpoints of the minor axis at $(0, 6)$ and $(10, 6)$.

52. Vertices at $(\pm 5, 0)$, $e = \dfrac{3}{5}$.

53. Vertices at $(0, \pm 8)$, $e = \dfrac{1}{2}$.

Given that the area A of the ellipse: $\dfrac{x^2}{a^2} + \dfrac{y^2}{b^2} = 1$ *is* $A = \pi \cdot a \cdot b$ *and* $a + b = 30$, *find the equation or function to satisfy each of the following.*

54. Write the area of the ellipse as a function of a.

55. Find the equation of an ellipse with an area of 200π square inches.

Graph the following parabolas and determine the focus and directrix of each.

56. $(x - 2)^2 = 2(y + 1)$

57. $y^2 + 2y = 4x - 1$

58. $(y + 1)^2 = 16(x + 2)$

59. $y^2 - 3x = 0$

60. $x^2 = 4y$

61. $x + \dfrac{1}{4} y^2 = 0$

62. $2y + 4x^2 = 8$

63. $x^2 + 12y = 0$

64. $x^2 + 2x + 6y + 17 = 0$

65. $y^2 - 4y + 2x + 24 = 0$

Find the equation, in standard form, for the parabola with the given properties.

66. Focus at $(-3, 2)$, directrix is the y-axis.

67. Focus at $(1, -2)$, directrix is the x-axis.

68. Vertex at $(4, -2)$, focus at $(4, 2)$.

69. Vertex at $(2,-1)$, directrix is the line $x = -2$.

70. Symmetric with respect to the x-axis, focus at $(-3, 0)$, and $p = 4$.

71. Symmetric with respect to the line $y = 2$, directrix is the line $x = 4$, and $p = -5$.

72. A parabolic archway is 12 meters high at the vertex. At a height of 10 meters the width of the archway is 8 meters. How wide is the archway at ground level?

Find the center, foci, and vertices of each of the following hyperbolas, and sketch the graph, using asymptotes as guides.

73. $x^2 - y^2 = 1$

74. $\dfrac{x^2}{9} - \dfrac{y^2}{16} = 1$

75. $\dfrac{y^2}{25} - \dfrac{x^2}{144} = 1$

76. $\dfrac{x^2}{36} - \dfrac{y^2}{4} = 1$

77. $\dfrac{(x+1)^2}{144} - \dfrac{(y+2)^2}{1} = 1$

78. $\dfrac{(y-1)^2}{\frac{1}{4}} - \dfrac{(x+3)^2}{\frac{1}{9}} = 1$

79. $9x^2 - y^2 - 36x - 6y + 18 = 0$

Find the center, foci, and vertices of each hyperbola that the equation describes.

80. $\dfrac{(x+1)^2}{4} - \dfrac{(y-2)^2}{25} = 1$

81. $16x^2 - 9y^2 = 144$

82. $(x+1)^2 - 4(y-2)^2 = 36$

83. $x^2 - 9y^2 + 36y - 72 = 0$

Find the equation, in standard form, for the hyperbola with the given properties.

84. Vertices at $(0, \pm 2)$ and foci at $(0, \pm 4)$.

85. Vertices at $(\pm 3, 0)$ and foci at $(\pm 5, 0)$.

86. Vertices at $(\pm 1, 0)$ and asymptotes at $y = \pm 3x$.

87. Vertices at $(3, 0)$ and $(3, 4)$ and asymptotes at $y = \dfrac{2}{3}x$ and $y = 4 - \dfrac{2}{3}x$.

chapter SEVEN

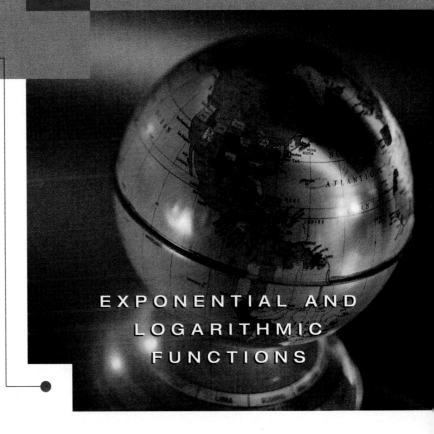

EXPONENTIAL AND LOGARITHMIC FUNCTIONS

OVERVIEW

7.1 Exponential Functions and Their Graphs

7.2 Applications of Exponential Functions

7.3 Logarithmic Functions and Their Graphs

7.4 Properties and Applications of Logarithms

7.5 Exponential and Logarithmic Equations

By the end of this chapter you should be able to estimate the behavior of growing populations, interest-earning accounts, and decaying elements. On page 573 you will find a problem regarding world population. Given information about current population and growth rate, you will be asked to calculate the time it will take to reach a population of 20 billion (disregarding complicated issues of limited resources to support such a population). You will master this type of problem, using tools such as the *Summary of Logarithmic Properties* on page 562.

chapter seven   Exponential and Logarithmic Functions

Introduction

This chapter introduces two entirely new classes of functions, both of which are of enormous importance in many natural and man-made contexts. As we will see, the two classes of functions are inverses of one another, though historically exponential and logarithmic functions were developed independently and for unrelated reasons.

We will begin with a study of exponential functions. These are functions in which the variable appears in the exponent while the base is a constant, just the opposite of what we have seen so often in the individual terms of polynomials. As with many mathematical concepts, the argument can be made that exponential functions exist in the natural world independently of mankind, and that consequently mathematicians have done nothing more than observe (and formalize) what there is to be seen. Exponential behavior is exhibited, for example, in the rate at which radioactive substances decay, in how the temperature of an object changes when placed in an environment held at a constant temperature, and in the fundamental principles of population growth. But exponential functions also arise in discussing such man-made phenomena as the growth of investment funds.

In fact, we will use the formula for compound interest to motivate the introduction of the most famous and useful base for exponential functions, the irrational constant e (the first few digits of which are 2.718281828459…). The Swiss mathematician Leonhard Euler (1707 - 1783), who identified many of this number's unique properties (such as the fact that $e = 1 + \frac{1}{1} + \frac{1}{1 \cdot 2} + \frac{1}{1 \cdot 2 \cdot 3} + \ldots$), was one of the first to recognize the fundamental importance of e, and in fact is responsible for the choice of the letter e as its symbol. The constant e also arises very naturally in the context of calculus, but that discussion must wait for a later course.

Napier

Logarithms are inverses, in the function sense, of exponentials, but historically their development was for very different reasons. Much of the development of logarithms is due to John Napier (1550 - 1617), a Scottish writer of political and religious tracts and an amateur mathematician and scientist. It was the goal of simplifying computations of large numbers that led him to devise early versions of logarithmic functions, to construct what today would be called tables of logarithms, and to design a prototype of what would eventually become a slide rule. Of course, today it is not necessary to resort to logarithms in order to carry out difficult computations, but the properties of logarithms make them invaluable in solving certain equations. Further, the fact that they are inverses of exponential functions means they have just as much a place in the natural world.

7.1 Exponential Functions and Their Graphs

TOPICS

1. Definition and classification of exponential functions
2. Graphing exponential functions
3. Solving elementary exponential equations

Topic 1: Definition and Classification of Exponential Functions

We have extensively studied simple functions such as $f(x) = x^2$ and $g(x) = x^3$, and we are familiar with radical functions such as $h(x) = x^{\frac{1}{3}}$. We have also explored more exotic species of functions, such as rational functions, step functions, and the like. To this point, though, we have not encountered a function in which the variable appears in the exponent. That will change in this chapter.

An *exponential function* is, informally, a function in which a constant is raised to a variable power. As a class, exponential functions are extremely important because of the large number of diverse situations in which they arise. Examples include radioactive decay, population growth, compound interest, spread of epidemics, rates of temperature change, and many others. The formal definition follows.

Exponential Functions

Let a be a fixed positive real number not equal to 1. The exponential function with base a is the function
$$f(x) = a^x.$$

chapter seven   Exponential and Logarithmic Functions

Note that for any such constant a, a^x is defined for all real numbers x; consequently, the domain of $f(x) = a^x$ is the set of real numbers. How about the range of f? This question requires a bit more thought to answer.

First, recall that if a is any non-zero number, a^0 is defined to be 1. This means the y-intercept of any exponential function, regardless of the base, is the point $(0, 1)$. Aside from this commonality, exponential functions fall into two classes, depending on whether a lies between 0 and 1 or if a is larger than 1. (Incidentally, the only reason we don't allow a to equal 1 is because the result is too boring: $1^x = 1$ for all x. And the reason we don't allow a to be negative is because a^x would not be real for some values of x; consider, for example, what would happen if $a = -1$ and $x = \frac{1}{2}$.)

Consider the following calculations for two sample exponential functions:

x	$f(x) = \left(\frac{1}{3}\right)^x$	$g(x) = 2^x$
-2	$f(-2) = 9$	$g(-2) = \frac{1}{4}$
-1	$f(-1) = 3$	$g(-1) = \frac{1}{2}$
1	$f(1) = \frac{1}{3}$	$g(1) = 2$
2	$f(2) = \frac{1}{9}$	$g(2) = 4$

Note that the values of f decrease as x increases while the values of g do just the opposite. Note also that for any x, both $f(x)$ and $g(x)$ are positive. A bit of experimentation will convince you that any exponential function with a base between 0 and 1 will behave like the function f above, while any exponential function with a base larger than 1 will behave like g. We have terminology to describe this: f is an example of a *decreasing* function and g is an example of an *increasing* function. If we plot a few more points to be sure of the behavior and then fill in the gaps, we get the graphs of f and g that appear in Figure 1.

section 7.1 Exponential Functions and Their Graphs

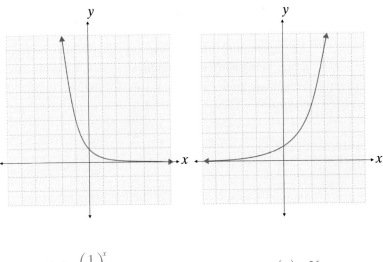

$$f(x) = \left(\frac{1}{3}\right)^x \qquad\qquad g(x) = 2^x$$

Figure 1: Two Exponential Functions

The above graphs suggest that the range of an exponential function is $(0, \infty)$, and that is indeed the case. Note that the base is immaterial: the range of a^x is the positive real numbers for any allowable base (that is, for any positive a not equal to 1).

The following summarizes what we have learned.

Behavior of Exponential Functions

Given a positive real number a not equal to 1, the function $f(x) = a^x$ is:

- a decreasing function if $0 < a < 1$, with $f(x) \to \infty$ as $x \to -\infty$ and $f(x) \to 0$ as $x \to \infty$
- an increasing function if $a > 1$, with $f(x) \to 0$ as $x \to -\infty$ and $f(x) \to \infty$ as $x \to \infty$

In either case, the point $(0, 1)$ lies on the graph of f, the domain of f is the set of real numbers, and the range of f is the set of positive real numbers.

chapter seven   Exponential and Logarithmic Functions

Topic 2: Graphing Exponential Functions

An exponential function, like any function, can be transformed in ways that result in the graph being shifted, reflected, stretched, or compressed. You may want to review the basic ideas underlying such transformations in Section 4.4 as you study the following examples.

example 1

Sketch the graphs of each of the following functions.

a. $f(x) = 2^{-x}$

b. $g(x) = -3^x + 1$

c. $r(x) = 3^{x-2} - 1$

d. $s(x) = \left(\dfrac{1}{2}\right)^{1-x}$

Solutions:

a.

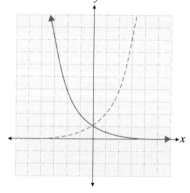

The graph of $f(x) = 2^{-x}$ is the reflection of the graph of $y = 2^x$ with respect to the y-axis, as x has been replaced by $-x$. In the picture at left, the dashed teal line is the graph of 2^x while the blue line, its reflection, is the graph of $f(x) = 2^{-x}$.

There is another way to think about the function $f(x) = 2^{-x}$. Note that $2^{-x} = \left(2^{-1}\right)^x = \left(\dfrac{1}{2}\right)^x$ so f is actually a decreasing exponential function.

b.

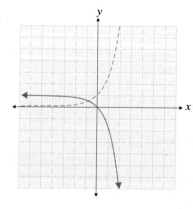

The graph of $y = 3^x$ is the dashed teal line at left (note that $(0, 1)$ and $(1, 3)$, two easily determined points, lie on the graph).

The effect of multiplying a function by -1 is, as always, the reflection of the graph with respect to the x-axis, and the effect of adding 1 to a function is a vertical upward shift of the graph. The blue line at left is the graph of $g(x) = -3^x + 1$.

section 7.1 Exponential Functions and Their Graphs

example 2

Solve the following exponential equations.

a. $25^x - 125 = 0$ **b.** $8^{y-1} = \dfrac{1}{2}$ **c.** $\left(\dfrac{2}{3}\right)^x = \dfrac{9}{4}$

Solutions:

a. $25^x - 125 = 0$

$25^x = 125$

$\left(5^2\right)^x = 5^3$

$5^{2x} = 5^3$

$2x = 3$

$x = \dfrac{3}{2}$

We may as well start by isolating the term with the variable on one side, with the hope that both sides can then be rewritten with the same base. Doing so requires finding a number for which both 25 and 125 are powers; the number 5 meets this criterion. The remainder of the work involves using the properties of exponents to write both sides with a base of 5, and then equating the powers. The result is a simple linear equation. Verify that $\dfrac{3}{2}$ does indeed solve the original equation.

b. $8^{y-1} = \dfrac{1}{2}$

$\left(2^3\right)^{y-1} = 2^{-1}$

$2^{3y-3} = 2^{-1}$

$3y - 3 = -1$

$3y = 2$

$y = \dfrac{2}{3}$

Again, we need to rewrite both sides so that they have the same base. Since 8 and $\dfrac{1}{2}$ can both be written as 2 raised to an appropriate power, we do so, writing 8 as 2^3 and $\dfrac{1}{2}$ as 2^{-1}.

After applying the properties of exponents to simplify both sides, we equate the exponents and solve for y. You should again verify that this solution works in the original equation.

c. $\left(\dfrac{2}{3}\right)^x = \dfrac{9}{4}$

$\left(\dfrac{2}{3}\right)^x = \left(\dfrac{3}{2}\right)^2$

$\left(\dfrac{2}{3}\right)^x = \left(\dfrac{2}{3}\right)^{-2}$

$x = -2$

The choice of base may not be as obvious in this problem, but 3 and 2 seem likely to appear in some way. If we initially write the right-hand side as shown in the second step at left, it is then clear that we can make the two bases equal if we make the change shown in the third step.

After making both bases the same, the solution is clear.

chapter seven Exponential and Logarithmic Functions

exercises

Sketch the graphs of the following functions. See Example 1.

1. $f(x) = 4^x$
2. $g(x) = (0.5)^x$
3. $s(x) = 3^{x-2}$
4. $f(x) = \left(\dfrac{1}{3}\right)^{x+1}$
5. $r(x) = 5^{x-2} + 3$
6. $h(x) = 1 - 2^{x+1}$
7. $f(x) = 2^{-x}$
8. $r(x) = 3^{2-x}$
9. $g(x) = 3\left(2^{-x}\right)$
10. $h(x) = 2^{2x}$
11. $s(x) = (0.2)^{-x}$
12. $f(x) = \dfrac{1}{2^x} + 1$
13. $g(x) = 3 - 2^{-x}$
14. $r(x) = \dfrac{1}{2^{3-x}}$
15. $h(x) = \left(\dfrac{1}{2}\right)^{5-x}$
16. $m(x) = 3^{2x+1}$
17. $p(x) = 2 - 4^{2-x}$
18. $q(x) = 5^{3-2x}$
19. $r(x) = \left(\dfrac{9}{2}\right)^{-x}$
20. $p(x) = \left(\dfrac{1}{3}\right)^{2-x}$
21. $r(x) = 1 - \left(\dfrac{15}{4}\right)^x$

Solve the following exponential equations. See Example 2.

22. $5^x = 125$
23. $3^{2x-1} = 27$
24. $9^{2x-5} = 27^{x-2}$
25. $10^x = 0.01$
26. $4^{-x} = 16$
27. $2^x = \left(\dfrac{1}{2}\right)^{13}$
28. $2^{x+1} = 64^3$
29. $\left(\dfrac{2}{3}\right)^{x+3} = \left(\dfrac{9}{4}\right)^{-x}$
30. $\left(\dfrac{1}{5}\right)^{x-4} = 625^{\frac{1}{2}}$
31. $4^{3x+2} = \left(\dfrac{1}{4}\right)^{-2x}$
32. $5^x = 0.2$
33. $7^{x^2+3x} = \dfrac{1}{49}$
34. $3^{x^2+4x} = 81^{-1}$
35. $\left(\dfrac{1}{2}\right)^{x-3} = \left(\dfrac{1}{4}\right)^{x-5}$
36. $64^{x+\frac{7}{6}} = 2$
37. $6^{2x} = 36^{2x-3}$
38. $4^{2x-5} = 8^{\frac{x}{2}}$
39. $\left(\dfrac{2}{5}\right)^{2x+4} = \left(\dfrac{4}{25}\right)^{11}$
40. $4^{4x-7} = \dfrac{1}{64}$
41. $-10^x = -0.001$
42. $3^x = 27^{x+4}$
43. $1000^{-x} = 10^{x-8}$
44. $1^{3x-7} = 4^{2-x}$
45. $5^{3x-1} = 625^x$
46. $\left(e^{x+2}\right)^3 = \left(e^x\right)\dfrac{1}{\left(e^{3x}\right)}$
47. $3^{2x-7} = 81^{\frac{x}{2}}$

section 7.1 Exponential Functions and Their Graphs

Match the graphs of the following functions to the appropriate equation.

48. $f(x) = 2^{3x}$ **49.** $h(x) = 5^x - 1$ **50.** $g(x) = 2(4^{x-1})$

51. $p(x) = 1 - 2^{-x}$ **52.** $f(x) = 6^{4-x}$ **53.** $r(x) = \dfrac{1}{3^x}$

54. $m(x) = -2 + 2^{-3x}$ **55.** $g(x) = \left(\dfrac{1}{4}\right)^{1+x}$ **56.** $h(x) = 3^{\frac{1}{2}x}$

57. $s(x) = 1^x - 4$

a.

b.

c.

d.

e.

f.

g.

h.

i.

j.

chapter seven Exponential and Logarithmic Functions

technology exercises

Graphing calculators and computer algebra systems generally produce very nice graphs of exponential functions, provided you specify the plotting "window" appropriately. The graphs below were generated using the simple `Plot` command in *Mathematica*.

Use a graphing calculator or a computer algebra system, if available, to check your graphs of the functions that follow.

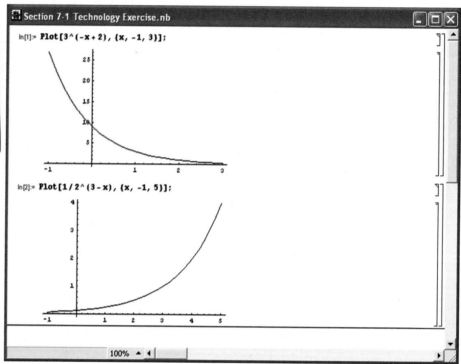

58. $f(x) = 2^{5-x}$

59. $g(x) = \dfrac{3}{2^{x-1}}$

60. $h(x) = 0.4^{x+5}$

61. $f(x) = 2.1^{3-x}$

62. $g(x) = \dfrac{1}{3^{x+2}}$

63. $h(x) = \left(\dfrac{2}{3}\right)^{x+2}$

64. $m(x) = \dfrac{1}{\left(\dfrac{7}{4}\right)^x} - 3$

65. $g(x) = \left(\dfrac{11}{4}\right)^{x-3}$

66. $f(x) = 1 + \left(\dfrac{1}{5}\right)^{4-x}$

67. $m(x) = 3 - \left(\dfrac{11}{4}\right)^x$

68. $n(x) = \left(\dfrac{19}{4}\right)^x$

69. $r(x) = 2^{3x+1}$

7.2 Applications of Exponential Functions

TOPICS

1. Models of population growth
2. Radioactive decay
3. Compound interest and the number *e*

Topic 1: Models of Population Growth

Exponential functions arise naturally in a wide array of situations. In this section, we will study a few such situations in some detail, beginning with population models.

Over the centuries many people, working in such areas as mathematics, biology, and sociology, have created mathematical models of population. The models have represented many types of populations, such as people in a given city or country, wolves in a wildlife habitat, or number of bacteria in a Petri dish. In general, such models can be quite complex, depending on factors like availability of food, space constraints, and effects of disease and predation. The length of time the model is intended to correspond to in reality also plays a large role in the mathematical details of the model. But at their most basic, many population models assume that population growth displays exponential behavior.

The reason for this is that the growth of a population usually depends to a large extent on the number of members capable of producing more members. For instance, in the simple case of bacteria in a Petri dish, the more bacteria there are the faster the population will grow, as there are more bacteria cloning themselves. (This assumes an abundant food supply and no constraints on population from lack of space, but at least initially this is often the case.) In any situation where the rate of growth of a population is proportional to the size of the population, the population will grow exponentially.

Example 1 illustrates this with a specific model of bacterial population growth.

chapter seven Exponential and Logarithmic Functions

example 1

A biologist is culturing bacteria in a large Petri dish. She begins with 1000 bacteria, and supplies sufficient food so that for the first five hours the bacteria population grows exponentially. If the population doubles every hour, how many bacteria are there two and a half hours after she begins?

Solution:

We are seeking a function that describes the population size. The function will depend on the variable t (for time), and the situation suggests we should measure t in hours. If we name the function P (for population), and if we assume that $t = 0$ corresponds to the moment when the biologist begins the culturing process, we can determine a few values of P: $P(0) = 1000$, $P(1) = 2000$, $P(2) = 4000$, and so on, up to $P(5) = 32000$. That is, the population at time 0 is 1000 bacteria, the population at time 1 (hour) is 2000 bacteria, and the population doubles every hour thereafter up to at least $t = 5$ (hours).

But we have been asked to calculate $P(2.5)$, and this is not so easily done. A good starting point, though, is the fact that since the population growth is exponential, we know

$$P(t) = P_0 a^t,$$

for some constant P_0 and some base a. In other words, the population function is an exponential function (with some as-yet-unknown base), possibly multiplied by a constant (labeled P_0 here). We just have to determine P_0 and a.

Note that $P(0) = P_0 a^0 = P_0$, and we already know that $P(0) = 1000$, so it must be the case that $P_0 = 1000$. In fact, this is the rationale for the name of this constant: P_0 is the population at time 0. Now we know that $P(t) = 1000a^t$, and we have only the task of determining a. But since we know $P(1) = 2000$, and from the above $P(1) = 1000a^1 = 1000a$, we have the equation $2000 = 1000a$, or $a = 2$. Thus $P(t) = 1000(2^t)$.

We can now use a calculator to determine that $P(2.5) = 1000(2^{2.5}) \approx 5657$.

section 7.2   Applications of Exponential Functions

Topic 2:   **Radioactive Decay**

In contrast to populations (at least healthy populations), radioactive substances diminish with time. To be exact, the mass of a radioactive substance decreases over time as the substance decays into other elements. Since exponential functions with a base between 0 and 1 are decreasing functions, this suggests that radioactive decay is modeled by

$$A(t) = A_0 a^t,$$

where $A(t)$ represents the amount of a given substance at time t, A_0 is the amount at time $t = 0$, and a is a number between 0 and 1.

The fact of radioactive decay is important (indirectly) in using radioactivity for power generation, and important again in working out the details of storing spent radioactive fuel rods. Radioactive decay also has other uses, one of which goes by the name of *radiocarbon dating*. This is a technique used by archaeologists, anthropologists, and others to estimate how long ago an organism died. The method depends on the fact that living organisms constantly absorb molecules of the radioactive substance carbon-14 while alive, but the intake of carbon-14 ceases once the organism dies. It is believed that the percentage of carbon-14 on Earth (relative to other isotopes of carbon) has been relatively constant over time, so the first step in the method is to determine the percentage of carbon-14 in the remains of a given organism. By comparing this (smaller) percentage to the percentage found in living tissue, an estimate of when the organism died can then be made.

The mathematics of the age estimation depends on the fact that half of a given mass of carbon-14 decays over a period of 5728 years. This is known as the *half-life* of carbon-14; every radioactive substance has a half-life, and the half-life is usually an important issue when working with such substances. In the case of carbon-14, the half-life of 5728 years means that if, for instance, an organism contained 12 grams of carbon-14 at death, it would contain 6 grams after 5728 years, 3 grams after another 5728 years, and so on. (In exponential growth, there is the related concept of *doubling-time*, which is the length of time needed for the function to double in value. The doubling-time for the bacteria population in *Example 1* is 1 hour.)

example 2

Determine the base, a, so that the function $A(t) = A_0 a^t$ accurately describes the decay of carbon-14 as a function of t years.

Solution:

Note that $A(0) = A_0 a^0 = A_0$, so A_0 represents the amount of carbon-14 at time $t = 0$. Since we are seeking a general formula, we don't know what A_0 is specifically; that is, the value of A_0 will vary depending on the details of the situation. But we can still determine the base constant, a.

What we know is that half of the original amount of carbon-14 decays over a period of 5728 years, so $A(5728)$ will be half of A_0. This gives us the equation

$$A_0 \left(a^{5728} \right) = \frac{A_0}{2}.$$

To solve this for a, we can first divide both sides by A_0; the fact that A_0 then cancels from the equation just emphasizes that its exact value is irrelevant for the task at hand. So we have the equation

$$a^{5728} = \frac{1}{2}.$$

At this point, a calculator is called for, as we need to take the 5728th root of both sides. This gives us the value for a that we seek:

$$a = \left(\frac{1}{2} \right)^{\frac{1}{5728}} \approx 0.999879$$

The function is thus $A(t) = A_0 (0.999879)^t$ (using our approximate value for a). We can now verify that the function behaves as expected by evaluating A at various multiples of the half-life for carbon-14, as shown.

$$A(1 \cdot 5728) = A_0 (0.999879)^{5728} \approx 0.5 A_0$$

$$A(2 \cdot 5728) = A_0 (0.999879)^{11456} \approx 0.25 A_0$$

$$A(3 \cdot 5728) = A_0 (0.999879)^{17184} \approx 0.125 A_0$$

This is our first encounter with radioactive decay and radio-carbon dating. We will come back to the topic in a later section and refine our understanding.

section 7.2 Applications of Exponential Functions

Topic 3: Compound Interest and the Number e

It is difficult to think of an application of exponential functions that is of more immediate importance to more people than compound interest. Virtually everyone living today eventually has some experience, positive or negative, with the notion of compound interest. The positive experiences derive from investing money in various ways, while the negative (but usually unavoidable) experiences derive from car loans, mortgages, credit card charges, and the like.

The basic compound interest formula can be easily understood by considering what happens when money is invested in a simple savings account. Typically, a savings account is set up to pay interest at an annual rate of r (which we will represent in decimal form) compounded n times a year. For instance, a bank may offer an annual interest rate of 5% (meaning $r = 0.05$) compounded monthly (so $n = 12$). Compounding, in this context, is the act of calculating the interest earned on an investment and adding that amount to the investment. An investment in a monthly-compounded account will have interest added to it twelve times over the course of a year.

Suppose an amount of P (for principal) dollars is invested in a savings account at an annual rate of r compounded n times per year. What we are seeking is a formula for the amount of money, which we will call A, in the account after t years. If we say that a period is the length of time between compoundings, interest is calculated at the rate of $\frac{r}{n}$ per period (for instance, if $r = 0.05$ and $n = 12$, interest is earned at a rate of $\frac{0.05}{12} \approx 0.00417$ per month). Table 1 illustrates how compounding increases the amount in the account over the course of several periods.

Period	Amount accumulated
0	$A = P$
1	$A = P\left(1 + \frac{r}{n}\right)$
2	$A = P\left(1 + \frac{r}{n}\right)\left(1 + \frac{r}{n}\right) = P\left(1 + \frac{r}{n}\right)^2$
3	$A = P\left(1 + \frac{r}{n}\right)^2\left(1 + \frac{r}{n}\right) = P\left(1 + \frac{r}{n}\right)^3$
$\vdots$	$\vdots$
k	$A = P\left(1 + \frac{r}{n}\right)^{k-1}\left(1 + \frac{r}{n}\right) = P\left(1 + \frac{r}{n}\right)^k$

Table 1: Effect of Compounding on an Investment of P Dollars

chapter seven   Exponential and Logarithmic Functions

Notice that the accumulation at the end of each period is the accumulation at the end of the preceding period multiplied by $\left(1+\frac{r}{n}\right)$; that is, the account has the amount from the end of the preceding period plus that amount multiplied by $\frac{r}{n}$. The amount A is then a function of t, the number of years the account stays active. Since the number of investment periods in t years is nt, we obtain the following formula.

Compound Interest Formula

An investment of P dollars at an annual interest rate of r compounded n times per year has an accumulated value after t years of

$$A(t) = P\left(1+\frac{r}{n}\right)^{nt}.$$

example 3

Sandy invests \$10,000 in a savings account earning 4.5% annual interest compounded quarterly. What is the value of her investment after three and a half years?

Solution:

The first step is to determine the values of the various pieces of the compound interest formula. We know that $P = 10{,}000$, $r = 0.045$ (remember to express the interest rate in decimal form), $n = 4$ (the account is compounded four times a year), and $t = 3.5$ (the length of time concerned, in years). Thus the accumulation after three and a half years is

$$A(3.5) = 10000\left(1+\frac{0.045}{4}\right)^{(4)(3.5)}$$

$$= 10000(1.01125)^{14}$$

$$\approx \$11{,}695.52.$$

The compound interest formula can also be used to determine the interest rate of an existing savings account, as demonstrated in *Example 4*.

section 7.2   Applications of Exponential Functions

example 4

Nine months after depositing $520.00 in a monthly-compounded savings account, Frank checks his balance and finds the account has $528.84. Being the forgetful type, he can't quite remember what the annual interest rate for his account is supposed to be, but he sees that the bank is advertising a rate of 2.5% for new accounts. Should he close out his existing account and open a new one?

Solution:

As in *Example 3*, we will begin by identifying the known quantities in the compound interest formula: note that $P = 520$, $n = 12$ (12 compoundings per year), and $t = 0.75$ (nine months is three-quarters of a year). Further, the accumulation, A, at the time in question is $528.84. This gives us the equation

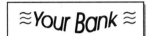

Introducing 2.5% Annual Interest Rate on all new accounts!

$$528.84 = 520\left(1 + \frac{r}{12}\right)^{(12)(0.75)}$$

to solve for r, the annual interest rate.

$$528.84 = 520\left(1 + \frac{r}{12}\right)^9$$

$$1.017 = \left(1 + \frac{r}{12}\right)^9$$

After dividing both sides by 520, we can use a calculator to take the 9th root of both sides and solve for r, as shown.

$$1.001875 \approx 1 + \frac{r}{12}$$

$$0.001875 \approx \frac{r}{12}$$

$$0.0225 \approx r$$

Thus his current savings account is paying an annual interest rate of 2.25%, and he would gain a slight advantage by switching to a new account.

The compound interest formula serves as a convenient means of introducing the extremely important irrational number e. Just as π is the symbol conventionally used for the irrational number obtained by dividing the circumference of a circle by its diameter, e is the symbol traditionally used to denote what is undoubtedly the most useful base for an exponential function, and it is consequently called the natural base.

chapter seven   Exponential and Logarithmic Functions

A complete understanding of the preceding statement requires a discussion of some concepts encountered in calculus, but we'll see many purely algebraic uses of e in this text. The first arises from considering what happens to the compound interest formula when the number of compoundings per year increases without bound. That is, we will examine the formula as $n \to \infty$.

In order to do this easily, we need to perform some algebraic manipulation of the formula, as follows:

$$A(t) = P\left(1 + \frac{r}{n}\right)^{nt}$$

$$= P\left(1 + \frac{1}{m}\right)^{rmt} \quad \text{we have replaced } \frac{r}{n} \text{ with } \frac{1}{m};$$
$$\text{therefore } n = rm$$

$$= P\left[\left(1 + \frac{1}{m}\right)^{m}\right]^{rt}$$

Although the manipulation may appear somewhat strange, it has accomplished the important task of isolating the part of the formula that changes as $n \to \infty$. First, note that the change of variables made in the second step implies $m = \frac{n}{r}$, and so letting $n \to \infty$ means that $m \to \infty$ as well. Since every other quantity in the formula remains fixed, we only have to understand what happens to

$$\left(1 + \frac{1}{m}\right)^{m}$$

as m grows without bound.

After entering the data for m, **L1** can be used to find the values of **L2** by highlighting the column name and entering L2=(1+1/L1)^L1 (L1 may be selected by pressing **2ND**).

Unfortunately, this is no trivial undertaking. You might think that letting m get larger and larger would make the expression grow larger and larger, as m is the exponent in the expression and the base, for any positive value of m, is larger than 1. But at the same time, the base approaches 1 as m increases without bound, and 1 raised to any power is simply 1. It turns out that these two effects balance one another out, with the result that

$$\left(1 + \frac{1}{m}\right)^{m} \to e \approx 2.7183$$

as $m \to \infty$. As far as compound interest is concerned, this means that if P dollars is invested in an account that is compounded continuously (the interest is calculated at every instant and added to the account), the accumulation in the account is determined by the following formula:

Continuous Compounding Formula

If P dollars is invested for t years in a continuously compounded account with an annual interest rate of r, the accumulation is:

$$A(t) = Pe^{rt}$$

example 5

If Sandy (last seen in *Example 3*) has the option of investing her $10,000 in a continuously compounded account earning 4.5% annual interest for three and a half years, what will her accumulation be?

Solution:

The relevant formula is applied:

$$A(3.5) = 10000e^{(0.045)(3.5)}$$
$$= 10000e^{0.1575}$$
$$\approx \$11705.81$$

Such an account thus earns $10.29 more than the quarterly compounded account in *Example 3*.

As we will see in Section 7.4, all exponential functions can be expressed with the natural base e (or any other legitimate base, for that matter). For instance, the formula for the radioactive decay of carbon-14, using the base e, is

$$A(t) = A_0 e^{-0.000121t}.$$

You should verify that this version of the decay formula does indeed give the same values for $A(t)$ as the version derived in *Example 2*.

chapter seven   Exponential and Logarithmic Functions

exercises

The following problems all involve working with exponential functions similar to those encountered in this section. See Examples 1 through 5 for guidance as you work through the problems.

1. A new virus has broken out in isolated parts of Africa and is spreading exponentially through tribal villages. The growth of this new virus can be mapped using the following formula where P stands for the number of people in a village and d stands for the number of days since the virus first appeared. According to this equation, how many people in a tribe of 300 will be infected after 5 days?

$$V = P\left(1 - e^{-0.18d}\right)$$

2. A new hybrid car is equipped with a battery that is meant to improve gas mileage by making the car run on electric power as well as gasoline. The batteries in these new cars must be changed out every so often to ensure proper operation. The power in the battery decreases according to the exponential equation below. After 30 days (d), how much power in watts (w) is left in the battery?

$$w(d) = 40e^{-0.06d}$$

3. A young economics student has come across a very profitable investment scheme in which his money will accrue interest according to the equation listed below. If this student invests $1250 into this lucrative endeavor, how much money will he have after 24 months? I represents the investment and m represents the number of months the money has been invested for.

$$C = Ie^{0.08m}$$

4. A family releases a couple of pet rabbits into the wild. Upon being released the rabbits begin to reproduce at an exponential rate, as shown in the formula below. After 2 years how large is the rabbit population where n stands for the initial rabbit population (2) and m stands for the number of months?

$$P = ne^{0.5m}$$

section 7.2 Applications of Exponential Functions

5. Inside a business network, an email worm was downloaded by an employee. This worm goes through the infected computer's address book and sends itself to all the listed email addresses. This worm very rapidly works its way through the network following the equation below where C is the number of computers in the network and W is the number of computers infected h hours after its discovery. After only 8 hours, how many computers has the worm infected if there are 150 computers in the network?

$$W = C\left(1 - e^{-0.12h}\right)$$

6. A construction crew has been assigned to build an apartment complex. The work of the crew can be modeled using the exponential formula below where A is the total number of apartments to be built, w is the number of weeks, and F is the number of finished apartments. Out of a total of 100 apartments, how many apartments have been finished after 4 weeks of work?

$$F = A\left(1 - e^{-0.1w}\right)$$

7. The half-life of radium is approximately 1600 years. **a.** Determine a so that $A(t) = A_0 a^t$ describes the amount of radium left after t years, where A_0 is the amount at time $t = 0$. **b.** How much of a 1-gram sample of radium would remain after 100 years? **c.** How much of a 1-gram sample of radium would remain after 1000 years?

8. The radioactive element Polonium-210 has a relatively short half-life of 140 days, and one way to model the amount of Polonium-210 remaining after t days is with the function $A(t) = A_0 e^{-0.004951t}$, where A_0 is the mass at time $t = 0$ (note that $A(140) = \frac{A_0}{2}$.) What percentage of the original mass of a sample of Polonium-210 remains after one year?

9. A certain species of fish is to be introduced into a new man-made lake, and wildlife experts estimate the population will grow according to $P(t) = (1000)2^{\frac{t}{3}}$, where t represents the number of years from the time of introduction. **a.** What is the doubling time for this population of fish? **b.** How long will it take for the population to reach 8000 fish, according to this model?

10. The population of a certain inner-city area is estimated to be declining according to the model $P(t) = 237{,}000e^{-0.018t}$, where t is the number of years from the present. What does this model predict the population will be in ten years?

11. In an effort to control vegetation overgrowth, 100 rabbits are released in an isolated area that is free of predators. After one year, it is estimated that the rabbit population has increased to 500. Assuming exponential population growth, what will the population be after another 6 months?

12. Assuming a current world population of 6 billion people, an annual growth rate of 1.9% per year, and a worst-case scenario of exponential growth, what will the world population be in: **a.** 10 years? **b.** 50 years?

13. Madiha has $3500 that she wants to invest in a simple savings account for two and a half years, at which time she plans to close out the account and use the money as a down payment on a car. She finds one local bank offering an annual interest rate of 2.75% compounded monthly, and another bank offering an annual interest rate of 2.7% compounded daily (365 times per year). Which bank should she choose?

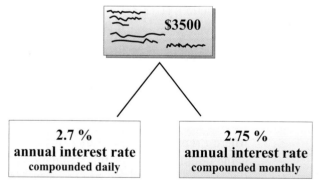

14. Madiha, from the last problem, does some more searching and finds an on-line bank offering an annual rate of 2.75% compounded continuously. How much more money will she earn over two and a half years if she chooses this bank rather than the local bank offering the same rate compounded monthly?

section 7.2   Applications of Exponential Functions

15. Tom hopes to earn $1000 in interest in three years time from $10,000 that he has available to invest. To decide if it's feasible to do this by investing in a simple monthly-compounded savings account, he needs to determine the annual interest rate such an account would have to offer for him to meet his goal. What would the annual rate of interest have to be?

16. An investment firm claims that its clients usually double their principal in five years time. What annual rate of interest would a savings account, compounded monthly, have to offer in order to match this claim?

17. The function $C(t) = C(1 + r)^t$ models the rise in the cost of a product that has a cost of C today, subject to an average yearly inflation rate of r for t years. If the average annual rate of inflation over the next decade is assumed to be 3%, what will the inflation-adjusted cost of a $100,000 house be in 10 years? Round your answer to the nearest dollar.

18. Given the inflation model $C(t) = C(1 + r)^t$ (see the previous problem), and given that a loaf of bread that currently sells for $3.60 sold for $3.10 six years ago, what has the average annual rate of inflation been for the past six years?

19. The function $N(t) = \dfrac{10,000}{1 + 999e^{-t}}$ models the number of people in a small town who have caught the flu t weeks after the initial outbreak. **a.** How many people were ill initially? **b.** How many people have caught the flu after 8 weeks? **c.** Describe what happens to the function $N(t)$ as $t \to \infty$.

20. The concentration, $C(t)$, of a certain drug in the bloodstream after t minutes is given by the formula $C(t) = 0.05(1 - e^{-0.2t})$. What is the concentration after 10 minutes?

21. Carbon-11 has a radioactive half-life of approximately 20 minutes, and is useful as a diagnostic tool in certain medical applications. Because of the relatively short half-life, time is a crucial factor when conducting experiments with this element. **a.** Determine a so that $A(t) = A_0 a^t$ describes the amount of Carbon-11 left after t minutes, where A_0 is the amount at time $t = 0$. **b.** How much of a 2-kg sample of Carbon-11 would be left after 30 minutes? **c.** How many milligrams of a 2-kg sample of Carbon-11 would be left after 6 hours?

22. Charles has recently inherited $8000 which he wants to deposit into a savings account. He has determined that his two best bets are an account that compounds annually at a rate of 3.20% and an account that compounds continuously at an annual rate of 3.15%. Which account would pay Charles more interest?

23. Marshall invests $1250 in a mutual fund which boasts a 5.7% annual return compounded semiannually (twice a year). After three and a half years, Marshall decides to withdraw his money. **a.** How much is in his account? **b.** How much has he made in interest from his investment?

24. Adam is working in a lab testing bacteria populations. After starting out with a population of 375 bacteria, he observes the change in population and notices that the population doubles every 27 minutes. Find **a.** the equation for the population P in terms of time t in minutes, rounding a to the nearest thousandth, and **b.** the population after 2 hours.

After 27 minutes

25. Your credit union offers a special interest rate of 10% compounded monthly for the first year for a student savings account opened in August if the student deposits $5000 or more. You received a total of $9000 for graduation, and you decide to deposit all of it in this special account. Assuming you open your account in August and make no withdrawals for the first year, how much money will you have in your account at the end of February (after 6 months)? How much will you have at the end of the following July (after one full year)?

26. You have a savings account of $3000 with an interest rate of 6.8%.
   a. How much interest would be earned in 2 years if the interest is compounded annually?
   b. How much interest would be earned in 2 years if the interest is compounded semiannually?
   c. In which case do you make more money on interest? Explain why this is so.

27. If $2500 is invested in a continuously compounded certificate of deposit with an annual interest rate of 4.2%, what would be the account balance at the end of 3 years?

section 7.2   Applications of Exponential Functions

28. The new furniture store in town boasts a special in which you can buy any set of furniture in their store and make no monthly payments for the first year. However, the fine print says that the interest rate of 7.25% is compounded quarterly beginning when you buy the furniture. You are considering buying a set of living room furniture for $4000, but know you can not save up more than $4500 in one year's time. Can you fully pay off your furniture on the one year anniversary of having bought the furniture? If so, how much money will you have left over? If not, how much more money will you need?

29. When Nicole was born, her grandmother was so excited about her birth that she opened a certificate of deposit in Nicole's honor to help send her to college. Now at age 18, Nicole's account has $81,262.93. How much did her grandmother originally invest if the interest rate has been 8.1% compounded annually?

30. Inflation is a relative measure of your purchasing power over time. The formula for inflation is the same as the compound interest formula, but with $n = 1$. Given the current prices below, what will the price of the following items be 10 years from now if inflation is at 6.4%?
    a. an SUV: $38,000
    b. a loaf of bread: $1.79
    c. a gallon of milk: $3.40
    d. your salary: $34,000

31. Depreciation is the decrease of an item's value and can be determined using a formula similar to that for compound interest:
$$V = P(1-r)^t, \text{ where } V \text{ is the new value}$$
If the particular car you buy upon graduation from college costs $17,500 and depreciates at a rate of 16% per year, what will the value of the car be in 5 years when you pay it off?

32. Assume the interest on your credit card is compounded continuously with an APR (annual percentage rate) of 19.8%. If you put your first term bill of $3984 on your credit card, but do not have to make payments until you graduate (4 years later), how much will you owe when you start making payments?

33. Suppose you deposit $5000 in an account for 5 years at a rate of 8.5%.
    a. What would be the ending account balance if the interest is continuously compounded?
    b. What would be the ending account balance if the interest is compounded daily?
    c. Are these two answers similar? Why or why not?

chapter seven Exponential and Logarithmic Functions

technology note

Scientific and graphing calculators usually have a button labeled e^x, to be used in evaluating or graphing exponential functions with the base e. Computer algebra systems such as *Mathematica* are similarly equipped. In *Mathematica*, the expression e^x can be entered as either `E^x` or `Exp[x]`. The example below shows how the function $f(x) = e^{3x}$ is defined, graphed, and then evaluated for two different values in *Mathematica*.

Window:
$x = [-3, 5]$
$y = [-6, 18]$

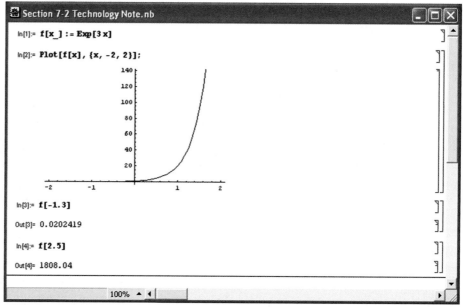

Use a graphing calculator or a computer algebra system, if available, to sketch the graphs of the following functions.

34. $m(x) = 1 - 3e^x$ **35.** $p(x) = e^{4x} - 2$ **36.** $b(x) = \dfrac{1}{e^{x-2}}$

37. $m(x) = e^{2x^2 - 3x + 1}$ **38.** $g(x) = e^{x+3} - 3$ **39.** $m(x) = 6e^{2x} - 2$

7.3 Logarithmic Functions and Their Graphs

TOPICS

1. Definition of logarithmic functions
2. Graphing logarithmic functions
3. Evaluating elementary logarithmic expressions
4. Common and natural logarithms

Topic 1: Definition of Logarithmic Functions

The need for logarithmic functions can be demonstrated with a few sample problems, some easily solved and some, at present, not.

Solvable	Not easily solvable yet
We can solve the equation $$2^x = 8$$ by writing 8 as 2^3 and equating exponents. This is an example of an elementary exponential equation such as we learned to solve in Section 7.1.	We cannot solve the equation $$2^x = 9$$ in the same way, although this equation is only slightly different. All we can say at the moment is that x must be a bit larger than 3.
If we know $A, P, n,$ and t, we can solve the compound interest equation $$A = P\left(1 + \frac{r}{n}\right)^{nt}$$ for the annual interest rate r, as we did in *Example 4* of Section 7.2.	If we know $A, P,$ and t, it is not so easy to determine the annual interest rate r from the equation $$A = Pe^{rt},$$ the continuous compounding equation.

chapter seven   Exponential and Logarithmic Functions

In both of the first two equations, the variable x appears in the exponent (making them exponential equations), but only in the first equation can x be extricated from the exponent with the tools we already have. In the second pair of equations, we are able to solve for the interest rate r in the first case, but it is difficult in the continuous-compounding case, and for the same reason: the variable is inconveniently "stuck" in the exponent. What is needed in such situations is a way of undoing the exponentiation.

This might seem familiar. We have "undone" functions before, and we have done so (whether we realized it at the time or not) by finding inverses of functions. In this case, we need to find the inverse of the generic exponential function $f(x) = a^x$. Happily, we know at the outset that $f(x) = a^x$ has an inverse, because the graph of f passes the horizontal line test (regardless of whether $0 < a < 1$ or $a > 1$).

If we apply the algorithm for finding inverses of functions, our work looks something like the following:

$f(x) = a^x$	The first step is to rewrite the function as an equation, simply by replacing $f(x)$ with y. We then interchange x and y and proceed to solve for y.
$y = a^x$	
$x = a^y$	
$y = ?$	At this point, we are stuck again. What is y?

No concept or notation that we have encountered up to this point allows us to solve the equation $x = a^y$ for the variable y, and this is ultimately the reason for introducing the new class of functions called *logarithms*.

Logarithmic Functions

Let a be a fixed positive real number not equal to 1. The **logarithmic function with base a** is defined to be the inverse of the exponential function with base a, and is denoted $\log_a x$. In symbols, if $f(x) = a^x$, then $f^{-1}(x) = \log_a x$.

In equation form, the definition of logarithm means that the equations

$$x = a^y \text{ and } y = \log_a x$$

are equivalent. Note that a is the base in both equations: either the base of the exponential function or the base of the logarithmic function.

section 7.3   Logarithmic Functions and Their Graphs

Since logarithmic functions are an entirely new class of functions, the first task is to study them and build an intuitive understanding of their behavior. For instance, we can determine that the domain of $\log_a x$ (for any allowable a) is the positive real numbers, because the positive real numbers constitute the range of a^x (remember that the domain of f^{-1} is the range of f). Similarly, the range of $\log_a x$ is the entire set of real numbers, because the real numbers make up the domain of a^x. But probably the best way to become familiar with logarithmic functions is to gain experience with their graphs.

Topic 2: Graphing Logarithmic Functions

Recall that the graphs of a function and its inverse are reflections of one another with respect to the line $y = x$. Since exponential functions come in two flavors ($0 < a < 1$ and $a > 1$), logarithmic functions fall into two similar categories. In both of the graphs in Figure 1, the dashed teal line is the graph of an exponential function representative of its class, and the solid blue line is the corresponding logarithmic function.

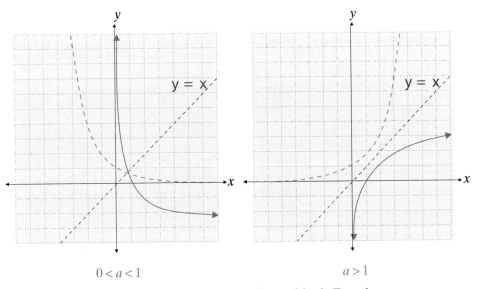

$0 < a < 1$ $a > 1$

Figure 1: The Two Classes of Logarithmic Functions

chapter seven   Exponential and Logarithmic Functions

Note that the domain of each of the logarithmic functions in Figure 1 is indeed $(0, \infty)$, and that the range in each case is $(-\infty, \infty)$. Also note that the y-axis is a vertical asymptote for both, and that neither has a horizontal asymptote.

Of course, we can shift, reflect, stretch, and compress the graphs of logarithms as we can any other function. The techniques we learned in Section 4.4 are all we need to do the work in *Example 1*.

example 1

Sketch the graphs of the following functions.

a. $f(x) = \log_3(x+2) + 1$ **b.** $g(x) = \log_2(-x-1)$ **c.** $h(x) = \log_{\frac{1}{2}}(x) - 2$

Solutions:

a.

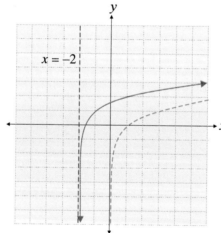

The dashed teal line at left is the graph of the function $y = \log_3 x$, the shape of which is the basis for the graph of f. We know several points on the graph of $\log_3 x$ exactly; for instance, $(1, 0)$ and $(3, 1)$ are on the graph of $\log_3 x$ because $(0, 1)$ and $(1, 3)$ are on the graph of 3^x.

To graph f, we then shift everything 2 units to the left and 1 unit up.

536

section 7.3 Logarithmic Functions and Their Graphs

b.

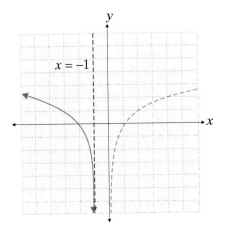

Again, we start with the basic shape of the graph, and then worry about the transformations. The basic shape of the graph of g is the same as the shape of $y = \log_2 x$, drawn as a dashed teal line at left. As in part **a**, some points on $\log_2 x$ are easily determined, such as $(1, 0)$ and $(2, 1)$.

To obtain g from $\log_2 x$, the variable x is replaced with $x - 1$, which shifts the graph 1 unit to the right, and then x is replaced by $-x$, which reflects the graph with respect to the y-axis.

c.

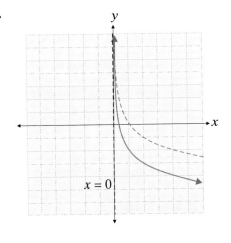

The basic logarithmic function $y = \log_{\frac{1}{2}} x$ is a decreasing function, as the base is between 0 and 1. Note that $(1, 0)$ and $\left(\frac{1}{2}, 1\right)$ are two points on the dashed teal line, the graph of $\log_{\frac{1}{2}} x$. If we then shift the graph 2 units down, we obtain the graph of the function h.

Topic 3: Evaluating Elementary Logarithmic Expressions

Now that we have graphed logarithmic functions, we can augment our understanding of their behavior with a few algebraic observations. These will enable us to evaluate some logarithmic expressions and even solve some elementary logarithmic equations.

First, our work in *Example 1* certainly suggests that the point $(1, 0)$ is always on the graph of $\log_a x$ for any allowable base a, and this is indeed the case. A similar observation is that $(a, 1)$ is always on the graph of $\log_a x$. These two facts are nothing more than restatements of two corresponding facts about exponential functions, and are a consequence of the definition of logarithms:

$$\log_a 1 = 0 \text{, because } a^0 = 1$$
$$\log_a a = 1 \text{, because } a^1 = a$$

More generally, we can use the fact that the functions $\log_a x$ and a^x are inverses of one another to write:

$$\log_a(a^x) = x \text{ and } a^{\log_a x} = x$$

Example 2 makes use of these facts in evaluating some specific logarithms.

example 2

Evaluate the following logarithmic expressions.

a. $\log_5 25$
b. $\log_{\frac{1}{2}} 2$
c. $\log_\pi \sqrt{\pi}$
d. $\log_{17} 1$
e. $\log_{16} 4$
f. $\log_{10}\left(\dfrac{1}{100}\right)$

Solutions:

a. $\log_5 25 = \log_5 5^2$
          $= 2$

Equivalent exponential equation: $25 = 5^2$

b. $\log_{\frac{1}{2}} 2 = \log_{\frac{1}{2}}\left(\dfrac{1}{2}\right)^{-1}$
          $= -1$

Equivalent exponential equation: $2 = \left(\dfrac{1}{2}\right)^{-1}$

section 7.3 Logarithmic Functions and Their Graphs

c. $\log_\pi \sqrt{\pi} = \log_\pi \pi^{\frac{1}{2}}$ Equivalent exponential equation: $\sqrt{\pi} = \pi^{\frac{1}{2}}$

$= \frac{1}{2}$

d. $\log_{17} 1 = 0$ Equivalent exponential equation: $1 = 17^0$

e. $\log_{16} 4 = \log_{16} 16^{\frac{1}{2}}$ Equivalent exponential equation: $4 = 16^{\frac{1}{2}}$

$= \frac{1}{2}$

f. $\log_{10}\left(\frac{1}{100}\right) = \log_{10} 10^{-2}$ Equivalent exponential equation: $\frac{1}{100} = 10^{-2}$

$= -2$

example 3

Use the elementary properties of logarithms to solve the following equations.

a. $\log_6(2x) = -1$ **b.** $3^{\log_{3x} 2} = 2$ **c.** $\log_2 8^x = 5$

Solutions:

a. $\log_6(2x) = -1$

$2x = 6^{-1}$

$2x = \frac{1}{6}$

$x = \frac{1}{12}$

One technique that we will find very fruitful for solving logarithmic equations is to convert the equation to the equivalent exponential form. The second equation at left means exactly the same thing as the first.

Note that $\log_6\left(\frac{2}{12}\right) = \log_6\left(6^{-1}\right) = -1$.

b. $3^{\log_{3x} 2} = 2$

$\log_{3x} 2 = \log_3 2$

$3x = 3$

$x = 1$

This time, we have converted from the exponential form to the logarithmic form and then equated the bases. We could also have noted from the start that the base of the logarithm must be the same as the base of the exponential function for the statement to be true.

continued on next page...

c. $\log_2 8^x = 5$
$8^x = 2^5$
$(2^3)^x = 2^5$
$2^{3x} = 2^5$
$3x = 5$
$x = \dfrac{5}{3}$

The exponential form of the equation, shown as the second line, is of a form that we have solved before. We know that if we can write both sides with the same base, the exponents must be equal.

Note that $\log_2\left(8^{\frac{5}{3}}\right) = \log_2\left(8^{\frac{1}{3}}\right)^5 = \log_2\left(2^5\right) = 5.$

Topic 4: Common and Natural Logarithms

We have already mentioned the fact that the number e, called the natural base, plays a fundamental role in many important real-world situations, so it is perhaps not surprising that the logarithmic function with base e is worthy of special attention. For historical reasons, the logarithmic function with base 10 is also singled out for further mention.

Common and Natural Logarithms

- The function $\log_{10} x$ is called the **common logarithm**, and is usually written as $\log x$.
- The function $\log_e x$ is called the **natural logarithm**, and is usually written as $\ln x$.

Another way in which these particular logarithms are special is that most calculators, if they are capable of calculating logarithms at all, are only equipped to evaluate common and natural logarithms. Such calculators normally have a button labeled **LOG** for the common logarithm and a button labeled **LN** for the natural logarithm.

section 7.3 Logarithmic Functions and Their Graphs

example 4

Evaluate the following logarithmic expressions.

a. $\ln \sqrt[3]{e}$ b. $\log 1000$ c. $\ln(4.78)$ d. $\log(10.5)$

Solutions:

a. $\ln \sqrt[3]{e} = \ln e^{\frac{1}{3}} = \frac{1}{3}$

No calculator is necessary for this problem, just an application of an elementary property of logarithms.

b. $\log 1000 = \log 10^3 = 3$

Again, no calculator is required.

c. $\ln(4.78) \approx 1.564$

This time, a calculator is needed, and only an approximate answer can be given. Be sure to use the correct logarithm.

d. $\log(10.5) \approx 1.021$

Again, we must use a calculator, though we can say beforehand that the answer should be only slightly larger than 1, as $\log 10 = 1$ and 10.5 is only slightly larger than 10.

exercises

Write the following equations in logarithmic terms.

1. $625 = 5^4$
2. $216 = 6^3$
3. $x^3 = 27$
4. $b^2 = 3.2$
5. $4.2^3 = C$
6. $1.3^2 = V$
7. $4^x = 31$
8. $16^{2x} = 215$
9. $(4x)^{\sqrt{3}} = 13$
10. $e^x = \pi$
11. $2^{e^x} = 11$
12. $4^e = N$

chapter seven   Exponential and Logarithmic Functions

Write the following logarithmic equations as exponential equations.

13. $\log_3 81 = 4$
14. $\log_2 \dfrac{1}{8} = -3$
15. $\log_b 4 = \dfrac{1}{2}$
16. $\log_y 9 = 2$
17. $\log_2 15 = b$
18. $\log_5 8 = d$
19. $\log_5 W = 12$
20. $\log_7 T = 6$
21. $\log_\pi 2x = 4$
22. $\log_{\sqrt{3}} 2\pi = x$
23. $\ln 2 = x$
24. $\ln 5x = 3$

Sketch the graphs of the following functions. See Example 1.

25. $f(x) = \log_3(x - 1)$
26. $g(x) = \log_5(x + 2) - 1$
27. $r(x) = \log_{\frac{1}{2}}(x - 3)$
28. $p(x) = 3 - \log_2(x + 1)$
29. $q(x) = \log_3(2 - x)$
30. $s(x) = \log_{\frac{1}{3}}(5 - x)$
31. $h(x) = \log_7(x - 3) + 3$
32. $m(x) = \log_{\frac{1}{2}}(1 - x)$
33. $f(x) = \log_3(6 - x)$
34. $p(x) = 4 - \log_{10}(x + 3)$
35. $s(x) = -\log_{\frac{1}{3}}(-x)$
36. $g(x) = \log_5(2x) - 1$

Match the graph of the appropriate equation to the logarithmic function.

37. $f(x) = \log_2 x - 1$
38. $f(x) = \log_2(2 - x)$
39. $f(x) = \log_2(-x)$
40. $f(x) = \log_2(x - 3)$
41. $f(x) = 1 - \log_2 x$
42. $f(x) = -\log_2 x$
43. $f(x) = -\log_2(-x)$
44. $f(x) = \log_2 x$
45. $f(x) = \log_2 x + 3$

a.

b.

c.
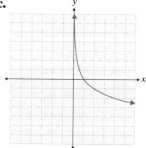

section 7.3 Logarithmic Functions and Their Graphs

d.
e.
f.

g.
h.
i.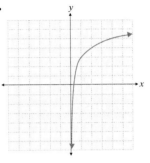

Evaluate the following logarithmic expressions without the use of a calculator. See Example 2.

46. $\log_7 \sqrt{7}$
47. $\log_{\frac{1}{2}} 4$
48. $\log_9 \left(\dfrac{1}{81}\right)$

49. $\log_3 27$
50. $\log_{27} 3$
51. $\log_3 (\log_{27} 3)$

52. $\ln e^{2.89}$
53. $\log 0.0001$
54. $\log_a a^{\frac{5}{3}}$

55. $\ln\left(\dfrac{1}{e}\right)$
56. $\log(\log(10^{10}))$
57. $\log_3 1$

58. $\ln \sqrt[5]{e}$
59. $\log_{\frac{1}{16}} 4$
60. $\log_8 4^{\log 1000}$

Use the elementary properties of logarithms to solve the following equations. See Example 3.

61. $\log_{16} x = \dfrac{3}{4}$
62. $\log_{16} x^{\frac{1}{2}} = \dfrac{3}{4}$
63. $\log_{16} x = -\dfrac{3}{4}$

64. $\log_5 5^{\log_3 x} = 2$
65. $\log_a a^{\log_b x} = 0$
66. $\log_3 9^{2x} = -2$

67. $\log_{\frac{1}{3}} 3^x = 2$
68. $\log_7 (3x) = -1$
69. $4^{\log_3 x} = 0$

70. $\log x^{10} = 10$
71. $\log_x \left(\log_{\frac{1}{4}} \dfrac{1}{4}\right) = 1$
72. $6^{\log_x e^2} = e$

chapter seven — Exponential and Logarithmic Functions

Solve the following logarithmic equations, using a calculator if necessary to evaluate the logarithms. See Example 4. Express your answer either as a fraction or a decimal rounded to the nearest hundredth.

73. $\log(3x) = 2.1$
74. $\log x^2 = -2$
75. $\ln(x + 1) = 3$
76. $\ln 2x = -1$
77. $\ln e^x = 5.6$
78. $\ln(\ln x^2) = 0$
79. $\log 19 = 3x$
80. $\log e^x = 5.6$
81. $\log 300^{\log x} = 9$
82. $\log_9(2x - 1) = 2$
83. $\log(\log(x - 2)) = 1$

technology exercises

As already mentioned, most calculators can only calculate natural logarithms and common logarithms directly. The computer algebra system *Mathematica*, in contrast, can evaluate logarithms of any allowable base. The command to evaluate $\log_2 15$, for example, is `Log[2., 15]` (the decimal point tells *Mathematica* to give us a decimal approximation of the answer). If only one number is supplied to the `Log` function, *Mathematica* assumes the base is e and calculates the natural logarithm of the given number. The illustration below shows the result of asking *Mathematica* to determine $\log_2 15$ and $\ln(9)$.

Use a computer algebra system, if available, to evaluate the expressions that follow. (We will see in the next section how a calculator can be used to evaluate such expressions.)

84. $\log_7 45$
85. $\log_{\frac{1}{2}} 9$
86. $\log_5 30$
87. $\log_{0.9} 2$
88. $\log_{\frac{1}{10}}\left(\frac{1}{9}\right)$
89. $\log_{4.5} 12$
90. $\log_{2.3} 10$
91. $\log_{19} 5$

7.4 Properties and Applications of Logarithms

TOPICS

1. Properties of logarithms
2. The change of base formula
3. Applications of logarithmic functions

Topic 1: Properties of Logarithms

In Section 7.3 we introduced logarithmic functions and made note of some of their elementary properties. Our motivation was our inability, at that time, to solve certain equations. Let us reconsider the two sample problems that initiated our discussion of logarithms and see if we have made progress. We'll begin with the continuously compounding interest problem.

example 1

Anne reads an ad in the paper for a new bank in town. The bank is advertising "continuously compounded savings accounts!" in an attempt to attract customers, but fails to mention the annual interest rate. Curious, she goes to the bank and is told by an account agent that if she were to invest, say, $10,000 in an account, her money would grow to $10,202.01 in a year's time. But, strangely, the agent also refuses to divulge the yearly interest rate. What rate is the bank offering?

Solution:

We need to solve the equation $A = Pe^{rt}$ for r, given that $A = 10202.01$, $P = 10000$, and $t = 1$. We are now able to do so:

$$10202.01 = 10000e^{r(1)}$$
$$1.020201 = e^r \quad \text{Divide by 10000.}$$
$$\ln(1.020201) = r \quad \text{Convert to logarithmic form.}$$
$$r = 0.0199 \approx 0.02 \quad \text{Use a calculator to determine } r.$$

continued on next page...

chapter seven   Exponential and Logarithmic Functions

To solve the equation, we convert the exponential equation to its logarithmic form. Note that we use the natural logarithm since the base of the exponential function is e. We can then use a calculator to evaluate r, and discover that the annual interest rate is 2%.

example 2

Solve the equation $2^x = 9$.

Solution:

To solve this problem, we convert the equation to logarithmic form to obtain the solution $x = \log_2 9$. The only problem with this answer is that we still don't know anything about x in decimal form, other than that it is bound to be slightly more than 3.

As *Example 2* shows, our ability to work with logarithms leaves something to be desired. In this section we will derive some more important properties of logarithms that will allow us to solve more complicated equations, as well as provide a decimal approximation to the solution of $2^x = 9$.

The following three properties of logarithmic functions are a consequence of corresponding properties of exponential functions; this is not surprising considering how logarithms are defined.

Properties of Logarithms

In the following, assume that a (the logarithmic base) is a fixed positive real number not equal to 1, that x and y represent positive real numbers, and that r represents any real number.

1. $\log_a(xy) = \log_a x + \log_a y$ ("the log of a product is the sum of the logs")

2. $\log_a\left(\dfrac{x}{y}\right) = \log_a x - \log_a y$ ("the log of a quotient is the difference of the logs")

3. $\log_a(x^r) = r \log_a x$ ("the log of something raised to a power is the power times the log")

section 7.4 Properties and Applications of Logarithms

We will illustrate the correspondence between these properties and the more familiar properties of exponents by proving the first one. The proofs of the second and third will be left as exercises.

Proof: We will start with the right-hand side of the statement, and convert the expressions to exponential form. To do this, let $m = \log_a x$ and $n = \log_a y$. The equivalent exponential form of these two equations is:

$$x = a^m \text{ and } y = a^n$$

Since we are interested in the product xy, note that

$$xy = a^m a^n = a^{m+n}.$$

The statement $xy = a^{m+n}$ can be converted to logarithmic form, giving us

$$\log_a (xy) = m + n.$$

If we now refer back to the original definition of m and n, we have achieved our goal:

$$\log_a (xy) = \log_a x + \log_a y$$

The properties of logarithms may appear strange at first, but with time they will come to seem as natural as the properties of exponents. In fact, they *are* the properties of exponents, simply restated in logarithmic form.

caution

Errors in working with logarithms often arise from imperfect recall of the above properties. It is common to be misled, for example, into making mistakes of the sort shown on the next page. Note carefully the differences (and tempting similarities) between the incorrect and correct statements.

chapter seven   Exponential and Logarithmic Functions

Incorrect Statements	Correct Statements
$\log_a(x+y) = \log_a x + \log_a y$	$\log_a(xy) = \log_a x + \log_a y$
$\log_a(xy) = (\log_a x)(\log_a y)$	$\log_a(xy) = \log_a x + \log_a y$
$\dfrac{\log_a x}{\log_a y} = \log_a x - \log_a y$	$\log_a\left(\dfrac{x}{y}\right) = \log_a x - \log_a y$
$\dfrac{\log_a x}{\log_a y} = \log_a\left(\dfrac{x}{y}\right)$	$\log_a\left(\dfrac{x}{y}\right) = \log_a x - \log_a y$
$\dfrac{\log_a(xz)}{\log_a(yz)} = \dfrac{\log_a(x)}{\log_a(y)}$	$\dfrac{\log_a(xz)}{\log_a(yz)} = \dfrac{\log_a x + \log_a z}{\log_a y + \log_a z}$

In some situations, we will find it useful to use properties of logarithms to decompose a complicated expression into a sum or difference of simpler expressions, while in other situations we will do just the reverse, combining a sum or a difference of logarithms into one logarithm. *Examples 3* and *4* illustrate the process.

example 3

Use the properties of logarithms to expand the following expressions as much as possible (that is, decompose the expressions into sums or differences of the simplest possible terms).

a. $\log_4\left(64x^3\sqrt{y}\right)$ **b.** $\log_a \sqrt[3]{\dfrac{xy^2}{z^4}}$ **c.** $\log\left(\dfrac{2.7 \times 10^4}{x^{-2}}\right)$

Solutions:

a. $\log_4\left(64x^3\sqrt{y}\right) = \log_4 64 + \log_4 x^3 + \log_4 \sqrt{y}$

$\qquad\qquad\qquad\qquad = \log_4 4^3 + \log_4 x^3 + \log_4 y^{\frac{1}{2}}$

$\qquad\qquad\qquad\qquad = 3 + 3\log_4 x + \dfrac{1}{2}\log_4 y$

We can initially write the expression as a sum of three log terms. We evaluate the first, and the second and third can be rewritten by bringing the exponent down in front.

b. $\log_a \sqrt[3]{\dfrac{xy^2}{z^4}} = \log_a \left(\dfrac{xy^2}{z^4}\right)^{\frac{1}{3}}$

$\qquad\qquad = \dfrac{1}{3}\log_a\left(\dfrac{xy^2}{z^4}\right)$

$\qquad\qquad = \dfrac{1}{3}\left(\log_a x + \log_a y^2 - \log_a z^4\right)$

$\qquad\qquad = \dfrac{1}{3}\left(\log_a x + 2\log_a y - 4\log_a z\right)$

Before we can decompose the fraction, we must apply the third property of logarithms and eliminate the radical. Once the exponent of $\dfrac{1}{3}$ has been brought to the front, the fraction can be written in simpler form as shown. Note that the base is immaterial: a can represent any legitimate logarithmic base.

section 7.4 Properties and Applications of Logarithms

c. $\log\left(\dfrac{2.7 \times 10^4}{x^{-2}}\right) = \log(2.7) + \log(10^4) - \log x^{-2}$

$= \log(2.7) + 4 + 2\log x$

$\approx 4.43 + 2\log x$

Recall that if the base is not explicitly written, it is assumed to be 10. This is a convenient base when dealing with scientific notation. Note that we can use a calculator to approximate $\log(2.7)$.

example 4

Use the properties of logarithms to condense the following expressions as much as possible (that is, rewrite the expressions as a sum or difference of as few logarithms as possible).

a. $2\log_3\left(\dfrac{x}{3}\right) - \log_3\left(\dfrac{1}{y}\right)$ **b.** $\ln x^2 - \dfrac{1}{2}\ln y + \ln 2$ **c.** $\log_b 5 + 2\log_b x^{-1}$

Solutions:

a. $2\log_3\left(\dfrac{x}{3}\right) - \log_3\left(\dfrac{1}{y}\right) = \log_3\left(\dfrac{x}{3}\right)^2 + \log_3\left(\dfrac{1}{y}\right)^{-1}$

$= \log_3\left(\dfrac{x^2}{9}\right) + \log_3 y$

$= \log_3\left(\dfrac{x^2 y}{9}\right)$

Note that before a sum or difference of log terms can be combined, they have to have the same coefficient. This is usually straightforward to arrange, as the coefficient can be moved up into the exponent, as shown. Note also that the order in which the properties are applied can vary, though all lead to the final answer.

b. $\ln x^2 - \dfrac{1}{2}\ln y + \ln 2 = \ln x^2 - \ln y^{\frac{1}{2}} + \ln 2$

$= \ln\left(\dfrac{x^2}{y^{\frac{1}{2}}}\right) + \ln 2$

$= \ln\left(\dfrac{2x^2}{y^{\frac{1}{2}}}\right)$ or $\ln\left(\dfrac{2x^2}{\sqrt{y}}\right)$

Again, we begin by rewriting each term so that its coefficient is 1 or −1; the sums or differences that result can then be combined.

The final answer can be written in several different ways, two of which are shown at left.

continued on next page...

c. $\log_b 5 + 2\log_b x^{-1} = \log_b 5 + \log_b x^{-2}$

$\qquad\qquad\qquad\qquad\quad = \log_b 5x^{-2}$ or $\log_b\left(\dfrac{5}{x^2}\right)$

The base is immaterial; the properties of logarithms hold for any legitimate base.

Topic 2: The Change of Base Formula

The properties we just derived can be used to provide an answer to a question about logarithms that has been left hanging to this point. A specific illustration of the question arose in *Example 2*: how do we determine the decimal form of a number like $\log_2 9$?

Paradoxically enough, to answer this question we will undo our work in *Example 2*. In the following, we assign a variable to the quantity $\log_2 9$, convert the resulting logarithmic equation into exponential form, take the natural logarithm of both sides, and then solve for the variable.

$$x = \log_2 9 \qquad \text{Let } x \text{ stand for the number } \log_2 9.$$
$$2^x = 9 \qquad \text{Convert to exponential form.}$$
$$\ln(2^x) = \ln 9 \qquad \text{Take the natural log of both sides.}$$
$$x \ln 2 = \ln 9 \qquad \text{Apply a property of logarithms.}$$
$$x = \dfrac{\ln 9}{\ln 2} \qquad \text{Solve for } x.$$
$$\approx 3.17$$

Of course, there is nothing special about the natural logarithm, at least as far as this problem is concerned. However, if a calculator is to be used to approximate the number $\log_2 9$, there are (for most calculators) only two good choices: the natural log and the common log. If we had done the work above with the common logarithm, the final answer would have been the same. That is,

$$\dfrac{\log 9}{\log 2} \approx 3.17.$$

More generally, a logarithm with base b can be converted to a logarithm with base a through the same reasoning, as summarized on the following page.

section 7.4 Properties and Applications of Logarithms

Change of Base Formula

Let a and b both be positive real numbers, neither of them equal to 1, and let x be a positive real number. Then

$$\log_b x = \frac{\log_a x}{\log_a b}.$$

example 5

Evaluate the following logarithmic expressions, using the base of your choice.

a. $\log_7 15$ **b.** $\log_{\frac{1}{2}} 3$ **c.** $\log_\pi 5$

Solutions:

a. $\log_7 15 = \dfrac{\ln 15}{\ln 7}$

≈ 1.392

The calculation at left uses the natural logarithm, but the common logarithm achieves the same result. Use a calculator to verify that $7^{1.392} \approx 15$.

b. $\log_{\frac{1}{2}} 3 = \dfrac{\log 3}{\log\left(\dfrac{1}{2}\right)}$

≈ -1.585

Since the base of the original logarithm is between 0 and 1, we expect the answer to be negative, as it is.

c. $\log_\pi 5 = \dfrac{\ln 5}{\ln \pi}$

≈ 1.406

Although a base of π is not often encountered, the original expression is legitimate, and the change of base formula allows us to evaluate it.

chapter seven Exponential and Logarithmic Functions

Topic 3: Applications of Logarithmic Functions

Logarithms appear in many different contexts and have a wide variety of uses. This is due partly to the fact that logarithmic functions are the inverses of exponential functions, and partly to the logarithmic properties we have discussed. In fact, the one individual who can be most credited for "inventing" logarithms, John Napier (1550–1617) of Scotland, was inspired in his work by the very utility of what we now call logarithmic properties.

Computationally, logarithms are useful because they relocate exponents as coefficients, thus making them more easily worked with. Consider, for example, a very large number such as 3×10^{17} or a very small number such as 6×10^{-9}. The common logarithm (used because it has a base of 10) expresses these numbers on a more comfortable scale:

$$\log(3 \times 10^{17}) = \log 3 + \log(10^{17}) = 17 + \log 3 \approx 17.477$$

$$\log(6 \times 10^{-9}) = \log 6 + \log(10^{-9}) = -9 + \log 6 \approx -8.222$$

Napier, working long before the advent of electronic calculating devices, devised logarithms in order to take advantage of this property. Although his particular motivation is not as meaningful today, it is still convenient in many instances to express quantities in terms of their logarithms. We will illustrate this with the next three examples.

example 6

In chemistry, the concentration of hydronium ions in a solution determines its acidity. Since the concentrations are small numbers, but can vary over many orders of magnitude, it is convenient to express acidity in terms of the pH scale, as follows.

> The **pH** of a solution is defined to be $-\log[H_3O^+]$, where $[H_3O^+]$ is the concentration of hydronium ions in moles/liter. Solutions with a pH less than 7 are said to be *acidic*, while those with a pH greater than 7 are *basic*.

If a sample of orange juice is determined to have a $[H_3O^+]$ concentration of 1.58×10^{-4} moles/liter, what is its pH?

section 7.4   Properties and Applications of Logarithms

1.58×10^{-4} moles / liter

Solution:

Applying the above formula (and using a calculator), the pH is equal to

$$\text{pH} = -\log(1.58 \times 10^{-4}) = -(-3.80) = 3.8.$$

After doing one calculation, the reason for the minus sign in the formula is more apparent. By multiplying the log of the concentration by -1, the pH of a solution is necessarily positive, and convenient for comparative purposes.

example 7

The energy released during earthquakes can vary greatly, but logarithms provide a convenient way to analyze and compare the intensity of earthquakes.

Earthquake intensity is measured on the *Richter scale* (named for the American seismologist Charles Richter, 1900 – 1985). In the formula that follows, I_0 is the intensity of a just-discernable earthquake, I is the intensity of an earthquake being analyzed, and R is its ranking on the Richter scale:

$$R = \log\left(\frac{I}{I_0}\right)$$

By this measure, earthquakes range from a classification of small $(R < 4.5)$, to moderate $(4.5 \leq R < 5.5)$, to large $(5.5 \leq R < 6.5)$, to major $(6.5 \leq R < 7.5)$, and finally to greatest $(7.5 \leq R)$.

Note that the base 10 logarithm means that every increase of 1 unit on the Richter scale corresponds to an increase by a factor of 10 in the intensity, and a barely discernable earthquake has a rank of 0 $(\log(1) = 0.)$

continued on next page...

The January 2001 earthquake in the state of Gujarat in India was 80,000,000 times as intense as a 0-level earthquake. What was the Richter ranking of this devastating event?

Solution:

If we let I denote the intensity of the Gujarat earthquake, then $I = 80000000 I_0$, so

$$R = \log\left(\frac{80000000 I_0}{I_0}\right)$$
$$= \log(8 \times 10^7)$$
$$= \log(8) + 7$$
$$\approx 7.9.$$

The Gujarat earthquake thus fell in the category of greatest on the Richter scale.

example 8

Sound intensity is another quantity that varies greatly, and the measure of how the human ear perceives intensity, in units called *decibels*, is very similar to the measure of earthquake intensity.

In the **decibel** formula that follows, I_0 is the intensity of a just-discernable sound, I is the intensity of the sound being analyzed, and D is its decibel level:

$$D = 10 \log\left(\frac{I}{I_0}\right)$$

Decibel levels range from 0 for a barely discernable sound, to 40 for the level of normal conversation, to 80 for heavy traffic, to 120 for a loud rock concert, and finally (as far as humans are concerned) to around 160, at which point the eardrum is likely to rupture.

section 7.4 Properties and Applications of Logarithms

Given that $I_0 = 10^{-12}$ watts/meter^2, what is the decibel level of jet airliner's engines at a distance of 45 meters, for which the sound intensity is 50 watts/meter^2?

Solution:

$$D = 10 \log\left(\frac{50}{10^{-12}}\right)$$
$$= 10 \log\left(5 \times 10^{13}\right)$$
$$= 10 \left(13 + \log 5\right)$$
$$\approx 137$$

In other words, the sound level would probably not be literally ear-splitting, but it would be very painful.

exercises

Use the properties of logarithms to expand the following expressions as much as possible. Simplify any numerical expressions that can be evaluated without a calculator. See Example 3.

1. $\log_5\left(125x^3\right)$
2. $\ln\left(\dfrac{x^2 y}{3}\right)$
3. $\ln\left(\dfrac{e^2 p}{q^3}\right)$
4. $\log(100x)$
5. $\log_9 9xy^{-3}$
6. $\log_6 \sqrt[3]{\dfrac{p^2}{q}}$
7. $\ln\left(\dfrac{\sqrt{x^3}\, pq^5}{e^7}\right)$
8. $\log_a \sqrt[5]{\dfrac{a^4 b}{c^2}}$
9. $\log\left(\log\left(100x^3\right)\right)$
10. $\log_3\left(9x + 27y\right)$
11. $\log\left(\dfrac{10}{\sqrt{x+y}}\right)$
12. $\ln\left(\ln\left(e^{ex}\right)\right)$

13. $\log_2\left(\dfrac{y^2+z}{16x^4}\right)$ 14. $\log(\log(100{,}000^{2x}))$ 15. $\log_b\sqrt{\dfrac{x^4 y}{z^2}}$

16. $\ln(7x^2 - 42x + 63)$ 17. $\log_b ab^2 c^b$ 18. $\ln(\ln(e^{e^x}))$

Use the properties of logarithms to condense the following expressions as much as possible, writing each answer as a single term with a coefficient of 1. See Example 4.

19. $\log x - \log y$
20. $\log_5 x - 2\log_5 y$

21. $\log_5(x^2 - 25) - \log_5(x - 5)$
22. $\ln(x^2 y) - \ln y - \ln x$

23. $\dfrac{1}{3}\log_2 x + \log_2(x+3)$
24. $\dfrac{1}{5}\left(\log_7(x^2) - \log_7(pq)\right)$

25. $\ln 3 + \ln p - 2\ln q$
26. $2\left(\log_5 \sqrt{x} - \log_5 y\right)$

27. $\log(x - 10) - \log x$
28. $2\log a^2 b - \log\dfrac{1}{b} + \log\dfrac{1}{a}$

29. $3\left(\ln\sqrt[3]{e^2} - \ln xy\right)$
30. $\log_2(4x) - \log_2 x$

31. $\log_5 20 - \log_5 5$
32. $\log 30 - \log 2 - \log 5$

33. $\ln 15 + \ln 3$
34. $\ln 8 - \ln 4 + \ln 3$

35. $0.5\log_3 16 - \log_3 4$
36. $3\log_7 2 - 2\log_7 4$

37. $0.25 \ln 81 + \ln 4$
38. $2(\log 4 - \log 1 + \log 2)$

39. $\log 11 + 0.5\log 9 - \log 3$
40. $3\log_4(x^2) + \log_4(x^6)$

41. $\log_8(2x^2 - 2y) - 0.25\log_8 16$
42. $\log_{3x} x^2 + \log_{3x} 18 - \log_{3x} 6$

Use the properties of logarithms to write each of the following as a single term that does not contain a logarithm.

43. $5^{2\log_5 x}$
44. $10^{\log y^2 - 3\log x}$
45. $e^{2 - \ln x + \ln p}$

46. $e^{5\left(\ln\sqrt[5]{3} + \ln x\right)}$
47. $10^{\log x^3 - 4\log y}$
48. $a^{\log_a b + 4\log_a \sqrt{a}}$

49. $10^{2\log x}$
50. $10^{4\log x - 2\log x}$
51. $\log_4 16 \cdot \log_x x^2$

52. $e^{\ln x + 2 + \ln x^2}$
53. $4^{\log_4(3x) + 0.5\log_4(16x^2)}$
54. $4^{2\log_2 6 - \log_2 9}$

section 7.4 Properties and Applications of Logarithms

Evaluate the following logarithmic expressions. See Example 5.

55. $\log_4 17$
56. $2\log_{\frac{1}{3}} 5$
57. $\log_9 8$

58. $\log_2 0.01$
59. $\log_{12} 10.5$
60. $\log(\ln 2)$

61. $\log_6 3^4$
62. $\log_7 14.3$
63. $\log_{\frac{1}{2}} \pi^{-2}$

64. $\log_{\frac{1}{5}} 626$
65. $\ln(\log 123)$
66. $\log_{17} 0.041$

67. $\log 16$
68. $\log_3 9$
69. $\log_5 20$

70. $\log_8 26$
71. $\log_4 0.25$
72. $\log_{1.8} 9$

73. $\log_{2.5} 34$
74. $\log_{0.5} 10$
75. $\log_4 2.9$

76. $\log_{0.4} 14$
77. $\log_{0.2} 17$
78. $\log_{0.16} 2.8$

Without using a calculator, evaluate the following expressions.

79. $\log_4 16$
80. $\log_5 25^3$
81. $\ln e^4 + \ln e^3$

82. $\log_4 \frac{1}{64}$
83. $\ln e^{1.5} - \log_4 2$
84. $\log_2 8^{(2\log_2 4 - \log_2 4)}$

Find the value of x in each of the following equations. Express your answer as exact as possible, or as a decimal rounded to the nearest hundredth.

85. $\log_x 1024 = 4$
86. $\log_6 729 = x$
87. $\log_2 529 = x$

88. $\log_4 625 = x$
89. $\log_x 729 = 9$
90. $\log_4 x = 8$

91. $\log_{12} x = 1$
92. $\log_x 16807 = 7$
93. $\log_4 x = 10$

chapter seven   Exponential and Logarithmic Functions

Use your knowledge of logarithms to answer the following questions. See Examples 6, 7, and 8.

94. A certain brand of tomato juice has a $\left[H_3O^+\right]$ concentration of 3.16×10^{-6} moles/liter. What is the pH of this brand?

3.16×10^{-6} moles/liter

95. One type of detergent, when added to neutral water with a pH of 7, results in a solution with a $\left[H_3O^+\right]$ concentration that is 5.62×10^{-4} times weaker than that of the water. What is the pH of the solution?

96. What is the concentration of $\left[H_3O^+\right]$ in lemon juice with a pH of 3.2?

97. The 1994 Northridge, California earthquake measured 6.7 on the Richter scale. What was the intensity, relative to a 0-level earthquake, of this event?

98. How much stronger was the 2001 Gujarat earthquake (7.9 on the Richter scale) than the 1994 Northridge earthquake described in Exercise 97?

99. A construction worker operating a jackhammer would experience noise with an intensity of 20 watts/meter² if it weren't for ear-protection. Given that $I_0 = 10^{-12}$ watts/meter², what is the decibel level for such noise?

section 7.4   Properties and Applications of Logarithms

100. A microphone picks up the sound of a thunderclap and measures its decibel level as 105. Given that $I_0 = 10^{-12}$ watts/meter^2, with what sound intensity did the thunderclap reach the microphone?

101. Matt, a lifeguard, has to make sure that the pH of the swimming pool stays between 7.2 and 7.6. If the pH is out of this range, he has to add chemicals which alter the pH level of the pool. If Matt measures the $\left[H_3O^+\right]$ concentration in the swimming pool to be 2.40×10^{-8} moles/liter, what is the pH? Does he need to change the pH by adding chemicals to the water?

102. The intensity of a cat's soft purring is measured to be 2.19×10^{-11}. Given that $I_0 = 10^{-12}$ watts/meter^2, what is the decibel level of this noise?

chapter seven Exponential and Logarithmic Functions

103. Newton's Law of Cooling states that the rate at which an object cools is proportional to the difference between the temperature of the object and the surrounding temperature. If C denotes the surrounding temperature and T_0 denotes the temperature at time $t = 0$, the temperature of an object at time t is given by $T(t) = C + (T_0 - C)e^{-kt}$, where k is a constant that depends on the particular object under discussion.

   a. You are having friends over for tea and want to know how long after boiling the water it will be drinkable. If the temperature of your kitchen stays around 74°F and you found online that the constant k for tea is approximately 0.049, how many minutes after boiling the water will the tea be drinkable (you prefer your tea no warmer than 140°F)? Recall that water boils at 212°F.

   b. As you intern for your local crime scene investigation department, you are asked to determine at what time a victim died. If you are told k is approximately 0.1947 for a human body and the body's temperature was 72°F at 1:00 a.m., and the body has been in a storage building at a constant 60°F, approximately what time did the victim die? Recall the average temperature for a human body is 98.6°F. Note in this situation, t is measured in hours.

   c. When helping your father cook the Christmas turkey, you were told to remove the turkey when the thickest part had reached 180°F. If you remove the turkey and place it on the table in a room that is 72°F, and it cools to 155°F in 20 minutes, what will the temperature of the turkey be at lunch time (an hour and 15 minutes after the turkey is removed from the oven)? Should you warm the turkey before eating?

7.5 Exponential and Logarithmic Equations

TOPICS

1. Converting between exponential and logarithmic forms
2. Further applications of exponential and logarithmic equations
3. Interlude: analysis of a stock market investment

Topic 1: Converting Between Exponential and Logarithmic Forms

At this point, we have all the tools we need to solve the most common sorts of exponential and logarithmic equations. All that is left is to develop our skill in using the tools.

We have already solved many exponential and logarithmic equations, using the more elementary facts about exponential and logarithmic functions to quickly obtain solutions. But many equations require a bit more work to solve. While there is no algorithm to follow in dealing with more complicated equations, you may find it useful to keep the following general approach in mind: If a given equation doesn't yield a solution easily, try converting it from exponential form to logarithmic form or vice versa, as the occasion warrants.

Of course, all of the well-known properties of exponents and their logarithmic counterparts are of great use too. For reference, the logarithmic properties that we have noted throughout Sections 7.3 and 7.4 are reproduced here, in brief form.

chapter seven   Exponential and Logarithmic Functions

Summary of Logarithmic Properties

1. The equations $x = a^y$ and $y = \log_a x$ are equivalent, and are, respectively, the exponential form and the logarithmic form of the same statement.
2. The inverse of the function $f(x) = a^x$ is $f^{-1}(x) = \log_a x$, and vice versa.
3. A consequence of the last point is that $\log_a(a^x) = x$ and $a^{\log_a x} = x$. In particular, $\log_a 1 = 0$ and $\log_a a = 1$.
4. $\log_a(xy) = \log_a x + \log_a y$ ("the log of a product is the sum of the logs")
5. $\log_a\left(\dfrac{x}{y}\right) = \log_a x - \log_a y$ ("the log of a quotient is the difference of the logs")
6. $\log_a(x^r) = r \log_a x$ ("the log of something raised to a power is the power times the log")

The next several examples illustrate typical uses of the properties, and the desirability of converting between the exponential and the logarithmic forms of an equation.

example 1

Solve the equation $3^{2-5x} = 11$. Express the answer exactly and as a decimal approximation.

Solution:

This problem is a good illustration of an exponential equation that is not easily solved in its original form (because the two sides of the equation do not have the same base), but is less complex as a logarithmic equation. The work on the next page begins with the conversion of the equation into logarithmic form and then proceeds with the solution of the resulting linear equation.

section 7.5 Exponential and Logarithmic Equations

$$3^{2-5x} = 11$$
$$\ln(3^{2-5x}) = \ln 11 \qquad \text{Take the log of both sides (any base).}$$
$$(2-5x)\ln 3 = \ln 11 \qquad \text{Bring the exponent down in front.}$$
$$2 - 5x = \frac{\ln 11}{\ln 3} \qquad \text{Solve the linear equation.}$$
$$-5x = \frac{\ln 11}{\ln 3} - 2$$
$$x = -\frac{\ln 11}{5\ln 3} + \frac{2}{5} \qquad \text{An exact form of the answer.}$$

The exact form of the answer is not unique. For one thing, a logarithm of any base may be used in the second line above, as any logarithm allows the variable to be brought down from the exponent. And if elegance is the only goal, using a logarithm with base 3 is undeniably best, as shown:

$$3^{2-5x} = 11$$
$$2 - 5x = \log_3 11$$
$$-5x = \log_3 11 - 2$$
$$x = \frac{2 - \log_3 11}{5}$$

But for most purposes, either the natural or common logarithm is an efficient choice, as the answer can then be approximated with a calculator:

$$x = -\frac{\ln 11}{5\ln 3} + \frac{2}{5} \approx -0.037$$

It can be verified with a calculator that this is indeed the solution.

example 2

Solve the equation $5^{3x-1} = 2^{x+3}$. Express the answer exactly and as a decimal approximation.

Solution:

As in the first example, taking a logarithm of both sides is the key. For variety, we will use the common logarithm this time, but the natural log would work as well.

continued on next page...

chapter seven Exponential and Logarithmic Functions

$$\log(5^{3x-1}) = \log(2^{x+3})$$ Take the log of both sides.
$$(3x-1)\log 5 = (x+3)\log 2$$ Use a property of logarithms.
$$3x\log 5 - \log 5 = x\log 2 + 3\log 2$$ Multiply out.
$$3x\log 5 - x\log 2 = 3\log 2 + \log 5$$ Collect terms with x on one side.
$$x(3\log 5 - \log 2) = \log 8 + \log 5$$ Factor out x.
$$x = \frac{\log 40}{\log 125 - \log 2}$$ Proceed to simplify the answer.
$$x = \frac{\log 40}{\log\left(\frac{125}{2}\right)} \approx 0.892$$

Again, the exact form of the answer could appear in many different forms, depending on the base of the logarithm chosen and the logarithmic properties used in simplifying the answer.

example 3

Solve the equation $\log_5 x = \log_5(2x+3) - \log_5(2x-3)$.

Solution:

This is an example of a logarithmic equation that is not easily solved in its logarithmic form. But once a few properties of logarithms have been exploited, the equation can be rewritten in a very familiar form.

$$\log_5 x = \log_5(2x+3) - \log_5(2x-3)$$
$$\log_5 x = \log_5\left(\frac{2x+3}{2x-3}\right)$$
$$x = \frac{2x+3}{2x-3}$$
$$x(2x-3) = 2x+3$$
$$2x^2 - 3x - 2x - 3 = 0$$
$$2x^2 - 5x - 3 = 0$$

section 7.5 Exponential and Logarithmic Equations

The second equation on the previous page is derived from the first by applying a property of logarithms, and the third equation can be obtained from the second either by applying the function $f(x) = 5^x$ to both sides (thus "undoing" $\log_5$) or by noting that logarithms are one-to-one functions, so $\log_5 A = \log_5 B \Rightarrow A = B$. The resulting rational equation is then turned into a quadratic equation, which we now proceed to solve.

$$2x^2 - 5x - 3 = 0$$

$$(2x + 1)(x - 3) = 0$$

$$x = -\frac{1}{2} \text{ or } x = 3$$

But and crucial step remains. While the two solutions above unquestionably solve the quadratic equation, remember that the goal is to solve the initial logarithmic equation. As we have seen with some other types of equations, the process of solving logarithmic equations can introduce extraneous solutions. If we check our two potential solutions in the original equation, we quickly discover that only one of them is valid.

$$\log_5\left(-\frac{1}{2}\right) \stackrel{?}{=} \log_5\left(2\left(-\frac{1}{2}\right) + 3\right) - \log_5\left(2\left(-\frac{1}{2}\right) - 3\right)$$

$-\frac{1}{2}$ is not a solution, as $\log_5\left(-\frac{1}{2}\right)$ is not defined.

$$\log_5(3) \stackrel{?}{=} \log_5(2(3) + 3) - \log_5(2(3) - 3)$$

$$\log_5 3 \stackrel{?}{=} \log_5 9 - \log_5 3$$

$$\log_5 3 \stackrel{?}{=} \log_5\left(\frac{9}{3}\right)$$

$\log_5 3 = \log_5 3$ The solution set of the equation is $\{3\}$.

The last example illustrates again the importance of verifying that potential solutions actually solve the original problem. In the case of logarithmic equations, always keep in mind that logarithms of non-positive numbers are not defined.

example 4

Solve the equation $\log_7(3x-2) = 2$.

Solution:

This equation is more quickly solved than the previous one. Once the equation is rewritten in its exponential form, the solution is apparent. We could also verify that the resulting solution actually solves the original equation.

$$\log_7(3x-2) = 2$$
$$3x - 2 = 7^2$$
$$3x = 51$$
$$x = 17$$

Topic 2: Further Applications of Exponential and Logarithmic Equations

We will conclude our discussion of exponential and logarithmic equations by revisiting some important applications.

example 5

Rita is saving up money for a down payment on a new car. She currently has $5500 but she knows she can get a loan at a lower interest rate if she can put down $6000. If she invests her $5500 in a money market account that earns 4.8% annually compounded monthly, how long will it take her to accumulate the $6000?

Solution:

We need to solve the compound interest formula, which in this case takes the form

$$6000 = 5500\left(1 + \frac{0.048}{12}\right)^{12t}.$$

section 7.5 Exponential and Logarithmic Equations

Since we are looking for a numerical value for t, the number of years Rita needs to invest her money, we should apply either the natural log or the common log to both sides of the equation, as follows.

$$\frac{6000}{5500} = \left(1 + \frac{0.048}{12}\right)^{12t}$$

$$\ln\left(\frac{6000}{5500}\right) = \ln\left(1 + \frac{0.048}{12}\right)^{12t}$$

$$\ln\left(\frac{6000}{5500}\right) = (12t)\ln(1.004)$$

$$\frac{\ln\left(\frac{6000}{5500}\right)}{12\ln(1.004)} = t$$

$$t \approx 1.82 \text{ years}$$

In other words, it will take a bit less than a year and 10 months for the $5500 to grow to $6000 (meaning that Rita will probably want to speed up the process by investing additional principal as she is able).

example 6

We have already discussed how the radioactive decay of carbon-14 can be used to arrive at age estimates of carbon-based fossils, and we constructed an exponential function describing the rate of decay. A much more common form of the function is

$$A(t) = A_0 e^{-0.000121t},$$

where $A(t)$ is the mass of carbon-14 remaining after t years, and A_0 is the mass of carbon-14 initially. Use this formula to determine:

a. The half-life of carbon-14 (that is, the number of years required for half of a given mass of carbon-14 to decay).
b. The number of years necessary for 2.3 grams of carbon-14 to decay to 1.5 grams.

continued on next page...

chapter seven   Exponential and Logarithmic Functions

Solutions:

a. $$\frac{A_0}{2} = A_0 e^{-0.000121t}$$

$$\ln\left(\frac{1}{2}\right) = -0.000121t$$

$$\frac{\ln\left(\frac{1}{2}\right)}{-0.000121} = t$$

$$t \approx 5728 \text{ years}$$

We are looking for the value of t for which $A(t)$ is half of A_0, so we are led to the first equation at left. As the base of the exponential function is e, the natural logarithm is the natural one to apply.

After solving for t, we have the desired half-life.

b. $$1.5 = 2.3 e^{-0.000121t}$$

$$\ln\left(\frac{1.5}{2.3}\right) = -0.000121t$$

$$t \approx 3533 \text{ years}$$

We would be led to solve the equation at left if it could be determined that a given organism had contained 2.3 grams of carbon-14 at death, and that it now contains 1.5 grams. This information would mean the organism died approximately 3533 years ago.

Topic 3: Interlude: Analysis of a Stock Market Investment

Exponential growth is a recurring topic in any discussion of investing and in any long-term observation of stock and bond markets. Claims of the relevance of exponential functions as models of market behavior range from the fairly simple and defensible to the very complex and speculative. At this point, we have learned enough algebra to answer one simple but fundamental question that investors should periodically ask themselves: How well (or poorly) are my investments performing?

We will answer this question as it applies to an investment technique that many Americans currently use, called *dollar-cost averaging*. The technique calls for an investor to commit P dollars on a periodic basis in a chosen stock, mutual fund, bond, or other investment vehicle. The amount P and the period of investment are, of course, up to the investor, but once chosen should remain fixed. For instance, an investor might choose to invest $100 on the first of every month in stock of a given company. The argument for dollar-cost averaging is that as the stock price fluctuates over time, the fixed $100 per month investment buys fewer high-priced shares in the company and more lower-priced shares, thus hopefully increasing the investor's profits when he or she eventually decides to sell the shares.

section 7.5   Exponential and Logarithmic Equations

One slight disadvantage of dollar-cost averaging is that it is not so easy to monitor the percentage gain (or loss) on the investment over time. The simplest sort of investment to analyze is a one-time investment; at any point it is possible to compare the current value of the investment to the purchase price and calculate the percentage change (even on an annual basis, if desired). And the performance of fixed periodic investments in a given stock is certainly harder to keep track of than, say, the performance of a savings account, which is often reported as an annual rate compounded monthly.

This points the way toward one possible solution. Suppose we invest P dollars on the first of January in Company XYZ, and suppose that one month later the value of our investment is A dollars. Then we could determine the annual rate r at which our investment has changed, assuming monthly compounding, by solving the equation

$$A = P\left(1 + \frac{r}{12}\right)$$

for r. We could then compare r to the rate for a typical monthly-compounded savings account to see how Company XYZ is doing. Of course, on the first of February we invest P dollars again. One month later our total investment has a new value A, and we need to solve the equation

$$A = \left(P\left(1 + \frac{r}{12}\right) + P\right)\left(1 + \frac{r}{12}\right) = P\left[\left(1 + \frac{r}{12}\right)^2 + \left(1 + \frac{r}{12}\right)\right]$$

for r if we want to compare the performance to a savings account again. In general, after n months (and just before our $(n+1)^{\text{th}}$ investment), we need to solve

$$A = P\left[\left(1 + \frac{r}{12}\right)^n + \left(1 + \frac{r}{12}\right)^{n-1} + \ldots + \left(1 + \frac{r}{12}\right)\right]$$

for r in order to keep track of our investment relative to a savings account.

This is a formidable task, even though at the moment we are only wondering if this can be done in theory, not in practice. Note that we are actually giving ourselves an nth degree polynomial equation in r to solve. To make this clearer, let $x = 1 + \frac{r}{12}$. Then the equation becomes

$$A = P\left(x^n + x^{n-1} + \ldots + x\right).$$

569

chapter seven Exponential and Logarithmic Functions

If we can solve this equation for x, we can then easily determine r. But from Chapter 5, we know that an nth degree polynomial equation has potentially n different solutions, only one of which has any meaning for us as far as this particular application is concerned. Is there any hope that we can identify that one solution?

Thankfully, yes. We will first make one more change to the equation, in order to eliminate the inconvenient dots. Notice that if we multiply both sides of the equation by $(x - 1)$ and expand the polynomial product, most of the terms on the right-hand side cancel:

$$A(x-1) = P(x^n + x^{n-1} + \ldots + x)(x-1)$$

$$Ax - A = P\left[\left(x^{n+1} + x^n + \ldots x^2\right) - \left(x^n + x^{n-1} + \ldots + x\right)\right]$$

$$Ax - A = P(x^{n+1} - x)$$

If we then put all the terms on one side, we obtain the final equation

$$x^{n+1} - \left(\frac{A}{P} + 1\right)x + \frac{A}{P} = 0.$$

This last equation has two changes of sign (note that $\frac{A}{P}$ is guaranteed to be positive), so Descartes' Rule of Signs tells us the equation has either 2 or 0 positive real solutions. Further, we already know a positive solution: $x = 1$ is certainly a solution, which shouldn't surprise us as we obtained the $(n + 1)^{\text{th}}$ degree equation from the nth degree equation by multiplying both sides by the factor $x - 1$. This means there must be exactly 2 positive real solutions, and the other positive solution is the one we care about.

Why do we only care about positive solutions? Note that if the r we are ultimately solving for is negative (an unpleasant but unfortunately real possibility), $x = 1 + \frac{r}{12}$ will be less than 1 but larger than 0. (The reason x can't be negative is for financial reasons, not mathematical reasons. The only way x could be negative is if r is so negative that the value A of our investment has actually fallen below 0, which can't happen.) If r is positive, then $x = 1 + \frac{r}{12}$ is larger than 1. In either event, the second positive real solution of our $(n + 1)^{\text{th}}$ degree polynomial equation leads to a value for r and an understanding of how the stock is performing.

section 7.5 Exponential and Logarithmic Equations

exercises

Solve the following exponential and logarithmic equations. When appropriate, write the answer as both an exact expression and as a decimal approximation (round your answer to the nearest hundredth). See Examples 1 through 4.

1. $3e^{5x} = 11$
2. $4^{3-2x} = 7$
3. $11^{\frac{3}{x}} = 10$
4. $8^{3x+2} = 7^{2x+3}$
5. $e^{15-3x} = 28$
6. $10^{\frac{5}{x}} = 150$
7. $10^{2x+5} = e$
8. $e^{8x+e} = 8^{ex+8}$
9. $6^{x-7} = 7$
10. $2e^{3x} = 145$
11. $2^{6-x} = 10$
12. $e^{3x-6} = 10^{x+2}$
13. $e^{-4x-2} = 12$
14. $10^{x-9} = 2001$
15. $e^{x-4} = 4^{\frac{2x}{3}}$
16. $5^{x-2} = 20$
17. $8^{x^2+1} = 23$
18. $3^{\frac{4}{x}} = 15$
19. $6^{3x-4} = 36^{2x+4}$
20. $81^x = 3^{2x+16}$
21. $e^{2x} = 14$
22. $e^{4x} = e^{3x+14}$
23. $5^{5x-7} = 10^{2x}$
24. $10^{6x} = 3^{3x+4}$
25. $\log_5 x = 3$
26. $\log_2 x = 4$
27. $\log x + \log(4x) = 2$
28. $\log_4 x^2 - \log_4 x = 2$
29. $\ln 2x - \ln 4 = 3$
30. $\ln 15x - \ln 3 = 6$
31. $\log_4(x-3) + \log_4 2 = 3$
32. $\log_3 24 - \log_3 4 = x - 5$
33. $\log_5(8x) - \log_5 3 = 2$
34. $e^{2\ln x} = 4\log 10$
35. $9^{\log_3 x} = 16$
36. $3\log_8 512^{x^2} = 36$
37. $\log_3 6x - 2\log_3 6x = 3$
38. $\ln(3e) = \log x$
39. $\ln 2^{4e^x} = \ln 16^e$
40. $\log(x-2) + \log(x+2) = 2$
41. $\log(x-3) + \log(x+3) = 4$
42. $\log_2(7x-4) = \log_2(16-3x)$

43. $\log_\pi(x-5) + \log_\pi(x+3) = \log_\pi(1-2x)$

44. $\log_3(x+3) + \log_3(x-5) = 2$ **45.** $\log x + \log(x-3) = 1$

46. $\log_7(3x+2) - \log_7 x = \log_7 4$ **47.** $\log_2 x + \log_2(x-7) = 3$

48. $\log_{12}(x-2) + \log_{12}(x-1) = 1$ **49.** $\log_3(x+1) - \log_3(x-4) = 2$

50. $\ln(x+1) + \ln(x-2) = \ln(x+6)$

51. $\log_4(x-3) + \log_4(x-2) = \log_4(x+1)$

52. $2\ln(x+3) = \ln(12x)$

53. $\log_5(x-1) + \log_5(x+4) = \log_5(x-5)$

54. $\log_{255}(2x+3) + \log_{255}(2x+1) = 1$ **55.** $\log_2(x-5) + \log_2(x+2) = 3$

56. $\log_6(x+1) + \log_6(x-4) = 2$ **57.** $\ln(x+2) + \ln(x) = 0$

58. $e^{2x} - 3e^x - 10 = 0$ (Hint: first solve for e^x.)

59. $2^{2x} - 12(2^x) + 32 = 0$ (Hint: first solve for 2^x.)

60. $e^{2x} + 2e^x - 8 = 0$ **61.** $3^{2x} - 12(3^x) + 27 = 0$

Using the properties of logarithmic functions, simplify the following functions as much as possible. Write each function as a single term with a coefficient of 1, if possible.

62. $f(x) = 0.5 \ln x^2$ **63.** $f(x) = 0.25 \log 16x^8$ **64.** $f(x) = 4 \ln \sqrt{5x}$

65. $f(x) = 8 \ln \sqrt[4]{3x}$ **66.** $f(x) = 3\ln e^x - 3$ **67.** $f(x) = 10^{2x \log 16}$

68. $f(x) = 2\ln x^3 + \ln x^6$ **69.** $f(x) = 2\ln x^3 - \ln x^6$ **70.** $f(x) = \ln(x^2 + x) - \ln x$

71. $f(x) = 2\ln 5^{x \log_{20}(2\sqrt{5})}$ **72.** $f(x) = e^{\ln(\log x^e - 1)}$ **73.** $f(x) = 2\ln 5^{\log_4 2}$

section 7.5   **Exponential and Logarithmic Equations**

Use your knowledge of exponential and logarithmic functions to answer the following questions. See Examples 5 and 6.

74. Assuming that there are currently 6 billion people on Earth and a growth rate of 1.9% per year, how long will it take for the Earth's population to reach 20 billion?

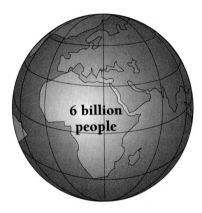

75. How long does it take for an investment to double in value if: **a.** the investment is in a monthly-compounded savings account earning 4% a year? **b.** the investment is in a continuously-compounded account earning 7% a year?

76. Assuming a half-life of 5728 years, how long would it take for 3 grams of carbon-14 to decay to 1 gram?

77. Suppose a population of bacteria in a Petri dish has a doubling time of one and a half hours. How long will it take for an initial population of 10,000 bacteria to reach 100,000?

78. According to Newton's Law of Cooling, the temperature $T(t)$ of a hot object, at time t after being placed in an environment with a constant temperature C, is given by $T(t) = C + (T_0 - C)e^{-kt}$, where T_0 is the temperature of the object at time $t = 0$ and k is a constant that depends on the object.

If a hot cup of coffee, initially at $190°F$, cools to $125°F$ in 5 minutes when placed in a room with a constant temperature of $75°F$, how long will it take for the coffee to reach $100°F$?

79. Wayne has $12,500 in a high interest savings account at 3.66% annual interest compounded monthly. Assuming he makes no deposits or withdrawals, how long will it take for his investment to grow to $15,000?

80. Ben and Casey both open money market accounts with 4.9% annual interest compounded continuously. Ben opens his account with $8700 while Casey opens her account with $3100. **a.** How long will it take Ben's account to reach $10,000? **b.** How long will it take Casey's account to reach $10,000? **c.** How much money will be in Ben's account after the time found in part b?

81. Cesium-137 has a half-life of approximately 30 years. How long would it take for 160 grams of cesium-137 to decay to 159 grams?

82. A chemist, running tests on an unknown sample from an illegal waste dump, isolates 50 grams of what he suspects is a radioactive element. In order to help identify the element, he would like to know its half-life. He determines that after 40 days only 44 grams of the original element remains. What is the half-life of this mystery element?

section 7.5 Exponential and Logarithmic Equations

technology exercises

The interlude of this section concludes with a polynomial equation of degree $n+1$, where n is the number of months during which an investor makes a fixed periodic investment. In practice, n is typically large enough to make solving the polynomial equation a nontrivial task, and a computer algebra system such as *Mathematica* is very useful.

Suppose, for instance, that an investor makes a periodic investment of $100.00 per month and has an accumulation of $1300.00 after one year. Since only $1200.00 was invested over the 12-month period, the investor clearly made money. But at what rate? As described in the Interlude, it would be nice to determine the equivalent annual rate of interest, compounded monthly, that would have generated the same return. The *Mathematica* commands below illustrate how this can be done. First the constants A and P, representing the total accumulation and the monthly investment respectively, are defined. Next, the command `NSolve` is used to solve the 13th degree polynomial equation. `NSolve` is used in the same way that `Solve` is used, but returns only numerical approximations to the solutions of the given equation and is a better choice in this case because of the high degree of the equation.

We know, from our study of polynomial functions in Chapter 5, that a 13th degree polynomial equation can have, potentially, 13 distinct solutions. The particular equation in this example does indeed have 13 different solutions, only three of which are real numbers. *Mathematica* returns a list of all 13 solutions, with the single negative solution listed first, the ten complex solutions next, and the two positive solutions listed last. As expected, $x = 1$ is one solution, and the second positive real solution is slightly larger than 1. When we use this second positive solution to solve for r, we obtain an annual monthly-compounded interest rate of 14.7%.

```
Section 7-5 Technology Exercise.nb

In[1]:= A = 1300
Out[1]= 1300

In[2]:= P = 100
Out[2]= 100

In[3]:= NSolve[x^13 - (A/P + 1)*x + A/P == 0, x]
Out[3]= {{x -> -1.30311}, {x -> -1.13687 - 0.630111 i}, {x -> -1.13687 + 0.630111 i},
        {x -> -0.683116 - 1.09374 i}, {x -> -0.683116 + 1.09374 i}, {x -> -0.064897 - 1.26924 i},
        {x -> -0.064897 + 1.26924 i}, {x -> 0.54846 - 1.11259 i}, {x -> 0.54846 + 1.11259 i},
        {x -> 0.981856 - 0.668899 i}, {x -> 0.981856 + 0.668899 i}, {x -> 1.}, {x -> 1.01225}}

In[4]:= Solve[1 + r/12 == 1.01225, r]
Out[4]= {{r -> 0.147}}
```

continued on next page...

chapter seven   Exponential and Logarithmic Functions

What if the accumulation at the end of one year had been $1100.00 instead of $1300.00? This means the investment lost money so the equivalent annual interest rate should be negative. The *Mathematica* output below confirms this; note that again there are only two positive real solutions to the equation, but this time we use the one slightly less than 1 to determine that $r = -16.2\%$.

```
Section 7-5 Technology Exercise b.nb

In[1]:= A = 1100
Out[1]= 1100

In[2]:= P = 100
Out[2]= 100

In[3]:= NSolve[x^13 - (A/P + 1)*x + A/P == 0, x]
Out[3]= {{x -> -1.28648}, {x -> -1.12236 - 0.622068 i}, {x -> -1.12236 + 0.622068 i},
         {x -> -0.674394 - 1.07978 i}, {x -> -0.674394 + 1.07978 i}, {x -> -0.064066 - 1.25304 i},
         {x -> -0.064066 + 1.25304 i}, {x -> 0.541462 - 1.09839 i}, {x -> 0.541462 + 1.09839 i},
         {x -> 0.969328 - 0.660355 i}, {x -> 0.969328 + 0.660355 i}, {x -> 0.986537}, {x -> 1.}}

In[4]:= Solve[1 + r/12 == 0.986537, r]
Out[4]= {{r -> -0.161556}}
```

Use a computer algebra system, if available, to determine the equivalent monthly-compounded interest rate in the following scenarios. In each problem, P represents the principal invested each month and A is the accumulated value at the end of n months.

83. $n = 24$, $P = \$50.00$, and $A = \$1275.00$
84. $n = 120$, $P = \$100.00$, and $A = \$15{,}000.00$
85. $n = 12$, $P = \$50.00$, and $A = \$590.00$
86. $n = 24$, $P = \$75.00$, and $A = \$2000.00$

chapter seven project

Exponential Functions

Computer viruses have cost U.S. companies billions of dollars in damages and lost revenues over the last few years. One factor that makes computer viruses so devastating is the rate at which they spread. A virus can potentially spread across the world in a matter of hours depending on its characteristics and whom it attacks.

Consider the growth of the following virus. A new virus has been created and is distributed to 100 computers in a company via a corporate e-mail. From these workstations the virus continues to spread. Let $t = 0$ be the time of the first 100 infections, and at $t = 17$ minutes the population of infected computers grows to 200. Assume the anti-virus companies are not able to identify the virus or slow its progress for 24 hours, allowing the virus to grow exponentially.

1. What will the population of the infected computers be after 1 hour?
2. What will the population be after 1 hour 30 minutes?
3. What will the population be after a full 24 hours?

Suppose another virus is developed and released on the same 100 computers. This virus grows according to $P(t) = (100)2^{\frac{t}{2}}$, where t represents the number of hours from the time of introduction.

4. What is the doubling time for this virus?
5. How long will it take for the virus to infect 2000 computers, according to this model?

chapter seven — Exponential and Logarithmic Functions

CHAPTER SUMMARY

A summary of concepts and skills follows each chapter. Refer to these summaries to make sure you feel comfortable with the material in the chapter. The concepts and skills are organized according to the section title and topic title in which the material is first discussed.

7.1 Exponential Functions and Their Graphs

topics	pages
Definition and Classification of Exponential Functions • The definition of *exponential functions* • The two classes of exponential functions, and their behavior	p. 507 – 509
Graphing Exponential Functions • Graphing transformations of exponential functions	p. 510 – 511
Solving Elementary Exponential Equations • The definition of an exponential equation • Using the one-to-one nature of exponential functions to solve simple exponential equations	p. 511 – 513

7.2 Applications of Exponential Functions

topics	pages
Models of Population Growth • Modeling population growth with exponential functions • Using population information to construct a corresponding exponential function	p. 517 – 518
Radioactive Decay • Modeling radioactive decay with exponential functions • The use of exponential functions in *radiocarbon* dating • Using half-life measurements and other data to construct exponential functions that model radioactive decay	p. 519 – 520

chapter seven summary

topics (continued)

Compound Interest and the Number e — p. 521 – 525
- The meaning of *compounding* in investment contexts
- The use of the basic *compound interest formula*
- Extending the compound interest formula to the *continuous compounding* case
- The meaning of the *natural base e*, and its approximate value

7.3 Logarithmic Functions and Their Graphs

topics

Definition of Logarithmic Functions — p. 533 – 535
- Understanding the need for *logarithmic functions*
- The connection between a^x and $\log_a x$

Graphing Logarithmic Functions — p. 535 – 537
- The two classes of logarithmic functions, and their behavior

Evaluating Elementary Logarithmic Expressions — p. 538 – 540
- Evaluating simple logarithmic expressions manually

Common and Natural Logarithms — p. 540 – 541
- The definition of the *common* and *natural* logarithms
- Using a calculator to evaluate common and natural logarithms

7.4 Properties and Applications of Logarithms

topics

Properties of Logarithms — p. 545 – 550
- Basic properties of logarithmic functions, and how they are derived from properties of exponents
- Using the properties of logarithms to expand a single complicated logarithmic expression into a sum of simpler logarithmic expressions
- Using the properties of logarithms to combine sums of logarithmic expressions into a single logarithmic expression

chapter seven — Exponential and Logarithmic Functions

topics (continued)	pages
The Change of Base Formula • The derivation of the *change of base formula* and its use	p. 550 – 551
Applications of Logarithmic Functions • The historical use of logarithms in simplifying computation • The use of logarithms in defining pH, earthquake intensity, and noise levels	p. 552 – 555

7.5 Exponential and Logarithmic Equations

topics	pages
Converting Between Exponential and Logarithmic Forms • Using the definition of logarithms and their properties to convert between the exponential and logarithmic forms of a given equation	p. 561 – 566
Further Applications of Exponential and Logarithmic Equations • Using logarithms to solve exponential equations	p. 566 – 568
Interlude: Analysis of a Stock Market Investment • The meaning of *dollar-cost averaging*, and how to determine the equivalent annual rate of interest for an investment	p. 568 – 570

chapter seven review

chapter review

Section 7.1

Sketch the graphs of the following functions.

1. $f(x) = \left(\dfrac{1}{2}\right)^{x-1} + 3$

2. $r(x) = 2^{-x+4} - 2$

Solve the following elementary exponential equations.

3. $3^{3x-5} = 81$

4. $\left(\dfrac{2}{5}\right)^{-4x} = \left(\dfrac{25}{4}\right)^{x-1}$

5. $10{,}000^x = 10^{-2x-1}$

Section 7.2

Solve the following problems involving exponential functions.

6. Melissa has recently inherited $15,000 which she wants to deposit into a savings account for 10 years. She has determined that her two best bets are an account that compounds annually at a rate of 3.95% and an account that compounds continuously at an annual rate of 3.85%. Which account would pay Melissa more interest?

7. Bill has come upon a 37 gram sample of iodine-131. He isolates the sample and waits for 2 weeks. After this time period, only 11 grams of iodine-131 remains. What is the half-life of this isotope?

8. Katherine is working in a lab testing bacteria populations. Starting out with a population of 870 bacteria, she notices that the population doubles every 22 minutes. Find **a.** the equation for the population P in terms of time t in minutes, and **b.** the time it would take for the population to reach 7500 bacteria.

Section 7.3

Evaluate the following logarithmic expressions without the use of a calculator.

9. $\log_{27} 9^{\log 1000}$

10. $\log_{\frac{1}{3}} 9$

11. $\log_4 \left(\dfrac{1}{64}\right)$

Use the elementary properties of logarithms to solve the following equations.

12. $\log_6 6^{\log_5 x} = 3$

13. $\log_9 x^{\frac{1}{2}} = \frac{3}{4}$

14. $\log_x \left(\log_{\frac{1}{2}} \frac{1}{16} \right) = 2$

Section 7.4

Use properties of logarithms to expand the following expressions as much as possible. Simplify any numerical expressions that can be evaluated without a calculator.

15. $\log \sqrt{\dfrac{x^3}{4\pi^5}}$

16. $\ln \left(\dfrac{\sqrt{a^5 mn^2}}{e^5} \right)$

Use properties of logarithms to condense the following expressions as much as possible, writing each answer as a single term with a coefficient of 1.

17. $\dfrac{1}{3} \left(\log_2 (a^5) - \log_2 (bc^3) \right)$

18. $\ln 4 - \ln x^2 - 7 \ln y$

Section 7.5

Solve the following exponential and logarithmic equations. When appropriate, write the answer as both an exact expression and as a decimal approximation.

19. $e^{8-5x} = 16$

20. $10^{\frac{6}{x}} = 321$

21. $\ln(x+1) + \ln(x-1) = \ln(x+5)$

22. $\log_2 (x+3) + \log_2 (x+4) = \log_2 (3x+8)$

Use your knowledge of exponential and logarithmic functions to answer the following questions.

23. Rick puts $6500 in a high interest money market account at 4.36% annual interest compounded monthly. Assuming he makes no deposits or withdrawals, how long will it take for his investment to grow to $7000?

24. Sodium-24 has a half-life of approximately 15 hours. How long would it take for 350 grams of sodium-24 to decay to 12 grams?

chapter seven test

chapter test

Sketch the graphs of the following functions.

1. $f(x) = 3^x$
2. $g(x) = 2^{-x}$
3. $f(x) = 1 - 2^{-x}$
4. $f(x) = 4^{1-2x}$

Solve the following exponential equations.

5. $5^x = \dfrac{1}{625}$
6. $9^{x-1} = 27^{-x+2}$
7. $2^{-x} + 1 = 17$
8. $\left(\dfrac{1}{3}\right)^{x-1} = 81^{\frac{1}{2}}$
9. $\left(\dfrac{2}{3}\right)^{x-3} = \left(\dfrac{9}{4}\right)^{x+1}$
10. $5^{3x-6} = 1$

11. Suppose you invest a total of $12,000 at an annual interest rate of 9%. Find the balance to the nearest dollar after 5 years if it is compounded:

    a. Quarterly
    b. Continuously

12. The number of fruit flies in an experimental population after t hours is given by $Q(t) = 20e^{0.03t}$, $t \geq 0$.

    a. How large is the population of fruit flies after 72 hours?
    b. Find the initial number of fruit flies in the population.

13. In 1986 a nuclear reactor accident occured in Chernobyl in what was then the Soviet Union. The explosion spread radioactive chemicals over hundreds of square miles and the government evacuated the city and surrounding areas. To see why the city is now uninhabited, consider the following model.

    $$P = 10e^{-0.00002845t}$$

    This model represents the amount of plutonium that remains (from an initial amount of 10 pounds) after t years. Sketch the graph of this function over the interval from $t = 0$ to $t = 100{,}000$. How much of the 10 pounds will remain after 100,000 years?

14. Let Q represent a mass, in grams. of carbon-14 (^{14}C) whose half-life is 5730 years. The quantity of carbon 14 after t years is given by $Q = 10\left(\dfrac{1}{2}\right)^{\frac{t}{5730}}$.

    a. Determine the initial quantity (when $t = 0$).
    b. Determine the quantity present after 2000 years.

chapter seven   Exponential and Logarithmic Functions

Write the following equations in logarithmic terms.

15. $125 = 5^x$

16. $3^x = 8$

17. $e^x = 3\pi$

18. $(3a)^{\sqrt{2}} = 10$

Write the following logarithmic equations as exponential equations.

19. $\log_4 64 = x$

20. $\log_3 \dfrac{1}{27} = -3$

21. $\ln 6x = 2$

22. $\log_5 a = 10$

Sketch the graphs of the following functions.

23. $f(x) = \log_2(x-1)$

24. $f(x) = -\log_2 x + 1$

25. $f(x) = -\log_3(-x)$

26. $f(x) = \log_{\frac{1}{2}} x$

Evaluate the following logarithmic expressions or equations.

27. $\log_{\frac{1}{2}} 8$

28. $\log_{\sqrt{3}} \dfrac{1}{3}$

29. $\log_2 [\log_3 81]$

30. $\ln \sqrt[3]{e^2}$

31. $\log 0.0001$

32. $\log_3 x = \dfrac{-1}{2}$

33. $\log_4 16^{2x} = -1$

34. $\log_2 [\log_3(x+1)] = 1$

35. $3^{\log_2(x-1)} = 0$

36. $\log_b b^{\log_3 x} = \dfrac{1}{3}$

Use the properties of logarithms to expand the following expressions as much as possible. Simplify any numerical expressions.

37. $\log_3(27a^3)$

38. $\ln\left(\dfrac{2a^2 b}{3c}\right)$

39. $\log_4(12x + 20y)$

40. $\ln\left(\ln e^{2ex}\right)$

Use the properties of logarithms to condense the following expressions as much as possible, writing each answer as a single term with a coefficient of 1.

41. $\log x + \log y - \log c$

42. $\log_2(x^2 - 9) - \log_2(x+3)$

43. $2\log a + 3\log b - \dfrac{1}{2}\log c - \log d$

44. $\log_3(x-2) + \log_2 x - \log_2(x^2 + 4)$

45. $1 + \log_5 x - \log_5 y - \log_5(y-1) + \dfrac{1}{3}\log_5 8$

chapter seven test

Solve the following exponential and logarithmic equations, if possible.

46. $2e^{3x} = 5$

47. $5^{1-2x} = 7$

48. $9^{2x-1} = 4^{x-2}$

49. $e^{5x-2} = 40$

50. $e^{x-2} = 10^{2x+1}$

51. $2^{\frac{3}{x}} = 5$

52. $\log_4(2x-1) = 2$

53. $\log_5(x-2) + \log_5 2 = 1$

54. $\log_3 4x - \log_3(x+1) = 2$

55. $\log_{\sqrt{3}} x + \log_{\sqrt{3}} 2x = 4$

56. $4^x - 2^x = 12$

57. $(\log_2 x)^2 - 2(\log_2 x) - 8 = 0$

58. $e^{2x} + e^x = 2$

59. $3^{2\log_3(x-1)} = 81$

60. $\log(x-1) = \log(x-2) - \log(x+2)$

61. $1 - 2\log_{10}(x+1) = -1$

62. $x\log_2 x + x - 2\log_2 x - 2 = 0$

63. The relationship between the number of decibels D and the intensity of sound I in watts per square meter is given by $D = 10 \cdot \log_{10}\left(\dfrac{I}{10^{-12}}\right)$. Determine the intensity of sound in watts per square meter if the decibel level is 125.

64. On the Richter scale, the magnitude R on a earthquake of intensity I is given by $R = \log_{10} \dfrac{I}{I_0}$, where $I_0 = 1$ is the minimium intensity used for comparision. Find the intensity per unit of area for the following values of R.

   a. $R = 8.4$ **b.** $R = 6.85$ **c.** $R = 9.1$

Find the inverses of each of the following functions.

65. $f(x) = 3^{2x} - 1$

66. $f(x) = e^{x-2} + 3$

67. $f(x) = 2\ln(x+1) - 1$

68. $f(x) = 3\log_2(x-3)$

chapter EIGHT

OVERVIEW

8.1 Solving Systems by Substitution and Elimination

8.2 Matrix Notation and Gaussian Elimination

8.3 Determinants and Cramer's Rule

8.4 The Algebra of Matrices

8.5 Inverses of Matrices

8.6 Linear Programming

8.7 Nonlinear Systems of Equations

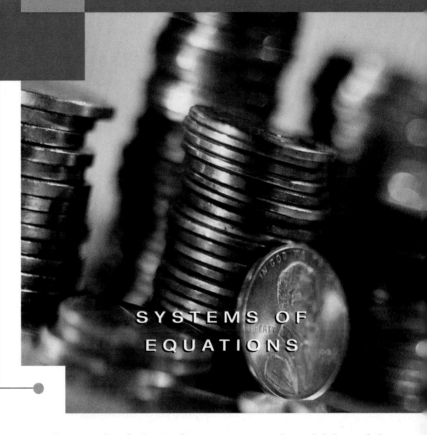

SYSTEMS OF EQUATIONS

By the end of this chapter you should be able to solve both linear and nonlinear systems of equations. Given two or more equations in the same variables, you will find solutions for the variables involved that solve all the equations simultaneously. On page 603 you will find a problem involving an application of such a system in which a variety of loose coins is found. You will be given the total value of the coins and asked to calculate how many of the total number are nickels, and how many are pennies. You will master this type of problem, using tools such as the method for *Solving Systems by Elimination* as shown in *Example 3* on page 594.

chapter eight   Systems of Equations

Introduction

In this chapter, we return to the study of linear equations, arguably the simplest class of equations. But as we will soon see, our understanding of linear equations has room for growth in many directions. In fact, the material in this chapter serves as a good illustration of how, given incentive and opportunity, mathematicians extend ideas and techniques beyond the familiar.

In the particular case of linear equations, this extension has taken the form of:

1. Considering equations containing more variables (we have already seen the first step in this process when we moved from linear equations in one variable to linear equations in two).

2. Trying to find solutions that satisfy more than one linear equation at a time; such sets of equations are called *systems*.

3. Making use of elementary methods to solve systems if possible, and developing entirely new methods if necessary or desirable.

The third point above constitutes the bulk of this chapter. We start off, in our quest to solve systems of linear equations, by using some intuitive methods that work admirably if the equations contain only a few variables and if the system has only a few equations. Several centuries ago, however, mathematicians began to realize the limitations of such simple methods. As the problems that people wanted to solve led to systems of many variables and equations, refinements were made to the elementary methods of solution, and eventually, entirely new techniques were developed. The two refinements we will study are called *Gaussian elimination* and *Gauss-Jordan elimination*. Some of the new techniques that were developed for solving large systems of equations make use of *determinants of matrices* and *matrix inverses*, two concepts that can serve as an introduction to higher mathematics.

Gauss

The notion of the inverse of a matrix (a *matrix* is a rectangular array of numbers) closes out our discussion of systems of linear equations, but in fact it brings us full circle and allows us to write systems of linear equations as single matrix equations. This illustrates how mathematicians, in extending our reach in terms of problem-solving and in constructing new mathematics, are guided by the achievements of the past.

8.1 Solving Systems by Substitution and Elimination

TOPICS

1. Definition and classification of linear systems of equations
2. Solving systems by substitution
3. Solving systems by elimination
4. Larger systems of equations
5. Applications of systems of equations

Topic 1: Definition and Classification of Linear Systems of Equations

Many problems of a mathematical nature are most naturally described by two or more equations in two or more variables. When the equations are all linear, such a collection of equations is called a *linear system of equations*, or sometimes *simultaneous linear equations*. The word *simultaneous* refers to the goal of identifying all the associated values – if any – of all the variables that solve all of the equations simultaneously.

We have some incidental familiarity with linear systems of equations from our work in graphing solutions of multiple linear inequalities in Section 3.5. In that section, we were concerned with identifying the regions of the plane that simultaneously satisfy collections of inequalities joined by the word "and" or the word "or," but the boundary line for the solution set of a given linear inequality in two variables is the graph of a linear equation, so we were effectively graphing solutions to systems of equations at the same time. This point is illustrated by the graphs in Figure 1, which pictorially identify the three possible varieties of systems of two linear equations in two variables.

chapter eight   Systems of Equations

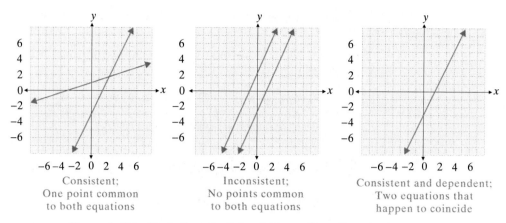

Figure 1: The Varieties of a Pair of Linear Equations in the Plane

In the first graph of Figure 1, the two lines intersect in exactly one point, and the ordered pair associated with that point is consequently a solution to both linear equations at the same time (of course, each equation individually has an infinite number of solutions). In the second graph, the two lines are parallel, and the corresponding system of equations has no solution as there is no point lying on both lines. In the final graph, the two lines actually coincide and appear as one, meaning that any ordered pair that solves one of the equations in the system solves the other as well, so the system has an infinite number of solutions.

We will soon encounter systems consisting of more than two equations and/or more than two variables, but larger systems will have one important similarity to the two-variable, two-equation systems in Figure 1: every linear system will have exactly one solution, no solution, or an infinite number of solutions. This fact is worthy of a definition.

Varieties of Linear Systems

- A linear system of equations with no solution is said to be **inconsistent**.
- A linear system of equations that has at least one solution is, naturally enough, called **consistent**. Any linear system with more than one solution must have an infinite number, and is said to be **dependent**.

section 8.1   Solving Systems by Substitution and Elimination

Of course, the goal is to develop a systematic and effective method of solving linear systems. Because linear systems of equations arise in so many different contexts and are of great importance both theoretically and practically, many solution methods have been devised. In fact, the foundation of the branch of mathematics called *linear algebra* is the study of linear systems and their solutions. In this text, we will use solution methods that fall into four broad classes, the first two of which are fairly easy to apply when the number of equations and variables is small.

Topic 2: ## Solving Systems by Substitution

The solution method of substitution hinges on solving one equation in a system for one of the variables, and substituting the result for that variable in the remaining equations. This can be a time-consuming process if the system is large (meaning more than a few equations and more than a few variables), and in fact the task may have to be repeated many times. But it is a very natural method to use for some small systems, as illustrated in *Example 1*.

example 1

Use the method of substitution to solve the system $\begin{cases} 2x - y = 1 \\ x + y = 5 \end{cases}$.

Solution:

Either equation can be solved for either variable, and the choice doesn't affect the final answer. One option is to solve the second equation for x. In the work that follows, we do exactly that, and then substitute the result in the first equation, giving us one equation in the variable y. Once we have solved for y, we can then go back and determine x.

$x = -y + 5$ Solve the second equation for x.
$2(-y + 5) - y = 1$ Substitute the result in the first equation.
$-2y + 10 - y = 1$ Distribute 2.
$-3y = -9$ Proceed to solve for y.
$y = 3$, so $x = -3 + 5$ Use y to then solve for x.
Solution is $(2, 3)$ The ordered pair $(2, 3)$ solves both equations.

Notice that the y-values are both "3" at $x=2$ after entering the two equations in [Y=] and selecting [2ND] [GRAPH] (Table).

Although it is not a necessary step in the solution process, it is worthwhile to graph a few systems and their solutions. Note how the following graph corresponds to the system and the ordered pair that we have found.

continued on next page...

chapter eight   Systems of Equations

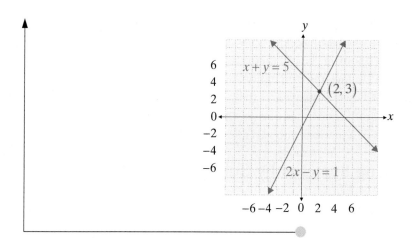

technology note

Computer software and graphing calculators are convenient for graphing systems of equations, and can be used to quickly get a sense of where (or whether) to expect a solution. In *Mathematica*, any number of equations (linear or otherwise) may be plotted on the same graph. The first step is to execute the command `<<Graphics`ImplicitPlot`` as shown in the figure below (execute this command just once per *Mathematica* session). The syntax for plotting the two equations $4x - y = 3$ and $x + 3y = 2$ is then shown in the second command. The part reading `{x,-5,5}` tells *Mathematica* to display the portion of the plane from –5 to 5 on the *x*-axis, and the part reading `{y,-5,5}` does the same thing for the *y*-axis.

Remember to first solve for *y* before entering the equations in [Y=] and selecting [GRAPH].

592

section 8.1   Solving Systems by Substitution and Elimination

As Figure 1 illustrates, though, a single solution to a system is just one of three possible outcomes. A natural question, therefore, is: How does the solution process differ when the system is inconsistent or dependent?

example 2

Use the method of substitution to solve the system $\begin{cases} -2x+6y=6 \\ x+3=3y \end{cases}$.

Solution:

$$x = 3y - 3 \quad \text{Solve the second equation for } x.$$
$$-2(3y-3) + 6y = 6 \quad \text{Substitute the result in the first equation.}$$
$$-6y + 6 + 6y = 6 \quad \text{Distribute } -2 \text{ and proceed to simplify.}$$
$$0 = 0 \quad \text{True, and trivial, statement.}$$

The last equation, obtained from equation 1 of the system after replacing x with (according to equation 2) an equivalent expression, is always true. Another way of saying this is that for any value of y, the corresponding value $3y - 3$ for x results in an ordered pair (x, y) that solves both equations. Since there are an infinite number of solutions, then, the system is dependent; in the two-variable, two-equation case, this means the graphs of the two equations coincide exactly. In fact, the first equation is simply a rearranged multiple of the second equation.

The question of how to describe the infinite number of solutions remains. One way is with a picture: since the graphs of the two equations are the same line in the plane, the graph of either one represents all the ordered pairs that solve the system. A more accurate method is to describe, with a formula, all the ordered pairs that make up the solution set. From the discussion above, we know one such description is

$$\{(3y-3, y) \mid y \in \mathbb{R}\},$$

or simply $\{(3y-3, y)\}$. If we had solved either equation for y instead of x, we would have obtained the alternative but equivalent set

$$\left\{ \left(x, \frac{x+3}{3}\right) \;\middle|\; x \in \mathbb{R} \right\}.$$

We will soon encounter an example of the third possibility, a system with no solution.

chapter eight   Systems of Equations

Topic 3:   Solving Systems by Elimination

In some systems, the expressions obtained by solving one equation for one variable are all somewhat unwieldy. (In practice, this often means that awkward fractions appear, no matter which variable is solved for.) In such cases, the solution method of elimination may be a more efficient choice. The elimination method is often a better choice for larger systems as well.

The method of elimination is based on the goal of eliminating one variable in one equation by adding two equations together, and in fact, some texts call this the *addition method*. The phrase "adding two equations together" means making use of the fact that if $A = B$ and $C = D$, then $A + C = B + D$. If the equation $A + C = B + D$ has fewer variables than either of the original two, then progress has been made. In fact, if the system is of the two-variable, two-equation variety, the new equation is ready to be solved for its remaining variable, and the solution of the system is then straightforward to find. The following examples illustrate the use of this method.

example 3

Use the method of elimination to solve the system $\begin{cases} 5x + 3y = -7 \\ 7x - 6y = -20 \end{cases}$.

Solution:

Although the method of substitution could indeed be used, solving for either x or y in either equation introduces some fractions that it would be nice to avoid, if possible. Note, though, that the coefficient of y in the second equation is an integer multiple of the coefficient of y in the first. In particular, if we multiply all the terms in the first equation by 2, the coefficients of y will be negatives of one another; adding the two equations will then produce a third equation in which y does not appear.

In order to keep track of the steps we use, it is helpful to annotate our work with reminders of how, exactly, we are modifying the appearance of the system. Each step must result, of course, in an equivalent system: we are not changing the problem, only the way it looks. (The ultimate goal is to change the appearance of the system so that we can "read off" the answer.) In the work that follows, the arrow and notation in teal indicates that we have modified the system by multiplying each term in equation 1 by the constant 2.

section 8.1 Solving Systems by Substitution and Elimination

Notice that the y-values are both "1" at $x=-2$ after entering the two equations in and selecting **2ND** **GRAPH** (Table).

$$\begin{cases} 5x + 3y = -7 \\ 7x - 6y = -20 \end{cases} \xrightarrow{2E_1} \begin{cases} 10x + 6y = -14 \\ 7x - 6y = -20 \end{cases}$$

If we now add the left-hand sides of the two equations, and equate this with the sum of the right-hand sides, we get

$$17x = -34,$$
$$\text{or } x = -2.$$

We can now return to any of the other equations to determine y. For instance, equation 1 of the original system tells us that $5(-2) + 3y = -7$, or $y = 1$. The ordered pair $(-2, 1)$ is thus the solution of the system. (Note that using the second equation gives the same y-value and is a good way to check our work.)

example 4

Use the method of elimination to solve the system $\begin{cases} 2x - 3y = 3 \\ 3x - \dfrac{9}{2}y = 5 \end{cases}$.

Solution:

Instead of multiplying just one equation by a constant, it is probably easier to eliminate x from the system by modifying both equations, as shown:

$$\begin{cases} 2x - 3y = 3 \\ 3x - \dfrac{9}{2}y = 5 \end{cases} \xrightarrow[-2E_2]{3E_1} \begin{cases} 6x - 9y = 9 \\ -6x + 9y = -10 \end{cases}$$

Although the intent of the above operation was to obtain coefficients of x that were negatives of one another, we have achieved the same thing for y. If we now add the equations, the result is $0 = -1$, a false statement. This means that no ordered pair solves both equations, and the system is inconsistent. Graphically, the two lines defined by the equations are parallel.

chapter eight   Systems of Equations

Topic 4:   Larger Systems of Equations

Algebraically, larger systems of equations can be dealt with much the same way as the two-variable, two-equation systems that we have studied (though the number of steps needed to obtain a solution might increase). Geometrically, however, larger systems mean something quite different.

For example, if an equation contains three variables, say x, y, and z, a given solution of the equation must consist of an *ordered triple* of numbers, not an ordered pair. A graphical representation of the ordered triple requires three coordinate axes, as opposed to two. This leads to the concept of three-dimensional space and to the association of points in space with ordered triples. Figure 2 is an illustration of the way in which the positive x, y, and z axes are typically represented on a two-dimensional surface, such as a piece of paper, a computer monitor, or a blackboard. The negative portion of each of the axes is not drawn, and the three axes meet at the origin (the point with $(0, 0, 0)$ as its coordinates) at mutual right angles. As an illustration of how ordered triples appear in $\mathbb{R}^3$, the point $(1, 2, 4)$ is plotted (the thin colored lines are drawn merely for reference and are not part of the plot).

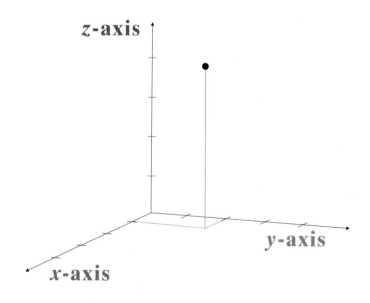

Figure 2: A Depiction of $\mathbb{R}^3$ and the Point $(1, 2, 4)$

section 8.1 Solving Systems by Substitution and Elimination

For our purposes, we are interested in how the graph of a linear equation in three variables appears in $\mathbb{R}^3$. The answer is that any equation of the form

$$Ax + By + Cz = D,$$

where not all of A, B, and C are 0, depicts a plane in three-dimensional space. A linear system of equations in three variables will thus describe a collection of planes in $\mathbb{R}^3$, one plane per equation. If a linear system of three variables contains three equations, then, it is possible for the three planes to intersect in exactly one point. It is also possible, however, for two of the planes to intersect in a line while the third plane contains no point of that line, for the three planes to intersect in a common line, for two or three of the planes to be parallel, or for two or three of the planes to coincide. In other words, the addition of another variable and another equation leads to many more possibilities than we have seen thus far. However, it is still the case that a linear system has no solution, exactly one solution, or an infinite number of solutions. Figure 3 illustrates some of the possible configurations of a three-variable, three-equation linear system.

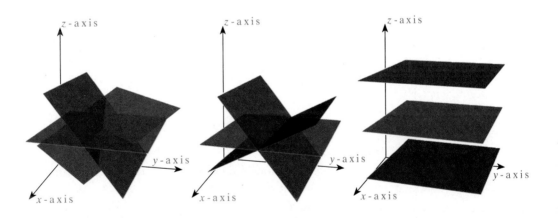

Figure 3: Some Possible 3-Variable, 3-Equation Systems

chapter eight   Systems of Equations

In *Example 5*, we will use the method of elimination to find the solution of a linear system in three variables.

example 5

Solve the system $\begin{cases} 3x - 5y + z = -10 \\ -x + 2y - 3z = -7 \\ x - y - 5z = -24 \end{cases}$.

Solution:

In general, a good approach to solving a large system of equations is to try to eliminate a variable and obtain a smaller system. In this problem, we know how to handle any 2-variable system that results from eliminating one of the original three variables.

There are many possible ways to proceed. One option is to use the second equation (or a multiple of it) to eliminate x when we add it to the first and third equations. This is what we have done below in order to obtain the two equations in teal.

$$\begin{cases} 3x - 5y + z = -10 \\ -x + 2y - 3z = -7 \end{cases} \xrightarrow{3E_2} \begin{cases} 3x - 5y + z = -10 \\ -3x + 6y - 9z = -21 \end{cases}$$
$$\overline{\quad y - 8z = -31\quad}$$

$$\begin{cases} -x + 2y - 3z = -7 \\ x - y - 5z = -24 \end{cases}$$
$$\overline{\quad y - 8z = -31\quad}$$

The two new equations are identical, and tell us that if an ordered triple (x, y, z) is to solve the system, it must be the case that $y = 8z - 31$. We can now use any equation that contains x to determine the relation between x and z. For instance, the third equation in the system tells us that $x = y + 5z - 24$, or

$$x = (8z - 31) + 5z - 24 = 13z - 55.$$

One description of the solution set is thus $\{(13z - 55, 8z - 31, z) \mid z \in \mathbb{R}\}$. This is a one-dimensional figure in $\mathbb{R}^3$, indicating that the three planes in the system intersect in a line.

We will soon see an example of a 3-variable system whose solution is a single point.

section 8.1   Solving Systems by Substitution and Elimination

technology note

Computer software and many calculators are able to solve systems of equations, and can be used to check your understanding of the material in this chapter. The command in *Mathematica* for solving a system of equations is the **Solve** command, as illustrated below. In the first instance, *Mathematica* has been used to verify our solution from *Example 5*; the warning in blue indicates that one of the variables (z in this case) is free to take on any value. In the second instance, a very similar system of equations has a solution consisting of the single point $(-2, 3, 1)$.

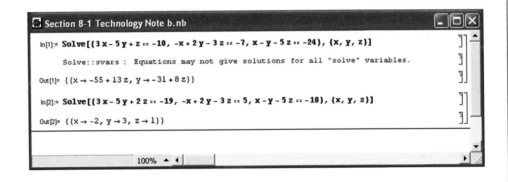

Topic 5: Applications of Systems of Equations

Many applications that we have previously analyzed mathematically are more naturally stated in terms of two or more equations. Consider, for example, the following mixture problem.

example 6

A foundry needs to produce 75 tons of an alloy that is 34% copper. It has supplies of 9% copper alloy and 84% copper alloy. How many tons of each must be mixed to obtain the desired result?

Solution:

To solve this problem without using a system of equations, it is necessary to arrive at one equation in one variable. But the work in doing this involves, essentially, stating the problem in system form.

continued on next page...

If we let x represent the number of tons of 9% copper alloy needed, and y the number of tons of 84% copper alloy needed, one immediate fact is that $x + y = 75$; that is, we require 75 tons of alloy in the end. Another fact is that 9% of the x tons and 84% of the y tons represent the total amount of copper, and this amount must equal 34% of 75 tons. The second equation is thus $0.09x + 0.84y = 0.34(75)$. This gives us a system that can be solved by elimination:

$$\begin{cases} x+y=75 \\ 0.09x+0.84y=0.34(75) \end{cases} \xrightarrow[100E_2]{-9E_1} \begin{cases} -9x-9y=-675 \\ 9x+84y=2550 \end{cases}$$

Adding these two equations gives us the equation $75y = 1875$, or $y = 25$. From this, we deduce that $x = 50$, so 50 tons of 9% alloy and 25 tons of 84% alloy are needed.

The last example involves three variables and is definitely most easily stated and solved in system form.

example 7

If the ages of three girls, Xenia, Yolanda, and Zsa Zsa, are added, the result is 30. The sum of Xenia's and Yolanda's ages is Zsa Zsa's age, while Xenia's age subtracted from Yolanda's is half of Zsa Zsa's age a year ago. How old are the three?

Solution:

Conveniently enough, x, y, and z are natural names for Xenia's, Yolanda's, and Zsa Zsa's ages. The first sentence tells us that $x+y+z=30$, the second that $x+y=z$, and the third that $y-x = \dfrac{(z-1)}{2}$. To make the work easier, these equations can be modified as shown:

$$\begin{cases} x+y+z=30 \\ x+y=z \\ y-x=\dfrac{z-1}{2} \end{cases} \xrightarrow{2E_3} \begin{cases} x+y+z=30 \\ x+y-z=0 \\ -2x+2y-z=-1 \end{cases}$$

section 8.1 Solving Systems by Substitution and Elimination

From this, we see that the sum of the first and second equations results in $2x + 2y = 30$, or $x + y = 15$, and the sum of the first and third equations is $-x + 3y = 29$. The method of elimination is the best choice for solving the new system

$$\begin{cases} x + y = 15 \\ -x + 3y = 29, \end{cases}$$

as the sum of the two equations gives us $4y = 44$, or $y = 11$. With this knowledge, the first of the last two equations tells us $x = 4$, and then any equation that includes z can be used to determine that $z = 15$. If this solution is interpreted geometrically, it means that the ordered triple (4, 11, 15) lies on the three planes described by the three equations. In context, it means that Xenia is 4, Yolanda is 11, and Zsa Zsa is 15.

exercises

Use the method of substitution to solve the following systems of equations. If a system is dependent, express the solution set in terms of one of the variables. See Examples 1 and 2.

1. $\begin{cases} 2x - y = -12 \\ 3x + y = -13 \end{cases}$
2. $\begin{cases} 2x - 4y = -6 \\ 3x - y = -4 \end{cases}$
3. $\begin{cases} 3y = 9 \\ x + 2y = 11 \end{cases}$

4. $\begin{cases} -3x - y = 2 \\ 9x + 3y = -6 \end{cases}$
5. $\begin{cases} 2x + y = -2 \\ -4x - 2y = 5 \end{cases}$
6. $\begin{cases} 5x - y = -21 \\ 9x + 2y = -34 \end{cases}$

7. $\begin{cases} 2x - y = -3 \\ -4x + 2y = 6 \end{cases}$
8. $\begin{cases} 3x + 6y = -12 \\ 2x + 4y = -8 \end{cases}$
9. $\begin{cases} 2x + 5y = 33 \\ 3x = -3 \end{cases}$

10. $\begin{cases} 5x + 2y = 8 \\ 2x + y = 6 \end{cases}$
11. $\begin{cases} -2x + y = 5 \\ 9x - 2y = 5 \end{cases}$
12. $\begin{cases} 3x + y = 4 \\ -2x + 3y = 1 \end{cases}$

13. $\begin{cases} 4x - y = -1 \\ -8x + 2y = 2 \end{cases}$
14. $\begin{cases} 4x - 2y = 3 \\ -2x + y = -7 \end{cases}$
15. $\begin{cases} 9x - y = -1 \\ 3x + 2y = 44 \end{cases}$

Use the method of elimination to solve the following systems of equations. If a system is dependent, express the solution set in terms of one of the variables. See Examples 3 and 4.

16. $\begin{cases} 2x - 3y = 8 \\ 8x + 5y = -2 \end{cases}$
17. $\begin{cases} -2x + 3y = 13 \\ 4x + 2y = -18 \end{cases}$
18. $\begin{cases} 5x + 7y = 1 \\ -2x + 3y = -12 \end{cases}$

19. $\begin{cases} x + 2y = 17 \\ 3x + 4y = 39 \end{cases}$
20. $\begin{cases} 5x - 10y = 9 \\ -x + 2y = -3 \end{cases}$
21. $\begin{cases} -2x - 2y = 4 \\ 3x + 3y = -6 \end{cases}$

22. $\begin{cases} 4x+y=11 \\ 3x-2y=0 \end{cases}$
23. $\begin{cases} 7x+8y=-3 \\ -5x-4y=9 \end{cases}$
24. $\begin{cases} -2x-y=9 \\ 4x+2y=1 \end{cases}$

25. $\begin{cases} -2x+4y=6 \\ 3x-y=-4 \end{cases}$
26. $\begin{cases} 5x-6y=-1 \\ -4x+3y=-10 \end{cases}$
27. $\begin{cases} \dfrac{2}{3}x+y=-3 \\ 3x+\dfrac{5}{2}y=-\dfrac{7}{2} \end{cases}$

28. $\begin{cases} \dfrac{x}{5}-y=-\dfrac{11}{5} \\ \dfrac{x}{4}+y=4 \end{cases}$
29. $\begin{cases} \dfrac{2}{3}x+2y=1 \\ x+3y=0 \end{cases}$
30. $\begin{cases} -x-5y=-6 \\ \dfrac{3}{5}x+3y=1 \end{cases}$

Use any convenient method to solve the following systems of equations. If a system is dependent, express the solution set in terms of one or more of the variables, as appropriate. See Example 5.

31. $\begin{cases} x-y+4z=-4 \\ 4x+y-2z=-1 \\ -y+2z=-3 \end{cases}$
32. $\begin{cases} x+2y=-1 \\ y+3z=7 \\ 2x+5z=21 \end{cases}$
33. $\begin{cases} x+y=4 \\ y+3z=-1 \\ 2x-2y+5z=-5 \end{cases}$

34. $\begin{cases} 2x-y=0 \\ 5x-3y-3z=5 \\ 2x+6z=-10 \end{cases}$
35. $\begin{cases} 3x-y+z=2 \\ -6x+2y-2z=-4 \\ -3x+y-z=-2 \end{cases}$
36. $\begin{cases} 2x-3y=-2 \\ x-4y+3z=0 \\ -2x+7y-5z=0 \end{cases}$

37. $\begin{cases} 3x-y+z=2 \\ -6x+2y-2z=1 \\ 5x+2y-3z=2 \end{cases}$
38. $\begin{cases} 4x-y+5z=6 \\ 4x-3y-5z=-14 \\ -2x-5z=-8 \end{cases}$
39. $\begin{cases} 3x+8z=3 \\ -3x-7z=-3 \\ x+3z=1 \end{cases}$

40. $\begin{cases} x+2y+z=8 \\ 2x-3y-4z=-16 \\ x-5y+5z=6 \end{cases}$
41. $\begin{cases} 2x-7y-4z=7 \\ -x+4y+2z=-3 \\ 3y-4z=-1 \end{cases}$
42. $\begin{cases} 4x+4y-2z=6 \\ x-5y+3z=-2 \\ -2x-2y+z=3 \end{cases}$

43. $\begin{cases} 2x+3y+4z=1 \\ 3x-4y+5z=-5 \\ 4x+5y+6z=5 \end{cases}$
44. $\begin{cases} x-4y+2z=-1 \\ 2x+y-3z=10 \\ -3x+12y-6z=3 \end{cases}$
45. $\begin{cases} x+2y+3z=29 \\ 2x-y-z=-2 \\ 3x+2y-6z=-8 \end{cases}$

46. $\begin{cases} 5x-2y+z=14 \\ 8x+4y=12 \\ 9x=18 \end{cases}$
47. $\begin{cases} 2x+5y=6 \\ 3y+8z=-6 \\ x+4y=-5 \end{cases}$
48. $\begin{cases} 4x+3y+4z=5 \\ 5x-6y-2z=-12 \\ 5z=20 \end{cases}$

49. $\begin{cases} 9x+4y-8z=-4 \\ -6x+3y-9z=-9 \\ 8y-3z=18 \end{cases}$
50. $\begin{cases} 21x-7y+51z=141 \\ 13x+9y-5z=-19 \\ 19x-8y+23z=30 \end{cases}$
51. $\begin{cases} 14x-23y+11z=-551 \\ 17x+19y+27z=1595 \\ 32x+26y-4z=-16 \end{cases}$

section 8.1 Solving Systems by Substitution and Elimination

Solve each of the following systems of equations using a graphing calculator.

52. $\begin{cases} 98x + 43y - 82z = -784 \\ -65x + 34y = 3032 \\ 28y - 13z = 966 \end{cases}$

53. $\begin{cases} 7.5x + 5.2y - 9.3z = -23.971 \\ -6.8x + 4.4y = 2.708 \\ 0.9x - 1.88y = -2.0194 \end{cases}$

54. $\begin{cases} -5x + 2y - 20z = 14 \\ 2x - 3y + 10z = -19 \\ 7x + 4y - 7z = -7 \end{cases}$

55. $\begin{cases} -5.5x + 2.2y - 5.1z = 11.29 \\ 1.8x + 4.9y - 0.5z = 7.066 \\ 3.9x - 2.6y + 6.3z = -3.698 \end{cases}$

56. $\begin{cases} 5x - 10y + 11z = 19 \\ 27x + 9y + 7z = -44 \\ 2x + 19y - 4z = -3 \end{cases}$

57. $\begin{cases} -23x + 17y - 7z = -51 \\ -13x + 25y - 11z = 45 \\ 51x - 21y - 28z = -58 \end{cases}$

Use a system of equations to solve the following problems. See Examples 6 and 7.

58. Karen empties out her purse and finds 45 loose coins, consisting entirely of nickels and pennies. If the total value of the coins is $1.37, how many nickels and how many pennies does she have?

59. What choice of *a*, *b*, and *c* will force the graph of the polynomial $f(x) = ax^2 + bx + c$ to have a y-intercept of 5 and to pass through the points (1, 3) and (2, 0)?

60. A tour organizer is planning on taking a group of 40 people to a musical. Balcony tickets cost $29.95 and regular tickets cost $19.95, and she collects a total of $1048.00 from her group to buy the tickets. How many people chose to sit in the balcony?

61. How many ounces each of a 12% alcohol solution and a 30% alcohol solution must be combined to obtain 60 ounces of an 18% solution?

62. Eliza's mother is 20 years older than Eliza, but 3 years younger than Eliza's father. Eliza's father is 7 years younger than three times Eliza's age. How old is Eliza?

63. An investor decides at the beginning of the year to invest some of his cash in an account paying 8% annual interest, and to put the rest in a stock fund that ends up earning 15% over the course of the year. He puts $2000 more in the first account than in the stock fund, and at the end of the year he finds he has earned $1310 in interest. How much money was invested at each of the two rates?

64. Jack and Tyler went shopping for summer clothes. Shirts were $12.47 each, including tax, and shorts were $17.23 per pair, including tax. Jack and Tyler spent a total of $156.21 on 11 items. How many shirts and pairs of shorts did they buy?

65. Three years ago, Bob was twice as old as Marla. Fifteen years ago, Bob was three times as old as Marla. How old is Bob?

66. Deyanira empties her pockets and finds 42 coins consisting of quarters, dimes, and pennies. There appear to be twice as many pennies as dimes and quarters total. If the total value of the coins is $2.13, how many coins of each denomination does she have?

67. If an investor has invested $1000 in stocks and bonds, how much has he invested in stocks if he invested four times more in stocks than in bonds?

68. Twelve years ago, Jim was twice as old as Kristin. Sixteen years ago, Jim was three times older than Kristin. How old is Jim?

69. A movie brought in $740 in ticket sales in one day. Tickets during the day cost $5 and tickets at night cost $7. If 120 tickets were sold, how many were sold during the day?

70. A computer has 24 screws in its case. If there are 7 times more slotted screws than thumb screws, how many thumb screws are in the computer?

71. Jael has $10,000 she would like to invest. She has narrowed her options down to a certificate of deposit paying 5% annually, bonds paying 4% annually, and stocks with an expected annual rate of return of 13.5%. If she wants to invest twice as much in the stocks as in the certificate of deposit and she wants to earn $1000 in interest by the end of the year, how much should she invest in each type of investment?

72. Lea ordered fruit baskets for three of her coworkers. One containing 5 apples, 2 oranges, and 1 mango cost $6.81. Another containing 2 mangos, 8 oranges, and 3 apples cost $11.88. The third contained 4 apples, 4 oranges, and 4 mangos and cost $11.04. How much did each type of fruit cost?

8.2 Matrix Notation and Gaussian Elimination

TOPICS

1. Linear systems, matrices, and augmented matrices
2. Gaussian elimination and row echelon form
3. Gauss-Jordan elimination and reduced row echelon form

Topic 1: Linear Systems, Matrices, and Augmented Matrices

The scheduling of planes and passengers at an airport and the modeling of materials at the atomic level, to name just two, are applications of linear systems with literally thousands of variables. The solution methods of substitution and elimination, learned in the last section, are inadequate for dealing with such large systems. More sophisticated and powerful methods, several of which we will learn in this text, are all based on matrices of numbers associated with a given system. We will begin the discussion with a definition of the relevant terms.

Matrices and Matrix Notation

A **matrix** is a rectangular array of numbers, called **elements** or **entries** of the matrix. As the numbers are in a rectangular array, they naturally form **rows** and **columns**. It is often important to determine the size of a given matrix; we say that a matrix with m rows and n columns is an $m \times n$ matrix (read "m by n"), or of *order* $m \times n$. By convention, the number of rows is always stated first.

Matrices are often labeled with capital letters. The same letter in lower-case, with a pair of subscripts attached, is usually used to refer to its individual elements. For instance, if A is a matrix, a_{ij} refers to the element in the ith row and the jth column (again, rows are stated before columns).

chapter eight — Systems of Equations

example 1

Given the matrix $A = \begin{bmatrix} -27 & 0 & 1 \\ 5 & -\pi & 13 \end{bmatrix}$, determine:

a. The order of A. **b.** The value of a_{13}. **c.** The value of a_{21}.

Solution:

a. A has 2 rows and 3 columns, and is thus a 2×3 matrix.
b. The value of the entry in the first row and third column is 1.
c. The value of the entry in the second row and first column is 5.

technology note

To enter matrices, select **2ND** x^{-1} (Matrix). Use the right arrow to highlight Edit, and press **1** for matrix A. Enter the desired dimensions and values.

Computer algebra systems and many calculators understand matrix notation and operations. In *Mathematica*, for instance, the matrix A from *Example 1* is defined as shown below. Note that A is entered as a list consisting of two sub-lists; each sub-list corresponds to a row in the matrix, so in this example there are two sub-lists of three elements each. The command **MatrixForm** is used to view the matrix as a rectangular array of numbers. The last two commands below illustrate how the values a_{13} and a_{21} are referred to in *Mathematica*.

```
Section 8-2 Technology Note.nb

In[1]:= A = {{-27, 0, 1}, {5, -Pi, 13}}

Out[1]= {{-27, 0, 1}, {5, -π, 13}}

In[2]:= MatrixForm[A]

Out[2]//MatrixForm=
        ( -27  0   1  )
        (  5  -π  13  )

In[3]:= A[[1, 3]]

Out[3]= 1

In[4]:= A[[2, 1]]

Out[4]= 5
```

section 8.2 Matrix Notation and Gaussian Elimination

The following application of matrices to solving linear systems is based on a slight refinement of the method of elimination. The first step is to construct the augmented matrix associated with a given system.

Augmented Matrices

The **augmented matrix** of a linear system of equations is a matrix consisting of the coefficients of the variables (retaining the appropriate order), with an adjoined column consisting of the constants from the right-hand side of the system, assuming the equations have been written in standard form. (Informally, an equation is in *standard form* when it has been simplified and all variables appear on the left of the equation and the constant appears on the right.) The matrix of coefficients and the column of constants are customarily separated by a vertical bar.

For example, the augmented matrix for the system $\begin{cases} 3x - 7y = -4 \\ -x + 4y = 9 \end{cases}$ is $\begin{bmatrix} 3 & -7 & | & -4 \\ -1 & 4 & | & 9 \end{bmatrix}$.

Of course, it may be necessary to rearrange the terms of a system in order to construct its augmented matrix.

example 2

Construct the augmented matrix for the linear system $\begin{cases} \dfrac{2x - 6y}{2} = 3 - z \\ z - x + 5y = 12 \\ x + 3y - 2 = 2z \end{cases}$.

Solution:

The first step is to write each equation in standard form:

$$\begin{cases} \dfrac{2x - 6y}{2} = 3 - z \\ z - x + 5y = 12 \\ x + 3y - 2 = 2z \end{cases} \longrightarrow \begin{cases} x - 3y + z = 3 \\ -x + 5y + z = 12 \\ x + 3y - 2z = 2 \end{cases}$$

continued on next page...

chapter eight Systems of Equations

To create this screen, use the **Matrix** operation as described on page 606.

We can now read off the coefficients and constants to construct the augmented matrix:

$$\begin{bmatrix} 1 & -3 & 1 & | & 3 \\ -1 & 5 & 1 & | & 12 \\ 1 & 3 & -2 & | & 2 \end{bmatrix}$$

Topic 2: Gaussian Elimination and Row Echelon Form

Consider the following augmented matrix:

$$\begin{bmatrix} 1 & 2 & -2 & | & 11 \\ 0 & 1 & -1 & | & 3 \\ 0 & 0 & 1 & | & -1 \end{bmatrix}$$

We can translate this back into system form to obtain

$$\begin{cases} x+2y-2z=11 \\ y-z=3 \\ z=-1 \end{cases},$$

and we note that this system can be solved when we *back-substitute* the last equation $z=-1$ into the second equation to obtain $y-(-1)=3$, or $y=2$, and then back-substitute again in the first equation to obtain $x+2(2)-2(-1)=11$, or $x=5$.

Of course, we have no reason to expect systems to be so singularly nice. The point of *Gaussian elimination* (named after the ubiquitous Carl Friedrich Gauss) is that it transforms an arbitrary augmented matrix into a form like the one above, and the solution of the corresponding system then solves the original system as well. The technical name for an augmented matrix in the form shown above is *row echelon form*.

Row Echelon Form

A matrix is in **row echelon form** if:

1. The first non-zero entry in each row is 1.
2. Every entry below each such 1 (called a **leading 1**) is 0, and each leading 1 appears farther to the right than leading 1's in previous rows.
3. All rows consisting entirely of 0's (if there are any) appear at the bottom.

The operations that make up Gaussian elimination are called *elementary row operations*. There are three such matrix operations, all based on very familiar facts concerning equations. Namely, that

- interchanging the order of two equations in a system has no effect on the system,
- all the terms in an equation can be multiplied by a non-zero constant, and finally
- two equations in a system can be added together.

The matrix form of these facts, along with the symbols we will use to annotate our work, appear below.

Elementary Row Operations

Assume A is an augmented matrix corresponding to a given system of equations. Each of the following operations on A results in the augmented matrix of an equivalent system. In the notation, R_i refers to row i of the matrix A.

1. Rows i and j can be interchanged. (Denoted $R_i \leftrightarrow R_j$)
2. Each entry in row i can be multiplied by a non-zero constant c. (Denoted cR_i)
3. Row j can be replaced with the sum of itself and a constant multiple of row i. (Denoted $cR_i + R_j$)

In summary, then, Gaussian elimination refines the method of elimination by removing the need to write variables at each step and by providing a framework for systematic application of row operations. *Examples 3* and *4* illustrate its use in practice.

chapter eight — Systems of Equations

example 3

Use Gaussian elimination to solve the system $\begin{cases} -2x+y-5z=-6 \\ x+2y-z=-8 \\ 3x-y+2z=2 \end{cases}$.

Solution:

We will first construct the augmented matrix for the system, and then transform it into row echelon form. A good rule of thumb is to proceed by columns: after getting a leading 1 in the first row, use row operations to obtain 0's below it, and then repeat the process with each successive column.

$$\begin{bmatrix} -2 & 1 & -5 & | & -6 \\ 1 & 2 & -1 & | & -8 \\ 3 & -1 & 2 & | & 2 \end{bmatrix} \xrightarrow{R_1 \leftrightarrow R_2} \begin{bmatrix} 1 & 2 & -1 & | & -8 \\ -2 & 1 & -5 & | & -6 \\ 3 & -1 & 2 & | & 2 \end{bmatrix} \xrightarrow[-3R_1+R_3]{2R_1+R_2} \begin{bmatrix} 1 & 2 & -1 & | & -8 \\ 0 & 5 & -7 & | & -22 \\ 0 & -7 & 5 & | & 26 \end{bmatrix}$$

$$\xrightarrow{\frac{1}{5}R_2} \begin{bmatrix} 1 & 2 & -1 & | & -8 \\ 0 & 1 & -\frac{7}{5} & | & -\frac{22}{5} \\ 0 & -7 & 5 & | & 26 \end{bmatrix} \xrightarrow{7R_2+R_3} \begin{bmatrix} 1 & 2 & -1 & | & -8 \\ 0 & 1 & -\frac{7}{5} & | & -\frac{22}{5} \\ 0 & 0 & -\frac{24}{5} & | & -\frac{24}{5} \end{bmatrix} \xrightarrow{-\frac{5}{24}R_3} \begin{bmatrix} 1 & 2 & -1 & | & -8 \\ 0 & 1 & -\frac{7}{5} & | & -\frac{22}{5} \\ 0 & 0 & 1 & | & 1 \end{bmatrix}$$

The third row in the last matrix tells us that $z = 1$. When we back-substitute this in the second row, we obtain

$$y - \frac{7}{5}(1) = -\frac{22}{5},$$

so $y = -3$. From the first row, then, we obtain

$$x + 2(-3) - 1(1) = -8,$$

or $x = -1$. Thus the ordered triple $(-1, -3, 1)$ solves the system of equations.

section 8.2 — Matrix Notation and Gaussian Elimination

example 4

Use Gaussian elimination to solve the system $\begin{cases} x - 3y + 2z = 1 \\ 2x + y - z = 0 \\ -3x - 5y + 4z = 3 \end{cases}$.

Solution:

We construct the system's augmented matrix and proceed as before.

$$\begin{bmatrix} 1 & -3 & 2 & | & 1 \\ 2 & 1 & -1 & | & 0 \\ -3 & -5 & 4 & | & 3 \end{bmatrix} \xrightarrow[3R_1+R_3]{-2R_1+R_2} \begin{bmatrix} 1 & -3 & 2 & | & 1 \\ 0 & 7 & -5 & | & -2 \\ 0 & -14 & 10 & | & 6 \end{bmatrix}$$

We might, however, notice at this point that the −14 in the third row, second column, is most easily changed to 0 if we add 2 times the second row to the third row. Of course, we eventually want a 1 in the second row, second column, but if we wish to avoid fractions as long as possible, we can delay, for one step, dividing each entry in the second row by 7.

$$\begin{bmatrix} 1 & -3 & 2 & | & 1 \\ 0 & 7 & -5 & | & -2 \\ 0 & -14 & 10 & | & 6 \end{bmatrix} \xrightarrow{2R_2+R_3} \begin{bmatrix} 1 & -3 & 2 & | & 1 \\ 0 & 7 & -5 & | & -2 \\ 0 & 0 & 0 & | & 2 \end{bmatrix}$$

In fact, we see now that we can stop at this point. Although the matrix is not yet in row echelon form, we already know there is no solution to the system. The third row of the matrix corresponds to the equation $0 = 2$, and we know that a false statement of this sort means the solution set of the system is the empty set.

Topic 3: Gauss-Jordan Elimination and Reduced Row Echelon Form

The process of back-substitution, once a matrix has been put in row echelon form, amounts to eliminating all but one variable per equation, beginning with the last equation and working backward. Once this is realized, it is natural to ask: Why not accomplish this while the system is still in matrix form?

Gauss-Jordan elimination is the answer to this question. Just as Gaussian elimination is a refinement of the method of elimination, Gauss-Jordan elimination, co-named for Wilhelm Jordan (1842 - 1899), is a refinement of Gaussian elimination. The goal in the method is to put a given matrix into *reduced row echelon form*.

chapter eight Systems of Equations

Reduced Row Echelon Form

A matrix is said to be in **reduced row echelon form** if:

1. It is in row echelon form.
2. Each entry **above** a leading 1 is also 0.

Consider, for instance, the last matrix obtained in *Example 3*. If we begin with that matrix and eliminate entries (that is, convert to 0) above the leading 1 in the third column, and then do the same in the second column, we obtain a matrix in reduced row echelon form:

$$\begin{bmatrix} 1 & 2 & -1 & | & -8 \\ 0 & 1 & -\frac{7}{5} & | & -\frac{22}{5} \\ 0 & 0 & 1 & | & 1 \end{bmatrix} \xrightarrow[R_3+R_1]{\frac{7}{5}R_3+R_2} \begin{bmatrix} 1 & 2 & 0 & | & -7 \\ 0 & 1 & 0 & | & -3 \\ 0 & 0 & 1 & | & 1 \end{bmatrix} \xrightarrow{-2R_2+R_1} \begin{bmatrix} 1 & 0 & 0 & | & -1 \\ 0 & 1 & 0 & | & -3 \\ 0 & 0 & 1 & | & 1 \end{bmatrix}$$

If we now write this in system form, we have

$$\begin{cases} x = -1 \\ y = -3, \\ z = 1 \end{cases}$$

which is equivalent to the original system, but in a form that tells us the solution of the system.

example 5

Use Gauss-Jordan elimination to solve the system $\begin{cases} x - 2y + 3z = -5 \\ 2x + 3y - z = 1 \\ -x - 5y + 4z = -6 \end{cases}$.

Solution:

$$\begin{bmatrix} 1 & -2 & 3 & | & -5 \\ 2 & 3 & -1 & | & 1 \\ -1 & -5 & 4 & | & -6 \end{bmatrix} \xrightarrow[R_1+R_3]{-2R_1+R_2} \begin{bmatrix} 1 & -2 & 3 & | & -5 \\ 0 & 7 & -7 & | & 11 \\ 0 & -7 & 7 & | & -11 \end{bmatrix} \xrightarrow{R_2+R_3} \begin{bmatrix} 1 & -2 & 3 & | & -5 \\ 0 & 7 & -7 & | & 11 \\ 0 & 0 & 0 & | & 0 \end{bmatrix}$$

section 8.2   Matrix Notation and Gaussian Elimination

We can pause at this intermediate stage and note that the row of 0's at the bottom corresponds to the true statement $0 = 0$, indicating that this system has an infinite number of solutions. But we don't yet have a good way of describing the infinite number of ordered triples that make up the solution set, so we continue:

$$\begin{bmatrix} 1 & -2 & 3 & | & -5 \\ 0 & 7 & -7 & | & 11 \\ 0 & 0 & 0 & | & 0 \end{bmatrix} \xrightarrow{\frac{1}{7}R_2} \begin{bmatrix} 1 & -2 & 3 & | & -5 \\ 0 & 1 & -1 & | & \frac{11}{7} \\ 0 & 0 & 0 & | & 0 \end{bmatrix} \xrightarrow{2R_2+R_1} \begin{bmatrix} 1 & 0 & 1 & | & -\frac{13}{7} \\ 0 & 1 & -1 & | & \frac{11}{7} \\ 0 & 0 & 0 & | & 0 \end{bmatrix}$$

The last matrix is in reduced row echelon form, as every leading 1 has 0's both below and above it. The corresponding system of equations is

$$\begin{cases} x + z = -\dfrac{13}{7} \\ y - z = \dfrac{11}{7} \end{cases},$$

To find the reduced row echelon form of a matrix, enter it using the **Matrix** operation as described on pg. 606. Then **Quit** (**2ND** **MODE**), select **Matrix**, and use the right arrow to highlight Math. Select B:rref. This enters the command; to use it on the matrix you created, select **Matrix** then **1**.

and we can thus describe the solution set as $\left\{ \left(-z - \dfrac{13}{7}, z + \dfrac{11}{7}, z \right) \middle| z \in \mathbb{R} \right\}$.

exercises

1. Let $A = \begin{bmatrix} 4 & -1 \\ 0 & 3 \\ 9 & -5 \end{bmatrix}$. Determine the following, if possible:

   a. The order of A **b.** The value of a_{12} **c.** The value of a_{23}

2. Let $B = \begin{bmatrix} -7 & 2 & 11 \end{bmatrix}$. Determine the following, if possible:

   a. The order of B **b.** The value of b_{12} **c.** The value of b_{31}

chapter eight — Systems of Equations

3. Let $C = \begin{bmatrix} 1 & 0 \\ 5 & -3 \\ 2 & 9 \\ \pi & e \\ 10 & -7 \end{bmatrix}$. Determine the following, if possible:

   a. The order of C **b.** The value of c_{23} **c.** The value of c_{51}

4. Let $D = \begin{bmatrix} -8 & 13 & -1 \\ 0 & 6 & 3 \\ 0 & -9 & 0 \end{bmatrix}$. Determine the following, if possible:

   a. The order of D **b.** The value of d_{23} **c.** The value of d_{33}

5. Let $E = \begin{bmatrix} -443 & 951 & 165 & 274 \\ 286 & -653 & 812 & -330 \\ 909 & 377 & 429 & -298 \end{bmatrix}$. Determine the following, if possible:

   a. The order of E **b.** The value of e_{42} **c.** The value of e_{21}

6. Let $A = \begin{bmatrix} 9 & 5 & 0 \\ 7 & 4 & 2 \end{bmatrix}$. Determine the following, if possible:

   a. The order of A **b.** The value of a_{22} **c.** The value of a_{13}

7. Let $B = \begin{bmatrix} 8 & 1 \\ 3 & 0 \\ 6 & 7 \end{bmatrix}$. Determine the following, if possible:

   a. The order of B **b.** The value of b_{12} **c.** The value of b_{13}

8. Let $C = \begin{bmatrix} 65 & 32 & 91 & 45 \\ 23 & 18 & 75 & 47 \\ 8 & 63 & 28 & 31 \end{bmatrix}$. Determine the following, if possible:

   a. The order of C **b.** The value of c_{43} **c.** The value of c_{23}

9. Let $D = \begin{bmatrix} 4 & 9 & 7 & 1 & 8 \\ 5 & 3 & 0 & 2 & 6 \end{bmatrix}$. Determine the following, if possible:

   a. The order of D **b.** The value of d_{21} **c.** The value of d_{24}

section 8.2   Matrix Notation and Gaussian Elimination

Construct the augmented matrix that corresponds to each of the following systems of equations. See Example 2. (Answers may appear in slightly different, but equivalent, form.)

10. $\begin{cases} 4x+5y-3z=8 \\ 7x-2y+9=3 \\ 5x-6y+3z=0 \end{cases}$

11. $\begin{cases} y-2z+4=3x \\ \dfrac{x}{2}-4y-1=z \\ 3(-y+z)-1=0 \end{cases}$

12. $\begin{cases} 5x+\dfrac{y-z}{2}=3 \\ 7(z-x)+y-2=0 \\ x-(4-z)=y \end{cases}$

13. $\begin{cases} \dfrac{2-3x}{2}=y \\ 3z+2(x+y)=0 \\ 2x-y=2(x-3z) \end{cases}$

14. $\begin{cases} 2(z+3)-x+y=z \\ -3(x-2y)-1=5z \\ \dfrac{x}{3}-(y-2z)=x \end{cases}$

15. $\begin{cases} \dfrac{12x-1}{5}+\dfrac{y}{2}=\dfrac{3z}{2} \\ y-(x+3z)=-(1-y) \\ 2x-2-z-2y=7x \end{cases}$

16. $\begin{cases} \dfrac{3x+4y}{2}-3z=6 \\ 3(x-2y+9z)=0 \\ 2x+6y=3-z \end{cases}$

17. $\begin{cases} \dfrac{2x-4y}{3}=2z \\ 8x=2(y-3z)+7 \\ 3x=2y \end{cases}$

18. $\begin{cases} \dfrac{2(2x-y)}{3}+z=7 \\ 4=\dfrac{3}{-x+y+3z} \\ 4x-8y+4=9x \end{cases}$

19. $\begin{cases} 0.5x-14y=\dfrac{z}{4}-8 \\ \dfrac{x}{5}-y+\dfrac{z}{4}=\dfrac{y}{6}-3 \\ \dfrac{2}{3}\left(\dfrac{4}{y-x-1}\right)=\dfrac{5}{z} \end{cases}$

Construct the system of equations that corresponds to each of the following matrices.

20. $\begin{bmatrix} 5 & 3 & | & 9 \\ 1 & 4 & | & 12 \end{bmatrix}$

21. $\begin{bmatrix} 1 & 0 & | & 8 \\ 0 & 1 & | & 3 \end{bmatrix}$

22. $\begin{bmatrix} 14 & 0 & 1 & | & 16 \\ 3 & 6 & 4 & | & 0 \\ 8 & 2 & 5 & | & 21 \end{bmatrix}$

23. $\begin{bmatrix} 1 & 3 & 6 & | & 16 \\ 0 & 1 & 2 & | & 9 \\ 0 & 0 & 1 & | & 4 \end{bmatrix}$

24. $\begin{bmatrix} 2 & 1 & 1 & | & 22 \\ 1 & 3 & 1 & | & 17 \\ 1 & 1 & 4 & | & 8 \end{bmatrix}$

25. $\begin{bmatrix} 0 & 9 & 13 & | & 27 \\ 2 & 0 & 21 & | & 19 \\ 7 & 18 & 0 & | & 32 \end{bmatrix}$

chapter eight   Systems of Equations

Fill in the blanks by performing the indicated row operations.

26. $\begin{bmatrix} 3 & 2 & | & -7 \\ 1 & 3 & | & 5 \end{bmatrix} \xrightarrow{-3R_2+R_1} ?$

27. $\begin{bmatrix} 2 & -5 & | & 3 \\ -4 & 3 & | & -1 \end{bmatrix} \xrightarrow{2R_1+R_2} ?$

28. $\begin{bmatrix} 4 & 2 & | & -8 \\ 3 & -9 & | & 0 \end{bmatrix} \xrightarrow[\frac{1}{3}R_2]{\frac{1}{2}R_1} ?$

29. $\begin{bmatrix} 9 & -2 & | & 7 \\ 1 & 3 & | & -2 \end{bmatrix} \xrightarrow{R_1 \leftrightarrow R_2} ?$

30. $\begin{bmatrix} 4 & 1 & | & 5 \\ 3 & 6 & | & 0 \end{bmatrix} \xrightarrow{2R_1} ?$

31. $\begin{bmatrix} 8 & -2 & | & -4 \\ 3 & -1 & | & 7 \end{bmatrix} \xrightarrow{-2R_2} ?$

32. $\begin{bmatrix} 9 & 12 & | & -6 \\ 15 & -3 & | & 0 \end{bmatrix} \xrightarrow{-\frac{1}{3}R_1} ?$

33. $\begin{bmatrix} 4 & 12 & | & -6 \\ 7 & 3 & | & 9 \end{bmatrix} \xrightarrow{\frac{1}{2}R_1+R_2} ?$

34. $\begin{bmatrix} 3 & 0 & | & 1 \\ 5 & 7 & | & -2 \end{bmatrix} \xrightarrow{3R_1+R_2} ?$

35. $\begin{bmatrix} 8 & -2 & | & 10 \\ 9 & -3 & | & 0 \end{bmatrix} \xrightarrow[-\frac{2}{3}R_2]{\frac{1}{2}R_1} ?$

36. $\begin{bmatrix} 5 & 2 & 9 & | & 7 \\ 1 & 3 & -5 & | & 0 \\ 2 & -4 & 1 & | & 8 \end{bmatrix} \xrightarrow[-R_1+R_3]{2R_2} ?$

37. $\begin{bmatrix} 6 & -2 & 5 & | & 14 \\ -7 & 19 & 2 & | & 3 \\ -9 & 11 & -4 & | & 7 \end{bmatrix} \xrightarrow[0.5R_3]{3R_1} ?$

38. $\begin{bmatrix} 5 & 3 & 13 & | & 15 \\ 17 & 9 & -8 & | & -14 \\ 4 & -11 & 19 & | & 8 \end{bmatrix} \xrightarrow{-2R_2+R_3} ?$

39. $\begin{bmatrix} 8 & 11 & 18 & | & 2 \\ 14 & 33 & -3 & | & -5 \\ -9 & 21 & 12 & | & 9 \end{bmatrix} \xrightarrow[-2R_3+R_2]{\frac{1}{3}R_3+R_1} ?$

40. $\begin{bmatrix} 1 & 3 & -2 & | & 4 \\ 3 & -1 & 8 & | & 2 \\ -5 & 0 & 2 & | & 7 \end{bmatrix} \xrightarrow[5R_1+R_3]{-3R_1+R_2} ?$

41. $\begin{bmatrix} 2 & 3 & -3 & | & 5 \\ 1 & 1 & 3 & | & 4 \\ 3 & 3 & 9 & | & 12 \end{bmatrix} \xrightarrow[-3R_2+R_3]{-2R_2+R_1} ?$

42. $\begin{bmatrix} -3 & 2 & | & 2 \\ 5 & -4 & | & 1 \end{bmatrix} \xrightarrow{2R_1+R_2} ?$

43. $\begin{bmatrix} -5 & 20 & | & -15 \\ 2 & -12 & | & 5 \end{bmatrix} \xrightarrow[\frac{1}{2}R_2]{\frac{1}{5}R_1} ?$

44. $\begin{bmatrix} 2 & 2 & 3 & | & 7 \\ -3 & 2 & 8 & | & -2 \\ 1 & 5 & 2 & | & 6 \end{bmatrix} \xrightarrow[3R_3+R_2]{-2R_3+R_1} ?$

45. $\begin{bmatrix} 1 & 5 & -9 & | & 11 \\ 1 & 4 & -1 & | & 4 \\ 4 & 3 & 5 & | & 45 \end{bmatrix} \xrightarrow[-4R_1+R_3]{-R_1+R_2} ?$

For each matrix below, determine if it is in row echelon form, reduced row echelon form, or neither.

46. $\begin{bmatrix} 1 & 5 & | & 4 \\ 0 & 1 & | & 3 \end{bmatrix}$

47. $\begin{bmatrix} 1 & 2 & 0 & | & 9 \\ 0 & 1 & 3 & | & 4 \\ 0 & 1 & 1 & | & 12 \end{bmatrix}$

48. $\begin{bmatrix} 1 & 0 & 0 & | & 4 \\ 0 & 1 & 0 & | & 1 \\ 0 & 0 & 1 & | & 8 \end{bmatrix}$

49. $\begin{bmatrix} 1 & 0 & 0 & | & 7 \\ 5 & 1 & 0 & | & 14 \\ 3 & 4 & 1 & | & -16 \end{bmatrix}$ **50.** $\begin{bmatrix} 1 & 2 & 5 & | & 0 \\ 0 & 1 & 9 & | & 3 \\ 0 & 0 & 0 & | & 1 \end{bmatrix}$ **51.** $\begin{bmatrix} 0 & 1 & | & 3 \\ 1 & 0 & | & 6 \end{bmatrix}$

Use Gaussian elimination and back-substitution to solve the following systems of equations. See Examples 3 and 4.

52. $\begin{cases} 2x - 4y = -6 \\ 3x - y = -4 \end{cases}$ **53.** $\begin{cases} 2x - 5y = 11 \\ 3x + 2y = 7 \end{cases}$ **54.** $\begin{cases} 5x - y = -21 \\ 9x + 2y = -34 \end{cases}$

55. $\begin{cases} x - 4y = -11 \\ 7x - y = 4 \end{cases}$ **56.** $\begin{cases} x + 2y = 17 \\ 3x + 4y = 39 \end{cases}$ **57.** $\begin{cases} 2x + 6y = 4 \\ -4x - 7y = 7 \end{cases}$

58. $\begin{cases} 3x - 2y = 5 \\ -5x + 4y = -3 \end{cases}$ **59.** $\begin{cases} 2x + y = -2 \\ -4x - 2y = 5 \end{cases}$ **60.** $\begin{cases} 6x - 16y = 10 \\ -3x + 8y = 4 \end{cases}$

61. $\begin{cases} 2x - 3y = 0 \\ 5x + y = 17 \end{cases}$ **62.** $\begin{cases} 6x + 3y = 3 \\ x + y = 3 \end{cases}$ **63.** $\begin{cases} 3x + 6y = -12 \\ 2x + 4y = -8 \end{cases}$

64. $\begin{cases} 4x + 5y = 9 \\ 8x + 3y = -17 \end{cases}$ **65.** $\begin{cases} \frac{2}{3}x + 2y = 1 \\ x + 3y = 0 \end{cases}$ **66.** $\begin{cases} 13x - 17y = -3 \\ -19x + 15y = -35 \end{cases}$

67. $\begin{cases} 3x - 9y - 7z = -9 \\ 5x + 11y - z = 17 \\ -4x - 8y + 7z = 5 \end{cases}$ **68.** $\begin{cases} 8x - y + 5z = -8 \\ 11x - 2y + 9z = -9 \\ 7x - 3y + 13z = 4 \end{cases}$ **69.** $\begin{cases} 17x + 13y + 8z = 46 \\ -12x + 3y + 28z = -19 \\ 14x + 5y - 15z = -15 \end{cases}$

Use Gauss-Jordan elimination to solve the following systems of equations. See Example 5.

70. $\begin{cases} 2x - 3y = 8 \\ 8x + 5y = -2 \end{cases}$ **71.** $\begin{cases} \frac{2}{3}x + y = -3 \\ 3x + \frac{5}{2}y = -\frac{7}{2} \end{cases}$ **72.** $\begin{cases} 3y = 9 \\ x + 2y = 11 \end{cases}$

73. $\begin{cases} 6x + 2y = -4 \\ -9x - 3y = 6 \end{cases}$ **74.** $\begin{cases} 3y = 6 \\ 5x + 2y = 4 \end{cases}$ **75.** $\begin{cases} 3x + 8y = -4 \\ x + 2y = -2 \end{cases}$

76. $\begin{cases} -3x + 2y = 5 \\ 5x - 2y = 1 \end{cases}$ **77.** $\begin{cases} 9x - 11y = 10 \\ -4x + 3y = -12 \end{cases}$ **78.** $\begin{cases} 9x - 15y = -6 \\ -3x + 11y = -10 \end{cases}$

chapter eight   **Systems of Equations**

79. $\begin{cases} 3x - 8y = 7 \\ 18x - 35y = -23 \end{cases}$

80. $\begin{cases} 4x + y - 3z = -9 \\ 2x - 3z = -19 \\ 7x - y - 4z = -29 \end{cases}$

81. $\begin{cases} -5x + 9y + 3z = 1 \\ 3x + 2y - 6z = 9 \\ x + 4y - z = 16 \end{cases}$

82. $\begin{cases} 2x - y = 0 \\ 5x - 3y - 3z = 5 \\ 2x + 6z = -10 \end{cases}$

83. $\begin{cases} x + y = 4 \\ y + 3z = -1 \\ 2x - 2y + 5z = -5 \end{cases}$

84. $\begin{cases} 2x - 3y = -2 \\ x - 4y + 3z = 0 \\ -2x + 7y - 5z = 0 \end{cases}$

85. $\begin{cases} 3x + 8z = 3 \\ -3x - 7z = -3 \\ x + 3z = 1 \end{cases}$

86. $\begin{cases} 3x - y + z = 2 \\ -6x + 2y - 2z = 1 \\ 5x + 2y - 3z = 2 \end{cases}$

87. $\begin{cases} x + 2y = -1 \\ y + 3z = 7 \\ 2x + 5z = 21 \end{cases}$

88. $\begin{cases} 2x + 8y - z = -5 \\ -5x + 3y + 4z = -6 \\ x - 4y - 5z = -8 \end{cases}$

89. $\begin{cases} 7x - 8y + 2z = -2 \\ 5x - 3y - z = -3 \\ 8x + y - 3z = 7 \end{cases}$

90. $\begin{cases} 8x + 14y - 3z = 3 \\ -6x + 2y + 7z = -13 \\ 8x + 19y + 3z = 11 \end{cases}$

91. $\begin{cases} 8x + 5y + 3z = -2 \\ 12x - y - 18z = 1 \\ 7x + 6y + 10z = 19 \end{cases}$

92. $\begin{cases} 4x + 8y + 7z = 27 \\ -2x + 9y - 8z = -15 \\ 9x + 13y + 7z = -33 \end{cases}$

93. $\begin{cases} w - x + 2z = 9 \\ 2w + 3y = -1 \\ -2w - 5y - z = 0 \\ x + 2y = -4 \end{cases}$

94. $\begin{cases} 3w - x + 5y + 3z = 2 \\ -4w - 10y - 2z = 10 \\ w - x + 2z = 7 \\ 4w - 2x + 5y + 5z = 9 \end{cases}$

Use your knowledge of systems of equations and Gaussian elimination to solve the following application problems.

95. The sum of three integers is 155. The first integer is sixteen more than the second. The third integer is seven less than the sum of the first integer and twice the second. What are the three integers?

96. Mario bought a pound of bacon, a dozen eggs, and a loaf of bread to make breakfast for his family. The total cost was $7.42. The bacon cost $0.03 more than twice the price of the bread and the eggs cost $0.03 less than half the price of the bread. Find the price of each item.

97. The Pizza House sells three sizes of pizzas: small, medium, large. The prices of the pizzas are $9.00, $12.00, and $15.00, respectively. In one day, they sold 82 pizzas for a total of $1098.00. If the number of large pizzas sold was twice the number of medium pizzas sold, how many of each size pizza did the Pizza House sell?

8.3 Determinants and Cramer's Rule

TOPICS

1. Determinants and their evaluation
2. Using Cramer's rule to solve linear systems

Topic 1: Determinants and Their Evaluation

As far as solving linear systems is concerned, Gaussian elimination is certainly an improvement over the methods of substitution and elimination. This improvement is due primarily to the systematic approach that Gaussian elimination, and in particular Gauss-Jordan elimination, encourages, and not to any significantly deeper understanding of systems. Cramer's rule, named after the Swiss mathematician Gabriel Cramer (1704 - 1752), is the first of two solution methods we will study that truly brings something new to the discussion.

Cramer's rule relies on the computation of a number, called the *determinant*, associated with every square (that is, $n \times n$) matrix, and we accordingly begin with a definition of determinant.

Determinant of a 2 × 2 Matrix

The **determinant** of the matrix $A = \begin{bmatrix} a_{11} & a_{12} \\ a_{21} & a_{22} \end{bmatrix}$, denoted $|A|$, is $a_{11}a_{22} - a_{21}a_{12}$.

caution

In the context of matrices, $|A|$ stands for the determinant of A, not the absolute value of A.

chapter eight Systems of Equations

Of course, this only defines determinants for matrices of one size, and unfortunately the simplicity of the definition does not carry over to larger sizes. The good news, however, is that determinants of 3×3, 4×4, and larger matrices can all be related, ultimately, to determinants of 2×2 matrices. We will do a few calculations before considering larger matrices.

example 1

Evaluate the determinants of each of the following matrices:

a. $A = \begin{bmatrix} -2 & 3 \\ -1 & 3 \end{bmatrix}$ **b.** $B = \begin{bmatrix} 2 & -3 \\ -4 & 6 \end{bmatrix}$

Solution:

a. $|A| = (-2)(3) - (-1)(3) = -6 + 3 = -3$
b. $|B| = (2)(6) - (-4)(-3) = 12 - 12 = 0$

To find the determinant of a matrix, enter it using the **Matrix** operation (pg. 606). Then Quit (**2ND** **MODE**), select **Matrix**, and use the right arrow to highlight Math. Select 1:det. This enters the command; to use it on the matrix you created, select **Matrix**, then .

The most elementary evaluation of a 3×3 determinant depends on first calculating three 2×2 determinants, and then combining these three numbers in a certain way. Similarly, the most basic evaluation of a 4×4 determinant requires the evaluation of four 3×3 determinants, each of which in turn depends on three 2×2 determinants. Fortunately, as we will soon see, there are some useful properties of determinants that allow us to reduce the required number of computations, but we need a few more definitions first.

Minors and Cofactors

Let A be an $n \times n$ matrix, and let i and j be two fixed numbers each between 1 and n.

- The **minor** of the element a_{ij} is the determinant of the $(n-1) \times (n-1)$ matrix formed from A by deleting its ith row and j^{th} column.
- The **cofactor** of the element a_{ij} is $(-1)^{i+j}$ times the minor of a_{ij}. Thus, if $i+j$ is even, the cofactor of a_{ij} is the same as the minor of a_{ij}; if $i+j$ is odd, the cofactor is the negative of the minor.

620

section 8.3 Determinants and Cramer's Rule

The following matrix of signs may help you remember which minors to multiply by −1: if a_{ij} is in a position occupied by a − sign, change the sign of the minor of a_{ij} to obtain the cofactor.

$$\begin{bmatrix} + & - & + & \cdots \\ - & + & - & \cdots \\ + & - & + & \cdots \\ \vdots & \vdots & \vdots & \ddots \end{bmatrix}$$

example 2

Evaluate the minor of a_{12} and cofactor of a_{23} of the matrix $A = \begin{bmatrix} -5 & 3 & 2 \\ 1 & 0 & -1 \\ -3 & 1 & 0 \end{bmatrix}$.

Solution:

The minor of $a_{12} = \begin{vmatrix} 1 & -1 \\ -3 & 0 \end{vmatrix} = (1)(0) - (-1)(-3) = 0 - 3 = -3$.

The cofactor of $a_{23} = (-1)^{2+3} \begin{vmatrix} -5 & 3 \\ -3 & 1 \end{vmatrix} = (-1)^5 [(-5)(1) - (-3)(3)]$

$= (-1)[-5 + 9] = (-1)(4) = -4$.

Determinant of an $n \times n$ Matrix

Evaluation of an $n \times n$ determinant is accomplished by what is called **expansion** along a fixed row or column. The end result does not depend on which row or column is chosen.

- To expand along the ith row, each element of that row is multiplied by its cofactor, and the n products are then added.
- To expand along the jth column, each element of that column is multiplied by its cofactor, and the n products are then added.

Note that, in either case, a total of n $(n-1) \times (n-1)$ determinants must potentially be computed. A wise choice of row or column will often reduce the number of determinants that must actually be evaluated, however.

chapter eight Systems of Equations

caution

Note that the definition of determinant only applies to square matrices; non-square matrices do not have determinants.

It's time to put the definitions and techniques to use.

example 3

Evaluate the determinants of the following matrices. In each case, minimize the required number of computations by choosing carefully a row or column to expand along.

a. $A = \begin{bmatrix} -1 & 3 & 2 \\ -2 & 0 & 0 \\ 4 & 1 & 5 \end{bmatrix}$ **b.** $B = \begin{bmatrix} 4 & -2 & 3 & 0 \\ 2 & 1 & -1 & 3 \\ 3 & 0 & 1 & 1 \\ 2 & -2 & 0 & 0 \end{bmatrix}$

Solution:

a. When calculating determinants, a row or column with many zeros is generally a good choice for the expansion, as we know beforehand the result of multiplying zero by any number. Thus we will expand along the second row of A. In the computations below, each entry of the second row appears in black, while its corresponding cofactor appears in teal. Note the minus sign that appears in front of the first and third 2×2 determinant; the change in sign is part of the definition of the cofactor. (The cofactors corresponding to the zero entries are written for reference only; we won't bother to compute them.) Make sure you understand how each 2×2 matrix is generated by deleting the appropriate row and column of the original 3×3 matrix.

$$\begin{vmatrix} -1 & 3 & 2 \\ -2 & 0 & 0 \\ 4 & 1 & 5 \end{vmatrix} = (-2)\left(-\begin{vmatrix} 3 & 2 \\ 1 & 5 \end{vmatrix}\right) + (0)\left(\begin{vmatrix} -1 & 2 \\ 4 & 5 \end{vmatrix}\right) + (0)\left(-\begin{vmatrix} -1 & 3 \\ 4 & 1 \end{vmatrix}\right)$$

$$= (-2)(-13) + 0 + 0 = 26$$

We have the answer then: $|A| = 26$. But to illustrate the fact that the initial choice of row or column to expand along doesn't affect the answer, it might be good to evaluate the determinant a second way. If we expand along the first column, our work appears as follows:

$$\begin{vmatrix} -1 & 3 & 2 \\ -2 & 0 & 0 \\ 4 & 1 & 5 \end{vmatrix} = (-1)\left(\begin{vmatrix} 0 & 0 \\ 1 & 5 \end{vmatrix}\right) + (-2)\left(-\begin{vmatrix} 3 & 2 \\ 1 & 5 \end{vmatrix}\right) + (4)\left(\begin{vmatrix} 3 & 2 \\ 0 & 0 \end{vmatrix}\right)$$

$$= (-1)(0) + (-2)(-13) + (4)(0) = 26$$

section 8.3 Determinants and Cramer's Rule

The screen below was produced using the Matrix operation as described on page 606 and the det feature as described on page 620.

b. The fourth row and the fourth column both contain two zeros, so either is a good choice as the basis for the expansion. If we use the fourth column, we obtain:

$$|B| = (0)\left(-\begin{vmatrix} 2 & 1 & -1 \\ 3 & 0 & 1 \\ 2 & -2 & 0 \end{vmatrix}\right) + (3)\begin{vmatrix} 4 & -2 & 3 \\ 3 & 0 & 1 \\ 2 & -2 & 0 \end{vmatrix} + (1)\left(-\begin{vmatrix} 4 & -2 & 3 \\ 2 & 1 & -1 \\ 2 & -2 & 0 \end{vmatrix}\right) + (0)\begin{vmatrix} 4 & -2 & 3 \\ 2 & 1 & -1 \\ 3 & 0 & 1 \end{vmatrix}$$

We now have two 3×3 determinants to evaluate, so our work continues:

$$|B| = (3)\begin{pmatrix} \begin{vmatrix} 4 & -2 & 3 \\ 3 & 0 & 1 \\ 2 & -2 & 0 \end{vmatrix} \end{pmatrix} + (1)\begin{pmatrix} -\begin{vmatrix} 4 & -2 & 3 \\ 2 & 1 & -1 \\ 2 & -2 & 0 \end{vmatrix} \end{pmatrix}$$

$$= (3)\left((2)\begin{pmatrix}\begin{vmatrix} -2 & 3 \\ 0 & 1 \end{vmatrix}\end{pmatrix} + (-2)\begin{pmatrix}-\begin{vmatrix} 4 & 3 \\ 3 & 1 \end{vmatrix}\end{pmatrix}\right) + (-1)\left((2)\begin{pmatrix}\begin{vmatrix} -2 & 3 \\ 1 & -1 \end{vmatrix}\end{pmatrix} + (-2)\begin{pmatrix}-\begin{vmatrix} 4 & 3 \\ 2 & -1 \end{vmatrix}\end{pmatrix}\right)$$

$$= (3)\big((2)(-2) + (-2)(5)\big) + (-1)\big((2)(-1) + (-2)(10)\big)$$

$$= (3)(-14) + (-1)(-22)$$

$$= -20$$

(Note that in both cases, the 3×3 determinants were evaluated by expanding along the third row, and those cofactors that were to be multiplied by zero were omitted.)

The last example shows that any tricks that might reduce the number of computations necessary would certainly be welcome. There are three such tricks, each of them due to a property of determinants.

chapter eight Systems of Equations

Properties of Determinants

1. A constant can be factored out of each of the terms in a given row or column when computing determinants. For example,

$$\begin{vmatrix} 2 & -1 \\ 15 & 5 \end{vmatrix} = 5 \begin{vmatrix} 2 & -1 \\ 3 & 1 \end{vmatrix} \text{ and } \begin{vmatrix} 4 & 7 \\ 12 & 9 \end{vmatrix} = 4 \begin{vmatrix} 1 & 7 \\ 3 & 9 \end{vmatrix}.$$

2. Interchanging two rows or two columns changes the determinant by a factor of -1. For example,

$$\begin{vmatrix} 2 & -1 \\ 15 & 5 \end{vmatrix} = - \begin{vmatrix} 15 & 5 \\ 2 & -1 \end{vmatrix} \text{ and } \begin{vmatrix} 3 & -2 \\ 7 & 1 \end{vmatrix} = - \begin{vmatrix} -2 & 3 \\ 1 & 7 \end{vmatrix}.$$

3. The determinant is unchanged by adding a multiple of one row (or column) to another row (or column). For example,

$$\begin{vmatrix} 3 & -2 \\ 1 & -1 \end{vmatrix} \stackrel{-3R_2+R_1}{=} \begin{vmatrix} 0 & 1 \\ 1 & -1 \end{vmatrix}.$$

It's a good idea to annotate, as shown in teal above, applications of this last property.

To demonstrate the utility of these properties, in particular the third one, let us revisit the 4×4 determinant from the last example.

example 4

Evaluate the determinant of the matrix $B = \begin{bmatrix} 4 & -2 & 3 & 0 \\ 2 & 1 & -1 & 3 \\ 3 & 0 & 1 & 1 \\ 2 & -2 & 0 & 0 \end{bmatrix}.$

Solution:

We will make use of the properties of determinants to try to obtain rows or columns with as many zeros as possible. In this problem, one application of the third property results in a column with only one non-zero entry, as shown:

$$\begin{vmatrix} 4 & -2 & 3 & 0 \\ 2 & 1 & -1 & 3 \\ 3 & 0 & 1 & 1 \\ 2 & -2 & 0 & 0 \end{vmatrix} \overset{-3R_3+R_2}{=} \begin{vmatrix} 4 & -2 & 3 & 0 \\ -7 & 1 & -4 & 0 \\ 3 & 0 & 1 & 1 \\ 2 & -2 & 0 & 0 \end{vmatrix}$$

We proceed to expand along the fourth column (don't forget the minus sign as part of the cofactor):

$$\begin{vmatrix} 4 & -2 & 3 & 0 \\ -7 & 1 & -4 & 0 \\ 3 & 0 & 1 & 1 \\ 2 & -2 & 0 & 0 \end{vmatrix} = (1)\left(-\begin{vmatrix} 4 & -2 & 3 \\ -7 & 1 & -4 \\ 2 & -2 & 0 \end{vmatrix} \right)$$

We can continue to apply the third property of determinants to simplify the evaluation of the 3×3 determinant. In what follows, the second column is replaced by the sum of the first and second columns, and then the determinant is evaluated by expanding around the third row.

$$|B| = -\begin{vmatrix} 4 & -2 & 3 \\ -7 & 1 & -4 \\ 2 & -2 & 0 \end{vmatrix} \overset{C_1+C_2}{=} -\begin{vmatrix} 4 & 2 & 3 \\ -7 & -6 & -4 \\ 2 & 0 & 0 \end{vmatrix}$$

$$= -(2)\left(\begin{vmatrix} 2 & 3 \\ -6 & -4 \end{vmatrix} \right)$$

$$= (-2)(10)$$

$$= -20$$

Notice that we have reduced the work to the evaluation of only one 3×3 determinant, which in turn involved evaluating only one 2×2 determinant.

chapter eight   Systems of Equations

technology note

The screens below were produced using the **Matrix** operation as described on page 606 and the **det** feature as described on page 620.

Clearly, evaluating determinants of large matrices can be a time-consuming and error-prone process, and technology offers welcome assistance. Computer algebra systems and many calculators are able to calculate determinants of matrices up to a certain size (the exact size depends on the technology being used). In *Mathematica*, the command to calculate the determinant of a matrix is **Det**. In the illustration below, the matrix B from *Example 3* is first defined (refer to the Technology Note in Section 8.2 for another example of defining matrices). The second command simply returns the matrix B in the familiar rectangular form, and allows us to verify that we defined B correctly. The third command then verifies the value for the determinant that we found manually.

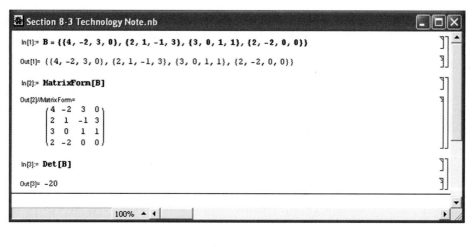

Topic 2:   Using Cramer's Rule to Solve Linear Systems

To motivate the form of Cramer's rule, let us solve a 2-variable, 2-equation linear system once and for all. To do this, we note that any such system can be put into the form

$$\begin{cases} ax + by = e \\ cx + dy = f \end{cases},$$

where $a, b, c, d, e,$ and f are all constants. If we can solve this system for x and y, then we will have a formula for the solution of any such system.

The methods we have at our disposal, at the moment, are substitution, elimination, Gaussian elimination, and Gauss-Jordan elimination. Any method can be used; if we choose the method of elimination, we can obtain an equation in x alone as follows:

section 8.3   Determinants and Cramer's Rule

$$\begin{cases} ax+by=e \\ cx+dy=f \end{cases} \xrightarrow[-bE_2]{dE_1} \begin{cases} adx+bdy=ed \\ -bcx-bdy=-bf \end{cases}$$
$$(ad-bc)x=ed-bf$$

This equation can then be solved for x to obtain

$$x = \frac{ed-bf}{ad-bc}.$$

Similarly, the system can be solved for y to obtain

$$y = \frac{af-ce}{ad-bc}.$$

These formulas are worthwhile on their own, but their true value becomes apparent when we express them using determinants. Note that the above formulas are equivalent to:

$$x = \frac{\begin{vmatrix} e & b \\ f & d \end{vmatrix}}{\begin{vmatrix} a & b \\ c & d \end{vmatrix}} \text{ and } y = \frac{\begin{vmatrix} a & e \\ c & f \end{vmatrix}}{\begin{vmatrix} a & b \\ c & d \end{vmatrix}}$$

The denominator in both formulas is the determinant of the coefficient matrix, the square matrix consisting of the coefficients of the variables. The numerators differ, but there is an easily remembered pattern that describes them. If we let D_x and D_y represent the numerators of the formulas for x and y, respectively, then D_x is the determinant of the coefficient matrix with the first column (the x-column) replaced by the column of constants, and D_y is the determinant of the coefficient matrix with the second column replaced by the column of constants. If we then define D to be the determinant of the coefficient matrix, we obtain Cramer's rule for the 2-variable, 2-equation case:

$$x = \frac{D_x}{D} \text{ and } y = \frac{D_y}{D}$$

Before looking at an application of the rule and before generalizing the rule to larger systems, we should make one more observation. Whenever a fraction appears in our work, it's appropriate to pause and ask if the expression in the denominator might ever be zero, and what it means if in fact this happens. In this case, there are certainly numbers a, b, c, and d for which $D = 0$; clearly, in such a system, we can't solve for x and y by Cramer's rule. The meaning of $D = 0$ is that the system is either dependent or inconsistent. If both D_x and D_y are also zero, the system is dependent. If at least one of D_x and D_y is non-zero, the system has no solution.

chapter eight   Systems of Equations

example 5

Use Cramer's rule to solve the following systems.

a. $\begin{cases} 4x - 5y = 3 \\ -3x + 7y = 1 \end{cases}$ **b.** $\begin{cases} -x + 2y = -1 \\ 3x - 6y = 3 \end{cases}$

Solution:

a. $D = \begin{vmatrix} 4 & -5 \\ -3 & 7 \end{vmatrix} = 28 - 15 = 13$

$D_x = \begin{vmatrix} 3 & -5 \\ 1 & 7 \end{vmatrix} = 21 - (-5) = 26$

$D_y = \begin{vmatrix} 4 & 3 \\ -3 & 1 \end{vmatrix} = 4 - (-9) = 13$

$x = \dfrac{26}{13} = 2$ and $y = \dfrac{13}{13} = 1$

so the single solution is $(2, 1)$.

We will learn something about the system merely by calculating D, so that is what we do first. The fact that D is non-zero tells us there is a single solution to the system; we then find D_x and D_y in order to determine the solution.

Cramer's rule gives us the ordered pair $(2, 1)$ as the single point that solves both equations.

b. $D = \begin{vmatrix} -1 & 2 \\ 3 & -6 \end{vmatrix} = 6 - 6 = 0$

$D_x = \begin{vmatrix} -1 & 2 \\ 3 & -6 \end{vmatrix} = 6 - 6 = 0$

$D_y = \begin{vmatrix} -1 & -1 \\ 3 & 3 \end{vmatrix} = -3 - (-3) = 0$

Solution set is $\{(2y + 1, y) | y \in \mathbb{R}\}$.

Again, we calculate D first. The fact that $D = 0$ means either the system has no solution, or it has an infinite number of solutions. Since both D_x and D_y also turn out to be zero, we know the system is dependent.

The infinite number of ordered pairs that solve the system can be described by solving either equation for either variable, as has been done at left.

The screens above were produced using the **Matrix** operation as described on page 606 and the `:rref` feature as described on page 613.

Cramer's rule can be extended to solve any linear system of n equations in n variables, as shown in the box below. This makes it appear to be an extremely powerful method, and indeed Cramer's rule is remarkable for its succinctness. Keep in mind, however, that using Cramer's rule to solve an n-equation, n-variable system entails calculating $(n+1)$ $n \times n$ determinants, so it is important to make liberal use of the labor-saving properties of determinants.

Cramer's Rule

A linear system of n equations in the n variables $x_1, x_2, \ldots, x_n$ can be written in the form

$$\begin{cases} a_{11}x_1 + a_{12}x_2 + \ldots + a_{1n}x_n = b_1 \\ a_{21}x_1 + a_{22}x_2 + \ldots + a_{2n}x_n = b_2 \\ \vdots \\ a_{n1}x_1 + a_{n2}x_2 + \ldots + a_{nn}x_n = b_n \end{cases}$$

The solution of the system is given by the n formulas $x_1 = \dfrac{D_{x_1}}{D}$, $x_2 = \dfrac{D_{x_2}}{D}$, $\ldots$ $x_n = \dfrac{D_{x_n}}{D}$, where D is the determinant of the coefficient matrix and D_{x_i} is the determinant of the same matrix with the ith column replaced by the column of constants $b_1, b_2, \ldots, b_n$. If $D = 0$ and if each $D_{x_i} = 0$ as well, for $i = 1, \ldots, n$, the system is dependent and has an infinite number of solutions. If $D = 0$ and at least one of the D_{x_i}'s is non-zero, the system has no solution.

example 6

Use Cramer's rule to solve the system $\begin{cases} 3x - 2y - 2z = -1 \\ 3y + z = -7 \\ x + y + 2z = 0 \end{cases}$.

Solution:

Note how the properties of determinants are used to simplify each calculation. In each case, the row or column used for expansion is written in teal.

$$D = \begin{vmatrix} 3 & -2 & -2 \\ 0 & 3 & 1 \\ 1 & 1 & 2 \end{vmatrix} \stackrel{-3R_3+R_1}{=} \begin{vmatrix} 0 & -5 & -8 \\ 0 & 3 & 1 \\ 1 & 1 & 2 \end{vmatrix} = (1)\left(\begin{vmatrix} -5 & -8 \\ 3 & 1 \end{vmatrix}\right) = 19$$

$$D_x = \begin{vmatrix} -1 & -2 & -2 \\ -7 & 3 & 1 \\ 0 & 1 & 2 \end{vmatrix} \stackrel{-2C_2+C_3}{=} \begin{vmatrix} -1 & -2 & 2 \\ -7 & 3 & -5 \\ 0 & 1 & 0 \end{vmatrix} = (1)\left(-\begin{vmatrix} -1 & 2 \\ -7 & -5 \end{vmatrix}\right) = -19$$

continued on next page...

chapter eight   Systems of Equations

$$D_y = \begin{vmatrix} 3 & -1 & -2 \\ 0 & -7 & 1 \\ 1 & 0 & 2 \end{vmatrix} \overset{-2C_1+C_3}{=} \begin{vmatrix} 3 & -1 & -8 \\ 0 & -7 & 1 \\ 1 & 0 & 0 \end{vmatrix} = (1)\left(\begin{vmatrix} -1 & -8 \\ -7 & 1 \end{vmatrix}\right) = -57$$

$$D_z = \begin{vmatrix} 3 & -2 & -1 \\ 0 & 3 & -7 \\ 1 & 1 & 0 \end{vmatrix} \overset{-C_1+C_2}{=} \begin{vmatrix} 3 & -5 & -1 \\ 0 & 3 & -7 \\ 1 & 0 & 0 \end{vmatrix} = (1)\left(\begin{vmatrix} -5 & -1 \\ 3 & -7 \end{vmatrix}\right) = 38$$

We know immediately upon calculating D that the system has a unique solution, as $D \neq 0$. And after determining D_x, D_y, and D_z, we know the solution is the single ordered triple

$$(x, y, z) = \left(\frac{-19}{19}, \frac{-57}{19}, \frac{38}{19}\right) = (-1, -3, 2).$$

Why does Cramer's rule work? Once we have the formulas, we can use the properties of determinants (specifically, the first and third properties) to verify that they work. Consider again the generic n-equation, n-variable system

$$\begin{cases} a_{11}x_1 + a_{12}x_2 + \ldots + a_{1n}x_n = b_1 \\ a_{21}x_1 + a_{22}x_2 + \ldots + a_{2n}x_n = b_2 \\ \vdots \\ a_{n1}x_1 + a_{n2}x_2 + \ldots + a_{nn}x_n = b_n \end{cases}.$$

If we assume that the system has a unique solution, which we will denote $(x_1, x_2, \ldots, x_n)$, then we can rewrite D_{x_1} as follows:

$$D_{x_1} = \begin{vmatrix} b_1 & a_{12} & \ldots & a_{1n} \\ b_2 & a_{22} & \ldots & a_{2n} \\ \vdots & \vdots & \ldots & \vdots \\ b_n & a_{n2} & \ldots & a_{nn} \end{vmatrix} = \begin{vmatrix} a_{11}x_1 + a_{12}x_2 + \ldots + a_{1n}x_n & a_{12} & \ldots & a_{1n} \\ a_{21}x_1 + a_{22}x_2 + \ldots + a_{2n}x_n & a_{22} & \ldots & a_{2n} \\ \vdots & \vdots & \ldots & \vdots \\ a_{n1}x_1 + a_{n2}x_2 + \ldots + a_{nn}x_n & a_{n2} & \ldots & a_{nn} \end{vmatrix}$$

We can now use the third property of determinants $n - 1$ times to simplify the first column. We do this by multiplying the second column by $-x_2$ and adding it to the first, multiplying the third column by $-x_3$ and adding it to the first, and so on. After doing this, we obtain

$$D_{x_1} = \begin{vmatrix} a_{11}x_1 & a_{12} & \ldots & a_{1n} \\ a_{21}x_1 & a_{22} & \ldots & a_{2n} \\ \vdots & \vdots & \ldots & \vdots \\ a_{n1}x_1 & a_{n2} & \ldots & a_{nn} \end{vmatrix} = x_1 \begin{vmatrix} a_{11} & a_{12} & \ldots & a_{1n} \\ a_{21} & a_{22} & \ldots & a_{2n} \\ \vdots & \vdots & \ldots & \vdots \\ a_{n1} & a_{n2} & \ldots & a_{nn} \end{vmatrix} = x_1 D.$$

section 8.3 Determinants and Cramer's Rule

Note that we have used the first property of determinants to factor x_1 out of the determinant. We can now solve this equation for x_1 (since $D \neq 0$) to obtain

$$x_1 = \frac{D_{x_1}}{D},$$

and we can then repeat the process for the remaining variables $x_2, x_3, \ldots, x_n$.

exercises

Evaluate the following determinants. See Example 1.

1. $\begin{vmatrix} 4 & -3 \\ 1 & 2 \end{vmatrix}$
2. $\begin{vmatrix} 5 & -2 \\ 5 & -2 \end{vmatrix}$
3. $\begin{vmatrix} 0 & 3 \\ -5 & 2 \end{vmatrix}$
4. $\begin{vmatrix} 34 & -2 \\ 17 & -1 \end{vmatrix}$
5. $\begin{vmatrix} a & x \\ x & b \end{vmatrix}$
6. $\begin{vmatrix} 5x & 2 \\ -x & 1 \end{vmatrix}$

7. $\begin{vmatrix} -2 & 2 \\ -2 & -2 \end{vmatrix}$
8. $\begin{vmatrix} ac & 2ad \\ bc & db \end{vmatrix}$
9. $\begin{vmatrix} -1 & 2 \\ 3 & 4 \end{vmatrix}$
10. $\begin{vmatrix} w & x \\ y & z \end{vmatrix}$
11. $\begin{vmatrix} -2 & 9 \\ 5 & -3 \end{vmatrix}$
12. $\begin{vmatrix} 2y & 3x \\ y-1 & x^2 \end{vmatrix}$

Solve for x by finding the determinant.

13. $\begin{vmatrix} x-2 & 2 \\ 2 & x+1 \end{vmatrix} = 0$
14. $\begin{vmatrix} x+7 & -2 \\ 9 & x-2 \end{vmatrix} = 0$
15. $\begin{vmatrix} x+1 & 8 \\ 1 & x+3 \end{vmatrix} = 0$

16. $\begin{vmatrix} x-8 & 11 \\ -2 & x+5 \end{vmatrix} = 0$
17. $\begin{vmatrix} x+6 & 2 \\ -1 & x+3 \end{vmatrix} = 0$
18. $\begin{vmatrix} x-4 & -4 \\ 3 & x+9 \end{vmatrix} = 0$

19. $\begin{vmatrix} x+5 & 3 \\ 3 & x-3 \end{vmatrix} = 0$
20. $\begin{vmatrix} x+3 & 6 \\ 5 & x+7 \end{vmatrix} = 0$
21. $\begin{vmatrix} x-3 & 2 \\ 1 & x-4 \end{vmatrix} = 0$

Use the matrix $A = \begin{bmatrix} 2 & -1 & 5 \\ 0 & 1 & 3 \\ 1 & 0 & -2 \end{bmatrix}$ to evaluate the following. See Example 2.

22. The minor of a_{12}
23. The cofactor of a_{12}
24. The minor of a_{22}

25. The cofactor of a_{22}
26. The cofactor of a_{32}
27. The cofactor of a_{33}

28. The minor of a_{13}
29. The cofactor of a_{21}
30. The cofactor of a_{31}

chapter eight   Systems of Equations

Find the determinant of the matrix by the method of expansion by cofactors along the given row or column.

31. $\begin{vmatrix} 4 & 5 & 3 \\ -1 & 2 & 7 \\ 11 & 6 & 2 \end{vmatrix}$ Expand along Row 3

32. $\begin{vmatrix} 8 & 2 & 0 \\ 3 & 4 & 7 \\ 1 & 0 & 2 \end{vmatrix}$ Expand along Column 3

33. $\begin{vmatrix} 5 & 8 & 5 \\ 0 & -6 & 3 \\ 2 & 4 & -1 \end{vmatrix}$ Expand along Row 2

34. $\begin{vmatrix} -4 & 2 & 1 \\ 9 & 12 & 8 \\ 0 & 6 & -3 \end{vmatrix}$ Expand along Column 1

35. $\begin{vmatrix} 13 & 0 & -7 \\ 4 & 2 & 3 \\ 1 & 4 & 0 \end{vmatrix}$ Expand along Row 2

36. $\begin{vmatrix} 7 & 0 & 1 \\ 2 & 5 & 3 \\ 8 & 6 & 2 \end{vmatrix}$ Expand along Column 3

37. $\begin{vmatrix} 8 & 0 & -7 & 5 \\ 4 & -2 & 3 & 3 \\ -1 & 1 & 0 & 2 \\ 2 & 0 & 6 & 0 \end{vmatrix}$ Expand along Row 4

38. $\begin{vmatrix} 4 & -2 & 9 & 2 \\ 7 & 0 & 1 & 7 \\ -6 & 3 & 0 & 1 \\ 3 & 1 & 2 & 0 \end{vmatrix}$ Expand along Column 2

Evaluate the following determinants. In each case, minimize the required number of computations by carefully choosing a row or column to expand along, and use the properties of determinants to simplify the process. See Examples 3 and 4.

39. $\begin{vmatrix} 2 & 0 & 1 \\ -5 & 1 & 0 \\ 3 & -1 & 1 \end{vmatrix}$

40. $\begin{vmatrix} 12 & 3 & 1 \\ 1 & 1 & -1 \\ 0 & 2 & 0 \end{vmatrix}$

41. $\begin{vmatrix} 12 & 3 & 6 \\ 2 & 2 & -4 \\ 0 & 2 & 0 \end{vmatrix}$

42. $\begin{vmatrix} 1 & 2 & 3 \\ 4 & 5 & 6 \\ 7 & 8 & 9 \end{vmatrix}$

43. $\begin{vmatrix} 2 & 1 & -3 & 0 \\ 1 & -2 & 1 & 0 \\ 0 & 1 & 0 & 1 \\ 2 & 0 & 1 & 1 \end{vmatrix}$

44. $\begin{vmatrix} x & 0 & 0 & 0 \\ 0 & x & 0 & 0 \\ 0 & 0 & x & 0 \\ 0 & 0 & 0 & x \end{vmatrix}$

45. $\begin{vmatrix} x & x & x & x \\ 0 & x & x & x \\ 0 & 0 & x & x \\ 0 & 0 & 0 & x \end{vmatrix}$

46. $\begin{vmatrix} 0 & 2 & 0 & 0 \\ -2 & -4 & 5 & 9 \\ 1 & 3 & -1 & 1 \\ 0 & 7 & 0 & 2 \end{vmatrix}$

47. $\begin{vmatrix} x & x & 0 & 0 \\ yz & x^3 & z & x^4 \\ z & xy & x & 0 \\ x^2 & 0 & 0 & 0 \end{vmatrix}$

Determinants and Cramer's Rule

Using a graphing calculator or other computational aid, find the determinant of the matrix.

48. $\begin{vmatrix} 0.1 & 0.4 & -0.7 \\ 0.3 & -0.1 & 0.2 \\ 0.5 & -0.2 & 0.3 \end{vmatrix}$
49. $\begin{vmatrix} 0.1 & 0.3 & 0.1 \\ 0.2 & -0.2 & -0.1 \\ -0.1 & -0.4 & 0.5 \end{vmatrix}$
50. $\begin{vmatrix} 2.2 & 0.3 & -1.7 \\ 0.4 & -0.2 & 0.1 \\ 0.2 & 0.3 & -1.6 \end{vmatrix}$

51. $\begin{vmatrix} 3.1 & 0.6 & -1.1 \\ 1.2 & 5.2 & -7.3 \\ -0.1 & -4.1 & 6.5 \end{vmatrix}$
52. $\begin{vmatrix} 13 & 23 & -21 \\ 17 & -32 & 14 \\ 15 & 12 & -16 \end{vmatrix}$
53. $\begin{vmatrix} 25 & 32 & 17 \\ -13 & 14 & -24 \\ 16 & 26 & 36 \end{vmatrix}$

Use Cramer's rule to solve the following systems. See Examples 5 and 6.

54. $\begin{cases} 2x - 3y = 8 \\ 8x + 5y = -2 \end{cases}$
55. $\begin{cases} 5x + 7y = 9 \\ 2x + 3y = -7 \end{cases}$
56. $\begin{cases} 5x - 10y = 9 \\ -x + 2y = -3 \end{cases}$

57. $\begin{cases} -2x - 2y = 4 \\ 3x + 3y = -6 \end{cases}$
58. $\begin{cases} \dfrac{2}{3}x + y = -3 \\ 3x + \dfrac{5}{2}y = -\dfrac{7}{2} \end{cases}$
59. $\begin{cases} \dfrac{2}{3}x + 2y = 1 \\ x + 3y = 0 \end{cases}$

60. $\begin{cases} x + 2y = -1 \\ y + 3z = 7 \\ 2x + 5z = 21 \end{cases}$
61. $\begin{cases} 2x - y = 0 \\ 5x - 3y - 3z = 5 \\ 2x + 6z = -10 \end{cases}$
62. $\begin{cases} 3x + 8z = 3 \\ -3x - 7z = -3 \\ x + 3z = 1 \end{cases}$

63. $\begin{cases} 3w - x + 5y + 3z = 2 \\ -4w - 10y - 2z = 10 \\ w - x + 2z = 7 \\ 4w - 2x + 5y + 5z = 9 \end{cases}$
64. $\begin{cases} 2w + x - 3y = 3 \\ w - 2x + y = 1 \\ x + z = -2 \\ y + z = 0 \end{cases}$
65. $\begin{cases} 3w - 2x + y - 5z = -1 \\ w + x - y + 4z = 2 \\ 4w - x - z = 1 \\ 5w - x = 9 \end{cases}$

66. $\begin{cases} -4x + y = 1 \\ 7x + 2y = 407 \end{cases}$
67. $\begin{cases} 5x - 4y = -49 \\ 24x - 19y = 179 \end{cases}$
68. $\begin{cases} 2w - 3x + 4y - z = 21 \\ w + 5x = 2 \\ -2x + 3y + z = 12 \\ -3w + 4z = -5 \end{cases}$

69. $\begin{cases} -5x + 10y = 3 \\ \dfrac{7}{2}x - 7y = 20 \end{cases}$
70. $\begin{cases} 23x + 21y = -4 \\ x - 3y = -8 \end{cases}$
71. $\begin{cases} w - x + y - z = 2 \\ 2w - x + 3y = -5 \\ x - 2z = 7 \\ 3w + 4x = -13 \end{cases}$

chapter eight   Systems of Equations

Use a graphing calculator or other computational aid and Cramer's rule to solve the systems of equations.

72. $\begin{cases} x - 2y + 3z = 9 \\ -x + 3y = -4 \\ 2x - 5y + 5z = 17 \end{cases}$

73. $\begin{cases} 2x + 4y + z = 1 \\ x - 2y - 3z = 2 \\ x + y - z = -1 \end{cases}$

74. $\begin{cases} w + x + y + z = 6 \\ -w + 2x + 3y = 0 \\ 2w - 3x + 4y + z = 4 \\ w + x + 2y - z = 0 \end{cases}$

Use the information below to set up systems of linear equations and then solve each system by using Cramer's Rule.

75. The three sides of a triangle are related as follows: the perimeter is 43 feet, the second side is 5 feet more than twice the first side, and the third side is 3 feet less than the sum of the other two sides. Find the lengths of the three sides of the triangle.

76. Eric's favorite candy bar and ice cream flavor have fat and calorie contents as follows: each candy bar has 5 grams of fat and 280 calories; each serving of ice cream has 10 grams of fat and 150 calories. How many candy bars and servings of ice cream did he eat during the weekend he consumed 85 grams of fat and 2300 calories from these two treats?

77. A farmer plants soybeans, corn, and wheat and rotates the planting each year on his 500-acre farm. In a particular year, the profits of her crops were $120 per acre of soybeans, $100 per acre of corn, and $80 per acre of wheat. She planted twice as many acres with corn as with soybeans. How many acres did she plant with each crop the year she made a total profit of $51,800?

8.4 The Algebra of Matrices

TOPICS

1. Matrix addition
2. Scalar multiplication
3. Matrix multiplication
4. Interlude: transition matrices

Topic 1: Matrix Addition

As you continue on in mathematics, you will encounter many sets of objects that can be combined in ways similar to the ways real numbers can be combined. We have, in fact, seen several examples already in this text: addition, subtraction, multiplication, and division are all defined on the set of complex numbers and on real-valued functions. Loosely speaking, an *algebra* is a collection of objects together with one or more methods of combining the objects. In this section, we will see that a very natural algebra exists for sets of matrices. We start by defining what it means to add two matrices.

Matrix Addition

> Two matrices A and B can be added to form the new matrix $A + B$ only if A and B are of the same order. The addition is performed by adding corresponding entries of the two matrices together; that is, the element in the ith row and the jth column of $A + B$ is given by $a_{ij} + b_{ij}$.

It is very important to note the restriction on the order of the two matrices in the above definition; a matrix with m rows and n columns can only be added to another matrix with m rows and n columns. It is, in fact, a hallmark of matrix algebra that the orders of the matrices under discussion must always be considered. This is true even for the concept of *matrix equality*, which we now define so we can use the equality symbol in our work with matrices.

chapter eight   Systems of Equations

Matrix Equality

Two matrices A and B are said to be **equal**, denoted $A = B$, if they are of the same order and corresponding entries of A and B are equal.

example 1

To add matrices, use the **Matrix** operation (pg. 606) to define matrices A and B. Then **Quit** (**2ND** **MODE**); enter A by selecting **Matrix 1**, enter "+", and then select **Matrix 2** (B).

Perform the indicated addition, if possible.

a. $\begin{bmatrix} -3 & 2 \\ 0 & -5 \\ 11 & -9 \end{bmatrix} + \begin{bmatrix} 3 & 17 \\ 5 & 4 \\ -10 & 4 \end{bmatrix}$

b. $\begin{bmatrix} 2 & -5 \\ 1 & 0 \\ 0 & 3 \\ -7 & 10 \end{bmatrix} + \begin{bmatrix} 2 & 1 & 0 & -7 \\ -5 & 0 & 3 & 10 \end{bmatrix}$

Solutions:

a. Both matrices are 3×2, so the sum is defined and

$$\begin{bmatrix} -3 & 2 \\ 0 & -5 \\ 11 & -9 \end{bmatrix} + \begin{bmatrix} 3 & 17 \\ 5 & 4 \\ -10 & 4 \end{bmatrix} = \begin{bmatrix} 0 & 19 \\ 5 & -1 \\ 1 & -5 \end{bmatrix}.$$

b. The first matrix is 4×2 and the second is 2×4, so the sum cannot be performed.

Of course, the entries in matrices do not all have to be constants. In many applications of matrices, we will find it convenient to represent some of the entries initially as variables, with the intent of eventually solving for the variables. We can solve some examples of such *matrix equations* now, with nothing other than matrix addition and matrix equality defined.

example 2

Determine the values of the variables that will make each of the following statements true.

a. $\begin{bmatrix} -3 & a & b \\ -2 & a+b & 5 \end{bmatrix} = \begin{bmatrix} c & 3 & 7 \\ -2 & d & 5 \end{bmatrix}$ **b.** $\begin{bmatrix} 3x \\ 4 \end{bmatrix} + \begin{bmatrix} -y \\ 2x \end{bmatrix} = \begin{bmatrix} 13 \\ 7y \end{bmatrix}$

Solutions:

a. Solving for the four variables $a, b, c,$ and d in the equation

$$\begin{bmatrix} -3 & a & b \\ -2 & a+b & 5 \end{bmatrix} = \begin{bmatrix} c & 3 & 7 \\ -2 & d & 5 \end{bmatrix}$$

is just a matter of comparing the entries one-by-one. Beginning with the first rows, we note that $c = -3$, $a = 3$, and $b = 7$. Of course, we have no control over the first and third entries in the second row of each matrix; if the constants in these two positions were not equal, there would be no way to make the matrix equality true. The only variable left to solve for is d, and since we have already determined a and b, it must be the case that $d = 10$.

b. Each matrix consists of only two entries, and after performing the matrix addition on the left and comparing corresponding entries, we arrive at the system of equations

$$\begin{cases} 3x - y = 13 \\ 4 + 2x = 7y \end{cases}.$$

We know a number of ways of solving such a system. If we choose to use Cramer's rule, we first rewrite the system as

$$\begin{cases} 3x - y = 13 \\ 2x - 7y = -4 \end{cases},$$

from which we can determine

$$D = \begin{vmatrix} 3 & -1 \\ 2 & -7 \end{vmatrix} = (3)(-7) - (2)(-1) = -19,$$

$$D_x = \begin{vmatrix} 13 & -1 \\ -4 & -7 \end{vmatrix} = (13)(-7) - (-4)(-1) = -95, \text{ and}$$

$$D_y = \begin{vmatrix} 3 & 13 \\ 2 & -4 \end{vmatrix} = (3)(-4) - (2)(13) = -38,$$

so $x = \dfrac{D_x}{D} = \dfrac{-95}{-19} = 5$ and $y = \dfrac{D_y}{D} = \dfrac{-38}{-19} = 2$.

chapter eight   Systems of Equations

Topic 2: Scalar Multiplication

In the context of matrix algebra, a *scalar* is a real number (or, more generally, a complex number), and *scalar multiplication* refers to the product of a real number and a matrix. At first glance, it may seem strange to combine two very different sorts of objects (a scalar and a matrix), but the notion arises naturally from considering such matrix sums as

$$\begin{bmatrix} -5 & 2 \\ 1 & -3 \\ -2 & 7 \end{bmatrix} + \begin{bmatrix} -5 & 2 \\ 1 & -3 \\ -2 & 7 \end{bmatrix} = \begin{bmatrix} -10 & 4 \\ 2 & -6 \\ -4 & 14 \end{bmatrix}.$$

Since the result of adding a matrix to itself has the effect of doubling each entry, we are led to write

$$2\begin{bmatrix} -5 & 2 \\ 1 & -3 \\ -2 & 7 \end{bmatrix} = \begin{bmatrix} -10 & 4 \\ 2 & -6 \\ -4 & 14 \end{bmatrix}.$$

Extending the idea to arbitrary scalars leads to the following definition.

Scalar Multiplication

If A is an $m \times n$ matrix and c is a scalar, cA stands for the $m \times n$ matrix for which each entry is c times the corresponding entry of A. In other words, the entry in the ith row and j^{th} column of cA is ca_{ij}.

example 3

Given the matrices

$$A = \begin{bmatrix} -1 & 6 & 2 \\ -8 & 0 & 1 \end{bmatrix} \text{ and } B = \begin{bmatrix} 0 & -3 & 4 \\ 1 & -2 & 6 \end{bmatrix},$$

write $-3A + 2B$ as a single matrix.

Solution:

Our work consists of two tasks.

section 8.4 The Algebra of Matrices

The first is to multiply each entry of either A or B by its corresponding scalar, and the second is to then add the resulting matrices together:

$$-3A+2B = -3\begin{bmatrix} -1 & 6 & 2 \\ -8 & 0 & 1 \end{bmatrix} + 2\begin{bmatrix} 0 & -3 & 4 \\ 1 & -2 & 6 \end{bmatrix}$$

$$= \begin{bmatrix} 3 & -18 & -6 \\ 24 & 0 & -3 \end{bmatrix} + \begin{bmatrix} 0 & -6 & 8 \\ 2 & -4 & 12 \end{bmatrix}$$

$$= \begin{bmatrix} 3 & -24 & 2 \\ 26 & -4 & 9 \end{bmatrix}$$

Enter the matrices in A and B using the **Matrix** command. Then **Quit** and enter the expression, selecting each matrix as desired through the **Matrix** menu.

Remember that our overall goal is to develop an algebra of matrices, and we can use the algebra of real numbers as a model. Since subtraction of real numbers is defined in terms of addition $(a-b = a+(-b))$, we can define matrix subtraction in the same way: if A and B are two matrices of the same order, then

$$A - B = A + (-B).$$

example 4

Perform the indicated subtraction: $\begin{bmatrix} 3 & -5 & 2 \end{bmatrix} - \begin{bmatrix} -2 & -5 & 3 \end{bmatrix}$.

Solution:

As both matrices are of order 1×3, we know the subtraction is possible. The computation consists of subtracting entries in the second matrix from the corresponding entries in the first:

$$\begin{bmatrix} 3 & -5 & 2 \end{bmatrix} - \begin{bmatrix} -2 & -5 & 3 \end{bmatrix} = \begin{bmatrix} 3-(-2) & -5-(-5) & 2-3 \end{bmatrix} = \begin{bmatrix} 5 & 0 & -1 \end{bmatrix}$$

Topic 3: Matrix Multiplication

Unlike the definitions of matrix addition and scalar multiplication, both of which probably seem fairly natural and expected, the definition of matrix multiplication may at first be confusing and appear unnecessarily complicated. For one thing, matrix multiplication does *not* refer to multiplying corresponding entries of two matrices together. Such a definition would be undeniably simpler than the actual definition, but matrix multiplication is defined the way it is because of the demands of applications of matrices.

chapter eight   Systems of Equations

One of the best ways of motivating the definition of matrix multiplication requires a bit of preamble. To begin with, it is useful to consider matrices as functions. Note that the system of equations

$$\begin{cases} x' = ax + by \\ y' = cx + dy \end{cases}$$

can be viewed as a function that transforms the ordered pair (x, y) into the ordered pair (x', y'), and that the function is characterized entirely by the matrix

$$A = \begin{bmatrix} a & b \\ c & d \end{bmatrix}.$$

Similarly, the system

$$\begin{cases} x' = ex + fy \\ y' = gx + hy \end{cases}$$

is a function that transforms ordered pairs into ordered pairs, and it is associated with the matrix

$$B = \begin{bmatrix} e & f \\ g & h \end{bmatrix}.$$

Consider now the result of plugging the output of the first function into the second function (that is, of *composing* the two functions together so that B acts on the result of A). If we let (x', y') denote the output of the first function, we obtain the new ordered pair (x'', y'') given by

$$\begin{cases} x'' = ex' + fy' \\ y'' = gx' + hy' \end{cases} \text{ or } \begin{cases} x'' = e(ax+by) + f(cx+dy) \\ y'' = g(ax+by) + h(cx+dy) \end{cases}.$$

We can change the way the last system is written to obtain

$$\begin{cases} x'' = (ea + fc)x + (eb + fd)y \\ y'' = (ga + hc)x + (gb + hd)y \end{cases}$$

so the composition of the two functions, in the order BA, is characterized entirely by the matrix

$$\begin{bmatrix} ea + fc & eb + fd \\ ga + hc & gb + hd \end{bmatrix}.$$

Note that the entries of this last matrix are constructed by a combination of products and sums, where the products in each sum are between elements of the rows of B with elements of the columns of A. This is the basis of our formal definition of matrix multiplication.

section 8.4   The Algebra of Matrices

Matrix Multiplication

Two matrices A and B can be multiplied together, resulting in a new matrix denoted AB, only if the number of columns of A (the matrix on the left) is the same as the number of rows of B (the matrix on the right). Thus, if A is of order $m \times n$, the product AB is only defined if B is of order $n \times p$. The order of AB in this case will be $m \times p$.

If we let c_{ij} denote the entry in the ith row and j^{th} column of AB, c_{ij} is obtained from the ith row of A and the jth column of B by the formula
$$c_{ij} = a_{i1}b_{1j} + a_{i2}b_{2j} + \ldots + a_{in}b_{nj}.$$

caution

Unlike numerical multiplication, matrix multiplication is not commutative. That is, given two matrices A and B, AB *in general* is not equal to BA. As an illustration of this fact, suppose A is a 3×4 matrix and B is a 4×2 matrix. Then AB is defined (and is of order 3×2), but BA doesn't even exist. We will see many more examples of the non-commutativity of matrix multiplication.

example 5

Given the two matrices $A = \begin{bmatrix} 2 & -3 \\ 4 & -1 \\ 1 & 0 \end{bmatrix}$ and $B = \begin{bmatrix} 5 & 0 & -2 \\ -4 & 1 & 3 \end{bmatrix}$, find the following products.

a. AB **b.** BA

Solutions:

a. A is of order 3×2 and B is of order 2×3, so AB is defined and will be of order 3×3. Remember that each entry of AB is formed from a row of A and a column of B. For instance, the first-row, first-column entry of AB is calculated as follows:

$$AB = \begin{bmatrix} 2 & -3 \\ 4 & -1 \\ 1 & 0 \end{bmatrix} \begin{bmatrix} 5 & 0 & -2 \\ -4 & 1 & 3 \end{bmatrix} = \begin{bmatrix} (2)(5)+(-3)(-4)=22 & ? & ? \\ ? & ? & ? \\ ? & ? & ? \end{bmatrix}$$

continued on next page...

641

Similarly, the entry in the third row and second column of AB comes from the third row of A and the second column of B:

$$AB = \begin{bmatrix} 2 & -3 \\ 4 & -1 \\ 1 & 0 \end{bmatrix} \begin{bmatrix} 5 & 0 & -2 \\ -4 & 1 & 3 \end{bmatrix} = \begin{bmatrix} ? & ? & ? \\ ? & ? & ? \\ ? & (1)(0)+(0)(1)=0 & ? \end{bmatrix}$$

Continuing the process, we obtain

$$AB = \begin{bmatrix} 2 & -3 \\ 4 & -1 \\ 1 & 0 \end{bmatrix} \begin{bmatrix} 5 & 0 & -2 \\ -4 & 1 & 3 \end{bmatrix} = \begin{bmatrix} 22 & -3 & -13 \\ 24 & -1 & -11 \\ 5 & 0 & -2 \end{bmatrix}.$$

b. From the orders of the two matrices, we know BA will be a 2×2 matrix. Note that each entry of BA is a sum of three products, the first of which is

$$BA = \begin{bmatrix} 5 & 0 & -2 \\ -4 & 1 & 3 \end{bmatrix} \begin{bmatrix} 2 & -3 \\ 4 & -1 \\ 1 & 0 \end{bmatrix} = \begin{bmatrix} (5)(2)+(0)(4)+(-2)(1)=8 & ? \\ ? & ? \end{bmatrix}.$$

Continuing, the final answer is

$$BA = \begin{bmatrix} 5 & 0 & -2 \\ -4 & 1 & 3 \end{bmatrix} \begin{bmatrix} 2 & -3 \\ 4 & -1 \\ 1 & 0 \end{bmatrix} = \begin{bmatrix} 8 & -15 \\ -1 & 11 \end{bmatrix}.$$

section 8.4   **The Algebra of Matrices**

technology note

Computer algebra systems and sufficiently-advanced calculators are all capable of performing matrix addition, scalar multiplication, and matrix multiplication, though the exact notation used varies from one technology to another. In the illustration below, the matrices A and B from *Example 5*, along with a third matrix M, are defined in *Mathematica*. The expression $3B - 2M$ is then evaluated (note that B and M are the same size, so this expression makes sense), and then the matrix products AB and BA are evaluated. The notation for matrix multiplication in *Mathematica* is a period between the two matrices.

Now that matrix multiplication has been defined, it can be used in another way to illustrate the fact that matrices can be considered as functions. As already mentioned, the matrix

$$A = \begin{bmatrix} a & b \\ c & d \end{bmatrix}$$

completely characterizes the function that transforms (x, y) into (x', y') in the system

$$\begin{cases} x' = ax + by \\ y' = cx + dy \end{cases}.$$

What we mean by this is that the four numbers a, b, c, and d, along with knowledge of their placement in the matrix, suffice to describe the function. This is even more clear if we associate the ordered pair (x, y) with the 2×1 matrix

$$\begin{bmatrix} x \\ y \end{bmatrix}.$$

Then the matrix product

$$\begin{bmatrix} a & b \\ c & d \end{bmatrix} \begin{bmatrix} x \\ y \end{bmatrix},$$

which has the "look" of a function acting on its argument, gives us the expressions in the system. Verify for yourself that

$$\begin{bmatrix} a & b \\ c & d \end{bmatrix} \begin{bmatrix} x \\ y \end{bmatrix} = \begin{bmatrix} ax + by \\ cx + dy \end{bmatrix}.$$

Topic 4: Interlude: Transition Matrices

In a variety of important applications, matrices can be used to model how the "state" of a situation changes over time. To illustrate the idea, we will consider one such situation in some detail.

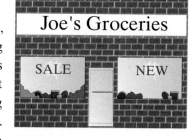

Suppose a new grocery store opens in a small town, with the intent of competing with the one existing store. The new store, which we will call store A, begins an aggressive advertising campaign, with the result that every month 45% of the customers of the existing store (store B) decide to start shopping at store A. Store B, however, responds with a strong appeal to win back customers and every month 30% of the customers of store A return to store B. In this highly dynamic situation, a number of questions may arise in the minds of the managers of the two stores, among them:

1. Given the known number of customers one month, how many customers can be expected the following month?
2. Can a certain percentage of the total number of the town's shoppers be expected in the long run? If so, what is that percentage?

To answer the first question, let a denote the number of customers of store A in a given month, and let b denote the number of customers of store B in the same month. In this simple example, we will assume that $a + b$, the total number of grocery store customers in the town, remains fixed over time.

section 8.4 The Algebra of Matrices

In the following month, store A gains $0.45b$ customers (45% of its competitor's customers) and retains $0.7a$ of its existing customers. At the same time, store B gains $0.3a$ new customers (the 30% of A's clientele that switch back to store B), and retains $0.55b$ existing customers. In equation form, the number of customers of each store the following month are given by the following two expressions:

$$\begin{cases} 0.7a + 0.45b \\ 0.3a + 0.55b \end{cases}$$

Matrix multiplication allows us to write this information as

$$\begin{bmatrix} 0.7 & 0.45 \\ 0.3 & 0.55 \end{bmatrix} \begin{bmatrix} a \\ b \end{bmatrix},$$

and if we let P stand for the 2×2 matrix of percentages, we can think of P as a function that transforms one month's distribution of customers into the next month's. For instance, if store A has 360 customers and store B has 640 customers one month, then the following month their respective number of customers will be 540 and 460, as

$$\begin{bmatrix} 0.7 & 0.45 \\ 0.3 & 0.55 \end{bmatrix} \begin{bmatrix} 360 \\ 640 \end{bmatrix} = \begin{bmatrix} 540 \\ 460 \end{bmatrix}.$$

In such situations, a matrix like P is called a **transition matrix**, as it characterizes the transition of the system from one state to the next. Transition matrices are characterized by the fact that all of their entries are positive, and the sum of the entries in each column is 1.

To answer the second question, we might continue the above specific example and ask how the 540 and 460 customers of the respective stores realign themselves one month later. Note that

$$\begin{bmatrix} 0.7 & 0.45 \\ 0.3 & 0.55 \end{bmatrix} \begin{bmatrix} 540 \\ 460 \end{bmatrix} = \begin{bmatrix} 585 \\ 415 \end{bmatrix},$$

meaning that at the end of the second month, store A has 585 customers and store B has 415. But we can obtain the same result by applying the matrix P^2 to the original distribution of 360 and 640 customers, as

$$P^2 \begin{bmatrix} 360 \\ 640 \end{bmatrix} = P\left(P\begin{bmatrix} 360 \\ 640 \end{bmatrix}\right) = P\begin{bmatrix} 540 \\ 460 \end{bmatrix}.$$

In other words, we can form the composition of P with itself to obtain a function that transforms one month's distribution of customers into the distribution two months later. Note that

$$P^2 = \begin{bmatrix} 0.625 & 0.5625 \\ 0.375 & 0.4375 \end{bmatrix}$$

and that

$$\begin{bmatrix} 0.625 & 0.5625 \\ 0.375 & 0.4375 \end{bmatrix} \begin{bmatrix} 360 \\ 640 \end{bmatrix} = \begin{bmatrix} 585 \\ 415 \end{bmatrix}.$$

chapter eight   Systems of Equations

At this point, a calculator or computer software with matrix capability may be useful. We can use either aid to calculate high powers of P to see if there is a long-term trend in the distribution of the customers. For instance, to six decimal places,

$$P^5 = \begin{bmatrix} 0.600391 & 0.599414 \\ 0.399609 & 0.400586 \end{bmatrix},$$

and after a certain point calculators and software will round off the entries and give us a result of

$$P^n = \begin{bmatrix} 0.6 & 0.6 \\ 0.4 & 0.4 \end{bmatrix}$$

for large n (the value for n at which this happens will vary depending on the technology used). This means that, after a few months, store A can count on roughly 60% of the town's customers and store B can count on roughly 40% (the actual identities of the customers will, of course, keep changing from month to month, but the relative proportions will have stabilized). We can verify that the situation is stable by applying the transition matrix to an assumed 1000 customers split 60/40:

$$\begin{bmatrix} 0.7 & 0.45 \\ 0.3 & 0.55 \end{bmatrix} \begin{bmatrix} 600 \\ 400 \end{bmatrix} = \begin{bmatrix} 600 \\ 400 \end{bmatrix}.$$

You will see another approach to determining the stable long-term state in the exercises.

exercises

Given $A = \begin{bmatrix} 3 & -2 \\ 1 & 0 \\ 0 & 5 \end{bmatrix}$, $B = \begin{bmatrix} 4 & -5 \\ 3 & 0 \\ -2 & 2 \end{bmatrix}$, $C = \begin{bmatrix} 2 & -1 \\ 6 & 10 \\ -3 & 7 \end{bmatrix}$, and $D = \begin{bmatrix} 3 & 2 & 5 \\ -2 & -4 & 1 \end{bmatrix}$, determine the following, if possible. See Examples 1 through 4.

1. $3A - B$ 2. $B - 2D$ 3. $3C$ 4. $\frac{1}{2}D$ 5. $3D + C$ 6. $A + B + C$

7. $2A + 2B$ 8. $\frac{3}{2}B + \frac{1}{2}C$ 9. $C - 3A$ 10. $3C - A$ 11. $4A - 3D$ 12. $2(A - 3B)$

Determine values of the variables that will make the following equations true, if possible. See Examples 1 through 4.

13. $[a \quad 2b \quad c] + 3[a \quad 2 \quad -c] = [8 \quad 2 \quad 2]$

14. $\begin{bmatrix} w & 5x \\ 2y & z \end{bmatrix} - 5\begin{bmatrix} w & x \\ y & -z \end{bmatrix} = \begin{bmatrix} w+5 & 0 \\ 6 & 1 \end{bmatrix}$

15. $\begin{bmatrix} 3x \\ 2y \end{bmatrix} + \begin{bmatrix} x \\ -y \\ z \end{bmatrix} = \begin{bmatrix} 4 \\ 0 \\ 2 \end{bmatrix}$

16. $[2a \quad 3b \quad c] = \begin{bmatrix} 4 \\ 3 \\ 0 \end{bmatrix}$

17. $\begin{bmatrix} x \\ 3x \end{bmatrix} - \begin{bmatrix} y \\ 2y \end{bmatrix} = \begin{bmatrix} 5 \\ 20 \end{bmatrix}$

18. $7\begin{bmatrix} -1 \\ y \end{bmatrix} = \begin{bmatrix} 2x \\ 5x \end{bmatrix} + 3\begin{bmatrix} y \\ 1 \end{bmatrix}$

19. $2\begin{bmatrix} x \\ 2y \end{bmatrix} - 3\begin{bmatrix} 5y \\ -3x \end{bmatrix} = \begin{bmatrix} -9 \\ 31 \end{bmatrix}$

20. $2[3r \quad s \quad 2t] - [r \quad s \quad t] = [15 \quad 3 \quad 9]$

21. $2\begin{bmatrix} 2x^2 & x \\ 7x & 4 \end{bmatrix} - \begin{bmatrix} 5x & 0 \\ x-2 & \end{bmatrix} = \begin{bmatrix} 2x & 0 \\ 6 & x^2 \end{bmatrix}$

22. $\begin{bmatrix} -x \\ 3 \end{bmatrix} - 5\begin{bmatrix} 2 \\ y \end{bmatrix} = \begin{bmatrix} -2y \\ 3x \end{bmatrix}$

23. $3\begin{bmatrix} 2a \\ -a \end{bmatrix} - 3\begin{bmatrix} 3b \\ 2b \end{bmatrix} = \begin{bmatrix} 3 \\ -54 \end{bmatrix}$

24. $2\begin{bmatrix} -s \\ -7 \end{bmatrix} + 2\begin{bmatrix} -2r \\ r \end{bmatrix} = -2\begin{bmatrix} 8 \\ s \end{bmatrix}$

Evaluate the following matrix products, if possible. See Example 5.

25. $[3 \quad -2 \quad 1]\begin{bmatrix} 5 & -1 \\ 0 & 3 \\ 9 & 4 \end{bmatrix}$

26. $\begin{bmatrix} 0 & -8 \\ 5 & 6 \end{bmatrix}[3 \quad 7]$

27. $[3 \quad 7]\begin{bmatrix} 0 & -8 \\ 5 & 6 \end{bmatrix}$

28. $[5 \quad 0 \quad -3]\begin{bmatrix} 4 \\ 2 \\ -6 \end{bmatrix}$

29. $\begin{bmatrix} 3 & 9 & -4 \\ 0 & 0 & 2 \\ 5 & -2 & 7 \end{bmatrix}\begin{bmatrix} 3 & 2 \\ 2 & 1 \end{bmatrix}$

30. $\begin{bmatrix} 4 \\ 2 \\ -6 \end{bmatrix}[5 \quad 0 \quad -3]$

31. $[-3 \quad -6 \quad -3]\begin{bmatrix} 6 & 9 \\ 6 & -8 \\ -8 & 8 \end{bmatrix}$

32. $\begin{bmatrix} 4 & -5 \\ 7 & -9 \end{bmatrix}[-8 \quad 3]$

33. $\begin{bmatrix} -3 \\ -5 \\ -6 \end{bmatrix}[-5 \quad 1 \quad 8]$

chapter eight Systems of Equations

Given $A = \begin{bmatrix} -3 & 1 \\ 2 & 3 \end{bmatrix}$, $B = \begin{bmatrix} 8 & -5 \end{bmatrix}$, $C = \begin{bmatrix} 4 \\ 7 \\ -2 \end{bmatrix}$, and $D = \begin{bmatrix} -5 & 4 \\ -1 & -1 \end{bmatrix}$, determine the following, if possible. See Example 5.

34. AB **35.** BA **36.** $BA + B$ **37.** A^2 **38.** C^2 **39.** CB

40. D^2 **41.** $CD + C$ **42.** DA **43.** AD **44.** DB **45.** $(BD)A$

Given $A = \begin{bmatrix} 3.8 & -1.2 & 4.6 \end{bmatrix}$, $B = \begin{bmatrix} -8.2 & -4.9 \\ 7.4 & -1.3 \\ 3.5 & -2.1 \end{bmatrix}$, $C = \begin{bmatrix} 6.3 \\ 5.7 \end{bmatrix}$, and $D = \begin{bmatrix} 2.8 & -7.1 \\ -5.4 & 6.6 \end{bmatrix}$, use a graphing calculator to determine the following, if possible.

46. BD **47.** CA **48.** D^2

49. AB **50.** DC **51.** BC

The following questions refer to the discussion of transition matrices in this section.

52. Suppose that each month 20% of store B's customers switch to store A, and 10% of store A's customers switch back to store B. At the start of January, store A has 300 customers and store B has 700. How many customers can each store expect at the start of February? At the start of March?

53. Given the percentages stated in the last problem, what long-term proportion of the town's customers can each store expect? (A technological aid may be used to compute high powers of the transition matrix, or you can use the method described in the following exercise.)

section 8.4   **The Algebra of Matrices**

54. Suppose P is a 2×2 transition matrix, and we want to determine the effect of applying high powers of P to the matrix

$$\begin{bmatrix} x \\ y \end{bmatrix},$$

where $x + y$ is a fixed constant, say c. (In our competing store situation, $x + y = 1000$.) If the long-term behavior approaches a steady state, as in our two-store example, then there is some value for x and some value for y such that $x + y = c$ and

$$P \begin{bmatrix} x \\ y \end{bmatrix} = \begin{bmatrix} x \\ y \end{bmatrix}.$$

In other words, once the steady state has been reached, applying the matrix P to it has no effect on the state.

We can use this fact to actually solve for x and y as follows. Given the matrix

$$P = \begin{bmatrix} 0.7 & 0.45 \\ 0.3 & 0.55 \end{bmatrix},$$

write the equation

$$P \begin{bmatrix} x \\ y \end{bmatrix} = \begin{bmatrix} x \\ y \end{bmatrix}$$

in system form. You should find that the two equations that result are actually identical. But if we now also use the fact that $x + y = 1000$, we can solve for the variables and find that $x = 600$ and $y = 400$. Verify that this is indeed the case.

55. Your friend Jared is having trouble with matrices, so you offer to study with him. Check his solution of the following problem. If the solution is incorrect, explain the error that has been made.

$$2 \begin{bmatrix} -8 & -9 & -1 \\ -8 & 1 & 5 \end{bmatrix} - 2 \begin{bmatrix} -2 & 3 \\ 5 & -7 \\ -8 & -1 \end{bmatrix}$$

$$= \begin{bmatrix} -16 & -18 & -2 \\ -16 & 2 & 10 \end{bmatrix} + \begin{bmatrix} 4 & -6 \\ -10 & 14 \\ 16 & 2 \end{bmatrix}$$

$$= \begin{bmatrix} -16+4-18-10-2+16 & -16-6-18+14-2+2 \\ -16+4+2-10+10+2 & -16-6+2+14+10+2 \end{bmatrix}$$

$$= \begin{bmatrix} -26 & -26 \\ -8 & 6 \end{bmatrix}$$

649

chapter eight   Systems of Equations

8.5 Inverses of Matrices

TOPICS

1. The matrix form of a linear system
2. Finding the inverse of a matrix
3. Using matrix inverses to solve linear systems

Topic 1: The Matrix Form of a Linear System

In the last section, we noted that matrix multiplication allows us to think of matrices as functions. For example, if we express the ordered pair (x, y) as a 2×1 matrix, then the linear system

$$\begin{cases} ax + by = e \\ cx + dy = f \end{cases}$$

can be written as

$$\begin{bmatrix} a & b \\ c & d \end{bmatrix} \begin{bmatrix} x \\ y \end{bmatrix} = \begin{bmatrix} e \\ f \end{bmatrix}.$$

While it can be verified (after multiplying out the left-hand side) that the matrix equation is equivalent to the system of equations above it, this is not, by any means, a trivial observation. For one thing, the matrix form of a linear system is more efficient: it converts a system of any number of equations into a single matrix equation. More importantly, it uses the function interpretation of a matrix to express a system of equations in a form reminiscent of linear equations of a single variable. Since the generic linear equation $ax = b$ can be solved by dividing both sides by a,

$$ax = b \Leftrightarrow x = \frac{b}{a} \quad (\text{assuming } a \neq 0)$$

it is tempting to solve

$$\begin{bmatrix} a & b \\ c & d \end{bmatrix} \begin{bmatrix} x \\ y \end{bmatrix} = \begin{bmatrix} e \\ f \end{bmatrix}$$

by "dividing" both sides by the 2×2 matrix of coefficients. Unfortunately, while we can add, subtract, and multiply matrices (at least when the dimensions match up appropriately), we don't have any way yet of making sense of "matrix division."

We will return to this thought soon, but first we will see how some specific linear systems appear in matrix form.

example 1

Write each linear system as a matrix equation.

a. $\begin{cases} -3x + 5y = 2 \\ x - 4y = -1 \end{cases}$
b. $\begin{cases} 3y - x = -2 \\ 4 - z + y = 5 \\ z - 3x + 3 = y - x \end{cases}$

Solutions:

a. Since the system is in standard form, we can just read off the coefficients of x and y to form the equation

$$\begin{bmatrix} -3 & 5 \\ 1 & -4 \end{bmatrix} \begin{bmatrix} x \\ y \end{bmatrix} = \begin{bmatrix} 2 \\ -1 \end{bmatrix}.$$

Verify for yourself that $\begin{bmatrix} -3 & 5 \\ 1 & -4 \end{bmatrix} \begin{bmatrix} x \\ y \end{bmatrix} = \begin{bmatrix} -3x + 5y \\ x - 4y \end{bmatrix}.$

b. While we could, with care, determine by inspection what the coefficient matrix of the system is, writing each equation in standard form first is probably wise:

$$\begin{cases} 3y - x = -2 \\ 4 - z + y = 5 \\ z - 3x + 3 = y - x \end{cases} = \begin{cases} -x + 3y = -2 \\ y - z = 1 \\ -2x - y + z = -3 \end{cases}$$

Now we can read off the coefficients to form the matrix equation

$$\begin{bmatrix} -1 & 3 & 0 \\ 0 & 1 & -1 \\ -2 & -1 & 1 \end{bmatrix} \begin{bmatrix} x \\ y \\ z \end{bmatrix} = \begin{bmatrix} -2 \\ 1 \\ -3 \end{bmatrix}.$$

chapter eight Systems of Equations

Topic 2: Finding the Inverse of a Matrix

In order to solve matrix equations like the two we obtained in *Example 1*, we need a way to "undo" the matrix of coefficients that appears in front of the column of variables. To get a better understanding of what, exactly, we are seeking, consider the following matrix products:

$$\begin{bmatrix} 1 & 0 \\ 0 & 1 \end{bmatrix} \begin{bmatrix} x \\ y \end{bmatrix} \text{ and } \begin{bmatrix} 1 & 0 & 0 \\ 0 & 1 & 0 \\ 0 & 0 & 1 \end{bmatrix} \begin{bmatrix} x \\ y \\ z \end{bmatrix}$$

If these matrix products appeared as the left-hand side of matrix equations, the equations would correspond to solutions of linear systems:

$$\begin{bmatrix} 1 & 0 \\ 0 & 1 \end{bmatrix} \begin{bmatrix} x \\ y \end{bmatrix} = \begin{bmatrix} a \\ b \end{bmatrix} \text{ corresponds to } \begin{cases} x = a \\ y = b \end{cases}$$

and

$$\begin{bmatrix} 1 & 0 & 0 \\ 0 & 1 & 0 \\ 0 & 0 & 1 \end{bmatrix} \begin{bmatrix} x \\ y \\ z \end{bmatrix} = \begin{bmatrix} a \\ b \\ c \end{bmatrix} \text{ corresponds to } \begin{cases} x = a \\ y = b \\ z = c \end{cases}$$

Such matrices, composed of 0's and 1's, deserve a name.

Identity Matrices

The $n \times n$ **identity matrix**, denoted I_n (or by just I when there is no possibility of confusion), is the $n \times n$ matrix consisting of 1's on the *main diagonal* and 0's everywhere else. The **main diagonal** consists of those entries in the first row-first column, the second row-second column, and so on down to the nth row-n^{th} column. Every identity matrix, then, has the form

$$I = \begin{bmatrix} 1 & 0 & 0 & \cdots & 0 \\ 0 & 1 & 0 & \cdots & 0 \\ 0 & 0 & 1 & \cdots & 0 \\ \vdots & \vdots & \vdots & \ddots & \vdots \\ 0 & 0 & 0 & \cdots & 1 \end{bmatrix}.$$

The identity matrix (of a certain size) serves as the multiplicative identity on the set of appropriate matrices. This means that if matrices A and B are such that the matrix products are defined, then $AI = A$ and $IB = B$. In this sense, I serves the same purpose as the number 1 in the set of real numbers.

section 8.5 Inverses of Matrices

(Although we have no immediate use for it, every set of matrices of a certain fixed order contains an *additive identity*. It is the matrix of the same order consisting entirely of 0's, and it is easy to see that if it is added to any matrix A, the result is A. The 0 matrices, then, play the same role as the number 0 in the set of real numbers.)

We know that a linear system of n equations and n variables can be expressed as a matrix equation $AX = B$, where A is an $n \times n$ matrix composed of coefficients, X is an $n \times 1$ matrix composed of the n variables, and B is an $n \times 1$ matrix composed of the constants from the right-hand sides of the equations. If we could find a matrix, which we will label A^{-1}, with the property that $A^{-1}A = I$, then we could use A^{-1} to "undo" the matrix A as desired. We call the matrix A^{-1}, if it exists, the *inverse of A*; the situation is completely analogous to the fact that $\frac{1}{a}$ is the (multiplicative) inverse of the real number a. The formal definition follows.

The Inverse of a Matrix

Let A be an $n \times n$ matrix. If there exists an $n \times n$ matrix A^{-1} such that

$$A^{-1}A = I_n \text{ and } AA^{-1} = I_n,$$

we call A^{-1} the **inverse** of A.

As the following example illustrates, we already possess the knowledge of how to find the inverse of a matrix.

example 2

Find the inverse of the matrix $A = \begin{bmatrix} 2 & -3 \\ -1 & 2 \end{bmatrix}$.

Solution:

If we let $A^{-1} = \begin{bmatrix} w & x \\ y & z \end{bmatrix}$, then one equation that must be satisfied is $AA^{-1} = I$, or

$$\begin{bmatrix} 2 & -3 \\ -1 & 2 \end{bmatrix} \begin{bmatrix} w & x \\ y & z \end{bmatrix} = \begin{bmatrix} 1 & 0 \\ 0 & 1 \end{bmatrix}.$$

continued on next page...

If we carry out the matrix product, we see that we need to solve the equation

$$\begin{bmatrix} 2w-3y & 2x-3z \\ -w+2y & -x+2z \end{bmatrix} = \begin{bmatrix} 1 & 0 \\ 0 & 1 \end{bmatrix},$$

which, if we equate columns on each side, means we need to solve the two linear systems

$$\begin{cases} 2w-3y=1 \\ -w+2y=0 \end{cases} \text{ and } \begin{cases} 2x-3z=0 \\ -x+2z=1 \end{cases}.$$

Of course, we have many methods for solving such systems. One efficient approach is to use Gauss-Jordan elimination, and to solve both systems at the same time by forming a new kind of augmented matrix. Since the coefficients in front of the variables are the same, if we put the matrix

$$\left[\begin{array}{cc|cc} 2 & -3 & 1 & 0 \\ -1 & 2 & 0 & 1 \end{array}\right]$$

Enter the matrix using the **Matrix** operation (pg. 606). Then **Quit**, enter matrix *A* through the **Matrix** menu, and select x^{-1}. Be careful... calculators may find the "inverse" of non-invertible matrices!

into reduced row-echelon form, the numbers that appear in the third column will solve the first system and the numbers that appear in the fourth column will solve the second system.

$$\left[\begin{array}{cc|cc} 2 & -3 & 1 & 0 \\ -1 & 2 & 0 & 1 \end{array}\right] \xrightarrow{R_1 \leftrightarrow R_2} \left[\begin{array}{cc|cc} -1 & 2 & 0 & 1 \\ 2 & -3 & 1 & 0 \end{array}\right] \xrightarrow{2R_1+R_2} \left[\begin{array}{cc|cc} -1 & 2 & 0 & 1 \\ 0 & 1 & 1 & 2 \end{array}\right]$$

$$\xrightarrow{-R_1} \left[\begin{array}{cc|cc} 1 & -2 & 0 & -1 \\ 0 & 1 & 1 & 2 \end{array}\right] \xrightarrow{2R_2+R_1} \left[\begin{array}{cc|cc} 1 & 0 & 2 & 3 \\ 0 & 1 & 1 & 2 \end{array}\right]$$

This tells us that $w=2$ and $y=1$ (from the third column) and $x=3$ and $z=2$ (from the fourth column). So

$$A^{-1} = \begin{bmatrix} 2 & 3 \\ 1 & 2 \end{bmatrix}.$$

We can now verify that

$$\begin{bmatrix} 2 & -3 \\ -1 & 2 \end{bmatrix}\begin{bmatrix} 2 & 3 \\ 1 & 2 \end{bmatrix} = \begin{bmatrix} 1 & 0 \\ 0 & 1 \end{bmatrix} \text{ and also } \begin{bmatrix} 2 & 3 \\ 1 & 2 \end{bmatrix}\begin{bmatrix} 2 & -3 \\ -1 & 2 \end{bmatrix} = \begin{bmatrix} 1 & 0 \\ 0 & 1 \end{bmatrix},$$

so we have indeed found A^{-1}.

Once we understand how the above method works, it becomes clear that we can omit the intermediate step of constructing the systems of equations, and skip to the process of putting the appropriate augmented matrix into reduced row-echelon form.

Finding the Inverse of a Matrix

> Let A be an $n \times n$ matrix. The inverse of A can be found by:
>
> 1. Forming the augmented matrix $[A \mid I]$, where I is the $n \times n$ identity matrix.
> 2. Using Gauss-Jordan elimination to put $[A \mid I]$ into the form $[I \mid B]$, if possible.
> 3. Defining A^{-1} to be B.
>
> If it is not possible to put $[A \mid I]$ into reduced row echelon form, then A doesn't have an inverse, and we say A is not *invertible*.

We will find the inverses of several more matrices soon, but the last sentence in the above box merits further discussion. Since it is certainly possible that a linear system has either no solution or an infinite number of solutions, we shouldn't be surprised by the possibility that some matrices have no inverses. In such cases, fortunately, it becomes obvious during the process of trying to find the inverse that there isn't one. We will examine such a situation in *Example 3*.

One last comment before we look at another example: there is a short-cut for finding inverses of 2×2 matrices that saves significant time. The short-cut also quickly identifies those 2×2 matrices that are not invertible.

Inverse of a 2 × 2 Matrix

> Let $A = \begin{bmatrix} a & b \\ c & d \end{bmatrix}$. Then $A^{-1} = \dfrac{1}{|A|} \begin{bmatrix} d & -b \\ -c & a \end{bmatrix}$, where $|A| = ad - bc$ is the determinant of A. Since $|A|$ appears in the denominator of a fraction, A^{-1} fails to exist if $|A| = 0$.

Finding the inverse of a 3×3 or larger matrix requires more work than the above short-cut, but there is one similarity. Regardless of the order of a matrix, if its determinant is zero, the matrix has no inverse.

chapter eight   Systems of Equations

example 3

Find the inverses of the following matrices, if possible.

a. $A = \begin{bmatrix} 3 & -5 \\ 2 & 1 \end{bmatrix}$

b. $B = \begin{bmatrix} 2 & 4 & 2 \\ -1 & 5 & -1 \\ 3 & 1 & 3 \end{bmatrix}$

Solutions:

a. Using the short-cut,

$$A^{-1} = \frac{1}{3-(-10)} \begin{bmatrix} 1 & 5 \\ -2 & 3 \end{bmatrix} = \begin{bmatrix} \frac{1}{13} & \frac{5}{13} \\ \frac{-2}{13} & \frac{3}{13} \end{bmatrix}.$$

We can verify our work as follows:

$$\begin{bmatrix} 3 & -5 \\ 2 & 1 \end{bmatrix} \begin{bmatrix} \frac{1}{13} & \frac{5}{13} \\ \frac{-2}{13} & \frac{3}{13} \end{bmatrix} = \begin{bmatrix} 1 & 0 \\ 0 & 1 \end{bmatrix} = \begin{bmatrix} \frac{1}{13} & \frac{5}{13} \\ \frac{-2}{13} & \frac{3}{13} \end{bmatrix} \begin{bmatrix} 3 & -5 \\ 2 & 1 \end{bmatrix}$$

b. A quick examination of B reveals two identical columns, so from the properties of determinants we know $|B| = 0$. Based on the comment immediately preceding this example, then, we expect that B is not invertible, but we will begin the process of inverting B anyway to see what happens.

$$\begin{bmatrix} 2 & 4 & 2 & | & 1 & 0 & 0 \\ -1 & 5 & -1 & | & 0 & 1 & 0 \\ 3 & 1 & 3 & | & 0 & 0 & 1 \end{bmatrix} \xrightarrow[3R_2+R_3]{2R_2+R_1} \begin{bmatrix} 0 & 14 & 0 & | & 1 & 2 & 0 \\ -1 & 5 & -1 & | & 0 & 1 & 0 \\ 0 & 16 & 0 & | & 0 & 3 & 1 \end{bmatrix}$$

$$\xrightarrow{R_1 \leftrightarrow R_2} \begin{bmatrix} -1 & 5 & -1 & | & 0 & 1 & 0 \\ 0 & 14 & 0 & | & 1 & 2 & 0 \\ 0 & 16 & 0 & | & 0 & 3 & 1 \end{bmatrix} \xrightarrow[\frac{1}{14}R_2]{-R_1} \begin{bmatrix} 1 & -5 & 1 & | & 0 & -1 & 0 \\ 0 & 1 & 0 & | & \frac{1}{14} & \frac{1}{7} & 0 \\ 0 & 16 & 0 & | & 0 & 3 & 1 \end{bmatrix}$$

$$\xrightarrow{-16R_2+R_3} \begin{bmatrix} 1 & -5 & 1 & | & 0 & -1 & 0 \\ 0 & 1 & 0 & | & \frac{1}{14} & \frac{1}{7} & 0 \\ 0 & 0 & 0 & | & -\frac{8}{7} & \frac{5}{7} & 1 \end{bmatrix}$$

At this point, we can stop. Once the first three entries of any row are 0 in the 3×3 matrix, there is no way to put the matrix into reduced row-echelon form. Thus, B has no inverse.

section 8.5   Inverses of Matrices

technology note

Computer algebra systems and sufficiently-advanced calculators are able to determine inverses of matrices. In *Mathematica*, the command to do so is `Inverse`. The illustration below shows first the definition of a 3×3 matrix A and then the evaluation of its inverse. Note that the output of the `Inverse` command is a list of sub-lists, which is the way that *Mathematica* works with matrices internally. To see A^{-1} in rectangular form, the command `MatrixForm` is used along with the `Inverse` command.

Enter the matrix and select **x⁻¹** as described on page 654. To show the answer in fraction form, select **MATH** and 1 (Frac.)

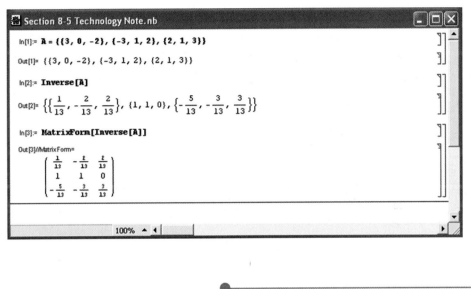

Topic 3: Using Matrix Inverses to Solve Linear Systems

We have now assembled all the tools we need to solve linear systems by the inverse matrix method.

Given a linear system of n equations in n variables, the first step is to write the system in matrix form as $AX = B$, where A is the $n \times n$ matrix of coefficients, X is the $n \times 1$ matrix of variables, and B is the $n \times 1$ matrix of constants. The next step is to construct A^{-1}, if A^{-1} exists. There are now two ways of interpreting the meaning of the final step, though the two interpretations entail identical computations.

657

chapter eight Systems of Equations

Interpretation 1: As we know, multiplying A by A^{-1} results in the $n \times n$ identity matrix, just as multiplying a non-zero number a by $\frac{1}{a}$ gives us the multiplicative identity 1. So if we multiply both sides of the equation $AX = B$ by A^{-1}, we obtain

$$A^{-1}AX = A^{-1}B$$
$$IX = A^{-1}B$$
$$X = A^{-1}B$$

The entries in the $n \times 1$ matrix $A^{-1}B$ are thus the sought-after values for the variables listed in the $n \times 1$ matrix X.

Interpretation 2: The more sophisticated interpretation is to use the fact that A^{-1} is the inverse function of the function A to rewrite the equation $AX = B$ in the equivalent form $X = A^{-1}B$. This is an illustration of the generic statement that $f(x) = y$ if and only if $x = f^{-1}(y)$, if f indeed has an inverse function. The advantage of interpreting the solution process this way is that it emphasizes the point that a given matrix A may not have an inverse. The ramifications of this are explored in more detail in such classes as linear algebra.

example 4

Solve the following systems by the inverse matrix method.

a. $\begin{cases} 4x - 5y = 3 \\ -3x + 7y = 1 \end{cases}$

b. $\begin{cases} -x + 2y = 3 \\ 3x - 6y = -5 \end{cases}$

Solutions:

a. $\begin{cases} 4x - 5y = 3 \\ -3x + 7y = 1 \end{cases}$

$$\begin{bmatrix} 4 & -5 \\ -3 & 7 \end{bmatrix} \begin{bmatrix} x \\ y \end{bmatrix} = \begin{bmatrix} 3 \\ 1 \end{bmatrix}$$

$$\begin{bmatrix} 4 & -5 \\ -3 & 7 \end{bmatrix}^{-1} = \frac{1}{28 - 15} \begin{bmatrix} 7 & 5 \\ 3 & 4 \end{bmatrix}$$

$$= \begin{bmatrix} \frac{7}{13} & \frac{5}{13} \\ \frac{3}{13} & \frac{4}{13} \end{bmatrix}$$

The first step is to write the system in matrix form. Since both equations are in standard form, this is a straightforward task.

The main step in the inverse matrix method is to find the inverse of the coefficient matrix A. Since the matrix is of order 2×2 in this problem, we can use the shortcut to obtain the inverse quickly. Of course, we can check our work at this stage by making sure that the product of the original matrix and its inverse is the identity matrix.

$$\begin{bmatrix} x \\ y \end{bmatrix} = \begin{bmatrix} \dfrac{7}{13} & \dfrac{5}{13} \\ \dfrac{3}{13} & \dfrac{4}{13} \end{bmatrix} \begin{bmatrix} 3 \\ 1 \end{bmatrix}$$

$$= \begin{bmatrix} \dfrac{21}{13} + \dfrac{5}{13} \\ \dfrac{9}{13} + \dfrac{4}{13} \end{bmatrix}$$

$$= \begin{bmatrix} 2 \\ 1 \end{bmatrix}$$

We can now rewrite the equation $AX = B$ in the form $X = A^{-1}B$, and proceed to carry out the matrix multiplication on the right to get the answer. Once we have $x = 2$ and $y = 1$, we can verify that this is indeed the solution of the system.

Solution is $(2, 1)$.

b. $\begin{cases} -x + 2y = 3 \\ 3x - 6y = -5 \end{cases}$

$$\begin{bmatrix} -1 & 2 \\ 3 & -6 \end{bmatrix} \begin{bmatrix} x \\ y \end{bmatrix} = \begin{bmatrix} 3 \\ -5 \end{bmatrix}$$

Again, we start by writing the system in matrix form.

$$\begin{bmatrix} -1 & 2 \\ 3 & -6 \end{bmatrix}^{-1} = \dfrac{1}{6-6} \begin{bmatrix} -6 & -2 \\ -3 & -1 \end{bmatrix}$$

In this problem, we note immediately that the coefficient matrix has no inverse, so we know that either the system has no solution or else it has an infinite number of solutions.

$$D_x = \begin{vmatrix} 3 & 2 \\ -5 & -6 \end{vmatrix} = -18 - (-10) \neq 0$$

Since $D_x \neq 0$, Cramer's rule tells us that this system has no solution.

No Solution.

We have now reached the end of our list of solution methods for linear systems; there is more to say on the topic, but further discussion is best left for a more advanced course. One question that might have occurred to you, though, has not yet been adequately addressed. That question is: How should the method of solution be chosen when faced with a specific system of equations?

Unfortunately, there is no simple answer. To begin with, if the system is small, or if it looks especially simple, then the method of substitution or the method of elimination may be the quickest route to a solution. Generally speaking, the larger and/or the more complicated the system, the likelier it is that one of the more advanced methods will be preferable. But if you find that your first choice of method leads to a computational headache, don't be reluctant to stop and try another method.

chapter eight   Systems of Equations

In practical applications, it is often the case that a number of closely related systems must be solved, and the only difference in the systems lies in the constants on the right-hand sides of the equations. That is, the coefficient matrix is the same for all the systems. In this case, the inverse matrix method is almost certainly the best choice, as illustrated below.

example 5

Solve the following three linear systems.

$$\begin{cases} x-4y+z=-6 \\ -2x+y=5 \\ 3y-z=3 \end{cases} \quad \begin{cases} x-4y+z=-17 \\ -2x+y=2 \\ 3y-z=14 \end{cases} \quad \begin{cases} x-4y+z=3 \\ -2x+y=5 \\ 3y-z=-5 \end{cases}$$

Solution:

The coefficient matrix is the same for the three systems, and this means we can avoid unnecessary computation if we use the inverse matrix method. We begin by finding the inverse of the coefficient matrix.

$$\begin{bmatrix} 1 & -4 & 1 & | & 1 & 0 & 0 \\ -2 & 1 & 0 & | & 0 & 1 & 0 \\ 0 & 3 & -1 & | & 0 & 0 & 1 \end{bmatrix} \xrightarrow{2R_1+R_2} \begin{bmatrix} 1 & -4 & 1 & | & 1 & 0 & 0 \\ 0 & -7 & 2 & | & 2 & 1 & 0 \\ 0 & 3 & -1 & | & 0 & 0 & 1 \end{bmatrix}$$

$$\xrightarrow{\frac{1}{3}R_3} \begin{bmatrix} 1 & -4 & 1 & | & 1 & 0 & 0 \\ 0 & -7 & 2 & | & 2 & 1 & 0 \\ 0 & 1 & -\frac{1}{3} & | & 0 & 0 & \frac{1}{3} \end{bmatrix} \xrightarrow[7R_3+R_2]{4R_3+R_1} \begin{bmatrix} 1 & 0 & -\frac{1}{3} & | & 1 & 0 & \frac{4}{3} \\ 0 & 0 & -\frac{1}{3} & | & 2 & 1 & \frac{7}{3} \\ 0 & 1 & -\frac{1}{3} & | & 0 & 0 & \frac{1}{3} \end{bmatrix}$$

$$\xrightarrow{-3R_2} \begin{bmatrix} 1 & 0 & -\frac{1}{3} & | & 1 & 0 & \frac{4}{3} \\ 0 & 0 & 1 & | & -6 & -3 & -7 \\ 0 & 1 & -\frac{1}{3} & | & 0 & 0 & \frac{1}{3} \end{bmatrix} \xrightarrow{R_2 \leftrightarrow R_3} \begin{bmatrix} 1 & 0 & -\frac{1}{3} & | & 1 & 0 & \frac{4}{3} \\ 0 & 1 & -\frac{1}{3} & | & 0 & 0 & \frac{1}{3} \\ 0 & 0 & 1 & | & -6 & -3 & -7 \end{bmatrix}$$

$$\xrightarrow[\frac{1}{3}R_3+R_2]{\frac{1}{3}R_3+R_1} \begin{bmatrix} 1 & 0 & 0 & | & -1 & -1 & -1 \\ 0 & 1 & 0 & | & -2 & -1 & -2 \\ 0 & 0 & 1 & | & -6 & -3 & -7 \end{bmatrix}$$

$$A^{-1} = \begin{bmatrix} -1 & -1 & -1 \\ -2 & -1 & -2 \\ -6 & -3 & -7 \end{bmatrix}$$

The majority of the work has now been done, and we can solve the three systems quickly by multiplying by the three matrices consisting of the right-hand-side constants:

$$\begin{bmatrix} x \\ y \\ z \end{bmatrix} = \begin{bmatrix} -1 & -1 & -1 \\ -2 & -1 & -2 \\ -6 & -3 & -7 \end{bmatrix} \begin{bmatrix} -6 \\ 5 \\ 3 \end{bmatrix} = \begin{bmatrix} -2 \\ 1 \\ 0 \end{bmatrix}$$

$$\begin{bmatrix} x \\ y \\ z \end{bmatrix} = \begin{bmatrix} -1 & -1 & -1 \\ -2 & -1 & -2 \\ -6 & -3 & -7 \end{bmatrix} \begin{bmatrix} -17 \\ 2 \\ 14 \end{bmatrix} = \begin{bmatrix} 1 \\ 4 \\ -2 \end{bmatrix}$$

$$\begin{bmatrix} x \\ y \\ z \end{bmatrix} = \begin{bmatrix} -1 & -1 & -1 \\ -2 & -1 & -2 \\ -6 & -3 & -7 \end{bmatrix} \begin{bmatrix} 3 \\ 5 \\ -5 \end{bmatrix} = \begin{bmatrix} -3 \\ -1 \\ 2 \end{bmatrix}$$

exercises

Write each of the following systems of equations as a single matrix equation. See Example 1.

1. $\begin{cases} 14x - 5y = 7 \\ x + 9y = 2 \end{cases}$

2. $\begin{cases} x - 5 = 9y \\ 3y - 2x = 8 \end{cases}$

3. $\begin{cases} \dfrac{3x - 8y}{5} = 2 \\ y - 2 = 0 \end{cases}$

4. $\begin{cases} x - y = 5 \\ 2 - z = x \\ z - 3y = 4 \end{cases}$

5. $\begin{cases} 3x_1 - 7x_2 + x_3 = -4 \\ x_1 - x_2 = 2 \\ 8x_2 + 5x_3 = -3 \end{cases}$

6. $\begin{cases} x_3 = x_2 \\ x_2 = x_1 \\ x_1 = x_3 \end{cases}$

7. $\begin{cases} -6 - 2y = x \\ 9x + 14 = 3y \end{cases}$

8. $\begin{cases} x - 7y = 5 \\ \dfrac{6 + x}{2} = 3y - 2 \end{cases}$

9. $\begin{cases} 4x = 3y - 9 \\ 13 - 2x = -4y \end{cases}$

10. $\begin{cases} -\dfrac{7}{3}y = \dfrac{5 - x}{6} \\ x - 5(y - 3) = -2 \end{cases}$

11. $\begin{cases} 2x - y = -3z \\ y - x = 17 \\ 2 + z + 4x = 5y \end{cases}$

12. $\begin{cases} 2x_1 - 3x_3 = 7 \\ x_2 - 10x_3 = 0 \\ 2x_1 - x_2 + x_3 = 1 \end{cases}$

chapter eight Systems of Equations

Find the inverse of each of the following matrices, if possible. See Examples 2 and 3.

13. $\begin{bmatrix} 0 & 4 \\ -5 & -1 \end{bmatrix}$
14. $\begin{bmatrix} -2 & -2 \\ -1 & 2 \end{bmatrix}$
15. $\begin{bmatrix} 3 & 4 \\ -4 & -5 \end{bmatrix}$

16. $\begin{bmatrix} -1 & -1 \\ -\frac{1}{4} & -\frac{1}{2} \end{bmatrix}$
17. $\begin{bmatrix} -\frac{1}{5} & 0 \\ \frac{1}{5} & \frac{1}{2} \end{bmatrix}$
18. $\begin{bmatrix} -7 & 2 \\ 7 & -2 \end{bmatrix}$

19. $\begin{bmatrix} -2 & -4 & -2 \\ 1 & -4 & 1 \\ 4 & -3 & 4 \end{bmatrix}$
20. $\begin{bmatrix} -3 & 0 & -4 \\ 2 & 5 & 4 \\ 1 & -5 & -2 \end{bmatrix}$
21. $\begin{bmatrix} -\frac{5}{11} & -\frac{8}{11} & 1 \\ \frac{13}{11} & \frac{12}{11} & -2 \\ -\frac{2}{11} & -\frac{1}{11} & 0 \end{bmatrix}$

22. $-\frac{1}{31}\begin{bmatrix} 17 & -8 & -2 \\ 1 & 5 & 9 \\ -6 & 1 & 8 \end{bmatrix}$
23. $\begin{bmatrix} -1 & 2 & -1 \\ 0 & 3 & -1 \\ 0 & 4 & -1 \end{bmatrix}$
24. $\begin{bmatrix} -1 & 0 & -1 \\ -\frac{3}{2} & \frac{1}{2} & -\frac{3}{2} \\ -\frac{1}{2} & 0 & -\frac{1}{4} \end{bmatrix}$

25. $\begin{bmatrix} -\frac{6}{5} & -\frac{2}{5} & -1 \\ \frac{3}{5} & \frac{1}{5} & 1 \\ 1 & 0 & 1 \end{bmatrix}$
26. $\begin{bmatrix} 2 & -2 & 1 \\ -2 & 2 & -3 \\ 1 & 0 & 2 \end{bmatrix}$
27. $\begin{bmatrix} 0 & 1 & 1 \\ 1 & 1 & 0 \\ 0 & 1 & 2 \end{bmatrix}$

28. $\begin{bmatrix} 9 & 8 & 7 \\ 6 & 5 & 4 \\ 3 & 2 & 1 \end{bmatrix}$
29. $\begin{bmatrix} \frac{2}{3} & \frac{8}{9} & \frac{1}{9} \\ -\frac{1}{3} & \frac{2}{9} & -\frac{2}{9} \\ -\frac{1}{3} & -\frac{7}{9} & -\frac{2}{9} \end{bmatrix}$
30. $\begin{bmatrix} -3 & -3 & -4 \\ 0 & \frac{1}{4} & \frac{1}{2} \\ 2 & 2 & 3 \end{bmatrix}$

For each set of matrices, determine if either matrix is the inverse of the other.

31. $\begin{bmatrix} -5 & -2 \\ -7 & 4 \end{bmatrix}, \begin{bmatrix} 10 & 4 \\ 14 & -8 \end{bmatrix}$
32. $\begin{bmatrix} 9 & -18 \\ 3 & 12 \end{bmatrix}, \begin{bmatrix} -3 & -6 \\ -1 & -4 \end{bmatrix}$

33. $\begin{bmatrix} -6 & -1 & 1 \\ 4 & -1 & -2 \\ 1 & -1 & -1 \end{bmatrix}, \begin{bmatrix} -1 & -2 & 3 \\ 2 & 5 & -8 \\ -3 & -7 & 10 \end{bmatrix}$
34. $\begin{bmatrix} -1 & 4 & 5 \\ 3 & -11 & -17 \\ 4 & -17 & -19 \end{bmatrix}, \begin{bmatrix} -80 & -9 & -13 \\ -11 & -1 & -2 \\ -7 & -1 & -1 \end{bmatrix}$

35. $\begin{bmatrix} 2 & 0 & -1 \\ 3 & 4 & 2 \\ 1 & 1 & -3 \end{bmatrix}, \begin{bmatrix} 4 & 0 & -2 \\ 6 & 8 & 4 \\ 2 & 2 & -6 \end{bmatrix}$
36. $\begin{bmatrix} -7 & 0 & -2 \\ -10 & -1 & -2 \\ -7 & -1 & -1 \end{bmatrix}, \begin{bmatrix} -1 & 2 & -2 \\ 4 & -7 & 6 \\ 3 & -7 & 7 \end{bmatrix}$

Inverses of Matrices

Using a graphing calculator, find the inverse of each of the following matrices, if possible. Round answers to the nearest thousandth, when necessary.

37. $\begin{bmatrix} -7 & 3 \\ -1 & 2 \end{bmatrix}$

38. $\begin{bmatrix} -6 & 2 \\ -5 & 5 \end{bmatrix}$

39. $\begin{bmatrix} -2 & 0 & 2 \\ 2 & -3 & 1 \\ 1 & -2 & 3 \end{bmatrix}$

40. $\begin{bmatrix} 2.3 & 7.8 \\ -3.4 & 1.6 \end{bmatrix}$

41. $\begin{bmatrix} 4.5 & -9.4 & 6.9 \\ 8.6 & -2.8 & 1.2 \\ 3.1 & 0.3 & -7.0 \end{bmatrix}$

42. $\begin{bmatrix} 38 & -44 & 72 \\ -93 & 16 & 29 \\ 65 & 23 & -19 \end{bmatrix}$

Solve the following systems by the inverse matrix method, if possible. If the inverse matrix method doesn't apply, use any other method to determine if the system is inconsistent or dependent. See Example 4.

43. $\begin{cases} -2x - 2y = 9 \\ -x + 2y = -3 \end{cases}$

44. $\begin{cases} 3x + 4y = -2 \\ -4x - 5y = 9 \end{cases}$

45. $\begin{cases} -2x + 3y = 1 \\ 4x - 6y = -2 \end{cases}$

46. $\begin{cases} -2x + 4y = 5 \\ x - 4y = -3 \end{cases}$

47. $\begin{cases} -5x = 10 \\ 2x + 2y = -4 \end{cases}$

48. $\begin{cases} -3x + y = 2 \\ 9x - 3y = 5 \end{cases}$

49. $\begin{cases} -x + 2y - z = 6 \\ 3y - z = -1 \\ 4y - z = 5 \end{cases}$

50. $\begin{cases} x - 4z = 8 \\ -3x + 2y = -2 \\ -2x + 4z = 4 \end{cases}$

51. $\begin{cases} -2x - 4y - 2z = 1 \\ x - 4y + z = 0 \\ 4x - 3y + 4z = -1 \end{cases}$

52. $\begin{cases} 4y + 3z = -254 \\ 2x - 2y - z = 100 \\ -x + y - 2z = 155 \end{cases}$

53. $\begin{cases} 2x - y - 3z = -10 \\ 2y - z = 11 \\ -x + 4z = 0 \end{cases}$

54. $\begin{cases} 3y - 4z = 15 \\ x + 2y - 3z = 9 \\ -x - y + 2z = -5 \end{cases}$

Solve the following sets of systems by the inverse matrix method. See Example 5.

55. $\begin{cases} x + 2y - z = 2 \\ 3x + 3y - z = -5 \\ 4x + 4y - z = 1 \end{cases}$ $\begin{cases} x + 2y - z = 1 \\ 3x + 3y - z = 1 \\ 4x + 4y - z = 1 \end{cases}$ $\begin{cases} x + 2y - z = 0 \\ 3x + 3y - z = 1 \\ 4x + 4y - z = 1 \end{cases}$

56. $\begin{cases} -x - y - 2z = 4 \\ x + 3y + 3z = 0 \\ -3y - 2z = 9 \end{cases}$ $\begin{cases} -x - y - 2z = 1 \\ x + 3y + 3z = 0 \\ -3y - 2z = 0 \end{cases}$ $\begin{cases} -x - y - 2z = -2 \\ x + 3y + 3z = -3 \\ -3y - 2z = 1 \end{cases}$

57. $\begin{cases} -x + z = 6 \\ -x + 3y + 2z = -11 \\ 2x - 4y - 3z = 13 \end{cases}$ $\begin{cases} -x + z = -2 \\ -x + 3y + 2z = 2 \\ 2x - 4y - 3z = -1 \end{cases}$ $\begin{cases} -x + z = -4 \\ -x + 3y + 2z = 2 \\ 2x - 4y - 3z = 0 \end{cases}$

chapter eight — Systems of Equations

8.6 Linear Programming

TOPICS

1. Planar feasible regions
2. Linear programming in two variables

Topic 1: Planar Feasible Regions

Many real-world applications of mathematics call for identifying the values that a set of variables can assume. If a particular application imposes limitations on the values that two variables can assume, the collection of all allowable values forms a portion of the plane called the **feasible region** or **region of constraint**. For instance, the constraints $1 \leq x \leq 5$ and $2 \leq y \leq 4$ define the rectangular feasible region shown below.

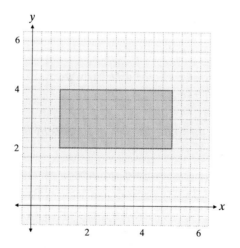

Typically, constraint inequalities involve both variables and the corresponding feasible region is a bit more interesting than a rectangle. The collection of all constraints in a problem comprises a system of inequalities, and the techniques learned in this chapter will be very useful in extracting certain information about the feasible region described by such systems.

We have already seen, in *Example 4* of Section 3.5, an illustration of how a feasible region may typically be described in an application. It is included again here for reference.

section 8.6   Linear Programming

example 1

A family orchard is in the business of selling peaches and nectarines. The owners know that to prevent a certain pest infestation, the number of nectarine trees in the orchard cannot exceed the number of peach trees. Also, because of the space requirements of each type of tree, the number of nectarine trees plus twice the number of peach trees cannot exceed 100 trees. Construct the constraints and graph the feasible region for this situation.

Solution:

If p represents the number of peach trees and n the number of nectarine trees, the paragraph above gives us various constraints on p and n. For instance, the second sentence translates into the inequality $n \leq p$ and the third sentence translates into the inequality $n + 2p \leq 100$. In addition, we can assume that $n \geq 0$ and $p \geq 0$, as it is not possible to have a negative number of trees. Letting the horizontal axis represent p and the vertical axis n, the portion of the plane satisfying all four of these inequalities appears as shown below (refer to Section 3.5 for a review of graphing inequalities in two variables).

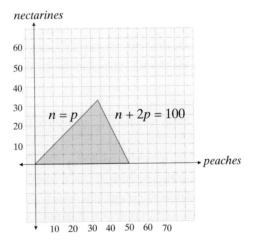

Any point (with integer coordinates) in the shaded feasible region corresponds to a certain number of peach trees and nectarine trees, and all such points meet the stated conditions. The question of how to choose a particular ordered pair then arises; we will soon see how the existence of one more condition resolves this question.

chapter eight   Systems of Equations

Topic 2: Linear Programming in Two Variables

Real-world applications involving the construction of feasible regions often also possess a quantity that must be either maximized or minimized. Two extremely important examples come from business: profit is a quantity that people usually wish to maximize, and cost is a quantity best kept to a minimum. When the quantity to be optimized (that is, either maximized or minimized) is a linear function of the variables concerned, the collection of mathematical tools known as **linear programming** comes into play.

Linear programming is used in a wide variety of applications, some involving thousands of variables, and a complete study of its techniques is beyond the scope of this text. But the basic concepts are simple and easily applied, especially when restricted to applications calling for just two variables. We begin our study with a description of the steps taken in linear programming, followed by an example that will illustrate why linear programming works.

Linear Programming Method (Two-Variable Case)

1. Identify the variables to be considered in the problem, determine all the constraints on the variables imposed by the problem, and sketch the feasible region described by the constraints.

2. Determine the function that is to be either maximized or minimized. Such a function is called the **objective function**.

3. Evaluate the function at each of the vertices of the feasible region and compare the values found. If the feasible region is *bounded*, the optimum value of the function will occur at a vertex. If the feasible region is *unbounded*, the optimum value of the function may not exist, but if it does it will occur at a vertex.

(A **bounded** feasible region is one in which all the points lie within some fixed distance of the origin. An **unbounded** region has points that are arbitrarily far away from the origin.)

example 2

Find the maximum and minimum values of $f(x,y) = 3x + 2y$ subject to the constraints

$$\begin{cases} x \geq 0 \\ y \geq 0 \\ y \geq 3 - 3x \\ 2x + y \leq 10 \\ x + y \leq 7 \\ y \geq x - 2 \end{cases}.$$

Solution:

The two variables are given to us, as are the constraints; all we must do to complete the first step is sketch the region defined by the constraints. The figure below contains the sketch, with selected boundaries identified by their corresponding equations.

The function to be optimized is also given to us in this problem, so step two is complete. According to the linear programming method, then, the only remaining task is to evaluate the function at the vertices of the feasible region. Why should we expect the maximum and minimum values of f to occur at vertices, and not in the interior of the region?

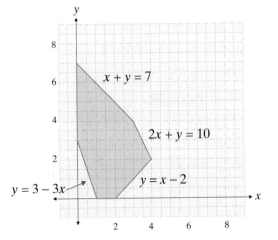

Another diagram will answer this question. The function $f(x,y)$ will, of course, take on a variety of values at different ordered pairs (x,y). But for any particular value, say k, there will be a line of ordered pairs in the plane such that $f(x,y) = k$; such lines, for different values of k, will be parallel, as shown here. Two particular values for k are labeled in the diagram, and these values are the minimum and maximum values of f.

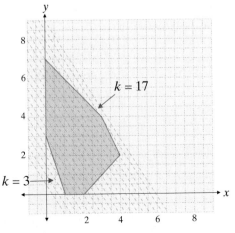

continued on next page...

chapter eight Systems of Equations

The extreme values of a linear function over a planar feasible region bounded by straight lines will thus occur where a line of constant value just touches the region, and this will be at a vertex or one of the edges of the region, if the edge is parallel to the line of constant value. In any event, evaluating the function at the vertices of the feasible region will suffice to identify the extreme values.

It is in the third step that the solution methods of this chapter are useful. Each vertex of a feasible region is the intersection of two edges, so the coordinates of each vertex are found by solving a system of two equations. The six vertices of the region in this problem are determined by the following six systems:

$$\begin{cases} x+y=7 \\ 2x+y=10 \end{cases} \begin{cases} 2x+y=10 \\ y=x-2 \end{cases} \begin{cases} y=x-2 \\ y=0 \end{cases} \begin{cases} y=0 \\ y=3-3x \end{cases} \begin{cases} y=3-3x \\ x=0 \end{cases} \begin{cases} x=0 \\ x+y=7 \end{cases}.$$

Solving each of these systems, we find the vertices to be $(3, 4)$, $(4, 2)$, $(2, 0)$, $(1, 0)$, $(0, 3)$, and $(0, 7)$. Substituting each of these ordered pairs into the objective function we find the maximum value of 17 occurs at the vertex $(3, 4)$, and the minimum value of 3 occurs at the vertex $(1, 0)$. The verification of these facts is left to the reader.

example 3

Find the maximum and minimum values of $f(x, y) = x + 2y$ subject to the constraints

$$\begin{cases} x \geq 0 \\ y \geq 3-x \\ x-y \leq 1 \\ y \leq 2x+4 \end{cases}.$$

Solution:

The feasible region defined by the constraints is unbounded; a portion of its sketch appears to the right.

Because the region is unbounded, the function f may not have a maximum or a minimum value over the region. In this case, f has no maximum value. One way to see this is to note that ordered pairs of the form (x, x) lie in the region for all $x \geq \frac{3}{2}$, and $f(x, x) = x + 2x = 3x$ grows without bound as x increases.

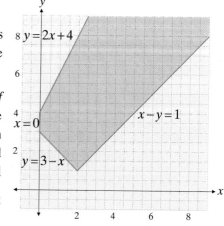

On the other hand, f does have a minimum value over the feasible region. The three vertices are $(0, 4)$, $(0, 3)$, and $(2, 1)$. Evaluating the function at these points, we obtain $f(0, 4) = 8$, $f(0, 3) = 6$, and $f(2, 1) = 4$, so the minimum value is 4.

Our last example illustrates how linear programming can be applied in a typical business application.

example 4

A manufacturer makes two models of minidisc players. Model A requires 3 minutes to assemble, 4 minutes to test, and 1 minute to package. Model B requires 4 minutes to assemble, 3 minutes to test, and 6 minutes to package. The manufacturer can allot 7400 minutes total for assembly, 8000 minutes for testing, and 9000 minutes for packaging. Model A generates a profit of $7.00 and model B generates a profit of $8.00. How many of each model should be made in order to maximize profit?

Solution:

If we let x denote the number of Model A players made and y the number of Model B players made, the constraint inequalities are

$$\begin{cases} x \geq 0, y \geq 0 \\ 3x + 4y \leq 7400 \\ 4x + 3y \leq 8000 \\ x + 6y \leq 9000 \end{cases}$$

and the sketch of the feasible region appears below.

The profit function in this situation is $f(x, y) = 7x + 8y$, and our goal is to maximize this function. The vertices of the feasible region are: $(0, 1500)$, $(600, 1400)$, $(1400, 800)$, $(2000, 0)$, and $(0, 0)$. Of course, manufacturing 0 units of Model A and Model B is not going to be profitable, so we really only need to evaluate f at four vertices:

$f(0, 1500) = 12{,}000$
$f(600, 1400) = 15{,}400$
$f(1400, 800) = 16{,}200$
$f(2000, 0) = 14{,}000$

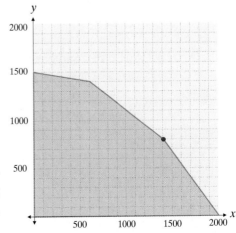

From these calculations, we conclude that the maximum profit is generated from making 1400 units of Model A and 800 units of Model B. The maximum profit would be $16,200.

chapter eight   Systems of Equations

exercises

Construct the constraints and graph the feasible regions for the following situations. See Example 1.

1. A plane carrying relief food and water can carry a maximum of 50,000 pounds, and is limited in space to carrying no more than 6000 cubic feet. Each container of water weighs 60 pounds and takes up 1 cubic foot, and each container of food weighs 50 pounds and takes up 10 cubic feet. What is the region of constraint for the number of containers of food and water that the plane can carry?

2. A furniture company makes two kinds of sofas, the Standard model and the Deluxe model. The Standard model requires 40 hours of labor to build, and the Deluxe model requires 60 hours of labor to build. The finish of the Deluxe model, however, uses both teak and fabric, while the Standard uses only fabric, with the result that each Deluxe sofa requires 5 square yards of fabric and each Standard sofa requires 8 square yards of fabric. Given that the company can use 200 hours of labor and 25 square yards of fabric per week building sofas, what is the region of constraint for the number of Deluxe and Standard sofas the company can make per week?

3. Sarah is looking through a clothing catalog and she is willing to spend up to $80 on clothes and $10 for shipping. Shirts cost $12 each plus $2 shipping, and a pair of pants costs $32 plus $3 shipping. What is the region of constraint for the number of shirts and pairs of pants Sarah can buy?

4. Suppose you inherit $75,000 from a previously unknown (and highly eccentric) uncle, and that the inheritance comes with certain stipulations regarding investments. First, the dollar amount invested in bonds must not exceed the dollar amount invested in stocks. Second, a minimum of $10,000 must be invested in stocks and a minimum of $5000 must be invested in bonds. Finally, a maximum of $40,000 can be invested in stocks. What is the region of constraint for the dollar amount that can be invested in the two categories of stocks and bonds?

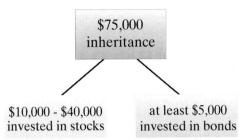

Find the minimum and maximum values of the given functions, subject to the given constraints. See Examples 2 and 3.

5. Objective Function
   $f(x, y) = 2x + 3y$
   Constraints
   $x \geq 0$
   $y \geq 0$
   $x + y \leq 7$

6. Objective Function
   $f(x, y) = 4x + y$
   Constraints
   $x \geq 0$
   $y \geq 0$
   $x + y \leq 3$

7. Objective Function
   $f(x, y) = 2x + 5y$
   Constraints
   $x \geq 0$
   $y \geq 0$
   $2x + y \leq 6$

8. Objective Function
   $f(x, y) = 7x + 4y$
   Constraints
   $x \geq 0$
   $y \geq 0$
   $3x + y \leq 3$

9. Objective Function
   $f(x, y) = 5x + 6y$
   Constraints
   $0 \leq x \leq 7$
   $0 \leq y \leq 10$
   $8x + 5y \leq 40$

10. Objective Function
    $f(x, y) = 9x + 7y$
    Constraints
    $0 \leq x \leq 20$
    $0 \leq y \leq 10$
    $6x + 12y \leq 140$

11. Objective Function
$f(x, y) = 6x + 4y$
Constraints
$0 \leq x \leq 4$
$0 \leq y \leq 5$
$4x + 3y \leq 10$

12. Objective Function
$f(x, y) = 3x + 7y$
Constraints
$0 \leq x \leq 8$
$0 \leq y \leq 6$
$7x + 10y \leq 50$

13. Objective Function
$f(x, y) = 6x + 8y$
Constraints
$x \geq 0; y \geq 0$
$4x + y \leq 16$
$x + 3y \leq 15$

14. Objective Function
$f(x, y) = x + 2y$
Constraints
$x \geq 0; y \geq 0$
$3x + y \leq 45$
$x + 3y \leq 24$

15. Objective Function
$f(x, y) = 6x + y$
Constraints
$x \geq 0; y \geq 0$
$3x + 4y \leq 24$
$3x + 4y \leq 48$

16. Objective Function
$f(x, y) = 15x + 30y$
Constraints
$x \geq 0; y \geq 0$
$5x + 7y \leq 70$
$5x + 7y \leq 140$

17. Objective Function
$f(x, y) = 3x + 10y$
Constraints
$x \geq 0$
$2x + 4y \geq 8$
$5x - y \leq 10$
$x + 3y \leq 40$

18. Objective Function
$f(x, y) = 20x + 30y$
Constraints
$x \geq 0$
$12x + 6y \geq 120$
$9x - 6y \leq 144$
$x + 4y \leq 12$

Use Linear Programming to answer the following questions. See Example 4.

19. A manufacturer produces two models of computers. The times (in hours) required for assembling, testing and packaging each model are listed in the table below.

Process	Model X	Model Y
Assemble	2.5	3
Test	2	1
Package	0.75	1.25

The total times available for assembling, testing and packaging are 4000 hours, 2500 hours, and 1500 hours, respectively. The profits per unit are $50 for Model X and $52 for Model Y. How many of each type should be produced to maximize profit? What is the maximum profit?

20. A manufacturer produces two types of fans. The time (in minutes) required for assembling, packaging, and shipping each type are listed in the table below.

Process	Type X	Type Y
Assemble	20	25
Package	40	10
Ship	10	7.5

The total times available for assembling, packaging, and shipping are 4000 minutes, 4800 minutes, and 1500 minutes, respectively. The profits per unit are $4.50 for Type X and $3.75 for Type Y. How many of each type should be produced to maximize profit? What is the maximum profit?

Type X   Type Y

21. Ashley is making a set of patchwork curtains for her apartment. She needs a minimum of 16 yards of the solid material, at least 5 yards of the striped material, and at least 20 yards of the flowered material. She can choose between two sets of precut bundles. The olive-based bundle costs $10 per bundle and contains 8 yards of the solid material, 1 yard of the striped material, and 2 yards of the flowered material. The cranberry-based bundle costs $20 per bundle and includes 2 yards of the solid material, 1 yard of the striped material, and 7 yards of the flowered material. How many of each bundle should Ashley buy to minimize her cost and yet buy enough material to complete the curtains? What is her minimum cost?

22. A volunteer has been asked to drop off some supplies at a facility housing victims of a hurricane evacuation. The volunteer would like to bring at least 60 bottles of water, 45 first aid kits, and 30 security blankets on his visit. The relief organization has a standing agreement with two companies that provide victim packages. Company

A can provide packages of 5 water bottles, 3 first aid kits, and 4 security blankets at a cost of $1.50. Company B can provide packages of 2 water bottles, 2 first aid kits, and 1 security blanket at a cost of $1.00. How many of each package should the volunteer pick up to minimize the cost? What total amount does the relief organization pay?

chapter eight   Systems of Equations

23. On your birthday your grandmother gave you $25,000, but told you she would like you to invest the money for 10 years before you used any of it. Since you wish to respect your grandmother's wishes, you seek out the advice of a financial adviser. She suggests you invest at least $15,000 in municipal bonds yielding 6% and no more than $5,000 in Treasury bills yielding 9%. How much should be placed in each investment so that income is maximized?

24. An independent cell phone manufacturer produces two models: a flip phone and a camera phone. The manufacturer's quota per day is to produce at least 100 flip phones and 80 camera phones. No more than 200 flip phones and 170 camera phones can be produced per day due to limitations on production. A total of at least 200 phones must be shipped every day.
    a. If the production cost is $5 for a flip phone and $7 for a camera phone, how many of each model should be produced on a daily basis to minimize cost? What is the minimum cost to produce the minimum amount?
    b. If each flip phone results in a $2 loss but each camera phone results in a $5 gain, how many of each model should be manufactured daily to maximize profit? What is the maximum profit if this number of phones were produced?

section 8.7 — **Nonlinear Systems of Equations**

8.7 Nonlinear Systems of Equations

TOPICS

1. Approximating solutions by graphing
2. Solving nonlinear systems algebraically

Topic 1: Approximating Solutions by Graphing

In this last section of the chapter, we deal with systems of equations in which one or more of the equations is nonlinear. For instance, many of the systems we will solve contain at least one equation with at least one second-degree term.

Because the systems are nonlinear, the matrix-based methods that we have learned do not apply. In fact, the theory of nonlinear systems is much more complex than that of linear systems, and for the most part we will rely on only very general guiding principles to solve them. What this means in practice is that you shouldn't be surprised or discouraged if your first approach fails to yield a solution; when this happens, simply try another angle of attack.

The first general guiding principle is to think about the equations that make up the system and to try to get some idea for what real solutions to expect. As we have learned how to graph many classes of equations, we can put our graphing knowledge to good use in applying this principle. Keep in mind, however, that graphing will only help us identify the *real* solutions; as we will see, some nonlinear systems also have solutions containing non-zero imaginary parts.

example 1

Use graphing to guess the real solution(s) of the following system, and then verify that your answer is correct.

$$\begin{cases} x + y = 2 \\ x^2 + y^2 = 4 \end{cases}$$

continued on next page...

chapter eight — Systems of Equations

Solution:

> Display two graphs simultaneously by entering multiple equations in **Y=**. When solving for *y* in this case, enter an equation for both the positive and negative square roots. After selecting **GRAPH**, **TRACE** was selected to view specific coordinates.

The first equation describes a line in the plane with an *x*-intercept at $(2, 0)$ and a *y*-intercept of $(0, 2)$, and the second describes a circle of radius two centered at the origin. As is always the case when given a system of two equations, we are interested in whether any ordered pairs of numbers satisfy both equations simultaneously. Graphically, we are looking for the points of intersection of the two graphs. The following graph of the system certainly suggests a non-empty solution set.

In fact, the graph suggests that $(2, 0)$ and $(0, 2)$ make up the real solutions of the system. We already know these two points lie on the line, and it can be verified that they also solve the second equation:

$$2^2 + 0^2 = 4 \text{ and } 0^2 + 2^2 = 4.$$

This simple example points out another very significant difference between linear and nonlinear systems. Recall that the solution set of any linear system is guaranteed to be empty, to contain exactly one element, or to be infinite. As we have now seen, nonlinear systems may have a different number of solutions.

Before we look at another example, verify for yourself (by graphing) that the following systems have, respectively, one real solution and no real solution:

$$\begin{cases} x + y = 4 \\ x^2 + y^2 = 8 \end{cases} \quad \begin{cases} x + y = 5 \\ x^2 + y^2 = 8 \end{cases}$$

example 2

Use graphing to guess the real solution(s) of the following system, and then verify that your answer is correct.

$$\begin{cases} x + 2 = y^2 \\ x^2 + y^2 = 4 \end{cases}$$

Nonlinear Systems of Equations

Solution:

From our work in Section 6.3, we know the first equation represents a horizontally-oriented parabola opening to the right, with vertex at $(-2, 0)$. The second equation is again a circle of radius two centered at the origin. The graph of the system is shown below.

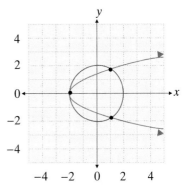

This time, the system appears to have three real solutions. The point $(-2, 0)$ is easily verified to be one of them, but the other two cannot be so readily found just by looking at the graph. However, the x-coordinate of the other two points appears to be 1, and if we replace x with 1 in the two equations we obtain the equation $y^2 = 3$ in *both* cases, meaning that $(1, \sqrt{3})$ and $(1, -\sqrt{3})$ are the other two solutions of the system. (Checking these solutions verifies that both ordered pairs solve both equations.)

Topic 2: Solving Nonlinear Systems Algebraically

In *Example 2*, we found it necessary to combine some graphical knowledge with some algebraic work in order to solve the system. The picture corresponding to the system suggested strongly that there were three real solutions, and for that example we can be fairly confident that we found all of them. But graphs can be misleading, especially if they are not drawn accurately. Consider, for example, the following very similar system.

example 3

Solve the following nonlinear system.

$$\begin{cases} x + 4 = y^2 \\ x^2 + y^2 = 4 \end{cases}$$

continued on next page...

chapter eight   Systems of Equations

Solution:

The only difference between this system and the one in *Example 2* is that the vertex of the parabola is at $(-4, 0)$, so the picture of the system is as shown.

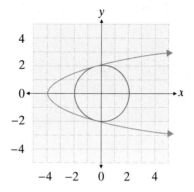

It isn't clear, from the graphs of the two equations, exactly how many intersections exist or where those intersections are. A reasonable guess might be the two points $(0, 2)$ and $(0, -2)$, and it can be easily verified that these two points do indeed lie on both equations. But if we were to conclude that these are the only two solutions, we would be wrong.

We need a more reliable method for this example, and algebra comes to the rescue. Although matrix-based solution methods don't apply to nonlinear systems, the simpler methods of substitution and elimination do. For the system

$$\begin{cases} x + 4 = y^2 \\ x^2 + y^2 = 4 \end{cases},$$

the method of substitution works well if we solve both equations for y^2:

$$\begin{cases} x + 4 = y^2 \\ y^2 = 4 - x^2 \end{cases}$$

The two expressions equal to y^2 both contain only the variable x, so if we equate the expressions we will have a single equation in one variable. The resulting equation is solved as shown.

$$x + 4 = 4 - x^2$$
$$x = -x^2$$
$$x^2 + x = 0$$
$$x(x + 1) = 0$$
$$x = 0, -1$$

We already know that 0 is one possible value for x, as we have verified that $(0, 2)$ and $(0, -2)$ are solutions of the system. But the value of -1 is new, and if we substitute $x = -1$ into the equations, we find that $\left(-1, \sqrt{3}\right)$ and $\left(-1, -\sqrt{3}\right)$ are also solutions. So the system has a total of four solutions, appearing as shown in the following figure.

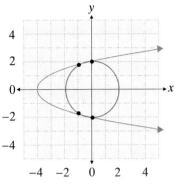

The next example of a nonlinear system is readily solved by the method of elimination.

example 4

Solve the following nonlinear system.

$$\begin{cases} y+1=(x-2)^2 \\ 3-y=(x-1)^2 \end{cases}$$

Solution:

Before solving the system algebraically, you may want to graph these two vertically-oriented parabolas to get some idea of what to expect.

We can eliminate the variable y by adding the two equations, leaving the quadratic equation

$$4 = (x-2)^2 + (x-1)^2.$$

We know very well how to solve such single-variable equations:

$$4 = (x-2)^2 + (x-1)^2$$
$$4 = x^2 - 4x + 4 + x^2 - 2x + 1$$
$$0 = 2x^2 - 6x + 1$$
$$x = \frac{6 \pm \sqrt{36-8}}{4}$$
$$x = \frac{3 \pm \sqrt{7}}{2}$$

continued on next page...

chapter eight — Systems of Equations

For each of the two values of x, we need to determine the corresponding values of y, and we can use either equation to do so. In the work below, we have used the first equation.

For $x = \dfrac{3+\sqrt{7}}{2}$:

$$y = \left(\dfrac{3+\sqrt{7}}{2} - 2\right)^2 - 1$$

$$y = \left(\dfrac{-1+\sqrt{7}}{2}\right)^2 - 1$$

$$y = \dfrac{8-2\sqrt{7}}{4} - 1$$

$$y = 2 - \dfrac{\sqrt{7}}{2} - 1$$

$$y = 1 - \dfrac{\sqrt{7}}{2}$$

For $x = \dfrac{3-\sqrt{7}}{2}$:

$$y = \left(\dfrac{3-\sqrt{7}}{2} - 2\right)^2 - 1$$

$$y = \left(\dfrac{-1-\sqrt{7}}{2}\right)^2 - 1$$

$$y = \dfrac{8+2\sqrt{7}}{4} - 1$$

$$y = 2 + \dfrac{\sqrt{7}}{2} - 1$$

$$y = 1 + \dfrac{\sqrt{7}}{2}$$

The two solutions of the system are thus

$$\left(\dfrac{3+\sqrt{7}}{2},\, 1-\dfrac{\sqrt{7}}{2}\right) \text{ and } \left(\dfrac{3-\sqrt{7}}{2},\, 1+\dfrac{\sqrt{7}}{2}\right).$$

We will conclude with an example that illustrates the last type of complexity we might encounter with nonlinear systems.

example 5

Solve the following nonlinear system.

$$\begin{cases} \dfrac{(x+1)^2}{9} + y^2 = 1 \\ y^2 = x + 2 \end{cases}$$

Solution:

The first equation describes an ellipse, and the second describes a parabola opening to the right, and a graph of the system indicates we should expect two real solutions:

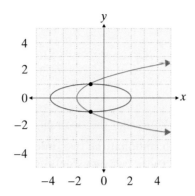

But if we solve this system algebraically, we discover another aspect to the problem. The method of substitution suggests itself, since the second equation is already solved for y^2. When we make the appropriate substitution in the first equation, we obtain

$$\frac{(x+1)^2}{9} + x + 2 = 1.$$

We solve this second-degree equation in x the usual way:

$$\frac{(x+1)^2}{9} + x + 2 = 1$$

$$\frac{(x+1)^2}{9} + x + 1 = 0$$

$$(x+1)^2 + 9x + 9 = 0$$

$$x^2 + 11x + 10 = 0$$

$$(x+10)(x+1) = 0$$

$$x = -10, -1$$

The solution $x = -1$ isn't surprising, since our graph certainly suggested this value, but the solution $x = -10$ doesn't seem to relate to the picture at all. But remember, these are only the x-coordinates of the ordered pairs that solve the system, and we still need to determine the corresponding y-coordinates. We will use the second equation to do this.

continued on next page...

chapter eight Systems of Equations

$$\begin{array}{ll} \text{For } x = -10 & \text{For } x = -1 \\ y^2 = -10 + 2 & y^2 = -1 + 2 \\ y^2 = -8 & y^2 = 1 \\ y = \pm 2i\sqrt{2} & y = \pm 1 \end{array}$$

Now we know that the two real solutions are $(-1, -1)$ and $(-1, 1)$, as was suggested by our graph, and that the two ordered pairs $(-10, -2i\sqrt{2})$ and $(-10, 2i\sqrt{2})$ also solve the system. These last two solutions don't appear on the graph because of their non-real second coordinates.

technology note

As we have seen, computer algebra systems and some calculators are capable of solving systems of equations. The command in *Mathematica* for solving nonlinear systems of equations is again **Solve**. The illustration below shows the syntax used in *Mathematica* for solving the nonlinear system from *Example 5*; note how *Mathematica* presents the solution that we found manually.

section 8.7 Nonlinear Systems of Equations

exercises

Use graphing to guess the real solution(s) of the following systems, and then verify your answer algebraically. See Examples 1 and 2.

1. $\begin{cases} 3x - 2y = 6 \\ \dfrac{x^2}{4} + \dfrac{y^2}{9} = 1 \end{cases}$
2. $\begin{cases} x + 2y = 2 \\ \dfrac{x^2}{4} + y^2 = 1 \end{cases}$
3. $\begin{cases} x^2 + 4y^2 = 5 \\ x^2 + y^2 = 2 \end{cases}$

4. $\begin{cases} 4x^2 + y^2 = 5 \\ 4(x-2)^2 + y^2 = 5 \end{cases}$
5. $\begin{cases} y = x^2 \\ 2 - y = x^2 \end{cases}$
6. $\begin{cases} x^2 + (y-1)^2 = 4 \\ (x-3)^2 + (y-1)^2 = 1 \end{cases}$

7. $\begin{cases} y - x^2 = 1 \\ y + 2 = 4x^2 \end{cases}$
8. $\begin{cases} x^2 + y^2 = 10 \\ x^2 + y = -2 \end{cases}$
9. $\begin{cases} x = y^2 - 3 \\ x^2 + 4y^2 = 4 \end{cases}$

10. $\begin{cases} (x-1)^2 + (y-6)^2 = 9 \\ (x-1)^2 + (y+1)^2 = 16 \end{cases}$
11. $\begin{cases} (x+1)^2 + (y-1)^2 = 4 \\ (x+1)^2 + 4(y-1)^2 = 4 \end{cases}$
12. $\begin{cases} x^2 + y^2 = 1 \\ y = x^2 - 1 \end{cases}$

13. $\begin{cases} (x-2)^2 + y^2 = 4 \\ (x+2)^2 + y^2 = 4 \end{cases}$
14. $\begin{cases} x^2 + y^2 = 9 \\ x^2 + y^2 - 2x - 3 = 1 \end{cases}$
15. $\begin{cases} x^2 + y^2 = 9 \\ \dfrac{x^2}{9} + \dfrac{y^2}{25} = 1 \end{cases}$

16. $\begin{cases} x^2 + y^2 = 4 \\ -x^2 = 2y - 1 \end{cases}$
17. $\begin{cases} x^2 + \dfrac{y^2}{4} = 1 \\ y = 0 \end{cases}$
18. $\begin{cases} y = x^2 + 1 \\ y - 1 = x^3 \end{cases}$

19. $\begin{cases} 2y^2 - 3x^2 = 6 \\ 2y^2 + x^2 = 22 \end{cases}$
20. $\begin{cases} y = x^3 \\ y = \sqrt[3]{x} \end{cases}$
21. $\begin{cases} x = y^2 - 4 \\ x + 13 = 6y \end{cases}$

22. $\begin{cases} y = 2x^2 - 3 \\ y = -x^2 \end{cases}$
23. $\begin{cases} 2y = x^2 - 4 \\ x^2 + y^2 = 4 \end{cases}$
24. $\begin{cases} x^2 - y^2 = 5 \\ \dfrac{x^2}{25} + \dfrac{4y^2}{25} = 1 \end{cases}$

chapter eight Systems of Equations

Solve the following nonlinear systems algebraically. Be sure to check for non-real solutions. See Examples 3, 4, and 5.

25. $\begin{cases} x^2 + y^2 = 30 \\ x^2 = y \end{cases}$

26. $\begin{cases} 3x^2 + 2y^2 = 12 \\ x^2 + 2y^2 = 4 \end{cases}$

27. $\begin{cases} x^2 - 1 = y \\ 4x + y = -5 \end{cases}$

28. $\begin{cases} x^2 + y^2 = 4 \\ 3x^2 + 4y^2 = 24 \end{cases}$

29. $\begin{cases} y = \dfrac{4}{x} \\ 2x^2 + y^2 = 18 \end{cases}$

30. $\begin{cases} xy = 5 \\ x^2 + y^2 = 10 \end{cases}$

31. $\begin{cases} y - x^2 = 4 \\ x^2 + y^2 = 16 \end{cases}$

32. $\begin{cases} y - x^2 = 6x \\ y = 4x \end{cases}$

33. $\begin{cases} 2x^2 + 3y^2 = 6 \\ x^2 + 3y^2 = 3 \end{cases}$

34. $\begin{cases} x^2 + y^2 = 4 \\ \dfrac{x^2}{4} - \dfrac{y^2}{8} = 1 \end{cases}$

35. $\begin{cases} 3x^2 - y = 3 \\ 9x^2 + y^2 = 27 \end{cases}$

36. $\begin{cases} \dfrac{1}{x} + \dfrac{1}{y} = 5 \\ \dfrac{1}{x} - \dfrac{1}{y} = -3 \end{cases}$

37. $\begin{cases} x + y^2 = 2 \\ 2x^2 - y^2 = 1 \end{cases}$

38. $\begin{cases} y - 2 = (x+3)^2 \\ \dfrac{1}{3}y = (x-1)^2 \end{cases}$

39. $\begin{cases} y - 2 = (x-2)^2 \\ y + 2 = (x-1)^2 \end{cases}$

40. $\begin{cases} y^2 + 2 = 2x^2 \\ y^2 = x^2 - 6 \end{cases}$

41. $\begin{cases} (x+1)^2 + y^2 = 10 \\ \dfrac{(x-2)^2}{4} + y^2 = 1 \end{cases}$

42. $\begin{cases} x^2 + y^2 = 10 \\ x^2 + y = 8 \end{cases}$

43. $\begin{cases} 2x = y - 1 \\ \dfrac{x^2}{25} + y^2 = 1 \end{cases}$

44. $\begin{cases} 2x^2 + y^2 = 4 \\ 2(x-1)^2 + y^2 = 3 \end{cases}$

45. $\begin{cases} x^2 + 7y^2 = 14 \\ x^2 + y^2 = 3 \end{cases}$

46. $\begin{cases} x^2 + y^2 = 25 \\ y^2 = x - 5 \end{cases}$

47. $\begin{cases} y = x^3 + 8x^2 + 17x + 10 \\ -y = x^3 + 8x^2 + 17x + 10 \end{cases}$

48. $\begin{cases} \dfrac{x^2}{25} + \dfrac{y^2}{16} = 1 \\ x^2 + y^2 = 16 \end{cases}$

49. $\begin{cases} xy = 6 \\ (x-2)^2 + (y-2)^2 = 1 \end{cases}$

50. $\begin{cases} y^2 = x + 1 \\ \dfrac{x^2}{5} + \dfrac{y^2}{6} = 1 \end{cases}$

51. $\begin{cases} y = x^3 - 1 \\ 3y = 2x - 3 \end{cases}$

52. $\begin{cases} y + 5 = (x+1)^2 \\ y - 3 = (x-3)^2 \end{cases}$

53. $\begin{cases} xy - y = 4 \\ (x-1)^2 + y^2 = 10 \end{cases}$

54. $\begin{cases} 2x^2 + 5y^2 = 16 \\ 4x^2 + 3y^2 = 4 \end{cases}$

55. $\begin{cases} y = \sqrt{x-4} + 1 \\ (x-3)^2 + (y-1)^2 = 1 \end{cases}$

56. $\begin{cases} y = \sqrt[3]{x} \\ \sqrt{y} = x \end{cases}$

57. $\begin{cases} y^2 - y - 12 = x - x^2 \\ y - 1 + \dfrac{2x - 12}{y} = 0 \end{cases}$

58. $\begin{cases} y = 7x^2 + 1 \\ x^2 + y^2 = 1 \end{cases}$

59. $\begin{cases} \dfrac{(y+2)^2}{(x+y)} = 1 \\ x = y^2 + 5y + 4 \end{cases}$

60. $\begin{cases} x = \sqrt{6y + 1} \\ y = \sqrt{\dfrac{x^2 + 7}{2}} \end{cases}$

61. $\begin{cases} \dfrac{-2}{x^2} + \dfrac{1}{y^2} = 8 \\ \dfrac{9}{x^2} - \dfrac{2}{y^2} = 4 \end{cases}$

62. $\begin{cases} x^2 + 3x - 2y^2 = 5 \\ -4x^2 + 6y^2 = 3 \end{cases}$

Use a graphing calculator to solve the following systems, and then verify your answer algebraically.

63. $\begin{cases} \dfrac{xy}{2} = \dfrac{71}{5} \\ \dfrac{6x^2}{7} + \dfrac{3y^2}{8} = \dfrac{39}{11} \end{cases}$

64. $\begin{cases} 2.4x^2 + 1.6y^2 = 18.9 \\ 2.1(x+1)^2 + 3.7y^2 = 16.3 \end{cases}$

65. $\begin{cases} y = 3.9(x+1)^2 \\ y + 8.4 = (x-1)^2 \end{cases}$

66. $\begin{cases} \dfrac{7}{9}x^2 + \dfrac{5}{6}y^2 = \dfrac{29}{5} \\ \dfrac{8}{3}x^2 + \dfrac{7}{4}y^2 = \dfrac{24}{7} \end{cases}$

67. $\begin{cases} 2.7x^2 - 10.3 = 4.1y \\ 6.5x + 0.4y = -19.8 \end{cases}$

68. $\begin{cases} x^2 + y^2 = 28.09 \\ y + 14.6 = x^2 \end{cases}$

Use the methods of this section to solve the following problems.

69. The area of a certain rectangle is 45 square inches, and its perimeter is 28 inches. Find the dimensions of the rectangle.

70. The product of two positive integers is 88, and their sum is 19. What are the integers?

chapter eight   Systems of Equations

71. Jack takes half an hour longer than his wife does to make the 210-mile drive between two cities. His wife drives 10 miles an hour faster. How fast do the two drive?

72. To construct the two garden beds shown below, 48.5 meters of fencing are needed. The combined area of the beds is 95 square meters. There are two possibilities for the overall dimensions of the two beds. What are they?

73. The product of two integers is -84, and their sum is -5. What are the integers?

74. Paul and Maria were driving the same 24 mile route, and they departed at the same time. After 20 minutes, Maria was 4 miles ahead of Paul. If it took Paul 10 minutes longer to reach their destination, how fast were they each driving?

75. The surface area of a certain right circular cylinder is 54π cm^2 and the volume is 54π cm^3. Find the height h and the radius r of this cylinder.

(**Hint:** The formulas for the volume and surface area of a right circular cylinder are as follows: $V = \pi r^2 h$ and $A = 2\pi rh + 2\pi r^2$.)

chapter eight project

Market Share Matrix

Assume you are the sales and marketing director for Joe's Java, a coffee shop located on a crowded city street corner. There are two competing coffee shops on this block – Buck's Café and Tweak's Coffee. The management has asked you to develop a marketing campaign to increase your market share from 25% to at least 35% within 6 months. With the resulting plan to meet this goal, you predict that each month:

a. You will retain 93% of your customers, 4% will go to Buck's Café, and 3% will go to Tweak's Coffee.
b. Buck's Café will retain 91% of their customers, 6% will come to Joe's Java, and 3% will go to Tweak's Coffee.
c. Tweak's Coffee will retain 92% of their customers, 3% will come to Joe's Java and 5% will go to Buck's Café.

The current percentage of the market is shown in this matrix:
$$x_0 = \begin{bmatrix} 0.25 \\ 0.45 \\ 0.30 \end{bmatrix} \begin{matrix} Joe's \\ Buck's \\ Tweak's \end{matrix}$$

After one month the shares of the coffee shops are:
$$x_1 = Px_0 = \begin{bmatrix} 0.93 & 0.06 & 0.03 \\ 0.04 & 0.91 & 0.05 \\ 0.03 & 0.03 & 0.92 \end{bmatrix} \begin{bmatrix} 0.25 \\ 0.45 \\ 0.30 \end{bmatrix} = \begin{bmatrix} 0.2685 \\ 0.4345 \\ 0.2970 \end{bmatrix}$$

1. Construct a table that lists the market share for all of the coffee shops at the end of each of the first 6 months.

2. Will your campaign be successful based on this model (will you reach 35% market share in 6 months)?

3. What actions do you think Buck's Café and Tweak's Coffee will take as your market share changes?

4. What effect could their actions make on the market?

chapter eight   **Systems of Equations**

CHAPTER SUMMARY

A summary of concepts and skills follows each chapter. Refer to these summaries to make sure you feel comfortable with the material in the chapter. The concepts and skills are organized according to the section title and topic title in which the material is first discussed.

8.1 Solving Systems by Substitution and Elimination

topics	pages
Definition and Classification of Linear Systems of Equations • The geometric and algebraic meaning of *linear* (or *simultaneous*) *systems of equations* • The varieties of linear systems of equations, and the meaning of the terms *inconsistent, consistent,* and *dependent*	p. 589 – 591
Solving Systems by Substitution • The use of the *substitution method* to solve linear systems of equations • Alternatives for describing solutions that consist of more than one point	p. 591 – 593
Solving Systems by Elimination • The use of the *elimination method* to solve linear systems of equations • The use of arrow notation to keep track of the steps used in solving a system	p. 594 – 595
Larger Systems of Equations • The geometric and algebraic meaning of larger linear systems of equations • How the solution process of larger systems relates to that of smaller systems	p. 596 – 599
Applications of Systems of Equations • Using systems of equations to solve previously-encountered application problems	p. 599 – 601

8.2 Matrix Notation and Gaussian Elimination

topics	pages
Linear Systems, Matrices, and Augmented Matrices • The definition of a *matrix*, the meaning of *rows* and *columns*, and the definition of *matrix order* • Notation commonly used to define and refer to matrices and their elements • The meaning of an *augmented matrix*	p. 605 – 608

chapter eight summary

topics (continued)	pages
Gaussian Elimination and Row Echelon Form • The meaning of *Gaussian elimination* as a solution strategy • The definition of *row echelon form*, and how it relates to solving systems of equations • The definition, notation, and use of the three *elementary row operations*	p. 608 – 611
Gauss-Jordan Elimination and Reduced Row Echelon Form • The meaning of *Gauss-Jordan elimination* as a solution strategy • The definition of *reduced row echelon form*	p. 611 – 613

8.3 Determinants and Cramer's Rule

topics	pages
Determinants and Their Evaluation • The definition of the *determinant* of a matrix • The definition of a matrix element's *minor* and *cofactor*, and the use of minors and cofactors in evaluating determinants • The use of properties of determinants and elementary row and column operations to simplify the computation of determinants	p. 619 – 626
Using Cramer's Rule to Solve Linear Systems • The formulas that go by the name Cramer's rule, and how these formulas can be used to solve linear systems of equations	p. 626 – 631

8.4 The Algebra of Matrices

topics	pages
Matrix Addition • The definition of *matrix equality* • The meaning of *matrix addition*, and how to perform it	p. 635 – 637
Scalar Multiplication • The meaning of *scalar multiplication*, and how to perform it	p. 638 – 639
Matrix Multiplication • The meaning of *matrix multiplication*, and how it relates to composition of functions	p. 639 – 644

chapter eight   Systems of Equations

topics (continued)	pages
Interlude: Transition Matrices • The definition of a *transition matrix* • How to determine the steady state of a system in transition	p. 644 – 646

8.5 Inverses of Matrices

topics	pages
The Matrix Form of a Linear System • How to write a linear system of equations as a single matrix equation	p. 650 – 651
Finding the Inverse of a Matrix • The meaning of the *identity matrix*, and the effect of multiplying a given square matrix by the appropriate identity • The meaning of the *inverse* of a matrix, and why the inverse of a matrix is useful in solving a given system • The algorithm used to find the inverse of a matrix	p. 652 – 657
Using Matrix Inverses to Solve Linear Systems • The different interpretations that can be put on solving a system by the inverse matrix method • Why the inverse matrix method is particularly efficient in some cases	p. 657 – 661

8.6 Linear Programming

topics	pages
Planar Feasible Regions • The definition of feasible region or region of constraints • How to construct graphs of the feasible region	p. 664 – 665
Linear Programming in Two Variables • How to use the linear programming method • How to find the objective function	p. 666 – 669

8.7 Nonlinear Systems of Equations

topics	pages
Approximating Solutions by Graphing • Using graphs of systems of nonlinear equations to gain an approximate understanding of the real-number solutions to expect	p. 675 – 677
Solving Nonlinear Systems Algebraically • Using algebra to determine all the solutions of nonlinear systems of equations	p. 677 – 682

chapter eight review

chapter review

Section 8.1

Use any convenient method to solve the following systems of equations. If a system is dependent, express the solution set in terms of one or more of the variables, as appropriate.

1. $\begin{cases} 3x - y + z = 2 \\ -x + y - 2z = -4 \\ -6x + 2y - 2z = -7 \end{cases}$

2. $\begin{cases} 2x - y = 13 \\ 5x - 2y - z = 25 \\ 7x - 6z = -2 \end{cases}$

3. $\begin{cases} x + y - z = 1 \\ 3x - 4y - 5z = -1 \\ 6x - 3y + z = 20 \end{cases}$

4. $\begin{cases} 6x - 5y = 17 \\ -4x + 9y = -17 \end{cases}$

Section 8.2

Construct the augmented matrix that corresponds to each of the following systems of equations.

5. $\begin{cases} 2x + (y - z) = 3 \\ 2(y - x) + y - 2 = z \\ 3x - \dfrac{3 - z}{2} = 4y \end{cases}$

6. $\begin{cases} z - 4x = 5y \\ 14z + 7(x + 3y) = 21 \\ 8x - y = -2(x - 3z) \end{cases}$

Fill in the blanks by performing the indicated elementary row operations.

7. $\begin{bmatrix} 3 & 1 & | & -2 \\ 1 & 2 & | & 3 \end{bmatrix} \xrightarrow{-3R_2 + R_1} \underline{\ ?\ }$

8. $\begin{bmatrix} 2 & 3 & | & 5 \\ -4 & -1 & | & 2 \end{bmatrix} \xrightarrow{2R_1 + R_2} \underline{\ ?\ }$

Section 8.3

Evaluate the following determinants.

9. $\begin{vmatrix} x^3 & -x^2 \\ x^2 & x \end{vmatrix}$

10. $\begin{vmatrix} -1 & 3 & 1 \\ 1 & -4 & 0 \\ 0 & 2 & 3 \end{vmatrix}$

11. $\begin{vmatrix} -2 & -1 & -3 & 0 \\ 3 & 3 & 1 & 5 \\ 4 & 0 & 0 & 1 \\ 2 & 0 & 1 & 0 \end{vmatrix}$

12. $\begin{vmatrix} x^4 & x & x & 2x \\ 0 & x & x^3 & x \\ 0 & 0 & x & x \\ 0 & 0 & 0 & x^2 \end{vmatrix}$

chapter eight — Systems of Equations

Use Cramer's rule to solve the following systems.

13. $\begin{cases} x + 6y = 2 \\ 3x - y = -13 \end{cases}$

14. $\begin{cases} x + 2y - 3z = -3 \\ -5x - y + 4z = -5 \\ 3x + y + z = 6 \end{cases}$

Section 8.4

Determine values of the variables that will make the following equations true, if possible.

15. $\begin{bmatrix} w & 5x \\ 2y & z \end{bmatrix} - 3 \begin{bmatrix} w & x \\ 2 & -z \end{bmatrix} = \begin{bmatrix} 4 & 2 \\ y-3 & -16 \end{bmatrix}$

16. $\begin{bmatrix} 4x & 2y^2 & z \end{bmatrix} = \begin{bmatrix} 12 \\ 18 \\ -2 \end{bmatrix}$

Evaluate the following matrix products, if possible.

17. $\begin{bmatrix} 7 & 1 & -1 \end{bmatrix} \begin{bmatrix} 1 & 6 \\ 2 & 1 \\ -3 & -3 \end{bmatrix}$

18. $\begin{bmatrix} 4 \\ 5 \\ 6 \end{bmatrix} \begin{bmatrix} -3 & 2 & 3 \end{bmatrix}$

Section 8.5

Write each of the following systems of equations as a single matrix equation.

19. $\begin{cases} \dfrac{x - 8y}{3} = -1 \\ -2y - 3 = 4x \end{cases}$

20. $\begin{cases} x_1 - x_2 + 2x_3 = -4 \\ 2x_1 - 3x_2 - x_3 = 1 \\ -3x_1 + 6x_3 = 5 \end{cases}$

Find the inverse of each of the following matrices, if possible.

21. $\begin{bmatrix} 4 & -2 \\ 2 & 3 \end{bmatrix}$

22. $\begin{bmatrix} 2 & 2 \\ \dfrac{1}{2} & 1 \end{bmatrix}$

Section 8.6

Construct the constraints and graph the feasible regions for the following situations.

23. Each bag of nuts contains peanuts and cashews. The total number of nuts in the bag cannot exceed 60. There must be at least 20 peanuts and 10 cashews per bag. There can be no more than 40 peanuts or 40 cashews per bag. What is the region of constraint for the number of nuts per bag?

24. You wish to study at least 15 hours (over a 4 day span) for your upcoming Statistics and Biology tests. You need to study a minimum of 6 hours for each test. The maximum you wish to study for Statistics is 10 hours and for Biology is 8 hours. What is the region of constraint for the number of hours you should study for each test?

Find the minimum and maximum values of the given functions, subject to the given constraints.

25. Objective Function
$f(x, y) = 6x + 10y$
Constraints
$x \geq 0, y \geq 0, 2x + 5y \leq 10$

26. Objective Function
$f(x, y) = 2x + y$
Constraints
$x \geq 0, x + 4y \leq 16, 3x - y \leq 9, 2x + 3y \geq 6$

27. Objective Function
$f(x, y) = 5x + 2y$
Constraints
$x \geq 0, y \geq 0, x + y \leq 10,$
$x + 2y \geq 10, 2x + y \geq 10$

28. Objective Function
$f(x, y) = 5x + 4y$
Constraints
$x \geq 0, y \geq 0, 2x + 3y \leq 12,$
$3x + y \leq 12, x + y \geq 2$

29. Objective Function
$f(x, y) = 70x + 82y$
Constraints
$x \geq 0, y \geq 0, x \leq 10, y \leq 20,$
$x + y \geq 5, x + 2y \leq 18$

30. Objective Function
$f(x, y) = 2x + 4y$
Constraints
$x \geq 0, y \geq 0, 2x + 2y \leq 21,$
$x + 4y \leq 20, x + y \leq 18$

Use Linear Programming to answer the following questions.

31. Krueger's Pottery manufactures two kinds of hand-painted pottery: a vase and a pitcher. There are three processes to create the pottery: throwing (forming the pottery on the potter's wheel), baking, and painting. No more than 90 hours are available per day for throwing, only 120 hours are available per day for baking, and no more than 60 hours per day are available for painting. The vase requires 3 hours for throwing, 6 hours for baking, and 2 hours for painting. The pitcher requires 3 hours for throwing, 4 hours for baking, and 3 hours for painting. The profit for each vase is $25 and the profit for each pitcher is $30. How many of each piece of pottery should be produced a day to maximize profit? What would the maximum profit be if Krueger's produced this amount?

32. Pranas produces bionic arms and legs for those that are missing a limb. Pranas can produce at least 20, but no more than 60 arms in a week due to the lab limitations. They can produce at least 15, but no more than 40 legs in a week. To keep their research grant, the company must produce at least 50 limbs. It cost $450 to produce the bionic arm and $550 to produce the bionic leg. How many of each should be produced per week to minimize the cost? What would the minimum cost be if Pranas produced this amount?

Section 8.7

Solve the following nonlinear systems algebraically. Be sure to check for non-real solutions.

33. $\begin{cases} x^2 + 2y^2 = 1 \\ x^2 = y \end{cases}$

34. $\begin{cases} 2x^2 + y^2 = 6 \\ x^2 + 3y^2 = 4 \end{cases}$

35. $\begin{cases} x^2 - 2 = y \\ 3x - 5 = y \end{cases}$

Solve the following problems involving nonlinear systems of equations.

36. The product of two positive integers is 144, and their sum is 25. What are the integers?

37. Stephen and Scott were driving the same 72 mile route, and they departed at the same time. After 30 minutes, Stephen was 6 miles ahead of Scott. If it took Scott one more hour than Stephen to reach their destination, how fast were they each driving?

chapter test

Use the method of substitution to solve the following system of equation. If a system is dependent, express the solution set in terms of one of the variables.

1. $\begin{cases} x + 2y = 4 \\ 2x - y = 3 \end{cases}$

2. $\begin{cases} 3x - y = 2 \\ -6x + 2y = 5 \end{cases}$

3. $\begin{cases} 2x - y + 1 = 0 \\ \dfrac{-x}{2} + \dfrac{y}{2} = \dfrac{1}{2} \end{cases}$

4. $\begin{cases} 3x - 2y = -10 \\ x + 2y = 2 \end{cases}$

Use the method of elimination to solve the following systems of equations. If a system is dependent, express the solution set in terms of one of the variables.

5. $\begin{cases} 3x + 2y = 5 \\ 2x - 3y = -4 \end{cases}$

6. $\begin{cases} \dfrac{x}{3} + y - 1 = 0 \\ x + 3y = 3 \end{cases}$

7. $\begin{cases} 3x - 2y = \dfrac{1}{2} \\ 6x - 4y = 3 \end{cases}$

8. $\begin{cases} \dfrac{x}{3} - \dfrac{y+1}{2} = 1 \\ \dfrac{x}{2} - \dfrac{y}{4} = \dfrac{3}{4} \end{cases}$

Use any convenient method to solve the following systems of equations. If a system is dependent, express the solution set in terms of one or more of the variables, as appropriate.

9. $\begin{cases} x + y = 5 \\ 2x - y = 4 \\ 5x + y = 17 \end{cases}$

10. $\begin{cases} x - 2y = 6 \\ 3x + y = 11 \\ 2x - 2y = -10 \end{cases}$

11. $\begin{cases} x - 2y + z = 2 \\ 2x + y - 2z = -2 \\ -x + 3z = 8 \end{cases}$

12. $\begin{cases} 2x + 3y - 4z = -7 \\ x - y + 4z = 6 \\ x + y + z = 2 \end{cases}$

13. $\begin{cases} 3x - 2y + z = 10 \\ x + y + z = 30 \\ 2x - y - z = -6 \end{cases}$

14. $\begin{cases} 3x - 2y - 2z = -8 \\ x - y - z = -5 \\ x + y + z = -3 \end{cases}$

15. Rose's age is half of what John's will be in 3 years and half of what Bob's was 3 years ago. If Bob's age is $\dfrac{11}{9}$ of John's age, how old is Rose?

16. Find a system of linear equations that has the solution $\left(\frac{2}{3}, -2\right)$. Answers may vary.

17. Find the equation of a parabola $y = ax^2 + bx + c$, passing through the points $(1, 0)$, $(-4, 9)$, and $(-1, 2)$.

18. Find the equation of a circle $x^2 + y^2 + Dx + Ey + F = 0$ that passes through the points $(2, 1)$, $(-1, -2)$, and $(5, -2)$.

19. Find a and b such that the following system of equations has an infinite number of solutions.
$$\begin{cases} 6x - 4y = a \\ 9x - 6y = b \end{cases}$$

20. Two jets start from the same airport and fly in opposite directions. The second jet departs half an hour after the first but its speed is 40 miles per hour faster. Find the speed of each jet, if 2 hours after the first jet departs they are 1600 miles apart.

21. Let $X = \begin{bmatrix} 3 & -2 \\ -1 & 0 \\ -4 & 1 \\ 2 & 3 \end{bmatrix}$. Determine the following, if possible.

    a. The order of X b. The value of x_{42} c. The value of x_{23}

Construct the augmented matrix that corresponds to each of the following systems of equations.

22. $\begin{cases} 3x - y + 5z = 1 \\ x + 2y + 4 = 5 \\ 3x - 4y + z = -1 \end{cases}$

23. $\begin{cases} y - 2z = 1 \\ 3x + y - z = 4 \\ 1 - 4y + 6z = 0 \end{cases}$

24. $\begin{cases} \dfrac{x}{2} - \dfrac{y-z}{3} = 4 \\ \dfrac{x-2y}{3} - \dfrac{1}{4} = \dfrac{z}{6} \\ 2x - y + \dfrac{z-1}{3} = 4 \end{cases}$

25. $\begin{cases} \dfrac{x}{5} - \dfrac{y}{2} = \dfrac{z+4}{10} \\ \dfrac{2}{x-y+z} = \dfrac{1}{3} \\ -x = 1 - \dfrac{z-3y}{4} \end{cases}$

Fill in the blanks by performing the indicated row operations

26. $\begin{bmatrix} 1 & -4 & | & -4 \\ 3 & -1 & | & 3 \end{bmatrix} \xrightarrow{-2R_1 + R_2}$ _____

27. $\begin{bmatrix} 8 & -6 & | & -4 \\ 5 & -10 & | & 15 \end{bmatrix} \xrightarrow[\frac{1}{5}R_2]{\frac{1}{2}R_1}$ _____

28. $\begin{bmatrix} -1 & 0 & 2 & | & -6 \\ 1 & -3 & 4 & | & 1 \\ -2 & -1 & -3 & | & 0 \end{bmatrix} \xrightarrow[-R_1+R_3]{2R_2}$ _____

29. $\begin{bmatrix} -8 & 4 & 6 & | & 2 \\ 3 & -6 & 0 & | & -1 \\ -1 & 2 & 4 & | & 3 \end{bmatrix} \xrightarrow[-2R_1+R_2]{R_3+R_1}$ _____

Use Gaussian elimination and back substitution to solve the following systems of equations.

30. $\begin{cases} 3x - y = 7 \\ x - 4y = 6 \end{cases}$

31. $\begin{cases} \dfrac{x}{2} - \dfrac{y}{3} = 1 \\ \dfrac{2x}{5} - \dfrac{5y}{3} = 5 \end{cases}$

32. $\begin{cases} \dfrac{x}{5} - \dfrac{y}{3} = 2 \\ -6x + 5y = 20 \end{cases}$

33. $\begin{cases} 2x - 5y + 3z = 1 \\ x + y - 5z = -12 \\ -4x + 6y - 3z = -1 \end{cases}$

34. $\begin{cases} x - y + 3z = -1 \\ -5x + 6y - 4z = -8 \\ 3x - 5z = 5 \end{cases}$

35. $\begin{cases} \dfrac{x}{2} - y + \dfrac{z}{3} = \dfrac{-1}{6} \\ 2x - \dfrac{y-3}{2} + \dfrac{z}{4} = \dfrac{13}{4} \\ x + y - 2z = 0 \end{cases}$

Evaluate the following determinants.

36. $\begin{vmatrix} 4 & -2 \\ 3 & -3 \end{vmatrix}$

37. $\begin{vmatrix} 2+i & 3 \\ -1 & 2-i \end{vmatrix}$

38. $\begin{vmatrix} \dfrac{3}{2} & \dfrac{-1}{3} \\ -6 & \dfrac{2}{5} \end{vmatrix}$

39. $\begin{vmatrix} xy & -axy \\ bx & -xab \end{vmatrix}$

Solve for x by finding the determinant.

40. $\begin{vmatrix} x-1 & 5 \\ 3 & x+1 \end{vmatrix} = 0$

41. $\begin{vmatrix} x+2 & 2 \\ 7 & x-3 \end{vmatrix} = 0$

42. $\begin{vmatrix} 2-x & 3 \\ 4 & x+2 \end{vmatrix} = 0$

Use Cramer's rule to solve the following systems.

43. $\begin{cases} 3x - 4y = 14 \\ x + 3y = -4 \end{cases}$

44. $\begin{cases} -4x + 2y = 3 \\ 2x - y = 4 \end{cases}$

chapter eight   Systems of Equations

45. $\begin{cases} x+y-3z=-1 \\ 2x+2y+z=5 \\ -x+3y-2z=-8 \end{cases}$

46. $\begin{cases} x-2y=0 \\ x+y+z=6 \\ 3x-y-4z=10 \end{cases}$

Determine values of the variables that will make the following equations true, if possible. Answers may vary.

47. $\begin{bmatrix} x \\ 2y \end{bmatrix} + \begin{bmatrix} x \\ -y \\ z \end{bmatrix} = \begin{bmatrix} 1 \\ 2 \\ -1 \end{bmatrix}$

48. $2\begin{bmatrix} x \\ -3y \end{bmatrix} - \begin{bmatrix} y \\ 2x \end{bmatrix} = \begin{bmatrix} 7 \\ 14 \end{bmatrix}$

49. $\begin{bmatrix} 3x \\ 5y \end{bmatrix} - \begin{bmatrix} y \\ -2x \end{bmatrix} = \begin{bmatrix} 4 \\ -3 \end{bmatrix}$

50. $3\begin{bmatrix} 2r & -s & t \end{bmatrix} - \begin{bmatrix} -r & 2s & 3t \end{bmatrix} = \begin{bmatrix} 21 & 20 & 0 \end{bmatrix}$

Evaluate the following matrix products, if possible.

51. $\begin{bmatrix} 1 & -2 & 3 & 4 \end{bmatrix} \begin{bmatrix} -2 & 0 \\ 3 & -1 \\ 2 & 4 \\ -1 & 1 \end{bmatrix}$

52. $\begin{bmatrix} -2 & 1 \\ 0 & 3 \\ -4 & 6 \end{bmatrix} \begin{bmatrix} 1 & 0 & 3 \\ 2 & -1 & 1 \end{bmatrix}$

53. $\begin{bmatrix} 3 & 0 \\ 1 & 1 \\ -1 & 4 \\ 2 & 3 \end{bmatrix} \begin{bmatrix} 3 & 2 \\ 1 & -1 \\ 4 & 0 \end{bmatrix}$

54. $A = \begin{bmatrix} 1 & 2 \\ -1 & 3 \end{bmatrix}; A^2 = ?$

Given $A = \begin{bmatrix} 4 & -1 \\ -2 & 3 \end{bmatrix}$, $B = \begin{bmatrix} -4 & 6 \end{bmatrix}$, $C = \begin{bmatrix} 4 \\ -1 \\ 3 \end{bmatrix}$, and $D = \begin{bmatrix} -4 & 2 \\ 1 & 0 \end{bmatrix}$, *determine the following, if possible.*

55. AB 56. BA 57. A^2 58. B^2 59. $D+A$ 60. $CD+C$ 61. BD

Write each of the following systems as a single matrix equation.

62. $\begin{cases} 2x-y=4 \\ x+y=0 \end{cases}$

63. $\begin{cases} \dfrac{x-y}{3}=4 \\ x+1=0 \end{cases}$

64. $\begin{cases} 3x-y+z=4 \\ 2x-5z=1 \\ 4x+3y-6=0 \end{cases}$

65. $\begin{cases} 3x-2y+z=1 \\ x-y=6 \\ y-z=4 \end{cases}$

chapter eight test

Find the inverse of each of the following matrices, if possible.

66. $\begin{bmatrix} 4 & 12 \\ 3 & 9 \end{bmatrix}$

67. $\begin{bmatrix} \frac{1}{2} & \frac{-3}{2} \\ -\frac{5}{2} & \frac{-7}{2} \end{bmatrix}$

68. $\begin{bmatrix} -1 & 2 & 3 \\ 1 & -1 & 4 \\ 2 & 0 & -2 \end{bmatrix}$

69. $\begin{bmatrix} 3 & -1 & 4 \\ 0 & 2 & -2 \\ -3 & 1 & -4 \end{bmatrix}$

Solve the following nonlinear systems algebraically.

70. $\begin{cases} x^2 + y^2 = 25 \\ 2x^2 - y^2 = 23 \end{cases}$

71. $\begin{cases} x^2 + y^2 = 5 \\ x^2 + y = 5 \end{cases}$

72. $\begin{cases} y - 5 = 2x \\ 4x^2 + y^2 = 13 \end{cases}$

73. $\begin{cases} y = (x-1)^2 \\ y + 8 = (x+1)^2 \end{cases}$

74. $\begin{cases} xy = 4 \\ x^2 + y^2 = 17 \end{cases}$

75. $\begin{cases} x^2 + y^2 = 8 \\ 2x^2 + 3y^2 = 25 \end{cases}$

chapter NINE

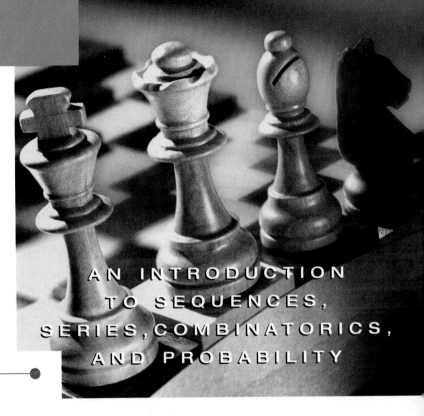

AN INTRODUCTION TO SEQUENCES, SERIES, COMBINATORICS, AND PROBABILITY

OVERVIEW

9.1 Sequences and Series

9.2 Arithmetic Sequences and Series

9.3 Geometric Sequences and Series

9.4 Mathematical Induction

9.5 An Introduction to Combinatorics

9.6 An Introduction to Probability

By the end of this chapter you should be able to articulate both individual terms and general formulas for sequences and series, in addition to finding sums for finite and infinite sequences and series. On page 748 you will find a problem in which the inventor of the game of chess receives a reward for his creation. As his reward, he requests a chess board filled with grains of wheat, with each square on the board having twice the number of grains as the previous square. You will be asked to use tools such as the formula for the *Partial Sum of a Geometric Sequence* on page 737 to find the total number of grains he collects.

chapter nine   An Introduction to Sequences, Series, Combinatorics, and Probability

Introduction

This last chapter introduces a variety of mathematical topics which, depending on your major, you are likely to encounter and study in greater detail in later courses.

Nearly everyone will, at some point, find it convenient to know the basic facts about sequences and series presented here. (This is not to say that everyone *will* use this material, but those who don't will be forced to find someone who does when faced with a sequence or series problem.) For instance, sequences and series arise in the financial world in the guise of retirement planning, portfolio analysis, and loan repayment; in sociology in the construction of population models; in the study of computer algorithms; and in medical models of epidemic spread. The counting techniques of *combinatorics*, a branch of mathematics, are immensely useful in a large number of surprisingly diverse situations; the fields of statistics and computer science, to name just two, make heavy use of the methods studied here. Finally, the science of *probability* finds applications ranging from the deadly serious (actuarial mortality tables) to the purely entertaining (the modern "gaming" industry).

Probability, in particular, has a long, convoluted, and fascinating history. People have pondered questions of chance, and the possibility of analyzing or predicting fate, for all of recorded history (and longer, of course). Mathematicians, as a subset of the general population, have been no exception to this rule, but their understanding of probability has had no particularly mathematical aspect until relatively recently. While Italian mathematicians of the Renaissance made some small progress in the area, mathematical probability is usually said to have been born in France in the year 1654. It was in that year that a member of the French nobility, Antoine Gombaud, Chevalier de Mere, posed several questions born from his gambling experience to the mathematician Blaise Pascal (1623 - 1662). Pascal, intrigued by the questions, consulted with Pierre de Fermat (1601 - 1665), another mathematician famous today largely for his so-called "Last Theorem." In a series of letters back and forth, Pascal and Fermat laid the foundation of modern mathematical probability, a field with profound applications inside and outside of the world of gambling.

Pascal

The material in this chapter can be viewed as a bridge between algebra and further study in mathematics. All of the topics introduced in this chapter rely on the mathematical maturity you have gained by studying algebra, but of course there are many more areas of math that are not explored at all in this book. Calculus, in particular, is one such area; many of you will soon be taking at least one semester of calculus, a field of mathematics that extends the static world of algebra to the dynamic world in which we live.

9.1 Sequences and Series

TOPICS

1. Recursively and explicitly defined sequences
2. Summation notation and a few formulas
3. Partial sums and series
4. Interlude: Fibonacci sequences

Topic 1: Recursively and Explicitly Defined Sequences

In attempting to explain – or at least model – many phenomena, it is often useful and very natural to construct a list of associated numbers. For instance, the growth of an isolated population of rabbits over the course of many months might be described with a simple list of the number of rabbits born each month. Or the number of steps necessary for a computer algorithm to sort a certain number of elements might be described with a list of numbers, where the nth number in the list is the number of steps needed to sort n elements.

Such lists are called *sequences*, and they arise often enough that specialized notation is warranted.

Sequences

> An **infinite sequence** (or just **sequence**) is a function whose domain is the set of natural numbers $\{1, 2, 3, \ldots\}$. Instead of giving the function a name such as f and referring to the values of the sequence as $f(1), f(2), f(3), \ldots$, we commonly use subscripts and refer to the elements of the sequence as $a_1, a_2, a_3, \ldots$. In other words, a_n stands for the nth term in the sequence. **Finite sequences** are defined similarly, but the domain of a finite sequence is a set of the form $\{1, 2, 3, \ldots, k\}$ for some natural number k.

chapter nine An Introduction to Sequences, Series, Combinatorics, and Probability

Note that a consequence of sequence notation is that we usually don't have a label, such as f, to attach to a sequence; instead, a given sequence is normally identified by some formula for determining a_n, the nth term of the sequence, and the entire sequence may be referred to as $\{a_n\}$ or $\{a_n\}_{n=1}^{\infty}$. If it is necessary to discuss several sequences at once, we use different letters to denote the different terms. For instance, a discussion may refer to two sequences $\{a_n\}$ and $\{b_n\}$.

example 1

Determine the first five terms of the sequences whose n^{th} terms are as follows:

a. $a_n = 5n - 2$ **b.** $b_n = \dfrac{(-1)^n + 1}{2}$ **c.** $c_n = \dfrac{n}{n+1}$

Solutions:

a. Replacing n in the formula for the general term with the first five positive integers, we obtain

$$a_1 = 5(1) - 2, \quad a_2 = 5(2) - 2, \quad a_3 = 5(3) - 2, \quad a_4 = 5(4) - 2, \quad a_5 = 5(5) - 2,$$

so the sequence starts out as 3, 8, 13, 18, 23, ...

b. Again, we replace n with the first five positive integers to determine the first five terms of the sequence. Note that $(-1)^n$ is equal to -1 if n is odd and is equal to 1 if n is even. When we add 1 to $(-1)^n$ and divide the result by 2, we obtain either 0 or 1, so the sequence begins 0, 1, 0, 1, 0, ... This example illustrates the fact that a given value may appear more than once in a sequence.

c. Replacing n in the formula for the general term with the first five positive integers, we obtain

$$c_1 = \dfrac{1}{1+1}, \quad c_2 = \dfrac{2}{2+1}, \quad c_3 = \dfrac{3}{3+1}, \quad c_4 = \dfrac{4}{4+1}, \quad c_5 = \dfrac{5}{5+1},$$

so the sequence starts out as $\dfrac{1}{2}, \dfrac{2}{3}, \dfrac{3}{4}, \dfrac{4}{5}, \dfrac{5}{6}, ...$

Enter the sequence in **Y=** as Y_1. Then select **VARS** and use the right arrow to highlight Y-Vars. Press **ENTER** and select Y_1; then use parentheses and enter the desired value for n.

section 9.1 Sequences and Series

The three formulas for the general n^{th} term in *Example 1* were all examples of **explicit** formulas, so-named because they provide a very direct and easily applied rule for calculating any term in the sequence. In many cases, though, an explicit formula for the general term cannot be found, or at least is not as easily determined as some other sort of formula. In such cases, the terms of a sequence are often defined *recursively*. A **recursive** formula is one that refers to one or more of the terms preceding a_n in the definition for a_n. For instance, if the first term of a sequence is -5, and if it is known that each of the remaining terms of the sequence is 7 more than the term preceding it, the sequence can be defined by the rules $a_1 = -5$ and $a_n = a_{n-1} + 7$ for $n \geq 2$.

example 2

Determine the first five terms of the following recursively defined sequences.

a. $a_1 = 3$ and $a_n = a_{n-1} + 5$ for $n \geq 2$ **b.** $a_1 = 2$ and $a_n = 3a_{n-1} + 1$ for $n \geq 2$

Solutions:

a. We find the first five terms by replacing n with the first five positive integers, just as in *Example 1*. Note, though, that in using the recursive definition given, we must determine the elements of the sequence in order; that is, to determine a_5, we need to first know a_4. And to determine a_4, we need to first know a_3, and so on back to a_1.

$$a_1 = 3$$
$$a_2 = a_1 + 5 = 3 + 5 = 8$$
$$a_3 = a_2 + 5 = 8 + 5 = 13$$
$$a_4 = a_3 + 5 = 13 + 5 = 18$$
$$a_5 = a_4 + 5 = 18 + 5 = 23$$

Thus the sequence starts out as 3, 8, 13, 18, 23, ... Note that the sequence defined by this recursive definition appears to be the same as the first sequence in *Example 1*, defined explicitly by $a_n = 5n - 2$. This is no illusion, and illustrates the fact that a given sequence can often be defined several different ways.

continued on next page...

b. Using the recursive formula $a_1 = 2$ and $a_n = 3a_{n-1} + 1$ for $n \geq 2$, we obtain

$$a_1 = 2$$
$$a_2 = 3a_1 + 1 = 3(2) + 1 = 7$$
$$a_3 = 3a_2 + 1 = 3(7) + 1 = 22$$
$$a_4 = 3a_3 + 1 = 3(22) + 1 = 67$$
$$a_5 = 3a_4 + 1 = 3(67) + 1 = 202$$

Thus the sequence starts out as 2, 7, 22, 67, 202, ...

technology note

Computer algebra systems such as *Mathematica* allow sequences to be defined both explicitly and recursively. The illustration below shows how the sequence seen in *Examples 1a* and *2a* can be defined either way; note that `a[5]` has the same value in both cases. *See Appendix A for more guidance on defining functions (explicitly or recursively) in Mathematica.*

Y_1 was defined as $5n - 2$ for this screen.

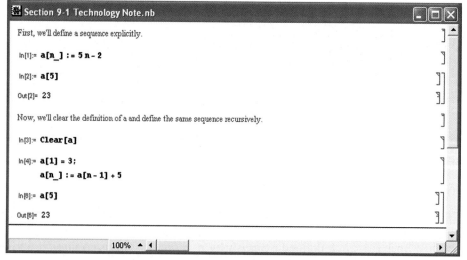

section 9.1 Sequences and Series

A comparison of *Examples 1* and *2* demonstrates that, in general, an explicit formula is far more useful when finding the terms of a sequence than a recursive formula. To convince yourself of this fact, consider the task of calculating a_{100} based on the formula $a_n = 5n - 2$, versus the same task given the formula $a_1 = 3$ and $a_n = a_{n-1} + 5$ for $n \geq 2$.

To a certain extent, the problems in *Examples 1* and *2* can be turned around, so that the question becomes one of finding a formula for the general n^{th} term of a sequence given the first few terms of the sequence. In fact, this is often the challenge in modeling situations, where the desire is to extrapolate the behavior of a sequence of numbers by finding a formula that produces the first few terms of the sequence. The catch is that there is always more than one formula that will produce identical terms of a sequence up to a certain point and then differ beyond that. To see why this must be so, consider the following two explicit formulas:

$$a_n = 3n \text{ and } b_n = 3n + (n-1)(n-2)(n-3)(n-4)(n-5)n^2$$

Do you see why these two formulas will produce identical results for the first five terms, but different results from then on? For this reason, the instructions in *Example 3* ask for just possible formulas for the given sequences.

example 3

Find a possible formula for the general n^{th} term of the sequences that begin as follows.

a. $-3, 9, -27, 81, -243, \ldots$ b. $1, 3, 6, 10, 15, \ldots$

Solutions:

a. As a rule, there is no particular method to use to find a formula for the terms of a sequence, other than the very general method of observation. If a formula for a given sequence does not come quickly to mind, it may help to associate each term of the sequence with its place in the sequence, as shown below:

$$\begin{array}{cccccc} 1 & 2 & 3 & 4 & 5 & \ldots \\ \updownarrow & \updownarrow & \updownarrow & \updownarrow & \updownarrow & \updownarrow \\ -3 & 9 & -27 & 81 & -243 & \ldots \end{array}$$

continued on next page...

chapter nine An Introduction to Sequences, Series, Combinatorics, and Probability

If a pattern is still not apparent, try rewriting the terms of the sequence in different ways. For instance, factoring the terms in this case leads to:

$$
\begin{array}{cccccc}
1 & 2 & 3 & 4 & 5 & \cdots \\
\updownarrow & \updownarrow & \updownarrow & \updownarrow & \updownarrow & \updownarrow \\
-3 & 3^2 & -3^3 & 3^4 & -3^5 & \cdots
\end{array}
$$

At this point, it is clear that the nth term of the sequence is the nth power of 3, multiplied by -1 if n is odd. One way of expressing alternating signs in a sequence is to multiply the nth term by $(-1)^n$ (if the odd terms are negative) or by $(-1)^{n+1}$ (if the even terms are negative). In this case, a possible formula for the general n^{th} term is $a_n = (-1)^n (3)^n$, or $a_n = (-3)^n$.

Note that we might also have come up with the recursive formula $a_1 = -3$ and $a_n = -3a_{n-1}$ for $n \geq 2$.

b. A good way to begin would be to again associate each term with its place in the sequence, as follows:

$$
\begin{array}{cccccc}
1 & 2 & 3 & 4 & 5 & \cdots \\
\updownarrow & \updownarrow & \updownarrow & \updownarrow & \updownarrow & \updownarrow \\
1 & 3 & 6 & 10 & 15 & \cdots
\end{array}
$$

In this case, factoring the terms does not seem to help in identifying a pattern, but thinking about the difference between successive terms does:

$$
\begin{array}{cccccccc}
1 & & 2 & & 3 & & 4 & & 5 & \cdots \\
\updownarrow & & \updownarrow & & \updownarrow & & \updownarrow & & \updownarrow & \updownarrow \\
1 & & 3=1+2 & & 6=3+3 & & 10=6+4 & & 15=10+5 & \cdots
\end{array}
$$

This observation leads to the recursive formula $a_1 = 1$ and $a_n = a_{n-1} + n$ for $n \geq 2$. There is an explicit formula that defines the sequence as well, and we will return to this problem soon.

section 9.1   Sequences and Series

Topic 2: Summation Notation and a Few Formulas

One very common use of sequence notation is to define terms that are to be added together. If the first n terms of a given sequence are to be added, one way of denoting the sum is $a_1 + a_2 + \ldots + a_n$, and in some cases this notation is sufficient. But there are many occasions when the use of ellipses (which in this context means the "dot dot dot", not the conic sections of the same name) is too vague. For instance, there is nothing in the expression $a_1 + a_2 + \ldots + a_n$ to indicate what, exactly, the terms in the sequence look like.

Summation notation (also known as *sigma notation*) is the response to this shortcoming. Summation notation borrows the capital Greek letter Σ ("sigma") to denote the operation of summation, as described below.

Summation Notation

The sum $a_1 + a_2 + \ldots + a_n$ is expressed in **summation notation** as $\sum_{i=1}^{n} a_i$.

When this notation is used, the letter i is called the **index of summation**, and a_i often appears as the formula for the ith term of a sequence. In the sum above, all the terms of the sequence beginning with a_1 and ending with a_n are to be added. The notation can be modified to indicate a different first term or last term of the sum.

example 4

Rewrite the following sums in expanded form, and evaluate them.

a. $\sum_{i=1}^{4}(3i-2)$ **b.** $\sum_{i=3}^{5} i^2$

continued on next page...

chapter nine An Introduction to Sequences, Series, Combinatorics, and Probability

Solutions:

a. $\displaystyle\sum_{i=1}^{4}(3i-2) = (3\cdot 1-2)+(3\cdot 2-2)+(3\cdot 3-2)+(3\cdot 4-2)$
$= 1+4+7+10$
$= 22$

In order to expand the sum, we replace the index i with the numbers 1 through 4. Note how the formula for the ith term is used to generate the terms of the sum.

b. $\displaystyle\sum_{i=3}^{5} i^2 = 3^2+4^2+5^2$
$= 9+16+25$
$= 50$

This sum differs from the first sum in the formula for the terms and in the beginning and ending point of the sum.

Keep in mind that sigma notation represents nothing more than addition. Thus the following properties of Σ are nothing more than restatements of familiar properties.

Properties of Sigma

The following properties refer to two sequences $\{a_i\}$ and $\{b_i\}$ and a constant c.

1) $\displaystyle\sum_{i=1}^{n}(a_i+b_i) = \sum_{i=1}^{n}a_i + \sum_{i=1}^{n}b_i$ (terms of a sum can be rearranged)

2) $\displaystyle\sum_{i=1}^{n}ca_i = c\sum_{i=1}^{n}a_i$ (constants can be factored out of a sum)

3) $\displaystyle\sum_{i=1}^{n}a_i = \sum_{i=1}^{k}a_i + \sum_{i=k+1}^{n}a_i$ for any $1 \le k \le n-1$, (a sum can be broken apart into two smaller sums)

section 9.1   Sequences and Series

caution

There are many similar looking statements that are not true. For instance, it is not generally true that $\sum_{i=1}^{n} a_i b_i = \sum_{i=1}^{n} a_i \cdot \sum_{i=1}^{n} b_i$. As an example, consider this statement for $n = 5$, $a_i = b_i = 1$, for $1 \le i \le 5$.

In addition to the properties listed above, there are formulas for many sums that occur frequently. We will list only a few in this section, and prove only two of them.

Four Summation Formulas

1) $\sum_{i=1}^{n} 1 = n$

2) $\sum_{i=1}^{n} i = \dfrac{n(n+1)}{2}$

3) $\sum_{i=1}^{n} i^2 = \dfrac{n(n+1)(2n+1)}{6}$

4) $\sum_{i=1}^{n} i^3 = \dfrac{n^2(n+1)^2}{4}$

The first formula really requires no proof at all, as it is clearly true if we write the sum in expanded form:

$$\sum_{i=1}^{n} 1 = \underbrace{1+1+\ldots+1}_{n \text{ terms}} = n.$$

The second formula requires a bit more in the way of justification, and several different proofs are possible. One proof begins by letting S stand for the sum; that is,

$$S = \sum_{i=1}^{n} i = 1 + 2 + \ldots + n.$$

Then it is also certainly true, since addition can be performed in any order, that

$$S = n + (n-1) + \ldots + 1.$$

Note, now, the result of adding these two equations:

$$\begin{array}{rllll} S &=& 1 & +2 & +\ldots & +n \\ S &=& n & +(n-1) & +\ldots & +1 \\ \hline 2S &=& (n+1) & +(n+1) & +\ldots & +(n+1) \end{array}$$

chapter nine An Introduction to Sequences, Series, Combinatorics, and Probability

Since the term $n + 1$ appears n times on the right-hand side of the last equation, we have $2S = n(n+1)$ or, after dividing both sides by 2,

$$\sum_{i=1}^{n} i = \frac{n(n+1)}{2}.$$

Note that this provides an explicit formula for the nth term of the sequence in *Example 3b*. Since the nth term is the sum $1 + 2 + \ldots + n$, the formula is $a_n = \frac{n(n+1)}{2}$.

example 5

Use the above properties and formulas to evaluate the following sums.

    a. $\sum_{i=1}^{9}(7i-3)$ **b.** $\sum_{i=4}^{6} 3i^2$

Solutions:

To find the sum of a list, select **2ND** **STAT** (List), use the right arrow to highlight Math, and select 5:sum. To enter the sequence, select **2ND** **STAT** (List), highlight Ops, and select 5:seq. In the parenthesis that appears, enter the desired equation, the name of the variable (X), beginning value, ending value, and increment of 1, separated by commas and followed by an end parenthesis.

a.
$$\sum_{i=1}^{9}(7i-3) = \sum_{i=1}^{9} 7i - \sum_{i=1}^{9} 3$$
$$= 7\sum_{i=1}^{9} i - 3\sum_{i=1}^{9} 1$$
$$= 7\left(\frac{9 \cdot 10}{2}\right) - 3 \cdot 9$$
$$= 315 - 27$$
$$= 288$$

The first of the three properties of sigma allows us to split the given sum into two sums, and the second property allows us to factor the constants out of each sum. (Note that we have factored a 3 from the second sum in order to apply one of the formulas.)

The first two of the four summation formulas then provide the final answer.

b.
$$\sum_{i=4}^{6} 3i^2 = 3\sum_{i=4}^{6} i^2$$
$$= 3\sum_{i=1}^{6} i^2 - 3\sum_{i=1}^{3} i^2$$
$$= 3\left(\frac{6 \cdot 7 \cdot 13}{6}\right) - 3\left(\frac{3 \cdot 4 \cdot 7}{6}\right)$$
$$= 273 - 42$$
$$= 231$$

We can again factor out a constant, but in order to apply the formula for the sum of squares of integers, one further step is necessary. In the second line, we have used the third property of sigma to obtain two sums that begin with $i = 1$, allowing us to use the third summation formula as shown.

Topic 3: Partial Sums and Series

Given a sequence $\{a_i\}$, many applications call for an understanding of sums of the form $a_1 + a_2 + \ldots + a_n$ or for the sum of *all* the terms, which we might express as $a_1 + a_2 + a_3 + \ldots$. The following definition formalizes these two notions.

Partial Sums and Series

Given the infinite sequence $\{a_i\}$, the nth **partial sum** is $S_n = \sum_{i=1}^{n} a_i$. S_n is an example of a **finite series**. The **infinite series** associated with $\{a_i\}$ is the sum $\sum_{i=1}^{\infty} a_i$. Note that the adjective *infinite* refers to the infinite number of terms that appear in the sum.

Many students, when first encountering the definition of an infinite series, have the initial reaction that a sum of an infinite number of terms can't possibly be a finite number, especially if all the terms are positive. But in fact, many simple examples of infinite series with finite sums exist. Consider, for instance, the fractions of an inch as marked on a typical ruler.

Figure 1: Visualizing an Infinite Series

If half of an inch and a quarter of an inch are added, the result is three-quarters of an inch. If an additional eighth of an inch is added, the result is seven-eighths of an inch. If the ruler is of sufficient precision, a sixteenth of an inch can be added, resulting in a total of fifteen-sixteenths of an inch.

chapter nine   An Introduction to Sequences, Series, Combinatorics, and Probability

In principle, further appropriate fractions can be added, with the result that the sum of the fractions at every stage is closer to, but never exceeds, one inch. In the language of partial sums and series, what we have informally deduced is

$$S_n = \frac{1}{2} + \frac{1}{4} + \ldots + \frac{1}{2^n} = \frac{2^n - 1}{2^n}$$

and

$$S = \frac{1}{2} + \frac{1}{4} + \frac{1}{8} + \ldots = \sum_{i=1}^{\infty} \frac{1}{2^i} = 1.$$

We will explore such infinite series further when we define and study geometric series.

We say that an infinite series $\sum_{i=1}^{\infty} a_i$ **converges** if the sequence of partial sums $S_n = \sum_{i=1}^{n} a_i$ approaches some fixed real number S, and in this case we write $S = \sum_{i=1}^{\infty} a_i$. If the sequence of partial sums does not approach some fixed real number (either by getting larger and larger in magnitude or by "bouncing around"), we say the series **diverges**. The next example illustrates both possible outcomes.

example 6

Examine the partial sums associated with each infinite series to determine if the series converges or diverges.

a. $\displaystyle\sum_{i=1}^{\infty}\left(\frac{1}{i} - \frac{1}{i+1}\right)$ **b.** $\displaystyle\sum_{i=1}^{\infty}(-1)^i$

Solutions:

a. $S_1 = 1 - \dfrac{1}{2}$

$S_2 = \left(1 - \dfrac{\cancel{1}}{\cancel{2}}\right) + \left(\dfrac{\cancel{1}}{\cancel{2}} - \dfrac{1}{3}\right) = 1 - \dfrac{1}{3}$

$S_3 = \left(1 - \dfrac{\cancel{1}}{\cancel{2}}\right) + \left(\dfrac{\cancel{1}}{\cancel{2}} - \dfrac{\cancel{1}}{\cancel{3}}\right) + \left(\dfrac{\cancel{1}}{\cancel{3}} - \dfrac{1}{4}\right) = 1 - \dfrac{1}{4}$

$\vdots$

$S_n = 1 - \dfrac{1}{n+1}$

The partial sums of this series exhibit very convenient behavior that allows us to easily evaluate each one. When most of the terms that appear in a partial sum cancel, as shown at left, the series is called a *telescoping series*.

The partial sums approach 1 as n gets larger, so we say the series converges. In fact,

$$\sum_{i=1}^{\infty}\left(\frac{1}{i} - \frac{1}{i+1}\right) = 1.$$

b. $S_1 = -1$

$S_2 = -1 + 1 = 0$

$S_3 = -1 + 1 - 1 = -1$

$\vdots$

$S_n = \begin{cases} -1 \text{ if } n \text{ is odd} \\ 0 \text{ if } n \text{ is even} \end{cases}$

The partial sums of this series oscillate between -1 and 0, and do not approach a fixed number as n gets larger. Thus this series diverges.

Topic 4: Interlude: Fibonacci Sequences

One of the most famous sequences in mathematics is one that has a long and colorful history.

Leonardo Fibonacci (meaning "Leonardo, son of Bonaccio") was born in Pisa, *ca.* 1175, and was one of the first of many famous Italian mathematicians of the Middle Ages. As a child and young adult, he traveled with his father (a merchant) to ports in northern Africa and the Middle East where Arab scholars had collected, preserved, and expanded the mathematics of many different cultures. Among other achievements, Leonardo is known for the book he wrote after returning to Italy, the *Liber abaci*, which exposed European scholars to Hindu-Arabic notation and the mathematics he had acquired, as well as to his own contributions.

One of the problems in the *Liber abaci* gives rise to what we now call the Fibonacci sequence: 1, 1, 2, 3, 5, 8, 13, 21, ... The problem, loosely translated, asks

> *How many pairs of rabbits can a single pair produce, if every month each pair of rabbits can beget a new pair and if each pair becomes productive in the second month of existence?*

We are to assume that in the first month we begin with a single pair of rabbits. In the second month, the rabbits are too young to reproduce, so the total number of pairs remains 1. But in the third month, the pair produces a new pair of rabbits, and so we have both the original pair and the new pair, for a total of 2 pairs. In the fourth month, we have the 2 pairs that we begin the month with, plus another pair is produced from the 1st pair of rabbits that is old enough to do so, so the total number of pairs is 3. Similarly, in the fifth month we have 3 pairs plus 2 more pairs from those old enough to be productive, for a total of 5 pairs.

chapter nine An Introduction to Sequences, Series, Combinatorics, and Probability

If we analyze the pattern that is emerging, we are led to the recursive definition

$$a_1 = 1 \quad \text{and} \quad a_2 = 1 \quad \text{and} \quad a_n = a_{n-2} + a_{n-1} \quad \text{for } n \geq 3.$$

Generalized Fibonacci sequences are defined similarly, with the first two terms given and each successive term defined as the sum of the previous two.

exercises

Determine if each of the sequences given is finite or infinite.

1. the sequence of odd numbers
2. 2, 4, 6, 8, 10, 12…
3. 1, 3, 5, 7, 9, 11
4. 3, 9, 27, 81, 243…
5. the sequence of the days of the week
6. 1, 0, 1, 0, 1, 0, 1, 0…
7. 1, 2, 3, 4, 5, 6, 7…
8. the sequence of letters in the alphabet
9. the sequence of the number of ants in a colony recorded daily.

section 9.1 Sequences and Series

Determine the first five terms of the sequences whose n^{th} terms are defined as follows. See Examples 1 and 2.

10. $a_n = 7n - 3$
11. $a_n = -3n + 5$
12. $a_n = (-2)^n$
13. $a_n = \dfrac{3n}{n+2}$
14. $a_n = \dfrac{(-1)^n}{n^2}$
15. $a_n = \dfrac{(-1)^{n+1} 2^n}{3^n}$
16. $a_n = \left(-\dfrac{1}{3}\right)^{n-1}$
17. $a_n = \dfrac{n^2}{n+1}$
18. $a_n = \dfrac{(n-1)^2}{(n+1)^2}$
19. $a_n = (-2)^n + n$
20. $a_n = (-n+4)^3 - 1$
21. $a_n = \dfrac{2n^2}{3n-2}$
22. $a_n = (-1)^n \sqrt{n}$
23. $a_n = \dfrac{2^n}{n^2}$
24. $a_n = 4n - 3$
25. $a_n = -5n + 15$
26. $a_n = 2^{n-2}$
27. $a_n = 3^{-n-2}$
28. $a_n = (3n)^n$
29. $a_n = \sqrt[2n]{64}$
30. $a_n = \dfrac{5n}{n+3}$
31. $a_n = \dfrac{n^2}{n+2}$
32. $a_n = \dfrac{n^2 + n}{2}$
33. $a_n = (-1)^n n$
34. $a_n = \dfrac{(n+1)^2}{(n-1)^2}$
35. $a_n = n^2 + n$
36. $a_n = \dfrac{2n-1}{3n}$
37. $a_n = \sqrt{3n} + 1$
38. $a_n = -(n-1)^2$
39. $a_n = (n-1)(n+2)(n-3)$
40. $a_1 = 2$ and $a_n = (a_{n-1})^2$ for $n \geq 2$
41. $a_1 = -2$ and $a_n = 7a_{n-1} + 3$ for $n \geq 2$
42. $a_1 = 1$ and $a_n = na_{n-1}$ for $n \geq 2$
43. $a_1 = -1$ and $a_n = -a_{n-1} - 1$ for $n \geq 2$
44. $a_1 = 2$ and $a_n = \sqrt{(a_{n-1})^2 + 1}$ for $n \geq 2$
45. $a_1 = 3$, $a_2 = 1$, and $a_n = (a_{n-2})^{a_{n-1}}$ for $n \geq 3$

Find a possible formula for the general n^{th} term of the sequences that begin as follows. Answers may vary. See Example 3.

46. $5, 12, 19, 26, 33, \ldots$
47. $-2, 4, -8, 16, -32, \ldots$
48. $-1, 2, -6, 24, -120, \ldots$
49. $\dfrac{1}{3}, \dfrac{2}{4}, \dfrac{3}{5}, \dfrac{4}{6}, \dfrac{5}{7}, \ldots$
50. $1, \dfrac{1}{4}, \dfrac{1}{9}, \dfrac{1}{16}, \dfrac{1}{25}, \ldots$
51. $1, \dfrac{1}{2}, \dfrac{1}{6}, \dfrac{1}{24}, \dfrac{1}{120}, \ldots$

chapter nine An Introduction to Sequences, Series, Combinatorics, and Probability

52. $-34, -25, -16, -7, 2, \ldots$ 53. $\dfrac{3}{14}, \dfrac{2}{15}, \dfrac{1}{16}, 0, -\dfrac{1}{18}, \ldots$ 54. $\dfrac{1}{4}, \dfrac{1}{2}, 1, 2, 4, \ldots$

55. $-1, -6, -11, -16, -21, \ldots$ 56. $\dfrac{1}{2}, \dfrac{1}{2}, \dfrac{3}{8}, \dfrac{1}{4}, \dfrac{5}{32}, \ldots$ 57. $1, 4, 15, 64, 325, \ldots$

Translate each expanded sum that follows into summation notation, and vice versa. Then use the formulas and properties from this section to evaluate the sums. See Examples 4 and 5.

58. $\displaystyle\sum_{i=1}^{7}(3i-5)$ 59. $\displaystyle\sum_{i=1}^{5}-3i^2$ 60. $1+8+27+\ldots+216$

61. $1+4+7+\ldots+22$ 62. $\displaystyle\sum_{i=3}^{10}5i^2$ 63. $9+16+25+\ldots+81$

64. $\displaystyle\sum_{i=1}^{6}-3(2)^i$ 65. $\displaystyle\sum_{i=6}^{13}(i+3)(i-10)$ 66. $9+27+81+\ldots+19683$

Find a formula for the nth partial sum S_n of each of the following series. If the series is finite, determine the sum. If the series is infinite, determine if it converges or diverges, and if it converges, determine the sum. See Example 6.

67. $\displaystyle\sum_{i=1}^{100}\left(\dfrac{1}{i+3}-\dfrac{1}{i+4}\right)$ 68. $\displaystyle\sum_{i=1}^{\infty}\left(\dfrac{1}{i+3}-\dfrac{1}{i+4}\right)$ 69. $\displaystyle\sum_{i=1}^{\infty}\left(2^i-2^{i-1}\right)$

70. $\displaystyle\sum_{i=1}^{15}\left(2^i-2^{i-1}\right)$ 71. $\displaystyle\sum_{i=1}^{49}\left(\dfrac{1}{2i}-\dfrac{1}{2i+2}\right)$ 72. $\displaystyle\sum_{i=1}^{\infty}\left(\dfrac{1}{2i}-\dfrac{1}{2i+2}\right)$

73. $\displaystyle\sum_{i=1}^{100}\ln\left(\dfrac{i}{i+1}\right)$ (Hint: make use of a property of logarithms to rewrite the sum.)

74. $\displaystyle\sum_{i=1}^{\infty}\ln\left(\dfrac{i}{i+1}\right)$ (Hint: make use of a property of logarithms to rewrite the sum.)

75. $\displaystyle\sum_{i=1}^{30}\left(\dfrac{1}{2i+5}-\dfrac{1}{2i+7}\right)$ 76. $\displaystyle\sum_{i=1}^{\infty}\left(\dfrac{1}{3i+1}-\dfrac{1}{3i+4}\right)$ 77. $\displaystyle\sum_{i=1}^{65}\ln\left(\dfrac{i}{i+1}\right)$

Determine the first five terms of the following generalized Fibonacci sequences.

78. $a_1=4$, $a_2=7$, and $a_n=a_{n-2}+a_{n-1}$ for $n\geq 3$

79. $a_1=-9$, $a_2=1$, and $a_n=a_{n-2}+a_{n-1}$ for $n\geq 3$

80. $a_1=10$, $a_2=20$, and $a_n=a_{n-2}+a_{n-1}$ for $n\geq 3$

81. $a_1 = -17$, $a_2 = 13$, and $a_n = a_{n-2} + a_{n-1}$ for $n \geq 3$

82. $a_1 = 13$, $a_2 = -17$, and $a_n = a_{n-2} + a_{n-1}$ for $n \geq 3$

Determine the first five terms of the following recursively defined sequences.

83. $a_1 = 2$, $a_2 = -3$, and $a_n = 3a_{n-1} + a_{n-2}$ for $n \geq 3$

84. $a_1 = 1$, $a_2 = -3$, and $a_n = a_{n-1} a_{n-2}$ for $n \geq 3$

Use your knowledge of sequences and series to answer the following questions.

85. Suppose you buy one cow and a number of bulls. In year one, your cow gives birth to a female calf, and continues to bear another female calf every year for the rest of her life. Assuming every calf born is female, that each cow begins calving in her third year (at age two), and that your cows never die, determine the number of cows (do not count the bulls) you will have at the end of the 14th year.

86. Beginning with yourself, create a sequence describing the number of biological predecessors you have in each of the past 7 generations of your family.

87. You borrow $638 to buy a new car stereo. You plan to pay this sum back with monthly payments of $74. The interest rate on your loan is 6% compounded monthly (recall that's 0.5% per month). Let A_n be the amount you owe at the end of the nth month. Find a recursive sequence to represent A_n. Use this sequence to find the amount owed after 4 months and the amount owed after 6 months. How many months will it take to pay off your loan?

88. The Fibonacci sequence is quite prevalent in nature. Do some research on your own to find an occurrence in nature (other than population growth) of the Fibonacci sequence.

chapter nine An Introduction to Sequences, Series, Combinatorics, and Probability

9.2 Arithmetic Sequences and Series

TOPICS

1. Characteristics of arithmetic sequences and series
2. The formula for the general term of an arithmetic sequence
3. Evaluating partial sums of arithmetic sequences

Topic 1: Characteristics of Arithmetic Sequences and Series

An *arithmetic sequence* $\{a_n\}$ is a sequence with the property that the difference between any two consecutive terms is the same fixed constant. Arithmetic sequences arise in many different situations. Suppose, for example, that the parents of a ten-year-old child decide to increase her \$1.00/week allowance by \$0.50/week with the start of each new year. The sequence describing her weekly allowance, beginning with age ten, is then

$$1.00,\ 1.50,\ 2.00,\ 2.50,\ 3.00,\ 3.50,\ldots$$

By the informal definition given above, this is an arithmetic sequence with a difference of 0.50 between consecutive terms. And as a mechanism for increasing a child's allowance, it is not a bad one. As we will see in the next section, though, adult workers would probably not benefit from having yearly raises determined in this manner.

Arithmetic Sequences

A sequence $\{a_n\}$ is an **arithmetic sequence** (sometimes also called an **arithmetic progression**) if there is some constant d such that $a_{n+1} - a_n = d$ for each $n = 1, 2, 3, \ldots$ The constant d is called the **common difference** of the sequence.

section 9.2   Arithmetic Sequences and Series

As we saw in the last section, every sequence $\{a_n\}$ can be used to determine an associated series $a_1 + a_2 + a_3 + ...$, so arithmetic series follow naturally from arithmetic sequences. But the first question that must be asked of any series is whether it converges, and it doesn't take long to realize that any non-trivial arithmetic sequence gives rise to a series that diverges.

Our reasoning goes something like the following. If we denote the first term of the sequence by a_1, the second term is then $a_1 + d$ (where d is the common difference), the third term is $(a_1 + d) + d = a_1 + 2d$, and so on. So if we add up the first n terms of the sequence (that is, if we find the nth partial sum), we have

$$S_n = a_1 + (a_1 + d) + (a_1 + 2d) + ... + (a_1 + (n-1)d)$$

$$= \sum_{i=1}^{n} (a_1 + (i-1)d)$$

$$= \sum_{i=1}^{n} a_1 + \sum_{i=1}^{n} (i-1)d \quad \text{(note the use of a property of } \sum \text{)}$$

For our present purposes, this might already be enough to convince you that the partial sums are getting larger and larger in magnitude as n grows, since S_n consists of a_1 added to itself n times, plus a sum of multiples of d. But we will soon find it useful to know more about S_n, so we continue the analysis:

$$S_n = \sum_{i=1}^{n} a_1 + \sum_{i=1}^{n} (i-1)d$$

$$= a_1 \sum_{i=1}^{n} 1 + d \sum_{i=1}^{n} (i-1)$$

$$= na_1 + d(0 + 1 + 2 + ... + (n-1))$$

$$= na_1 + d \sum_{i=1}^{n-1} i$$

We have now also used the distributive property of sigma in order to factor out the constants a_1 and d, and we have used the simple formula that tells us adding up n 1's gives us n. We also temporarily expanded the second of the two sums above in order to find a more useful way of writing it, and we can now apply another formula to complete our task.

$$S_n = na_1 + d \sum_{i=1}^{n-1} i$$

$$= na_1 + d\left(\frac{(n-1)n}{2}\right)$$

chapter nine   An Introduction to Sequences, Series, Combinatorics, and Probability

It is now very clear that the sequence of partial sums is not going to approach some fixed number S, except in the trivial case when $a_1 = 0$ and $d = 0$. That is, except for the series $0 + 0 + 0...$, every arithmetic series diverges. But we will have further use for the above formula for partial sums of arithmetic sequences.

Topic 2: The Formula for the General Term of an Arithmetic Sequence

Another formula we will find useful for arithmetic sequences, as with all sequences, is the explicit formula for the general n^{th} term of the sequence. And the thinking that went into the analysis above has all but given us the formula.

We have already noted that if a_1 is the first term of an arithmetic sequence, and if the common difference is d, then $a_1 + d$ is the second term, $a_1 + 2d$ is the third term, and so on. This pattern is summarized in the box below.

General Term of an Arithmetic Sequence

The explicit formula for the general n^{th} term of an arithmetic sequence is

$$a_n = a_1 + (n-1)d,$$

where d is the common difference for the sequence.

example 1

Find the formula for the general n^{th} term of the following arithmetic sequences.

    a. $-3, 2, 7, 12,...$ **b.** $9, 1, -7, -15,...$

Solutions:

a. $d = 2 - (-3) = 5$

$a_1 = -3$

$a_n = -3 + 5(n-1)$

To construct the formula, we just need to determine d and the first term. Note that we could have determined d by taking the difference between any two consecutive terms given.

section 9.2   Arithmetic Sequences and Series

b. $d = 1 - 9 = -8$
$a_1 = 9$
$a_n = 9 - 8(n-1)$

This sequence is also arithmetic, though it is an example of a decreasing sequence (the numbers diminish as n gets larger). This means that d is a negative constant.

In modeling situations the mathematical task may not quite take the form seen in *Example 1*.

example 2

A demographer decides to model the population growth of a small town as an arithmetic progression. He knows that the population in 1992 was 12,790 and that in 1995 the population was 13,150. He wants to treat the population in 1990 as the first term of the arithmetic progression. What is the sought-after formula?

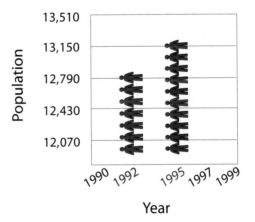

Solution:

Although we do not yet know its value, we know that a_1 represents the population in 1990. Similarly, a_2 represents the population in 1991 and a_3 represents the population in 1992. From what we have been given, we know that $a_3 = 12{,}790$. We also know that $a_6 = 13{,}150$ (the population in 1995 is 13,150). If we let d represent the common difference, the population must have grown by $3d$ between 1992 and 1995, so

$$3d = 13150 - 12790 = 360,$$
$$\text{or } d = 120.$$

continued on next page...

chapter nine An Introduction to Sequences, Series, Combinatorics, and Probability

Now we can use the fact that $a_3 = 12{,}790$ to determine a_1, as follows:

$$a_3 = a_1 + 2d$$
$$12{,}790 = a_1 + 2(120)$$
$$12{,}790 - 240 = a_1$$
$$a_1 = 12{,}550$$

Our formula, then, is $a_n = 12{,}550 + 120(n-1)$. Note that to apply the formula, we need to remember $n = 1$ corresponds to 1990, $n = 2$ corresponds to 1991, and so on. We might also find it convenient to simplify the expression and write

$$a_n = 12{,}550 + 120(n-1) = 12{,}430 + 120n.$$

Topic 3: Evaluating Partial Sums of Arithmetic Sequences

To begin our study of partial sums of arithmetic sequences, we will return to the ten-year-old girl whose weekly allowance grows in an arithmetic progression.

example 3

Melinda, the ten-year-old in question, knows that her parents plan to increase her weekly allowance by $0.50/week with the start of each new year, and she currently receives $1.00/week. Being fiscally minded, she naturally wonders how much money she will have been given by the time she is 18 years old. Assuming that Melinda receives an allowance every week from the current year up through the year she turns 18, what is her total take?

section 9.2   Arithmetic Sequences and Series

Solution:

In the current year, Melinda will receive $(52)(\$1.00) = \$52.00$. Similarly, she will receive $(52)(\$1.50) = \$78.00$ in the next year, so the sum of these two amounts is her total allowance during her 10th and 11th years. We need to extend this reasoning up through Melinda's 18th year, and we can make use of the partial sum formula we have already derived:

$$S_n = na_1 + d\left(\frac{(n-1)n}{2}\right)$$

In order to use this, we note that $a_1 = 52$ (her total allowance in year 1), that $d = 26$ (since her yearly income changes by $(52)(\$0.50) = \$26.00$ from year-to-year), and that S_9 is the sum we are trying to determine (since S_1 corresponds to her income when she is 10, S_9 will be her income up through age 18). So applying the formula, we obtain

$$S_9 = 9 \cdot 52 + (26)\left(\frac{(9-1)\cdot 9}{2}\right) = 468 + 936 = 1404.$$

Melinda will thus receive a total of $1404 in allowance money from age 10 through age 18.

We can use the partial sum formula that we have to derive a slightly different formula that is useful in other contexts. Recall that the formula for the general term of an arithmetic sequence is

$$a_n = a_1 + (n-1)d.$$

From this we see that $(n-1)d = a_n - a_1$, and we can use this fact in our partial sum formula to eliminate d, as follows:

$$S_n = na_1 + d\left(\frac{(n-1)n}{2}\right)$$

$$= na_1 + \left(\frac{n}{2}\right)(n-1)d$$

$$= na_1 + \left(\frac{n}{2}\right)(a_n - a_1)$$

$$= \left(\frac{n}{2}\right)(a_1 + a_n)$$

chapter nine An Introduction to Sequences, Series, Combinatorics, and Probability

The box below summarizes these two forms for the partial sums of an arithmetic sequence.

Partial Sums of Arithmetic Sequences

Let $\{a_n\}$ be an arithmetic sequence with common difference d. The sum of the first n terms (the nth partial sum) is given by both

$$S_n = na_1 + d\left(\frac{(n-1)n}{2}\right)$$

and

$$S_n = \left(\frac{n}{2}\right)(a_1 + a_n).$$

In practice, the choice of formula is based on whether it is easier to determine d or the last term, a_n, in the sum to be evaluated.

example 4

Determine the sum of the first 100 positive odd integers.

Solution:

First, note that the nth positive odd integer is $2n - 1$, so the 100th positive odd integer is 199. We can use either partial sum formula to evaluate $1 + 3 + \ldots + 199$. If we choose to use the second formula, we obtain

$$S_{100} = \left(\frac{100}{2}\right)(1 + 199) = 10,000.$$

section 9.2 Arithmetic Sequences and Series

Note that an alternative method is to make use of the summation formulas from Section 9.1, as follows:

$$\sum_{n=1}^{100}(2n-1) = \sum_{n=1}^{100} 2n - \sum_{n=1}^{100} 1$$
$$= 2\sum_{n=1}^{100} n - \sum_{n=1}^{100} 1$$
$$= 2\left(\frac{(100)(101)}{2}\right) - 100$$
$$= (100)(101) - 100$$
$$= 10,000$$

exercises

Find the explicit formula for the general n^{th} term of the arithmetic sequences described below. See Examples 1 and 2.

1. $-2, 1, 4, 7, 10, \ldots$
2. $5, 7, 9, 11, 13, \ldots$
3. $7, 5, 3, 1, -1, \ldots$
4. $a_2 = 14$ and $a_3 = 19$
5. $a_1 = 5$ and $a_5 = 41$
6. $a_2 = 13$ and $a_4 = 21$
7. $a_3 = -9$ and $d = -6$
8. $a_{12} = 43$ and $d = 3$
9. $a_5 = 100$ and $d = 19$
10. $-37, -20, -3, 14, 31, \ldots$
11. $\frac{7}{2}, \frac{9}{2}, \frac{11}{2}, \frac{13}{2}, \frac{15}{2}, \ldots$
12. $15, 11, 7, 3, -1, \ldots$
13. $a_1 = 12$ and $a_3 = -7$
14. $a_{73} = 224$ and $a_{75} = 230$
15. $a_1 = -1$ and $a_6 = -11$
16. $a_5 = -\frac{5}{2}$ and $d = \frac{3}{2}$
17. $a_4 = 17$ and $d = -4$
18. $a_{34} = -71$ and $d = -2$

Determine if each of the following sequences is arithmetic. If so, find the common difference.

19. the sequence of even numbers
20. $1, 2, 4, 7, 11, 16, \ldots$
21. $1, 2, 3, 4, 5, 6, 7, \ldots$
22. the Fibonacci sequence

chapter nine An Introduction to Sequences, Series, Combinatorics, and Probability

23. 1, 2, 4, 8, 16, 32, ...

24. 42, 38, 34, 30, 26, 22, ...

25. 0, 1, 0, 2, 0, 3, 0, 4, ...

26. 12, 12, 12, 12, 12, ...

Given the initial term and the common difference, find the value of the 7th term of the arithmetic sequences described below. See Examples 1 and 2.

27. $a_1 = 1$ and $d = 2$

28. $a_1 = 4$ and $d = -3$

29. $a_1 = 0$ and $d = \frac{1}{3}$

30. $a_1 = 3$ and $d = \pi$

31. $a_1 = 8$ and $d = -1$

32. $a_1 = \frac{1}{2}$ and $d = 3$

Given two terms of an arithmetic sequence, find the common difference and the first five terms of the sequence.

33. $a_1 = 5$ and $a_2 = 7.5$

34. $a_6 = 27$ and $a_9 = 42$

35. $a_7 = 49$ and $a_{11} = 77$

36. $a_4 = 76$ and $a_8 = 156$

37. $a_5 = -26$ and $a_9 = 10$

38. $a_8 = 45$ and $a_{10} = 53$

Find the common difference of the given sequence. See Examples 1 and 2.

39. $\{5n - 3\}$

40. $\{3n - \frac{1}{2}\}$

41. $\{n + 6\}$

42. $\{1 - 4n\}$

43. $\{\sqrt{2} - 2n\}$

44. $\{n\sqrt{3} + 5\}$

Use the given information about each arithmetic sequence to answer the question. See Examples 1 and 2.

45. Given that $a_1 = -3$ and $a_5 = 5$, what is a_{100}?

46. In the sequence 24, 43, 62, ..., which term is 955?

47. In the sequence $1, \frac{4}{3}, \frac{5}{3}, ...$, which term is 25?

48. Given that $a_5 = -\frac{5}{3}$ and $a_9 = 1$, what is a_{62}?

49. In the sequence $-16, -9, -2, ...$, what is a_{20}?

50. In the sequence $\frac{1}{4}, \frac{7}{16}, \frac{5}{8}, ...$, which term is $\frac{35}{8}$?

section 9.2 Arithmetic Sequences and Series

51. In the sequence 2, 5, 8, 11, ..., what is the 9th term?

52. In the sequence 1, 3, 5, 7, ..., what is the 6th term?

53. In the sequence 16, 12, 8, 4, ..., what is the 7th term?

54. In the sequence $\frac{1}{2}, 2, \frac{7}{2}, 5, ...$, what is the 8th term?

55. In the sequence −2, 1, 4, 7, ..., what is the 6th term?

56. In the sequence 9, 6, 3, 0, ..., what is the 10th term?

57. In the sequence 5, 10, 15, 20, ..., what is the 11th term?

58. In the sequence $2\sqrt{2}, 4\sqrt{2}, 6\sqrt{2}, 8\sqrt{2}, ...$, what is the 7th term?

Each of the following sums is a partial sum of an arithmetic sequence; use either formula to determine its value. See Examples 3 and 4.

59. $\sum_{i=1}^{100}(3i-8)$

60. $\sum_{i=1}^{50}(-2i+5)$

61. $\sum_{i=5}^{90}(4i+9)$

62. $3+11+...+795$

63. $25+18+...+(-143)$

64. $-12+2+...+674$

65. $\sum_{i=1}^{37}\left(-\frac{3}{5}i-6\right)$

66. $\sum_{i=100}^{200}(3i+57)$

67. $\sum_{i=2}^{42}(2i-22)$

68. $-90+(-77)+...+92$

69. $7+3+...+-101$

70. $4+\frac{81}{20}+...+900$

Use a graphing calculator or other technology to evaluate each of the following sums.

71. $1.2+2.8+4.4+6.0+7.6+...+28.4$

72. $5+\frac{5}{2}+0+\frac{-5}{2}+...+\frac{-45}{2}$

73. $\sum_{i=1}^{25} 89.47-7.35i$

74. $\sum_{i=1}^{32} 4.12i+17.54$

75. $\sum_{i=49}^{83}(4.37i+8.21)$

76. $\sum_{i=23}^{79}\left(\frac{256i}{397}+\frac{57}{481}\right)$

chapter nine An Introduction to Sequences, Series, Combinatorics, and Probability

Each of the following can be answered by finding the partial sum of an arithmetic sequence. See Examples 3 and 4.

77. Cynthia borrows $21,000, interest-free, from her parents to help pay for her college education, and promises that upon graduation she will pay back the sum beginning with $1000 the first year and increasing the amount by $1000 with each successive year. How many years will it take for her to repay the entire $21,000?

78. A certain theatre is shaped so that the first row has 30 seats and, moving toward the back, each successive row has two seats more than the previous one. If there are 40 rows, how many seats does the last row contain? How many seats are there altogether?

79. A brick-mason spends a morning moving a pile of bricks from his truck to the work-site by wheelbarrow. Each brick weighs two pounds, and on his first trip he transports 100 pounds. On each successive trip, as he tires, he decides to move one less brick. How many pounds of bricks has he transported after 20 trips?

80. The manager of a grocery store decides to create a display of soup cans by placing cans in a row on the floor and then stacking successive rows so that each level of the tower has one less can than the one below it. The manager wants the top row to have 5 cans, and the store has 290 cans that can be used for the display. If all of the cans are used, how many rows will the display have?

section 9.2   Arithmetic Sequences and Series

81. A man decides to lease a car and is told that his payment to the car dealership will be $50 in the first month. He is also told that every month thereafter, for the next 60 months, his payments will increase by $25. How much is his monthly payment after two years? How much has he paid in total after the first two years?

82. Your grandmother, who lived through the Depression, doesn't trust banks, so she decided to save for your college education by periodically adding money to a mason jar buried in her flower garden. She began the practice with $65 and added $15 every time she got her monthly paycheck. If she continued this routine for 18 years, how much money did she manage to save for you?

chapter nine An Introduction to Sequences, Series, Combinatorics, and Probability

9.3 Geometric Sequences and Series

TOPICS

1. Characteristics of geometric sequences
2. The formula for the general term of a geometric sequence
3. Evaluating partial sums of geometric sequences
4. Evaluating infinite geometric series
5. Interlude: Zeno's paradoxes

Topic 1: Characteristics of Geometric Sequences

We began the last section with a discussion of Melinda, a young girl whose weekly allowance was to increase by a fixed amount ($0.50/week) with the start of each new year, and we noted that the resulting sequence constituted an arithmetic progression. In this section, we will examine a similar situation as it might apply to Melinda's mother, Marilyn.

Suppose Marilyn is nearing the end of a job-hunt, and has a choice of two positions. Both offer a starting salary of $40,000 per year, but they differ in their projected future salaries. Employer A has a well-defined tier structure for its employees, and offers a yearly increase in grade with a corresponding increase in salary of $1250/year. Employer B has a flatter hierarchy, and simply offers an increase of 3%/year for all its employees. Assuming the positions are equally desirable in all other respects, which one should Marilyn choose?

As you might expect, real life being what it is, this question is not as easily answered as, say, a problem in a college algebra textbook. One important consideration for Marilyn is the number of years she anticipates working for her next employer. A table comparing the two projected salaries over the next decade or so will help us determine the answer, and the table will be most easily constructed if we can find formulas for the salaries in the nth year.

section 9.3   Geometric Sequences and Series

We know how to find such a formula for Employer A, as the sequence of yearly salaries forms an arithmetic progression with a common difference of $1250; this means the nth term of the sequence is given by

$$a_n = 40000 + (n-1)1250.$$

The nth term of the sequence of salaries for Employer B is of a different form. We can begin by noting that since each successive salary is 3% greater than the preceding one, the recursive formula

$$b_1 = 40{,}000 \text{ and } b_n = b_{n-1} + (0.03)b_{n-1} \text{ for } n \geq 2$$

describes the sequence of salaries. We can then simplify the formula by writing $b_n = (1.03)b_{n-1}$ for $n \geq 2$. With these two formulas in hand, we can now construct the table comparing salaries seen in Figure 1 (all salaries are rounded off to the nearest dollar).

	1	2	3	4	5	6	7	8	9	10
a_n	40,000	41,250	42,500	43,750	45,000	46,250	47,500	48,750	50,000	51,250
b_n	40,000	41,200	42,436	43,709	45,020	46,371	47,762	49,195	50,671	52,191

Figure 1: Comparison of Salaries of Employers A and B in Year n

The first thing we notice is that Employer A offers a higher yearly salary up through the fourth year, but that thereafter Employer B pays more. So if Marilyn anticipates staying in her next job for five years or more, she may want to go with Employer B. This reasoning doesn't consider the *accumulated* salary through year n; for a more accurate comparison of the two jobs, we really need to compare the partial sums of the two sequences and determine when that of Employer B overtakes that of Employer A. We will return to the question of partial sums later in this section.

The salary sequence for Employer B is an example of a *geometric sequence*, and its identifying characteristic is that the *ratio* of consecutive terms in the sequence is a fixed constant. As we noted above, the ratio of any term in the sequence to the preceding term is 1.03. (Remember that for arithmetic sequences, the *difference* between consecutive terms is fixed.)

chapter nine An Introduction to Sequences, Series, Combinatorics, and Probability

Geometric Sequences

A sequence $\{a_n\}$ is a **geometric sequence** (sometimes also called a **geometric progression**) if there is some constant $r \neq 0$ so that $\frac{a_{n+1}}{a_n} = r$ for each $n = 1, 2, 3, \ldots$. The constant r is called the **common ratio** of the sequence.

Topic 2: The Formula for the General Term of a Geometric Sequence

Our goal is to develop an explicit formula for the general n^{th} term of a geometric sequence. This can be done if we relate each term to a_1, the first term of the sequence $\{a_n\}$, just as we did for arithmetic sequences.

Since the sequence $\{a_n\}$ is geometric, there is some fixed number r such that $\frac{a_{n+1}}{a_n} = r$ for each $n = 1, 2, 3, \ldots$. In particular,

$$\frac{a_2}{a_1} = r, \text{ or } a_2 = a_1 r.$$

Similarly,

$$\frac{a_3}{a_2} = r, \text{ so } a_3 = a_2 r = (a_1 r) r = a_1 r^2.$$

This pattern persists, leading to the following observation.

General Term of a Geometric Sequence

The explicit formula for the general n^{th} term of a geometric sequence is

$$a_n = a_1 r^{n-1},$$

where r is the common ratio for the sequence.

section 9.3 Geometric Sequences and Series

Given this information, we now know that the yearly salary offered by Employer B above is given by the formula

$$b_n = (40{,}000)(1.03)^{n-1}.$$

example 1

Find the formula for the general term of the following geometric sequences.

a. $\dfrac{1}{3}, \dfrac{1}{9}, \dfrac{1}{27}, \ldots$ b. $3, -6, 12, \ldots$

Solutions:

a. $r = \dfrac{\frac{1}{9}}{\frac{1}{3}} = \dfrac{3}{9} = \dfrac{1}{3}$

$a_1 = \dfrac{1}{3}$

$a_n = \left(\dfrac{1}{3}\right)\left(\dfrac{1}{3}\right)^{n-1} = \left(\dfrac{1}{3}\right)^n$

Since we have been told the sequence is geometric, we could use any two consecutive terms to determine the common ratio r. Once we have done this, the formula for the general term follows. Note that in this case we can simplify the formula a bit, as shown.

b. $r = \dfrac{-6}{3} = -2$

$a_1 = 3$

$a_n = (3)(-2)^{n-1}$

In this sequence, the common ratio is negative. It will always be the case that a geometric sequence consists of terms alternating in sign if and only if the common ratio is negative.

We can also determine the formula for the general term when given other information about a geometric sequence.

chapter nine An Introduction to Sequences, Series, Combinatorics, and Probability

example 2

Given that the second term of a geometric sequence is -6 and the fifth term is 162, what is the fourth term?

Solution:

First, the fact that the second term and the fifth term alternate in sign tells us that the common ratio must be negative. (But be careful with this type of reasoning; if, say, the second term and the fourth term have the same sign, the common ratio might be positive or negative.) Given that the sequence is geometric, we know that

$$-6 = a_1 r \quad \text{and} \quad 162 = a_1 r^4.$$

We can eliminate a_1 from this pair of equations with the following observation:

$$\frac{162}{-6} = \frac{a_1 r^4}{a_1 r},$$

or

$$-27 = r^3.$$

This tells us $r = -3$, a negative number as expected, and we can then determine that $a_1 = 2$. So the general term is given by $a_n = 2(-3)^{n-1}$, and the fourth term is

$$a_4 = 2(-3)^{4-1} = -54.$$

Topic 3: Evaluating Partial Sums of Geometric Sequences

Given a geometric sequence $\{a_n\}$, many applications require us to compute the partial sum $S_n = a_1 + a_2 + \ldots + a_n$ for some fixed n. The ideal situation is to have an explicit formula that tells us what the nth partial sum is for any given geometric sequence.

We can come up with such a formula with a few simple, though not necessarily obvious, steps. The first is to note that if we define $S_n = a_1 + a_2 + \ldots + a_n$, then the formula for the general term of a geometric sequence allows us to write

$$S_n = a_1 + a_1 r + a_1 r^2 + \ldots + a_1 r^{n-1}.$$

section 9.3 Geometric Sequences and Series

If we multiply both sides of the previous equation by r, we obtain

$$rS_n = a_1r + a_1r^2 + a_1r^3 + \ldots + a_1r^n.$$

If we now subtract the second equation from the first, most of the terms from the right-hand sides of the equations cancel, as shown below:

$$S_n - rS_n = \left(a_1 + \cancel{a_1r} + \ldots + \cancel{a_1r^{n-1}}\right) - \left(\cancel{a_1r} + \cancel{a_1r^2} + \ldots + a_1r^n\right)$$

$$= a_1 - a_1r^n$$

Of course, the goal is to solve for S_n, and we can do this now by factoring the left and right sides of the equation and dividing:

$$S_n(1-r) = a_1(1-r^n)$$

$$S_n = \frac{a_1(1-r^n)}{1-r}$$

This formula is reproduced in the following box for reference.

Partial Sum of a Geometric Sequence

Let $\{a_n\}$ be a geometric sequence with the common ratio r, and assume r is neither 0 nor 1. Then the nth partial sum of the sequence, $S_n = a_1 + a_1r + a_1r^2 + \ldots + a_1r^{n-1}$, is given by

$$S_n = \frac{a_1(1-r^n)}{1-r}.$$

(Note that the condition $r \neq 1$ is necessary to prevent division by zero in the formula, but it isn't difficult to determine the partial sum without the formula if r happens to be 1. In this case, $S_n = a_1 + a_1 + \ldots + a_1 = na_1$.)

With this formula in hand, along with one of the two partial sum formulas for arithmetic sequences, let us return to the question with which we began this section.

chapter nine An Introduction to Sequences, Series, Combinatorics, and Probability

example 3

Recall that Marilyn has the option of taking a job with Employer A, who offers a yearly salary increase of $1250, or Employer B, who offers a yearly salary increase of 3%. Given that the starting salary is $40,000 for both, in what year does the accumulated salary paid by Employer B overtake the accumulated salary paid by Employer A?

Solution:

We have already determined that Employer B pays a higher salary beginning with the fifth year of employment, but we need to compare the partial sums of the salaries paid by the two employers up through year n to answer the question above. Recall that the sequence of yearly salaries paid by Employer A is defined by

$$a_n = 40{,}000 + (n-1)1250$$

and that the sequence of yearly salaries paid by Employer B is defined by

$$b_n = 40{,}000(1.03)^{n-1}.$$

Recall also that one partial sum formula for an arithmetic sequence is

$$S_n = na_1 + d\left(\frac{(n-1)n}{2}\right),$$

where d is the common difference for the sequence.

In order to keep the partial sums straight, let A_n denote the sum of the salaries paid by Employer A through year n, and let B_n be the same for Employer B. Then

$$A_n = n(40{,}000) + 1250\left(\frac{(n-1)n}{2}\right) = 40{,}000n + 625(n^2 - n) = 625n^2 + 39{,}375n$$

and

$$B_n = \frac{40{,}000(1 - 1.03^n)}{1 - 1.03} = \left(\frac{40{,}000}{0.03}\right)(1.03^n - 1).$$

We can now use these formulas to compute the accumulated salaries paid by the two employers up through year n, as shown in Figure 2.

	1	2	3	4	5	6	7	8
A_n	40,000	81,250	123,750	167,500	212,500	258,750	306,250	355,000
B_n	40,000	81,200	123,636	167,345	212,365	258,736	306,498	355,693

Figure 2: Accumulated Salaries through Year 8

Comparing values in the table indicates that the accumulated salary paid by Employer B overtakes that of Employer A in the seventh year.

We can also use the geometric partial sum formula to evaluate certain expressions defined with the sigma notation, as shown in the next example.

example 4

Evaluate $\sum_{i=2}^{7} 5\left(\frac{-1}{2}\right)^i$.

Solution:

The given sum represents a partial sum of a geometric sequence, but as it is written the first term and the common ratio of the sequence are not apparent. So a good way to begin is to write the sum in expanded form:

continued on next page...

chapter nine An Introduction to Sequences, Series, Combinatorics, and Probability

The above screen was performed using the summation and sequence commands as described on page 712. To find the answer in fraction form, select **MATH** and 1.

$$\sum_{i=2}^{7} 5\left(\frac{-1}{2}\right)^i = 5\left(\frac{-1}{2}\right)^2 + 5\left(\frac{-1}{2}\right)^3 + \ldots + 5\left(\frac{-1}{2}\right)^7$$

$$= \frac{5}{4} - \frac{5}{8} + \ldots - \frac{5}{128}$$

From this, we can see that $a_1 = \frac{5}{4}$ and that the common ratio is $r = -\frac{1}{2}$. We also need to observe that there are a total of *six* terms in the sum, so to use the partial sum formula we let $n = 6$. Putting this all together, we have

$$S_6 = \frac{\left(\frac{5}{4}\right)\left(1-\left(-\frac{1}{2}\right)^6\right)}{1-\left(-\frac{1}{2}\right)} = \frac{\left(\frac{5}{4}\right)\left(1-\frac{1}{64}\right)}{\frac{3}{2}} = \frac{\left(\frac{5}{4}\right)\left(\frac{63}{64}\right)}{\frac{3}{2}} = \frac{105}{128}.$$

Topic 4: Evaluating Infinite Geometric Series

In Section 9.1, we used a fairly intuitive approach to think about the result of adding half an inch to a quarter of an inch to an eighth of an inch, and so on. We decided then that if the process were continued indefinitely, the sum of the fractions would be 1. That is,

$$\frac{1}{2} + \frac{1}{4} + \frac{1}{8} + \ldots = 1, \text{ or } \sum_{i=1}^{\infty}\left(\frac{1}{2}\right)^i = 1.$$

Our rationale for this conclusion was that the sequence of partial sums of the sequence

$$\left\{\left(\frac{1}{2}\right)^n\right\}$$

never exceeded 1, yet got closer and closer to 1 as more terms were added.

We now have the machinery to analyze infinite series like the one above more rigorously and more efficiently. For one thing, we know that the sequence

$$\frac{1}{2}, \frac{1}{4}, \frac{1}{8}, \ldots$$

is a geometric sequence, and that both the first term and the common ratio are $\frac{1}{2}$. So we can use the partial sum formula for geometric sequences to determine that

$$S_n = \frac{1}{2} + \frac{1}{4} + \ldots + \frac{1}{2^n} = \frac{\left(\frac{1}{2}\right)\left(1-\left(\frac{1}{2}\right)^n\right)}{\left(1-\frac{1}{2}\right)} = 1-\left(\frac{1}{2}\right)^n$$

where we have made the substitutions

$$a_1 = \frac{1}{2} \quad \text{and} \quad r = \frac{1}{2}.$$

If we were to write S_n as a single fraction, we would obtain

$$S_n = \frac{2^n - 1}{2^n},$$

the formula we intuitively derived in Section 9.1. In either form, we can see that S_n approaches 1 as $n \to \infty$, but we can now generalize this observation for all convergent geometric series.

Infinite Geometric Series

If $|r| < 1$, the infinite geometric series

$$\sum_{n=1}^{\infty} a_1 r^{n-1} = a_1 + a_1 r + a_1 r^2 + a_1 r^3 + \ldots$$

converges, and the sum of the series is given by $S = \dfrac{a_1}{1-r}$. Alternatively, we can use sigma notation with the index beginning at 0 to write the same series; in this form,

$$\sum_{n=0}^{\infty} a_1 r^n = a_1 + a_1 r + a_1 r^2 + a_1 r^3 + \ldots = \frac{a_1}{1-r}.$$

chapter nine An Introduction to Sequences, Series, Combinatorics, and Probability

The proof of this result follows from the fact that if $|r|<1$, then $r^n \to 0$ as $n \to \infty$, so

$$S_n = \frac{a_1(1-r^n)}{1-r} \to \frac{a_1}{1-r}, \text{ as } n \to \infty.$$

That is, the sequence of partial sums of such geometric sequences approaches the given formula as n gets larger and larger.

In the case of the series $\frac{1}{2} + \frac{1}{4} + \frac{1}{8} + \ldots$, where $a_1 = \frac{1}{2}$ and $r = \frac{1}{2}$, the sum is

$$S = \frac{\frac{1}{2}}{1-\frac{1}{2}} = \frac{\frac{1}{2}}{\frac{1}{2}} = 1.$$

example 5

Find the sums of the following series.

a. $\sum_{n=1}^{\infty} 5\left(\frac{-1}{2}\right)^{n-1}$

b. $\sum_{n=1}^{\infty} 3\left(\frac{1}{10}\right)^n$

Solutions:

a. In the case of the first series, it is easy to identify a_1 and r from the way the series is written:

$$a_1 = 5 \text{ and } r = -\frac{1}{2}$$

Since the common ratio r is less than 1 in magnitude, we can proceed to apply the formula for the sum to determine that

$$\sum_{n=1}^{\infty} 5\left(\frac{-1}{2}\right)^{n-1} = \frac{5}{1-\left(-\frac{1}{2}\right)} = \frac{5}{\frac{3}{2}} = \frac{10}{3}$$

b. Be careful to note exactly how a series is written. In this case, the form of the series doesn't quite match either of the forms to which the formula $\frac{a_1}{1-r}$ applies. But if we write out the first few terms, we can easily identify the first term of the series and the common ratio:

$$\sum_{n=1}^{\infty} 3\left(\frac{1}{10}\right)^n = \frac{3}{10} + \frac{3}{100} + \frac{3}{1000} + \ldots$$

From this, we can see that $a_1 = \frac{3}{10}$ and $r = \frac{1}{10}$ (determine r by dividing any term by the preceding term). So

$$\sum_{n=1}^{\infty} 3\left(\frac{1}{10}\right)^n = \frac{\frac{3}{10}}{1-\frac{1}{10}} = \left(\frac{\frac{3}{10}}{1-\frac{1}{10}}\right)\left(\frac{10}{10}\right) = \frac{3}{10-1} = \frac{1}{3}.$$

The last example illustrates an important relation between geometric series and the decimal system we use to write real numbers. Note that what we have really shown is that

$$\sum_{n=1}^{\infty} 3\left(\frac{1}{10}\right)^n = \frac{3}{10} + \frac{3}{100} + \frac{3}{1000} + \ldots = 0.\overline{333}$$

is the decimal representation of the fraction $\frac{1}{3}$. Similarly,

$$0.\overline{52} = \frac{52}{100} + \frac{52}{10,000} + \frac{52}{1,000,000} + \ldots = \sum_{n=1}^{\infty} 52\left(\frac{1}{100}\right)^n$$

can be written in fractional form by noting that for this series $a_1 = \frac{52}{100}$ and $r = \frac{1}{100}$, so

$$0.\overline{52} = \frac{\frac{52}{100}}{1-\frac{1}{100}} = \frac{52}{99}.$$

Topic 5: Interlude: Zeno's Paradoxes

The fact that the sum of an infinite number of positive quantities may in fact be finite has made people uneasy, at least at first glance, for millennia. The Greek philosopher Zeno (*ca.* 450 BC), a member of a community of thinkers on the island of Elea, used this uneasiness as the basis for a number of paradoxes. His purpose in creating these paradoxes was to point out complexities in the concept of infinity.

One of the most famous of Zeno's paradoxes concerns a race between the Greek hero Achilles and a tortoise. Because Achilles is so much faster, the tortoise is given a head start in the race. The paradox is that, once the race has begun, it will take a certain amount of time for Achilles to reach the tortoise's starting place, and that in this time the tortoise will have been able to move ahead some (small) distance to a new point.

chapter nine An Introduction to Sequences, Series, Combinatorics, and Probability

By the time Achilles reaches this new point, the tortoise will have again been able to move ahead, and Achilles will have to repeat the process. Because this argument can be continued indefinitely, Zeno argued, Achilles can never actually catch the tortoise, let alone get ahead of him.

Our knowledge of infinite geometric series now allows us to resolve the paradox. To be specific, let us suppose that Achilles can run 100 feet per second (remember, he's a mythical hero), and that the tortoise begins with a 1000 foot head start. Let us also suppose Achilles can run 10 times as fast as the tortoise (this means the tortoise runs at 10 feet per second, so it's fairly heroic, too). Once the race starts, it takes Achilles 10 seconds to get to the tortoise's starting place, and in this 10 seconds the tortoise has managed to crawl another 100 feet. It takes Achilles an additional 1 second to cover this 100 feet, but during that 1 second the tortoise is able to crawl another 10 feet. If we continue the process, the total amount of time required for Achilles to catch the tortoise is then

$$T = 10 + 1 + \frac{1}{10} + \frac{1}{100} + \ldots$$

This sum is a geometric series with a first term of 10 and a common ratio of $\frac{1}{10}$, so

$$T = \frac{10}{1 - \frac{1}{10}} = \frac{100}{9} = 11.\overline{1} \text{ seconds.}$$

We can also determine that Achilles has run $1111.\overline{1}$ feet in this time, so the tortoise actually only manages to run $111.\overline{1}$ feet before being caught.

section 9.3   Geometric Sequences and Series

exercises

Find the explicit formula for the general term of the geometric sequences described below. See Examples 1 and 2.

1. $-3, -6, -12, -24, -48, \ldots$
2. $7, \dfrac{7}{2}, \dfrac{7}{4}, \dfrac{7}{8}, \dfrac{7}{16}, \ldots$
3. $2, -\dfrac{2}{3}, \dfrac{2}{9}, -\dfrac{2}{27}, \dfrac{2}{81}, \ldots$
4. $a_1 = 5$ and $a_4 = 40$
5. $a_2 = -\dfrac{1}{4}$ and $a_5 = \dfrac{1}{256}$
6. $a_1 = 1$ and $a_4 = -0.001$
7. $a_2 = \dfrac{1}{7}$ and $r = \dfrac{1}{7}$
8. $a_3 = \dfrac{9}{16}$ and $r = -\dfrac{3}{4}$
9. $a_3 = 9$, $a_5 = 81$, and $r < 0$
10. $-3, 9, -27, 81, -243, \ldots$
11. $3, 2, \dfrac{4}{3}, \dfrac{8}{9}, \dfrac{16}{27}, \ldots$
12. $-5, \dfrac{5}{4}, -\dfrac{5}{16}, \dfrac{5}{64}, -\dfrac{5}{256}, \ldots$
13. $a_3 = 28$ and $a_6 = -224$
14. $a_2 = -24$ and $a_5 = -81$
15. $a_5 = 1$ and $a_6 = 2$
16. $a_4 = \dfrac{343}{3}$ and $r = 7$
17. $a_2 = \dfrac{13}{17}$ and $r = \dfrac{4}{3}$
18. $a_4 = 8$, $a_8 = 128$, and $r > 0$

Determine if each of the following sequences is geometric. If so, find the common ratio.

19. the sequence of odd numbers
20. $4, 4, 4, 4, 4, 4, \ldots$
21. $100, 50, 25, 12.5, 6.25, \ldots$
22. $2, 5, 11, 23, 47, \ldots$
23. $\dfrac{7}{8}, \dfrac{7}{4}, \dfrac{7}{2}, 7, 14, \ldots$
24. the sequence of numbers called out at a Bingo game
25. $7, 49, 343, 2401, \ldots$
26. $10, 15, 22.5, 33.75, \ldots$

Given the two terms of a geometric sequence, find the common ratio and first five terms of the sequence.

27. $a_1 = 8$ and $a_2 = 24$
28. $a_6 = \dfrac{1}{2}$ and $a_9 = \dfrac{1}{54}$
29. $a_7 = 16$ and $a_{11} = 256$
30. $a_4 = 108$ and $a_8 = 8748$
31. $a_5 = 100$ and $a_9 = \dfrac{4}{25}$
32. $a_8 = 100$ and $a_{10} = 1$

chapter nine An Introduction to Sequences, Series, Combinatorics, and Probability

Use the given information about each geometric sequence to answer the question.

33. Given that $a_2 = -\dfrac{5}{2}$ and $a_5 = \dfrac{5}{16}$, what is a_{15}?

34. Given that $a_1 = 1$ and $a_4 = \dfrac{8}{27}$, what is the common ratio r?

35. Given that $a_3 = -2$ and $a_4 = -16$, what is a_{13}?

36. Given that $a_2 = 24$ and $a_5 = 375$, what is the common ratio r?

37. Given that $a_1 = -1$ and $a_3 = -4$, what is the common ratio r?

38. Given that $a_3 = 108$ and $a_4 = -648$, what is the common ratio r?

39. Given that $a_3 = \dfrac{-4}{25}$ and $a_7 = \dfrac{-4}{15625}$, what is the common ratio r?

Each of the following sums is a partial sum of a geometric sequence. Use this fact to evaluate the sums. See Examples 3 and 4.

40. $\displaystyle\sum_{i=1}^{10} 3\left(-\dfrac{1}{2}\right)^i$

41. $\displaystyle\sum_{i=5}^{20} 5\left(\dfrac{3}{2}\right)^i$

42. $\displaystyle\sum_{i=10}^{40} 2^i$

43. $1 - \dfrac{1}{2} + \ldots + \dfrac{1}{16{,}384}$

44. $2 + 6 + \ldots + 39{,}366$

45. $5 - \dfrac{5}{3} + \ldots - \dfrac{5}{19{,}683}$

46. $1 - 3 + \ldots + 59{,}049$

47. $\displaystyle\sum_{i=4}^{15} 5(-2)^i$

48. $1 + \dfrac{3}{5} + \ldots + \dfrac{243}{3125}$

Determine if the following infinite geometric series converge. If a given sum converges, find the sum. See Example 5.

49. $\displaystyle\sum_{i=0}^{\infty} -\dfrac{1}{2}\left(\dfrac{2}{3}\right)^i$

50. $\displaystyle\sum_{i=1}^{\infty} \left(\dfrac{4}{5}\right)^i$

51. $\displaystyle\sum_{i=0}^{\infty} \left(-\dfrac{9}{8}\right)^i$

52. $\displaystyle\sum_{i=0}^{\infty} \left(-\dfrac{8}{9}\right)^i$

53. $\displaystyle\sum_{i=5}^{\infty} \left(\dfrac{19}{20}\right)^i$

54. $\displaystyle\sum_{i=0}^{\infty} (-1)^i$

55. $\displaystyle\sum_{i=1}^{\infty} \dfrac{1}{3}(2)^{i-1}$

56. $\displaystyle\sum_{i=0}^{\infty} 5\left(\dfrac{6}{11}\right)^i$

57. $\displaystyle\sum_{i=4}^{\infty} \left(\dfrac{13}{24}\right)^i$

section 9.3 Geometric Sequences and Series

Write each of the following repeating decimal numbers as fractions. See Example 5.

58. $1.\overline{65}$

59. $0.\overline{123}$

60. $-0.\overline{5}$

61. $-3.\overline{8}$

62. $0.\overline{029}$

63. $9.\overline{98}$

Use a graphing calculator or other technology to evaluate each of the following sums.

64. $1.2 + 1.44 + 1.728 + 2.0736 + \ldots + 38.3376$

65. $5 + \dfrac{5}{2} + \dfrac{5}{4} + \dfrac{5}{8} + \ldots + \dfrac{5}{256}$

66. $\sum_{i=53}^{92} (4.21)^i$

67. $\sum_{i=13}^{54} \left(\dfrac{19}{436}\right)^i$

68. $\sum_{i=23}^{79} \dfrac{25}{81}\left(\dfrac{57}{397}\right)^i$

69. $\sum_{i=16}^{71} 3.42(5.26)^i$

70. A rubber ball is dropped from a height of 10 feet, and on each bounce it rebounds up to 80% of its previous height. How far has it traveled vertically at the moment when it hits the ground for the tenth time? If we assume it bounces indefinitely, what is the total vertical distance traveled?

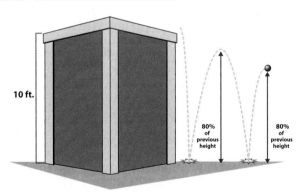

71. If $10,000 is invested in a simple savings account with an annual interest rate of 4% compounded once a year, what is the value of the account after ten years?

72. If $10,000 is invested in a simple savings account with an annual interest rate of 4% compounded once a month, what is the value of the account after ten years?

73. An ancient story about the game of chess tells of a king who offered to grant the inventor of the game a wish. The inventor replied "Place a grain of wheat on the first square of the board, 2 grains on the second square, 4 grains on the third, and so on. The wheat will be my reward." How many grains of wheat would the king have had to come up with? (There are 64 squares on a chessboard.)

74. An isosceles right triangle is divided into two similar triangles, one of the new triangles is divided into two similar triangles, and this process is continued without end. If the shading pattern seen below is continued indefinitely, what fraction of the original triangle is shaded?

75. Each year, the university admissions committee accepts 3% more students than they accepted in the previous year. If 2130 students were admitted in the first year of this trend, how many total students will have been admitted after 6 years?

76. On the day you were born, your parents deposited $15,000 in a simple savings account for your college education. If the annual interest rate is 6.8%, compounded quarterly, how much money will be in the account when you begin college at the age of 18? What is the common ratio of this series?

77. Spamway, an internet advertising agency, uses email forwards to collect addresses to which they send their advertisements. They begin with an email chain letter that they send to 10 people. According to the letter, each of those 10 people has to forward the email to 10 more people or they will have 7 years of bad luck. Assuming the emails are received and forwarded only once a day and all the recipients are superstitious and follow the rules, how many email addresses will Spamway have collected after 30 days? Is this series geometric? If so, find the common ratio.

78. Last summer you were a camp counselor for Sunny Days day camp. An arts and crafts project required you to distribute pieces of string (of different lengths) among your kids. If you began with a piece of string 18 feet long, and gave each child exactly half of your remaining string as you distributed, how long (in inches) was the seventh child's string?

79. Is it possible for a geometric sequence to also be an arithmetic sequence? If so, give an example and if not, explain your reasoning.

chapter nine   An Introduction to Sequences, Series, Combinatorics, and Probability

9.4 Mathematical Induction

TOPICS

1. The role of induction
2. Proofs by mathematical induction

Topic 1: The Role of Induction

The first three sections of this chapter have focused on concepts built on the foundation of the natural numbers: sequences are functions whose domain is the natural numbers, and series are sums based on sequences. This section is similarly focused. While the word "induction" has many meanings in everyday language, the *Principle of Mathematical Induction* has a precise mathematical meaning that depends on properties possessed by the natural numbers.

As a principle, mathematical induction is most often used to prove statements that involve natural numbers in some way. To be precise, suppose that $P(n)$ is a statement about each natural number n. If the statement $P(n)$ is actually true for each n, mathematical induction can be used to prove that fact. As an illustration, consider the statement

$$P(n): \text{The sum of the first } n \text{ positive integers is } \frac{n(n+1)}{2}.$$

We already know this statement is true for each n, as we proved it using a non-inductive technique in Section 9.1. However, induction provides a more powerful and more general method of proof, and we will see how it can be used to prove similar statements. In particular, we will use induction to prove one of the other summation formulas from Section 9.1.

Before we do that, though, it is important to realize how induction relates to some of the other concepts introduced in this chapter. By now, we have seen several examples of recursive formulas, for instance, $a_1 = 1$ and $a_n = a_{n-1} + n$ for $n \geq 2$ is a recursive formula that generates the sequence of sums $1 + 2 + \cdots + n$. As we know, this recursive formula is inconvenient for calculating, say, a_{100}. Its use for this purpose would require us to first calculate a_{99}, which would in turn require us to calculate a_{98}, and so on back to a_1. In contrast, the explicit formula

$$a_n = \frac{n(n+1)}{2}$$

allows us to instantly determine a_{100} or any other such sum. As we will see, the Principle of Mathematical Induction is a very powerful tool for making the transition from a recursive formula to an explicit formula.

Topic 2: Proofs by Mathematical Induction

Consider again the statement

$$P(n)\text{: The sum of the first } n \text{ positive integers is } \frac{n(n+1)}{2}.$$

Suppose we have been able to demonstrate that **if** the statement is true for an integer k, **then** the statement is also true for the next integer, $k + 1$. We can easily show that $P(1)$ is true – that is, the sum of the first 1 positive integers, namely 1, is indeed $\frac{(1)(1+1)}{2}$. So since $P(1)$ is true, we can conclude from our demonstration that $P(2)$ is also true. And since $P(2)$ is true, it follows that $P(3)$ is true. We don't have to literally show that $P(2)$ being true implies that $P(3)$ is true, because we've already demonstrated that, in general, the validity of $P(k)$ implies the validity of $P(k + 1)$. This is the essence of a Proof by Mathematical Induction; the process is summarized below.

Proof by Mathematical Induction

Assume that $P(n)$ is a statement about each natural number n. Suppose that the following two conditions hold:

1. $P(1)$ is true.

2. For each natural number k, if $P(k)$ is true, then $P(k + 1)$ is true.

Then the statement $P(n)$ is true for every natural number n.

The construction of an inductive proof thus consists of two steps. The first task, which we will call the **basic step**, is generally very easy; given a statement $P(n)$ simply verify that $P(1)$ is true. The second task, called the **induction step**, normally calls for a bit more work. It is important to realize, though, that the induction step begins with a very powerful assumption: under the *assumption* that $P(k)$ is true, our goal is to prove that $P(k + 1)$ is also true. The remainder of this section will be devoted to illustrating how inductive proofs are constructed.

chapter nine An Introduction to Sequences, Series, Combinatorics, and Probability

> **example 1**

Prove that for each natural number n, $\sum_{i=1}^{n} i^2 = \dfrac{n(n+1)(2n+1)}{6}$.

Solution:

This statement is another of the summation formulas first seen in Section 9.1, but it has not yet been proved. In less formal terms, we wish to prove the statement

$$P(n): 1^2 + 2^2 + \cdots + n^2 = \frac{n(n+1)(2n+1)}{6}.$$

Basic Step: Note that $P(1)$ is the statement $1^2 = \dfrac{(1)(1+1)(2+1)}{6}$ which is clearly true.

Induction Step: We begin this step with the assumption that the statement is true for some integer k, i.e., that $1^2 + 2^2 + \ldots + k^2 = \dfrac{k(k+1)(2k+1)}{6}$. Our task is to show that $P(k + 1)$ is then also true. To do so, consider the left-hand side of $P(k + 1)$:

$$1^2 + 2^2 + \cdots + k^2 + (k+1)^2$$

To show that $P(k + 1)$ is true, we need to demonstrate that the sum above is equal to the right-hand side of $P(k + 1)$, namely

$$\frac{(k+1)(k+1+1)(2(k+1)+1)}{6},$$

better written as

$$\frac{(k+1)(k+2)(2k+3)}{6}.$$

We can do this by using the inductive assumption to rewrite all but the last term of the left-hand side; some simple algebra then gives us the desired result:

$$1^2 + 2^2 + \cdots + k^2 + (k+1)^2 = \frac{k(k+1)(2k+1)}{6} + (k+1)^2 \quad \text{Use the inductive assumption to rewrite all but the last term.}$$

$$= \frac{k(k+1)(2k+1)}{6} + \frac{6(k+1)^2}{6} \quad \text{Add the two fractions.}$$

$$= \frac{(k+1)}{6}\left[k(2k+1)+6(k+1)\right] \qquad \text{Factor out the common factors.}$$

$$= \frac{(k+1)}{6}\left[2k^2+k+6k+6\right] \qquad \text{Expand the remaining terms.}$$

$$= \frac{(k+1)}{6}\left[2k^2+7k+6\right]$$

$$= \frac{(k+1)}{6}\left[(k+2)(2k+3)\right] \qquad \text{Factor the resulting quadratic.}$$

$$= \frac{(k+1)(k+2)(2k+3)}{6} \qquad \text{Rewrite in the desired form.}$$

Every inductive proof is similar in spirit. The single most important step, and the step that is most unfamiliar initially, is the one in which the inductive assumption is used to simplify part of the statement $P(k + 1)$ in order to prove that $P(k + 1)$ is true. Be sure to note how the inductive assumption is used in the following examples as well.

example 2

Prove that for each natural number n, $1+3+5+\cdots+(2n-1)=n^2$.

Solution:

Basic Step: The statement $P(1)$ is the equation $1=1^2$, which is clearly true.

Induction Step: As always, we begin this step with the assumption that $P(k)$ is true; that is, $1+3+5+\cdots+(2k-1)=k^2$. As in *Example 1*, we then work with the left-hand side of $P(k + 1)$, with the goal of transforming it into the right-hand side:

$$1+3+5+\cdots+(2k-1)+(2(k+1)-1)=k^2+(2(k+1)-1) \qquad \text{Use the inductive assumption to rewrite all but the last term.}$$

$$=k^2+(2k+1)$$

$$=(k+1)^2 \qquad \text{Factoring gives us the desired result.}$$

Induction can be used to prove statements other than equations, of course. The next example illustrates the proof of an inequality.

chapter nine An Introduction to Sequences, Series, Combinatorics, and Probability

example 3

Prove that for each natural number n, $2n \leq 2^n$.

Solution:

Basic Step: The statement $P(1)$ is the inequality $2(1) \leq 2^1$, or $2 \leq 2$. This is clearly true.

Induction Step: Assume that $2k \leq 2^k$. Then:

$$2(k+1) = 2k + 2 \quad \text{Begin by writing the left-hand side of } P(k+1).$$
$$\leq 2^k + 2 \quad \text{Use the inductive assumption.}$$
$$\leq 2^k + 2^k \quad \text{With the right-hand side of } P(k+1) \text{ as the goal,}$$
$$= 2(2^k) \quad \text{we use the fact that } k \geq 1 \text{ implies } 2 \leq 2^k.$$
$$= 2^{k+1} \quad \text{Algebra leads to the desired goal.}$$

We will finish this section with an example that uses the Principle of Mathematical Induction to prove a divisibility fact.

example 4

Prove that for each natural number n, $8^n - 3^n$ is divisible by 5.

Solution:

Basic Step: Clearly, $8^1 - 3^1$ is divisible by 5, so $P(1)$ is true.

Induction Step: Assume that $8^k - 3^k$ is divisible by 5. Then there is some integer p for which $8^k - 3^k = 5p$. Note that

$$8^{k+1} - 3^{k+1} = 8^k \cdot 8 - 3^k \cdot 3$$
$$= 8^k \cdot (5+3) - 3^k \cdot 3$$
$$= 5 \cdot 8^k + 3 \cdot 8^k - 3 \cdot 3^k$$
$$= 5 \cdot 8^k + 3 \cdot (8^k - 3^k)$$
$$= 5 \cdot 8^k + 3 \cdot 5p$$
$$= 5(8^k + 3p).$$

This shows that $8^{k+1} - 3^{k+1}$ is a product of 5 and the integer $8^k + 3p$, so we have shown that $8^{k+1} - 3^{k+1}$ is divisible by 5.

exercises

Find S_{k+1} for the given S_k.

1. $S_k = \dfrac{1}{3(k+2)}$

2. $S_k = \dfrac{k^2}{k(k-1)}$

3. $S_k = \dfrac{k(k+1)(2k+1)}{4}$

4. $S_k = \dfrac{1}{(2k-1)(2k+1)}$

Use the Principle of Mathematical Induction to prove the following statements. See Examples 1 through 4.

5. $1+2+3+4+\cdots+n = \dfrac{n(n+1)}{2}$

6. $\dfrac{1}{2}+\dfrac{1}{2^2}+\dfrac{1}{2^3}+\cdots+\dfrac{1}{2^n} = 1-\dfrac{1}{2^n}$

7. $2+4+6+8+\cdots+2n = n(n+1)$

8. $\displaystyle\sum_{i=1}^{n}\dfrac{1}{(2i-1)(2i+1)} = \dfrac{n}{2n+1}$

9. $4^0+4^1+4^2+\cdots+4^{n-1} = \dfrac{4^n-1}{3}$

10. Prove that $2^n > n^2$ for all $n \geq 5$.

11. $\dfrac{1}{1\cdot 4}+\dfrac{1}{4\cdot 7}+\dfrac{1}{7\cdot 10}+\cdots+\dfrac{1}{(3n-2)(3n+1)} = \dfrac{n}{3n+1}$

chapter nine An Introduction to Sequences, Series, Combinatorics, and Probability

12. $5^0 + 5^1 + 5^2 + \cdots + 5^{n-1} = \dfrac{5^n - 1}{4}$

13. $5 + 10 + 15 + \cdots + 5n = \dfrac{5n(n+1)}{2}$

14. Prove that $n^2 \geq 100n$ for all $n \geq 100$.

15. $\left(1 + \dfrac{1}{1}\right)\left(1 + \dfrac{1}{2}\right)\left(1 + \dfrac{1}{3}\right)\cdots\left(1 + \dfrac{1}{n}\right) = n + 1$

16. $3 + 5 + 7 + \cdots + (2n+1) = n(n+2)$

17. Prove that $n! > 2^n$ for all $n \geq 4$.

18. $1 + 4 + 7 + 10 + \cdots + (3n-2) = \dfrac{n}{2}(3n-1)$

19. $-2 - 3 - 4 - \cdots - (n+1) = -\dfrac{1}{2}(n^2 + 3n)$

20. Prove that $3^n > 2n + 1$ for all $n \geq 2$.

21. Prove that for all natural numbers n, $2^n > n$.

22. $1^3 + 2^3 + 3^3 + 4^3 + \cdots + n^3 = \dfrac{n^2(n+1)^2}{4}$

23. $1\cdot 2+2\cdot 3+3\cdot 4+\cdots+n(n+1)=\dfrac{n(n+1)(n+2)}{3}.$

24. Prove that if $a>1$, then $a^n>1$.

25. Prove that $2^n>4n$ for all $n\geq 5$.

26. $1^4+2^4+3^4+4^4+\cdots+n^4=\dfrac{n(n+1)(2n+1)(3n^2+3n-1)}{30}.$

27. $1^5+2^5+3^5+4^5+\cdots+n^5=\dfrac{n^2(n+1)^2(2n^2+2n-1)}{12}$

28. $\dfrac{1}{\sqrt{1}}+\dfrac{1}{\sqrt{2}}+\dfrac{1}{\sqrt{3}}+\cdots+\dfrac{1}{\sqrt{n}}>\sqrt{n},$ for all $n\geq 2$

29. $1+3+5+7+\cdots+(2n-1)=n^2$

Use the Principle of Mathematical Induction to prove the given properties.

30. $(ab)^n = a^n b^n$, for all positive integers n. (Assume a and b are constant.)

31. $(a^m)^n = a^{mn}$, for all positive integers m and n. (Assume a and m are constant.)

32. If $x_1 > 0$, $x_2 > 0$, ..., $x_n > 0$ then
$\ln(x_1 \cdot x_2 \cdot x_3 \cdot \ldots \cdot x_n) = \ln x_1 + \ln x_2 + \ln x_3 + \cdots + \ln x_n$.

33. 5 is a factor of $\left(2^{2n-1} + 3^{2n-1}\right)$.

34. 64 is a factor of $\left(9^n - 8n - 1\right)$ for all $n \geq 2$.

35. 3 is a factor of $\left(n^3 + 3n^2 + 2n\right)$.

36. Prove that for all natural numbers n, $n^3 - n + 3$ is divisible by 3.

37. Prove that for all natural numbers n, $5^n - 1$ is divisible by 4.

38. Prove that for all natural numbers n, $n(n+1)(n+2)$ is divisible by 6.

section 9.4 — Mathematical Induction

Use the Principle of Mathematical Induction to prove the following.

39. In the 19th century a mathematical puzzle was published telling of a mythical monastery in Benares, India with three crystal towers holding 64 disks made of gold. The disks are each of a different size and have holes in the middle so that they slide over the towers and sit in a stack with the largest on the bottom and the smallest on the top. The monks of the monastery were instructed to move all of the disks to the third tower following these three rules:
  1. Each disk sits over a tower except when it is being moved.
  2. No disk may ever rest on a smaller disk.
  3. Only one disk at a time may be moved.

According to the puzzle, when the monks complete their task, the world would end! To move n disks requires $H(n) = 2^n - 1$ moves. Prove this is true through mathematical induction.

40. If there are N people in a room, and every person shakes hands with every other person exactly once, then exactly $\dfrac{n(n-1)}{2}$ handshakes will occur. Prove this is true through mathematical induction.

41. Any monetary value of 4 cents or higher can be composed of twopence (a British two-cent coin) and nickels. Your basic step would be 4 cents = twopence + twopence. Use the fact that $k = 2t + 5n$ where k is the total monetary value, t is the number of twopence, and n is the number of nickels, to prove $P(k + 1)$. (**Hint:** There are 3 induction steps to prove.)

42. What is wrong with this "proof" by induction?
Proposition: All horses are the same color. (In any set of n horses, all horses are the same color.)
Basic Step: If you have only one horse in a group, then all of the horses in that group have the same color.
Induction Step: Assume that in any group of n horses, all horses are the same color. Now take any group of $n + 1$ horses. Remove the first horse from this group and the remaining n horses must be of the same color because of the hypothesis. Now replace the first horse and remove the last horse. Once again, the remaining n horses must be the same color because of the hypothesis. Since the two groups overlap, all $n + 1$ horses must be the same color.
Thus by induction, any group of n horses are the same color.

chapter nine An Introduction to Sequences, Series, Combinatorics, and Probability

9.5 An Introduction to Combinatorics

TOPICS

1. The multiplication principle of counting
2. Permutations
3. Combinations
4. The binomial and multinomial theorems

Topic 1: The Multiplication Principle of Counting

Combinatorics can be informally defined as "the science of counting." Such a simple definition is somewhat misleading, however, and runs the risk of trivializing the topic. A better – though less pithy – definition of combinatorics is "the study of techniques used to determine the sizes, or cardinalities, of sets." Even this may sound deceptively simple until a few examples are studied. As we will see, there are many occasions when the cardinality (number of elements) of a perfectly well-defined set may be difficult to determine at first glance. We will also see that in many cases the size of a set is of more importance than the actual elements of the set.

One of the joys of combinatorics is that there are many problems that can be solved with nothing more than a few fairly intuitive ideas, the most basic of which is the *Multiplication Principle of Counting*. Before formally stating the principle, we will use it to solve a problem.

example 1

In the United States, telephone numbers consist of a 3-digit area code followed by a 7-digit local number. Neither the first digit of the area code nor the first digit of the local number can be 0 or 1. How many such phone numbers are there if:

a. There is a further restriction that the middle digit of the area code must be 0 or 1?
b. There are no further restrictions?

section 9.5 — An Introduction to Combinatorics

Solutions:

a. Notice that this problem, in mathematical terms, concerns finding the cardinality of the set of all phone numbers of a certain form, but the words "set" and "cardinality" don't appear in the statement of the problem. This is very typical. Most of the combinatorics problems we will see use informal language to define a set, and then ask us, with equally informal language, to determine its size.

The method we will use to solve this problem is also very typical. We will count all the possible phone numbers by considering how we could go about constructing them. Every such phone number consists of a string of ten digits, with the restrictions that the first and fourth digits (reading from left to right) can't be either 0 or 1, but the second digit must be either 0 or 1. Since there are ten digits in all (0 through 9), this means there are eight possible ways to choose a digit for the first and fourth "slots", only two possible ways to choose a digit for the second slot, and ten possible ways to choose digits for all the remaining slots. This is illustrated in Figure 1.

___ ___ ___ ___ ___ ___ ___ ___ ___ ___
8 possible 2 possible 10 possible 8 possible 10 possible 10 possible 10 possible 10 possible 10 possible 10 possible
digits digits digits digits digits digits digits digits digits digits

Figure 1: Constructing Allowable Phone Numbers

For the moment, consider how just the first two slots can be filled. Any of the eight allowable digits for the first slot can be paired up with either of the two allowable digits for the second slot, meaning that there are 8×2 ways of filling the first two slots altogether. In fact, all of the sixteen possible choices for the first two slots can be easily listed:

$$\underbrace{20, 30, 40, \ldots, 90}_{8 \text{ ending in } 0}, \underbrace{21, 31, 41, \ldots, 91}_{8 \text{ ending in } 1}$$

Now, any of these sixteen possible choices for the first two slots can be matched with any of the ten possible digits for the third slot, giving us a total of 160 ways of filling the first three slots. This pattern continues, so that the total number of phone numbers of the required form is

$$8 \times 2 \times 10 \times 8 \times 10 \times 10 \times 10 \times 10 \times 10 \times 10 = 1,280,000,000.$$

continued on next page...

chapter nine An Introduction to Sequences, Series, Combinatorics, and Probability

b. The same sort of argument applies if there is no restriction on the middle digit of the area code. In this case, the total number of allowable phone numbers is

$$8 \times 10 \times 10 \times 8 \times 10 \times 10 \times 10 \times 10 \times 10 \times 10 = 6,400,000,000.$$

The generalization of the reasoning we used in *Example 1* is often stated in terms of a sequence of events.

The Multiplication Principle of Counting

Suppose $E_1, E_2, \ldots, E_n$ is a sequence of events, each of which has a certain number of possible outcomes. Suppose event E_1 has m_1 possible outcomes, and that after event E_1 has occurred, event E_2 has m_2 possible outcomes. Similarly, after event E_2, event E_3 has m_3 possible outcomes, and so on. Then the total number of ways that all n events can occur is $m_1 \cdot m_2 \cdot \ldots \cdot m_n$.

An alternative interpretation of the principle is to think of the events as a sequence of tasks to be completed. In *Example 1*, for instance, each task consists of selecting a digit for a given slot. There are ten tasks in all, and the product of the number of ways each task can be performed gives us the total number of phone numbers.

example 2

A certain state specifies that all non-personalized license plates consist of two letters followed by four digits, and that the letter O (which could be mistaken for the digit 0) cannot be used. How many such license plates are there?

Solution:

Generating such a license plate is a matter of choosing six characters, the first two of which can be any of 25 letters and the last four of which can be any of 10 digits. The total number of such license plates is thus $25 \times 25 \times 10 \times 10 \times 10 \times 10 = 6,250,000$.

Topic 2: Permutations

A **permutation** of a set of objects is a linear arrangement of the objects. In any such arrangement, one of the objects is first, another is second, and so on, so the construction of a permutation of n objects consists of "filling in" n slots. We can thus use the Multiplication Principle of Counting to determine the number of permutations of a given set of objects.

example 3

One brand of combination lock for a door consists of five buttons labeled A, B, C, D, and E, and the installer of the lock can set the combination to be any permutation of the five letters. How many such permutations are there?

Solution:

The difference between this problem and the first two examples is that once a letter has been chosen for a given slot, it can't be reused. So in constructing a combination code, there are five choices for the first letter but only four choices for the second letter, since whatever letter was used first cannot be used again.

continued on next page...

Similarly, there are only three choices for the third slot and only two choices for the fourth slot. Finally, whichever of the five letters is left *must* be used for the fifth slot, so there is only one choice. Figure 2 illustrates the slot-filling process.

$$\underbrace{}_{5\text{ choices}}\ \underbrace{}_{4\text{ choices}}\ \underbrace{}_{3\text{ choices}}\ \underbrace{}_{2\text{ choices}}\ \underbrace{}_{1\text{ choice}}$$

Figure 2: Constructing a Permutation of Five Characters

We can now use the Multiplication Principle of Counting to determine that there are $5 \times 4 \times 3 \times 2 \times 1 = 120$ such combinations.

Products of the form $n \times (n-1) \times (n-2) \times \ldots \times 2 \times 1$ occur so frequently that it makes sense to assign some shorthand notation to them, as follows.

Factorial Notation

If n is a positive integer, we use the notation $n!$ (which is read "***n* factorial**") to stand for the product of all the integers from 1 to n. That is,

$$n! = n \times (n-1) \times (n-2) \times \ldots \times 2 \times 1$$

In addition, we define $0! = 1$.

caution

Note that the factorial operation is defined only for positive integers and, as a special case, 0. You may encounter, in a later math class, a related function called the gamma function that extends the factorial idea to all positive real numbers.

If all permutation problems involved nothing more than a straightforward application of the multiplication principle, as in *Example 3*, there would be little left to say. But typically, the solution of a permutation problem requires counting the number of ways that a linear arrangement of k objects can be made from a collection of n objects, where $k \leq n$. Consider the following problem.

section 9.5 An Introduction to Combinatorics

example 4

How many different five-letter combination codes are possible if every letter of the alphabet can be used, but no letter may be repeated?

Solution:

The difference between this problem and the one in *Example 3* is that there are now 26 choices for the first letter, 25 choices for the second, and so on. The corresponding "slot diagram" describing the number of ways each slot can be filled appears in Figure 3.

$$\underline{\quad\quad}_{26\text{ choices}} \quad \underline{\quad\quad}_{25\text{ choices}} \quad \underline{\quad\quad}_{24\text{ choices}} \quad \underline{\quad\quad}_{23\text{ choices}} \quad \underline{\quad\quad}_{22\text{ choices}}$$

Figure: 3 Five-Letter Permutations Chosen from 26 Characters

The total number of such combination codes is thus $26 \times 25 \times 24 \times 23 \times 22 = 7,893,600$.

Products like the one in *Example 4* are also very common, and deserve special mention. To motivate the following formula, note that we can state the answer to *Example 4* in terms of factorials as follows:

$$26 \times 25 \times 24 \times 23 \times 22 = \frac{(26 \times 25 \times 24 \times 23 \times 22) \times 21!}{21!} = \frac{26!}{21!}$$

If we generalize this, we obtain a formula for the number of permutations of length k that can be formed from a collection of n objects, usually expressed as the number of permutations of n objects taken k at a time.

Permutation Formula

The number of permutations of n objects taken k at a time is

$$_nP_k = \frac{n!}{(n-k)!}.$$

This is a conveniently short way of expressing the product

$$_nP_k = n \times (n-1) \times \ldots \times (n-k+1).$$

chapter nine An Introduction to Sequences, Series, Combinatorics, and Probability

The formula for $_nP_k$ is especially useful when the number of factors to be multiplied is large, as in the next example, assuming you have access to a calculator or computer with the factorial function.

example 5

A magician is preparing to demonstrate a card trick that involves 20 cards chosen at random from a standard deck of cards. Once chosen, the 20 cards are arranged in a row, and the order of the cards plays a role in the trick. How many such orderings are possible?

To find the number of permutations, enter the first number, *n* (52); select **MATH** and use the right arrow to highlight `Prb` (probability). Then select `2:nPr` and enter the second number, *r* (20).

Solution:

A standard deck of cards contains 52 distinct cards (13 cards in each of 4 suits), and this problem asks for the number of permutations of 52 cards taken 20 at a time. One way to determine this would be to evaluate $52 \times 51 \times \ldots \times 33$ (note that this is a product of 20 numbers), but a far faster way is to use the permutation formula:

$$_{52}P_{20} = \frac{52!}{(52-20)!} = \frac{52!}{32!} \approx \frac{8.07 \times 10^{67}}{2.63 \times 10^{35}} \approx 3.07 \times 10^{32}$$

The two factorials that appear in the formula above have been determined by a calculator. Note that $52!$, $32!$, and the final answer are all very large numbers, so it is convenient to use scientific notation.

Topic 3: **Combinations**

In contrast to permutations, where the order of the objects under consideration is of great importance, combinations are simply collections of objects. To be specific, a **combination** of n objects taken k at a time is one of the ways of forming a subset of size k from a set of size n, where again $k \leq n$. Combination problems typically ask us to determine the number of different subsets of size k that can be formed.

To emphasize the difference between permutations and combinations, and to begin to understand the combination formula that we will soon derive, consider the following problem.

example 6

Let $S = \{a, b, c, d\}$.

a. How many permutations of size 3 can be formed from the set S?
b. How many combinations of size 3 can be formed from the set S?

Solutions:

a. The number of permutations of 4 objects taken 3 at a time is $_4P_3 = \dfrac{4!}{(4-3)!} = \dfrac{4!}{1!} = 24$.

To enter the factorial sign, select **MATH**, highlight Prb, and select 4:!.

b. We have already determined there are 24 permutations of size 3 that can be formed. For the purpose of determining the corresponding number of combinations, it might be useful to actually list all the permutations:

abc	acb	bac	bca	cab	cba
abd	adb	bad	bda	dab	dba
acd	adc	cad	cda	dac	dca
bcd	bdc	cbd	cdb	dbc	dcb

If we now view these collections of objects as sets, all six permutations in the first row describe the single set (a, b, c). Similarly, the six permutations in the second row are simply six different ways of describing the set (a, b, d). The third row describes the set (a, c, d) and the fourth row describes the set (b, c, d). In all, there are only four combinations of 4 objects taken 3 at a time.

chapter nine An Introduction to Sequences, Series, Combinatorics, and Probability

Although we don't yet have the formula we seek, let $_nC_k$ denote the number of combinations of n objects taken k at a time. Taking a cue from *Example 6*, we might work toward the desired formula by noting that each of the size k combinations formed from S gives rise to $k!$ permutations of size k, and we know that there are $_nP_k$ permutations taken k at a time overall. This means that

$$(k!)(_nC_k) = {_nP_k}, \text{ or } {_nC_k} = \frac{_nP_k}{k!}.$$

If we now replace $_nP_k$ with $\dfrac{n!}{(n-k)!}$, we have the following formula.

Combination Formula

The number of combinations of n objects taken k at a time is

$$_nC_k = \frac{n!}{(k!)(n-k)!}.$$

At this point, we have seen all of the counting techniques we will need. There is much more to the subject of combinatorics, but the three ideas we have discussed – the Multiplication Principle of Counting, the permutation formula, and the combination formula – are sufficient to answer many, many questions. This is especially true when the techniques are combined in various ways, as the next few examples show.

example 7

Suppose a Senate committee consists of 11 Democrats, 10 Republicans, and 1 Independent. The chair of the committee wants to form a sub-committee to be charged with researching a particular issue, and decides to appoint 3 Democrats, 2 Republicans, and the 1 Independent to the sub-committee. How many different sub-committees are possible?

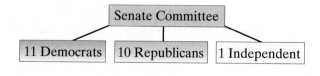

section 9.5 An Introduction to Combinatorics

Solution:

This is in large part a combination problem, as the chair needs to form a subset of size 3 from the 11 Democrats, a subset of size 2 from the 10 Republicans, and a subset of size 1 from the 1 Independent. (Note that the order of those chosen is irrelevant, which is why this is a combination problem and not a permutation problem.) The combination formula tells us that these tasks can be done in, respectively, $_{11}C_3$, $_{10}C_2$, and $_1C_1$ ways (of course, there is only 1 way to choose 1 member from a set of 1). These numbers are:

$$_{11}C_3 = \frac{11!}{3!8!} = \frac{11 \times 10 \times 9 \times 8!}{3! \, 8!} = \frac{990}{6} = 165$$

$$_{10}C_2 = \frac{10!}{2!8!} = \frac{10 \times 9 \times 8!}{2! \, 8!} = \frac{90}{2} = 45$$

$$_1C_1 = \frac{1!}{1!0!} = \frac{1}{1} = 1 \text{ (remember that } 0! = 1\text{)}$$

Once the appropriate number of people from each party have been chosen, any of the 165 possible groups of 3 Democrats can be matched up with any of the 45 possible groups of 2 Republicans, and then further matched up with the 1 Independent. The Multiplication Principle thus tells us there are $165 \times 45 \times 1 = 7425$ ways of forming the desired sub-committee (any one of which, inevitably, is bound to offend one party or another).

> To find the number of combinations, enter the first number, n (11); select **MATH** and use the right arrow to highlight Prb (probability). Then select 3:nCr and enter the second number, r (3).

example 8

How many different arrangements are there of the letters in the following words?

    a. STIPEND   **b.** SALAAM   **c.** MISSISSIPPI

Solutions:

a. The goal is to count the total number of "words" that can be formed from the letters in the word STIPEND, using each letter once and only once (of course, most of the arrangements will not actually be legitimate English words). Since the order of the letters is important, and since STIPEND contains 7 distinct letters, the answer is

$$_7P_7 = \frac{7!}{0!} = 7! = 5040.$$

continued on next page...

b. The word SALAAM contains 6 letters, but only 4 distinct letters. That is, the 3 A's are indistinguishable, so the answer is not simply $_6P_6$. In fact, 6! overcounts the total number of arrangements by a factor of 3!, because any one arrangement of the 6 letters in SALAAM is equivalent to 5 more arrangements. This is because the 3 A's can be permuted in 3! = 6 ways, as is evident if we temporarily color the A's differently:

$$M\text{A}ALSA, M\text{A}ALSA, MA\text{A}LSA, MA\text{A}LSA, MAA\text{L}SA, MAA\text{L}SA$$

This means the total number of arrangements is actually

$$\frac{6!}{3!} = \frac{720}{6} = 120.$$

c. MISSISSIPPI contains 11 characters, but 4 of them are S's, 4 of them are I's, 2 of them are P's, and the remaining 1 character is M. If the 11 characters were all distinct, the total number of arrangements of the letters would be 11!, but we need to divide this number by 4! to account for the indistinguishable S's, and then divide again by 4! to account for the I's, and then again by 2! to account for the P's. This gives us a total of

$$\frac{11!}{4!4!2!} = \frac{39,916,800}{(24)(24)(2)} = 34,650.$$

Topic 4: The Binomial and Multinomial Theorems

Consider the expressions $(x+y)^7$ and $(a+b+c)^3$. The first is a binomial raised to a power, and the second is a trinomial raised to a power. Expressions like these (or worse) can occur in the course of solving an algebra problem, and we often need to expand them in order to work toward the solution of the problem.

We could use only elementary methods to expand such expressions. For example, we could expand $(x+y)^7$ as follows:

$$(x+y)^7 = (x+y)(x+y)(x+y)^5$$
$$= (x^2 + 2xy + y^2)(x+y)^5$$
$$= (x^2 + 2xy + y^2)(x+y)(x+y)^4$$
$$= \ldots$$

But such work can be extremely tedious and error-prone if the exponent is larger than, say, 3 or 4. We can use the combinatorics methods we have seen to develop two formulas that greatly simplify the process.

When we expand an expression like $(x+y)^7$, we are really making sure we account for all the possible products that result from taking one term from each factor. For instance, when we multiply the terms that are boxed below, we obtain x^3y^4:

$$(\boxed{x}+y)(x+\boxed{y})(x+\boxed{y})(\boxed{x}+y)(x+\boxed{y})(\boxed{x}+y)(x+\boxed{y})$$

But there are many other choices of terms that also lead to x^3y^4, such as:

$$(x+\boxed{y})(x+\boxed{y})(\boxed{x}+y)(\boxed{x}+y)(\boxed{x}+y)(x+\boxed{y})(x+\boxed{y})$$

In all, there are 35 different ways of boxing 3 x's and 4 y's, meaning that there is a coefficient of 35 in front of x^3y^4 in the expansion of $(x+y)^7$.

Where did this number of 35 come from? We can relate the expansion of binomials (or, more generally, multinomials) to the "word" problems we studied in *Example 8*. For instance, the coefficient of x^3y^4 is the number of different arrangements of the letters *xxxyyyy*. Since these 7 letters consist of 3 x's and 4 y's, the total number of such arrangements is

$$\frac{7!}{3!4!} = \frac{5040}{(6)(24)} = 35.$$

(Note that the boxed letters above correspond to the "words" *xyyxyxy* and *yyxxxyy*.)

There is some special notation and language for the type of work we have just done.

Binomial Coefficients

Given non-negative integers n and k, with $k \leq n$, we define

$$\binom{n}{k} = \frac{n!}{k!(n-k)!}.$$

The expression $\binom{n}{k}$ is called a **binomial coefficient**, as it corresponds to the coefficient of $x^k y^{n-k}$ in the expansion of $(x+y)^n$.

We have already seen the formula for binomial coefficients in another context. Note that

$$\binom{n}{k} = {}_nC_k.$$

chapter nine An Introduction to Sequences, Series, Combinatorics, and Probability

The last step in expanding a binomial, now that we know how to find the coefficients, is to put all the pieces together. Consider again the expression $(x+y)^7$. When this is expanded, there will be terms containing y^7, xy^6, x^2y^5, x^3y^4, x^4y^3, x^5y^2, x^6y, and x^7 each multiplied by the appropriate binomial coefficient. Note that the sum of the exponents in each term is 7, as this is the total number of x's and y's in each term.

example 9

Expand the expression $(x+y)^7$.

Solution:

Based on the above reasoning, and the binomial coefficient formula,

$$(x+y)^7$$
$$= \binom{7}{0}x^0y^7 + \binom{7}{1}x^1y^6 + \binom{7}{2}x^2y^5 + \binom{7}{3}x^3y^4 + \binom{7}{4}x^4y^3 + \binom{7}{5}x^5y^2 + \binom{7}{6}x^6y^1 + \binom{7}{7}x^7y^0$$
$$= y^7 + 7xy^6 + 21x^2y^5 + 35x^3y^4 + 35x^4y^3 + 21x^5y^2 + 7x^6y + x^7.$$

The generalization of our work is called the Binomial Theorem.

The Binomial Theorem

Given the positive integer n, and any two expressions A and B,

$$(A+B)^n = \sum_{k=0}^{n} \binom{n}{k} A^k B^{n-k}$$
$$= \binom{n}{0}A^0B^n + \binom{n}{1}A^1B^{n-1} + \binom{n}{2}A^2B^{n-2} + \ldots + \binom{n}{n-1}A^{n-1}B^1 + \binom{n}{n}A^nB^0.$$

Note that A and B can be more complicated than we have considered so far.

example 10

Expand the expression $(2x-y)^4$.

Solution:

To use the Binomial Theorem, note that $A = 2x$ and $B = -y$. The theorem thus tells us that

$$(2x-y)^4$$
$$= \binom{4}{0}(2x)^0(-y)^4 + \binom{4}{1}(2x)^1(-y)^3 + \binom{4}{2}(2x)^2(-y)^2 + \binom{4}{3}(2x)^3(-y)^1 + \binom{4}{4}(2x)^4(-y)^0$$
$$= (-y)^4 + 4(2x)(-y)^3 + 6(2x)^2(-y)^2 + 4(2x)^3(-y) + (2x)^4$$
$$= y^4 - 8xy^3 + 24x^2y^2 - 32x^3y + 16x^4.$$

We will finish with one last definition and formula. To expand an expression like $(a+b+c)^3$ we need to determine the coefficients of terms of the form $a^{k_1}b^{k_2}c^{k_3}$, where $k_1 + k_2 + k_3 = 3$. Reasoning similar to that preceding the Binomial Theorem leads to the following definition.

Multinomial Coefficients

Let n be a positive integer, and let $k_1, k_2, ..., k_r$ be non-negative integers such that $k_1 + k_2 + ... + k_r = n$. We define

$$\binom{n}{k_1, k_2, ..., k_r} = \frac{n!}{k_1! k_2! \cdots k_r!}.$$

Such an expression is called a multinomial coefficient. It is the coefficient of the term $A_1^{k_1} A_2^{k_2} \cdots A_r^{k_r}$ in the expansion of $(A_1 + A_2 + ... + A_r)^n$.

To expand something of the form $(A_1 + A_2 + ... + A_r)^n$, we identify all the terms of the form $A_1^{k_1} A_2^{k_2} \cdots A_r^{k_r}$, multiply each one by its corresponding multinomial coefficient, and add them up. This is called the Multinomial Theorem.

chapter nine An Introduction to Sequences, Series, Combinatorics, and Probability

The Multinomial Theorem

Given the positive integer n and expressions $A_1, A_2, \ldots, A_r$,

$$(A_1 + A_2 + \ldots + A_r)^n = \sum_{k_1+k_2+\ldots+k_r=n} \binom{n}{k_1, k_2, \ldots, k_r} A_1^{k_1} A_2^{k_2} \cdots A_r^{k_r}.$$

example 11

Expand the expression $(a+b+c)^3$.

Solution:

We will start by noting that the expansion will have terms containing a^3, a^2b, a^2c, ab^2, ac^2, abc, b^3, b^2c, bc^2, and c^3. (Note that these are all the ways that the three non-negative integer exponents can add up to 3.) Each of these terms must be multiplied by its corresponding multinomial coefficient. For instance, the coefficient of a^2c is

$$\binom{3}{2,0,1} = \frac{3!}{2!0!1!} = \frac{6}{2} = 3.$$

Similarly, the coefficient of b^3 is

$$\binom{3}{0,3,0} = \frac{3!}{0!3!0!} = \frac{6}{6} = 1.$$

Altogether, we obtain

$$(a+b+c)^3 = a^3 + 3a^2b + 3a^2c + 3ab^2 + 3ac^2 + 6abc + b^3 + 3b^2c + 3bc^2 + c^3.$$

section 9.5   An Introduction to Combinatorics

exercises

Consider each of the following situations and determine if each is a combination or permutation.

1. double scoop options from 29 ice cream flavors

2. a poker hand from a standard deck

3. a board committee chosen from 15 candidates

4. a seating chart for 24 students

Use the Multiplication Principle of Counting to answer the following questions. See Examples 1 and 2.

5. Suppose you write down someone's phone number on a piece of paper, but then accidentally wash it along with your laundry. Upon drying the paper, all you can make out of the number is 42? – 3?7?. How many different phone numbers fit this pattern?

6. How many different 7-digit phone numbers contain no odd digits? (Ignore the fact that certain 7-digit sequences are disallowed as phone numbers.)

7. How many different 7-digit phone numbers do not contain the digit 9? (Ignore the fact that certain 7-digit sequences are disallowed as phone numbers.)

8. A certain combination lock allows the buyer to set any combination of five letters, with repetition allowed, but each of the letters must be A, B, C, D, E, or F. How many combinations are possible?

9. In how many different orders can 15 runners finish a race, assuming there are no ties?

775

10. How many different 4-letter radio-station names can be made, assuming the first letter must be a *K* or a *W*? Assume repetition of letters is allowed.

11. How many different 4-letter radio-station names can be made from the call-letters *K*, *N*, *I*, *T*, assuming the letter *K* must appear first? Each of the four letters can be used only once.

12. Three men and three women line up in a row for a photograph, and decide men and women should alternate. In how many different ways can this be done? (Don't forget that the left-most person can be a man or a woman.)

13. How many different ways can a 10-question multiple choice test be answered, assuming every question has 5 possible answers?

14. How many different ways can your 12 favorite novels be arranged in a row?

15. How many different 6-character license plates can be formed if all 26 letters and 10 numerical digits can be used with repetition?

16. How many different 6-character license plates can be formed if all 26 letters and 10 numerical digits can be used without repetition?

17. How many different 6-character license plates can be formed if the first 3 places must be letters and the last 3 places must be numerical digits? (Assume repetition is not allowed.)

18. A box of crayons comes with 8 different colored crayons arranged in a single row. How many different ways can the crayons be ordered in the box?

section 9.5 An Introduction to Combinatorics

Express the answer to the following permutation problems using permutation notation $({}_nP_k)$ and numerically. See Examples 3, 4, and 5.

19. Suppose you have a collection of 30 cherished math books. How many different ways can you choose 12 of them to arrange in a row?

20. In how many different ways can first-place, second-place, and third-place be decided in a 15-person race?

21. Suppose you need to select a user-ID for a computer account, and the system administrator requires that each ID consist of 8 characters with no repetition allowed. The characters you may choose from are the 26 letters of the alphabet (with no distinction between uppercase and lowercase) and the 10 digits. How many choices for a user-ID do you have?

22. How many different 5-letter "words" (they don't have to be actual English words) can be formed from the letters in the word *PLASTIC*?

23. Seven children rush into a room in which six chairs are lined up in a row. How many different ways can six of the seven children choose a chair to sit in? (The seventh remains standing.) How does the answer differ if there are seven chairs in the room?

24. At a meeting of 17 people, a president, vice president, secretary, and treasurer are to be chosen. How many different ways can these positions be filled?

25. Given 26 building blocks, each with a different letter of the alphabet printed on it, how many different 3-letter "words" can be formed?

Solve the following permutation expressions. See Examples 5 and 6.

26. ${}_4P_2$

27. ${}_{15}P_2$

28. ${}_6P_5$

29. ${}_{19}P_{17}$

chapter nine An Introduction to Sequences, Series, Combinatorics, and Probability

Solve the following combination expressions. See Example 7.

30. $_6C_4$

31. $_4C_2$

32. $_{12}C_5$

33. $_{21}C_{14}$

Express the answer to the following combination problems using combination notation $(_nC_k)$ and numerically. See Examples 7 and 8.

34. In many countries, it is not uncommon for quite a few political parties to have their representatives in power. Suppose a committee composed of 10 Conservatives, 13 Liberals, 6 Greens, and 4 Socialists decides to form a sub-committee consisting of 3 Conservatives, 4 Liberals, 2 Greens and 1 Socialist. How many different such sub-committees can be formed?

35. A trade-union asks its members to select 3 people, from a slate of 7, to serve as representatives at a national meeting. How many different sets of 3 can be chosen?

36. Many lottery games are set up so that players select a subset of numbers from a larger set and the winner is the person whose selection matches that chosen by some random mechanism. The order of the numbers is irrelevant. How many choices of six numbers can be made from the numbers 1 through 49?

37. How many different lines can be drawn through a set of nine points in the plane, assuming that no three of the points are collinear? (Points are said to be *collinear* if a single line containing them can be drawn.)

38. Suppose you are taking a 10-question True-False test, and you are guessing that the professor has arranged it so that five of the answers are True and five are False. How many different ways are there of marking the test with five True answers and five False answers?

39. A caller in a Bingo game draws 5 marked ping pong balls from a basket of 75 and calls the numbers out to the players. How many different combinations are possible assuming that the order is irrelevant?

B	I	N	G	O
10	17	39	49	64
12	21	36	55	62
14	25	FREE!	52	70
7	19	32	56	68
5	24	34	54	71

Determine the number of distinguishable different "words" that can be formed by rearranging the letters in each of the following words.

40. ABYSS   **41.** BANANA   **42.** COLLEGE

43. ALGEBRA   **44.** MATHEMATICS   **45.** FIBONACCI

Use a combination of techniques seen in this section, if necessary, to answer the following questions. See Examples 1 through 8.

46. How many different ways are there of choosing five cards from a standard 52-card deck and arranging them in a row? How many different five-card hands can be dealt from a standard 52-card deck?

47. Suppose you have 10 Physics texts, 8 Computer Science texts, and 13 Math texts. How many different ways can you select 4 of each to take with you on vacation?

48. Suppose you have 10 Physics texts, 8 Computer Science texts, and 13 Math texts. How many different ways can you select 4 of each and then arrange them in a row on a shelf, so that the books are grouped by discipline?

49. A certain ice cream store has four different kinds of cones and 28 different flavors of ice cream. How many different single-scoop ice cream cones is it possible to order at this ice cream store?

50. If a local pizza shop has three different types of crust, two different kinds of sauce, and five different toppings, how many different one topping pizzas can be ordered?

51. A man has 8 different shirts, 4 different shorts, and 3 different pairs of shoes. How many different outfits can the man choose from?

chapter nine An Introduction to Sequences, Series, Combinatorics, and Probability

52. A couple wants to have three children. They want to know the different possible gender outcomes for birth order. How many different birth orders are possible?

53. A student has to make out his schedule for classes next fall. He has to take a math class, a science class, an elective, a history class, and an English class. There are three math classes to choose from, two science classes, four electives, three history classes, and four English classes. How many different schedules could the student have?

54. A basketball team has 12 different people on the team. The team consists of three point guards, two shooting guards, one weak forward, three power forwards, and three centers. How many different starting line-ups are possible? (The starting line-up will consist of one player in each of the 5 positions.)

55. How many 5-letter strings can be formed using the letters V, W, X, Y, and Z, if the same letter cannot be repeated?

56. If at the racetrack nine greyhounds are racing against each other, how many different first, second, and third place finishes are possible?

57. A basketball tryout has four distinct positions available on the team. If 25 people show up for tryouts, how many different ways can the four positions be filled?

58. If a trumpet player is practicing eight different pieces of music, in how many different orders can he play his pieces of music?

59. A pizza place has 12 total toppings to choose from. How many different 4-topping pizzas can be ordered?

60. How many different 5-digit numbers can be formed using each of the numbers 6, 8, 1, 9, and 4?

61. If eight cards are chosen randomly from a deck of 52, how many possible groups of eight can be chosen?

62. A baseball team has 15 players on the roster and a batting line-up consists of 9 players. How many different batting line-ups are possible?

63. How many different ways can two red balls, one orange ball, one black ball, and three yellow balls be arranged?

Use the Binomial and Multinomial Theorems in the following problems. See Examples 9, 10, and 11.

64. Expand the expression $(3x+y)^5$.

65. Expand the expression $(x-2y)^7$.

66. Expand the expression $(x-3)^4$.

67. Expand the expression $(x^2-y^3)^4$.

68. Expand the expression $(6x^2+y)^5$.

69. Expand the expression $(4x+5y^2)^6$.

70. Expand the expression $(7x^2+8y^2)^4$.

71. Expand the expression $(x^3-y^2)^5$.

72. Expand the expression $(x+y+z)^2$.

73. Expand the expression $(a-2b+c)^3$.

74. Expand the expression $(2x+5)^6$.

75. Expand the expression $(2x+3y-z)^3$.

76. What is the coefficient of the term containing x^3y in the expansion of $(2x+y)^4$?

77. What is the coefficient of the term containing x^4y^3 in the expansion of $(x^2-2y)^5$?

78. Find the first four terms in the expansion of $(x+3y)^{16}$.

79. Find the first three terms in the expansion of $(2x+3)^{13}$.

80. Find the first two terms in the expansion of $\left(3x^{\frac{1}{4}}+5y\right)^{17}$.

81. Find the 11th term in the expansion of $(x+2)^{24}$.

82. Find the 17th term in the expansion of $(2x+1)^{21}$.

83. Find the 9th term in the expansion of $(x-6y)^{12}$.

Pascal's triangle is a triangular arrangement of binomial coefficients, the first few rows of which appear as follows:

$$\begin{array}{c}1\\ 1\quad 1\\ 1\quad 2\quad 1\\ 1\quad 3\quad 3\quad 1\\ 1\quad 4\quad 6\quad 4\quad 1\\ \vdots\end{array}$$

Each number (aside from those on the perimeter of the triangle) is the sum of the two numbers diagonally adjacent to it in the previous row. Pascal's triangle is a useful way of generating binomial coefficients, with the nth row containing the coefficients of a binomial raised to the $(n-1)^{\text{th}}$ power. It can also be used to suggest useful relationships between binomial coefficients. *Prove each of the following such relationships algebraically.*

84. $\binom{n}{k} = \binom{n-1}{k-1} + \binom{n-1}{k}$ (Note that this is a restatement of how Pascal's triangle is formed.)

85. $\binom{n}{k} = \binom{n}{n-k}$

86. $\binom{n}{0} = \binom{n}{n} = 1$

87. $\binom{n}{0} + \binom{n}{1} + \ldots + \binom{n}{n} = 2^n$ (Hint: use the Binomial Theorem on $(x+y)^n$ for a convenient choice of x and y.)

9.6 An Introduction to Probability

TOPICS

1. The language of probability
2. Using combinatorics to compute probabilities
3. Unions, intersections, and independent events

Topic 1: The Language of Probability

We all make use of the concept of *probability* nearly every day, and in a wide variety of ways. Most of the time, the use is informal; we might briefly wonder how likely it is we'll run out of gas as we drive to an appointment, or what the probability of rain is during an afternoon game of tennis, or what the chances are of striking it rich by winning the state lottery. In mathematics, probability is much more rigorously defined, and the results of probabilistic analysis are used in many important (and some less important, but entertaining) applications.

Mathematical probability, like combinatorics, is a huge subject, and this section will do no more than introduce the topic. Fittingly, most of the probability questions we will consider can be answered with the combinatorics techniques learned in the last section. To see why, we need to first define the terminology of probability.

An **experiment**, in probability theory, is any activity that results in well-defined **outcomes**. In a given problem, we are usually concerned with finding the probability that one or more of the outcomes will occur, and we use the word **event** to refer to a set of outcomes. (An event may be a set containing just one element; that is, an event may consist of a single outcome of an experiment.) The set of all possible outcomes of a given experiment is called the **sample space** of the experiment. This means that in the language of sets, an event is any subset of the sample space. (Note that the empty set thus meets the definition of an event, as does the entire sample space. We are usually interested in finding the probability of events that fall somewhere between being empty and being everything.)

chapter nine   An Introduction to Sequences, Series, Combinatorics, and Probability

For example, rolling a standard die qualifies as an experiment, and the outcomes (as in, the number showing on the face-up side when it stops rolling) can be reasonably described with the numbers 1, 2, 3, 4, 5, and 6. So the sample space of this experiment is the set $S = \{1, 2, 3, 4, 5, 6\}$, and the events consist of all the possible subsets of S. We might, for instance, be interested in the event E defined as "the number rolled is even." In terms of sets, $E = \{2, 4, 6\}$. Our intuition tells us that, if the die is a fair one, the probability of E occurring is one-half, as half the numbers are even and half are odd. We denote this by writing $P(E) = \dfrac{1}{2}$, which is read "the probability of E is one-half."

The intuition we used in this first simple experiment points the way toward the main tool we have for calculating probabilities.

Probabilities when Outcomes are Equally Likely

We say that the outcomes of an experiment are *equally likely* when they all have the same probability of occurring. If E is an event of such an experiment, and if S is the sample space of the experiment, then the **probability of E** is given by

$$P(E) = \frac{n(E)}{n(S)}$$

where $n(E)$ and $n(S)$ are, respectively, the cardinalities of the sets E and S.

It is important to note that the formula above has some restrictions. First, it only applies in equally likely situations, so we can't use it to analyze weighted coins, crooked roulette tables, tampered decks of cards, and so on. Secondly, and more subtly, it assumes that the size of the sample space is finite (otherwise, the formula makes no sense), so we can't use it to analyze experiments based on, say, choosing a real number at random.

The formula also tells us, indirectly, that probabilities are always going to be real numbers between 0 and 1, inclusive. The probability of an event E is 0 if E is the empty set, and 1 if E is the entire sample space S. In all other (finite) cases, the size of E is going to be something between 0 and the size of S, so the fraction will yield a number between 0 and 1.

example 1

A (fair) die is rolled, and the number on top noted. Determine the probability that the number is:

a. prime **b.** divisible by 5 **c.** 7

Solutions:

a. If we let E denote the event in question, then $E = \{2, 3, 5\}$. Since $n(E) = 3$ and the size of the sample space $S = \{1, 2, 3, 4, 5, 6\}$ is 6, $P(E) = \frac{1}{2}$.

b. If we let F denote the event, then $F = \{5\}$, as this is the only integer from 1 to 6 that is divisible by 5. So $P(F) = \frac{1}{6}$.

c. If we let G denote the event, then $G = \{\ \}$. In other words, G is just another name for the empty set, as there is no way for the top face to show a 7. So $P(G) = 0$.

Topic 2: Using Combinatorics to Compute Probabilities

Of course, most meaningful probability questions are not as basic as those in *Example 1*, and what makes a probability problem more interesting usually involves finding the size of an event and/or the sample space. This is where combinatorics comes into play. Consider the following problems.

example 2

A pair of dice is rolled, and the sum of the top faces noted. What is the probability that the sum is:

a. 2 **b.** 5 **c.** 7 or 11

Solutions:

The size of the sample space is the same for all three questions, so it makes sense to determine this first. If we want to use the single probability formula that we have, it is important to make sure that all the outcomes of the sample space are equally likely.

continued on next page...

chapter nine   An Introduction to Sequences, Series, Combinatorics, and Probability

For this reason, we do *not* want to define the sample space to be all integers between 2 (the smallest possible sum) and 12 (the largest possible sum). The reason for this is that these sums are not all equally likely. For instance, there is only one way for the sum to be 2: both top faces must show a 1. But there are two ways for the sum to be 3: one die (call it die A) shows a 1 and the other (die B) shows a 2, or else die A shows a 2 and die B shows a 1. Similarly, there are *three* ways for the sum of the top faces to be 4. In order to make sense of this situation, and in order to define the sample space properly, it may help to construct a table of ordered pairs. In Figure 1 below, the first number in each pair corresponds to the number showing on die A and the second number corresponds to the number showing on die B.

Number on die B

(1, 1)	(1, 2)	(1, 3)	(1, 4)	(1, 5)	(1, 6)
(2, 1)	(2, 2)	(2, 3)	(2, 4)	(2, 5)	(2, 6)
(3, 1)	(3, 2)	(3, 3)	(3, 4)	(3, 5)	(3, 6)
(4, 1)	(4, 2)	(4, 3)	(4, 4)	(4, 5)	(4, 6)
(5, 1)	(5, 2)	(5, 3)	(5, 4)	(5, 5)	(5, 6)
(6, 1)	(6, 2)	(6, 3)	(6, 4)	(6, 5)	(6, 6)

Number on die A

Figure 1: Possible Outcomes of a Pair of Dice

Each of these ordered pairs is equally likely to come up, as any of the numbers 1 through 6 are equally likely for the first slot (die A) and similarly for the second slot (die B). And referring to the positions in the ordered pairs as "slots" points to a quick way of determining the size of the sample space. Since there are 6 choices for each slot, the Multiplication Principle of Counting tells us there are 36 possible outcomes of this experiment.

We can now proceed to answer the three specific questions.

a. There is only one ordered pair corresponding to a sum of 2, namely (1, 1), so the probability of this event is $\frac{1}{36}$.

b. There are four ordered pairs corresponding to a sum of 5, namely the set

$$\{(1, 4), (2, 3), (3, 2), (4, 1)\},$$

so the probability of this event is $\frac{4}{36}$, or $\frac{1}{9}$.

c. A sum of 7 or 11 comes from any of the following ordered pairs:

$$\{(1, 6), (2, 5), (3, 4), (4, 3), (5, 2), (6, 1), (5, 6), (6, 5)\}$$

Since there are eight elements in this event, the probability of rolling a 7 or 11 is $\frac{8}{36}$, or $\frac{2}{9}$.

example 3

Suppose you are taking a 10-question True or False test, and that you are woefully unprepared for it. If you decide to guess on each question, what is the probability of getting 8 or more questions right?

Solution:

The sample space for this problem consists of all the possible sequences of 10 answers, each of which is True or False. Combinatorics tells us that there are $2^{10} = 1024$ such sequences (2 choices for each of 10 "slots"), and these 1024 possible sequences are equally likely if your choice of True or False on each question is truly random.

There is only one particular sequence of 10 answers that is completely correct, so your probability of randomly choosing this one sequence out of all 1024 possible sequences is $\frac{1}{1024}$. But there are more ways that you could guess the right answers on exactly 9 of the questions. You might, for instance, miss only question 3, or only question 8, to name two possibilities. This makes it clear, in fact, that there are 10 ways to miss exactly 1 question; stated another way, there are 10 ways to get exactly 9 questions right, so the probability of getting exactly 9 right is $\frac{10}{1024}$.

We now need to figure out the number of ways you could guess the right answer on exactly 8 questions. To do this, let us revisit the question of how you could get exactly 9 answers right. In the last paragraph, we resolved this by realizing that getting 9 right is equivalent to getting 1 wrong, and there are 10 ways of doing this. A more sophisticated way is to realize that there are $_{10}C_9$ ways of selecting 9 of the 10 questions (the 9 for which you guess the right answer), and then noting that

$$_{10}C_9 = \frac{10!}{9!1!} = 10.$$

Similarly, there are $_{10}C_8$ ways of selecting 8 of the 10 questions, so there are

$$_{10}C_8 = \frac{10!}{8!2!} = \frac{10 \times 9}{2} = 45$$

ways of getting exactly 8 questions right.

Altogether, there are $45 + 10 + 1 = 56$ ways of getting 8, 9, or 10 questions right, so the probability of this happening is

$$\frac{56}{1024} \approx 0.055.$$

In other words, you only have a 5.5% chance of scoring 8 or better by guessing, so studying ahead of time is the way to go (but you knew that already).

chapter nine An Introduction to Sequences, Series, Combinatorics, and Probability

Topic 3: ## Unions, Intersections, and Independent Events

Since mathematical probability is defined in terms of sets (in the form of sample spaces and events), it is natural to use terms like "union" and "intersection" to describe ways of combining events. The notion of independence of events, on the other hand, is unique to probability. We will look first at probabilities of unions and intersections of events. (For a review of set operations like union and intersection, see Section 1.2.)

If E and F are two subsets of the same sample space S, then $E \cup F$ and $E \cap F$ are also subsets and constitute events in their own right. It makes sense, then, to talk about the probability of the events $E \cup F$ and $E \cap F$, and it's reasonable to suspect that they bear some relation to $P(E)$ and $P(F)$. First, note that if all the outcomes in S are equally likely (as is the case with the problems we study in this section), then

$$P(E \cup F) = \frac{n(E \cup F)}{n(S)}$$

and

$$P(E \cap F) = \frac{n(E \cap F)}{n(S)}$$

so what we are really after are the relations between $n(E \cup F)$, $n(E \cap F)$, $n(E)$, and $n(F)$.

To understand the formula that follows, consider again the specific experiment that consists of rolling a die. Let E be the event "the number rolled is divisible by 2" and let F be the event "the number rolled is divisible by 3." For this small experiment, we can list the elements of each of the four events we are interested in, and then determine each event's cardinality:

$$E = \{2, 4, 6\}; \; n(E) = 3$$
$$F = \{3, 6\}; \; n(F) = 2$$
$$E \cup F = \{2, 3, 4, 6\}; \; n(E \cup F) = 4$$
$$E \cap F = \{6\}; \; n(E \cap F) = 1$$

Note that even though union is in some ways the set equivalent of numerical addition, the cardinality of the union of E and F in this example is not the sum of the cardinalities of E and F individually. The reason: both E and F contain the element 6, so if we simply add the sizes of E and F together to get the size of $E \cup F$, we wind up counting 6 twice. This happens in general when we try to determine the cardinality of a union of two sets, and we remedy the situation by subtracting the cardinality of the intersection from the sum of the individual cardinalities. By doing this, we count those elements that lie in both sets just once.

Cardinality of a Union of Sets

> Let E and F be two (finite) sets. Then
> $$n(E \cup F) = n(E) + n(F) - n(E \cap F).$$
> Note that if E and F are *disjoint*, meaning they have no elements in common, then $n(E \cap F) = 0$, so $n(E \cup F) = n(E) + n(F)$.

We can now use this knowledge about cardinalities of unions of sets to find a formula for the probability of the union of two events. Assuming E and F are two subsets of the same sample space S,

$$P(E \cup F) = \frac{n(E \cup F)}{n(S)}$$

$$= \frac{n(E) + n(F) - n(E \cap F)}{n(S)}$$

$$= \frac{n(E)}{n(S)} + \frac{n(F)}{n(S)} - \frac{n(E \cap F)}{n(S)}$$

$$= P(E) + P(F) - P(E \cap F).$$

(In fact, this statement about probabilities of unions is true in general, not just under the assumptions we have made, but discussion of this fact is best left for a later course in probability.)

Probability of a Union of Two Events

> Let E and F be two subsets of the same sample space. Then the probability of the event "E or F", denoted $P(E \cup F)$, is given by the formula
> $$P(E \cup F) = P(E) + P(F) - P(E \cap F).$$
> The term $P(E \cap F)$ represents the probability of both events E and F happening. If, as sets, events E and F are disjoint (so $E \cap F = \emptyset$), then E and F are said to be *mutually exclusive*. In this case, $E \cap F = \emptyset$ and $P(E \cup F) = P(E) + P(F)$.

chapter nine An Introduction to Sequences, Series, Combinatorics, and Probability

example 4

Assume a single die has been rolled and the number showing on top noted. Let E be the event "the number is divisible by 2" and let F be the event "the number is divisible by 3." Find the following probabilities:

a. $P(E \cap F)$ **b.** $P(E \cup F)$

Solutions:

We have already determined the sizes of all the relevant sets in the discussion above, so we are ready to note the answers or apply formulas, as necessary.

a. Since $n(E \cap F) = 1$, and since the sample space has six elements altogether,
$$P(E \cap F) = \frac{1}{6}.$$

b. Since $n(E) = 3$ and $n(F) = 2$,
$$P(E \cup F) = P(E) + P(F) - P(E \cap F)$$
$$= \frac{3}{6} + \frac{2}{6} - \frac{1}{6}$$
$$= \frac{2}{3}.$$

The formula for the probability of the union of three or more events can be complicated in general, and will be left for a later course. But if no two events have any elements in common, the formula is less complex. We say that events $E_1, E_2, ..., E_n$ are **pairwise disjoint** if $E_i \cap E_j = \emptyset$ whenever $i \neq j$. Another way of saying this is that every possible pair of events in the collection $E_1, E_2, ..., E_n$ is a mutually exclusive pair.

Unions of Mutually Exclusive Events

If $E_1, E_2, ..., E_n$ are pairwise disjoint, then
$$P(E_1 \cup E_2 \cup \cdots \cup E_n) = P(E_1) + P(E_2) + ... + P(E_n).$$

example 5

Suppose Jim has chosen a PIN of 8736 for use at his bank's ATM, and that all PIN's at the bank consist of four digits from 0 to 9. Using the ATM one day, he senses someone looking over his shoulder as he punches in the first two digits, and he decides to cancel the operation and leave. The next day Jim discovers his ATM card is missing, and he's not entirely sure he removed it from the machine the day before. In the worst case scenario, a stranger now has Jim's ATM card and the first two digits of his PIN. As he calls up the bank to cancel the card, he wonders what the chances are the unknown someone can guess the remaining digits in the three tries the ATM allows. What is the probability of this unfortunate event?

Solution:

If the first two digits are indeed known, the stranger has the task of filling in the last two digits, and there are 10 choices for each:

$$8 \quad 7 \quad \underbrace{}_{\text{10 choices}} \underbrace{}_{\text{10 choices}}$$

The size of the sample space is thus 100, and any guess at completing the PIN correctly has a probability equal to $\frac{1}{100}$ of being correct. Assuming the stranger tries three different ways of completing the PIN (so that the three events are pairwise disjoint), the probability of Jim's account being broken into is

$$\frac{1}{100} + \frac{1}{100} + \frac{1}{100} = \frac{3}{100}.$$

The last probability idea we will consider is that of *independence*. Most people have an intuitive understanding of what it means for two events to be independent, in terms of probability. Informally, we say that two events are independent if the occurrence of one of them has no effect on the occurrence of the other. Formally, independence of events is related to the probability of their intersection, as follows.

chapter nine An Introduction to Sequences, Series, Combinatorics, and Probability

Independent Events

Given two events E_1 and E_2 in the same sample space, we say E_1 and E_2 are **independent** if $P(E_1 \cap E_2) = P(E_1)P(E_2)$. More generally, a collection of events $E_1, E_2, ..., E_n$ is **independent** if for any subcollection $E_{n_1}, E_{n_2}, ..., E_{n_k}$ of $E_1, E_2, ..., E_n$, it is true that

$$P(E_{n_1} \cap E_{n_2} \cap \cdots \cap E_{n_k}) = P(E_{n_1})P(E_{n_2})\cdots P(E_{n_k}).$$

example 6

If a coin is flipped three times, what is the probability of it coming up tails each time?

Solution:

We actually have two good ways of answering this question, one using the notion of independence and one not.

Let E_i be the event "the coin comes up tails on the ith flip." We are interested, then, in the probability of $E_1 \cap E_2 \cap E_3$, the probability that we get tails each time. Since $P(E_i) = \frac{1}{2}$ for each $i = 1, 2, 3$ (remember, the coin is assumed fair), then

$$P(E_1 \cap E_2 \cap E_3) = P(E_1)P(E_2)P(E_3)$$
$$= \frac{1}{2} \cdot \frac{1}{2} \cdot \frac{1}{2}$$
$$= \frac{1}{8}.$$

The other way of obtaining the same answer is to consider the sample space made up of all possible three-toss sequences. Since each flip of the coin results in one of two possibilities, the Multiplication Principle of Counting tells us that there are $2^3 = 8$ possible sequences. Only one of these is the sequence consisting of three tails, so the probability of this event is $\frac{1}{8}$.

section 9.6 — An Introduction to Probability

exercises

Below is the given probability that an event will occur; find the probability that it will not occur.

1. $P = \dfrac{2}{5}$
2. $P = 0.72$
3. $P = \dfrac{4}{13}$
4. $P = 0.15$
5. $P = \dfrac{2}{3}$
6. $P = 0.49$

Apply the formulas for the probability of intersection or union to the following sets and determine **a.** $P(E \cap F)$ *and* **b.** $P(E \cup F)$. S = size of sample space.

7. $S = 8$, $E = \{2,5\}$, $F = \{3,7,9\}$
8. $S = 10$, $E = \{1,2,5\}$, $F = \{1,2,3,5\}$
9. $S = 5$, $E = \{4, B\}$, $F = \{3\}$
10. $S = 8$, $E = \{A\}$, $F = \{B, C, D, E\}$
11. $S = 4$, $E = \{1, \beta\}$, $F = \{\alpha, 2\}$
12. $S = 12$, $E = \{A, C, g, 5, n, 7, 8, t, L\}$, $F = \{n, 6\}$
13. $S = 16$, $E = \{1, 2, A, m, 13, Y, 8\}$, $F = \{1, 9, 11, m\}$
14. $S = 11$, $E = \{m, 7, D, 4, \theta\}$, $F = \{\phi, D, 3, 7, m, \Sigma\}$

Determine the sample space of each of the following experiments.

15. A coin is flipped four times, and the result recorded after each flip.

16. A card is drawn at random from the 13 hearts.

17. A coin is flipped, and a card is drawn at random from the 13 hearts.

18. A quadrant of the Cartesian plane is chosen at random.

19. A slot machine lever is pulled; there are 3 slots, each of which can hold 6 different values.

20. An individual die is rolled twice, and each of the two results is recorded.

21. At a casino, a roulette wheel spins until a ball comes to rest in one of the 38 pockets.

22. A lottery drawing consists of 6 randomly drawn numbers from 1 to 20; the order of the numbers matters in this case, and repetition is possible.

Answer the following probability questions. Be careful to properly identify the sample space and the appropriate event in each case. See Examples 1, 2, and 3.

23. An ordinary die is rolled. Find the probability of rolling:
    a. A 3 or higher.
    b. An even non-prime number.

24. A card is drawn from a standard 52-card deck. Find the probability of drawing:
    a. A face card (jack, queen, or king) in the suit of hearts.
    b. Anything but an ace.
    c. A black (clubs or spades) non-face card.

25. A coin is flipped three times. Find the probability of getting:
    a. Heads exactly twice.
    b. The sequence Heads, Tails, Heads.
    c. Two or more Heads.

26. A state lottery game is won by choosing the same six numbers (without repetition) as those selected by a mechanical device. The numbers are picked from the set $\{1, 2, \ldots, 49\}$, and the order of the numbers chosen is immaterial. What is the probability of winning?

27. What is the probability that a four-digit ATM PIN chosen at random ends in 7, 8, or 9?

28. Assume the probability of a newborn being male is one-half. What is the probability that a family with five children has exactly three boys?

29. What is the probability that a 9-digit driver's license number chosen at random will not have an 8 as a digit?

30. A roulette wheel in a casino has 38 pockets: 18 red, 18 black, and 2 green. Spinning the wheel causes a small ball to randomly drop into one of the pockets. All of the pockets are equally likely. The wheel is spun twice. Find the probability of getting:
    a. Green both times.
    b. Black at least once.
    c. Red exactly once.

31. There is a 25% chance of rain for each of the next 2 days. What is the probability that it will rain on one of the days but not the other?

The following problems all involve unions or intersections of events, or the notion of independent events. See Examples 4, 5, and 6.

32. A pair of dice is rolled. Find the probability that the sum of the top faces is:
    a. Seven.
    b. Seven or eleven.
    c. An even number or a number divisible by three.
    d. Ten or higher.

33. A card is drawn from a standard 52-card deck. Find the probability of drawing:
    a. A face card or a diamond.
    b. A face card but not a diamond.
    c. A red face card or a king.

34. A state lottery game is won by choosing the same six numbers (without repetition) as those selected by a mechanical device. The numbers are picked from the set $\{1, 2,...,49\}$, and the order of the numbers chosen is immaterial. What is the probability of winning if someone buys 1000 tickets? (Of course, no two of the tickets have the same set of six numbers.) How many tickets would have to be bought to raise the probability of winning to one half?

35. Two cards are drawn at random from a standard 52-card deck. What is the probability of them both being aces if:
    a. The first card is drawn, looked at, placed back in the deck, and the deck is then shuffled before the second card is drawn?
    b. The two cards are drawn at the same time?

36. What is the probability of being dealt a five-card hand (from a standard 52-card deck) that has four cards of the same rank?

37. The probability of rain today is 75%, and the probability that Bob will forget to put the top up on his convertible is 25%. What is the probability of the inside of his car getting wet?

38. What is the probability of drawing 3 face cards in a row, without replacement, from a 52-card deck?

39. Two dice are rolled, and the difference of the results is calculated by subtracting the smaller value from the larger value. Therefore, the result may range from 0 to 5. Find the probability of each of the following results:
   a. 0
   b. 1
   c. 4

40. A letter is randomly chosen from the word MISSISSIPPI. What is the probability of the letter being an S?

41. Mike works in a company of about 100 employees. If this year five people in the company are going to be randomly laid off, what is the probability that Mike will get laid off?

42. A pack of M&M's contains 10 yellow, 15 green, and 20 red pieces. What is the probability of choosing a green M&M out of the pack?

43. A jar of cookies has 3 sugar cookies, 4 chocolate chip cookies, and 2 peanut butter cookies. What is the probability of randomly choosing a peanut butter cookie out of the jar?

44. A big box of crayons contains 4 different blues, 3 different reds, 5 different greens, and 2 different yellows. What is the probability of randomly choosing a yellow crayon out of the box?

45. A bag of marbles contains 3 blue marbles, 2 red marbles, and 5 orange marbles. What is the probability of randomly picking a blue marble out of the bag?

46. Every week a teacher of a class of 25, randomly chooses a student to wash the blackboards. If there are 15 girls in the class, what is the probability that the student selected will be a boy?

47. If in a raffle 135 tickets are sold, how many tickets must be bought for an individual to have a 20% chance of winning?

48. Zach is running for student council. At Zach's school the student council is chosen randomly from all qualified candidates. If there are 42 candidates running, including Zach, and a total of three positions, what is the probability that Zach will be selected for the council?

chapter nine An Introduction to Sequences, Series, Combinatorics, and Probability

chapter nine project

Probability

You may be familiar with the casino game of roulette. But have you ever tried to compute the probability of winning on a given bet?

The roulette wheel has 38 total slots. The wheel turns in one direction and a ball is rolled in the opposite direction around the wheel until it comes to rest in one of the 38 slots. The slots are numbered 00, and 0 – 36. Eighteen of the slots between 1 and 36 are colored black and eighteen are colored red. The 0 and 00 slots are colored green and are considered neither even nor odd, and neither red nor black. These slots are the key to the house's advantage.

The following are some common bets in roulette:
A gambler may bet that the ball will land on a particular number, or a red slot, or a black slot, or an odd number, or an even number (not including 0 or 00). He or she could wager instead that the ball will land on a column (one of 12 specific numbers between 1 and 36), or on a street (one of 3 specific numbers between 1 and 36).

The payoffs for winning bets are:
1 to 1 on odd, even, red and black
2 to 1 on a column
11 to 1 on a street
35 to 1 any one number

1. Compute the probability of the ball landing on:
   a. A red slot
   b. An odd number
   c. The number 0
   d. A street (any of 3 specific numbers)
   e. The number 2
2. Based on playing each of the scenarios above (**a. – e.**) compute the winnings for each bet individually, if $5 is bet each time and all 5 scenarios lead to winnings.
3. If $1 is bet on hitting just one number, what would be the expected payoff? (**Hint:** expected payoff is [*probability of winning* times *payment for a win*] – [*probability of losing* times *payout for a loss*])
4. Given the information in question 3, would you like to play roulette on a regular basis? Why or why not? Why will the casino acquire more money in the long run?

chapter nine summary

CHAPTER SUMMARY

A summary of concepts and skills follows each chapter. Refer to these summaries to make sure you feel comfortable with the material in the chapter. The concepts and skills are organized according to the section title and topic title in which the material is first discussed.

9.1 Sequences and Series

topics	pages
Recursively and Explicitly Defined Sequences • The definition of *infinite* and *finite sequences*, and the notation used to define sequences • The meaning of *explicit* and *recursive* formulas, and how to use them • The process of contriving a formula that reproduces a given number of terms of a sequence	p. 703 – 708
Summation Notation and a Few Formulas • The meaning of *summation notation* using the Greek letter sigma • Converting between summation notation and expanded form • Properties of summations • The use of specific summation formulas	p. 709 – 712
Partial Sums and Series • The meaning of a *series*, and how series are related to sequences • The meaning of a *partial sum* of a series • *Finite series* and *infinite series* • *Convergence* and *divergence* of series	p. 713 – 715
Interlude: Fibonacci Sequences • Using recursive formulas to define Fibonacci sequences	p. 715 – 716

9.2 Arithmetic Sequences and Series

topics	pages
Characteristics of Arithmetic Sequences and Series • The definition of an *arithmetic sequence*, and the meaning of the *common difference* of an arithmetic sequence	p. 720 – 722

chapter nine An Introduction to Sequences, Series, Combinatorics, and Probability

topics (continued)	pages
The Formula for the General Term of an Arithmetic Sequence • The formula for the general term of an arithmetic sequence, and how to determine it	p. 722 – 724
Evaluating Partial Sums of Arithmetic Sequences • The two formulas for partial sums of arithmetic sequences	p. 724 – 727

9.3 Geometric Sequences and Series

topics	pages
Characteristics of Geometric Sequences • The definition of a *geometric sequence*, and the meaning of the *common ratio* of a geometric sequence	p. 732 – 734
The Formula for the General Term of a Geometric Sequence • The formula for the general term of a geometric sequence, and how to use it	p. 734 – 736
Evaluating Partial Sums of Geometric Sequences • The formula for the partial sum of a geometric sequence	p. 736 – 740
Evaluating Infinite Geometric Series • The extension of the partial sum formula to the case of infinite geometric series • How to determine convergence or divergence of infinite geometric series • The correspondence between geometric series and decimal notation	p. 740 – 743
Interlude: Zeno's Paradoxes • The use of geometric series to resolve apparent paradoxes	p. 743 – 744

9.4 Mathematical Induction

topics	pages
The Role of Induction • The mathematical meaning of *induction*	p. 750 – 751
Proofs by Mathematical Induction • Using mathematical induction to prove statements • The conditions for applying proof by mathematical induction • The two steps – *basic step* and *induction step* – for constructing an inductive proof	p. 751 – 754

chapter nine summary

9.5 An Introduction to Combinatorics

topics	pages
The Multiplication Principle of Counting • The meaning of the *Multiplication Principle of Counting*, and its use in counting problems	p. 760 – 763
Permutations • The application of the Multiplication Principle of Counting to count the number of *permutations* of a set of objects • The *Permutation Formula* and the meaning and use of the notation $_nP_k$	p. 763 – 766
Combinations • The connection between the Permutation Formula and the *Combination Formula* • The meaning and use of the notation $_nC_k$	p. 767 – 770
The Binomial and Multinomial Theorems • The formulas for *binomial* and *multinomial coefficients* • The *Binomial* and *Multinomial Theorems*, and their use	p. 770 – 774

9.6 An Introduction to Probability

topics	pages
The Language of Probability • The meaning of *experiment*, *outcome*, *event*, and *sample space* in probability theory • The calculation of the probability of an event when all outcomes are *equally likely*	p. 783 – 785
Using Combinatorics to Compute Probabilities • The use of combinatorics in calculating probabilities	p. 785 – 787
Unions, Intersections, and Independent Events • The meaning of *independence* of events • The meaning of *unions* and *intersections* of events, and the calculation of probabilities of unions of events • The meaning of *mutually exclusive events*	p. 788 – 792

chapter nine An Introduction to Sequences, Series, Combinatorics, and Probability

chapter review

Section 9.1

Find a possible formula for the general term of the sequences that begin as follows.

1. $-7, -1, 5, 11, 17, \ldots$
2. $\dfrac{1}{2}, \dfrac{3}{4}, \dfrac{9}{8}, \dfrac{27}{16}, \dfrac{81}{32}, \ldots$
3. $0, 3, 8, 15, 24, 35, \ldots$

Translate each expanded sum that follows into summation notation, and vice versa. Then use the formulas and properties from this section to evaluate the sums.

4. $\displaystyle\sum_{i=3}^{8}(-2i+3)$
5. $\displaystyle\sum_{i=2}^{7}(-2)^{i-1}$
6. $8+27+64+\ldots+343$

Section 9.2

Find the explicit formula for the general term of the arithmetic sequences described below.

7. $5, 2, -1, -4, -7, \ldots$
8. $a_2 = 14$ and $a_4 = 19$
9. $a_7 = -43$ and $d = -9$

Each of the following sums is a partial sum of an arithmetic sequence; use either formula to determine its value.

10. $\displaystyle\sum_{i=1}^{97}(2i-7)$
11. $\displaystyle\sum_{i=1}^{60}(-4i+3)$

Section 9.3

Each of the following sums is a partial sum of a geometric sequence. Use this fact to evaluate the sums.

12. $\displaystyle\sum_{i=3}^{9} 3\left(\dfrac{1}{2}\right)^i$
13. $5+10+\ldots+20{,}480$

Determine if the following infinite geometric series converge. If a given sum converges, find the sum.

14. $\displaystyle\sum_{i=0}^{\infty} -3\left(\dfrac{3}{4}\right)^i$
15. $\displaystyle\sum_{i=1}^{\infty}\left(-\dfrac{5}{4}\right)^i$
16. $\displaystyle\sum_{i=1}^{\infty} \dfrac{2}{5}\left(\dfrac{5}{7}\right)^i$

Section 9.4

Use the Principle of Mathematical Induction to prove the following statements.

17. $1+4+9+\cdots+n^2 = \dfrac{n(n+1)(2n+1)}{6}$

18. $\dfrac{1}{1\cdot 3}+\dfrac{1}{3\cdot 5}+\dfrac{1}{5\cdot 7}+\cdots+\dfrac{1}{(2n-1)(2n+1)} = \dfrac{n}{2n+1}$

19. $5+8+11+\cdots+(3n+2) = \dfrac{n(3n+7)}{2}$

20. $1\cdot 3+2\cdot 4+3\cdot 5+\cdots+n(n+2) = \dfrac{n(n+1)(2n+7)}{6}$

21. Prove that for all natural numbers n, $11^n - 7^n$ is divisible by 4.

22. Prove that for all natural numbers n, $7^n - 1$ is divisible by 3.

chapter nine An Introduction to Sequences, Series, Combinatorics, and Probability

Section 9.5

Use the Multiplication Principle of Counting and the Permutation and Combination formulas to answer the following questions.

23. A license plate must contain 4 numerical digits followed by 3 letters. If the first digit cannot be 0 or 1, how many different license plates can be created?

24. How many different 7-digit phone numbers do not contain the digits 6 or 7?

25. In how many different orders can the letters in the word "aardvark" be arranged?

26. In how many different ways can first-place, second-place, and third-place be awarded in a 10-person shot put competition?

27. At a meeting of 21 people, a president, vice president, secretary, treasurer, and recruitment officer are to be chosen. How many different ways can these positions be filled?

28. A college admissions committee selects 4 out of 12 scholarship finalists to receive merit-based financial aid. How many different sets of 4 recipients can be chosen?

29. Expand the expression $(x+2y+z)^3$.

Section 9.6

Answer the following probability questions. Be careful to properly identify the sample space and the appropriate event in each case.

30. A card is drawn from a standard 52-card deck. Find the probability of drawing:
    a. a seven or a club
    b. a face card but not a red queen
    c. a black three or a spade

31. What is the probability of being dealt a five-card hand (from a standard 52-card deck) that contains only face cards?

32. There is a 10% chance of rain each individual day for an entire week. What is the probability that it will rain at least once during this seven day period?

chapter nine test

chapter test

Determine the first five terms of the sequences whose n^{th} terms are defined as follows.

1. $a_n = -2n - 1$
2. $a_n = (-3)^n$
3. $a_n = -n^2$
4. $a_n = (-1)^n \sqrt[3]{n}$
5. $a_n = 1 + 2 + 3 + \ldots + n$
6. $a_1 = -3, a_{n-1} = a_n + 1$ for $n \geq 2$
7. $a_1 = 1$ and $a_n = \dfrac{a_{n-1}}{n}$, for $n \geq 2$
8. $a_n = \dfrac{n!}{n^n}$

Find a possible formula for the general n^{th} term of the sequences that begin as follows. Answers may vary.

9. $-1, 2, 7, 14, 23, \ldots$
10. $\dfrac{3}{2}, \dfrac{5}{3}, \dfrac{7}{4}, \dfrac{9}{5}, \dfrac{11}{6}, \ldots$
11. $-1, \dfrac{1}{4}, -\dfrac{1}{9}, \dfrac{1}{16}, -\dfrac{1}{25}, \ldots$

Translate each summation that follows into an expanded sum. Then use the formulas to evaluate the sums.

12. $\displaystyle\sum_{k=1}^{6}(2k-3)$
13. $\displaystyle\sum_{k=1}^{9}-2k^2$
14. $\displaystyle\sum_{i=1}^{5}-4(2^i)$
15. $\displaystyle\sum_{i=1}^{12}(1+i)(1-2i)$

Find a formula for the nth partial sum S_n of each of the following series, and determine the sum if possible.

16. $\displaystyle\sum_{n=1}^{100}\left(\dfrac{1}{n+1} - \dfrac{1}{n+2}\right)$
17. $\displaystyle\sum_{n=1}^{\infty}\left(\dfrac{1}{3n} - \dfrac{1}{3(n+1)}\right)$
18. $\displaystyle\sum_{n=1}^{80}\ln\left(\dfrac{n+1}{n+2}\right)$

19. Determine the first five terms of the following generalized Fibonacci sequence:
$a_1 = 3, a_2 = 6$ and $a_n = a_{n-1} + a_{n-2}$, for $n \geq 3$

20. Determine the first terms of the following recursively defined sequence:
$a_1 = 4, a_2 = -2$ and $a_n = 2a_{n-1} + a_{n-2}$, for $n \geq 3$

chapter nine An Introduction to Sequences, Series, Combinatorics, and Probability

Find the explicit formula for the general nth term of the arithmetic sequences described below.

21. $-1, -5, -9, -13, -17, \ldots$

22. $a_1 = 2, a_4 = 11$

23. $a_2 = 4, d = 8$

24. $a_9 = \dfrac{13}{2}, d = \dfrac{3}{4}$

Find the explicit formula for the general term of each geometric sequence described below.

25. $4, 12, 36, 108, 324, \ldots$

26. $a_1 = -3, r = 2$

27. $a_3 = 50, a_5 = 1250$

28. $a_2 = \dfrac{6}{5}, r = \dfrac{2}{3}$

Determine if the following geometric series converge. If a given sum converges find the sum.

29. $\displaystyle\sum_{n=0}^{\infty} \dfrac{1}{3}\left(\dfrac{1}{3}\right)^n$

30. $\displaystyle\sum_{n=0}^{\infty} \left(\dfrac{5}{2}\right)^n$

31. Given $\displaystyle\sum_{k=1}^{n} a_k = n^2 + 3n$, let $n = 6$ and solve.

Use the Principle of Mathematical Induction to prove the given property for all positive integers n.

32. If $x_1 \neq 0, x_2 \neq 0, \ldots, x_n \neq 0$, then $(x_1 \cdot x_2 \cdot x_3 \cdot \ldots \cdot x_n)^{-1} = x_1^{-1} \cdot x_2^{-1} \cdot x_3^{-1} \cdot \ldots \cdot x_n^{-1}$.

33. Prove that $n! > 2^n$ for all $n \geq 4$.

Prove the following formula for every positive integer n, using the Principle of Mathematical Induction.

34. $2 + 4 + 6 + 8 + \ldots + 2n = n(n+1)$

35. $2 + 7 + 12 + 17 + \ldots + (5n - 3) = \dfrac{n}{2}(5n - 1)$

chapter nine test

Use the Multiplication Principle of Counting to answer the following questions.

36. How many different 3-digit even numbers (digits are different from each other) can be written using digits 0, 1, 2, 3, 4, 5?

37. How many different 7-digit telephone numbers are possible within each area code in the USA? (Ignore the fact that certain 7-digit sequences are disallowed as phone numbers.)

38. The insurance form you are filling out online has a dropdown box containing every possible set of initials (first and last initials only). How many sets of initials does the dropdown box contain?

39. How many different 6-character license plates can be formed if the first 2 places must be letters and last 4 places must be digits? (Assume repetition is not allowed.)

40. How many different 7-letter "words" (they don't have to be actual English words) can be formed from the letters in the word MARMARA?

41. A restaurant serves 3 different salads, 10 different entrees, and 6 different desserts. How many different meals could be created, assuming each meal consists of all three courses?

42. There are 12 questions on a test and you only need to answer 10 questions. If you are required to answer the first 3 questions, how many different ways can you answer 10 questions on the test?

43. Expand the expression $(a-3b)^6$.

44. Expand the expression $(1-2j)^8$.

45. Expand the expression $(2b-2c)^3$.

46. Find the probability of obtaining at least one tail when a coin is tossed five times.

47. A man has five pairs of socks of which no two pairs are the same color. If he randomly selects two socks from a drawer, what is the probability that he gets a matched pair?

48. Are the chances of rolling a total of 3 with one die the same as the chances of rolling a total of 6 with two dice? If not, which has the higher probability?

chapter nine An Introduction to Sequences, Series, Combinatorics, and Probability

49. A sample of college students, faculty, and administration were asked whether they favored a proposed increase in the annual activity fee to enhance student life on campus. The results of the study are given in the following table.

	Students	Faculty	Admin	Total
Favor	237	37	18	292
Oppose	163	38	7	208
Total	400	75	25	500

A person is selected at random from the sample. Find each of the following specified probabilities.
   a. The person is not in favor of the proposal.
   b. The person is a student.
   c. The person is a faculty member and is in favor of the proposal.

50. There are 5 red, 4 black, and 3 yellow pencils in a box. Three pencils are selected without replacement at random from the box. Find each of the following specified probabilities.
   a. Each one is a different color.
   b. All three are the same color.
   c. All three are red.
   d. 2 are yellow, and 1 is black.

Appendix A: Fundamentals of *Mathematica*®

Mathematica is a very powerful and flexible software package with a wide variety of uses. To begin with, *Mathematica* (along with similar products such as *Maple*®, *Matlab*®, and *Derive*™) can be viewed as a sort of super-calculator. It understands the rules of algebra, has a huge number of built-in functions ranging from the trivial to the exotic, and is very good at generating high-quality graphs in one, two, and three dimensions. Beyond that, a package such as *Mathematica* is also a programming environment; it is this aspect of *Mathematica* that allows the user to extend its capabilities to suit specialized needs.

The applications in this text require only a basic understanding of *Mathematica*, and this appendix will serve as a quick guide to its use. It should thus be noted that a complete guide to the use of *Mathematica* can be found within *Mathematica* itself. Once it is installed and running on your computer, you have access to an electronic version of a very large *Mathematica* manual, all available by first clicking on the "Help" button located in the top toolbar (see Figure 1).

Figure 1: Getting On-Screen Help in *Mathematica*

After clicking on "Help" a drop-down menu appears. The full selection of Help categories appears by selecting "Help Browser," but a good place to start is with "Getting Started/Demos..." This leads to many useful examples of how Mathematica can be used to solve different sorts of problems.

At first, you will probably be making use of built-in *Mathematica* commands such as **Solve** and **Simplify** (as opposed to using your own user-defined commands). It is important to realize that *Mathematica* is case-sensitive, and that all built-in commands begin with a capital letter. Once a command has been typed in, you'll need to tell *Mathematica* to execute it. This can be done in one of two ways—either by pressing the "Shift" and "Enter" keys together (known as "Shift-Enter") or, if you are using an extended keyboard, by using the "Enter" key that appears in the numeric keypad area.

Pressing the normal "Enter" key alone will simply move the cursor to the next line and allow you to continue typing, but will not execute any commands.

appendix a Fundamentals of *Mathematica*

Each time you press "Shift-Enter" *Mathematica* will execute all the commands contained in a single cell. Different *Mathematica* cells are demarcated by blue bracketing symbols along the right-hand edge of the work area, and you can always start a new cell by positioning the mouse cursor over a blank part of the area and clicking the left mouse button once.

The remainder of this appendix contains examples of a few of the basic *Mathematica* commands used in this text, arranged in the order in which they appear. For instant on-screen help on any command, type the command into *Mathematica* and then press the "F1" key. Doing so will bring up the relevant help pages and, more often than not, provide more examples of how the command is used.

Simplify:

The **Simplify** command is used to simplify mathematical expressions according to the usual rules of algebra. The basic syntax is **Simplify[*expr*]**, where *expr* is the expression to be simplified. Note the examples in the two cells shown in Figure 2 below.

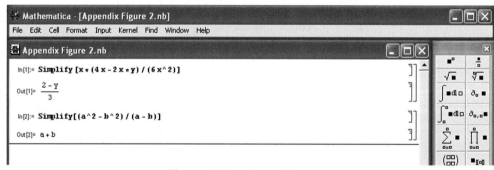

Figure 2: The Use of Simplify

Expand:

This command is used to multiply out factors in an expression. The syntax for the command is **Expand[*expr*]**. Figure 3 shows the use of the command in multiplying out the expression $(x-y)^5$.

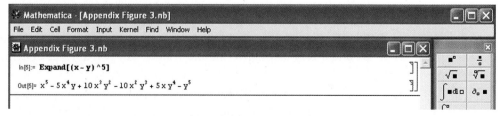

Figure 3: The Use of Expand

Factor:

The **Factor** command is the reverse of the **Expand** command when applied to polynomials. Its basic usage is **Factor**[*poly*], where *poly* is a polynomial expression to be factored.

Solve:

The **Solve** command is very powerful, and is used in several different ways in this text. Its basic usage is **Solve**[*eqns*, *vars*], where *eqns* represents one or more equations and *vars* represents one or more variables. If more than one equation is to be solved, the collection of equations must be enclosed in a set of braces, separated by commas. Similarly, if more than one variable is to be solved for, the variables must be enclosed in a set of braces. Figure 4 shows the use of **Solve** to first solve one equation for one variable, and then to solve a collection of three equations for all three variables. Note how *Mathematica* expresses the solution in each case.

It is important to note that equations in *Mathematica* are expressed with two "=" symbols, as seen in Figure 4. The use of just one "=" is reserved for assigning a permanent value to something. For instance, the expression **x = 3** assigns the value of 3 to the symbol *x*, while the expression **x = = 3** represents the equation $x = 3$ in *Mathematica*.

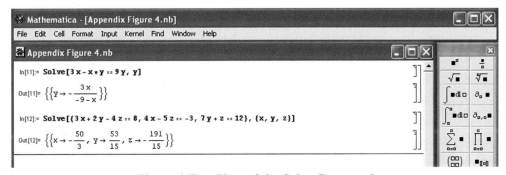

Figure 4: Two Uses of the Solve Command

Together:

The **Together** command is used primarily to combine two or more expressions into one with a common denominator, automatically canceling any common factors that may appear.

appendix a Fundamentals of *Mathematica*

Defining functions:

A few rules of syntax must be observed in order to define your own functions in *Mathematica*. The first is that each variable serving as a place-holder in the definition must be followed by the underscore symbol "_" when it appears on the left side of the definition and without the underscore when it appears on the right. The second rule is that ":=" is used in the definition, as opposed to "=" (see the on-screen *Mathematica* help for detailed explanations of these rules). Figure 5 illustrates the definition of the two functions $f(x) = x^2 + 5$ and $g(x, y) = 3x - 7y$, followed by an evaluation of each.

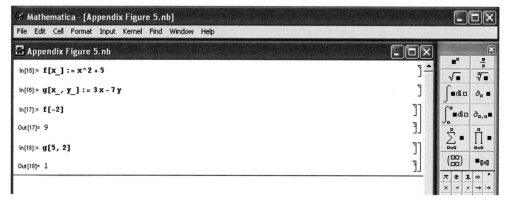

Figure 5: Defining Functions in *Mathematica*

Plot:

The **Plot** command is another very versatile command. Its basic usage is **Plot** [*f*, {*x*, *xmin*, *xmax*}], where *f* is an expression in *x* representing a function to be plotted and *xmin* and *xmax* define the endpoints of the interval on the *x*-axis over which *f* is to be graphed. However, the **Plot** command also recognizes many options that modify the details of the resulting picture; these options are best explored via the on-screen help.

ImplicitPlot:

The **ImplicitPlot** command (all one word) is used to graph equations in two variables. To use it, you must first execute the command **<<Graphics`ImplicitPlot`**, which extends the collection of built-in commands in *Mathematica* (execute this command only once per *Mathematica* session). Aside from this necessary first step, the usage of **ImplicitPlot** is similar to that of **Plot**: the syntax is **ImplicitPlot** [*eqn*, {*x*, *xmin*, *xmax*}], where *eqn* is an equation in *x* and *y*. Figure 6 illustrates its use.

appendix a Fundamentals of *Mathematica*

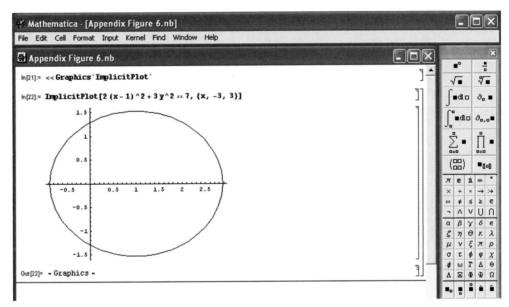

Figure 6: The ImplicitPlot Command

Defining and working with matrices:

A matrix in *Mathematica* is defined as a list consisting of a number of sub-lists. Each sub-list corresponds to a row of the matrix, so an $m \times n$ matrix appears in *Mathematica* as a list of m sub-lists, each of length n. The **MatrixForm** command can be used to view a matrix in a more familiar format. The command **Det** is used to evaluate the determinant of a matrix, and the command **Inverse** finds the inverse of a given matrix. The use of these commands is illustrated in Figure 7.

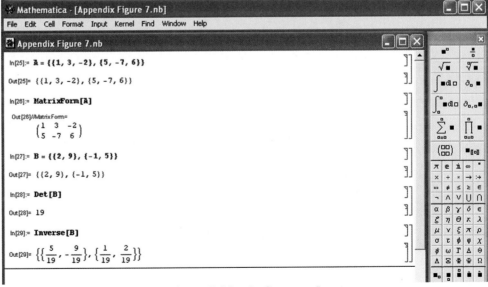

Figure 7: Matrix Commands

813

chapter one

section 1.1, pages 11-14

1. (a) 19, 2^5 **(b)** 19, $\frac{0}{15}$, 2^5 **(c)** 19, $\frac{0}{15}$, 2^5, -33 **(d)** 19, -4.3, $\frac{0}{15}$, 2^5, -33 **(e)** $-\sqrt{3}$ **(f)** all except $\frac{15}{0}$ **(g)** $\frac{15}{0}$
3. (a) $|-16|$, $\frac{12}{3}$, $\sqrt{4}$ **(b)** $|-16|$, $\frac{12}{3}$, 0, $\sqrt{4}$ **(c)** $|-16|$, $\frac{12}{3}$, 0, $\sqrt{4}$ **(d)** all **(e)** none **(f)** all **(g)** none
5. ←•—•—•→ at $-4.5, -1, 2.5$ **7.** ←•—•—•→ at $-24, 2, 15$ **9.** $>, \geq$ **11.** $\leq, \geq$ **13.** $>, \geq$ **15.** $\leq, \geq$ **17.** $<, \leq$ **19.** $2 \leq x$
21. $7 \leq 9$ **23.** $2c \leq 3d$ **25.** $6 + x \geq 4x$ **27.** $\{n \mid n \text{ is an integer and } 5 \leq n \leq 105\}$ **29.** $\{2^n \mid n \text{ is a whole number}\}$
31. $\{n \mid n \text{ is a whole number}\}$ **33.** $[-9, 6]$ **35.** $[-3, 19)$ **37.** $(0, \infty)$ **39.** $[1, 2]$ **41.** -11 **43.** -5 **45.** 2 **47.** $\sqrt{2}$
49. 35 **51.** $\sqrt{7} - 2$ **53.** 5 **55.** 0 **57.** 14 **59.** $\frac{7}{3}$ **61.** $\frac{41836}{99999}$ **63.** $-\frac{100}{99}$ **65.** $-3.\bar{6}$ **67.** $0.\overline{857142}$ **69.** JR > Freddie > Sarah > Aubrey > Elizabeth **71.** If age = x, $\{x \mid x < 2\} = [0, 2) \to$ free; $\{x \mid 2 \leq x < 12\} = [2, 12) \to \$3$; $\{x \mid 12 \leq x < 65\} = [12, 65) \to \$7$; $\{x \mid x \geq 65\} = [65, \infty) \to \$5$ **73.** No, because all natural numbers can be expressed as fractions. **75.** Answers may vary.

section 1.2, pages 23-27

1. $3x^2y^3$, $-2\sqrt{x+y}$, $7z$ **3.** $-2, \sqrt{x+y}$ **5.** $1, 8.5, -14$ **7.** $\frac{-5x}{2yz}$, $-8x^5y^3$, $6.9z$ **9.** $\frac{-5}{2}, \frac{1}{y}, \frac{1}{z}, x$ **11.** $6\pi + 8$ **13.** 1
15. -6 **17.** -20 **19.** -12 **21.** commutative **23.** distributive **25.** associative **27.** multiplicative inverse
29. associative **31.** multiplicative cancellation; $-\frac{1}{14}$ **33.** additive cancellation; $-2x + y$ **35.** zero-factor property **37.** zero-factor property **39.** additive cancellation; $\frac{-1}{4}y^9 + z$ **41.** -2 **43.** $-\frac{11}{2}$ **45.** $36 - 11\pi$
47. $-5 - 7\pi$ **49.** $\frac{-66}{59}$ **51.** $-9 - 2\pi$ **53.** $\frac{-154}{65}$ **55.** -2.50 **57.** -30.54 **59.** $\sqrt[4]{(-2)(-6+4)^3}$ **61.** $(-5, 4]$
63. $[3, 4]$ **65.** $\mathbb{Z}$ **67.** $[-\pi, 21)$ **69.** $(3, 9)$ **71.** $\mathbb{Z}$ **73.** $\$66$ **75.** $\$102$ **77.** 2.19 square meters **79.** It is the same number you began with. Explanations may vary. **81, 83.** Answers may vary.

section 1.3, pages 39-43

1. 16 **3.** -9 **5.** 81 **7.** 64 **9.** 1 **11.** $\frac{1}{7}$ **13.** x^3 **15.** $27s^{10}$ **17.** -2 **19.** x^3 **21.** $121x^7$ **23.** x **25.** $\frac{1}{x^2}$
27. x^3y^3 **29.** $\frac{16}{s^3}$ **31.** $-\frac{y^5}{3x^2}$ **33.** $\frac{1}{3y^2z}$ **35.** $27x^2y^4$ **37.** 1 **39.** $\frac{c^2}{9a^7b^3}$ **41.** $\frac{81y^3z^2}{2x^3}$ **43.** $\frac{64a^6}{b^{15}}$
45. $27x^9$ **47.** $\frac{1}{5z^6 - 81x^{12}}$ **49.** -0.0000176 **51.** 2.1×10^{-7} **53.** 5.1×10^3 **55.** 312.12 **57.** 2.587×10^{-8}
59. 3.1536×10^7 **61.** 6.75×10^5 **63.** 2.605×10^{-7} **65.** 1.2×10^{-9} **67.** 50 **69.** 3.2×10^{-6} **71.** 1.2×10^{13}
73. Answers may vary. **75.** $7s$ **77.** $\pi r^2 h$ **79.** 585 m^3 **81.** $81\pi d \text{ ft}^3$ **83.** $2\pi r^2$ **85.** Answers may vary.

section 1.4, pages 56-60

1. -3 **3.** not real **5.** -2 **7.** -5 **9.** not real **11.** $\dfrac{-3}{5}$ **13.** $-\dfrac{1}{2}$ **15.** 2 **17.** $\dfrac{2}{5}$ **19.** $3|x|$ **21.** $\dfrac{x^2|z|}{2}$ **23.** $x^2y^7z^3$
25. $\dfrac{ab^4}{3c^2}$ **27.** $\dfrac{|x^3|y^2}{2}$ **29.** $\dfrac{y^6 z^5}{2x^7}$ **31.** $\dfrac{\sqrt[3]{36x^2y^2}}{3y^2}$ **33.** $-\sqrt{2}-\sqrt{5}$ **35.** $\sqrt{6}+\sqrt{3}$ **37.** $\dfrac{x+\sqrt{2x}}{x-2}$ **39.** $\dfrac{x+2\sqrt{xy}+y}{x-y}$
41. $\dfrac{y-2\sqrt{y}}{y-4}$ **43.** $\dfrac{1}{\sqrt{5}+3}$ **45.** $\dfrac{9-y}{18-6\sqrt{y}}$ **47.** $\dfrac{1}{\sqrt{13}-\sqrt{t}}$ **49.** $\dfrac{6-y}{6+y-2\sqrt{6y}}$ **51.** $3x\sqrt[3]{2x}$ **53.** not possible
55. 0 **57.** $4z\sqrt[3]{2z}$ **59.** 0 **61.** $(3x^2-4)^2$ **63.** 27 **65.** n^2 **67.** $\dfrac{x^{\frac{4}{5}}}{y^{\frac{5}{3}}}$ **69.** $\dfrac{1}{125}$ **71.** $y\sqrt[3]{y^2}$ **73.** $\dfrac{1}{a}$ **75.** $a^{\frac{15}{4}}$ **77.** $x^{\frac{1}{4}}$
79. $6^{-\frac{1}{3}}$ **81.** $\sqrt[4]{125}$ **83.** $\sqrt[4]{|y|}$ **85.** x^3 **87.** $\sqrt[16]{16807}$ **89, 91.** Answers may vary. **93.** $3d^2\sqrt{3}$; 3.326 cm²
95. 1651 cm²; no **97.** 2.998×10^8 m/s **99.** Because a root is the same as a fractional exponent.

section 1.5, pages 71-75

1. not a polynomial **3.** degree 11 polynomial of four terms **5.** degree 0 monomial **7.** degree 4 binomial
9. degree 2 trinomial **11.** degree 5 binomial **13.** $\pi z^5 + 8z^2 - 2z + 1$ **(a)** 5 **(b)** π **15.** $2s^6 - 10s^5 + 4s^3$ **(a)** 6
(b) 2 **17.** $9y^6 - 3y^5 + y - 2$ **(a)** 6 **(b)** 9 **19.** $-x^{13} + 7x^{11} - 4x^{10} + 9$ **(a)** 13 **(b)** -1 **21.** $-4x^3y - 6y - x^2z$
23. $x^2y + xy^2 + 6x - 6y$ **25.** $-3ab$ **27.** $xy^2 - x^2y - y$ **29.** $3a^3b^3 + 21a^3b^2 + 2a^2b^2 + 14a^2b - 3ab^3 - 21ab^2$
31. $3a^2 - 2ab - 8b^2$ **33.** $6x^2 + 33xy - 18y^2$ **35.** $7y^4 - 34xy^2 - 5x^2$ **37.** $6x^3y^3 - 3x^3y + 36x^2y^3 + 4x^2y^2 - 18x^2y + 24xy^2$
39. $(3a^2b)(1+a-3b)$ **41.** $2x(x^5 - 7x^2 + 4)$ **43.** $3x^3y(1-3x+4y)$ **45.** $4y^2(3y^4 - 2 - 4y^3)$ **47.** $(a^2+b)(a-b)$
49. $z(1+z)(1+z^2)$ **51.** $(n-2)(x^2+y)$ **53.** $(a-5b)(x+5y)$ **55.** $(5x^2y-3)(5x^2y+3)$ **57.** $(x-10y)(x^2+10xy+100y^2)$
59. $(m^2+5n^3)(m^4-5m^2n^3+25n^6)$ **61.** $(3x^2-2y^4z)(9x^4+6x^2y^4z+4y^8z^2)$ **63.** $(4y^2z-3x^4)(4y^2z+3x^4)$
65. $(7y^3+3xz^2)(49y^6+21xy^3z^2+9x^2z^4)$ **67.** $(x+5)(x-3)$ **69.** $(x-1)^2$ **71.** $(x-2)^2$ **73.** $(y+7)^2$
75. $(x+11)(x+2)$ **77.** $(y-8)(y-1)$ **79.** $(5a+3)(a-8)$ **81.** $(x+6)(5x-3)$ **83.** $(16y-9)(y-1)$
85. $(4a-3)(2a+1)$ **87.** $(4y-5)(3y-1)$ **89.** $2x(2x-1)^{\frac{-3}{2}}$ **91.** $a^{-3}(7a^2-2b)$ **93.** $2y^{-5}(5y^3-x)$
95. $(5x+7)^{\frac{4}{3}}(5x+6)$ **97.** $y^{-4}(7y^3+5)$ **99.** No; polynomials can only have variables with positive exponents and having a variable in the denominator is the same as having a variable with negative exponents.
101. a. Yes; degree = 4; leading coefficient = 2; terms = 4 **b.** Yes; degree = 3; leading coefficient = 2; terms = 3

section 1.6, pages 83-84

1. $5i$ **3.** $-3i\sqrt{3}$ **5.** $4i\sqrt{2x}$ **7.** $i\sqrt{29}$ **9.** $1-3i$ **11.** $9+2i$ **13.** $8-6i$ **15.** -9 **17.** $-5+6i$ **19.** 10 **21.** 97 **23.** 6
25. 1 **27.** $-1-4i$ **29.** $7i$ **31.** $3+i$ **33.** $-i$ **35.** $-i$ **37.** $10-2i$ **39.** $\dfrac{14}{37}+\dfrac{10}{37}i$ **41.** $\dfrac{21}{17}-\dfrac{1}{17}i$ **43.** $-5+2i\sqrt{6}$
45. 8 **47.** $-\dfrac{7}{3}i$ **49.** $22+10i\sqrt{3}$ **51.** $6+3j$ ohms **53.** $11-2j$ ohms

chapter one review, pages 91-92

1. $\dfrac{23}{3}$ **2.** $-\dfrac{337}{165}$ **3.** -9 **4.** -7 **5.** $\sqrt{11}-\sqrt{5}$ **6.** 6 **7.** $\dfrac{4\pi}{3}-36$ **8.** multiplicative cancellation; $\dfrac{1}{4}$ **9.** zero-factor property **10.** $\dfrac{-t^9}{2s^7}$ **11.** $\dfrac{27z^3}{y^6}$ **12.** $\dfrac{a^2}{8b^4c^3}$ **13.** $(2x-5)(x+3)$ **14.** -0.0002004 **15.** 5.224×10^7 **16.** $\dfrac{2y\sqrt[3]{9x^2y}}{3}$ **17.** $\dfrac{|a|}{2}$ **18.** $-\sqrt{2}-\sqrt{6}$ **19.** $4|x|$ **20.** $\dfrac{-4y}{x^3}$ **21.** $\dfrac{a^2\sqrt[4]{a}}{3|b|}$ **22.** $(x-2y)(3m+n)$ **23.** $(6x^3+y)(6x^3-y)$ **24.** $(x+3)(x-4)$ **25.** $(2a+1)(3a-5)$ **26.** $(2x-5)(x+3)$ **27.** $(2a+3b^2)(2a-3b^2)$ **28.** $2+7i$ **29.** $5+9i$ **30.** $-30+10i$ **31.** $-\dfrac{7}{25}+\dfrac{24}{25}i$ **32.** $4+i$ **33.** $-\dfrac{1}{5}+\dfrac{3}{5}i$

chapter one test, pages 93-95

1. True; if $|x|=-x$ then $x<0$, so x minus any number would also be less than 0. **2.** $a=-5$ **3.** 511 **4.** -6000 **5.** $\dfrac{1}{5}$ **6.** 0 **7.** -129 **8.** $\dfrac{-1}{729}$ **9.** -8 **10.** $\dfrac{b}{512}$ **11.** $64x^8y^4$ **12.** 4.152×10^{12} **13.** -3.6×10^{-3} **14.** 2.0×10^{-8} **15.** 2.0×10^{-13} **16.** 4.0×10^{-4} **17.** $\sqrt{3}>\sqrt[3]{4}>\sqrt[6]{15}$ **18.** -11 **19.** -5 **20.** 5 **21.** $9i$ **22.** 4 **23.** 3 **24.** $\sqrt[6]{3}$ **25.** -16 **26.** 1 **27.** a^2+3b^2-5ab **28.** $5x+6y$ **29.** a^3+a^2-5a-3 **30.** y^2-y **31.** $-2a$ **32.** $\dfrac{\sqrt{2}\cdot\sqrt[3]{4}}{2}$ **33.** $\sqrt{5}-\sqrt{2}$ **34.** $\dfrac{2(\sqrt{a}+1)}{a-1}$ **35.** $-2\sqrt{3}$ **36.** $-2\sqrt{2}$ **37.** 1 **38.** $62-20\sqrt{6}$ **39.** $\sqrt[12]{2^7}\sqrt{3}$ **40.** $(2x-1)^2$ **41.** 1 **42.** $-12x^2+9x-15$ **43.** 3 **44.** $4x^2-6x+2$ **45.** $3xy(x-3y+2)$ **46.** $(a^2-b)(a^2-b+1)$ **47.** $2y^2(y^2-4y-5)$ **48.** $(x-y)(x+y)(x^4+y^2)$ **49.** $(m+1)^2(m-1)$ **50.** $(3b+2)(2a-1)$ **51.** $(4ab^2-3c)(4ab^2+3c)$ **52.** $4a(c-b)$ **53.** $y(y-2z)(y+2z)$ **54.** $(a-5b)(a^2+5ab+25b^2)$ **55.** $(4x^2+3y^3)(16x^4-12x^2y^3-9y^2)$ **56.** $(x-3)(x+2)$ **57.** $(x+6)(x-1)$ **58.** $(x+2)(x+4)$ **59.** $(x+5)^2$ **60.** $(x-9)(x+4)$ **61.** $-(x-7)^2$ **62.** $(x^2+4)(x^2-3)$ **63.** $(x^3-1)^2$ **64.** $(3b^2-2a)^2$ **65.** $(3x-2)(2x+3)$ **66.** $(2x-y)(x+y)$ **67.** $x(x-y)(2x+y)$ **68.** $a^{-\frac{3}{2}}(a-1)$ **69.** $3(a-1)^{-3}(3a-5)$ **70.** $(3x-2y)^{\frac{2}{3}}\left[(3x-2y)^{\frac{2}{3}}-1\right]$ **71.** $(2x-3y)^{-4}\left[2^{-2}(2x-3y)^2-1\right]$ **72.** 1 **73.** $-1+2i$ **74.** $-2\sqrt{2}i$ **75.** $-6i\sqrt{3x}$ **76.** $-3i$ **77.** $-108i$ **78.** $11-2i$ **79.** 5 **80.** $\dfrac{-4}{5}-\dfrac{3}{5}i$ **81.** $\dfrac{3}{13}-\dfrac{2}{13}i$

chapter two

section 2.1, pages 106-110

1. $t=-5$ **3.** $y=-1$ **5.** $w=-3$ **7.** $\mathbb{R}$ (Identity) **9.** $\varnothing$ (Contradiction) **11.** $m=7$ **13.** $x=3.7$ **15.** $x=1.05$ **17.** $y=-5$ **19.** $\mathbb{R}$ (Identity) **21.** $\mathbb{R}$ (Identity) **23.** $x=3$ **25.** $\varnothing$ (Contradiction) **27.** $y=-\dfrac{1}{3},-3$ **29.** $x=\dfrac{1}{3}$ **31.** $x=-311{,}420$ **33.** $x=-\dfrac{4}{5},2$ **35.** $\varnothing$ (Contradiction) **37.** $x=-2,2$ **39.** $x=5$ **41.** $x=-\dfrac{1}{2}$ **43.** $x=\dfrac{1}{4}$

section 2.1 answers

45. $x = \dfrac{1}{7}$ **47.** $r = \dfrac{C}{2\pi}$ **49.** $a = \dfrac{v^2 - v_0^2}{2x}$ **51.** $F = \dfrac{9}{5}C + 32$ **53.** $h = \dfrac{A - 2lw}{2w + 2l}$ **55.** $m = \dfrac{2K}{v^2}$ **57.** $x = -9.95$
59. $x = 764.82$ **61.** $\dfrac{19}{3}$ hours, or 6 hours and 20 minutes **63.** 13.5 miles **65.** $390 **67.** 7.5% **69.** 26 feet by 26 feet **71.** 53, 55, and 57 **73.** 36.4%

section 2.2, pages 118-121

1. $\{-9, 3.14, -2.83, 1, -3, 4\}$ **3.** $\{-2.83, 1, -3\}$ **5.** $(-\infty, -3]$ **7.** $(-\infty, 4.8)$
9. $(-\infty, 2.25)$ **11.** $\left(-\infty, \dfrac{3}{2}\right)$ **13.** $\left(-\infty, -\dfrac{3}{11}\right]$ **15.** $(7, \infty)$
17. $(35, \infty)$ **19.** $(-3, \infty)$ **21.** $(-0.11, \infty)$
23. $(1, 5]$ **25.** $(-10, 6]$ **27.** $[-8, -2)$
29. $(-25, -15]$ **31.** $\left[\dfrac{13}{2}, 16\right)$ **33.** $\left(-\dfrac{5}{3}, 1\right)$
35. $\left(-\infty, -\dfrac{7}{2}\right) \cup \left(\dfrac{15}{2}, \infty\right)$ **37.** $\left(-\infty, \dfrac{1}{2}\right) \cup \left(\dfrac{5}{2}, \infty\right)$ **39.** $\varnothing$
41. $(-\infty, 2) \cup (6, \infty)$ **43.** $\varnothing$ **45.** $\varnothing$ **47.** $[-4, 0]$ **49.** $(3, 15)$
51. $(-1, 3]$ **53.** $(-\infty, \infty)$ **55.** $[-2, 3)$
57. $[73, 113]$ for an A, $(113, 115]$ for an A+. **59.** $(1140, 1600]$

section 2.3, pages 132-135

1. $\left\{\dfrac{3}{2}, -1\right\}$ **3.** $\{7\}$ **5.** $\left\{\dfrac{-3}{2}, -3\right\}$ **7.** $\{-3, 1\}$ **9.** $\{2\}$ **11.** $\{3, 11\}$ **13.** $\{0, 6\}$ **15.** $\left\{\dfrac{3}{8}\right\}$ **17.** $\{17, 19\}$
19. $\left\{\dfrac{1}{2} \pm \sqrt{2}\right\}$ **21.** $\{-5, 9\}$ **23.** $\left\{\dfrac{6}{5}, 6\right\}$ **25.** $\{-5, -3\}$ **27.** $\left\{-5, \dfrac{3}{2}\right\}$ **29.** $\{-9, -1\}$ **31.** $\{-13, 5\}$ **33.** $\{-16, -6\}$
35. $\left\{-\dfrac{4}{3}, 1\right\}$ **37.** $\{0.17 \pm 0.86i\}$ **39.** $\left\{0, \dfrac{2}{3}\right\}$ **41.** $\left\{-\dfrac{17}{7}, 3\right\}$ **43.** $\{-4.5 \pm 4.5i\}$ **45.** $\{-14, -6\}$ **47.** $\{2, 14\}$
49. $\left\{\dfrac{-1 \pm \sqrt{7}}{2}\right\}$ **51.** $\{-3, 9\}$ **53.** $\{-16, 12\}$ **55.** $\{-7, -6\}$ **57.** $\{0, 6\}$ **59.** $\{-1, 2\}$ **61.** 3 seconds **63.** 3 seconds
65. 2.6 seconds **67.** $(3x - 1 - \sqrt{5})(3x - 1 + \sqrt{5})$ **69.** $(5x - 1 - i)(5x - 1 + i)$

section 2.4, pages 141-142

1. $\{-3, 4\}$ **3.** $\{8, 13\}$ **5.** $\{\pm\sqrt{2}, \pm i\sqrt{5}\}$ **7.** $\{1 \pm 2i, 1 \pm \sqrt{3}\}$ **9.** $\left\{\dfrac{1}{8}, 27\right\}$ **11.** $\{\pm 2i, \pm 3\}$ **13.** $\{-1, \pm 2, 3\}$
15. $\left\{1, -\dfrac{8}{27}\right\}$ **17.** $\{-1, -2, -3\}$ **19.** $\{\pm 1, 3\}$ **21.** $\left\{-\dfrac{5}{2}, 0, 3\right\}$ **23.** $\{\pm 2, \pm 5i\}$ **25.** $\left\{\pm 2, -\dfrac{6}{5}\right\}$ **27.** $\left\{\pm\dfrac{3}{2}, \pm\dfrac{3i}{2}\right\}$

29. $\left\{-\frac{5}{2}, 0, \frac{4}{7}\right\}$ **31.** $\left\{-\frac{4}{3}, \frac{2 \pm 2i\sqrt{3}}{3}\right\}$ **33.** $\left\{-3, \frac{3 \pm 3i\sqrt{3}}{2}\right\}$ **35.** $\left\{\frac{5}{2}\right\}$ **37.** $\left\{0, \frac{1}{2}, 2\right\}$ **39.** $\{0, 2, 3\}$ **41.** $\{0, 3\}$
43. $\left\{0, \frac{3}{5}, 8\right\}$ **45.** $\left\{-3, -\frac{13}{4}\right\}$ **47.** $h = -4, c = -12,$ and $d = 0$ **49.** $a = 1, c = 36,$ and $d = -144$ **51.** $a = 15,$ $b = -16,$ and $c = -5$

section 2.5, pages 152-156

1. $\frac{2x+1}{x-5}, x \neq -3, 5$ **3.** $x(x-1), x \neq -3$ **5.** $\frac{x+6}{x+5}, x \neq -5, 1$ **7.** $\frac{1}{x^2-x+1}, x \neq -1$ **9.** $2x+1, x \neq -5$
11. $2x-3, x \neq -7$ **13.** $\frac{x^3+9x^2+11x+19}{(x-3)(x+5)}$ **15.** $\frac{13x}{(x-3)(x+5)}$ **17.** $\frac{x^3+4x^2-7x+18}{(x+3)(x-3)}$ **19.** $\frac{x^2+11x+17}{x+3}$
21. $\frac{x+2}{x-6}$ **23.** $y-1$ **25.** $(x+2)(2x+3)$ **27.** $\frac{y-8}{y+8}$ **29.** $5y^2-2y-3$ **31.** -6 **33.** $\frac{x^2+9}{6x-3}$ **35.** $\frac{2x^2}{x+1}$ **37.** $\frac{s-r}{r^2s+s}$
39. $\frac{m+n}{mn}$ **41.** $\frac{x}{y}$ **43.** x^2y^2 **45.** $\frac{11x}{7y}$ **47.** $\frac{5z-3x}{z^2}$ **49.** $\{3 \pm \sqrt{10}\}$ **51.** $\{\pm i\}$ **53.** $(-\infty, -3) \cup (-3, 3) \cup (3, \infty)$
55. $\{-2, 2\}$ **57.** $(-\infty, -2) \cup (-2, 1) \cup (1, \infty)$ **59.** $\frac{x-2}{x+2}$ **61.** $\frac{(z^2-11z+54)(z-9)}{(z-2)}$ **63.** $\frac{2y^2+5y-4}{y+1}$ **65.** $\frac{35}{12}$
hours, or 2 hours and 55 minutes **67.** 7.5 hours **69.** 20 weeks **71.** 45 minutes

section 2.6, pages 162-163

1. $\{0\}$ **3.** $\emptyset$ **5.** $\{1\}$ **7.** $\left\{\frac{2}{3}\right\}$ **9.** $\emptyset$ **11.** $\left\{\frac{29}{8}\right\}$ **13.** $\{6\}$ **15.** $\emptyset$ **17.** $\{-2, 1\}$ **19.** $\{1\}$ **21.** $\{10\}$ **23.** $\{4\}$
25. $\{2\}$ **27.** $\{-32\}$ **29.** $\left\{\pm\frac{125}{343}\right\}$ **31.** $\{-2, 5\}$ **33.** $\{7, 10\}$ **35.** $a = \pm\sqrt{c^2-b^2}$ **37.** $m = \frac{k}{\omega^2}$ **39.** $v = \pm\sqrt{\frac{Fr}{m}}$
41. $h = \pm\sqrt{\frac{w}{23}}$ **43.** $c = \pm\sqrt{\frac{2gm}{r}}$ **45.** $b = \pm\sqrt{c^2-a^2}$ **47.** $a = \sqrt[3]{\frac{uP^2}{4\pi^2}}$

chapter two review, pages 169-170

1. $\{3, 4\}$ **2.** $\{1, 3\}$ **3.** $\left\{-\frac{10}{7}, 0\right\}$ **4.** $\{-3, 1\}$ **5.** 246.7 miles **6.** $85 **7.** $(-1, 7]$ ⟵—●————●—⟶ $_{-1}$ $_{7}$ **8.** $[-7, 4)$
⟵—●————●—⟶ $_{-7}$ $_{4}$ **9.** $(-5, -1)$ ⟵—○————○—⟶ $_{-5}$ $_{-1}$ **10.** $(-3, 13]$ ⟵—○————●—⟶ $_{-3}$ $_{13}$ **11.** $\left\{-4, \frac{5}{2}\right\}$ **12.** $\{2, 5\}$
13. $\left\{\frac{19 \pm \sqrt{701}}{17}\right\}$ **14.** $\{-3, -2\}$ **15.** $\{\pm 1, \pm\sqrt{2}\}$ **16.** $\{-27, 8\}$ **17.** $\{\pm\sqrt{2}, 4\}$ **18.** $\left\{0, \frac{1}{2}, 2\right\}$ **19.** $\frac{x+3}{3}$
20. $\frac{b-a}{4a+4b}$ **21.** $\left\{-\frac{1}{3}\right\}$ **22.** $\{-5\}$ **23.** $\{2\}$ **24.** $\{3, 5\}$ **25.** $\{3\}$ **26.** $\{-1, 8\}$

chapter 2 test   answers

chapter two test, pages 171-173

1. $\frac{7}{3}$ 2. No solution   3. $\frac{22}{9}$ 4. All real numbers   5. 6.25   6. $\{-3, 4\}$ 7. Ø   8. $\{0, -2\}$ 9. $\left\{\frac{-3}{2}\right\}$ 10. 7
11. $(-\infty, 3)$ 12. $(-\infty, 8]$ 13. $(-2, 4)$ 14. $(-2, 16]$ 15. $[-2, 8]$ 16. No solution   17. $\left(-\infty, \frac{-5}{2}\right) \cup \left(\frac{-1}{2}, \infty\right)$
18. $(-\infty, -6] \cup [8, \infty)$ 19. $(-\infty, \infty)$ 20. $\{-1, 2\}$ 21. $\{-1, 7\}$ 22. $\{-4, 2\}$
23. $\{0, 5\}$ 24. $\{-3, 3\}$ 25. $\{\pm 5i\}$ 26. $\{-3, 4\}$ 27. $\{-12, 16\}$ 28. $\{-3, 5\}$ 29. $\left\{\frac{-1 \pm i\sqrt{7}}{2}\right\}$ 30. $\left\{\frac{-1 \pm \sqrt{5}}{2}\right\}$
31. $\{4\}$ 32. $\{-6, 4\}$ 33. $\{\pm 2, \pm 3\}$ 34. $\{-1, \pm 2, 3\}$ 35. $\left\{\frac{1}{16}, 256\right\}$ 36. $\{\pm 1, 4\}$ 37. $\{1, \pm 2i\}$ 38. $\left\{\frac{5}{3}\right\}$
39. $\left\{\frac{-1}{2}, 1\right\}$ 40. $\{-1, 0, 4\}$ 41. $b=-4, c=1, d=6$ 42. 10   43. -5 44. $\frac{2x-1}{x-2}$ 45. $\frac{1}{x^2-2x+4}$ 46. $\frac{x+2}{3x+1}$
47. $\frac{x+3}{x-3}$ 48. $\frac{3x-1}{x^2-1}$ 49. $\frac{-2}{x}$ 50. 0   51. $\frac{a^2+a+2}{a+1}$ 52. 1   53. $\frac{x}{x-1}$ 54. 1   55. $\frac{1}{a+1}$ 56. $\frac{x^2}{(2x-1)(x+2)}$
57. $\frac{x-1}{x+1}$ 58. $\frac{xy}{x+y}$ 59. $\frac{x^2}{x+1}$ 60. 2   61. $\pm i$ 62. $\{0, 3\}$ 63. 1   64. No solution   65. All real numbers except 2 and -2 66. $\{4, 7\}$ 67. 3 hours 20 minutes   68. $\frac{24}{7}$ 69. $b_2 = \frac{2A}{h} - b_1$ 70. $y = \frac{xz}{2x-z}$ 71. 4   72. $\frac{7}{3}$

chapter three

section 3.1, pages 185-188

1.

3.

5.

7.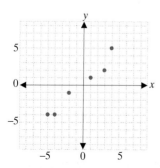

9. III   11. IV   13. positive x-axis   15. III   33. $\{(0, -3), (2, 0), \left(3, \frac{3}{2}\right), (4, 3)\}$
17. IV   19. II   21. IV   23. I   25. negative y-axis   27. $X = [-5, 6]; Y = [-8, 9]$
29. $X = [-3, 6]; Y = [-4, 5]$
31. $X = [-6, 8]; Y = [-9, 7]$

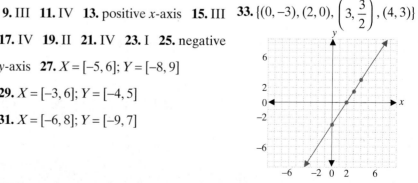

35. {(0,0), (1, ±1), (4, ±2), (9, ±3), (2, −√2)}

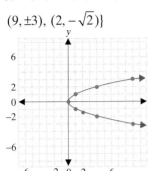

37. {(0, ±3), (±3, 0), (−1, ±2√2), (1, ±2√2), (±√5, 2)}

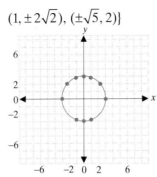

39. $\sqrt{34}, \left(\frac{-7}{2}, \frac{1}{2}\right)$ **41.** $\sqrt{58}, \left(\frac{3}{2}, \frac{7}{2}\right)$

43. $2\sqrt{2}, (-1,-1)$ **45.** $4\sqrt{34}, (3,-8)$

47. $10, (1,-6)$ **49.** $3\sqrt{13}, \left(2, \frac{1}{2}\right)$

51. $10\sqrt{2}, (3,3)$ **53.** $y = -12$ or 6

55. $2 + 3\sqrt{2} + \sqrt{34}$ **57.** $18 + \sqrt{226}$

59. 48 **61.** area = $\frac{15}{2}$ **63.** area = 25

65. area = 17 **67.** area = 48

69. 145 miles **71.** (6, −3), 3.54 kilometers **73.** (−1, 3) and $4\sqrt{5}$

section 3.2, pages 194-196

1. Yes **3.** No **5.** No **7.** No **9.** Yes **11.** Yes **13.** Yes **15.** No **17.** No **19.** No **21.** No **23.** Yes

25.   **27.**   **29.**

31.   **33.**   **35.**

37.   **39.**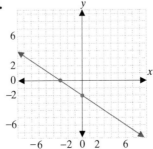

41. e **43.** c **45.** f

47. $a = P - b - c$

49. $j = 24000 + 9b$; Yes

$b = \dfrac{j - 24000}{9}$

section 3.3, pages 209-213

1. −4 **3.** 0 **5.** undefined **7.** $\dfrac{2}{3}$ **9.** −7 **11.** −6 **13.** 4 **15.** 3 **17.** 1 **19.** $\dfrac{1}{15}$ **21.** $-\dfrac{5}{6}$ **23.** −6

25.

27.

29.

31.

33.

35. $y = -3x + 5$ **37.** $y = 4x + 6$

39. $y = \dfrac{1}{2}x + 2$ **41.** $10x - y = 31$

43. $x = -2$ **45.** $2x + 7y = 19$

47. $x + 2y = -13$ **49.** $y = 3$

51. $3x - 10y = -47$

53. $30x - 49y = -9$ **55.** $x = 8$

57. $15x + 2y = 119$ **59.** $2x + 5y = -35$ **61.** f   **63.** a   **65.** b   **67.** $267,000   **69.** $V = -2766.67t + 51,500$; 18.61 yrs

71. a. $y = -0.93x + 10.14$ **73. a.** $y = 1.36x + 3.82$ **75. a.** $y = 1.20x - 0.85$

b.

b.

b.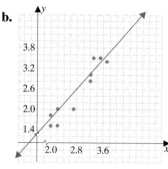

section 3.4, pages 219-222

1. $y = 4x + 9$ **3.** $y = 3x - 11$ **5.** $y = -9$ **7.** $y = x$ **9.** $y = \dfrac{7}{6}x + \dfrac{53}{6}$ **11.** Yes   **13.** Yes   **15.** Yes   **17.** No   **19.** No

21. Yes **23.** No **25.** Yes **27.** No **29.** $y = -\dfrac{1}{3}x - 1$ **31.** $y = 7$ **33.** $y = -\dfrac{1}{4}x - \dfrac{3}{4}$ **35.** $y = x + 3$ **37.** $y = -3x + 28$

39. No **41.** Yes **43.** Yes **45.** Yes **47.** No **49.** No **51.** No **53.** Yes **55.** $41\dfrac{2}{3}$ ft

section 3.5, pages 230-234

1.

3.

5.

7.

9.

11.

13.

15.

17.

19.

21.

23.

25.

27.

29.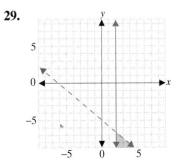

section 3.5 answers

31.   **33.**   **35.**

37.   **39.**   **41.**

43.   **45.**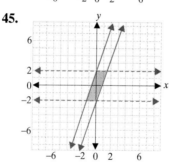

47. h **49.** b **51.** g **53.** c

55. $12x + 22y < 150$

57. $73x + 46y < 1750$

  59.   **61.**

section 3.6, pages 239-241

1. $(x+4)^2 + (y+3)^2 = 25$ **3.** $(x-7)^2 + (y+9)^2 = 9$ **5.** $x^2 + y^2 = 9$ **7.** $(x-\sqrt{5})^2 + (y-\sqrt{3})^2 = 16$ **9.** $(x-7)^2 + (y-2)^2 = 4$

11. $(x+3)^2 + (y-8)^2 = 2$ **13.** $(x-4)^2 + (y-8)^2 = 10$ **15.** $x^2 + y^2 = 85$ **17.** $\left(x+\frac{7}{2}\right)^2 + \left(y-\frac{17}{2}\right)^2 = \frac{53}{2}$

section 3.6   **answers**

19. $(x+6)^2 + \left(y - \frac{3}{2}\right)^2 = \frac{125}{4}$ **21.** $\left(x + \frac{13}{2}\right)^2 + (y+7)^2 = \frac{365}{4}$ **23.** $(x-4)^2 + (y-3)^2 = 25$ **25.** $(x-2)^2 + y^2 = 4$

27. $(x-2)^2 + (y-4)^2 = 49$ **29.** $(x+3)^2 + (y+2)^2 = 64$

31.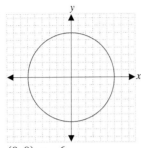
$(0,0), r = 6$

33.
$(0, 8), r = 3$

35.
$(8, 0), r = 2\sqrt{2}$

37.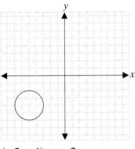
$(-5, -4), r = 2$

39.
$(5, -5), r = \sqrt{5}$

41.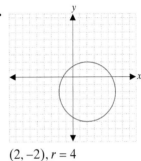
$(2, -2), r = 4$

43.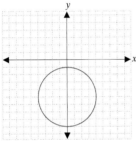
$(0, 5), r = 4$

45.
$(1, -3), r = 2\sqrt{2}$

47.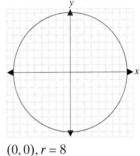
$(0, 0), r = 8$

49.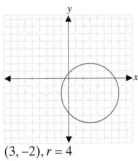
$(3, -2), r = 4$

51.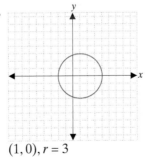
$(1, 0), r = 3$

53.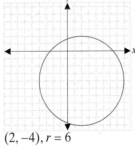
$(2, -4), r = 6$

55.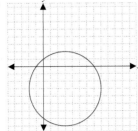
$(3, -3), r = 5$

chapter three review, pages 246-247

1. a. $\sqrt{2}$ **b.** $\left(\dfrac{5}{2}, -\dfrac{13}{2}\right)$ **2. a.** 10 **b.** $(0,-6)$ **3. a.** $2\sqrt{13}$ **b.** $(-5,3)$ **4.** $2+3\sqrt{2}+\sqrt{34}$ **5.** 24 **6.** No **7.** Yes
8. Yes **9.** 12 **10.** -4 **11.** undefined

12.   **13.**   **14.**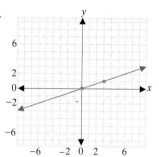

15. $x-y=5$ **16.** $3x-2y=-12$ **17.** $y=\dfrac{5}{9}x-2$ **18.** $y=-\dfrac{7}{3}x+9$ **19.** $9x-2y=31$ **20.** $2x+6y=9$
21. Neither **22.** Perpendicular **23.** Neither **24.** Yes **25.** No **26.** $y=-\dfrac{10}{3}x+\dfrac{5}{3}$ **27.** $y=\dfrac{7}{5}x-\dfrac{7}{5}$

28.   **29.**   **30.**

31. $(x+2)^2+(y-5)^2=18$ **32.** $r=4\ (h,k)=(-5,2)$ **33.** 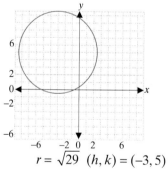 $r=\sqrt{29}\ (h,k)=(-3,5)$

chapter three test, pages 248-251

1. III quadrant **2.** negative y-axis **3.** origin (both axes) **4.** positive x-axis **5.** IV quadrant

6. 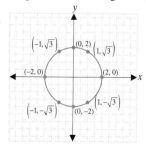 $(0,\pm 2),(\pm 2,0),$ $(-1,\pm\sqrt{3}),$ $(1,\pm\sqrt{3}),(0,2)$

7. 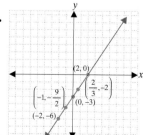 $(2,0),(0,-3),$ $\left(-1,-\dfrac{9}{2}\right),$ $\left(\dfrac{2}{3},-2\right),(-2,-6)$

8. 2   **9.** 5, −1   **10.** $\sqrt{26} + 2\sqrt{5} + \sqrt{10}$ **11.**

$(\sqrt{5})^2 + (\sqrt{80})^2 = (\sqrt{85})^2$;

Area $= \frac{1}{2}\sqrt{5}\sqrt{80} = 10$

12. 7   **13.** 1   **14.** No   **15.** Yes, $y \ne -1$

16. (0, −2) (6, 0)

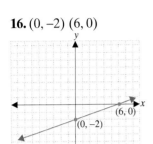

17. $-\frac{3}{4}$ **18.** undefined   **19.** 0   **20.** $-\frac{16}{3}$ **21.** undefined   **22.** $2y - x = -8$ **23.** $\frac{6}{5}$

24. $4x + 3y = 1$ **25.**

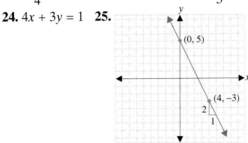

26. $55,000   **27.** $W = 0.08s + 2800$
28. 6   **29.** $y = 3x - 11$ **30.** $-\frac{8}{5}$
31. 4, −2   **32.** $y = 2x + 11$ **33.** $-\frac{1}{4}$
34. parallel   **35.** perpendicular
36. neither   **37.** perpendicular
38. No, if $m_1 \cdot m_2 = -1$, then m_1 and m_2 must be opposite in sign.

39.–41. Answers may vary.
42. (1, −4)

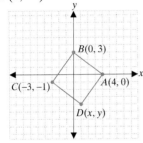

43. $m = 4$; $y = 1$

44. m is undefined; no y-intercept

45. $m = 0$; $y = 2$

46. $m = 1$; $y = -4$

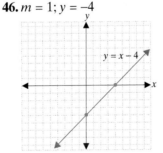

47. a. $-2x + y = -3$ **b.** $x + 2y = 4$
48. a. $x + y = -1$ **b.** $-x + y = -2$
49. a. $y = 0$ **b.** $x = -1$
50. a. $x = 2$ **b.** $y = 5$

51.

52.

53.

827

chapter 3 test   answers

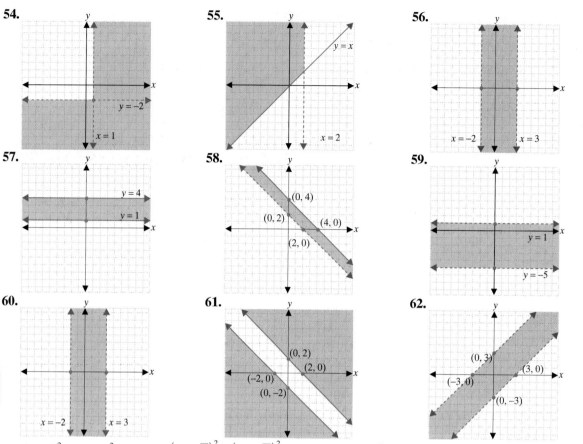

63. $(x+3)^2+(y+2)^2=5$ **64.** $(x-\sqrt{5})^2+(y+\sqrt{2})^2=16$ **65.** $(x+2)^2+y^2=18$ **66.** $(x-2)^2+(y+1)^2=20$
67. $x^2+y^2=169$ **68.** ±3 **69.** center: (−3, 1), radius: $2\sqrt{2}$ **70.** center: (−1, 3), radius: 4
71.   **72.** $B(6, 3)$ Equation: $(x-2)^2+(y-2)^2=17$

chapter four

section 4.1, pages 266-269

1. Dom = { −2 }, Ran = { 5, 3, 0, −9 } **3.** Dom = {π, −2π, 3, 1}, Ran = { 2, 4, 0, 7 } **5.** Dom = $\mathbb{Z}$, Ran = even integers **7.** Dom = $\mathbb{Z}$, Ran = {..., −2, 1, 4,...} **9.** Dom = Ran = $\mathbb{R}$ **11.** Dom = [0, ∞), Ran = $\mathbb{R}$ **13.** Dom = $\mathbb{R}$, Ran = { −1 } **15.** Dom = { 0 }, Ran = $\mathbb{R}$ **17.** Dom = [−3, 1], Ran = [0, 4] **19.** Dom = [0, 3], Ran = [1, 5] **21.** Dom = [−1, 3], Ran = [−4, 3] **23.** Dom = All males with siblings, Ran = All people who have brothers

section 4.2 answers

25. Not a function; $(-2,5)$ and $(-2,3)$ **27.** Function **29.** Not a function; $(-1,0)$ and $(-1,4)$ **31.** Function
33. Not a function $(6,-1)$ and $(6,4)$ **35.** Function **37.** $f(x) = -6x^2 + 2x, f(-1) = -8$
39. $f(x) = \dfrac{-x+10}{3}, f(-1) = \dfrac{11}{3}$ **41.** $f(x) = -2x - 10, f(-1) = -8$ **43.** $x^2 + x - 2, 2ax + 3a + a^2, x^4 + 3x^2$
45. $3x - 1, 3a, 3x^2 + 2$ **47.** $-6x + 16, -6a, -6x^2 + 10$ **49.** $\sqrt{2-x} - 3, \sqrt{1-x-a} - \sqrt{1-x}, \sqrt{1-x^2} - 3$
51. $2x + h - 5$ **53.** $\dfrac{-1}{(x+h+2)(x+2)}$ **55.** $5(2x+h)$ **57.** 2 **59.** $\dfrac{\sqrt{x+h} - \sqrt{x}}{h}$ **61.** $[1, \infty)$
63. $(-\infty, -2) \cup (-2, 3) \cup (3, \infty)$ **65.** $\mathbb{R}$ **67.** $\left(-\infty, \dfrac{1}{3}\right) \cup \left(\dfrac{1}{3}, \infty\right)$ **69.** $x \in [1, 2)$ **71.** $x \in [-6, \infty)$ **73.** $(-\infty, 0) \cup (0, \infty)$

section 4.2, pages 281–286

1.

3.

5.

7.

9.

11.

13.

15.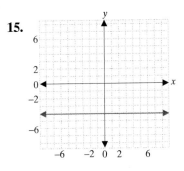

17. vertex: $(-2, -1)$, no x-int.

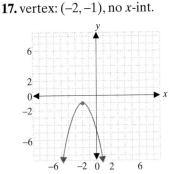

section 4.2   answers

19. vertex: (0, 2), no *x*-int.

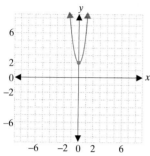

21. vertex: $\left(\frac{1}{2}, \frac{25}{2}\right)$, *x*-int. –2, 3

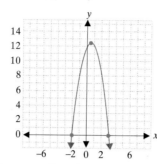

23. vertex: (0, –1), no *x*-int.

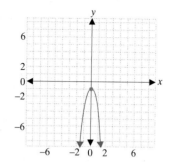

25. vertex: (–1, 3), no *x*-int.

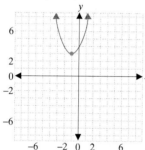

27. vertex: (1, –4), no *x*-int.

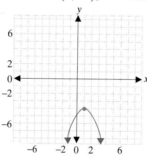

29. vertex: (1, –2), *x*-int. 0, 2

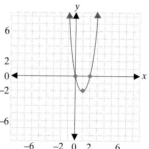

31. g   **33.** a   **35.** c   **37.** h   **39.** The total area, as a function of *x*, is $A(x) = -2x^2 + 200x$. This is maximized when $x = 50$, so the overall dimensions should be 50 feet by 100 feet.   **41.** The relation between length and width is $2l + 2w = 20$, or $l = 10 - w$. This leads to the area function $A(w) = w(10 - w)$, and this is maximized when both width and length are 5.   **43.** The minimum of the function $f(x) = x^2 + (-2x + 20)^2$ occurs when $x = 8$, with a corresponding *y* value of 4.   **45.** The two numbers are *x* and $16 - x$, so the product is $p(x) = -x^2 + 16x$. This function is maximized when $x = 8$, so the two numbers are 8 and 8.   **47.** The total area, as a function of *x*, is $A(x) = -2x^2 + 300x$. This is maximized when $x = 75$, so the area should be $75 \times 150 = 11{,}250$ square feet.   **49.** 500 rooms rented produces the maximum revenue.   **51.** 1500 cars should be produced to incur minimum cost.   **53.** The total area, as a function of *x*, is $A(x) = -2x^2 + 220x$. This is maximized when $x = 55$, so the area should be $55 \times 110 = 6050$ square feet.   **55.** The height function can be written $h(t) = -16\left(t - \frac{3}{2}\right)^2 + 180$, telling us that the maximum height is attained at $\frac{3}{2}$ seconds, and the maximum height is 180 feet.

57. vertex: (4, –1), *x*-int. $\dfrac{8 \pm \sqrt{2}}{2}$

59. vertex: (4, –36), *x*-int. –2, 10

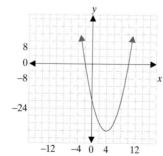

61. vertex: (0, 25), *x*-int. –5, 5

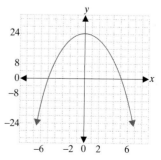

63. vertex: $(-1, 0)$, x-int. -1

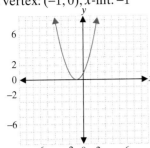

65. vertex: $(5, 21)$, x-int. $5 \pm \sqrt{21}$

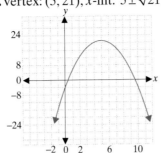

section 4.3, pages 299-303

1.

3.

5.

7.

9.

11.

13.

15.

17.

19.

21.

23.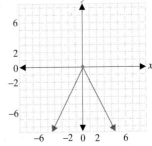

section 4.3 answers

25.
27.
29.
31.
33.
35.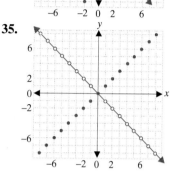

37. f **39.** g **41.** h **43.** b **45.** 256 feet **47.** $18\sqrt{5}$ **49.** 0.72 **51.** 315 pounds **53.** 20 amps **55.** 19.66 inches **57.** 210 pounds

section 4.4, pages 317-320

1. $f(x) = x^2$ **3.** $f(x) = \sqrt[3]{x}$ **5.** $f(x) = [\![x]\!]$ **7.** $f(x) = \sqrt{x}$ **9.** $f(x) = x^2$ **11.** $f(x) = |x|$

13.

Dom = Ran = $\mathbb{R}$

15.
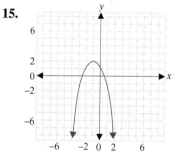
Dom = $\mathbb{R}$, Ran = $(-\infty, 2]$

17.
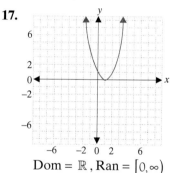
Dom = $\mathbb{R}$, Ran = $[0, \infty)$

19.
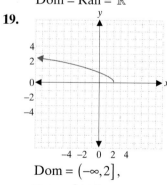
Dom = $(-\infty, 2]$, Ran = $[0, \infty)$

21.
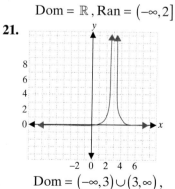
Dom = $(-\infty, 3) \cup (3, \infty)$, Ran = $(0, \infty)$

23.
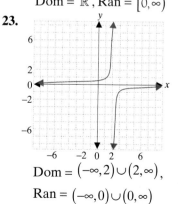
Dom = $(-\infty, 2) \cup (2, \infty)$, Ran = $(-\infty, 0) \cup (0, \infty)$

section 4.4 answers

25.
Dom = $\mathbb{R}$, Ran = $\mathbb{Z}$

27.
Dom = $\mathbb{R}$, Ran = $\mathbb{R}$

29.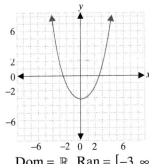
Dom = $\mathbb{R}$, Ran = $[-3, \infty)$

31.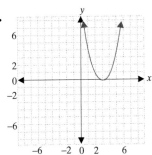
Dom = $\mathbb{R}$, Ran = $[0, \infty)$

33.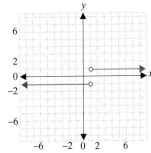
Dom = $(-\infty, 1) \cup (1, \infty)$, Ran = $\{-1, 1\}$

35.
Dom = $[1, \infty)$, Ran = $(-\infty, 2]$

37. $f(x) = (x-4)^2 + 2$ **39.** $f(x) = (x+2)^2$ **41.** $f(x) = (x-10)^3 + 4$ **43.** $f(x) = \sqrt{-x} - 3$ **45.** $f(x) = -|x-8| - 2$

47.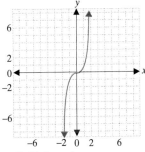
Odd function; origin symmetry

49.
Odd function; origin symmetry

51.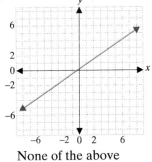
None of the above

53.
None of the above

55.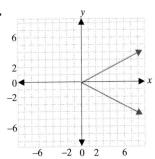
Not a function; x-axis symmetry

57.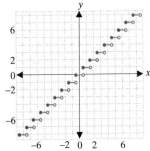
None of the above

59.
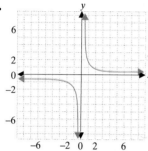
Odd function; Origin symmetry

61. Dec. on $(-\infty, -3)$, Inc. on $(-3, \infty)$ **63.** Dec. on $(-\infty, 1)$, Dec. on $(1, \infty)$
65. Inc. on $(-1, \infty)$ **67.** Inc. on $(-\infty, 1)$, Dec. on $(1, \infty)$
69. Inc. on $(-\infty, 7)$, Dec. on $(7, \infty)$
71. Dec. on $(-\infty, 1)$, Inc. on $(1, 3)$, Dec. on $(3, \infty)$

73.

Constant on $(0,1)\cup[1,2)\cup[2,3)$
$\cup[3,4)\cup[4,5)\cup[5,6)\cup[6,7)\cup$
$[7,8)\cup[8,9)\cup[9,10)$

75.

Dec. on $[0,3)\cup(3,\infty)$

83. $|x+4|-1$
85. $-\sqrt{6-x}+2$
87. $6-(x-3)^2$

section 4.5, pages 330-334

1. (a) 2 **(b)** -8 **(c)** -15 **(d)** $-\dfrac{3}{5}$ **3. (a)** -3 **(b)** -1 **(c)** 2 **(d)** 2   **5. (a)** 12 **(b)** 18 **(c)** -45 **(d)** -5 **7. (a)** 3 **(b)** 1 **(c)** 2 **(d)** 2   **9. (a)** 6 **(b)** 0 **(c)** 9 **(d)** 1   **11. (a)** 5 **(b)** -1 **(c)** 6 **(d)** $\dfrac{2}{3}$ **13. (a)** 3 **(b)** 5 **(c)** -4 **(d)** -4
15. (a) $|x|+\sqrt{x}$, Dom $= [0,\infty)$ **(b)** $\dfrac{|x|}{\sqrt{x}}$, Dom $= (0,\infty)$ **17. (a)** x^2+x-2, Dom $= \mathbb{R}$ **(b)** $\dfrac{1}{x+1}$, Dom $= (-\infty,-1)\cup(-1,1)\cup(1,\infty)$ **19. (a)** x^3+3x-8, Dom $= \mathbb{R}$ **(b)** $\dfrac{3x}{x^3-8}$, Dom $= (-\infty,2)\cup(2,\infty)$
21. (a) $-2x^2+[\![x+4]\!]$, Dom $= \mathbb{R}$ **(b)** $\dfrac{-2x^2}{[\![x+4]\!]}$, Dom $= (-\infty,-4)\cup[-3,\infty)$ **23.** 2   **25.** 0   **27.** 8   **29.** 3   **31.** 1
33. $\dfrac{1}{3}$ **35. (a)** $\dfrac{1}{x-1}$, Dom $= (-\infty,1)\cup(1,\infty)$ **(b)** $\dfrac{1}{x}-1$, Dom $= (-\infty,0)\cup(0,\infty)$ **37. (a)** $1-\sqrt{x}$, Dom $= [0,\infty)$
(b) $\sqrt{1-x}$, Dom $= (-\infty, 1]$ **39. (a)** x^2-4x+3, Dom $= \mathbb{R}$ **(b)** x^2+2x-3, Dom $= \mathbb{R}$ **41. (a)** $|x|^3+|x|^2-5|x|+3$, Dom $= \mathbb{R}$ **(b)** $|x^3+4x^2|-1$, Dom $= \mathbb{R}$ **43. (a)** $\left[\!\left[\dfrac{x^2+7}{2}\right]\!\right]$, Dom $= \mathbb{R}$ **(b)** $\dfrac{[\![x+2]\!]^2+3}{2}$, Dom $= \mathbb{R}$ **45.** $g(x)=\dfrac{2}{x}$,
$h(x)=5x-1$, $g(h(x))$ **47.** $g(x)=x+\sqrt{x}-7$, $h(x)=x+2$, $g(h(x))$ **49.** $g(x)=\dfrac{\sqrt{x}}{x^2}$, $h(x)=x-3$, $g(h(x))$
51. $h(x)=|x^2+3x|$, $g(x)=x-3$, $g(h(x))$ **53.** $g(x)=\sqrt{x+5}$ **55.** $g(x)=-x^3-7$ **57.** $V=3\pi r^3$
59. $V=\dfrac{1}{12}\pi r^2 t^2$ **61.** $f\circ g(x)=\sqrt[3]{\dfrac{-x^3}{3x^2-9}}$, $f\circ g(-x)=\sqrt[3]{\dfrac{x^3}{3x^2-9}}$ **63.** Yes   **65.** Yes   **67.** No   **69.** No   **71.** No

section 4.6, pages 344-348

1.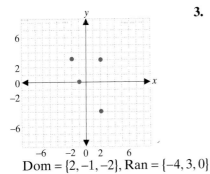
Dom = {2, −1, −2}, Ran = {−4, 3, 0}

3.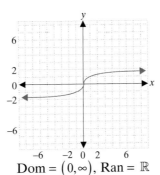
Dom = $(0, \infty)$, Ran = $\mathbb{R}$

5.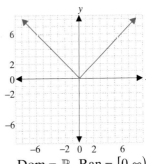
Dom = $\mathbb{R}$, Ran = $[0, \infty)$

7.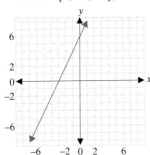
Dom = $\mathbb{R}$, Ran = $\mathbb{R}$

9.
Dom = $\mathbb{Z}$, Ran = $\mathbb{R}$

11.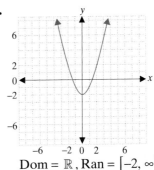
Dom = $\mathbb{R}$, Ran = $[-2, \infty)$

13.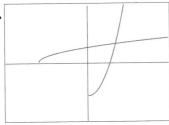

Dom = $\mathbb{R}$, Ran = $\mathbb{R}$

15. not a one-to-one function $f(-1) = f(1) = 1$ **17.** Restrict to $[0, \infty)$

19. Inverse exists **21.** Inverse exists **23.** Inverse exists

25. Restrict to $[2, \infty)$ **27.** Restrict to $[12, \infty)$ **29.** $f^{-1}(x) = (x+2)^3$

31. $r^{-1}(x) = \dfrac{-2x-1}{3x-1}$ **33.** $F^{-1}(x) = (x-2)^{\frac{1}{3}} + 5$ **35.** $V^{-1}(x) = 2x - 5$

37. $h^{-1}(x) = (x+2)^{\frac{5}{3}}$ **39.** $J^{-1}(x) = \dfrac{x-2}{3x}$ **41.** $h^{-1}(x) = (x-6)^{\frac{1}{7}}$

43. $r^{-1}(x) = \dfrac{x^5}{2}$ **45.** $f^{-1}(x) = \dfrac{x^3}{54}$ **47.– 55.** Answers may vary. **57.** b **59.** e **61.** a **63.** 73 1 53 13 97 73 29 57 17 73 **65.** FRISBEE VOLLEYBALL AND HORSESHOES **67.** CATCH A WAVE

69.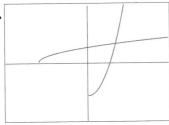
Dom = $[0, \infty)$, Ran = $[-5, \infty)$

71.
Dom = $(-\infty, 1) \cup (1, 2) \cup (2, \infty)$,
Ran = $(-\infty, 1) \cup (1, \infty)$

73.
Dom = $[0, \infty)$, Ran = $[4, \infty)$

chapter four review, pages 353-355

1. Dom = {−2, −3}, Ran = {9, −3, 2, −9}; No 2. Dom = $\mathbb{R}$, Ran = $\mathbb{R}$; Yes 3. Dom = $[-6, \infty)$, Ran = $\mathbb{R}$; No
4. Dom = $[0, \infty)$, Ran = $[4, \infty)$; Yes

5.
6.
7.
8.
9.
10.

11. 64 feet 12. about 1226 videos per month

13.
Dom = $\mathbb{R}$, Ran = $\mathbb{R}$

14.
Dom = $\mathbb{R}$, Ran = $[0, \infty)$

15. Dec. on $(-\infty, 2)$, Inc. on $(2, \infty)$

16. Dec. on $(-\infty, -2)$, Inc. on $(-2, -1)$, Dec. on $(-1, \infty)$

17. (a) $x^2 + \sqrt{x}$, Dom = $[0, \infty)$
(b) $x^{\frac{3}{2}}$, Dom = $(0, \infty)$

18. (a) $\frac{1}{x-2} + \sqrt[3]{x}$, Dom = $(-\infty, 2) \cup (2, \infty)$ (b) $\frac{1}{\sqrt[3]{x}(x-2)}$, Dom = $(-\infty, 0) \cup (0, 2) \cup (2, \infty)$ 19. 5 20. $-\frac{9}{2}$

21. $4x^3 + 7$ 22. $\frac{1}{\sqrt{x^2 + 4x}}$ 23. $r^{-1}(x) = \frac{x+2}{7x}$ 24. $g^{-1}(x) = \frac{3}{4-x}$ 25. $f^{-1}(x) = (x+6)^5$

chapter four test, pages 356-359

1. $-x^2 + 2(x-h) - h^2 + 2xh$ 2. $-2x - h + 2$ 3. $f(x) = 3x$; -6 4. $f(x) = \frac{x-2}{x+3}$; -4 5. $f(x) = \frac{8x}{4-3x}$; $\frac{-8}{5}$

6. $\left(-\infty, \frac{1}{2}\right]$ 7. $\mathbb{R}$ 8. $(-\infty, -2) \cup (-2, 2) \cup (2, \infty)$ 9. $[-3, 4)$ 10. $\mathbb{R}$

section 5.1 answers

11.   **12.**   **13.**

14. 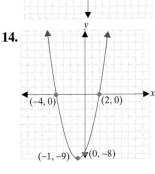   **15.** –15   **16.** –4   **17.**

18.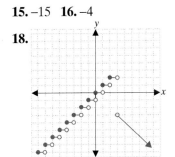

19. $f(x) = (x+2)^2 + 4$ **20.** $f(x) = (x-1)^2 - 2$ **21.** $y = -(x-3)^2$ **22.** (a) –2 (b) –6 (c) –8 (d) –2

23. (a) $\dfrac{-13}{2}$ (b) $\dfrac{-11}{2}$ (c) 3 (d) 12   **24.** 4   **25.** $1 - \sqrt[3]{-4}$ **26.** –1   **27.** –25   **28.** $-x^4 - 2x^2 - 2$ **29.** 0; –2

30. Answers may vary.   **31.** $(f \circ f)(x) = x$ **32.** $g(x) = \sqrt{x},\ h(x) = 2x^2 + 1,\ f(x) = (g \circ h)(x)$

33. $g(x) = |x| - 2,\ h(x) = x^2 - x,\ f(x) = (g \circ h)(x)$ **34.** $g(x) = \dfrac{3}{x},\ h(x) = 3x^2 + 1,\ f(x) = (g \circ h)(x)$

35. $g(x) = \dfrac{\sqrt{x}}{x^2},\ h(x) = x + 2,\ f(x) = (g \circ h)(x)$ **36.** $\dfrac{x+4}{6}$ **37.** $(x-3)^4$ **38.** $\pm\sqrt{4-x}$ **39.** $\dfrac{2}{x} + 1$ **40.** $\dfrac{x^7}{3}$

41. $\dfrac{-x-3}{x-2}$ **42.** $\dfrac{x+1}{2x-3}$ **43.** $(x+1)^3 - 2$ **44.** –3   **45.** $\dfrac{x-5}{2}$ **46.** –90   **47.** –3   **48.** 7   **49.** $\{-3,-1,0,3,4\}$ **50.** $r > 0$

51. y-axis   **52.** origin   **53.** none   **54.** origin   **55.** –4   **56.** 4   **57.** 4   **58.** $2x - 4$ **59.** $\dfrac{-dx+b}{cx-a}$ **60.** $\dfrac{4}{5}$ **61.** (a) –64

(b) 1   **62.** (a) undefined (b) 1   **63.** $\sqrt{x+h}$ **64.** $\dfrac{\sqrt{x+h}-\sqrt{x}}{x+h}$ **65.** $\sqrt[3]{(x+h)^2}$ **66.** $\dfrac{\sqrt[3]{(x+h)^2} - \sqrt[3]{x^2}}{h}$ **67.** $\mathbb{R}$

68. $(-\infty, 0] \cup (1, \infty)$ **69.** $\mathbb{R}$ **70.** $[-1, 1]$

chapter five

section 5.1, pages 372-376

1–17. Answers may vary.   **19.** Yes   **21.** Yes   **23.** Yes   **25.** $1 \pm 2i$ **27.** $-3, \dfrac{1}{2}$ **29.** $\pm\sqrt{3}, \pm\sqrt{5}$ **31.** $-\dfrac{5}{2}$

33. $0, 4 \pm 3i$ **35.** $\pm 1, \pm 2i\sqrt{2}$ **37.** 7th degree; lead coef = 4; $i(x) \to -\infty$ as $x \to -\infty$; $i(x) \to \infty$ as $x \to \infty$

39. 5th degree; lead coef = –6; $h(x) \to \infty$ as $x \to -\infty$; $h(x) \to -\infty$ as $x \to \infty$

41. 4th degree; lead coef = –2; $f(x) \to -\infty$ as $x \to -\infty$ and ∞

section 5.1 answers

43.
x-int: $(-4, 0), (-2, 0), (3, 0)$
y-int: $(0, 24)$

45.
x-int: $(-2, 0)$; *y*-int: $(0, -8)$

47.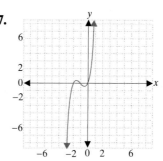
x-int: $(-2, 0), (-1, 0)$
$(0, 0)$; *y*-int: $(0, 0)$

49.
x-int: $(3, 0)$; *y*-int: $(0, -243)$

51. e **53.** a **55.** f **57.** d **59.** f **61.** b **63.** $(-\infty, -2) \cup (3, \infty)$

65. $(-\infty, -2) \cup (-1, 0)$ **67.** $[-2, 1] \cup [3, \infty)$ **69.** $[-5, -1] \cup [1, 4]$ **71.** $\left(-\frac{1}{2}, 2\right)$

73. $(-\infty, -4) \cup (2, 3)$ **75.** $5 \le x \le 27$ or all integers between 5 and 27, inclusive

77. $11 \le x \le 23$ or all integers between 11 and 23, inclusive **79.** ≈17.9 months

81. ≈141.4 weeks

section 5.2, pages 386-388

1. $3x^2 - x + 1 + \dfrac{5x-1}{2x^2+2}$ **3.** $x - 2 + \dfrac{-2}{x^2-4x+4}$ **5.** $4x^2 - 14x + 29 - \dfrac{65}{x+2}$ **7.** $x^3 + 6x^2 - 2x + 5 + \dfrac{2x+5}{3x^2-1}$

9. $2x^3 - 3x^2 + 2x - 5$ **11.** $x^3 + 3x^2 + 10x + 10 + \dfrac{22}{x-3}$ **13.** $3x^2 + 5x + 9 + \dfrac{45}{3x-5}$ **15.** $2x - 5 + \dfrac{7}{x+3}$

7. $x^2 - ix + 6 + \dfrac{1+i}{2x-i}$ **19.** $x^2 + 3$ **21.** $p(1) = 4$ **23.** k is a zero **25.** k is a zero **27.** $p(1) = 12$ **29.** k is a zero

31. k is a zero **33.** k is a zero **35.** k is a zero **37.** $p(5) = -2$ **39.** k is a zero **41.** $x^2 - 4x + 2 + \dfrac{-1}{x+5}$

43. $x^7 - 3x^2 + \dfrac{3}{x+1}$ **45.** $4x^2 - 4x + 2$ **47.** $x^4 - x^3 - x^2 - 7x - 14 - \dfrac{10}{x-2}$ **49.** $x^3 - x^2 + x$

51. $2x^2 - 4ix + 17 + \dfrac{8+48i}{x-3i}$ **53.** $f(x) = -x^2 - x + 12$ **55.** $f(x) = -x^2 + 4x - 13$

57. $f(x) = x^4 - 12x^3 + 54x^2 - 108x + 81$ **59.** $f(x) = 3x^4 + 9x^3 - 9x^2 - 21x + 18$

61. $SA = (x+5)(x+2) = x^2 + 7x + 10$

section 5.3, pages 401-404

1. $\pm\left\{\dfrac{1}{3}, \dfrac{2}{3}, 1, \dfrac{4}{3}, 2, \dfrac{8}{3}, 4, 8\right\}, \left\{-4, \dfrac{1}{3}, 2\right\}$ **3.** $\pm\{1, 2, 3, 4, 6, 8, 12, 24\}, \{\pm 2i, 2, 3\}$ **5.** $\pm\{1, 2, 7, 14\}, \{1, 2, 7\}$

7. $\pm\left\{\dfrac{1}{2}, 1, \dfrac{3}{2}, \dfrac{5}{2}, 3, 5, \dfrac{15}{2}, 15\right\}, \left\{-1, \dfrac{5}{2}, 3\right\}$ **9.** $\pm\left\{\dfrac{1}{3}, 1, 3\right\}, \{\pm 1, \pm i\}$ **11.** $\pm\{1, 11\}, \{-11, \pm 1\}$ **13.** $\pm\{1, i\}$ **15.** $\{-1, 2 \pm 3i\}$

838

17. $\{\pm 2i, 2, 3\}$ **19.** $\{4, 1\pm 2i\}$ **21.** $\{\pm 5i, -2, 1\}$ **23.** $\{-11, \pm 1\}$ **25.** 0 pos., 3 or 1 neg. **27.** 2 or 0 pos., 1 neg. **29.** 3 or 1 pos., 1 neg. **31.** 1 pos., 1 neg. **33.** 3 or 1 pos., 0 neg. **35.** 0 pos., 0 neg. **37.** $[-5, 1]$ **39.** $[-1, 6]$ **41.** $[-3, 6]$ **43.** $[-3, 3]$ **45.** $[-3, 6]$ **47.** $\{-4, \pm 1\}$ **49.** $\{-1, 2, 5\}$ **51.** $\{2, 3, \pm 2\sqrt{2}\}$ **53.** $\pm\{\sqrt{5}, i\sqrt{5}\}$ **55.** $\{-2, -\frac{1}{2}, 6\}$ **57.** $f(-3) = -84, f(-1) = 16$ **59.** $f(2) = -15, f(3) = 24$ **61.** $f(2) = 15, f(3) = -24$ **63.** Answers may vary. **65.** $\{-6, \frac{1}{4}, 1\}$ **67.** $\{-3, \pm\sqrt{3}, \frac{1}{2}\}$ **69.** $\{\frac{-9 \pm \sqrt{41}}{2}, \frac{1}{2}, 2\}$ **71.** $\{-4, -3, 1, 2\}$ **73.** $\{\pm 5, \pm 6, \pm 8\}$ **75.** $\{-3, 4, 5\}$ **77.** $\{-\frac{2}{3}, \frac{1}{2}, 7\}$ **79.** $\{-4, -3, 2\}$ **81.** $\{\pm i\sqrt{7}, 0, 3\}$ **83.** $\{\sqrt[3]{-5}, -\frac{3}{4}, 2\}$

section 5.4, pages 415-418

1.

3.

5.

7.

9.
$f(x) = (x+2)^3(x-1)^2$

11.
$s(x) = -(x-3)(x+1)$
$\times (x-\sqrt{5})(x+\sqrt{5})$

13.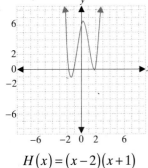
$H(x) = (x-2)(x+1)$
$\times (x-\sqrt{3})(x+\sqrt{3})$

15.
$f(x) = (x+2)^2(2x+3)$

17. $\{-2, 1\}$ **19.** $\{-2, 1, \pm i\sqrt{5}\}$
21. $\{-3, 1, \pm 2i\}$ **23.** $\{\pm 1, \pm 4i\}$
25. $(x-3+2i)(x-3-2i)(x-4)(x+1)$
27. $(x-1-3i)(x-2)(x+1)$
29. $(x+\sqrt{7})(x-\sqrt{7})(x-2+3i)(x-2-3i)$
31. $(x-2)(x+1)(x-1+2i)(x-1-2i)$
33. $x(x-3)(x-1)(x+11)$
35. $(x-2)(x-2+4i)(x-2-4i)$

section 5.4 answers

37. $f(x) = -2x^3 + 18x^2 - 32x - 52$ **39.** $f(x) = 2x^5 + 2x^4 - 10x^3 - 2x^2 + 16x - 8$
41. $f(x) = 3x^4 - 18x^3 + 12x^2 - 72x$ **43.** $-x^3 + 2x^2 + 14x - 40$ **45.** $-x^3 + 4x^2 + 15x - 68$
47. a. $V(x) = 2x(15 - 2x)(5 - x)$ **b.** $x = 0, x = 5, x = \dfrac{15}{2}$ **c.** $x = 0$ and $x = 5$ **49.** Answers may vary.

chapter five review, pages 423-425

1.- 4. Answers may vary. **5.** $\left[\dfrac{5}{2}, 3\right]$ **6.** $(-\infty, -1) \cup (-1, 3) \cup (3, \infty)$ **7.** $[-2, -1] \cup [1, 4]$ **8.** $4x^2 - 3x + 3 + \dfrac{7}{2x^2 - 1}$
9. $11x + 35 + \dfrac{100}{x - 3}$ **10.** $x^2 - 3x + 4 - \dfrac{5x + 16}{x^2 + 3x + 2}$ **11.** $p(1) = 90$ **12.** k is a zero **13.** $p\left(\dfrac{2}{3}\right) = -\dfrac{7}{3}$
14. $\pm\{1, 2, 3, 6\}, \{-3, -1, 2\}$ **15.** $\pm\left\{\dfrac{1}{2}, 1, \dfrac{3}{2}, 3, \dfrac{9}{2}, 9\right\}, \left\{1, \dfrac{3}{2}, 3\right\}$ **16.** 2 or 0 pos., 2 or 0 neg. **17.** 4, 2 or 0 pos., 2 or 0 neg.
18. $[-2, 6]$ **19.** $[-3, 7]$ **20.** $\left\{-1, \dfrac{5}{3}, \pm 2i\right\}$ **21.** $\left\{\pm\sqrt{3}, -\dfrac{1}{2}\right\}$ **22.** $\{\pm 2i, -3, \pm 1\}$ **23.** $(x - 2 + i)(x - 2 - i)(7x - 2)(2x - 7)$
24. $(x - 5i)(x + 5i)(x - 6)(x + 1)$ **25.** $(x - 1 - i)(x - 1 + i)(2x + 1)(x + 3)$ **26.** $f(x) = 2x^4 + 7x^3 - 18x^2 + 67x - 30$
27. $f(x) = x^5 + 3x^4 - 3x^3 - 17x^2 - 18x - 6$ **28.** $x^5 - 3x^4 + 8x^2 - 9x + 3$

chapter five test, pages 426-428

1.-2. Answers may vary **3.** $0, 5, -4$ **4.** $\dfrac{3 \pm i\sqrt{3}}{2}$ **5.** $\pm\sqrt{2}, \pm\sqrt{5}$ **6.** $0, \pm\sqrt{2}, \pm i$

7. x-int: $1, \pm 2$; y-int: -4
$f(x) \to \infty$ as $x \to -\infty$
$f(x) \to -\infty$ as $x \to \infty$

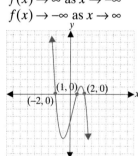

8. x-int: 2; y-int: -8
$f(x) \to -\infty$ as $x \to -\infty$;
$f(x) \to \infty$ as $x \to \infty$

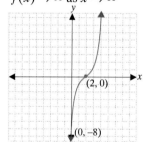

9. x-int: $-2, 1, 3$; y-int: 6
$f(x) \to -\infty$ as $x \to -\infty$;
$f(x) \to \infty$ as $x \to \infty$

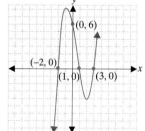

10. x-int: 2, -1; y-int: 4
$f(x) \to \infty$ as $x \to \pm\infty$

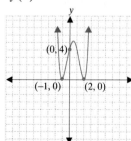

11. $(-6, 1)$ **12.** $[-2, 0] \cup [4, +\infty)$ **13.** All real numbers except $-1, 3$
14. $(-\infty, 0) \cup (0, 1) \cup (2, +\infty)$ **15.** All real numbers **16.** $\left(\dfrac{1}{3}, 2\right)$ **17.** $40 \le x \le 50$
18. $x^3 + x^2 + x + 1$ **19.** $2x^3 + 2x^2 - 2x - 3 + \dfrac{-2x - 2}{x^2 - x}$ **20.** $x^2 - 6 + \dfrac{-4 + 7i}{2x + i}$
21. $3x^2 i - 6x + i - 3$ **22.** $x^3 - 3x^2 + 8x - 16 + \dfrac{28}{x + 2}$ **23.** k is a zero
24. $p\left(\dfrac{1}{2}\right) = \dfrac{5}{2}$ **25.** $p(-2) = -41$ **26.** k is a zero **27.** k is a zero

section 6.1 answers

28. $y = (x + 2)(1 - x)$ **29.** $y = x(x + 3)(x - 2)$ **30.** $y = \frac{1}{2}(x+4)^4$ **31.** $y = 2(x^2 - 4)(x - 3)$ **32.** $-2, 1, 3$
33. $0, -2, \pm 2i$ **34.** $\pm 1, 2, 3$ **35.** 0 pos., 3 or 1 neg. **36.** 2 or 0 pos., 1 neg. **37.** 3 or 1 pos., 0 neg.
38. 3 or 1 pos., 1 neg. **39.** 15 **40.** -19 **41.** 8
42. $(2x-1)(x-2)(x+1)$ **43.** $(x^2+1)(x-3)$ **44.** $(x-2)^3(x+3)$

  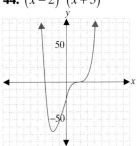

45. $(x+1)(x-1)(x-2)(x^2+x+1)$ **46.** $y = -2x^2 + 4x - 6$ **47.** $y = (x^2 - 2x + 2)(x - 2)$

48. $y = x^4 - 9x^3 + 28x^2 - 30x$ **49.** 1 **50.** No **51.** 18

52. $f(x) = \left(x + \frac{2}{3}\right)(15x^3 - 6x + 4) + \frac{34}{3}; f\left(\frac{-2}{3}\right) = \frac{34}{3}$

53. $f(x) = (x - \sqrt{2})\left(x^2 + (3+\sqrt{2})x + 3\sqrt{2}\right) - 8; f(\sqrt{2}) = -8$ **54.** Same **55.** Same

56. $r_1 + 5, r_2 + 5, r_3 + 5$ **57.** $\frac{r_1}{2}, \frac{r_2}{2}, \frac{r_3}{2}$ **58.** Cannot be determined

59. $-r_1, -r_2, -r_3$

chapter six

section 6.1, pages 443–447

1. $x = 1$ **3.** No vertical asymptote **5.** $x = 2$ **7.** $x = 0$ **9.** $x = -\frac{1}{2}$ **11.** No vertical asymptote **13.** $x = 7$
15. $x = -2$ **17.** $x = -2, x = 2$ **19.** $y = 0$ **21.** $y = x - 2$ **23.** $y = 0$ **25.** $y = 2$ **27.** $y = 3x + 6$ **29.** $y = 0$ **31.** $y = 0$
33. $y = x - 11$ **35.** $y = 5x + 4$

37.   **39.**   **41.**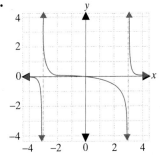

section 6.1   answers

43.   **45.**   **47.**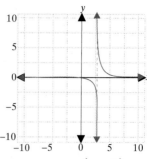

49. a. $x = -2$ **b.** $y = 0$ **c.** None **d.** None **e.** $(0, 5)$ **51. a.** $x = 9$ **b.** $y = 0$ **c.** None **d.** None **e.** $\left(0, -\dfrac{1}{3}\right)$

53. a. $x = -1, x = 1$ **b.** None **c.** $y = x$ **d.** $\left(\sqrt[3]{3}, 0\right)$ **e.** $(0, 3)$ **55. a.** $x = 1$ **b.** None **c.** $y = 3x$ **d.** None **e.** $(0, -3)$

57. $(-\infty, -2) \cup (-1, 1)$ **59.** $(-8, -2) \cup (2, \infty)$ **61.** $(-\infty, -2) \cup (-2, 3)$ **63.** $(0, 3)$ **65.** $(-2, -1) \cup (1, \infty)$

67. $(-\infty, -1) \cup \left[-\dfrac{1}{2}, 0\right)$ **69. a.**   **71. a.**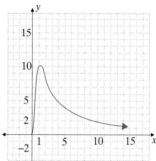

b. Joan's rabbit population reaches a maximum of 200 rabbits.   **b.** The concentration of the drug disappears in the long run.

section 6.2, pages 460-465

1. Center: $(5, 2)$; Foci: $\left(5, 2 \pm \sqrt{21}\right)$; Vertices: $(5, 7)$ $(5, -3)$ **3.** Center: $(-2, -5)$; Foci: $\left(-2 \pm \sqrt{6}, -5\right)$; Vertices: $(1, -5), (-5, -5)$ **5.** Center: $(-3, 2)$; Foci: $\left(-3 \pm \sqrt{2}, 2\right)$; Vertices: $(-1, 2), (-5, 2)$ **7.** Center: $(-5, 1)$; Foci: $\left(-5, 1 \pm 2\sqrt{3}\right)$; Vertices: $(-5, 5), (-5, -3)$ **9.** Center: $(-4, 2)$; Foci: $\left(-4 \pm 3\sqrt{2}, 2\right)$; Vertices: $\left(-4 \pm 3\sqrt{3}, 2\right)$ **11.** Center: $(2, 0)$; Foci: $(4, 0), (0, 0)$; Vertices: $\left(2 \pm \sqrt{5}, 0\right)$ **13.** e **15.** f **17.** c **19.** b

21.   **23.**   **25.**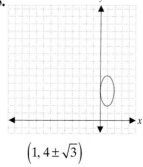

$\left(3 \pm 2\sqrt{2}, -1\right)$ $\left(3 \pm \sqrt{5}, 4\right)$ $\left(1, 4 \pm \sqrt{3}\right)$

section 6.3 **answers**

27.
$\left(-1\pm\sqrt{21},-5\right)$

29.
$\left(-2\pm\sqrt{7},-1\right)$

31.
$\left(-1\pm\sqrt{7},2\right)$

33.
$\left(-5, 3\pm\sqrt{15}\right)$

35.
$\left(-5\pm\sqrt{5},-5\right)$

37.
$\left(0,-2\pm\sqrt{3}\right)$

39. $\dfrac{x^2}{16}+\dfrac{y^2}{25}=1$ **41.** $(x-1)^2+\dfrac{(y-1)^2}{9}=1$ **43.** $\dfrac{(x-3)^2}{36}+\dfrac{y^2}{27}=1$ **45.** $(x+2)^2+\dfrac{(y+3)^2}{4}=1$

47. $\dfrac{(x-5)^2}{16}+\dfrac{(y-3)^2}{15}=1$ **49.** $\dfrac{(x-2)^2}{4}+\dfrac{(y+2)^2}{9}=1$ **51.** $\dfrac{(x-1)^2}{9}+\dfrac{y^2}{16}=1$ **53.** 3,253,954,454 miles

55. 185.93 million miles **57.** The string should be 5 cm long, and the tacks should be 4 cm apart.
59. $e = 0.9585$; 544.4079 inches

section 6.3, pages 474-477

1.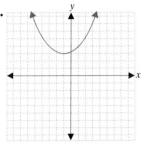
$(-1, 4)$, $y = 2$

3.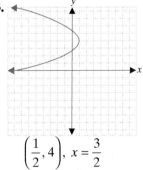
$\left(\dfrac{1}{2}, 4\right)$, $x = \dfrac{3}{2}$

5.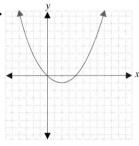
$(2, 0)$, $y = -2$

7.
 $\left(\dfrac{3}{2}, 0\right), x = -\dfrac{3}{2}$

9.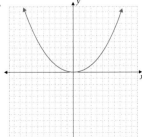
$\left(0, \dfrac{7}{4}\right), y = -\dfrac{7}{4}$

section 6.3   answers

11.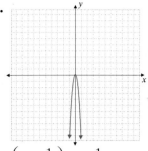
$\left(0, -\dfrac{1}{48}\right), y = \dfrac{1}{48}$

13.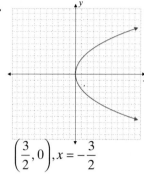
$\left(\dfrac{3}{2}, 0\right), x = -\dfrac{3}{2}$

15.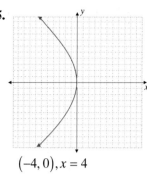
$(-4, 0), x = 4$

17.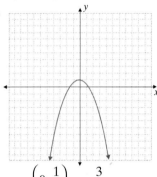
$\left(0, \dfrac{1}{2}\right), y = \dfrac{3}{2}$

19.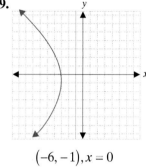
$(-6, -1), x = 0$

21.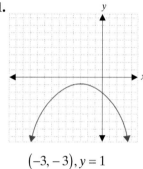
$(-3, -3), y = 1$

23.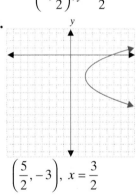
$\left(\dfrac{5}{2}, -3\right), x = \dfrac{3}{2}$

25. g **27.** b **29.** e **31.** d **33.** $(y-1)^2 = -4(x+1)$ **35.** $(x-3)^2 = 8(y+1)$

37. $(y+2)^2 = 24(x-3)$ **39.** $(x+3)^2 = -2(y+1)$ **41.** $(y-3)^2 = 10(x+4)$

43. $(y+1)^2 = -8(x-2)$ **45.** 2 feet **47.** 1.5 inches

section 6.4, pages 488–491

1.
$(-3 \pm \sqrt{13}, -1)$

3.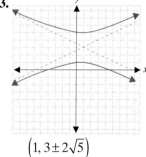
$(1, 3 \pm 2\sqrt{5})$

5.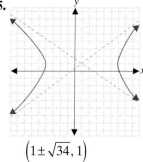
$(1 \pm \sqrt{34}, 1)$

7.
$\left(\pm 2\sqrt{5}, 2\right)$

9.
$\left(-2, 2\pm\sqrt{34}\right)$

11.
$(7,1), (-3,1)$

13.
$\left(-7, 2\pm\sqrt{113}\right)$

15.
$\left(-1\pm\sqrt{68}, -7\right)$

17. Center: $(-3, 2)$; Foci: $\left(-3\pm\sqrt{13}, 2\right)$; Vertices: $(-1, 2), (-5, 2)$ **19.** Center: $(1, -4)$; Foci: $\left(1\pm 2\sqrt{3}, -4\right)$; Vertices: $\left(1\pm\sqrt{3}, -4\right)$ **21.** Center: $(-2, 1)$; Foci: $\left(-2\pm\sqrt{30}, 1\right)$; Vertices: $(3, 1), (-7, 1)$ **23.** Center: $(-3, -1)$ Foci: $\left(-3\pm 2\sqrt{3}, -1\right)$ Vertices: $(-1, -1), (-5, -1)$

25. Center: $(1, 0)$; Foci: $\left(1\pm\dfrac{\sqrt{5}}{2}, 0\right)$; Vertices: $(2, 0), (0, 0)$ **27.** Center: $(8, 5)$; Foci: $\left(8\pm 4\sqrt{5}, 5\right)$; Vertices: $(12, 5), (4, 5)$ **29.** a **31.** b **33.** g **35.** h **37.** $\dfrac{x^2}{4}-\dfrac{y^2}{5}=1$ **39.** $y^2-4x^2=1$ **41.** $\dfrac{5x^2}{2}-\dfrac{5(y-4)^2}{18}=1$ **43.** $\dfrac{(x-6)^2}{9}-\dfrac{(y-5)^2}{7}=1$ **45.** $\dfrac{(x+4)^2}{4}-\dfrac{(y-3)^2}{16}=1$ **47.** $\dfrac{(x-1)^2}{9}-\dfrac{(y+1)^2}{4}=1$ **49.** $\dfrac{(y+4)^2}{16}-\dfrac{(x-3)^2}{25}=1$

51.

53.

55.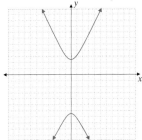

57. $\dfrac{x^2}{\left(6\times 10^7\right)^2}-\dfrac{y^2}{\left(7.2\times 10^7\right)^2}=1$

chapter six review, pages 498–499

1. $x=\dfrac{5}{2}$ **2.** no vertical asymptote **3.** $y=2x+9$ **4.** $y=\dfrac{1}{3}$

chapter 6 review answers

5.

6.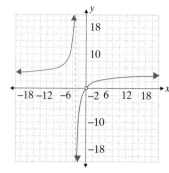

7. $\left(-3, \dfrac{7}{2}\right]$ **8.** $(-\infty, 2)$

9.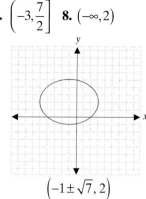
$(-1 \pm \sqrt{7}, 2)$

10.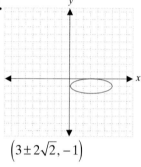
$(3 \pm 2\sqrt{2}, -1)$

11. $\dfrac{(x+1)^2}{9} + \dfrac{(y-4)^2}{16} = 1$

12. $\dfrac{(x-4)^2}{36} + \dfrac{(y-2)^2}{27} = 1$

13. $\dfrac{4(x-2)^2}{9} + \dfrac{4(y+1)^2}{9} = 1$

14. $\dfrac{(x-1)^2}{9} + \dfrac{(y+3)^2}{25} = 1$

15.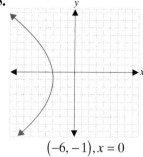
$(-6, -1), x = 0$

16.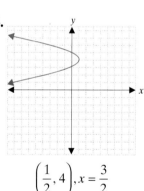
$\left(\dfrac{1}{2}, 4\right), x = \dfrac{3}{2}$

17. $(x+2)^2 = 4(y-3)$

18. $(x-5)^2 = 16(y+3)$

19. $(y+1)^2 = 2\left(x - \dfrac{5}{2}\right)$

20. $\dfrac{5}{4}$ inches

21.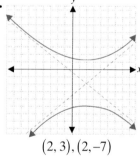
$(2, 3), (2, -7)$

22.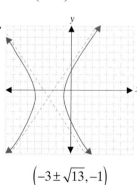
$(-3 \pm \sqrt{13}, -1)$

23. $\dfrac{(x-1)^2}{9} - \dfrac{(y+1)^2}{7} = 1$

24. $\dfrac{(x+1)^2}{4} - \dfrac{(y+2)^2}{25} = 1$

25. $\dfrac{(y-3)^2}{9} - \dfrac{(x+1)^2}{16} = 1$

26. $\dfrac{(y-2)^2}{9} - \dfrac{(x-2)^2}{1} = 1$

chapter six test, pages 500–503

1. $x = -1$ **2.** $x = -2$ **3.** none **4.** $x = 0$ **5.** none **6.** $x = \pm 2$ **7.** $x = 3$ **8.** none **9.** $y = 0$ **10.** $y = x - 2$
11. none **12.** $y = -1$ **13.** $y = -2$ **14.** none **15.** $y = \dfrac{-1}{3}$ **16.** $y = 0$

17.

18.

19.

20.

21.

22.

23.

24.

25.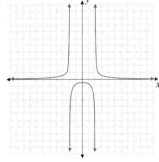

26. $x > 2$ **27.** $\dfrac{1}{2} \le x < 1$ **28.** $x < -2$ or $2 < x \le 4$ **29.** $x < -2$ or $-1 < x \le 1$ **30.** $-2 < x < 0$ or $x > 2$ **31.** $x < -4$ or $\dfrac{-7}{3} \le x < 1$ **32.** $-3 < x < \dfrac{8}{9}$ or $2 < x$ **33.** $x > -3$ **34.** $x < -8$ or $x > -3$ **35.** $1 < x \le \dfrac{5}{3}$ or $2 < x$ **36.** No; explanations may vary. **37.** $f(x) = \dfrac{x}{x^2 + 2x - 3}$ **38.** $f(x) = \dfrac{-2x^2 + 1}{x^2 + 2}$ **39.** $f(x) = \dfrac{-4x^2 + 1}{2x^2 - 3x}$ **40.** $f(x) = \dfrac{x + 3}{x^2 + x - 2}$
41. center: $(2, -1)$; vertices: $(2, 8), (2, -10)$; foci: $\left(2, -1 \pm \sqrt{65}\right)$ **42.** center: $(3, -1)$; vertices: $(7, -1), (-1, -1)$; foci: $\left(3 \pm 2\sqrt{3}, -1\right)$ **43.** center: $(0, 0)$; vertices: $\left(0, \pm 3\sqrt{3}\right)$; foci: $\left(0, \pm 3\sqrt{2}\right)$ **44.** center: $(-1, 2)$; vertices: $(-1, 0), (-1, 4)$; foci: $\left(-1, 2 \pm \dfrac{2\sqrt{5}}{3}\right)$ **45.** center: $(-3, 1)$; vertices: $(0, 1), (-6, 1)$; foci: $\left(-3 \pm \dfrac{3\sqrt{3}}{2}, 1\right)$ **46.** $\dfrac{(x-2)^2}{4} + (y-2)^2 = 1$
47. $\dfrac{(x-2)^2}{16} + \dfrac{y^2}{12} = 1$ **48.** $(x-2)^2 + \dfrac{(y+1)^2}{\frac{9}{4}} = 1$ **49.** $\dfrac{(x-3)^2}{32} + \dfrac{(y-2)^2}{36} = 1$ **50.** $\dfrac{x^2}{16} + \dfrac{(y-4)^2}{12} = 1$

51. $\dfrac{(x-5)^2}{25} + \dfrac{(y-6)^2}{36} = 1$ **52.** $\dfrac{x^2}{25} + \dfrac{y^2}{16} = 1$ **53.** $\dfrac{x^2}{48} + \dfrac{y^2}{64} = 1$ **54.** $30\pi a - \pi a^2$ **55.** $\dfrac{x^2}{400} + \dfrac{y^2}{100} = 1$

56.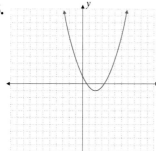
focus: $\left(2, \dfrac{-1}{2}\right)$; directrix: $y = \dfrac{-3}{2}$

57.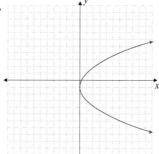
focus: $(1, -1)$; directrix: $x = -1$

58.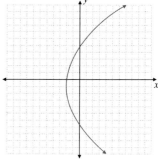
focus: $(2, -1)$; directrix: $x = -6$

59.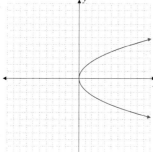
focus: $\left(\dfrac{3}{4}, 0\right)$; directrix: $x = \dfrac{-3}{4}$

60.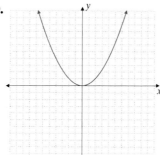
focus: $(0, 1)$; directrix: $y = -1$

61.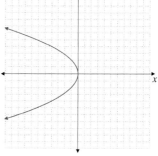
focus: $(-1, 0)$; directrix: $x = 1$

62.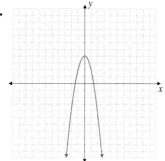
focus: $\left(0, \dfrac{31}{8}\right)$; directrix: $y = \dfrac{33}{8}$

63.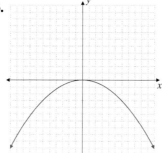
focus: $(0, -3)$; directrix: $x = 3$

64.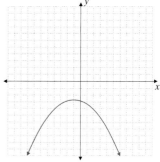
focus: $\left(-1, \dfrac{-25}{6}\right)$; directrix: $y = \dfrac{-7}{6}$

65.
focus: $\left(\dfrac{-21}{2}, 2\right)$; directrix: $x = \dfrac{-19}{2}$ **66.** $(y-2)^2 = -6\left(x + \dfrac{3}{2}\right)$

67. $(x-1)^2 = -4(y+1)$ **68.** $(x-4)^2 = 16(y+2)$ **69.** $(y+1)^2 = 16(x-2)$

70. $y^2 = 16(x+7)$ **71.** $(y-2)^2 = -20(x+1)$ **72.** $8\sqrt{6}$ m ≈ 19.6 m

73.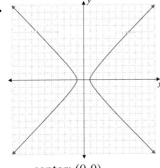
center: $(0,0)$
foci: $(\pm\sqrt{2},0)$
vertices: $(\pm 1,0)$

74.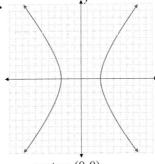
center: $(0,0)$
foci: $(\pm 5,0)$
vertices: $(\pm 3,0)$

75.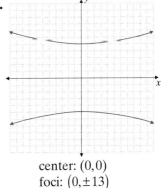
center: $(0,0)$
foci: $(0,\pm 13)$
vertices: $(0,\pm 5)$

76.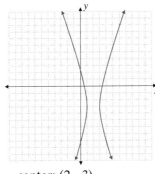
center: $(0,0)$
foci: $(\pm 2\sqrt{10},0)$
vertices: $(\pm 6,0)$

77.
center: $(-1,-2)$
foci: $(-1\pm\sqrt{145},-2)$
vertices: $(-13,-2),(11,-2)$

78.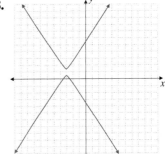
center: $(-3,1)$
foci: $\left(-3, 1\pm\dfrac{\sqrt{13}}{6}\right)$
vertices: $\left(-3,\dfrac{3}{2}\right),\left(-3,\dfrac{1}{2}\right)$

79.
center: $(2,-3)$
foci: $(2\pm\sqrt{10},-3)$
vertices: $(3,-3),(1,-3)$

80. center: $(-1,2)$
foci: $(-1\pm\sqrt{29},2)$
vertices: $(-3,2),(1,2)$

81. center: $(-1,-2)$,
foci: $(\pm 5,0)$
vertices: $(\pm 3,0)$

82. center: $(-1,2)$
foci: $(-1\pm 3\sqrt{5},2)$
vertices: $(-7,2),(5,2)$

83. center: $(0,2)$
foci: $(\pm 2\sqrt{10},2)$
vertices: $(\pm 6,2)$

84. $\dfrac{y^2}{4}-\dfrac{x^2}{12}=1$

85. $\dfrac{x^2}{9}-\dfrac{y^2}{16}=1$

86. $x^2-\dfrac{y^2}{9}=1$

87. $\dfrac{(y-2)^2}{4}-\dfrac{(x-3)^2}{36}=1$

chapter seven

section 7.1, pages 514-515

1.
3.
5.
7.
9.
11.
13.
15.
17.
19.
21.

23. $\{2\}$ **25.** $\{-2\}$ **27.** $\{-13\}$ **29.** $\{3\}$
31. $\{-2\}$ **33.** $\{-2,-1\}$ **35.** $\{7\}$ **37.** $\{3\}$
39. $\{9\}$ **41.** $\{-3\}$ **43.** $\{2\}$ **45.** $\{-1\}$
47. No Solution **49.** g **51.** b **53.** f
55. c **57.** j

section 7.2, pages 526-531

1. $V \approx 178$ people **3.** $C \approx \$8526.20$ **5.** $W \approx 93$ computers **7. (a)** 0.999567 **(b)** 0.958 grams **(c)** 0.648 grams **9. (a)** 3 years **(b)** 9 years **11.** 1118 rabbits **13.** The bank offering 2.75% and monthly compounding. **15.** Approx. 3.18% **17.** $134,392 **19. (a)** 10 **(b)** 7490 people **(c)** The function approaches 10,000 as time goes on. **21. (a)** 0.965936 **(b)** 0.707 kg **(c)** 7.628 mg **23. (a)** $1521.74 **(b)** $271.74 **25.** $9459.48; $9942.41 **27.** $2835.71 **29.** $20,000 **31.** $7318.71 **33. a.** $7647.95 **b.** $7647.57 **c.** yes; explanations may vary

section 7.3, pages 541-544

1. $4 = \log_5 625$ **3.** $3 = \log_x 27$ **5.** $3 = \log_{4.2} C$ **7.** $x = \log_4 31$ **9.** $\sqrt{3} = \log_{4x} 13$ **11.** $e^x = \log_2 11$ **13.** $81 = 3^4$
15. $4 = b^{\frac{1}{2}}$ **17.** $15 = 2^b$ **19.** $W = 5^{12}$ **21.** $2x = \pi^4$ **23.** $e^x = 2$

25.   **27.**   **29.**

31.   **33.**   **35.**

37. e **39.** b
41. h **43.** d
45. i **47.** −2
49. 3 **51.** −1
53. −4 **55.** −1

57. 0 **59.** $-\frac{1}{2}$ **61.** $\{8\}$ **63.** $\left\{\frac{1}{8}\right\}$ **65.** $\{1\}$ **67.** $\{-2\}$ **69.** No Solution **71.** $\{2\}$ **73.** $\{41.96\}$ **75.** $\{19.09\}$ **77.** $\{5.60\}$
79. $\{0.43\}$ **81.** $\{4297.83\}$ **83.** $\{10,000,000,002\}$

section 7.4, pages 555-560

1. $3 + 3\log_5 x$ **3.** $2 + \ln p - 3\ln q$ **5.** $1 + \log_9 x - 3\log_9 y$ **7.** $\frac{3}{2}\ln x + \ln p + 5\ln q - 7$ **9.** $\log(2 + 3\log x)$
11. $1 - \frac{1}{2}\log(x+y)$ **13.** $\log_2(y^2 + z) - 4\log_2 x - 4$ **15.** $2\log_b x + \frac{1}{2}\log_b y - \log_b z$ **17.** $2 + \log_b a + b\log_b c$
19. $\log\frac{x}{y}$ **21.** $\log_5(x+5)$ **23.** $\log_2\left(x^{\frac{4}{3}} + 3x^{\frac{1}{3}}\right)$ **25.** $\ln\left(\frac{3p}{q^2}\right)$ **27.** $\log\left(\frac{x-10}{x}\right)$ **29.** $\ln\left(\frac{e^2}{x^3 y^3}\right)$ **31.** $\log_5 4$
33. $\ln 45$ **35.** $\log_3 1 = 0$ **37.** $\ln 12$ **39.** $\log 11$ **41.** $\log_8(x^2 - y)$ **43.** x^2 **45.** $\frac{e^2 p}{x}$ **47.** $\frac{x^3}{y^4}$ **49.** x^2 **51.** 4
53. $12x^2$ **55.** 2.04 **57.** 0.95 **59.** 0.95 **61.** 2.45 **63.** 3.30 **65.** 0.74 **67.** 1.20 **69.** 1.86 **71.** −1 **73.** 3.85 **75.** 0.77
77. −1.76 **79.** 2 **81.** 7 **83.** 1 **85.** $4\sqrt{2} = 5.66$ **87.** 9.05 **89.** 2.08 **91.** 12 **93.** 1048576 **95.** 10.25
97. 5,011,872 times stronger **99.** 133 decibels **101.** 7.62, yes **103. a.** 15.05 minutes **b.** 7:00 pm **c.** 112°; no

section 7.5, pages 571-574

1. $x \approx 0.26$ **3.** $x \approx 3.12$ **5.** $x \approx 3.89$ **7.** $x \approx -2.28$ **9.** $x \approx 8.09$ **11.** $x \approx 2.68$ **13.** $x \approx -1.12$ **15.** $x \approx 52.77$ **17.** $x \approx \pm 0.71$
19. $x = -12$ **21.** $x \approx 1.32$ **23.** $x \approx 3.27$ **25.** $x = 125$ **27.** $x = 5$ **29.** $x \approx 40.17$ **31.** $x = 35$ **33.** $x \approx 9.38$ **35.** $x = 4$

section 7.5 answers

37. $x = \dfrac{1}{162}$ **39.** $x = 1$ **41.** $x \approx 100.04$ **43.** Undefined **45.** $\{5\}$ **47.** $\{8\}$ **49.** $\left\{\dfrac{37}{8}\right\}$ **51.** $\{5\}$ **53.** $\varnothing$ **55.** $\{6\}$
57. $\{\sqrt{2}-1\}$ **59.** $\{2,3\}$ **61.** $\{1,2\}$ **63.** $f(x) = \log 2x^2$ **65.** $f(x) = \ln 9x^2$ **67.** $f(x) = 256^x$ **69.** $f(x) = \ln 1 = 0$
71. $f(x) = \ln 5^x$ **73.** $f(x) = \ln 5$ **75. a.** 17.36 years **b.** 9.90 years **77.** 4.98 hours **79.** 4.99 years **81.** 0.271 years (about 99 days)

chapter seven review, pages 581-582

1.   **2.** 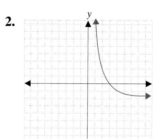   **3.** $\{3\}$ **4.** $\{-1\}$ **5.** $\{-2\}$
6. The account at 3.95%. **7.** 8 days
8. (a) $P(t) = 870(1.032^t)$
 (b) 68.37 minutes **9.** 2 **10.** -2 **11.** -3
12. $\{125\}$ **13.** $\{27\}$ **14.** $\{2\}$
15. $\dfrac{3}{2}\log x - \dfrac{5}{2}\log \pi - \log 2$ **16.** $\dfrac{5}{2}\ln a + \ln m + 2\ln n - 5$ **17.** $\log_2\left(\dfrac{a^{\frac{5}{3}}}{b^{\frac{1}{3}}c}\right)$ **18.** $\ln\left(\dfrac{4}{x^2 y^7}\right)$ **19.** $\dfrac{8 - \ln 16}{5} \approx 1.05$
20. $\dfrac{6}{\log 321} \approx 2.39$ **21.** $\{3\}$ **22.** $\{-2\}$ **23.** 20.4 months (1.7 years) **24.** 73.0 hours (about 3 days)

chapter seven test, pages 583-585

1.   **2.**   **3.**   **4.**

5. $x = -4$ **6.** $x = \dfrac{8}{5}$ **7.** $x = -4$ **8.** $x = -1$ **9.** $x = \dfrac{1}{3}$ **10.** $x = 2$ **11. a.** \$18,726 **b.** \$18,820 **12. a.** 173 flies **b.** 20 flies
13. 0.58 pounds; **14. a.** 10 grams **b.** 7.85 grams **15.** $\log_5 125 = x$ **16.** $x = \log_3 8$ **17.** $x = \ln 3\pi$
18. $\sqrt{2}\log 3a = 1$ **19.** $4^x = 64$ **20.** $3^{-3} = \dfrac{1}{27}$ **21.** $6x = e^2$ **22.** $a = 5^{10}$

23.   **24.**   **25.**

26.

27. -3 28. -2 29. 2 30. $\dfrac{2}{3}$ 31. -4 32. $\dfrac{1}{\sqrt{3}}$ 33. $x=\dfrac{-1}{4}$ 34. $x=8$

35. No solution 36. $3^{\frac{1}{3}}$ 37. $3+3\log_3 a$ 38. $\ln 2 + 2\ln a + \ln b - \ln 3 - \ln c$

39. $1+\log_4(3x+5y)$ 40. $1+\ln 2+\ln x$ 41. $\log\dfrac{xy}{c}$ 42. $\log_2(x-3)$

43. $\log\dfrac{a^2 b^3}{d\sqrt{c}}$ 44. $\log_3\dfrac{x^2-2x}{x^2+4}$ 45. $\log_5\dfrac{10x}{y^2-y}$ 46. $x=\dfrac{\ln 5-\ln 2}{3}$

47. $x=\dfrac{1-\log_5 7}{2}$ 48. $x=\dfrac{\ln 9-\ln 16}{\ln 81-\ln 4}$ 49. $x=\dfrac{2+\ln 40}{5}$ 50. $x=\dfrac{2+\ln 10}{1-2\ln 10}$ 51. $x=\dfrac{3}{\log_2 5}$ 52. $x=\dfrac{17}{2}$

53. $x=\dfrac{9}{2}$ 54. $x=\dfrac{-9}{5}$ 55. $x=\dfrac{3}{\sqrt{2}}$ 56. $x=2$ 57. $x=16, x=\dfrac{1}{4}$ 58. $x=0$ 59. $x=10$ 60. no solution

61. $x=9$ 62. $x=2, x=\dfrac{1}{2}$ 63. $I=3.16$ watts/m² 64. a. $251{,}188{,}643$ b. $7{,}079{,}458$ c. $1{,}258{,}925{,}412$

65. $f^{-1}(x)=\dfrac{\log_3(x+1)}{2}$ 66. $f^{-1}(x)=\ln(x-3)+2$ 67. $f^{-1}(x)=e^{\frac{x+1}{2}}-1$ 68. $f^{-1}(x)=2^{\frac{x}{3}}+3$

chapter eight

section 8.1, pages 601-604

1. $(-5,2)$ 3. $(5,3)$ 5. $\varnothing$ 7. $\left\{\left(\dfrac{y-3}{2},y\right)\bigg| y\in\mathbb{R}\right\}$ 9. $(-1,7)$ 11. $(3,11)$ 13. $\{(x,4x+1)|x\in\mathbb{R}\}$ 15. $(2,19)$

17. $(-5,1)$ 19. $(5,6)$ 21. $\{(-y-2,y)|y\in\mathbb{R}\}$ 23. $(-5,4)$ 25. $(-1,1)$ 27. $(3,-5)$ 29. $\varnothing$ 31. $(-1,3,0)$

33. $(2,2,-1)$ 35. $\left\{\left(\dfrac{y-z+2}{3},y,z\right)\bigg| y\in\mathbb{R}, z\in\mathbb{R}\right\}$ 37. $\varnothing$ 39. $\{(1,y,0)|y\in\mathbb{R}\}$ 41. $(9,1,1)$ 43. $(3,1,-2)$

45. $(4,5,5)$ 47. $\left(\dfrac{49}{3},\dfrac{-16}{3},\dfrac{5}{4}\right)$ 49. $(0,3,2)$ 51. $(-22,34,49)$ 53. $(0.43,1.28,3.64)$ 55. $(-3.42,2.98,2.76)$

57. $(6,8,7)$ 59. $a=-\dfrac{1}{2}, b=-\dfrac{3}{2}, c=5$ 61. 40 ounces of 12%, 20 ounces of 30% 63. $7000 at 8% and $5000 at 15% 65. Bob is 51 years old. 67. $800 in stocks 69. 50 tickets 71. Certificates of deposit: $3000, Stocks: $6000, Bonds: $1000

section 8.2, pages 613-618

1. a. 3×2 b. -1 c. none 3. a. 5×2 b. none c. 10 5. a. 3×4 b. none c. 286 7. a. 3×2 b. 1 c. none

9. a. 2×5 b. 5 c. 2 11. $\left[\begin{array}{ccc|c} -3 & 1 & -2 & -4 \\ \frac{1}{2} & -4 & -1 & 1 \\ 0 & -3 & 3 & 1 \end{array}\right]$ 13. $\left[\begin{array}{ccc|c} -\frac{3}{2} & -1 & 0 & -1 \\ 2 & 2 & 3 & 0 \\ 0 & -1 & 6 & 0 \end{array}\right]$ 15. $\left[\begin{array}{ccc|c} \frac{12}{5} & \frac{1}{2} & -\frac{3}{2} & \frac{1}{5} \\ 1 & 0 & 3 & 1 \\ 5 & 2 & 1 & -2 \end{array}\right]$ 17. $\left[\begin{array}{ccc|c} \frac{2}{3} & -\frac{4}{3} & -2 & 0 \\ 8 & -2 & 6 & 7 \\ 3 & -2 & 0 & 0 \end{array}\right]$

19. $\left[\begin{array}{ccc|c} \frac{1}{2} & -14 & -\frac{1}{4} & -8 \\ \frac{1}{5} & -\frac{7}{6} & \frac{1}{4} & -3 \\ 5 & -5 & \frac{8}{3} & -5 \end{array}\right]$ 21. $\begin{cases} x=8 \\ y=3 \end{cases}$ 23. $\begin{cases} x+3y+6z=16 \\ y+2z=9 \\ z=4 \end{cases}$ 25. $\begin{cases} 9y+13z=27 \\ 2x+21z=19 \\ 7x+18y=32 \end{cases}$ 27. $\left[\begin{array}{cc|c} 2 & -5 & 3 \\ 0 & -7 & 5 \end{array}\right]$

section 8.2 answers

29. $\begin{bmatrix} 1 & 3 & | & -2 \\ 9 & -2 & | & 7 \end{bmatrix}$ **31.** $\begin{bmatrix} 8 & -2 & | & -4 \\ -6 & 2 & | & -14 \end{bmatrix}$ **33.** $\begin{bmatrix} 4 & 12 & | & -6 \\ 9 & 9 & | & 6 \end{bmatrix}$ **35.** $\begin{bmatrix} 4 & -1 & | & 5 \\ -6 & 2 & | & 0 \end{bmatrix}$ **37.** $\begin{bmatrix} 18 & -6 & 15 & | & 42 \\ -7 & 19 & 2 & | & 3 \\ -4.5 & 5.5 & -2 & | & 3.5 \end{bmatrix}$

39. $\begin{bmatrix} 5 & 18 & 22 & | & 5 \\ 32 & -9 & -27 & | & -23 \\ -9 & 21 & 12 & | & 9 \end{bmatrix}$ **41.** $\begin{bmatrix} 0 & 1 & -9 & | & -3 \\ 1 & 1 & 3 & | & 4 \\ 0 & 0 & 0 & | & 0 \end{bmatrix}$ **43.** $\begin{bmatrix} -1 & 4 & | & -3 \\ 1 & -6 & | & \frac{5}{2} \end{bmatrix}$ **45.** $\begin{bmatrix} 1 & 5 & -9 & | & 11 \\ 0 & -1 & 8 & | & -7 \\ 0 & -17 & 41 & | & 1 \end{bmatrix}$ **47.** Neither **49.** Neither **51.** Neither

53. $(3,-1)$ **55.** $(1,3)$ **57.** $(-7,3)$ **59.** $\varnothing$ **61.** $(3,2)$ **63.** $\{(-2y-4, y) | y \in \mathbb{R}\}$ **65.** $\varnothing$ **67.** $(4,0,3)$ **69.** $(15,-21,8)$
71. $(3,-5)$ **73.** $\{(x, -3x-2) | x \in \mathbb{R}\}$ **75.** $(-4,1)$ **77.** $(6,4)$ **79.** $(-11,-5)$ **81.** $(7,3,3)$ **83.** $(2,2,-1)$
85. $\{(1, y, 0) | y \in \mathbb{R}\}$ **87.** $(3,-2,3)$ **89.** $(2,3,4)$ **91.** $(9,-19,7)$ **93.** $(1,-2,-1,3)$ **95.** $42, 26, 87$ **97.** small: 10, medium: 24, large: 48

section 8.3, pages 631-634

1. 11 **3.** 15 **5.** $ab - x^2$ **7.** 8 **9.** -10 **11.** -39 **13.** $\{-2,3\}$ **15.** $\{-5,1\}$ **17.** $\{-5,-4\}$ **19.** $\{-6,4\}$ **21.** $\{2,5\}$ **23.** 3 **25.** -9
27. 2 **29.** -2 **31.** 159 **33.** 78 **35.** -254 **37.** 404 **39.** 4 **41.** 120 **43.** 10 **45.** x^4 **47.** x^8 **49.** -0.051 **51.** 12.595
53. 21334 **55.** $(76,-53)$ **57.** $\{(-y-2, y) | y \in \mathbb{R}\}$ **59.** $\varnothing$ **61.** $\{(-3z-5, -6z-10, z) | z \in \mathbb{R}\}$
63. $\left\{\left(\dfrac{-5y-z-5}{2}, \dfrac{-5y+3z-19}{2}, y, z\right) \middle| y \in \mathbb{R}, z \in \mathbb{R}\right\}$ **65.** $\{(-z+8, -5z+31, -2z+37, z) | z \in \mathbb{R}\}$ **67.** $(1647, 2071)$
69. $\varnothing$ **71.** $(-3,-1,0,-4)$ **73.** $(5,-3,3)$ **75.** 6 ft, 17 ft, 20 ft **77.** soybeans: 147.5 acres, corn: 295 acres, wheat: 57.5 acres

section 8.4, pages 646-649

1. $\begin{bmatrix} 5 & -1 \\ 0 & 0 \\ 2 & 13 \end{bmatrix}$ **3.** $\begin{bmatrix} 6 & -3 \\ 18 & 30 \\ -9 & 21 \end{bmatrix}$ **5.** not possible **7.** $\begin{bmatrix} 14 & -14 \\ 8 & 0 \\ -4 & 14 \end{bmatrix}$ **9.** $\begin{bmatrix} -7 & 5 \\ 3 & 10 \\ -3 & -8 \end{bmatrix}$ **11.** not possible
13. $a=2, b=-2, c=-1$ **15.** not possible **17.** $x=10, y=5$ **19.** $x=3, y=1$ **21.** not possible **23.** $a=8, b=5$
25. $[24 \ -5]$ **27.** $[35 \ 18]$ **29.** not possible **31.** $[-30 \ -3]$ **33.** $\begin{bmatrix} 15 & -3 & -24 \\ 25 & -5 & -40 \\ 30 & -6 & -48 \end{bmatrix}$ **35.** $[-34 \ -7]$ **37.** $\begin{bmatrix} 11 & 0 \\ 0 & 11 \end{bmatrix}$

39. $\begin{bmatrix} 32 & -20 \\ 56 & -35 \\ -16 & 10 \end{bmatrix}$ **41.** not possible **43.** $\begin{bmatrix} 14 & -13 \\ -13 & 5 \end{bmatrix}$ **45.** $[179 \ 76]$ **47.** $\begin{bmatrix} 23.94 & -7.56 & 28.98 \\ 21.66 & -6.84 & 26.22 \end{bmatrix}$ **49.** $[-23.94 \ -26.72]$
51. $\begin{bmatrix} -79.59 \\ 39.21 \\ 10.08 \end{bmatrix}$ **53.** $\dfrac{2}{3}$ for store A; $\dfrac{1}{3}$ for store B **55.** Solution is incorrect. Explanations may vary.

section 8.5, pages 661-663

1. $\begin{bmatrix} 14 & -5 \\ 1 & 9 \end{bmatrix}\begin{bmatrix} x \\ y \end{bmatrix} = \begin{bmatrix} 7 \\ 2 \end{bmatrix}$ **3.** $\begin{bmatrix} \frac{3}{5} & -\frac{8}{5} \\ 0 & 1 \end{bmatrix}\begin{bmatrix} x \\ y \end{bmatrix} = \begin{bmatrix} 2 \\ 2 \end{bmatrix}$ **5.** $\begin{bmatrix} 3 & -7 & 1 \\ 1 & -1 & 0 \\ 0 & 8 & 5 \end{bmatrix}\begin{bmatrix} x_1 \\ x_2 \\ x_3 \end{bmatrix} = \begin{bmatrix} -4 \\ 2 \\ -3 \end{bmatrix}$ **7.** $\begin{bmatrix} 1 & 2 \\ 9 & -3 \end{bmatrix}\begin{bmatrix} x \\ y \end{bmatrix} = \begin{bmatrix} -6 \\ -14 \end{bmatrix}$

section 8.7 answers

9. $\begin{bmatrix} 4 & -3 \\ 2 & -4 \end{bmatrix} \begin{bmatrix} x \\ y \end{bmatrix} = \begin{bmatrix} -9 \\ 13 \end{bmatrix}$ **11.** $\begin{bmatrix} 2 & -1 & 3 \\ -1 & 1 & 0 \\ 4 & -5 & 1 \end{bmatrix} \begin{bmatrix} x \\ y \\ z \end{bmatrix} = \begin{bmatrix} 0 \\ 17 \\ -2 \end{bmatrix}$ **13.** $\begin{bmatrix} -\frac{1}{20} & -\frac{1}{5} \\ \frac{1}{4} & 0 \end{bmatrix}$ **15.** $\begin{bmatrix} -5 & -4 \\ 4 & 3 \end{bmatrix}$ **17.** $\begin{bmatrix} -5 & 0 \\ 2 & 2 \end{bmatrix}$

19. not invertible **21.** $\begin{bmatrix} 2 & 1 & -4 \\ -4 & -2 & -3 \\ -1 & -1 & -4 \end{bmatrix}$ **23.** $\begin{bmatrix} -1 & 2 & -1 \\ 0 & -1 & 1 \\ 0 & -4 & 3 \end{bmatrix}$ **25.** $\begin{bmatrix} -1 & -2 & 1 \\ -2 & 1 & -3 \\ 1 & 2 & 0 \end{bmatrix}$ **27.** $\begin{bmatrix} -2 & 1 & 1 \\ 2 & 0 & -1 \\ -1 & 0 & 1 \end{bmatrix}$

29. $\begin{bmatrix} 2 & -1 & 2 \\ 0 & 1 & -1 \\ -3 & -2 & -4 \end{bmatrix}$ **31.** No **33.** Yes **35.** No **37.** $\begin{bmatrix} -\frac{2}{11} & \frac{3}{11} \\ -\frac{1}{11} & \frac{7}{11} \end{bmatrix}$ **39.** $\begin{bmatrix} -\frac{7}{12} & -\frac{1}{3} & \frac{1}{2} \\ -\frac{5}{12} & -\frac{2}{3} & \frac{1}{2} \\ -\frac{1}{12} & -\frac{1}{3} & \frac{1}{2} \end{bmatrix}$ **41.** $\begin{bmatrix} -0.044 & 0.146 & -0.018 \\ -0.146 & 0.121 & -0.124 \\ -0.026 & 0.070 & -0.156 \end{bmatrix}$

43. $\left(-2, -\frac{5}{2}\right)$ **45.** $\left\{\left(\frac{3y-1}{2}, y\right) \middle| y \in \mathbb{R}\right\}$ **47.** $(-2, 0)$ **49.** $(-13, 6, 19)$ **51.** $\varnothing$ **53.** $(-4, 5, -1)$ **55.** $(-13, 19, 23)$ $(0, 0, -1)$ $(1, -1, -1)$ **57.** $(1, -8, 7)$ $(3, 1, 1)$ $(4, 2, 0)$

section 8.6, pages 670-674

1.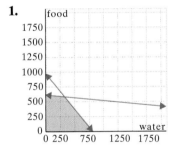
$60x + 50y \le 50{,}000;$
$x + 10y \le 6000;$
$x \ge 0;\ y \ge 0$

3.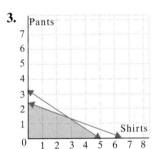
$12x + 32y \le 80;$
$2x + 3y \le 10$
$x \ge 0;\ y \ge 0$

5. Min = 0 at $(0, 0)$, Max = 21 at $(0, 7)$

7. Min = 0 at $(0, 0)$, Max = 30 at $(0, 6)$

9. Min = 0 at $(0, 0)$, Max = 48 at $(0, 8)$

11. Min = 0 at $(0, 0)$, Max = 15 at $(2.5, 0)$

13. Min = 0 at $(0, 0)$, Max = 50 at $(3, 4)$

15. Min = 0 at $(0, 0)$, Max = 48 at $(8, 0)$

17. Min = 15.64 at $(2.18, 0.91)$, Max = 133.33 at $(0, 13.33)$

19. Model X: 1000 units, Model Y: 500 units, Maximum Profit: $76,000 **21.** Ashley should buy 3 olive bundles and 2 cranberry bundles. The total cost to make the curtains is $70. **23.** $20,000 should be placed in municipal bonds, $5000 should be placed in Treasury bills.

section 8.7, pages 683-686

1. $\{(0, -3), (2, 0)\}$ **3.** $\{(-1, -1), (-1, 1), (1, -1), (1, 1)\}$ **5.** $\{(-1, 1), (1, 1)\}$

  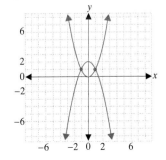

section 8.7   answers

7. $\{(-1, 2), (1, 2)\}$

9. no solution

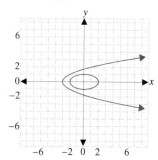

11. $\{(-3, 1), (1, 1)\}$

13. $\{(0, 0)\}$

15. $\{(\pm 3, 0)\}$

17. $\{(\pm 1, 0)\}$

19. $\{(2, \pm 3), (-2, \pm 3)\}$

21. $\{(5, 3)\}$

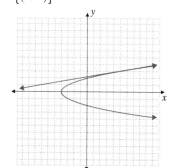

23. $\{(\pm 2, 0), (0, -2)\}$

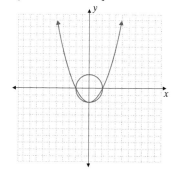

25. $\{(-i\sqrt{6}, -6), (i\sqrt{6}, -6), (-\sqrt{5}, 5), (\sqrt{5}, 5)\}$ **27.** $\{(-2, 3)\}$ **29.** $\{(-1, -4), (1, 4), (-2\sqrt{2}, -\sqrt{2}), (2\sqrt{2}, \sqrt{2})\}$

31. $\{(-3i, -5), (3i, -5), (0, 4)\}$ **33.** $\{(-\sqrt{3}, 0), (\sqrt{3}, 0)\}$ **35.** $\{(-i, -6), (i, -6), (-\sqrt{2}, 3), (\sqrt{2}, 3)\}$

37. $\left\{(1,1), (1,-1), \left(-\dfrac{3}{2}, \dfrac{\sqrt{14}}{2}\right), \left(-\dfrac{3}{2}, -\dfrac{\sqrt{14}}{2}\right)\right\}$ **39.** $\left\{\left(\dfrac{7}{2}, \dfrac{17}{4}\right)\right\}$ **41.** $\{(-6, -i\sqrt{15}), (-6, i\sqrt{15}), (2, -1), (2, 1)\}$

43. $\left\{(0, 1), \left(-\dfrac{100}{101}, -\dfrac{99}{101}\right)\right\}$ **45.** $\left\{\left(-\dfrac{\sqrt{42}}{6}, -\dfrac{\sqrt{66}}{6}\right), \left(-\dfrac{\sqrt{42}}{6}, \dfrac{\sqrt{66}}{6}\right), \left(\dfrac{\sqrt{42}}{6}, -\dfrac{\sqrt{66}}{6}\right), \left(\dfrac{\sqrt{42}}{6}, \dfrac{\sqrt{66}}{6}\right)\right\}$

47. $\{(-5, 0), (-2, 0), (-1, 0)\}$ **49.** $\left\{(3, 2), (2, 3), \left(\dfrac{-1 \pm i\sqrt{23}}{2}, \dfrac{12}{-1 \pm i\sqrt{23}}\right)\right\}$ **51.** $\left\{(0, -1), \left(\dfrac{\sqrt{6}}{3}, \dfrac{2\sqrt{6}}{9} - 1\right),\right.$

$\left.\left(-\dfrac{\sqrt{6}}{3}, \dfrac{-2\sqrt{6}}{9} - 1\right)\right\}$ **53.** $\{(\sqrt{2} + 1, 2\sqrt{2}), (-\sqrt{2} + 1, -2\sqrt{2}), (2\sqrt{2} + 1, \sqrt{2}), (-2\sqrt{2} + 1, -\sqrt{2})\}$ **55.** $\{(4, 1), (1, 1 + i\sqrt{3})\}$

57. $\{(0,4),(0,-3),(3,-2),(3,3)\}$ **59.** $\{(4,0)\}$ **61.** $\left\{\left(\pm\dfrac{1}{2}, \pm\dfrac{1}{4}\right)\right\}$ **63.** no solution **65.** no solution **67.** $\{(-3.345, 4.856)\}$ **69.** 9 inches by 5 inches **71.** 60 mph and 70 mph **73.** -12 and 7 **75.** $h = 6$ cm, $r = 3$ cm

chapter eight review, pages 691-694

1. $\varnothing$ **2.** $(4,-5,5)$ **3.** $(3,0,2)$ **4.** $(2,-1)$ **5.** $\begin{bmatrix} 2 & 1 & -1 & | & 3 \\ -2 & 3 & -1 & | & 2 \\ 6 & -8 & 1 & | & 3 \end{bmatrix}$ **6.** $\begin{bmatrix} 4 & 5 & -1 & | & 0 \\ 1 & 3 & 2 & | & 3 \\ 10 & -1 & -6 & | & 0 \end{bmatrix}$ **7.** $\begin{bmatrix} 0 & -5 & | & -11 \\ 1 & 2 & | & 3 \end{bmatrix}$

8. $\begin{bmatrix} 2 & 3 & | & 5 \\ 0 & 5 & | & 12 \end{bmatrix}$ **9.** $2x^4$ **10.** 5 **11.** 7 **12.** x^8 **13.** $(-4, 1)$ **14.** $(2,-1,1)$ **15.** $w=-2, x=1, y=3, z=-4$

16. not possible **17.** $[\,12\ \ 46\,]$ **18.** $\begin{bmatrix} -12 & 8 & 12 \\ -15 & 10 & 15 \\ -18 & 12 & 18 \end{bmatrix}$ **19.** $\begin{bmatrix} 1 & -8 \\ 4 & 2 \end{bmatrix}\begin{bmatrix} x \\ y \end{bmatrix} = \begin{bmatrix} -3 \\ -3 \end{bmatrix}$ **20.** $\begin{bmatrix} 1 & -1 & 2 \\ 2 & -3 & -1 \\ -3 & 0 & 6 \end{bmatrix}\begin{bmatrix} x_1 \\ x_2 \\ x_3 \end{bmatrix} = \begin{bmatrix} -4 \\ 1 \\ 5 \end{bmatrix}$

21. $\begin{bmatrix} \dfrac{3}{16} & \dfrac{1}{8} \\ -\dfrac{1}{8} & \dfrac{1}{4} \end{bmatrix}$ **22.** $\begin{bmatrix} 1 & -2 \\ -\dfrac{1}{2} & 2 \end{bmatrix}$

23. $x \geq 20,\ y \geq 10,\ x \leq 40,\ y \leq 40,\ x+y \leq 60$ **24.** $x \geq 6,\ y \geq 6,\ x \leq 10,\ y \leq 8,\ x+y \geq 15$

25. Min = 0 at $(0,0)$, Max = 30 at $(5,0)$ **26.** Min = 2 at $(0,2)$, Max = 11 at $(4,3)$ **27.** Min = 20 at $(0,10)$, Max = 50 at $(10,0)$ **28.** Min = 8 at $(0,2)$ Max = 24 at $(3.43, 1.71)$ **29.** Min = 350 at $(5,0)$ Max = 1028 at $(10,4)$ **30.** Min = 0 at $(0,0)$ Max = 27.33 at $(7.33, 3.16)$

31. 12 vases should be produced, 12 pitchers should be produced, Max profit: $660 **32.** 35 bionic arms, 15 bionic legs, Min cost: $24,000 **33.** $\left\{(i,-1), (-i,-1), \left(\dfrac{\sqrt{2}}{2}, \dfrac{1}{2}\right), \left(-\dfrac{\sqrt{2}}{2}, \dfrac{1}{2}\right)\right\}$

34. $\left\{\left(\dfrac{\sqrt{70}}{5}, \dfrac{\sqrt{10}}{5}\right), \left(\dfrac{\sqrt{70}}{5}, -\dfrac{\sqrt{10}}{5}\right), \left(-\dfrac{\sqrt{70}}{5}, \dfrac{\sqrt{10}}{5}\right), \left(-\dfrac{\sqrt{70}}{5}, -\dfrac{\sqrt{10}}{5}\right)\right\}$

35. $\left\{\left(\dfrac{3+i\sqrt{3}}{2}, \dfrac{-1+3i\sqrt{3}}{2}\right), \left(\dfrac{3-i\sqrt{3}}{2}, \dfrac{-1-3i\sqrt{3}}{2}\right)\right\}$ **36.** 9 and 16 **37.** 36 mph and 24 mph

chapter eight test, pages 695-699

1. $(2,1)$ **2.** $\varnothing$ **3.** $(0,1)$ **4.** $(-2,2)$ **5.** $\left(\dfrac{7}{13}, \dfrac{22}{13}\right)$ **6.** $\{(3-3y, y)\,|\,y \in \mathbb{R}\}$ **7.** $\varnothing$ **8.** $(0,-3)$ **9.** $(3,2)$ **10.** No solution

11. $\left(\dfrac{3}{2}, \dfrac{4}{3}, \dfrac{19}{6}\right)$ **12.** $(-1,1,2)$ **13.** $(8,12,10)$ **14.** $\varnothing$ **15.** 15 **16.** $\begin{cases} 3x - y = 4 \\ 6x + 2y = 0 \end{cases}$ **17.** $y = \dfrac{4}{15}x^2 - x + \dfrac{11}{15}$

18. $x^2 + y^2 - 4x + 4y - 1 = 0$ **19.** Answers may vary. **20.** 440 mph, 480 mph **21. a.** 4×2 **b.** 3 **c.** None

chapter 8 test answers

22. $\begin{bmatrix} 3 & -1 & 5 & | & 1 \\ 1 & 2 & 0 & | & 1 \\ 3 & -4 & 1 & | & -1 \end{bmatrix}$ **23.** $\begin{bmatrix} 0 & 1 & -2 & | & 1 \\ 3 & 1 & -1 & | & 4 \\ 0 & -4 & 6 & | & -1 \end{bmatrix}$ **24.** $\begin{bmatrix} 3 & -2 & 2 & | & 24 \\ 4 & -8 & -2 & | & 3 \\ 6 & -3 & 1 & | & 13 \end{bmatrix}$ **25.** $\begin{bmatrix} 2 & -5 & -1 & | & 4 \\ 1 & -1 & 1 & | & 6 \\ -4 & -3 & 1 & | & 4 \end{bmatrix}$ **26.** $\begin{bmatrix} 1 & -4 & | & -4 \\ 1 & 7 & | & 11 \end{bmatrix}$

27. $\begin{bmatrix} 4 & -3 & | & -2 \\ 1 & -2 & | & 3 \end{bmatrix}$ **28.** $\begin{bmatrix} -1 & 0 & 2 & | & -6 \\ 2 & -6 & 8 & | & 2 \\ -1 & -1 & -5 & | & 6 \end{bmatrix}$ **29.** $\begin{bmatrix} -9 & 6 & 10 & | & 5 \\ 5 & -10 & -8 & | & -7 \\ -1 & 2 & 4 & | & 3 \end{bmatrix}$ **30.** $(2,-1)$ **31.** $(0,-3)$ **32.** $\left(\dfrac{-50}{3},-16\right)$

33. $(1,2,3)$ **34.** $(0,-2,1)$ **35.** $(1,1,1)$ **36.** -6 **37.** 8 **38.** $\dfrac{-7}{5}$ **39.** 0 **40.** ± 4 **41.** $5,-4$ **42.** $\pm 2i\sqrt{2}$ **43.** $(2,-2)$

44. $\varnothing$ **45.** $(3,-1,1)$ **46.** $(4,2,0)$ **47.** Not possible **48.** $(2,-3)$ **49.** $(1,-1)$ **50.** $(3,-4,0)$ **51.** $\begin{bmatrix} -6 & 18 \end{bmatrix}$

52. $\begin{bmatrix} 0 & -1 & -5 \\ 6 & -3 & 3 \\ 8 & -6 & -6 \end{bmatrix}$ **53.** Not possible **54.** $\begin{bmatrix} -1 & 8 \\ -4 & 7 \end{bmatrix}$ **55.** Not possible **56.** $\begin{bmatrix} -28 & 22 \end{bmatrix}$ **57.** $\begin{bmatrix} 18 & -7 \\ -14 & 11 \end{bmatrix}$ **58.** Not possible **59.** $\begin{bmatrix} 0 & 1 \\ -1 & 3 \end{bmatrix}$ **60.** Not possible **61.** $\begin{bmatrix} 22 & -8 \end{bmatrix}$ **62.** $\begin{bmatrix} 2 & -1 \\ 1 & 1 \end{bmatrix}\begin{bmatrix} x \\ y \end{bmatrix}=\begin{bmatrix} 4 \\ 0 \end{bmatrix}$ **63.** $\begin{bmatrix} \dfrac{1}{3} & \dfrac{-1}{3} \\ 1 & 0 \end{bmatrix}\begin{bmatrix} x \\ y \end{bmatrix}=\begin{bmatrix} 4 \\ -1 \end{bmatrix}$

64. $\begin{bmatrix} 3 & -1 & 1 \\ 2 & 0 & -5 \\ 4 & 3 & 0 \end{bmatrix}\begin{bmatrix} x \\ y \\ z \end{bmatrix}=\begin{bmatrix} 4 \\ 1 \\ 6 \end{bmatrix}$ **65.** $\begin{bmatrix} 3 & -2 & 1 \\ 1 & -1 & 0 \\ 0 & 1 & -1 \end{bmatrix}\begin{bmatrix} x \\ y \\ z \end{bmatrix}=\begin{bmatrix} 1 \\ 6 \\ 4 \end{bmatrix}$ **66.** Not possible **67.** $\begin{bmatrix} \dfrac{7}{11} & \dfrac{-3}{11} \\ \dfrac{-5}{11} & \dfrac{-1}{11} \end{bmatrix}$ **68.** $\begin{bmatrix} \dfrac{1}{12} & \dfrac{1}{6} & \dfrac{11}{24} \\ \dfrac{5}{12} & \dfrac{-1}{6} & \dfrac{7}{24} \\ \dfrac{1}{12} & \dfrac{1}{6} & \dfrac{-1}{24} \end{bmatrix}$

69. Not possible **70.** $(4,3),(-4,3),(-4,-3),(4,-3)$ **71.** $(2,1),(-2,1),\left(-\sqrt{5},0\right),\left(\sqrt{5},0\right)$

72. $(-1,3),\left(\dfrac{-3}{2},2\right)$ **73.** $(2,1)$ **74.** $(1,4),(-1,-4),(4,1),(-4,-1)$ **75.** $(i,3),(i,-3),(-i,3),(-i,-3)$

chapter nine

section 9.1, pages 716-719

1. infinite **3.** finite **5.** finite **7.** infinite **9.** infinite **11.** $2,-1,-4,-7,-10$ **13.** $1,\dfrac{3}{2},\dfrac{9}{5},2,\dfrac{15}{7}$ **15.** $\dfrac{2}{3},\dfrac{-4}{9},\dfrac{8}{27},\dfrac{-16}{81},\dfrac{32}{243}$

17. $\dfrac{1}{2},\dfrac{4}{3},\dfrac{9}{4},\dfrac{16}{5},\dfrac{25}{6}$ **19.** $-1,6,-5,20,-27$ **21.** $2,2,\dfrac{18}{7},\dfrac{16}{5},\dfrac{50}{13}$ **23.** $2,1,\dfrac{8}{9},1,\dfrac{32}{25}$ **25.** $10,5,0,-5,-10$

27. $\dfrac{1}{27},\dfrac{1}{81},\dfrac{1}{243},\dfrac{1}{729},\dfrac{1}{2187}$ **29.** $8,2^{\frac{3}{2}},2^{\frac{3}{4}},2^{\frac{3}{8}},2^{\frac{3}{16}}$ **31.** $\dfrac{1}{3},1,\dfrac{9}{5},\dfrac{8}{3},\dfrac{25}{7}$ **33.** $-1,2,-3,4,-5$ **35.** $2,6,12,20,30$

37. $\sqrt{3}+1,\sqrt{6}+1,4,2\sqrt{3}+1,\sqrt{15}+1$ **39.** $0,-4,0,18,56$ **41.** $-2,-11,-74,-515,-3602$ **43.** $-1,0,-1,0,-1$

45. $3,1,3,1,3$ **47.** $a_n=(-2)^n$ **49.** $a_n=\dfrac{n}{n+2}$ **51.** $a_1=1,a_n=\dfrac{1}{n}a_{n-1},n\geq 2$ **53.** $a_n=\dfrac{4-n}{n+13}$ **55.** $a_n=4-5n$

57. $a_1=1,\ a_n=n+n(a_{n-1}),n\geq 2$ **59.** $-3-12-27-48-75=-165$ **61.** $\displaystyle\sum_{i=0}^{7}(1+3i)=92$ **63.** $\displaystyle\sum_{i=3}^{9}i^2=280$

65. $-36-30-22-12+0+14+30+48=-8$ **67.** $S_n=\dfrac{n}{4(n+4)},S_{100}=\dfrac{25}{104}$ **69.** $S_n=2^n-1$, sum does not converge **71.** $S_n=\dfrac{n}{2(n+1)},S_{49}=\dfrac{49}{100}$ **73.** $S_n=-\ln(n+1),S_{100}=-\ln 101$ **75.** $S_n=\dfrac{2n}{7(2n+7)},S_{30}=\dfrac{60}{469}$

77. $S_n = -\ln(n+1), S_{65} = -\ln(66)$ **79.** $-9, 1, -8, -7, -15$ **81.** $-17, 13, -4, 9, 5$ **83.** $2, -3, -7, -24, -79$ **85.** 987
87. $A_n = A_{n-1}(1.005) - 74; A_4 = 352.63; A_6 = 207.79$; 9 months

section 9.2, pages 727-731

1. $a_n = 3n - 5$ **3.** $a_n = -2n + 9$ **5.** $a_n = 9n - 4$ **7.** $a_n = -6n + 9$ **9.** $a_n = 19n + 5$ **11.** $a_n = n + \frac{5}{2}$
13. $a_n = -\frac{19}{2}n + \frac{43}{2}$ **15.** $a_n = -2n + 1$ **17.** $a_n = -4n + 33$ **19.** $d = 2$ **21.** $d = 1$ **23.** no **25.** no **27.** 13 **29.** 2
31. 2 **33.** $d = 2.5; 5, 7.5, 10, 12.5, 15$ **35.** $d = 7; 7, 14, 21, 28, 35$ **37.** $d = 9; -62, -53, -44, -35, -26$ **39.** 5 **41.** 1
43. -2 **45.** 195 **47.** a_{73} **49.** 117 **51.** 26 **53.** -8 **55.** 13 **57.** 55 **59.** 14,350 **61.** 17,114 **63.** -1475 **65.** $-\frac{3219}{5}$
67. 902 **69.** -1316 **71.** 266.4 **73.** -152 **75.** 10,382.05 **77.** 6 years **79.** 1620 pounds **81.** $625; $8100

section 9.3, pages 745-749

1. $a_n = -3(2)^{n-1}$ **3.** $a_n = 2\left(\frac{-1}{3}\right)^{n-1}$ **5.** $a_n = \left(\frac{-1}{4}\right)^{n-1}$ **7.** $a_n = \left(\frac{1}{7}\right)^{n-1}$ **9.** $a_n = (-3)^{n-1}$ **11.** $a_n = 3\left(\frac{2}{3}\right)^{n-1}$
13. $a_n = 7(-2)^{n-1}$ **15.** $a_n = \frac{1}{16}(2)^{n-1}$ **17.** $a_n = \frac{39}{68}\left(\frac{4}{3}\right)^{n-1}$ **19.** No **21.** $r = \frac{1}{2}$ **23.** $r = 2$ **25.** $r = 7$ **27.** $r = 3; 8,$
$24, 72, 216, 648$ **29.** $r = 2; \frac{1}{4}, \frac{1}{2}, 1, 2, 4$ **31.** $r = \frac{1}{5}; 62500, 12500, 2500, 500, 100$ **33.** $\frac{5}{16,384}$ **35.** $-2,147,483,648$
37. $r = \pm 2$ **39.** $r = \frac{1}{5}$ **41.** $\frac{52222139775}{1048576} \approx 49,802.9$ **43.** $\frac{10923}{16384} \approx 0.666687$ **45.** $\frac{73810}{19683} \approx 3.749936$
47. $-109,200$ **49.** $-\frac{3}{2}$ **51.** does not converge **53.** $\frac{2476099}{160000} \approx 15.475619$ **55.** does not converge
57. $\frac{28561}{152064} \approx 0.187822$ **59.** $\frac{123}{999}$ **61.** $-\frac{35}{9}$ **63.** $\frac{989}{99}$ **65.** 9.98 **67.** 2.137×10^{-18} **69.** 6.54×10^{51}
71. $14,802.44 **73.** 1.845×10^{19} **75.** approximately 13,778 students **77.** $S_{30} = 1.1 \times 10^{30}; r = 10$ **79.** Yes; explanations may vary (any example such that $r = 1$ and $d = 0$).

section 9.4, pages 755-759

1. $S_{k+1} = \frac{1}{3k+9}$ **3.** $S_{k+1} = \frac{(k+1)(k+2)(2k+3)}{4}$

5. Basic Step: $n = 1, 1 = 1$ and $\frac{1(1+1)}{2} = 1$;

Induction Step: If $1 + 2 + 3 + \cdots + k = \frac{k(k+1)}{2}$, then $(1 + 2 + 3 + \cdots + k) + (k+1) = \frac{k(k+1)}{2} + (k+1)$

$= \frac{k^2 + k + 2k + 2}{2} = \frac{(k+1)(k+2)}{2}$

section 9.4 answers

7. Basic Step: $n = 1$, $2(1) = 2$ and $1(1+1) = 2$;

Induction Step: If $2 + 4 + 6 + \cdots + 2k = k(k+1)$, then $(2 + 4 + 6 + \cdots + 2k) + 2(k+1) = k^2 + k + 2k + 2 = (k+1)(k+2)$

9. Basic Step: $n = 1$, $4^{1-1} = 1$ and $\dfrac{4^1 - 1}{3} = 1$;

Induction Step: If $4^0 + 4^1 + 4^2 + \cdots + 4^{k-1} = \dfrac{4^k - 1}{3}$, then $4^0 + 4^1 + 4^2 + \cdots + 4^{k-1} + 4^{k+1-1} = \dfrac{4^k - 1}{3} + 4^k$

$= \dfrac{4^k - 1 + 3 \cdot 4^k}{3} = \dfrac{4 \cdot 4^k - 1}{3} = \dfrac{4^{k+1} - 1}{3}$

11. Basic Step: $n = 1$, $\dfrac{1}{(3(1) - 2)(3(1) + 1)} = \dfrac{1}{4}$ and $\dfrac{1}{3(1) + 1} = \dfrac{1}{4}$;

Induction Step: If $\dfrac{1}{1 \cdot 4} + \dfrac{1}{4 \cdot 7} + \dfrac{1}{7 \cdot 10} + \cdots + \dfrac{1}{(3k-2)(3k+1)} = \dfrac{k}{3k+1}$, then $\dfrac{1}{1 \cdot 4} + \dfrac{1}{4 \cdot 7} + \dfrac{1}{7 \cdot 10} + \cdots$

$+ \dfrac{1}{(3k-2)(3k+1)} + \dfrac{1}{(3(k+1)-2)(3(k+1)+1)} = \left[\dfrac{1}{1 \cdot 4} + \dfrac{1}{4 \cdot 7} + \dfrac{1}{7 \cdot 10} + \cdots + \dfrac{1}{(3k-2)(3k+1)}\right] + \dfrac{1}{(3k+1)(3k+4)}$

$= \dfrac{k}{3k+1} + \dfrac{1}{(3k+1)(3k+4)} = \dfrac{3k^2 + 4k + 1}{(3k+1)(3k+4)} = \dfrac{(3k+1)(k+1)}{(3k+1)(3k+4)} = \dfrac{(k+1)}{(3(k+1)+1)}$

13. Basic Step: $n = 1$, $5(1) = 5$ and $\dfrac{5(1)(1+1)}{2} = 5$;

Induction Step: If $5 + 10 + 15 + \cdots + 5k = \dfrac{5k(k+1)}{2}$, then $5 + 10 + 15 + \cdots + 5k + 5(k+1)$

$= (5 + 10 + 15 + \cdots + 5k) + 5k + 5 = \dfrac{5k(k+1)}{2} + 5k + 5 = \dfrac{5k^2 + 15k + 10}{2} = \dfrac{5(k+1)(k+2)}{2} = \dfrac{5(k+1)[(k+1)+1]}{2}$

15. Basic Step: $n = 1$, $1 + \dfrac{1}{1} = 2$ and $1 + 1 = 2$

Induction Step: If $\left(1 + \dfrac{1}{1}\right)\left(1 + \dfrac{1}{2}\right)\left(1 + \dfrac{1}{3}\right) \cdots \left(1 + \dfrac{1}{k}\right) = k + 1$, then $\left(1 + \dfrac{1}{1}\right)\left(1 + \dfrac{1}{2}\right)\left(1 + \dfrac{1}{3}\right) \cdots \left(1 + \dfrac{1}{k}\right)\left(1 + \dfrac{1}{k+1}\right)$

$= (k+1)\left(1 + \dfrac{1}{k+1}\right) = k + 1 + \dfrac{k+1}{k+1} = (k+1) + 1$

17. Basic Step: $n = 4$, $4! = 24$ and $2^4 = 16$, so $4! > 2^4$;

Induction Step: If $k! > 2^k$, then $(k+1)! = k!(k+1) > 2^k(k+1) \geq 2^k(4+1) > 2^k \cdot 2 = 2^{k+1}$

19. Basic Step: $n = 1$, $-(1+1) = -2$ and $-\dfrac{1}{2}(1^2 + 3(1)) = -2$;

Induction Step: If $-2 - 3 - 4 - \cdots - (k+1) = -\dfrac{1}{2}(k^2 + 3k)$, then $-2 - 3 - 4 - \cdots - (k+1) - ((k+1)+1)$

$= -\dfrac{1}{2}(k^2 + 3k) - (k+2) = -\dfrac{k^2 + 3k + 2k + 4}{2} = -\dfrac{(k^2 + 2k + 1) + (3k + 3)}{2} = -\dfrac{1}{2}((k+1)^2 + 3(k+1))$

21. Basic Step: $n = 1$, since $2^1 = 2$, $2^1 > 1$;

Induction Step: If $2^k > k$, then $2^{k+1} = 2^1 \cdot 2^k > 2k = k + k \geq k + 1$

section 8.4

23. Basic Step: $n = 1$, $1(1+1) = 2$ and $\dfrac{1(1+1)(1+2)}{3} = 2$;

Induction Step: If $1 \cdot 2 + 2 \cdot 3 + 3 \cdot 4 + \cdots + k(k+1) = \dfrac{k(k+1)(k+2)}{3}$, then $[1 \cdot 2 + 2 \cdot 3 + 3 \cdot 4 + \cdots + k(k+1)]$

$+ (k+1)((k+1)+1) = \dfrac{k(k+1)(k+2)}{3} + (k+1)(k+2) = \dfrac{k(k+1)(k+2) + 3(k+1)(k+2)}{3}$

$= \dfrac{(k+1)(k+2)(k+3)}{3} = \dfrac{(k+1)((k+1)+1)((k+1)+2)}{3}$

25. Basic Step: $n = 5$, $2^5 = 32$ and $4 \cdot 5 = 20$, so $2^5 > 4 \cdot 5$;

Induction Step: If $2^k > 4k$, then $2^{k+1} = 2 \cdot 2^k = 2^k + 2^k > 4k + 4k > 4k + 4 = 4(k+1)$

27. Basic Step: $n = 1$, $1^5 = 1$ and $\dfrac{1^2(1+1)^2 \left(2(1)^2 + 2(1) - 1\right)}{12} = 1$;

Induction Step: If $1^5 + 2^5 + 3^5 + \cdots + k^5 = \dfrac{k^2(k+1)^2(2k^2 + 2k - 1)}{12}$, then $\left(1^5 + 2^5 + 3^5 + \cdots + k^5\right) + (k+1)^5$

$= \dfrac{k^2(k+1)^2(2k^2 + 2k - 1)}{12} + (k+1)^5 = \dfrac{(k+1)^2 \left[k^2(2k^2 + 2k - 1) + 12(k+1)^3\right]}{12}$

$= \dfrac{(k+1)^2(2k^4 + 14k^3 + 34k^2 + 36k + 12)}{12} = \dfrac{(k+1)^2(k+2)^2(2k^2 + 6k + 3)}{12} = \dfrac{(k+1)^2(k+2)^2 \left(2(k+1)^2 + 2(k+1) - 1\right)}{12}$

29. Basic Step: $n = 1$, $2(1) - 1 = 1$ and $1^2 = 1$; Induction Step: If $1 + 3 + 5 + 7 + \cdots + 2k - 1 = k^2$,

then $(1 + 3 + 5 + 7 + \cdots + 2k - 1) + (2(k+1) - 1) = k^2 + (2k+1) = (k+1)^2$

31. Basic Step: $n = 1$, $(a^m)^1 = a^m$ and $a^{m \cdot 1} = a^m$;

Induction Step: If $(a^m)^k = a^{mk}$, then $(a^m)^{k+1} = (a^m)^k \cdot (a^m)^1 = a^{mk} \cdot a^m = a^{mk+m} \cdot a^{m(k+1)}$

33. Basic Step: $n = 1$, $\left(2^{2(1)-1} + 3^{2(1)-1}\right) = 5$ so 5 is a factor;

Induction Step: If $\left(2^{2(k)-1} + 3^{2(k)-1}\right) = 5p$ for some integer p, then $\left(2^{2(k+1)-1} + 3^{2(k+1)-1}\right) = \left(2^{2k+1} + 3^{2k+1}\right)$

$= 2^2 \cdot 2^{2k-1} + 3^2 \cdot 3^{2k-1} = 4 \cdot 2^{2k-1} + 9 \cdot 3^{2k-1} = 4 \cdot 2^{2k-1} + (4+5) \cdot 3^{2k-1} = 4 \cdot 2^{2k-1} + 4 \cdot 3^{2k-1} + 5 \cdot 3^{2k-1}$

$= 4\left(2^{2k-1} + 3^{2k-1}\right) + 5 \cdot 3^{2k-1} = 4(5p) + 5 \cdot 3^{2k-1} = 5\left(4p + 3^{2k-1}\right)$

35. Basic Step: $n = 1$, $\left(1^3 + 3(1)^2 + 2(1)\right) = 6 = 3 \cdot 2$, so 3 is a factor;

Induction Step: If $\left(k^3 + 3k^2 + 2k\right) = 3p$ for some integer p, then $(k+1)^3 + 3(k+1)^2 + 2(k+1)$

$= k^3 + 6k^2 + 11k + 6 = \left(k^3 + 3k^2 + 2k\right) + \left(3k^2 + 9k + 6\right) = 3p + 3\left(k^2 + 3k + 2\right) = 3\left(p + k^2 + 3k + 2\right)$

37. Basic Step: $n = 1$, $5^1 - 1 = 4$, which is divisible by 4;

Induction Step: If $\dfrac{5^k - 1}{4} = p$ or $5^k - 1 = 4p$ for some integer p,

then $5^{k+1} - 1 = 5 \cdot 5^k - 5 + 4 = 5\left(5^k - 1\right) + 4 = 5(4p) + 4 = 4(5p + 1)$

section 9.4 answers

39. $1+2+4+8+\cdots+2^{n-1}=2^n-1$; Basic Step: $n=1$, $2^{1-1}=1$ and $2^1-1=1$;
Induction Step: If $1+2+4+8+\cdots+2^{k-1}=2^k-1$, then $(1+2+4+8+\cdots+2^{k-1})+2^{k+1-1}$
$=(2^k-1)+2^k=2\cdot 2^k-1=2^{k+1}-1$

41. Induction Steps: If $k=2t+5n$, then...
Case 1 $(t\geq 1, n\geq 1, \text{and } k=2t+5n)$: $k+1=2t+5n+1=2t+5n+(6-5)=(2t+6)+(5n-5)=2(t+3)+5(n-1)$
Case 2 $(t=0, n\geq 1, \text{and } k=5n)$: $k+1=5n+1=5n+(6-5)=6+(5n-5)=2(3)+5(n-1)$
Case 3 $(t\geq 2, n=0, \text{and } k=2t)$: $k+1=2t+1=2t+(5-4)=(2t-4)+5=2(t-2)+5(1)$

section 9.5, pages 775-782

1. Combination **3.** Combination **5.** $10^3=1000$ **7.** $9^7=4,782,969$ **9.** $15!\approx 1.308\times 10^{12}$ **11.** $3!=6$
13. $5^{10}=9,765,625$ **15.** $36^6=2,176,782,336$ **17.** $26\cdot 25\cdot 24\cdot 10\cdot 9\cdot 8=11,232,000$ **19.** $_{30}P_{12}\approx 4.143\times 10^{16}$
21. $_{36}P_8\approx 1.220\times 10^{12}$ **23.** $_7P_6=5040$ $_7P_7=5040$ as well. (Having a child remain standing is numerically equivalent to putting a seventh chair in the room.) **25.** $_{26}P_3=15,600$ **27.** 210 **29.** 6.0823×10^{16} **31.** 6
33. 116,280 **35.** $_7C_3=35$ **37.** $_9C_2=36$ **39.** $_{75}C_5=17,259,390$ **41.** $\dfrac{6!}{2!3!}=60$ **43.** $\dfrac{7!}{2!}=2520$
45. $\dfrac{9!}{2!2!}=90,720$ **47.** $_{10}C_4\cdot _8C_4\cdot _{13}C_4=10,510,500$ **49.** 112 cones **51.** 96 outfits **53.** 288 schedules **55.** 120 5-letter strings **57.** 303,600 ways **59.** 495 pizzas **61.** 752,538,150 groups **63.** 420 ways
65. $x^7-14x^6y+84x^5y^2-280x^4y^3+560x^3y^4-672x^2y^5+448xy^6-128y^7$ **67.** $x^8-4x^6y^3+6x^4y^6-4x^2y^9+y^{12}$
69. $4096x^6+30,720x^5y^2+96,000x^4y^4+160,000x^3y^6+150,000x^2y^8+75,000xy^{10}+15,625y^{12}$
71. $x^{15}-5x^{12}y^2+10x^9y^4-10x^6y^6+5x^3y^8-y^{10}$ **73.** $a^3-6a^2b+3a^2c+12ab^2-12abc+3ac^2-8b^3+12b^2c-6bc^2+c^3$
75. $8x^3+36x^2y-12x^2z+54xyz-36xyz+6xz^2+27y^3-27y^2z+9yz^2-z^3$ **77.** -80
79. $8192x^{13}+159,744x^{12}+1,437,696x^{11}$ **81.** $2,008,326,144x^{14}$ **83.** $831,409,920x^4y^8$
85. $\binom{n}{n-k}=\dfrac{n!}{(n-k)!(n-(n-k))!}=\dfrac{n!}{(n-k)!(n-n+k)!}=\dfrac{n!}{(n-k)!k!}=\binom{n}{k}$
87. $2^n=(1+1)^n=\sum_{k=0}^{n}\binom{n}{k}(1)^k(1)^{n-k}=\sum_{k=0}^{n}\binom{n}{k}=\binom{n}{0}+\binom{n}{1}+\cdots+\binom{n}{n}$

section 9.6, pages 793-797

1. $\dfrac{3}{5}$ **3.** $\dfrac{9}{13}$ **5.** $\dfrac{1}{3}$ **7. a.** 0 **b.** $\dfrac{5}{8}$ **9. a.** 0 **b.** $\dfrac{3}{5}$ **11. a.** 0 **b.** 1 **13. a.** $\dfrac{1}{8}$ **b.** $\dfrac{9}{16}$ **15.** The set of all ordered 4-tuples made up of H's and T's. There are 16 such 4-tuples. **17.** The set of all ordered pairs that have either an H or a T in the first slot and one of the 13 hearts in the second slot. There are 26 such ordered pairs. **19.** The set of all ordered triples with any of the 6 values in each slot. There are 216 such triples. **21.** The set of the 38 pockets. **23. a.** $\dfrac{2}{3}$ **b.** $\dfrac{1}{3}$ **25. a.** $\dfrac{3}{8}$ **b.** $\dfrac{1}{8}$ **c.** $\dfrac{1}{2}$ **27.** $\dfrac{3}{10}$ **29.** 0.3874 **31.** $\dfrac{3}{8}$ **33. a.** $\dfrac{11}{26}$ **b.** $\dfrac{9}{52}$ **c.** $\dfrac{2}{13}$ **35. a.** $\dfrac{1}{169}$ **b.** $\dfrac{1}{221}$ **37.** 18.75% **39. a.** $\dfrac{1}{6}$ **b.** $\dfrac{5}{18}$ **c.** $\dfrac{1}{9}$ **41.** $\dfrac{1}{20}$ **43.** $\dfrac{2}{9}$ **45.** $\dfrac{3}{10}$ **47.** 27 tickets

chapter nine review, pages 802-804

1. $a_n = 6n - 13$ **2.** $a_n = \dfrac{3^{n-1}}{2^n}$ **3.** $a_n = n^2 - 1$ **4.** $-3 - 5 - 7 - 9 - 11 - 13 = -48$ **5.** $-2 + 4 - 8 + 16 - 32 + 64 = 42$

6. $\displaystyle\sum_{i=2}^{7} i^3 = 783$ **7.** $a_n = -3n + 8$ **8.** $a_n = \dfrac{5}{2}n + 9$ **9.** $a_n = -9n + 20$ **10.** 8827 **11.** -7140 **12.** $\dfrac{381}{512}$ **13.** $40{,}955$

14. -12 **15.** does not converge **16.** 1

17. Basic Step: $n = 1$, $1^2 = 1$ and $\dfrac{1(1+1)(2(1)+1)}{6} = 1$; Induction Step: If $1 + 4 + 9 + \cdots + k^2 = \dfrac{k(k+1)(2k+1)}{6}$,

then $(1 + 4 + 9 + \cdots + k^2) + (k+1)^2 = \dfrac{k(k+1)(2k+1)}{6} + (k+1)^2 = \dfrac{(k+1)\left[2k^2 + k + 6(k+1)\right]}{6}$

$= \dfrac{(k+1)\left[2k^2 + 7k + 6\right]}{6} = \dfrac{(k+1)(k+2)(2k+3)}{6} = \dfrac{(k+1)((k+1)+1)(2(k+1)+1)}{6}$

18. Basic Step: $n = 1$, $\dfrac{1}{(2(1)-1)(2(1)+1)} = \dfrac{1}{3}$ and $\dfrac{1}{2(1)+1} = \dfrac{1}{3}$;

Induction Step: If $\dfrac{1}{1\cdot 3} + \dfrac{1}{3\cdot 5} + \dfrac{1}{5\cdot 7} + \cdots + \dfrac{1}{(2k-1)(2k+1)} = \dfrac{k}{2k+1}$, then $\dfrac{1}{1\cdot 3} + \dfrac{1}{3\cdot 5} + \dfrac{1}{5\cdot 7} +$

$\cdots + \dfrac{1}{(2k-1)(2k+1)} + \dfrac{1}{(2(k+1)-1)(2(k+1)+1)} = \dfrac{k}{2k+1} + \dfrac{1}{(2k+1)(2k+3)} = \dfrac{2k^2 + 3k + 1}{(2k+1)(2k+3)}$

$= \dfrac{(2k+1)(k+1)}{(2k+1)(2(k+1)+1)} = \dfrac{(k+1)}{(2(k+1)+1)}$

19. Basic Step: $n = 1$, $(3(1) + 2) = 5$ and $\dfrac{1(3(1)+7)}{2} = 5$; Induction Step: If $5 + 8 + 11 + \cdots + (3k+2) = \dfrac{k(3k+7)}{2}$, then

$5 + 8 + 11 + \cdots + (3k+2) + (3(k+1)+2) = \dfrac{k(3k+7)}{2} + (3k+5) = \dfrac{3k^2 + 7k + 6k + 10}{2} = \dfrac{(k+1)(3k+10)}{2} = \dfrac{(k+1)(3(k+1)+7)}{2}$

20. Basic Step: $n = 1$, $1(1+2) = 3$ and $\dfrac{1(1+1)(2(1)+7)}{6} = 3$;

Induction Step: If $1\cdot 3 + 2\cdot 4 + 3\cdot 5 + \cdots + k(k+2) = \dfrac{k(k+1)(2k+7)}{6}$, then $1\cdot 3 + 2\cdot 4 + 3\cdot 5 + \cdots + k(k+2)$

$+ (k+1)(k+1+2) = \dfrac{k(k+1)(2k+7)}{6} + (k+1)(k+3) = \dfrac{(k+1)\left[k(2k+7) + 6(k+3)\right]}{6} = \dfrac{(k+1)(k+2)(2k+9)}{6}$

$= \dfrac{(k+1)((k+1)+1)(2(k+1)+7)}{6}$

21. Basic Step: $n = 1$, $11^1 - 7^1 = 4$, which is divisible by 4;

Induction Step: If $\dfrac{11^k - 7^k}{4} = p$ or $11^k - 7^k = 4p$ for some integer p, then

$11^{k+1} - 7^{k+1} = 11\cdot 11^k - 7\cdot 7^k = 4\cdot 11^k + 7\cdot 11^k - 7\cdot 7^k = 4\cdot 11^k + 7(11^k - 7^k) = 4\cdot 11^k + 7(4p) = 4(11^k + 7p)$

22. Basic Step: $n = 1$, $7^1 - 1 = 6$, and $2\cdot 3 = 6$; Induction Step: If $\dfrac{7^k - 1}{3} = p$ or $7^k - 1 = 3p$

for some integer p, then $7^{k+1} - 1 = 7\cdot 7^k - 7 + 6 = 7(7^k - 1) + 6 = 7(3p) + 6 = 21p + 6 = 3(7p + 2)$

chapter 9 review answers

23. $8 \cdot 10^3 \cdot 26^3 = 140{,}608{,}000$ **24.** $8^7 = 2{,}097{,}152$ **25.** $\dfrac{8!}{3!2!} = 3{,}360$ **26.** $_{10}P_3 = 720$ **27.** $_{21}P_5 = 2{,}441{,}880$

28. $_{12}C_4 = 495$ **29.** $x^3 + 6x^2y + 3x^2z + 12xy^2 + 3xz^2 + 12xyz + 8y^3 + 12y^2z + 6yz^2 + z^3$ **30. a.** $\dfrac{4}{13}$ **b.** $\dfrac{5}{26}$ **c.** $\dfrac{7}{26}$

31. $\dfrac{33}{108{,}290}$ **32.** there is a 52.17% chance

chapter nine test, pages 805-808

1. $-3, -5, -7, -9, -11$ **2.** $-3, 9, -27, 81, -243$ **3.** $-1, -4, -9, -16, -25$ **4.** $-1, \sqrt[3]{2}, -\sqrt[3]{3}, \sqrt[3]{4}, -\sqrt[3]{5}$ **5.** $1, 3, 6, 10, 15$

6. $-3, -4, -5, -6, -7$ **7.** $1, \dfrac{1}{2}, \dfrac{1}{6}, \dfrac{1}{24}, \dfrac{1}{120}$ **8.** $1, \dfrac{1}{2}, \dfrac{2}{9}, \dfrac{3}{32}, \dfrac{24}{625}$ **9.** $a_n = n^2 - 2$ **10.** $a_n = \dfrac{2n+1}{n+1}$ **11.** $a_n = \dfrac{(-1)^n}{n^2}$

12. $-1 + 1 + 3 + 5 + 7 + 9 = 24$ **13.** $-2 - 8 - 18 - 32 - 50 - 72 - 98 - 128 - 162 = -570$ **14.** $-8 - 16 - 32 - 64 - 128 = -248$ **15.** $-2 - 9 - 20 - 35 - 54 - 77 - 104 - 135 - 170 - 209 - 252 - 299 = -1366$

16. $S_n = \dfrac{n}{2n+4}$, $S_{100} = \dfrac{25}{51}$ **17.** $S_n = \dfrac{1}{3}\left(\dfrac{n}{n+1}\right) \Sigma S_n = \dfrac{1}{3}$ **18.** $S_n = \ln\dfrac{2}{n+2}$, $S_{80} = \ln\dfrac{1}{41}$ **19.** $3, 6, 9, 15, 24$

20. $4, -2, 0, -2, -4$ **21.** $a_n = -4n + 3$ **22.** $a_n = 3n - 1$ **23.** $a_n = 8n - 12$ **24.** $a_n = \dfrac{3}{4}n - \dfrac{1}{4}$

25. $a_n = 4 \cdot 3^{n-1}$ **26.** $a_n = -3 \cdot 2^{n-1}$ **27.** $a_n = 2 \cdot 5^{n-1}$ **28.** $a_n = \dfrac{9}{5}\left(\dfrac{2}{3}\right)^{n-1}$ **29.** Yes, $S = \dfrac{1}{2}$ **30.** No **31.** $a_6 = 54$

32. Basic Step: $n = 1$, $(x_1)^{-1} = x_1^{-1}$;

Induction Step: If $(x_1 \cdot x_2 \cdot x_3 \cdot \ldots \cdot x_n)^{-1} = x_1^{-1} \cdot x_2^{-1} \cdot x_3^{-1} \cdot \ldots \cdot x_n^{-1}$ then $(x_1 \cdot x_2 \cdot x_3 \cdot \ldots \cdot x_n \cdot x_{n+1})^{-1}$
$= (x_1 \cdot x_2 \cdot x_3 \cdot \ldots \cdot x_n)^{-1} (x_{n+1})^{-1} = x_1^{-1} \cdot x_2^{-1} \cdot x_3^{-1} \cdot \ldots \cdot x_n^{-1} \cdot (x_{n+1})^{-1} = x_1^{-1} \cdot x_2^{-1} \cdot x_3^{-1} \cdot \ldots \cdot x_n^{-1} \cdot x_{n+1}^{-1}$.

33. Basic Step: $n = 4$, $4! > 2^4$; Induction Step: If $n! > 2^n$ then $(n+1)! = (n+1)n! > 2^n(n+1) > 2^n \cdot 2 = 2^{n+1}$.

34. Basic Step: $n = 1$, $2(1) = 2$ and $1(1+1) = 2$; Induction Step: If $2 + 4 + 6 + 8 + \ldots + 2n = n(n+1)$ then
$2 + 4 + 6 + 8 + \ldots + 2n + 2(n+1) = n(n+1) + 2(n+1) = n^2 + n + 2n + 2 = n^2 + 3n + 2 = (n+1)(n+2) = (n+1)((n+1)+1)$.

35. Basic Step: $n = 1$, $5(1) - 3 = 2$ and $\dfrac{1}{2}(5(1) - 1) = 2$;

Induction Step: If $2 + 7 + 12 + 17 + \ldots + (5n - 3) = \dfrac{n}{2}(5n - 1)$ then $2 + 7 + 12 + 17 + \ldots + (5n - 3) + (5(n+1) - 3)$
$= \dfrac{n}{2}(5n - 1) + (5(n+1) - 3) = \dfrac{5n^2 - n}{2} + 5n + 5 - 3 = \dfrac{5n^2 - n}{2} + \dfrac{10n - 4}{2} = \dfrac{5n^2 + 9n + 4}{2}$
$= \dfrac{(n+1)}{2}(5n + 4) = \left(\dfrac{n+1}{2}\right)(5n + 5 - 1) = \left(\dfrac{n+1}{2}\right)(5(n+1) - 1)$.

36. 60 **37.** 10,000,000 **38.** 676 **39.** 3,276,000 **40.** $\dfrac{7!}{2!3!2!} = 210$ **41.** 180 **42.** $\binom{9}{7} = \binom{9}{2} = 36$

43. $a^6 - 18a^5b + 135a^4b^2 - 540a^3b^3 + 1215a^2b^4 - 1458ab^5 + 729b^6$ **44.** $-527 - 336i$ **45.** $8b^3 - 24b^2c + 24bc^2 - 8c^3$

46. $\dfrac{31}{32}$ **47.** $\dfrac{1}{9}$ **48.** The probability of rolling a 3 with one die is higher. $\dfrac{1}{6}$ vs. $\dfrac{5}{36}$ **49. a.** $\dfrac{52}{125}$ **b.** $\dfrac{4}{5}$

c. $\dfrac{37}{500}$ **50. a.** $\dfrac{\binom{5}{3}}{\binom{12}{3}} + \dfrac{\binom{4}{3}}{\binom{12}{3}} + \dfrac{\binom{3}{3}}{\binom{12}{3}} = \dfrac{3}{44}$ **b.** $\dfrac{\binom{5}{1}\binom{4}{1}\binom{3}{1}}{\binom{12}{3}} = \dfrac{3}{11}$ **c.** $\dfrac{\binom{5}{3}}{\binom{12}{3}} = \dfrac{1}{22}$ **d.** $\dfrac{\binom{3}{2}\binom{4}{1}}{\binom{12}{3}} = \dfrac{3}{55}$

index of applications

applications

Note: **Examples are bold;** *chapter projects are italicized.*

biology and life science

Biological predecessors (sequences), 719
Body Mass Index (BMI), 26, (variation) 301
Concentration of a drug in the blood stream, 529, 447
Depth of Mariana Trench vs. Dead Sea, 196
Diameter of a white blood cell, 40
Family orchard (region of constraint), 229, 665
Fat and calorie intake, 634
Fencing a garden plot (area), 41, 110
Finding a body's surface area (DuBois Method), 26
Height of Mount Everest, 110
Human height (percentage), 109
Ideal Body Weight, 163
Maximize area of a garden plot, 280
Number of pairs of DNA in 46 chromosomes, 40
Number of people having the flu, 529
Ozone Layer, 349
Plutonium remains from nuclear reactor explosion in Chernobyl, 583
Pneumothorax, 75
Population
 of cows, 719
 of fruit flies, 583
 of mosquitos, 376
 of rabbits, 526, 446
 of sea lions, 376
Probability of rain, 795, 796, 804
Spread of a virus through African tribal villages, 526
Stocking a lake with fish (exponential function), 527
Temperature in a city (inequality), 117
Water level
 of a garden pool (intervals of monotonicity), 320
 of a river (intervals of monotonicity), 316
Weight of a person flying, 297

business applications

Amount of CDs sold (variation), 301
Constraints of furniture production, 670
Cost and revenue of producing and selling telescopes, 375
Cost function of producing widgets, 333
Defective units (inequality), 117
Depreciation of construction equipment, 212
Determining number of pizzas sold, 618
Expenses for an automobile manufacturer, 419
Finance, 108
Growth of a Computer Virus, 577
Market Share (matrices), 687
Maximizing
 area for parking lot enclosure, 285
 income for cruiseline, 284
 product, 283, 284
 profit
  for bionic arms and legs, 694
  for cell phone manufacturer, 674
  for pottery, 693
  of computer models, 672
  of fan production, 673
  of minidisc players, 669
 revenue
  for apartment rentals, 284
  for machinery rentals, 284
Minimizing cost
 for bionic arms and legs, 694
 for cell phone manufacturer, 674
 for manufacturing golf clubs, 284
 for producing cars, 284
Number
 of customers (transition matrix), 648
 of TV commercials compared to auto sales, 213
Profit
 function for MP3 players, 376
 of skateboards (polynomial equation), 375
Rate of furniture production (region of constraint), 233
Relationship
 between cost and dolls produced at a toy factory, 232
 between cost and production of canoes, 233
Salary with a linear growth pattern, 249
Sales increasing linearly, 212
Stacking cans for a display (arithmetic sequence), 730
Time function
 of a hotdog vendor, 320
 of a newspaper stand, 320
Total cost at a bottle manufacturer, 212
Video rental variation with price, 354
Revenue from the sale of
 compact refrigerators, 375
 cordless phones, 375
 fire extinguishers, 375

chemistry and physics

Acceleration of a falling object, 26
Acidity
 of a pool, 559
 of a solution, 552
 of detergent, 558
 of lemon juice, 558
 of tomato juice, 558
Angular frequency, 163
Bacteria growth, 518, 530, 573, 581

865

index of applications

Bouncing ball (geometric sequence), 747
Bungee jumping (quadratic equation), 135
Earthquake magnitude (logarithm), 553, 585
Einstein's Theory of Special Relativity, 60, 163
Force on an object in circular motion, 111, 163
Force to keep a car from skidding (variation), 302
Half-life
    of an unknown element, 574
    of Carbon-11, 529
    of Carbon-14, 567, 573, 583
    of Cesium-137, 574
    of Iodine-131, 581
    of Polonium-210, 527
    of Radium, 527
    of Sodium-24, 582
Ideal gas law, 108
Inductance of a circuit, 163
Kepler's Third Law, 163
Lateral acceleration, 163
Length of a spring (Hooke's Law), 298, 302
Magnitude of California earthquake (logarithm), 558
Mass of an electron (exponents), 36
Mechanical energy, 163
Mixture (systems of equations), 599, 603
Newton's Law of Cooling, 560, 573
Object's distance from a lens, 447
Orbit
    of Halley's Comet (eccentricity), 464
    of Pluto (eccentricity), 464
    of the Earth (Kepler's Law), 459, (eccentricity), 464
Orbital period of circular Keplerian orbits, 163
Path of a comet, 491
Period of a pendulum, 162
Radioactive decay of Carbon-14, 520
Schwarzschild Radius Formula (black hole), 163

Shooting a BB (quadratic equation), 134
Temperature conversions, 108
Tossing a ball from a building (quadratic equation), 134, 285
Tossing a ball in a stadium (quadratic equation), 131
Tossing a rock (quadratic equation), 134, 285
Trajectory of a reflected beam of light, 222
Velocity, 108
Volume of a gas in a pressurized container, 303
Weight of an astronaut in space, 303

combinatorics

Acceptance of University students, 748
Arithmetic growth of allowance, 724
Arithmetic growth of payments, 730, 731
Arrangement
    of books, 776, 777
    of letters, 769, 777, 779, 780, 804, 807
    of people for a photo, 776
Basketball line-up, 781
Bingo numbers, 778
Choices on an insurance form, 807
Choosing
    a cookie, 796
    a crayon, 796
    leaders of a meeting, 777, 804
Classroom seating chart, 775
Class schedule arrangement, 780
Clothing selections, 779
Creation of
    ice cream cones, 779
    pizzas, 779, 780
Combination lock codes, 763, 765, 775
Comparing salaries, 738
Computer user-ID account (combinatorics), 777
Crayon colors, 776
Determining a phone number, 775

Gender outcomes, 780
Impedance of a series circuit, 84
Kinetic energy of protons, 108
Lottery numbers, 778
Musical chairs, 777
Number of seats in a theater, 730
Order
    of playing cards, 766, 775, 779, 780
    of runners, 775, 777, 780
Ordering a meal, 807
Pascal's triangle, 782
Radio station names, 776
Representatives of a committee, 775, 778
Selecting ice cream flavors, 775
Terms of a geometric sequence, 736
Test questions, 776, 807
True/false questions, 778
Voltage of a series circuit, 84
Wheat on a chess board, 748
Winners of a shot put competition, 804

construction

Bulb of a motorcycle headlight (parabola), 499
Bulb of a spotlight (parabola), 477
Constructing a bridge (parabola, ellipse), 493-494
Constructing a watering trough (maximum), 283
Constructing garden beds (nonlinear systems), 684
Determining the length of cables (Pythagorean Theorem), 242
Enclosing a small garden, 232
Focus of a solar furnace (parabola), 477
Focus of the Hale Telescope (parabola), 473
Height of a bridge, 464
Maximizing area of a pasture, 284
Maximize dog-training arenas, 283
Maximum load supported by a beam, 302
Moving bricks (arithmetic sequence), 730

index of applications

Radius of a drainage pipe (variation), 302
Receiver of a satellite dish antenna (parabola), 477
Suspension cables on a bridge, 221
Work completed by a construction crew, 527

consumer applications

Balancing a non-interest checking account, 25
Buying clothing,
   (linear programming) 670
   (region of constraint) 234
   (systems of equations) 604
Cost function of a calling card (intervals of monotonicity), 320
Determining grocery prices from total, 618
Growth of savings with no interest, 731
Inflation model (exponential function), 529
Minimizing cost of patchwork curtains, 673
Original price of a product, 110
Paying off a loan (sequences), 719
Price of an Italian sportscar, 40
Product price increase, 529
Saving money for a gaming console, 26
State sales tax, 109
Zoo admission prices, 14

demographics

ATM pin numbers (probability), 791, 794
Driver's license number (probability), 794
License plate characters (combinatorics), 776, 804, 807
Population
   growth of a town, 723
   of an inner-city area, 528
   of Jamaica and Bahamas, 196
   of the Earth, 528, 573
Probability of sex of children, 794
Representatives of political parties (combinatorics), 778

State license plates (combinatorics), 762
Subcommittees of Senate (combinatorics), 768
Telephone numbers (combinatorics), 760, 775, 804, 807

geometry

Area
   of a picture frame, 75
   of a rectangle, 685
   of a regular hexagon, 55
   of a trapezoid, 75, 108
   of equilateral triangles, 59
Circumference of a circle, 108
Construction of open-top box, 418
Drawing
   an ellipse, 465
   lines through points, 778
Finding sides of a triangle, 634
Length of hypotenuse of a right triangle, 162
Perimeter
   and area of a rectangle, 283
   of a picture frame, 75
   of a rectangle, 104
   of a triangle, 196
Pythagorean Theorem, 163
Shading of a triangle (geometric sequence), 748
Standard form of a line, 196
Surface area
   of a birdbath, 43
   of a box, 37
   of a cube, 108
   of a cylinder, 111
   of a guinea pig home, 60
   of a prism, 59
   of a rectangular box, 196
   of a right circular cylinder, 104, 686
   of a right triangular cylinder, 60
   of a soup can (right circular cylinder), 37
   of a windsock, 333
   of an Egyptian pyramid, 59
Volume
   of a birdbath (sphere), 38

   of a box, 418
   of a circular pool, 42
   of a cone-shaped candle, 60
   of a gold ingot (right trapezoidal cylinder), 38
   of a prism, 41
   of a rectangular pool, 42
   of a right circular cone, 108
   of a right circular cylinder, 41, 333
   of a right trapezoidal cylinder, 42
   of a sphere, 163
   of a time capsule, 41
Wallpapering walls (area), 43

interest rate applications

Actual cash value of property, 111
Average change in stock value, 212
Average rate of return ($I=Prt$), 106, 109
Calculating investment outcome, 526
Comparing interest returns, 109, 169, 528, 529, 530, **556,** 573, 574, 581, 604
Constraints of investments, 671
Depreciation rate, 531
Equation for compound interest, 104
Inflation rate, 121, 531
Interest
   compounded annually, 530, 531, 619
   compounded continuously, 525, 545, 531, 583
   compounded daily, 528, 531
   compounded monthly, 523, 529, 530, 531, 573, 582, 747
   compounded quarterly, 522, 531, 583, 748
   compounded semi-annually, 530
Investment mixture (region of constraint), 234, 609, 674
Purchasing a new car, 164
U.S. national debt, 43

867

index of applications

miscellaneous

Arranging height for a photo-shoot, 13
British twopence, 759
Calculating student grades (polynomials), 86
Celsius to Fahrenheit temperature conversion, 25
Choosing a quadrant of the Cartesian plane, 793
Comparing ages (systems of equations), 600, 603, 604, 695
Consecutive integers, 110
Constraints
    of cargo for relief plane, 670
    for number of nuts in a bag, 692
    for study time, 693
Decorative arrangements of orchids and lillies, 232
Determining the number of crops planted, 634
Electric current (variation), 302
Filling a pool, 151
Finding integers, 686, 694
Height of a pebble shot in the air (maximum), 285
High school GPA compared to college GPA, 213
Locating runners in a marathon, 13
Magnitude
    of a cat's purr (logarithm), 559
    of a thunderclap (logarithm), 559
    of Gujarat earthquake (logarithm), 558
    of jackhammer noise (logarithm), 558
Maximizing
    cargo on a plane (region of constraint), 233
    income through investment, 674
Minimizing
    cost of relief supplies, 673
    distance from origin, 283
Mnemonic devices, 27
Monastic crystal towers puzzle, 759
Number of coins (systems of equations), 603, 604
Number of handshakes among people, 759
Number of tickets to a show (systems of equations), 603, 604
Racquetball games (inequality), 121
Recipe conversion, 14
SAT scores (inequality), 121
Seconds in a calendar year, 40
Sound intensity (logarithm), 554
Spread
    of an email worm, 527
    of forwarded emails, 749
String project, 749
Sum of geometric series, 111
Test Scores (inequality), 121
Types
    of fruits in a basket, 604
    of screws in a computer, 604
Wattage of a hybrid car battery, 526
Whispering Gallery, 465
Work-rate:
    Cleaning a house, 173
    Cleaning a yacht, 156
    Emptying a lake, 156
    Filling a sink, 155, 173
    Filling a tank, 156
    Filling/draining a pool, 152
    Flushing a radiator, 156
    Painting a room, 155
    Picking strawberries, 156
    Plowing a field, 156

probability

Buying raffle tickets, 797
Choices on an insurance form, 807
Choosing an M&M, 796
Creating sets with a poll, 27
Dealing five cards, 795, 804
Drawing
    a card, 793, 794, 795, 804
    three cards, 796
    two cards, 795
Flipping a coin, 743, 793, 794
Lay-offs, 796
Lottery numbers, 794, 795
Picking a marble, 797
Rolling a die (dice), 785, 790, 793, 794, 795, 796, 807
Roulette wheel, 794, *798*
Running for student council, 797
Selecting
    a letter at random, 796
    a student to wash the boards, 797
    from a sample, 808
    pencils, 808
    socks, 807
Slot machine values, 793
True/false questions, 787
Winning the lottery, 40

time and distance

Average speed of a boat, 108
Average speed of a car, 105
Direct distance between Eiffel Tower
    and Arc de Triomphe, 188
    and Notre Dame, 188
Distance, 108
    an object falls from rest, 246, 353
    on a theme park map, 188
    to the sun (exponents), 35
Driving distance
    from Shreveport, LA to Austin, TX, 105
    on a weekend roadtrip, 188
Location of radio transmitters (hyperbola), 409
Speed
    of light (exponents), 36
    of sound (hyperbola), 486
    of two jets traveling in opposite directions, 696
Straight-line distance between two trains, 14
Submarine position on a grid map, 188
Travel time
    for two cars, 109, 686, 694
    for two trains, 109, 169
    for two trucks, 109

technology index

technology

Calculating determinants, 626
Calculator commands,
    see TI-84 Plus
Circle, graphing, 241
Combining functions, 326, 335
Constructing polynomials, 389
Defining functions in *Mathematica*®, 263, 270
Evaluating expressions, 27
Evaluating functions, 263
Evaluating logarithms, 544
Finding zeros of polynomials, 400
Finding the equation for the line of best fit, 208-209
Graphing calculator,
    see TI-84 Plus
Graphing equations, 456-457, 466, 470, 477, 482, 492, 592
Graphing exponential functions, 516, 532
Graphing functions, 296, 303, 321, 400, 448
Inverse matrices, 657
Investment problems, 575-576
Mathematica® commands,
    (*see Appendix A, 809*)
    Det, 626, 815
    E^x, or Exp[x], 532
    Expand, 76, 85, 812
    Factor, 76, 448, 813
    ImplicitPlot, 241, 456-457, 592, 814
    Inverse, 657, 815
    Log, 544
    MatrixForm, 606, 643, 657, 815
    NSolve, 575
    Plot, 296, 448, 516, 814
    Simplify, 34, 44, 214, 812
    Solve, 111, 136, 214, 599, 682, 813
    Sqrt, 27
    Together, 149, 812
Matrix notation and operations, 606, 643
Multiplying polynomials, 76
Polynomial functions, 389
Sequences, 706
Simplifying
    an expression, 34, 44
    complex numbers, 85
    functions, 264
    rational expressions, 149, 157
Slope and y-intercept, 214, 222
Solving
    for a specific variable, 111, 136
    nonlinear systems of equations, 682
    systems of equations, 599
TI-84 Plus
    Absolute value, 293
    Combination $\left(_nC_r\right)$, 769
    Complex numbers, 80
    Composition of functions, 326
    Distance formula, 184
    Ellipse, graphing, 454, 456-457
    Equation solver, 101, 161
    Factorial, 767
    Fraction, 81, 657
    Graph
        of an ellipse, 454, 456-457
        of an equation, 193
        of a hyperbola, 482
        of a parabola, 470
    regression line, 209
    shading, 224
    shading multiple, 226
    viewing multiple, 215, 676
    viewing table of points, 180, 191
    Hyperbola, graphing, 482
    Inequalities, 370
    Log, 540
    Matrix
        addition of, 636
        determinant of, 620
        forming, 606
        inverse of, 654
        reduced row echelon form of, 613
        viewing, 608, 654
    Natural log, 540-541
    Parabola, graphing, 470
    Permutation $\left(_nP_r\right)$, 766
    Polynomial inequalities, 370
    Quadratic equation, 130, 139
    Regression line, 208-209
    Scatterplots, 208
    Scientific notation, 35
    Sequence
        finding the nth term, 704
        finding the sum of, 712
    Table, 144, 399, 524, 591, 595
    Trace, 279
    Variables, storing, 130
    Zoom, 400
Verifying truth of an inequality statement, 15

index

a

Abel, Niels, 362
Absolute value
    defined, 9
    function, 292
    inequalities, 116
    properties of, 10
Addition of matrices, 635
Addition of polynomials, 62
Algebraically complete, 77
Algebraic expressions, 16
Algorithm, division, 377
Apollonius, 430
Approximating solutions by graphing, 675
Archimedes, 430
Argument, 261
Arithmetic progression,
    see arithmetic sequence
Arithmetic sequence, 720
    common difference, 720
    general term, 722
    partial sum, 726
Associative property of real numbers, 17
Asymptotes
    equations for horizontal, 438
    equations for oblique, 438
    equations for vertical, 436
    horizontal, 433
    notation, 435
    oblique, 434
    of hyperbolas, 481
    vertical, 433
Augmented matrices, 605, 607
Axis of a parabola, 274

b

Back-substitution for matrices, 608
Base
    of exponents, 28
    of logarithmic functions, 534
Binomial
    coefficients, 771
    defined, 61
    factoring special, 67
    Theorem, 772
Boundary line, 223
Bounded feasible region, 666
Bounds of zeros, 396

c

Cancellation properties, 18
    for inequalities, 113
Cardinality, 760, 784, 788
    of a union of sets, 789
Cartesian
    coordinate system, 178
    plane, 178
Change of base formula, 551
Circle
    center of, 235–236
    graphing, 237–238
    radius of, 235–236
    standard form of, 235–236
Closed interval, 8
Closure property, of real numbers, 17
Coefficient
    binomial, 771
    defined, 61
    multinomial, 773
    of a polynomial, 61
Cofactor of a matrix, 620
Combination, 767
    formula, 768
Combinatorics
    binomial coefficient, 771
    Binomial Theorem, 772
    combination, 767
    combination formula, 768
    factorial notation, 764
    multinomial coefficients, 773
    Multinomial Theorem, 774
    multiplication principle of counting, 762
    permutation, 763
    permutation formula, 765
    using to compute probabilities, 785
Combining rational expressions, 52, 145
Common difference, 720
Common logarithms, 540
Common ratio, 734
Commutative property, of real numbers, 17
Completing the square, 126
Complex expressions, simplifying, 80
Complex numbers
    algebra of, 79
    conjugate, 81
    defined, 79
Complex rational expressions
    defined, 147
    simplifying, 147
Composition of functions, 325
Compound interest, 522
Compound linear inequalities, 114
Compressing, 308
Conditional equations, 99
Conic sections
    algebraic definition, 450
    graphical representation, 449
Conjugate
    radical expression, 50
    Roots Theorem, 411
Consistent system of linear equations, 590
Constant
    defined, 16
    functions, 273, 315
    of proportionality, 297
    term, 61
Constraint, region of, 228
Constructing polynomials with given zeros, 384
Continuous compounding interest, 525
Contradiction, 99
Converge, 714
Coordinate, 178
    on the real number line, 5
    system, Cartesian, 178
Counting, multiplication principle of, 762
Counting numbers, defined, 3
Cramer, Gabriel, 619
Cramer's Rule, 626, 629
Cube, perfect, 49
Cube root, 45, 46

d

Decibel, 554
Decomposition of functions, 327
Decreasing functions, 315
Dedekind, Richard, 2
Degree
    of a polynomial, 61
    of terms, 61
Denominator, rationalizing, 48
Dependent system of linear equations, 590
Dependent variable, 261
Descartes' Rule of Signs, 393
Descartes, René, 176, 430
Descending order, 62
Determinant(s)
    of a 2 by 2 matrix, 619
    of an n by n matrix, 621
    properties of, 624
Difference of
    two cubes, 67
    two squares, 67
Direct variation, 297
Directrix, 450
Dirichlet, Lejeune, 254
Disjoint, 789

index

Distance
    formula, 183
    on the real number line, 9
Distance-rate-time formula, 105
Distributive property, of real numbers, 17
Diverge, 714
Division
    algorithm, 377
    of functions, 322
Dollar-cost averaging, 568
Domain
    defined, 255
    implied, 265
    of rational functions, 431
    restriction of, 340
Double
    root, 124
    solution, 124
Doubling time, 519

e

e, 521, 540
Eccentricity, of an ellipse, 458
Einstein's Theory of Special Relativity, 60
Elimination, for solving systems of equations, 594
Ellipse
    algebraic definition, 450
    as a conic section, 449
    eccentricity of, 458
    foci of, 450
    major axis of, 454
    minor axis of, 454
    standard form of, 454, 455
    vertices of, 454
Empty set, 7
Equation
    conditional, 99
    contradiction, 99
    defined, 99
    equivalent, 100
    exponential, solving elementary, 512
    first-degree, 100
    graph of, 179
    horizontal asymptotes, 438
    identity, 99
    linear, 100
    linear, in two variables, 189
    oblique asymptotes, 438
    polynomial, 364
    polynomial-like, 140
    quadratic, 123
    quadratic, vertex form, 277
    quadratic-like, 137
    radical, 158
    rational, solving, 149
    second-degree, 122

    simultaneous linear, 589
    solving absolute value, 102
    solving radical, 158–159
    symmetry of, 312
    system of linear, 589
    two variable linear, in standard form, 189
    vertical asymptotes, 436
Equivalent equations, 100
Euclid, 430
Euler, Leonhard, 506
Evaluating
    elementary logarithmic expressions, 538
    partial sums of arithmetic sequences, 724
Even functions, 311
Event(s), 783
    independent, 792
    mutually exclusive, unions, 790
    probability of a union of, 789
Expansion along a row or column, 621
Experiment, 783
Explicit formula, 705
Exponent(s)
    defined, 28
    integer, 30
    natural number, 28
    negative integer, 31
    properties of, 32
    rational number, 53
    zero as an, 30
Exponential equations, solving elementary, 512
Exponential functions, 507
    behavior of, 509
    compound interest, 522
    continuous compounding interest, 525
    e, 521
    further applications of, 566
    graphing, 510
    population growth, 517
    radioactive decay, 519
Expressions
    algebraic, defined, 16
    logarithmic, evaluating elementary, 538
    rational, combining, 52, 145
    rational complex, defined, 147
    rational complex, simplifying, 147
    rational, defined, 143
    rational, simplifying, 143
Extraneous solution, 102

f

f^{-1}, finding a formula for, 341
Factorable, 65
Factorial notation, 764
Factoring
    defined, 64
    expressions containing fractional exponents, 70
    greatest common factor, 65
    grouping, 66
    quadratic equations, 122
    special binomials, 67
    trinomials, 68
Factors
    defined, 16
    linear, 378
    variable, defined, 16
Feasible region, 229, 664
    bounded, 666
    of a plane, 664
    unbounded, 666
    vertex of, 666
Fermat, Pierre de, 702
Fibonacci, Leonardo, 715
Fibonacci sequences, 715
Field properties (for real numbers), 17
Finite
    sequence, 703
    series, 713
First-degree
    equations, 100
    function, 271
Foci, 450
FOIL method, 64
Formula(s)
    change of base, 551
    combination, 768
    continuous compounding interest, 525
    distance, 105, 183
    distance-rate-time, 105
    explicit, 705
    geometric, 37
    interest, 105
    midpoint, 184
    permutation, 765
    quadratic, 129
    recursive, 705
    summation, 711
Function(s), 339
    absolute value, 292
    addition of, 322
    composition of, 325
    constant, 273, 315
    decomposing, 327
    decreasing, 315
    defined, 258
    division of, 322
    even, 311

871

exponential, 507
   behavior of, 509
   further applications of, 566
   graphing, 510
first-degree, 271
graphing rational, 439
greatest integer, 294
implied domain, 265
linear, 271
logarithmic,
   base of, 534
   defined, 534
   further applications of, 552
   graphing, 535
   multiplication of,, 322
   notation, 260
   objective, in linear programming, 666
   odd, 312
   of the form $\frac{a}{x^n}$, 289
   of the form $ax^{\frac{1}{n}}$, 291
   of the form ax^n, 287
   one-to-one, 339
   piecewise-defined, 295
   quadratic, 274
   rational, 431
   second-degree, 274
   subtraction of, 322
   vertical line test, 259
Fundamental Theorem of Algebra, 405

g

Galois, Evariste, 362
Gauss, Carl Friedrich, 362, 405, 430, 588, 608
Gauss-Jordan elimination, 611
Gaussian elimination, 608
General polynomial equations, solving by factoring, 138
General term
   arithmetic sequence, 722
   geometric sequence, 734
Geometric formulas, 37
Geometric progression,
   see geometric sequence
Geometric sequence, 734
   common ratio, 734
   general term, 734
   partial sum, 737
Geometric series, infinite, 741
Graph of an equation, 179
Graphing
   approximating solutions, 675
   circles, 237, 237–238
   exponential functions, 510

   factored polynomials, 366
   logarithmic functions, 535
   rational functions, 439
Graphs
   compressing, 308
   horizontal shift, 304
   reflection, 307
   stretching, 308
   vertical shift, 305
   y-axis symmetry, 311
Gravity, 130
Greater than, 6
Greatest common factor (GCF), 65
Greatest integer function, 294
Guidance systems, 484

h

Half-closed interval, 8
Half-life, 519
Half-open interval, 8
Half-plane, 223
Heron's Formula, 55
Hooke's Law, 298
Horizontal asymptotes
   defined, 433
   equations for, 438
Horizontal line, 192
Horizontal line test, 339
Horizontal shifting, 304
Hyperbola(s)
   algebraic definition, 450
   as a conic section, 449
   asymptotes of, 480, 481
   branches of, 478
   center of, 478
   foci of, 450
   standard form of, 480
   vertices, 478

i

i, 78
Identity matrix, 652
Identity of an equation, 99
Identity property, of real numbers, 17
Imaginary unit (i), 78
Implied domain of a function, 265
Inconsistent system of linear equations, 590
Increasing functions, 315
Independent events, 792
Independent variable, 261
Index
   of a radical, 46
   of summation, 709

Induction, mathematical, 751
   basic step, 751
   induction step, 751
   principle of, 750
   proofs by, 751
Inequalities
   absolute value, 116
   joined by "and" or "or", 226
   linear, 112
      joined by "and" or "or", 226
      two variables, 223–224
   non-strict, 6
   solving polynomial, 370
   solving rational, 441
   strict, 6
   symbols, 6
   translating, 117
   triangle, 10
Infinite
   geometric series, 741
   sequence, 703
   series, 713
Infinity symbol, 8
Integers
   defined, 3
   exponents, 30
Interest
   continuous compounding formula, 525
   formula, 105
Intermediate Value Theorem, 398
Intersection, of sets, 22
Interval(s)
   closed, 8
   defined, 7
   half-closed, 8
   half-open, 8
   notation, 8
   of monotonicity, 315
   open, 8
Inverse
   of a 2 by 2 matrix, 655
   of a matrix, 653
      finding, 655
      to solve linear systems, 657
   of a relation, 336
   property, of real numbers, 17
   variation, 297
Investment analysis, 568
Irrational numbers, defined, 4
Irreducible, 65
Iterate, 328

j

Jordan, Wilhelm, 611

index

k

Kepler, Johann, 430, 458

l

Lambert, Johann, 2
Larger systems of equations, 596
Laws of planetary motion
    Kepler, Johann, 459
Leading coefficient, 62
Least common denominator (LCD), 101, 145
Leibniz, Gottfried, 254
Less than, 6
Like
    radicals, 52
    terms, 63
Line(s)
    horizontal, 192
    parallel, 215
    perpendicular, 218
    point-slope form, 204
    slope-intercept form, 201
    slope of, 197
    standard form, 189
    vertical, 192
Linear
    equation(s), 100
        in two variables, 189
        simultaneous, 589
    factors, 378
    Factors Theorem, 406
    function, 271
    inequalities, 112
        compound, 114
        in two variables, 223
            solving, 224
        joined by "and" or "or", 226
    programming, 228, 666
        method for two-variable case, 666
    regression, 205
    system of equations, 589
    systems,
        Cramer's Rule, 626
        solving, matrix inverses, 657
        solving using matrices, 650
        varieties of, 590
Logarithmic expressions, evaluating, 538
Logarithmic functions
    applications of, 552
    base of, 534
    defined, 534
    further applications of, 552, 566
    graphing, 535
Logarithms
    change of base formula, 550–551
    common, 540
    natural, 540
    properties of, 546
    summary of properties, 562
Long division, of polynomials, 379
LORAN, 430, 484
Lower bounds of zeros, 396

m

Magnitude, 9
Main diagonal of a matrix, 652
Mandelbrot, Benoit, 329
Mathematical induction
    basic step, 751
    induction step, 751
    principle of, 750
    proofs by, 751
Matrices
    addition of, 635
    augmented, 605, 607
    back-substitution, 608
    cofactors of, 620
    columns of, 605
    defined, 605
    determinant, 619–621
    elementary row operations, 609
    elements of, 605
    entries, 605
    equality of, 636
    Gauss-Jordan elimination, 611
    Gaussian elimination, 608
    identity, 652
    inverse, 653
        finding, 655
        of a 2 by 2 matrix, 655
        to solve linear systems, 657
    main diagonal, 652
    minors, 620
    multiplication of, 641
    notation, 605
    properties of determinants, 624
    reduced row echelon form, 612
    rows of, 605
    row echelon form, 608–609
    scalar multiplication, 638
    solving linear systems using, 650
        finding, 655
        of a 2 by 2 matrix, 655
        to solve linear systems, 657
    transition, 644
Maximization, 279
Midpoint formula, 184
Minimization, 279
Minor of a matrix, 620
Models of population growth, 517
Monomial, 61
Monotonicity, intervals of, 315
Multinomial
    coefficients, 773
    Theorem, 774
Multiplication
    of functions, 322
    of matrices, 641
    principle of counting, 760, 762
    scalar, 638
Multiplicity, geometric meaning, 409
Mutually exclusive, 790

n

n by n matrix determinant, 621
n^{th} iterate, 328
n^{th} Roots, 46
Napier, John, 506, 552
Nappe, 449
Natural logarithms, 540
Natural number(s), 3
    exponents, 28
Negative integer exponents, 31
Newton, Sir Isaac, 430
Non-strict inequalities, 6
Nonlinear systems, solving,
    algebraically, 677
    graphically, 675
Notation
    asymptotes, 435
    factorial, 764
    function, 260
    interval, 7
    matrices, 605
    radical, 46
    scientific, 35
    set-builder, 7
    summation, 709
        index of summation, 709
        sigma, 709
Numbers
    complex, 79
    counting, 3
    integers, 3
    irrational, 4
    natural, 3
    rational, 3
    real, 4
    whole, 3
Number line, 5

o

Objective function, in linear programming, 666

index

Oblique asymptotes
    defined, 434
    equations for, 438
Odd functions, 312
One-to-one functions, 339
Open interval, 8
Order
    of operations, 20
    of transformations, 309
    on the real number line, 6
Ordered
    pair, 178
    triple, 596
Origin
    on the Cartesian plane, 178
    on the real number line, 5
    symmetry, 312
Outcomes, 783
Ozone layer, 349

p

Parabola, 274
    algebraic definition, 450
    as a conic section, 449
    axis of, 274
    directrix of, 450, 467
    focus of, 467
    standard form of, 469
    vertex of, 274
Parabolic mirrors, 472
Parallel lines, slope of, 216
Partial sums, 713
    arithmetic sequences, 726
    geometric sequences, 737
Pascal, Blaise, 702
Pascal's triangle, 782
Perfect cube, 49
Perfect square quadratic equations, 125
Permutation, 763
    formula, 765
Perpendicular lines, slope of, 218
pH, 552
Piecewise-defined function, 295
Pi (π), 2
Planar feasible regions, 664
Planetary motion
    Kepler, Johann, 459
Planetary Orbits, 458
Point-slope form of a line, 204
Polynomial(s)
    addition and subtraction of, 62
    construction with given zeros, 384
    equations, 364
        solving by factoring, 122
    graphing factored, 366
    inequalities, solving, 370
    long division, 379
    methods, summary of, 413
    of a single variable, 62
    roots of, 364
    solutions of, 364
    zeros of, 364
Polynomial-like equations
    solving by factoring, 140
Population growth, models of, 517
Positive rational exponents
    solving equations with, 161
Prime polynomial, 65
Principal square roots, 82
Principle of counting, multiplication, 762
Principle of mathematical induction, 750
Probability, 783
    event, 783
    experiment, 783
    independent events, 792
    of a union of two events, 789
    outcomes, 783
    sample space, 783
    unions of mutually exclusive events, 790
    using combinatorics, 785
    when outcomes are equally likely, 784
Progression, arithmetic,
    see sequence, arithmetic
Progression, geometric,
    see sequence, geometric
Programming, linear, 666
Proof by mathematical induction, 751
    basic step, 751
    induction step, 751
Properties
    cancellation, 18, 113
    field (real numbers), 17
    of absolute value, 10
    of determinants, 624
    of exponents, 32
    of logarithms, 546, 562
    of radicals, 48
    sigma, 710
    summary of logarithmic, 562
    zero-factor, 19, 123
Proportionality, constant of, 297
Pythagoras, 2
Pythagorean Theorem, 182

q

Quadrants, 178
Quadratic equation(s)
    defined, 123
    solving by completing the square, 126
    solving by factoring, 122
    solving perfect squares, 125
    vertex form, 277
Quadratic formula, 129
Quadratic function, 274
Quadratic, in form, 137
Quadratic-like equations, defined, 137

r

Radical(s)
    equations
        defined, 158
        solving, 158–159
    expression, conjugate, 50
    form, simplified, 47
    like, 52
    notation, 45, 46
    properties of, 48
    sign, 52
Radicand, 46
Radio-carbon dating, 519
Radioactive decay, 519
Radius, of a circle, 235–236
Range, defined, 255
Rational
    equations, solving, 149
    exponents, positive, 161
    expressions
        combining, 145
        complex, defined, 147
        complex, simplifying, 147
        defined, 143
        simplifying, 143
    functions
        defined, 431
        domain of, 431
        graphing, 439
    inequalities, solving, 441
    number
        defined, 3
        exponents, 53
Rational Zero Theorem, 391
Rationalizing the denominator, 48
Real number line, 5
    coordinate, 5
    distance on, 9
    order, 6
    origin, 5
Real numbers
    defined, 4
    types, 3–4
Reciprocal, 18
Rectangular coordinate system, 178
Recursion, 328
Recursive formula, 705
Reduced row echelon form, 612
Reflecting, with respect to the axes, 307
Region
    feasible, 229, 664
    of constraint, 228, 664

index

Relation
    defined, 255
    domain of, 255
    inverse of, 336
    range of, 255
Remainder Theorem, 377, 379
Restriction of domain, 340
Richter, Charles, 553
Richter scale, 553
Roots
    cube, 45
    double, 124
    n^{th}, 46
    of polynomials, 364
Row echelon form, 608–609
Row operations for matrices, 609

S

Sample space, 783
Scalar multiplication, 638
Scientific notation, 35
Second-degree
    equations, 122
    function, 274
Sequence, 703
    arithmetic
        common difference, 720
        general term, 722
        partial sum, 726
    explicitly defined, 703
    Fibonacci, 715
    finite, 703
    geometric
        common ratio, 734
        general term, 734
    infinite, 703
    partial sum, 713
    recursively defined, 703
    summation
        formulas, 711
        properties of, 710
    summation notation, 709
        index of summation, 709
        sigma, 709
Series
    converge, 714
    diverge, 714
    finite, 713
    geometric, infinite, 741
    infinite, 713
    telescoping, 714
Set(s)
    cardinality of a union of, 789
    empty, 7
    intersection of, 22
    union of, 21
Set-builder notation, 7
Shifting
    horizontal, 304
    vertical, 305
Sigma, 709
    properties of, 710
Similar terms, 63
Simplified radical form, 47
Simplifying
    complex expressions, 80
    complex rational expressions, 147
    rational expressions, 143
Simultaneous linear equations, 589
Slope, 197
    of parallel lines, 216
    of perpendicular lines, 218
Slope intercept form of a line, 201
Solution of an equation, 99
Solution set, 99
Solutions
    approximating by graphing, 675
    double, 124
    extraneous, 102
    of polynomials, 364
Solving
    absolute value inequalities, 116
    elementary exponential equations, 512
    equations with rational exponents, 161
    for a variable, 104
    compound, 114
    linear inequalities
        in two variables, 224
        joined by "and" or "or", 226
    linear systems
        Cramer's Rule, 626
        matrix inverses, 657
    nonlinear systems, algebraically, 677
    perfect square quadratic equations, 125
    polynomial equations by factoring, 138
    polynomial inequalities, 370
    polynomial-like equations by factoring, 140
    quadratic equations
        by completing the square, 126
        by factoring, 122
    quadratic-like equations, 137
    radical equations, 158–159
    rational equations, 149
    rational inequalities, 441
    systems of equations
        by elimination, 594
        by substitution, 591
Square roots
    of negative numbers, 78
    principal, 82
Standard form
    of a circle, 236
    of a hyperbola, 480
    of a parabola, 469
    of a two variable linear equation, 189, 607
    of an ellipse, 455
Stock market investment analysis, 568
Stretching, 308
Strict inequalities, 6
Substitution, for solving systems of equations, 591
Subtraction
    of functions, 322
    of polynomials, 62
Sum of two cubes, 67
Summary of logarithmic properties, 562
Summation
    formulas, 711
    properties of, 710
Summation notation, 709
    index of summation, 709
    sigma, 709
Symmetry
    of equations, 313
    origin, 312
    y-axis, 311
Synthetic division, 379
System(s) of equations
    applications of, 599
    consistent, 590
    dependent, 590
    inconsistent, 590
    independent, 590
    larger, 596
    linear, 589
        by elimination, 594
    solving
        by elimination, 594
        by substitution, 591
        using matrices, 605

T

Telescoping series, 714
Term(s)
    constant, 61
    defined, 16
    degree of, 61
    like, 63
    similar, 63
Test point, 224
Transition matrices, 644
Translating inequality phrases, 117
Triangle inequality, 10
Trinomial, 61
    factoring, 68
    factoring by grouping, 69
Turning point of a graph, 407
Two by two matrix determinant, 619

U

Unbounded feasible region, 666
Union
    of mutually exclusive events, 790
        cardinality, 789
    of sets, 21
        cardinality, 789
    of two events, probability, 789
Upper bounds of zeros, 396

V

Variable(s)
    defined, 16
    dependent, 261
    independent, 261
Variable factor, defined, 16
Variation
    constant of proportionality, 297
    direct, 297
    inverse, 297
Varieties of linear systems, 590
Venn diagram, 21
Vertex
    form of a quadratic equation, 277
    of a feasible region, 666
    of a parabola, 274
Vertical
    asymptotes
      defined, 433
      equations for, 436
    line, 192
    line test, 259
    shifting, 305

W

Whole numbers, defined, 3
Work-rate, 150

X

x-axis, 179
    symmetry, 312
x-coordinate, 179
x-intercept, 191

Y

y-axis, 179
    symmetry, 311
y-coordinate, 179
y-intercept, 191

Z

Zeno, 743
Zeno's paradoxes, 743
Zero, as an exponent, 30
Zero-Factor Property, 19, 123
Zeros, of polynomials, 364, 378

Properties of Absolute Value 1.1
For all real numbers a and b:
1. $|a| \geq 0$
2. $|-a| = |a|$
3. $a \leq |a|$
4. $|ab| = |a||b|$
5. $\left|\dfrac{a}{b}\right| = \dfrac{|a|}{|b|}$
6. $|a+b| \leq |a| + |b|$

Properties of Exponents 1.3
1. $a^m \cdot a^n = a^{m+n}$
2. $\dfrac{a^n}{a^m} = a^{n-m}$
3. $a^{-n} = \dfrac{1}{a^n}$
4. $(a^n)^m = a^{nm}$
5. $(ab)^n = a^n b^n$
6. $\left(\dfrac{a}{b}\right)^n = \dfrac{a^n}{b^n}$

Properties of Radicals 1.4
1. $\sqrt[n]{ab} = \sqrt[n]{a} \cdot \sqrt[n]{b}$
2. $\sqrt[n]{\dfrac{a}{b}} = \dfrac{\sqrt[n]{a}}{\sqrt[n]{b}}$
3. $\sqrt[m]{\sqrt[n]{a}} = \sqrt[mn]{a}$

Factoring Special Binomials 1.5
Difference of Two Squares:
$$A^2 - B^2 = (A-B)(A+B)$$
Difference of Two Cubes:
$$A^3 - B^3 = (A-B)(A^2 + AB + B^2)$$
Sum of Two Cubes:
$$A^3 + B^3 = (A+B)(A^2 - AB + B^2)$$

The Quadratic Formula 2.3
The solutions of the equation $ax^2 + bx + c = 0$ are:
$$x = \dfrac{-b \pm \sqrt{b^2 - 4ac}}{2a}$$

Pythagorean Theorem 3.1
Given a right triangle with legs a and b and hypotenuse c:
$$a^2 + b^2 = c^2$$

Distance Formula 3.1
$$d = \sqrt{(x_2 - x_1)^2 + (y_2 - y_1)^2}$$

Midpoint Formula 3.1
$$\left(\dfrac{x_1 + x_2}{2}, \dfrac{y_1 + y_2}{2}\right)$$

Slope of a Line 3.3-3.4
Given two points (x_1, y_1) and (x_2, y_2) on a line, the slope of the line is the ratio:
$$\dfrac{y_2 - y_1}{x_2 - x_1}$$
Horizontal lines have slope 0.
Vertical lines have undefined slope.
Given a line with slope m:
  slope of a parallel line $= m$
  slope of a perpendicular line $= -\dfrac{1}{m}$

Standard Form of a Circle 3.6
The standard form of the equation for a circle of radius r and center (h, k) is
$$(x-h)^2 + (y-k)^2 = r^2.$$

Transformations of Functions 4.4
Horizontal Shifting: The graph of $g(x) = f(x-h)$ is the same as the graph of f, but shifted h units to the right if $h > 0$ and h units to the left if $h < 0$.

Vertical Shifting: The graph of $g(x) = f(x) + k$ is the same as the graph of f, but shifted upward k units if $k > 0$ and downward k units if $k < 0$.

Reflecting with Respect to Axes:
1. The graph of the function $g(x) = -f(x)$ is the reflection of the graph of f with respect to the x-axis.
2. The graph of the function $g(x) = f(-x)$ is the reflection of the graph of f with respect to the y-axis.

Stretching and Compressing:
1. The graph of the function $g(x) = af(x)$ is stretched vertically compared to the graph of f if $a > 1$.
2. The graph of the function $g(x) = af(x)$ is compressed vertically compared to the graph of f if $0 < a < 1$.

Operations with Functions 4.5
Let f and g be functions.
1. $(f+g)(x) = f(x) + g(x)$
2. $(f-g)(x) = f(x) - g(x)$
3. $(fg)(x) = f(x)g(x)$
4. $\left(\dfrac{f}{g}\right)(x) = \dfrac{f(x)}{g(x)}$

Rational Zero Theorem 5.3
If $f(x) = a_n x^n + a_{n-1} x^{n-1} + \ldots + a_1 x + a_0$ is a polynomial with integer coefficients, then any rational zero of f must be of the form $\dfrac{p}{q}$, where p is a factor of the constant a_0 and q is a factor of a_n.

Fundamental Theorem of Algebra 5.4
If f is a polynomial of degree n with $n \geq 1$, then f has at least one zero. That is, $f(x) = 0$ has at least one root or solution (note, the solution may be a non-real complex number).

Compound Interest 7.2
An investment of P dollars at an annual interest rate of r, compounded n times per year for t years has an accumulated value of
$$A(t) = P\left(1 + \dfrac{r}{n}\right)^{nt}.$$
An investment compounded continuously has an accumulated value of $A(t) = Pe^{rt}$.

Properties of Logarithms 7.3, 7.4
$a > 0$, a is not equal to 1, $x, y > 0$ and r is a real number

$\log_a(x) = y$ and $x = a^y$ are equivalent

$\log_a(a) = 1$

$\log_a(1) = 0$

$\log_a(a^x) = x$

$a^{\log_a(x)} = x$

$\log_a(xy) = \log_a(x) + \log_a(y)$

$\log_a\left(\dfrac{x}{y}\right) = \log_a(x) - \log_a(y)$

$\log_a(x^r) = r\log_a(x)$

Change of Base Formula 7.4
$a, b, x > 0$; $a, b \neq 1$;
$$\log_b(x) = \dfrac{\log_a(x)}{\log_a(b)}$$

Matrices 8.3
The **minor** of the element a_{ij} is the determinant of the $(n-1)$ by $(n-1)$ matrix formed from A by deleting the ith row and the jth column.

The **cofactor** of the element a_{ij} is $(-1)^{i+j}$ times the minor of a_{ij}.

Find the **determinant** of an $n \times n$ matrix by expanding along a fixed row or column.
- To expand along the ith row, each element of that row is multiplied by its cofactor and the n products are then added.
- To expand along the jth column, each element of that column is multiplied by its cofactor and the n products are then added.

Cramer's Rule 8.3
A linear system of n equations $x_1, x_2, \ldots, x_n$ can be written in the form:
$$\begin{cases} a_{11}x_1 + a_{12}x_2 + \ldots + a_{1n}x_n = b_1 \\ a_{21}x_1 + a_{22}x_2 + \ldots + a_{2n}x_n = b_2 \\ \vdots \\ a_{n1}x_1 + a_{n2}x_2 + \ldots + a_{nn}x_n = b_n \end{cases}$$

The solution of the system is given by the n formulas $x_1 = \dfrac{D_{x_1}}{D}$, $x_2 = \dfrac{D_{x_2}}{D}$, ..., $x_n = \dfrac{D_{x_n}}{D}$, where D is the determinant of the coefficient matrix and D_{x_i} is the determinant of the same matrix with the ith column of constants replaced by the column of constants $b_1, b_2, \ldots, b_n$.